REALM OF THE UNIVERSE

Fifth Edition

GEORGE O. ABELL

Professor of Astronomy
University of California, Los Angeles

DAVID MORRISON

Chief, Space Science Division
NASA Ames Research Center

SIDNEY C. WOLFF

Director
National Optical Astronomy Observatories

SAUNDERS COLLEGE PUBLISHING

A Harcourt Brace Jovanovich College Publisher
Fort Worth   Philadelphia   San Diego   New York
Orlando   Austin   San Antonio   Toronto
Montreal   London   Sydney   Tokyo

Text Typeface: Times Roman
Compositor: General Graphic Services
Acquisitions Editor: John Vondeling
Developmental Editor: Lloyd W. Black
Managing Editor: Carol Field
Project Editor: Anne Gibby
Copy Editor: Bonnie Boehme
Manager of Art and Design: Carol Bleistine
Art Director: Christine Schueler
Art Assistant: Caroline McGowan
Text Designer: Gene Harris
Cover Designer: Lawrence R. Didona
Text Artwork: J & R Art Services Inc.
Layout Artist: Anne O'Donnell
Director of EDP: Tim Frelick
Production Manager: Charlene Squibb
Marketing Manager: Marjorie Waldron

Cover Credit: Interacting galaxies NGC 5194 (also known as M51, The Whirlpool Galaxy) and NGC 5195. Images in red, green, and blue were processed and combined into a true-color picture. The large spiral is classified Sbc while the smaller companion has been classified either Irr 2 or SB0. Note the strong red dust lanes at the point of contact between the spiral arm and the companion. *(National Optical Astronomy Observatories/T. Boroson, N. Sharp)*

Photo Opposite Contents Overview: A true-color picture of the Crab Nebula, which is the remnant of a supernova explosion. In this relatively short exposure, the pulsar is clearly visible (just below the center). The red filaments are tendrils of excited gas, emitting strong Hα radiation, still bearing mute testimony to the violence of the supernova explosion which created both the nebula and the pulsar. One of the more spectacular historical supernovae, it was recorded by Chinese astronomers in the year 1054 AD. *(National Optical Astronomy Observatories/W. Schoening/N. Sharp)*

Back Cover Credit: Artist's impression of the impact of an asteroid with the Earth. *(Painting by Don Davis)*

Printed in the United States of America

REALM OF THE UNIVERSE

ISBN: 0-03-074919-0

Library of Congress Catalog Card Number: 91-050655

2345 032 98765432

PREFACE

These are exciting times for science in general and for astronomy in particular. We have sent spacecraft to explore eight of the nine planets, to plunge into the heart of a comet, and to land on the surfaces of Venus, the Moon, and Mars. We have discovered the dying glow of the primeval fireball that began the expansion of the universe and have used theory to probe to within a fraction of a second of this cosmic beginning. We have identified gravitational lenses in space, probed deep into the stellar nurseries where stars and planets are born, and found convincing evidence for the existence of black holes. New telescopes on Earth and in space reveal a universe richer, more varied, and more violent than had been suspected by previous generations.

Public interest in the new frontiers of astronomy continues at a high level. A variety of publications interpret astronomy and other sciences for the lay person, and in the colleges our classes are crowded. Students and the public alike are becoming more sophisticated and more fervent in their desire to understand as much as they can of what we have learned about the cosmos. People seem to be aware of what astronomy can offer to the human perspective. What they are often not aware of, however, is the method of science—the exacting procedure and rigid rules that have permitted science its steady progress. They look for quick answers, not realizing that there are no final authorities, but only a process that leads toward a better but always inadequate understanding of nature.

That communication gap becomes especially obvious when we note that many people, thirsty for knowledge about new frontiers, have turned to all manner of unreliable sources for their information. Even in this era, when science and technology are crucial to our lives, the public is bombarded by misinformation and pseudoscience. Scientists have an obligation to the public to increase our efforts in presenting an honest view of our work. We try in this book to describe not only how the universe is but also to show how, by simple rational processes, we can probe its mysteries. We have also indicated that astronomy is a very human endeavor and have related it to those men and women who created our science.

Inevitably, we have had to be selective about the individuals mentioned by name, and we ask our colleagues' understanding if the choices often seem arbitrary.

Realm of the Universe is a textbook in astronomy—specifically for a comprehensive one-semester course for the liberal arts student. George Abell introduced this text in 1969, as a condensed version of his highly successful two-semester text *Exploration of the Universe*. Since that time the two texts, *Exploration* and *Realm*, have evolved separately. Both remain straightforward introductions to astronomy, covering both historical and modern topics with emphasis on the basic principles that underlie our concepts of the universe. While the mathematics prerequisites for *Realm of the Universe* are confined to simple high-school algebra, we believe that it is not possible to understand a subject like astronomy without the use of quantitative reasoning, and we therefore do not shy from the use of numbers in illustrating the facts and concepts in this book.

New Features of the Fifth Edition

The fifth edition of *Realm of the Universe* has been completely revised and updated, while maintaining the same approach and length of the previous edition. Important changes in the fifth edition include the following:

- The text is now printed in full color.

- Nearly one-third of the chapters have been either internally restructured or rearranged within the table of contents to present the material more logically. The discussion of tides has been moved to Chapter 3. Chapter 6 combines two former chapters in discussing astronomical observations at all wavelengths. Chapters 20 and 21 expand the coverage of stellar evolution; Chapter 20 also incorporates material on the search for extraterrestrial life from what had been the book's final chapter. Chapter 23 now combines the discussions of general relativity and black holes. A new chapter (27) has been added to discuss the exciting new discoveries and theories about the structure of the universe. And Chapters 4, 13 through 15, and 23 have been reordered within the contents.

• Recent discoveries from space-based as well as ground-based instruments are discussed. Included are dramatic images from the Magellan Venus probe, COBE, and the Hubble Space Telescope. Also presented are advances in infrared and submillimeter astronomy.

• New material has been added on solar astronomy and the structure and evolution of the universe.

• The end-of-chapter material now consists of review *and* thought questions as well as problems, some of which require the use of a hand calculator.

• New to this edition are interviews with five leading astronomers. They discuss their work, their attitudes toward science, and a little of the story of how they became involved in astronomical research so as to provide insight into the human elements and motivations of the people who are advancing our ideas of the cosmos.

• New to this edition are essays that provide suggestions for visualizing many of the concepts introduced in the book. Many of the essays also encourage students to make their own observations of the world around us.

• Jargon is held to a minimum, but where specialized vocabulary is needed, these words are introduced in boldface and defined in the text, repeated in boldface in the chapter summaries, and defined again in the Glossary at the end of the book.

Organization

Realm of the Universe is composed of 28 chapters organized in an Earth-outward approach. Chapters 1 through 6 present, respectively, the ancients' view of our world and the cosmos, an historical overview of major astronomical discoveries to the time of Newton, discussion of phenomena—seasons, time, calendar, phases of the Moon, eclipses, tides—tied to the motion of the Earth, a close look at the geology and chemistry of Earth, a discussion of the basic principles of light and energy, and coverage of the various types of telescopes used to explore the electromagnetic spectrum.

Chapters 7 through 12 cover the solar system and its occupants. Chapter 7 describes the cratered worlds of Mercury and the Moon. Chapter 8 compares Earth's nearest planetary neighbors, Venus and Mars. Based largely on the Voyager data, Chapter 9 compares and contrasts the jovian gas giants—Jupiter, Saturn, Uranus, Neptune. Chapter 10 looks at the rings, moons, and Pluto. Chapter 11 presents comets and asteroids, the debris of the solar system. Chapter 12 evaluates meteorites and their relation to the formation and evolution of the Solar System.

Moving outward, Chapters 13 through 23 discuss all aspects of stars. Chapter 13 begins by presenting the star we know best, our Sun. Chapter 14 discusses the special theory of relativity and its relationship to the special nature of the speed of light. Chapter 15 investigates the Sun as a nuclear powerhouse. Chapters 16 through 18 present, respectively, important details on determining the distances to and motions of stars, the brightness and color of stars, and stellar properties such as mass and diameter. Chapter 19 offers a look at the gas and dust of the interstellar medium. Chapters 20 to 22 deal with stellar birth, aging, and death. Chapter 23 connects Einstein's powerful general theory of relativity and the growing evidence for the existence of black holes.

Finally, we move to the realm of the galaxies and beyond. Chapter 24 describes the parts of and our place in the Milky Way Galaxy. Chapter 25 takes the reader beyond the Milky Way to discuss galaxies in general, the fundamental building block of the Universe. Chapter 26 looks at quasars and active galaxies. Chapter 27 presents exciting new discoveries and theories about large-scale structures in the universe and the creation and evolution of galaxies. Chapter 28 considers the ultimate question of cosmological creation by analyzing recent findings and current theories.

Ancillary Materials

Available free to all adopters, the *Instructor's Manual* that accompanies *Realm of the Universe* contains answers to thought questions and problems in the textbook.

Available free to qualified adopters are 100 *full-color* overhead transparency acetates composed of artwork and photographs taken from the textbook.

Also available to adopters is a printed *Test Bank* containing over 1,500 questions in multiple choice and other formats. Accompanying the Test Bank is the exclusive Saunders *ExaMaster*™ Computerized Test Bank for IBM and Macintosh personal computers. The Computerized Test Bank contains the same questions as the printed version and allows instructors to edit existing questions and add new ones of their own devising. Adopters may also choose the option of *RequesTest*™. By telephoning Software Support at (800) 447–9457, an adopter can request one or more tests prepared from the computerized test bank. Within 48 hours, a copy of these requested tests are then faxed or mailed back to the adopter.

Acknowledgments

We would particularly like to thank the readers of the manuscript of the fifth edition. Their comments and suggestions were helpful to us as we revised the text.

Robert H. Allen
University of Wisconsin, La Crosse

Laurence W. Fredrick
University of Virginia

Mark Littman
Starmaster Corporation

James C. LoPresto
Edinboro University of Pennsylvania

J. Scott Shaw
University of Georgia

Raymond E. White
University of Arizona

Arthur Young
San Diego State University

Many of our colleagues have helped to locate accurate and up-to-date material for this edition. We are grateful for their assistance.

Edward Bowell
Lowell Observatory

A. G. W. Cameron
Harvard University

Clark Chapman
Planetary Science Institute

Dale Cruikshank
NASA Ames Research Center

Alfred McEwen
U.S. Geological Survey

Stephen Ostro
Jet Propulsion Laboratory

Stephen Saunders
Jet Propulsion Laboratory

Eugene Shoemaker
U.S. Geological Survey

Anita Sohus
Jet Propulsion Laboratory

Ellen Stofan
Jet Propulsion Laboratory

Steven Squyres
Cornell University

Jill Tarter
University of California, Berkeley

Peter Thomas
Rand Corporation

Jurrie van der Woude
Jet Propulsion Laboratory

Don Yeomans
Jet Propulsion Laboratory

We are also especially grateful to Don Davis and Don Dixon for permission to reproduce their fine paintings. Janet Morrison offered many useful editorial suggestions. We also thank John Vondeling, Lloyd Black, and Anne Gibby, our editors at Saunders, who worked diligently with us to produce this book.

David Morrison
Sidney C. Wolff
Cupertino and Tucson
September 1991

CONTENTS OVERVIEW

CONTENTS

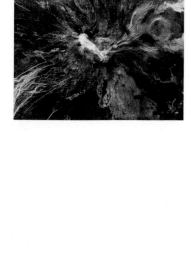

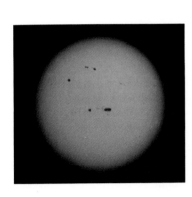

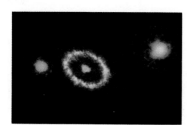

The Galileo spacecraft being launched from the Shuttle in Earth orbit, preparatory to its 5-year trip around the solar system, flying by Venus, Earth, and two asteroids on the way to Jupiter. *(MAX photo, courtesy of NASA/JPL)*

Galileo Galilei (1564–1642) advocated that we perform experiments or make observations to ask Nature her ways, rather than deciding how things must be on the basis of preconceived notions. When Galileo turned the telescope to the sky, he found that things are not as philosophers had supposed, discovering sunspots, the mountains of the Moon, the phases of Venus, and the four large satellites of Jupiter—worlds that are still called the Galilean satellites.

PROLOGUE

Astronomy is the oldest science. Since before the dawn of history, humans have tried to understand the progression of the seasons and the motions of the Moon and planets. They have also speculated on the origin of the Earth and of the cosmos in which it is embedded. Gradually they learned to advance from speculation and superstition to the systematic thinking about nature that we call science.

Science, unlike religion or philosophy, accepts nothing on faith. Of course, individual scientists may be religious, and a belief common to all scientists is that scientific laws—the rules of the game that nature plays—are truly universal, the same at all times and places. When you think of it, you realize that the universality of scientific laws is a marvelous concept.

The Scientific Method

Science itself is a *method* by which we attempt to understand nature and how it behaves. Specifically, we attempt to understand things in terms of *models* that correctly describe the behavior of nature. In their tentative stages these models are called *hypotheses*. Eventually, when a body of hypotheses has been checked out and pieced together into a self-consistent system, that system becomes a *theory*. Without theory, there is no science—just collections of unrelated facts.

Some models and theories—for example, general relativity and quantum mechanics—are quite mathematical and abstract. Others, such as the laws of planetary motion or the explanation of eclipses, are easier to visualize or even are susceptible to mechanical representation. No scientific theory, however, should be confused with absolute truth; such theories merely represent nature with descriptive models.

The scientific process is basically rather simple. To begin with, we must be aware of the results of work by previous generations of scientists, so that we do not make fools of ourselves by ignoring the vast body of pre-existing knowledge. That is why scientists spend years of study before undertaking original research. Usually, then, experiments or observations lead to the identification of a new phenomenon or of additional data on something already thought to be understood. One or more hypotheses are advanced that enable us to understand the phenomenon or experimental result in ways that do not violate other observations or experiments. Finally, a hypothesis must be susceptible to testing by further observations or experiments. This last point is crucial; if there is no possible way of testing a hypothesis, it does not belong in the realm of science. We will give a number of examples of this process as we recount some of the history of astronomy in later chapters of this book.

The real power of science, of course, is its success. If science does not give us absolute truth, it nevertheless can provide considerable insight into the work-

FIGURE P.1 Astronomy is primarily an observational science, with telescopes located on high mountains (such as Mauna Kea in Hawaii, shown here) playing a key role in the advancement of the science. *(David Morrison)*

ings of nature. It has also made possible an explosive development of technology. In fact, people have become so used to technological ''miracles'' that many believe that science can do anything and that everything is possible. Such is not the case. It is important, therefore, also to learn at least some of the limitations of science and to gain some understanding of what is possible and what is not.

One aspect of science that sets it off from most human activities is its *self-correcting* nature. The scientist learns never to accept results on faith. An experiment can always be improved upon, or a new theory developed to explain existing data. Research scientists spend a great deal of their time questioning and criticizing one another. No research project is funded, and no report published, without extensive ''peer review,'' which is just a way of saying that it is subjected to the criticism of colleagues. The refusal to accept authority is the essence of science.

That is why science progresses. A college undergraduate today knows more of science and mathematics than did Isaac Newton, who was among the most brilliant scientists who ever lived. Einstein's theory of relativity, a concept at the very forefront of human knowledge 75 years ago, is now the common fare of first-year graduate students in physics. Although the domain of science is limited, within that domain its progress has been glorious.

The laws of nature provide the interconnected threads that knit together all of the phenomena that we see, from the emission of light by a single atom to the large-scale motions of the universe. We cannot imagine a world without such laws. Sometimes people think that these laws might occasionally be suspended, but there is no evidence to support such beliefs. Such a fairy-tale world would not allow more possibilities; indeed, without the laws of nature, chaos would prevail and nothing would be possible!

Numbers in Astronomy

Astronomy, despite its great popular appeal, is difficult for most nonscientists because its subject matter is remote and unfamiliar. Astronomy also has a great deal of jargon and involves a lot of new ideas. When you first encounter these, there is a danger of your getting bogged down with details. To help prevent this from happening, we begin with a brief overview of what the universe is like. You may want to return to this thumbnail sketch from time to time, to keep things in perspective.

Even this overview might require you to imagine distances on a scale you have never thought about before and numbers larger than any you have encountered. For now, do not worry about precise figures, but strive to keep clear the distinction between such quantities as 10, 1000, 1 million, 1 billion, and 10^{15} (the figure 1 followed by 15 zeros). Many Americans think of any big number as a million, and even such an august group as the recent (1990) Presidential Advisory Committee on the Future of the U.S. Space Program issued an official report in which it mentioned that Uranus was 1.7 million miles from the Earth, whereas the correct value is 1.7 billion miles. But there is an incredible difference between 1 million and 1 billion!

In this text we adopt two approaches that make dealing with numbers easier. First, we use a consistent set of units—the *international metric system*. Metric units have been defined in a way that simplifies arithmetic, since all of the measures are interrelated and many calculations involve only multiples of ten. In the metric system you will never have to deal with such messy problems as remembering how many feet to a mile (5280) or converting from an ounce of weight to an ounce of volume. The metric system, which has been adopted by every major country in the world except the United States, is summarized in Appendix 4.

The second way that we try to make numbers simple is to use the *powers-of-ten notation* (also called *scientific notation*) for very large and very small quantities. Thus a million is 10^6, a billion is 10^9, a millionth is 10^{-6}, and so on. Arithmetic is very easy with this notation, since to multiply, for example, we need only to add the exponents: Thus $10^3 \times 10^9 = 10^{12}$.

As an example of the powers-of-ten notation, consider another area besides astronomy where big numbers are common. The national debt of the United States is now about 1 trillion (10^{12}) dollars, and there are about 100 million (10^8) taxpayers. Then we can calculate that this debt amounts to $10^{12}/10^8 = 10^4 = $10,000$ per taxpayer. If the annual interest on the national debt is 10 percent ($0.1 = 10^{-1}$), then the interest paid each year per taxpayer is $10^{12} \times 10^{-1}/10^8$ $= 10^{11}/10^8 = 10^3 = \$1000$. Other examples are given in Appendix 3.

From the Earth to the Stars

Now let us look at the sizes of some familiar things in the universe. Our Earth (Figure P.2) is approximately spherical and has a diameter of nearly 13,000 kilometers (km) (8000 miles). Light, at the speed of 300,000 km per second (3×10^8 m/s), can travel seven times the Earth's circumference in 1 second. A traveler by commercial airplane takes about two days to go once around the Earth, and an astronaut in orbit can do the same thing in about 100 minutes.

Our nearest astronomical neighbor is the Moon. We can draw the Earth and Moon to scale on the same diagram (Figure P.3). The Moon's distance from the Earth is about 30 times the Earth's diameter, or about 384,000 km (239,000 miles). Light takes about 1.3 seconds to make the journey.

Whereas the Moon revolves once about the Earth each month, the Earth revolves once about the Sun each year (3×10^7 seconds). The Sun is about 150 million km away, or about 400 times farther than the Moon. Light requires about 8 minutes to come from the Sun. Since the Sun is nearly 1.5 million km in diameter, it is also about 100 solar diameters distant. We could, with care, draw the Sun and Earth to scale on one diagram, but then the Earth would be only a point and even the orbit of the Moon could not be shown to scale unless the diagram was drawn on a very large sheet of paper.

Jupiter, the largest planet, is about five times the Earth's distance from the Sun. We call the average distance from the Earth to the Sun the astronomical unit (AU); Jupiter's orbit is thus about 5 AU in radius. Jupiter has ten times the Earth's diameter but only one-tenth the Sun's diameter. Light travels to the Earth from the planets in periods ranging from a few minutes to a few hours.

The Sun is a typical star. It is about a thousand times more massive than Jupiter and generates its energy by nuclear reactions taking place in its core, where the temperature is millions of degrees. The nearest star beyond the Sun is about 300,000 AU

FIGURE P.2   The Earth is a planet, as becomes evident when we view it from a distant perspective. *(NASA)*

away, or about 7000 times as far from the Sun as is Neptune, currently the outermost planet. On a scale map showing the orbits of the planets about the Sun (Figure P.4), the stars cannot be shown. If the Earth's orbit were represented as only 1 centimeter (cm) in radius, Pluto's orbit would be about 40 centimeters (cm) in radius and the nearest star would have to be placed 300 m (0.3 km) away. On this scale, most of the hundred or so nearest stars would be 1 to 2 km away. Remember that light takes 8 minutes to come from the Sun. It takes about four years to come from the nearest star. The distances to stars are thus often expressed in light years (LY), the time light takes to come from them to us.

Suppose we make a rough scale drawing to show the stars within 10 LY of the Sun. In Figure P.5a, the circle represents a sphere 10 LY in radius centered on the Sun. Roughly ten stars are included. Now we change scale. In Figure P.5b, the sphere of 10 LY is the small center circle, and the larger circle repre-

FIGURE P.3   The Earth and Moon, drawn to scale.

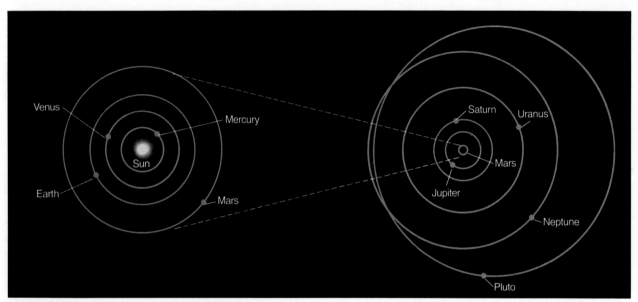

FIGURE P.4 The distances of the planets. (a) The inner planets. (b) The orbits of the outer planets (with a change of scale).

sents a sphere 100 LY in radius—ten times as large. In that sphere, we would have approximately 10,000 stars. Similarly, the sphere 100 LY in radius is the small circle in our next change of scale, Figure P.5c, and the larger circle represents a sphere 1000 LY in radius, within which there are 10 million (10^7) stars. In our next change of scale, Figure P.5d, the stars begin to thin out. The Sun is part of a wheel-shaped system

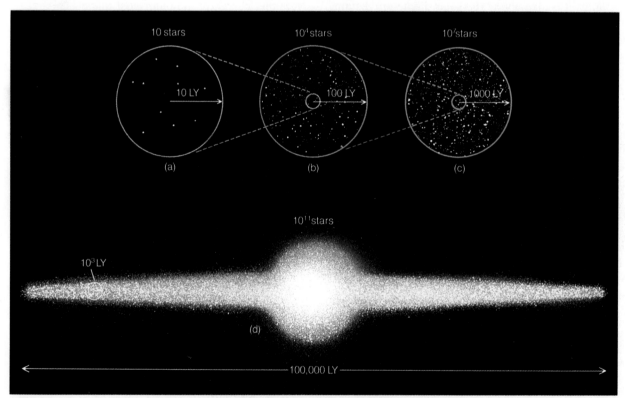

FIGURE P.5 The distribution of stars around the Sun within (a) 10 LY, (b) 100 LY, (c) 1000 LY, and (d) the Galaxy.

FIGURE P.6 The Sun is part of a wheel-shaped system of stars called the Galaxy. The galaxy shown here (called M83) is thought to be very like our own Galaxy. It is about 10 million LY distant. *(Cerro Tololo Inter-American Observatory, National Optical Astronomy Observatories [NOAO])*

Seen from outside, the Galaxy would probably look like the system shown in Figure P.6.

Perhaps many, possibly even a majority, of the stars in the Galaxy have planetary systems, as does the Sun. We do not know, for we cannot detect planets revolving about other stars—they are too insignificant to see, even with our greatest telescopes. When we look edge-on through our Galaxy, we find so many stars in our line of sight that the more remote ones give just a glow of light in a circular band around the sky—the Milky Way (Figure P.7). We do not see directly through the entire Galaxy to its far rim with ordinary light because the interstellar space is not completely empty. It contains a sparse distribution of gas (mostly hydrogen) intermixed with microscopic particles that we call interstellar dust. This material is so extremely sparse that interstellar space is a far, far better vacuum than any we can produce in terrestrial laboratories. Yet the dust, extending over thousands of light years, obscures the light from stars that lie far away in the Galaxy.

There are many double stars—pairs of stars revolving about each other—and even triple- and multiple-star systems. More than a thousand clusters of stars have been catalogued. These star clusters have anywhere from a few dozen to more than a hundred thousand member stars each and range in diameter

of stars (shown edge-on in Figure P.5d) at least 100,000 LY in diameter, containing more than 10^{11} stars, of which the Sun is typical. We call the system the Milky Way Galaxy or just simply our Galaxy.

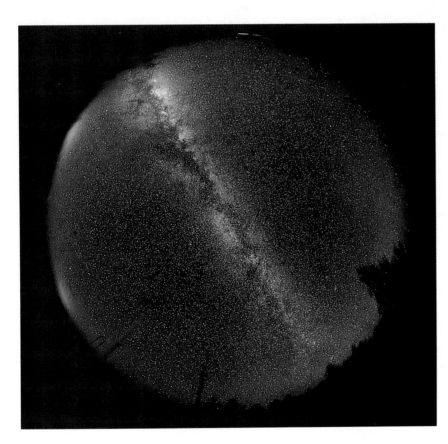

FIGURE P.7 The Milky Way Galaxy viewed from the inside, as seen in an all-sky photograph made at Mt. Graham in Arizona. *(University of Arizona)*

FIGURE P.8 Clusters of stars form within clouds of gas and dust. This photograph shows the Rosette Nebula. A cluster of hot, bright stars at the lower left has cleared away the gas and dust from the central region of the nebula. (Anglo-Australian Telescope)

from a few light years to several hundred light years (Figure P.8). The stars themselves evolve, sometimes ending their lives in cataclysmic explosions called supernovas. Nearly all of the elements that make the Earth and our own bodies were manufactured in the interiors of stars and then ejected into space by supernova explosions. We are literally made of "star dust."

The Universe on the Large Scale

Early in the 20th century, most astronomers thought our Galaxy constituted the entire universe. Now, however, we know that it is an insignificant part. Other comparable galaxies stretch as far in space as we can see—at least a billion (10^9) of them within the reach of our present telescopes. These galaxies tend to occur in clusters. Our own Galaxy is part of a small cluster of about 20 members extending over a region of space about 3 million LY in diameter; we call the cluster our Local Group. At 10 to 15 million LY from the Local Group are other small clusters, and at perhaps 60 million LY is the nearest rather important cluster, which contains at least 1000 member galaxies. Some 300 to 400 million LY away we find the first really great cluster, which has at least 10,000 galaxies as members. (The precise distances to remote clusters are still very uncertain, as we shall see later, so the possible range of distances is given here.)

At distances that are probably several times greater still, we find the quasars. Quasars are bright regions in the centers of galaxies otherwise too remote to see. The quasars emit enough light that we can detect them as faint points of light, even though many may have distances as great as 10 billion (10^{10}) LY.

Still beyond the most remote quasars lies the farthest part of the universe we can detect. We can see it only by means of feeble radio waves coming from all directions in space. Because light does not travel at an infinitely great speed, radiation arriving from great distances must have originated in the past. We see the remote quasars as they were when light left them billions of years ago. Still earlier, the universe itself was born in a gigantic fireball. The feeble radio radiation from the remote past is thought to be all that we can now observe of that primeval fireball that began the universe as we know it today.

The universe is thought to be infinite in extent, but we can see only a finite part of it, extending out to a distance of between 10 and 15 billion LY. If this observable universe were reduced to the size of the Earth, the billions of individual galaxies in it would be about the size of small villages, each separated from its neighbor villages by a kilometer or two. Individual stars on this scale are no larger than a single atom. Thus as an atom is to the Earth in size, a star is to the observable universe.

An Inner View of the Universe

The foregoing discussion should impress on you that the universe is extraordinarily large and extraordinarily empty. The universe, on the average, is 10,000 times as empty as (that is, less dense than) our own Galaxy. Yet, as we have seen, the Galaxy is mostly empty space. Recall the interstellar gas in our Galaxy; typically, there is about one atom of hydrogen in each cubic centimeter of space. In contrast, the air that we breathe has about 10^{19} times as much material per cubic centimeter, and we think of the air as quite empty stuff. Solid matter, such as our own bodies or this page, is tremendously dense in comparison.

Yet even the familiar solids are mostly space. If we could take such a solid apart, piece by piece, we would eventually reach the molecules of which it is

FIGURE P.9   A cluster of galaxies in Hercules. *(Palomar Observatory, Caltech)*

formed. Molecules are the smallest particles that matter can be divided into while still retaining its chemical properties. They are in turn composed of atoms, the building blocks of all matter. Nearly 100 different kinds of atoms (elements) exist in nature, but most of them are rare and only a handful account for more than 99 percent of everything with which we come in contact. The most cosmically abundant elements are listed in Table P.1.

All atoms consist of a central, positively charged nucleus surrounded by negatively charged electrons. The bulk of the matter of an atom is in the nucleus, which consists of a certain number of positive protons and a roughly equal number of electrically neutral neutrons all tightly bound together. In its regular condition, an atom has as many electrons around the nucleus as protons in the nucleus. Different kinds of atoms, responsible for the different kinds of chemical elements, are distinguished by the number of protons in their nuclei (or electrons outside). Thus the simplest kind of atom, hydrogen, has one proton and one electron. Helium has two of each, oxygen has eight of each, and so on.

Conclusion

The subject of this book is the astronomical universe—nature's grandest laboratory, where we can ask and sometimes partially answer the most profound questions about the laws of nature and the origin and evolution of galaxies, stars, planets, and ourselves. This book is organized as follows. The first

TABLE P.1   The Cosmically Abundant Elements

Element	Symbol	Number of Atoms per Million Hydrogen Atoms
Hydrogen	H	1,000,000
Helium	He	80,000
Carbon	C	450
Nitrogen	N	92
Oxygen	O	740
Neon	Ne	130
Magnesium	Mg	40
Silicon	Si	37
Sulfur	S	19
Iron	Fe	32

six chapters deal with a variety of topics, from the theory of gravitation to the nature of light and the construction of astronomical observatories. This material provides background for the chapters that follow. The next six chapters (Chapters 7 to 12) are devoted to the planets and smaller bodies in the solar system. Stellar astronomy (including the nearest star, our Sun) is the subject of Chapters 13 through 23, with emphasis on the evolution of stars, from formation through their sometimes catastrophic deaths. Finally, Chapters 24 through 28 take as their subject galaxies and the large-scale structure of the universe, as well as its origin and ultimate fate.

As the Earth rotates, the stars appear to move across the sky, as shown in this 10-hour exposure. The dome houses the Anglo-Australian Telescope. *(Copyright Anglo-Australian Telescope Board)*

Nicolaus Copernicus (1473–1543), cleric and scientist, played a leading role in the emergence of modern science. While he could not prove that the Earth revolves about the Sun, he presented such compelling arguments for this idea that he turned the tide of cosmological thought, laying the foundations upon which Galileo and Kepler so effectively built in the following century.

THE CELESTIAL CLOCKWORK

In antiquity it was discovered that there is a majestic regularity in the motions of the heavens. The Babylonians and Greeks, believing the planets to be gods, studied their motions in hopes of better understanding the presumed influences of those planet-gods on human affairs; thus developed the religion of astrology. At the same time, however, the ancient Greeks and Romans were laying the foundations of science. In astronomy, their crowning achievement was the cosmological scheme of Claudius Ptolemy in the second century. Ptolemy's system predicted the positions of planets with good precision for hundreds of years, and it was not substantially improved until the European Renaissance.

In the first two chapters of this text we reconstruct some of the history of our ancestors' ideas about the heavens. We will see how the great astronomers of the Renaissance—Copernicus, Tycho, Kepler, and Galileo—struggled to develop a better understanding of the skies based on experimental methods and new mathematical techniques. Their work led to the 17th-century synthesis of physics and astronomy—of things on Earth with things celestial—achieved by Isaac Newton. He showed that the force that makes apples fall to the ground and the force that makes the planets fall into perpetual orbits about the Sun are different manifestations of the same thing: gravitation.

Newton also developed, by his example, an approach to the study of nature that still guides us. This approach is based on the conviction that nature is rational and consistent and is governed by immutable laws. Further, it embodies a means with which to try to learn these laws, through systematic observation and experiment. Instead of accepting the traditions of superstition and supposed revelation, or exalting the powers of pure human reason as had many ancient philosophers, Newton and his successors found answers to their questions in the everyday world around them.

1.1 The Sky Above

Our senses suggest to us that the Earth is the center of the universe—the hub around which the heavens turn. This **geocentric** view was held almost universally until the European Renaissance. It is simple, logical, and seemingly self-evident. Further, the geocentric perspective reinforces philosophical and religious systems that teach the unique role of humans as the central focus of the cosmos. However, the geocentric view happens to be wrong. One of the great themes of intellectual history is the overthrow of the geocentric perspective and the successive steps by which we have re-evaluated the place of our world in the cosmic order.

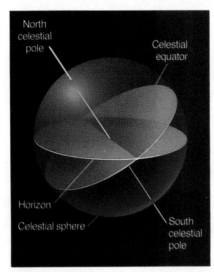

FIGURE 1.1   The celestial sphere showing celestial poles, celestial equator, and horizon.

The Celestial Sphere

Our study of astronomy begins with the view of the heavens above us, a view identical to that available to peoples all over the world before the invention of the telescope. If we look up on a clear night, we get the impression that the sky is a great hollow spherical shell with the Earth at the center. The early Greeks regarded the sky as just such a **celestial sphere** (Figure 1.1). Some apparently thought of it as an actual sphere of crystalline material, with the stars embedded in it like tiny jewels.

The Sun, Moon, and stars rise and set as the Earth turns within this imaginary sphere. In ancient times, of course, people did not realize that the Earth was a planet, spinning about its axis. Their concept of the Earth was restricted to the apparently flat world that they could see with their own eyes. It is easy to recreate this viewpoint just by finding a dark, quiet spot from which to look at the stars above. If we watch the sky for several hours, we see that the celestial sphere appears gradually to turn around us. Stars rise and set, moving completely across the vault of heaven in the course of the night. The ancients, unaware of the Earth's rotation, imagined that the celestial sphere rotated about an axis that passed through the Earth. As it turned, the celestial sphere carried the stars up in the east, across the sky, and down in the west.

Following along with us must be our **horizon,** that line in the distance at which the ground seems to dip out of sight, providing a demarcation between Earth and sky. (The horizon may, of course, be hidden from view by mountains, trees, buildings, or, in large cities, smog.) As our horizon tips down in the direction that the Earth's rotation carries us, stars hitherto hid-

den beyond it appear to rise. In the opposite direction the horizon tips up, and stars hitherto visible appear to set. Analogously, as we round a curve in a mountain road, new scenery comes into view while old scenery disappears behind us. The direction around the sky toward which the Earth's rotation carries us is (by definition) east; the opposite direction is west. The point on the celestial sphere that is directly overhead is called the **zenith.**

As the celestial sphere rotates, all of the objects in the sky maintain their positions with respect to one another. A grouping of stars like the Big Dipper has the same shape wherever we see it in the sky, although its apparent orientation with respect to the terrestrial foreground shifts during the night. Even objects that we know have their own motions, such as planets or the Moon, seem fixed relative to the stars over the period of a single night. Only the meteors—brief ''shooting stars'' that flash into view for just a few seconds—move appreciably with respect to the celestial sphere, and they are located within the atmosphere of the Earth.

Celestial Poles and Celestial Equator

The pole or point about which the celestial sphere appears to pivot lies along an extension of the line through the Earth's North and South Poles. As the Earth rotates about its polar axis, the sky appears to turn in the opposite direction about those **north** and **south celestial poles** (Figure 1.2). Halfway between them, and separating the sky into its northern and southern halves, is the **celestial equator**—just like the Earth's equator, which separates the Northern and Southern Hemispheres of our planet.

The apparent motion of the celestial sphere depends on the latitude of the observer. Someone at the North Pole of the Earth sees the north celestial pole directly overhead, at the position called the zenith. The stars would all circle about the sky parallel to the horizon, with none rising or setting. An observer at the Earth's equator, on the other hand, sees the celestial poles at the north and south points on the horizon. As the sky turns about these points, all the stars rise straight up from the eastern horizon and set straight down toward the west. For an observer in Europe or the United States, the north celestial pole appears above the northern horizon at an angular height, or altitude, equal to that observer's latitude (Figure 1.3).

Rising and Setting of the Sun

We have described the appearance of the night sky. The situation during the day is similar, except that the brilliance of the Sun renders the stars and planets

FIGURE 1.2 Time exposure showing trails left by stars as a consequence of the apparent rotation of the celestial sphere. The bright trail at top center was made by Polaris (the North Star), which is about 1° away from the north celestial pole. *(National Optical Astronomy Observatories)*

invisible. (The Moon is still easily seen in the daylight, however.) We can think of the Sun as being located at a position on the hypothetical celestial sphere. When the Sun rises—that is, when the rotation of the Earth carries the Sun above the horizon—sunlight scattered about by the molecules of the atmosphere produces the blue sky that hides the stars that are also above the horizon.

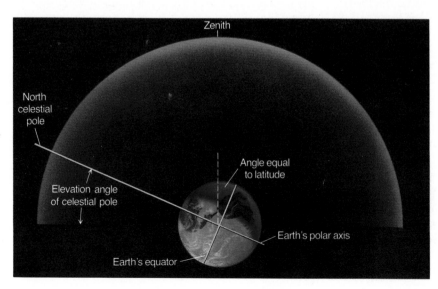

FIGURE 1.3 The altitude of the celestial pole is equal to the latitude of the observer.

ESSAY Observing The Planets

At almost any time of the night, and at any season, you can spot one or more bright planets visible in the sky. All five of the planets known to the ancients—Mercury, Venus, Mars, Jupiter, and Saturn—are more prominent than any but the brightest of the fixed stars, and they can be seen even from urban locations if you know where and when to look.

Venus, which appears either as an evening "star" in the west after sunset or as a morning "star" in the east before sunrise, is the brightest object in the sky after the Sun and Moon. It far outshines any real star, and under the most favorable circumstances it can even cast a visible shadow. Mars, with its distinctive red color, can be nearly as bright as Venus when it is close to the Earth, but normally it remains much less conspicuous. Jupiter is most often the second brightest planet, approximately equaling in brilliance the brightest of the stars. Saturn is dimmer, and it varies considerably in brightness, depending on whether its rings are seen nearly edge-on (faint) or more widely opened (bright). Finally, Mercury is quite bright, but few people ever notice it because it never moves very far from the Sun and is always seen against bright twilight skies. There is a story (probably apocryphal) that even the great Copernicus never saw the planet Mercury.

True to their name, the planets "wander" against the background of the fixed stars. Although their apparent motions are complex, they reflect an underlying order, which was the basis for the development of the heliocentric model of the solar system.

The two inner planets, Mercury and Venus, never appear far from the Sun, since their orbits lie inside the orbit of the Earth. Venus has the larger orbit, and it can achieve an angular separation from the Sun of about 45°. Mercury, with its smaller orbit, never ventures more than 28° from the Sun. As seen from the Earth, both planets appear to move back and forth with respect to the Sun, appearing alternately in the evening and morning sky.

The apparent motion of planets with orbits larger than that of the Earth is more complex. Since all the planets revolve about the Sun in the same direction (east to west), these exterior planets normally appear to move from east to west against the stellar background. To an observer on the Sun, they would always move from east to west. But from the moving Earth we get a somewhat different perspective. The Earth is also moving east to west, and at a faster pace, since it is nearer the Sun. Therefore we pass each planet from time to time, and when we do, the planet appears to move in a backward direction, from west to east. The effect is similar to that of overtaking a slower car on the freeway; as you pass the other car, it seems to move backward with respect to the distant landscape.

Today we understand how the complex apparent motion of each planet is a reflection of the relative motion of the Earth and the planet, each moving in its orbit about the Sun. To ancient peoples, however, these phenomena remained mysterious, and they were not really understood in detail until the time of Kepler, early in the 17th century.

For thousands of years, astronomers have been aware that the Sun gradually changes its position, moving each day about 1° to the east relative to the stars. The Sun's apparent path around the celestial sphere, which reflects the revolution of the Earth about the Sun, is called the **ecliptic.** Each day the Sun rises about 4 minutes later with respect to the stars; the Earth must make just a bit more than one complete rotation (with respect to the stars) to bring the Sun up again.

As we look at the Sun from different places in our orbit, we see it projected against different stars in the background, or we would, at least, if we could see the stars in the daytime. In practice, we must deduce what stars lie behind and beyond the Sun by observing the stars visible in the opposite direction at night. After a year, when the Earth has completed one trip around the Sun, the Sun will appear to have completed one circuit of the sky along the ecliptic. We have a similar experience if we walk around a campfire at night; we see the flames appear successively in front of each person seated about the fire.

The *year* is defined as the time required for the Sun to complete a full circuit of the celestial sphere. Of course, it is not really the Sun that moves; it is the Earth that revolves about the Sun, and an equivalent definition of the year is the period for the Earth to make one circuit about its orbit.

It was also known by the ancients that the ecliptic does not lie along the celestial equator but is inclined to it at an angle of about 23°. This inclination of the ecliptic explains why the Sun moves north and south in the sky as the seasons change. This angle is the same as the tilt of the Earth's axis of rotation, and it was measured with surprising accuracy by several ancient observers. In Chapter 3, we will discuss the progression of the seasons in more detail.

Fixed and Wandering Stars

The Sun is not the only object that moves among the fixed stars. The Moon and each of the five planets visible to the unaided eye—Mercury, Venus, Mars, Jupiter, and Saturn—also slowly change their positions from day to day. The Moon, being the Earth's nearest celestial neighbor, has the fastest apparent motion; it completes a trip around the sky in about one month. During a single day, of course, the Moon and planets all rise and set, as do the Sun and stars. But like the Sun, they have independent motions among the stars, superimposed on the daily rotation of the celestial sphere.

The Greeks of 2000 years ago distinguished between what they called the fixed stars, the stars that maintain fixed patterns among themselves throughout many generations, and the wandering stars or planets. The word planet means "wanderer" in Greek. Today, we do not regard the Sun and Moon as planets, but the ancients applied the term to all seven of the moving objects in the sky. Much of ancient astronomy was devoted to observing and predicting their motions. In the Romance languages the planets give the names for the seven days of our week, although modern English retains only Sunday (Sun), Monday (Moon), and Saturday (Saturn).

The individual paths of the Moon and planets in the sky all lie close to the ecliptic, although not exactly on it. The reason is that the paths of the planets about the Sun, and of the Moon about the Earth, are all in nearly the same plane, as if they were marbles rolling about on the top of a table. The planets and Moon are always found in the sky within a narrow belt 18° wide that is centered on the ecliptic, called the **zodiac.** The apparent motions of the planets in the sky result from a combination of their actual motions and the motion of the Earth about the Sun, and consequently they are somewhat complex. This very complexity fascinated and challenged ancient astronomers.

The famous philosopher Pythagoras (who lived 2500 years ago) pictured a series of concentric spheres in which each of the seven moving objects—the planets, the Sun, and the Moon—was carried by a sphere separate from the one that carried the stars, so that the motions of the planets resulted from independent rotations of the different spheres about the Earth. The friction between them gave rise to harmonious sounds, the music of the spheres, which only the most gifted ear could hear.

Constellations

The backdrop for the motions of the "wanderers" in the sky is the canopy of stars themselves. Like the Chinese and the Egyptians, the Greeks divided the sky into **constellations,** apparent configurations of stars (Figure 1.4). Modern astronomers still make use of these constellations to denote approximate locations in the sky, much as geographers use political areas to denote the locations of places on the Earth. The modern boundaries between the constellations are imaginary lines in the sky running north-south and east-west, so that each point in the sky falls in a specific constellation. The constellations are listed in Appendix 19.

Many of the 88 recognized constellations are of Greek origin and bear names that are Latin translations of those given them by the Greeks. Today, people are often puzzled because the constellations seldom resemble the people or animals for which they

FIGURE 1.4 The winter constellation of Orion, the hunter, as illustrated in the 17th-century atlas by Hevelius. *(J.M. Pasachoff and the Chapin Library)*

were named. In all likelihood, the Greeks themselves did not name groupings of stars because they looked like actual people or objects (any more than the outline of Washington State resembles George Washington). Rather, they named sections of the sky in honor of the characters in their mythology and then fitted the configurations of stars to the animals and people as best they could.

1.2   Greek and Roman Science

Let us now look briefly back into history. A great deal of modern Western civilization is derived in one way or another from the civilization of the ancient Greeks and Romans. This is true in astronomy as well, although most of the ideas have been superseded during the past three centuries.

Early Cosmology

Our concept of the cosmos—its basic structure and origin—is called **cosmology** (a Greek word). Before the invention of telescopes, humans had to depend on the simple evidence of their senses for a picture of the universe. The Greeks and Romans developed cosmologies that combined their direct view of the heavens with a rich variety of philosophical and religious symbolism.

At least 2000 years before Columbus, educated people in the eastern Mediterranean region knew the Earth was round. Belief in a spherical Earth may have stemmed from the time of Pythagoras. Even then, it was realized that the Moon shines by reflected sunlight, and that the curved shape of the line between the Moon's illuminated and dark portions showed that the Moon must be a sphere. The sphericity of the Earth might have seemed to follow by analogy.

The writings of Aristotle, the tutor of Alexander the Great, describe how the progression of the Moon's phases—its changing apparent shape—results from our seeing different portions of the Moon's illuminated hemisphere during the month. Only half of the spherical Moon is lit by the Sun at any time, and the apparent shape of the Moon in the sky depends simply on how much of that sunlit side we can see. Aristotle also knew that the Sun has to be more distant from the Earth than the Moon is because occasionally the Moon passes exactly between the Earth and Sun and temporarily hides the Sun from view (in a solar eclipse).

Aristotle also cited two convincing arguments that the Earth is round. First is the fact that during a lunar eclipse, as the Moon enters or emerges from the Earth's shadow, the shape of the shadow seen on the Moon is always round (Figure 1.5). Only a spherical object always produces a round shadow. If the Earth were a disk, for example, there would be some occasions when the sunlight would be striking the disk edge on, and its shadow on the Moon would be a line.

As a second argument, Aristotle explained that southbound travelers observe hitherto invisible stars to appear above the southern horizon. In the northern sky, the height of the Pole Star (North Star) decreases as a traveler moves south. The only possible explanation is that the travelers' horizons have tipped to the south, which indicates that they must have moved over a curved surface of the Earth.

Aristotle and most of the ancient Greek scholars rejected the idea that the Earth might itself be in motion. One of the reasons for their conclusion was the following: If the Earth moved about the Sun, we would be observing the stars from successively different places along our orbit, and their apparent directions in the sky would then change continually during the year. In a similar way, we see foreground objects appear to move against a more distant background whenever we move. Unconsciously, we use this phenomenon all of the time to estimate distances around us.

The apparent shift in the direction of an object as a result of a motion of the observer is called **parallax.** An annual shifting in the apparent directions of the stars that results from the Earth's orbital motion is called stellar parallax. For the nearer stars this shift is observable with modern telescopes (see Chapter 16), but it is impossible to measure with the unaided eye because of the great distances of even the nearest stars.

Measurement of the Earth by Eratosthenes

The first fairly accurate determination of the Earth's diameter was made about 200 B.C. by Eratosthenes, a Greek living in Alexandria in Egypt. His method was a geometrical one, based on observations of the Sun. The Sun is so distant from the Earth compared with its size that the Sun's light rays intercepted by all parts of the Earth approach it along essentially parallel lines. Imagine a light source near the Earth, say at position *A* in Figure 1.6. Its rays strike different parts of the Earth along diverging paths. From a light source at *B,* or at *C,* still farther away, the angle between rays that strike extreme parts of the Earth is smaller. The more distant the source, the smaller is the angle between the rays. For a source infinitely distant, the rays travel along parallel lines. The Sun is not, of course, infinitely far away, but light rays strik-

9:04 P.M.

11:32 P.M.

9:38 P.M.

12:04 A.M.

10:01 P.M.

12:23 A.M.

FIGURE 1.5 A lunar eclipse, with the Moon moving into the Earth's shadow. Note the curved shape of the shadow—evidence for a spherical Earth that has been recognized since antiquity. *(Yerkes Observatory)*

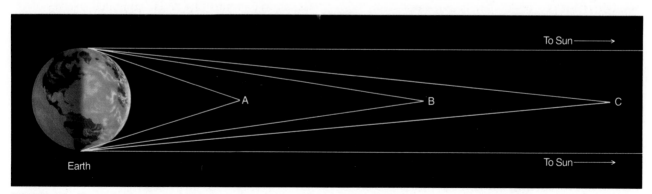

FIGURE 1.6 The more distant an object, the more nearly parallel are the rays of light coming from it.

ing the Earth from a point on the Sun diverge from one another by an angle far too small to be observed with the unaided eye. As a consequence, if people all over the Earth who could see the Sun were to point at it, their fingers would all be essentially parallel to one another.

Eratosthenes noticed that at Syene, Egypt (near modern Aswan), on the first day of summer, sunlight struck the bottom of a vertical well at noon, which indicated that Syene was on a direct line from the center of the Earth to the Sun. At the corresponding time and date in Alexandria, he observed that the Sun was not directly overhead but slightly south of the zenith, so that its rays made an angle with the vertical equal to 1/50 of a circle (about 7°). Since the Sun's rays striking the two cities are parallel to one another, Alexandria must be 1/50 of the Earth's circumference north of Syene (Figure 1.7). Alexandria had been measured to be 5000 stadia north of Syene (the stadium was a Greek unit of length; in another context, a length of one stadium, laid out as a racecourse, gave us our word for a place where games and competitions are held). Eratosthenes thus found that the Earth's circumference must be 50 × 5000, or 250,000 stadia.

It is not possible to evaluate precisely the accuracy of Eratosthenes' solution because there is doubt about which of the various kinds of Greek stadia he used. If it was the common Olympic stadium, his result was about 20 percent too large. According to another interpretation, he used a stadium equal to about 1/6 km, in which case his figure was within 1

percent of the correct value of 40,000 km. The diameter of the Earth is found from the circumference by dividing the latter by π (about 3.14).

Hipparchus

The greatest astronomer of pre-Christian antiquity was Hipparchus, who was born in Nicaea in Bithynia, in present-day Turkey. He erected an observatory on the island of Rhodes in the period around 150 B.C., when the Roman Republic was increasing its influence throughout the Mediterranean region. Here he measured as accurately as possible the directions of objects in the sky, compiling a star catalogue of about 850 entries. He designated for each star its celestial coordinates, that is, quantities analogous to latitude and longitude that specify its position (direction) in the sky. He also divided the stars according to their apparent brightness into six categories, or **magnitudes,** and specified the magnitude of each star.

In the course of his observations of the stars, and in comparing his data with older observations, he made one of his most remarkable discoveries: The position in the sky of the north celestial pole had altered over the previous century and a half. Hipparchus correctly deduced that the direction of the axis about which the celestial sphere appears to rotate continually changes. The real explanation for the phenomenon is that the direction of the Earth's rotational axis changes slowly because of the gravitational influence of the Moon and the Sun, much as a spinning top's axis describes a conical path as the Earth's gravitation tries to tumble the top over. This variation of the Earth's axis, called **precession,** requires about 26,000 years for one cycle (Figure 1.8).

Ptolemy

The last great astronomer of antiquity was Claudius Ptolemy (or Ptolemaeus), who flourished in Roman Alexandria about 140 A.D. He compiled a series of 13 volumes on astronomy known as the *Almagest*. The Almagest does not deal exclusively with Ptolemy's own work, for it includes a compilation of the astronomical achievements of the past, principally those of Hipparchus. In fact, it is our main source of information about Greek astronomy.

Ptolemy's most important contribution was a geometrical representation of the solar system that predicted the motions of the planets with considerable accuracy. Hipparchus, not having enough data on hand to solve the problem himself, instead amassed observational material for posterity to use. Ptolemy

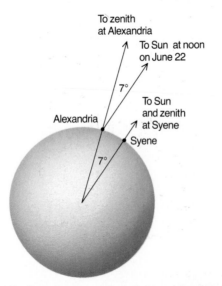

FIGURE 1.7 Eratosthenes' method of determining the size of the Earth.

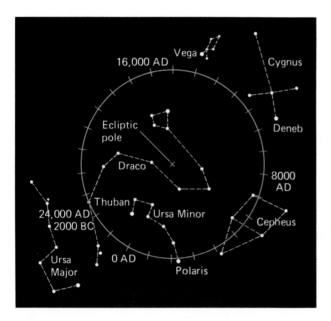

FIGURE 1.8 Path of the north celestial pole among the stars as caused by precession.

supplemented this material with new observations of his own and produced a cosmological model that endured more than a thousand years, until the time of Copernicus.

The complicating factor in the analysis of the planetary motions is that their apparent wanderings in the sky result from the combination of their own motions and the Earth's orbital revolution. In Figure 1.9a, notice the orbit of the Earth and the orbit of a hypothetical planet farther from the Sun than the Earth. The Earth travels around the Sun in the same direction as the planet and in nearly the same plane but has

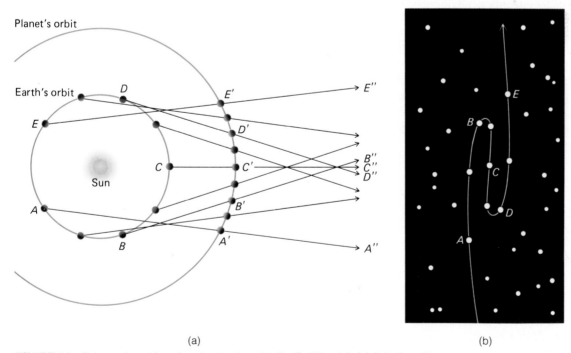

(a)

(b)

FIGURE 1.9 Retrograde motion of a planet external to the Earth's orbit. (a) Actual positions of the planet and the Earth. (b) The apparent path of the planet as seen from the moving Earth, against the background of stars.

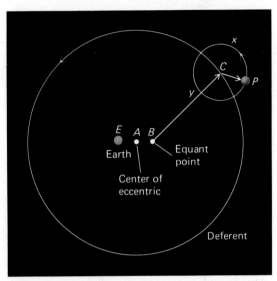

FIGURE 1.10 Ptolemy's cosmological system based on geocentric motion; the diagram shows the deferent, epicycle, eccentric, and equant.

a higher orbital speed. Consequently, it periodically overtakes the planet, like a faster race car on the inside track. The apparent directions of the planet, seen from the Earth, are shown at successive intervals of time along lines $A\,A'\,A''$, $B\,B'\,B''$, and so on. In Figure 1.9b, we see the resulting apparent path of the planet among the stars. From positions B to D, as the Earth passes the planet, it appears to drift backward, to the west in the sky, even though it is actually moving to the east. Similarly, a slowly moving car appears to drift backward with respect to the distant scenery when we pass it in a faster moving car.

As the Earth rounds its orbit toward position E, the planet again takes up its apparent eastward motion in the sky. The temporary apparent westward motion of a planet as the Earth swings between it and the Sun is called **retrograde motion.** Retrograde motion is easy to understand in the context of a moving Earth. But Ptolemy required a different explanation for retrograde motion, since he assumed that each planet is revolving about a stationary Earth.

Ptolemy solved the problem by having a planet P (Figure 1.10) revolve in a small orbit, called an **epicycle,** about C. The center of the epicycle C in turn revolved about the Earth. When the planet is at position x, it is moving in its epicycle orbit in the same direction as the movement of point C about the Earth, and the planet appears to be moving eastward. When the planet is at y, however, its epicyclic motion is in the opposite direction to the motion of C. By choosing the right combination of speeds and distances, Ptolemy succeeded in having the planet moving westward at the right speed at y and for the

correct interval of time. However, we shall see in the next chapter that the planets, like the Earth, travel about the Sun in orbits that are ellipses, not circles. Their actual behavior cannot be represented accurately by a scheme of uniform circular motions. In order to match the observed motions of the planets, Ptolemy had to introduce additional circles in his model, considerably complicating his scheme.

It is a tribute to the genius of Ptolemy as a mathematician that he was able to develop such a complex system to account successfully for the observations. His hypothesis, with some modifications, was accepted as absolute authority throughout the Middle Ages. Perhaps Ptolemy did not intend his cosmological model to describe reality, but rather to serve as a mathematical representation to predict the positions of the planets at any time.

1.3 Astrology: An Ancient Religion

The modern science of *astronomy* and the religion of *astrology* are unrelated, yet many people still confuse the two. For this reason, it is appropriate in an astronomy text to digress for a brief discussion of astrology.

Beginnings of Astrology

Modern research has shown that all matter in the universe is composed of atoms—and the same kinds of atoms. Our probes that landed on Mars and Venus and our telescopic studies of the light from the most remote galaxies indicate that these planets and galaxies are made of the same elements that make up our own bodies. It is a great discovery, for it suggests a beautiful unity of the universe.

Still, we cannot fault the ancients for assuming that the luminous orbs in the sky—the stars and planets—are made of "heavenly" substances and not of the "earthly" elements we find at home. The realization that celestial objects are actually worlds and not ethereal substances is relatively recent in the history of science. Small wonder, then, that the ancients regarded the planets (including the Sun and Moon), which alone moved about among the stars on the celestial sphere, as having special significance. The planets came to be associated with the gods of ancient mythologies; in some cases, they were themselves thought of as gods. Even in the comparatively sophisticated Greece of antiquity, the planets had the names of gods and were credited with having the same powers and influences as the gods whose names they bore. From such ideas grew the religion of astrology.

Astrology began in Babylonia a millennium or so before the Christian era. The Babylonians, believing that the planets and their motions influenced the fortunes of kings and nations, used their knowledge of astronomy to support their rulers. When the Babylonian culture was absorbed by the Greeks, astrology gradually influenced the entire Western world and eventually spread to the Orient as well.

By the second century B.C., the Greeks democratized astrology by developing the idea that the planets influence every individual. In particular, they believed that the configuration of the planets at the moment of a person's birth affected his or her personality and fortune, the doctrine called natal astrology. Natal astrology reached its acme with Ptolemy 400 years later. Ptolemy, as famous for his astrology as for his astronomy, compiled the *Tetrabiblos*, a four-volume treatise on astrology that remains the "bible" of the subject. It is this ancient religion, older than Christianity or Islam, that is practiced by today's astrologers.

The Horoscope

The key to natal astrology is the **horoscope,** a chart that shows the positions of the planets in the sky at the moment of an individual's birth. In charting a horoscope, the planets (including the Sun and Moon, classed as planets by the ancients) are located in the zodiac. For the purposes of astrology, the zodiac is divided into 12 sectors called signs, each 30° long. The signs have their origin at the place on the ecliptic where the Sun, in its annual journey about the sky, crosses from the south half to the north half of the sky at the beginning of spring (about March 21). The north and south halves of the celestial sphere are separated by the celestial equator, halfway between the north and south celestial poles, and the place where the Sun crosses it on the first day of spring is called the **vernal equinox.**

The first zodiacal sign is the 30° sector of the zodiac, centered on the ecliptic, immediately to the east of the vernal equinox; that first sign is called Aries, so the vernal equinox is also known as the first point of Aries. The subsequent 11 signs are, in order to the east (the direction of the Sun's annual motion), Taurus, Gemini, Cancer, Leo, Virgo, Libra, Scorpio, Sagittarius, Capricorn, Aquarius, and Pisces. The signs of the zodiac are named for the constellations that occupied the same parts of the sky in antiquity. Because of precession, the equinox and hence the signs of the zodiac slide westward along the ecliptic, going once around the sky in about 26,000 years. Thus today, after some 2000 years, the signs and constellations are out of step by about one sign, and the sign of Aries now occupies the constellation of Pisces.

A horoscope shows the position of each planet in the sky by indicating its position in the appropriate sign of the zodiac. However, as the celestial sphere turns (owing to the rotation of the Earth), the vernal equinox, and the entire zodiac with it, moves across the sky to the west, completing a circuit of the heavens each day. Thus the position in the sky (or "house" in astrology) must also be calculated. A complete horoscope is usually represented by a circle denoting the center of the zodiac (the ecliptic), with the 12 houses indicated as sectors inside the circle together with the positions of the seven planets. Figure 1.11 shows the horoscope of author Abell.

There are more or less standardized rules for the interpretation of the horoscope, most of which (at least in Western schools of astrology) are derived from the *Tetrabiblos* of Ptolemy. Each sign, each house, and each planet, the last supposedly acting as a center of force, is associated with particular matters. The detailed interpretation of a horoscope is a very complicated business, and although the rules may be standardized, how each rule is to be weighed and applied is a matter of judgment—and "art." It also means that it is very difficult to tie astrology down to specific predictions.

Astrology Today

Astrologers today use the same basic principles laid down by Ptolemy nearly 2000 years ago. They cast horoscopes (a process much simplified by the development of appropriate computer programs) and suggest interpretations. A modern variant of natal astrology is sun-sign astrology, which uses only one element of the horoscope, the sign occupied by the Sun at a person's birth. Although even professional astrologers do not place much trust in such a limited scheme, which tries to fit everyone into just 12 groups, sun-sign astrology is the mainstay of newspaper astrology columns and party games, and apparently many people take it quite seriously. In a recent poll of teenagers in the United States, more than half said they "believed in astrology."

Today, with our knowledge of the nature of the planets as physical bodies, as well as our understanding of human genetics, it is hard to imagine that the directions of these planets in the sky at the moment of one's birth could have anything to do with one's personality or future. There are no known forces, not gravity or anything else, that could cause such effects. Astrologers have to argue that there are un-

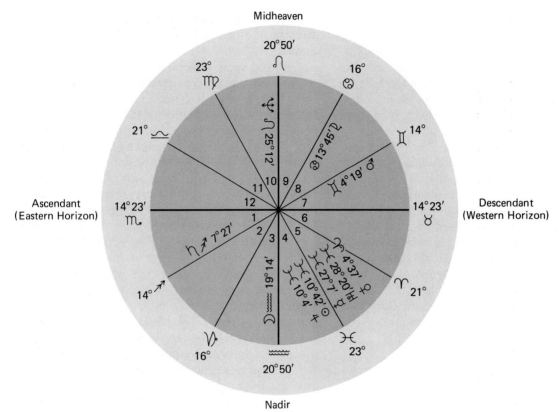

FIGURE 1.11   Natal horoscope of George O. Abell, who was born in Los Angeles, California, on March 1, 1927, at 10:50 P.M., PST. The 12 pie-shaped sectors represent the 12 houses, and the outer circular zone represents the zodiac. The definition of houses used in preparing this horoscope is that of Placedus, in which, as the rotating celestial sphere carries the planets around the sky, each place in the zodiac spends equal time in each of the six houses above the horizon (diurnal houses) and also in the six houses below the horizon (nocturnal houses). However, the time required for an object to pass through a diurnal house is not the same as that required for it to pass through a nocturnal house except for objects on the celestial equator. The boundaries between the houses (cusps) intersect the ecliptic in the zodiacal signs indicated by their symbols in the outer circular zone. The number beside each sign symbol is the angular distance of the cusp from the beginning of that sign. The position of each planet is shown in the house it occupied at the instant of birth. Beside the symbol for the planet is the symbol of the zodiacal sign it was also in at that time, and the angular distance of the planet from the beginning of that sign. The places where the horizon intersects the zodiac are shown, as well as the highest point of the ecliptic in the sky (midheaven) and its lowest point below the horizon (astrological nadir).

known forces exerted by the planets that depend on their configurations with respect to one another and with respect to arbitrary coordinate systems—forces for which there is not a whit of solid evidence.

Another curious aspect of astrology is its emphasis on the configurations of the planets at *birth*. What about the forces that might influence us at conception? Isn't our genetic makeup more important for determining our personality than the circumstances of our birth? Would we really be a different person if we had been born a few hours earlier or later, as astrology claims?

Actually, very few thinking people today would buy the claim that our entire lives are predetermined by astrological influences at our birth. But many people apparently believe that astrology has validity as an indicator of affinities and personality. A surprising number of Americans make judgments about people—who they will hire, associate with, and even marry—on the basis of astrological information. However, such claims can be tested.

During the past few years, a number of careful statistical tests have been carried out to assess astrology's predictive power. The simplest of these examine sun-sign astrology to determine whether some signs are more likely than others to be associated with such objective measures of success as winning Olympic medals, earning high corporate salaries, or achieving elective office or high military rank. You can make such a test yourself with, for example, the birthdates of all members of Congress or of all members of the U.S. Olympic Team. But more sophisticated studies have also been done, involving horoscopes calculated for thousands of individuals. (With modern computers, the once laborious process of calculating a horoscope is practically instantaneous). The results of all of these studies are the same: There is no evidence that natal astrology has any predictive power, even in a statistical sense. Astrology is not a science, it is not scientific in its methods, and it has failed every test so far applied to it.

1.4 The Birth of Modern Astronomy

Astronomy made no major advances in medieval Europe, where the prevailing philosophy was acceptance of dogmatic authority. Medieval cosmology combined the crystalline spheres of Pythagoras (as perpetuated by Aristotle) with the epicycles of Ptolemy. Astrology was widely practiced, however, and an interest in the motions of the planets was thus kept alive. Then came the Renaissance; in science, the rebirth was clearly embodied in Nicolaus Copernicus.

Copernicus

One of the most important events of the Renaissance was the displacement of the Earth from the center of the universe. This intellectual revolution was initiated by a Polish lay monk, Copernicus. Nicolus Copernicus (1473–1543) was born in Torun on the Vistula. His training was in law and medicine, but Copernicus' main interests were astronomy and mathematics. His great contribution to science was a critical reappraisal of the existing theories of planetary motion and the development of a new Sun-centered, or *heliocentric*, model of the solar system. Copernicus concluded that the Earth is a planet, and that the planets Mercury, Venus, Earth, Mars, Jupiter, and Saturn all circle the Sun. Only the Moon was left in orbit about the Earth (Figure 1.12).

His ideas were set forth in detail in his book *De Revolutionibus*, published in 1543, the year of his death. In this book, Copernicus described certain postulates from which he derived his system of planetary motions. His postulates include the assumptions that the universe is spherical and that the motions of the heavenly bodies must be made up of combinations of uniform circular motions; thus Copernicus was not free of all traditional prejudices. Yet he evidently found something orderly and pleasing in the heliocentric system, and his defense of it was elegant and persuasive. His ideas, although not widely accepted until more than a century after his death, never disappeared and were ultimately of immense influence.

One of the objections raised to the heliocentric theory was that if the Earth were moving, we would all sense or feel this motion. Solid objects would be ripped from the surface, a ball dropped from a great height would not strike the ground directly below, and the like. But a moving person is not necessarily aware of that motion. We have all experienced seeing an adjacent train, car, or ship appear to move, only to discover that it is we who are moving.

Copernicus argued that the apparent annual motion of the Sun about the Earth could be represented equally well by a motion of the Earth about the Sun. He also reasoned that the apparent rotation of the celestial sphere could be accounted for by assuming that the Earth rotates while the celestial sphere is stationary. To the objection that if the Earth rotated about an axis it would fly into pieces, Copernicus answered that if such motion would tear the Earth apart, the still faster motion (because of its great size) of the celestial sphere required by the alternative hypothesis would be even more devastating.

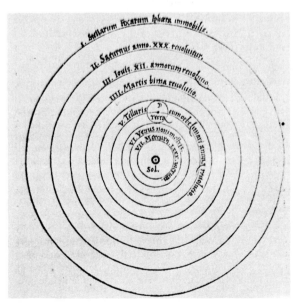

FIGURE 1.12 Heliocentric plan of the solar system in the first edition of Copernicus' *De Revolutionibus. (Crawford Collection, Royal Observatory, Edinburgh)*

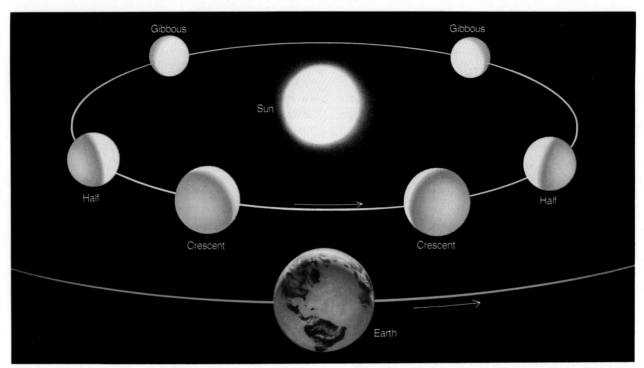

FIGURE 1.13 Phases of Venus according to the heliocentric theory.

The Heliocentric Model

The important assumption Copernicus made in *De Revolutionibus* is that the Earth is but one of six (then known) planets that revolve about the Sun. Given this, he was able to work out the correct general picture of the solar system. He placed the planets, starting nearest the Sun, in the order Mercury, Venus, Earth, Mars, Jupiter, and Saturn. Further, he deduced that the nearer a planet is to the Sun, the greater is its orbital speed. Thus the retrograde motions of the planets were easily understood without the necessity for epicycles. Also, Copernicus worked out the correct approximate scale of the solar system.

Copernicus did not prove that the Earth revolves about the Sun. In fact, with some adjustments, the old Ptolemaic system could have accounted as well for the motions of the planets in the sky. But the Ptolemaic cosmology was clumsy and lacked the beauty and symmetry of its successor.

In Copernicus' time, few people thought that ways existed to *prove* whether the heliocentric or the older geocentric system was correct. A long philosophical tradition, going back to the Greeks and defended by the Catholic Church, held that pure human thought combined with divine revelation represented the path to truth. Nature, as revealed by our senses, was suspect. In this environment, there was little motivation to carry out observations or experiments to try to distinguish between competing cosmological theories (or anything else). It should not surprise us, therefore, that the heliocentric idea was debated for more than half a century without any tests being applied to determine its validity. (In fact, the older geocentric system was still taught at Harvard University in the first years after it was founded in 1636.) Contrast this with the situation today, when scientists rush to test each new theory. For example, when two researchers at the University of Utah announced in 1989 that they had discovered a way to achieve nuclear fusion at room temperature, other scientists at more than 25 laboratories around the United States attempted to duplicate this feat within a few weeks.

When a new hypothesis or theory is proposed in science, it must first be checked for consistency with what is already known. Copernicus' heliocentric idea passed this test, for it allowed planetary positions to be calculated at least as well as did the geocentric theory. The next step is to see what predictions the new theory makes that differ from those of competing ideas. In the case of Copernicus, one example is the prediction that if Venus circles the Sun, it should go through the full range of phases just as the Moon does, whereas if it circles the Earth, it should not (Figure 1.13). But in those days, before the telescope, no one imagined testing this prediction.

FIGURE 1.14 The leaning bell-tower of the Cathedral at Pisa. While living in Pisa, Galileo carried out experiments to show that objects of different mass fall at the same rate. According to legend, one of his experiments consisted of dropping cannon balls of different weight from this tower. *(David Morrison)*

Today, the tests of competing theories are rarely as simple as this example of the phases of Venus. Since most predictions in science are quantitative, involving numerical values that can be calculated from mathematical equations, the checks usually require accurate measurements of the phenomena being studied. Often it is only through increasing the precision of the observations that a distinction between two competing theories can be made. For example, ordinary gravitational theory and Einstein's theory of general relativity make identical predictions about everyday experience, diverging only under extreme conditions, such as speeds near that of light.

Galileo and the Beginning of Modern Science

Many of the modern scientific concepts of observation, experimentation, and the testing of hypotheses through careful quantitative measurements were pioneered by a man who lived nearly a century after Copernicus. Galileo Galilei (1564–1642), a contemporary of Shakespeare, was born in Pisa (Figure 1.14). Like Copernicus, he began training for a medical career, but he had little interest in the subject and later switched to mathematics. He held faculty positions at the Universities of Pisa and Padua, and eventually he became mathematician to the Grand Duke of Tuscany in Florence.

Galileo's greatest contributions were in the field of mechanics, the study of motion and the actions of forces on bodies. It was familiar to all persons then, as it is to us now, that if a body is at rest, it tends to remain at rest and requires some outside influence to start it in motion. Rest was thus generally regarded as the natural state of matter. Galileo showed, however, that rest was no more natural than motion. If an object is slid along a rough horizontal floor, it soon comes to rest because friction between it and the floor acts as a retarding force. However, if the floor and object are both highly polished, the body, given the same initial speed, will slide farther before coming to rest. On a smooth layer of ice, it will slide farther still. Galileo reasoned that if all resisting effects could be removed, the object would continue in a steady state of motion indefinitely. In fact, he argued, a force is required not only to start an object moving from rest but also to slow down, stop, speed up, or change the direction of a moving object. You will appreciate this if you have ever tried to stop a rolling car by leaning against it or a moving boat by tugging on a line.

Galileo also studied the way bodies **accelerate,** or change their speed, as they fall freely or roll down inclined planes. He found that such bodies accelerate uniformly; that is, in equal intervals of time they gain equal increments in speed. Galileo formulated these newly found laws in precise mathematical terms that enabled one to predict, in future experiments, how far and how fast bodies would move in various lengths of time.

Sometime in the 1590s Galileo adopted the Copernican hypothesis of the solar system. In Roman Catholic Italy, this was not a popular philosophy, for the Church authorities still upheld the ideas of Aristotle and Ptolemy. It was primarily because of Galileo that in 1616 the Church issued a prohibition decree stating that the Copernican doctrine was "false and absurd" and was not to be held or defended.

Galileo's Astronomical Observations

It is not certain when the principle was first conceived of combining two or more pieces of glass to produce an instrument that enlarged images of distant objects, making them appear nearer. The first telescopes that attracted much notice were made by the Dutch spectacle maker Hans Lippershey in 1608. Galileo heard of the discovery, and without ever having seen an assembled telescope, he constructed one of his own with a three-power magnification, which made distant

FIGURE 1.15 Telescopes used by Galileo. The longer has a wooden tube covered with paper, a focal length of 1.33 m, and an aperture of 26 mm. *(Istituto e Museo di Storia della Scienza di Florenza)*

objects appear three times nearer and larger (Figure 1.15).

On August 25, 1609, Galileo demonstrated a telescope with a magnification of 9× to officials of the Venetian government. By a magnification of 9×, we mean that the linear dimensions of the object being viewed appeared nine times larger or, alternatively, that the objects appeared nine times closer than they really were. There were obvious military advantages associated with a device for seeing distant objects. For his invention Galileo's salary was nearly doubled, and he was granted lifetime tenure as a professor. His university colleagues were outraged, particularly since the invention was not even original.

Before using his telescope for astronomical observations, Galileo had to devise a stable mount, and he improved the optics to provide a magnification of 30×. Galileo also needed to acquire confidence in the telescope. At that time, the eyes were believed to be the final arbiter of truth about sizes, shapes, and colors. Lenses, mirrors, and prisms were known to distort distant images by enlarging them, reducing them, or even inverting them. Galileo undertook repeated experiments to convince himself that what he saw through the telescope was identical to what he saw up close. Only then could he begin to believe that the miraculous phenomena that were revealed in the heavens were real. While Galileo was convinced of the validity of what he saw, others were not. One unbelieving colleague said that he ''. . . tested this instrument of Galileo's in a thousand ways, both on things here below and on those above. Below, it works wonderfully; in the sky it deceives one, as some fixed stars are seen double.'' Another scholar refused even to look through the telescope because doing so gave him a headache.

FIGURE 1.16 The four Galilean satellites of Jupiter as photographed by the Voyager 1 spacecraft in 1979. *(NASA/ JPL)*

Beginning his astronomical work late in 1609, Galileo found that many stars too faint to be seen with the naked eye became visible with his telescope. In particular, he found that some nebulous blurs resolved into many stars and that the Milky Way was made up of multitudes of individual stars. He found four satellites or moons revolving about Jupiter, with periods ranging from just under 2 days to about 17 days. This discovery was particularly important because it showed that there could be centers of motion that in turn are in motion. Defenders of the geocentric view had argued that if the Earth were in motion the Moon would be left behind, because it could hardly keep up with a rapidly moving planet. Yet here were Jupiter's satellites doing exactly that!

With his telescope, Galileo was also able to carry out the test of the Copernican theory mentioned above, based on the phases of Venus. Within a few months he had found that Venus goes through phases like the Moon, showing that it must revolve about the Sun, so that we see different parts of its daylight side at different times (see Figure 1.11). These observations could not be reconciled with any model in which Venus circled about the Earth.

Galileo observed the Moon and saw craters, mountain ranges, valleys, and flat dark areas that he thought might be water (the dark maria, or "seas," on the Moon were thought to be water until long after Galileo's time). These discoveries show that the Moon might be not so dissimilar to the Earth, which suggested that the Earth, too, could belong to the realm of celestial bodies.

The discoveries of Copernicus and Galileo revolutionized our concept of the cosmos. Contrary to previous belief, they showed that space is vast and the Earth relatively small. They demonstrated that the Earth is a planet and raised the possibility that the other planets might be worlds themselves, perhaps even supporting life. They demoted the Earth (and humanity) from its position at the center of the universe. We take these things for granted, but four centuries ago such concepts were frightening and heretical for some, immensely stimulating for others. The pioneers of the Renaissance started the European world along the path toward science and technology that we still tread today. For them, Nature was rational and ultimately knowable, and science provided the means to reveal its secrets.

SUMMARY

1.1 The direct evidence of our senses supports a **geocentric** perspective, with the **celestial sphere** pivoting on the **celestial poles** and rotating about a stationary Earth. We see only half of this sphere at one time, limited by the **horizon;** the point directly overhead is the **zenith.** The Sun's annual path on the celestial sphere is the **ecliptic,** a line that runs through the center of the **zodiac,** the 18°-wide strip of sky within which are found the Moon and planets. The fixed stars are organized into 88 **constellations.**

1.2 Ancient Greeks such as Aristotle recognized that the Earth and Moon were spheres and understood the phases of the Moon, but because of their inability to detect stellar **parallax,** they rejected the idea that the Earth moved. Eratosthenes measured the size of the Earth with surprising precision, and Hipparchus carried out many astronomical observations, made a star catalogue, defined the system of stellar **magnitudes,** and discovered **precession** from the apparent shift of the position of the north celestial pole. Ptolemy of Alexandria summarized classical astronomy in his *Almagest*, explaining complex planetary motions, including **retrograde motion,** with high accuracy. This geocentric model, based on combinations of uniform circular motion using the **epicycle,** was accepted as authority for more than a thousand years.

1.3 The ancient religion of astrology began in Mesopotamia and reached its peak in the Greco-Roman world, especially as recorded in the *Tetrabiblos* of Ptolemy. Astrology uses a celestial coordinate system based on the ecliptic and the **vernal equinox** as they were two millennia ago. It is based on the assumption that the positions of the planets at the time of our birth, as described by a **horoscope,** determines our future. However, there is no evidence that this is so, even in a broad statistical sense.

1.4 The Pole Nicolus Copernicus (1473–1543) introduced the **heliocentric cosmology** to Renaissance Europe in his book *De Revolutionibus*. Although he retained the Aristotelian idea of uniform circular motion, Copernicus presented a convincing case that the Earth is a planet and that the planets all circle about the Sun, forever dethroning the Earth from its position at the center of the universe. The Italian Galileo Galilei (1564–1642) was the father of both modern experimental physics and telescopic astronomy. He studied the **acceleration** of moving objects, and in 1610 he began telescopic observations, discovering the nature of the Milky Way, the large-scale features of the Moon, the phases of Venus, and the four Galilean satellites of Jupiter. Although accused of heresy for his support of the heliocentric cosmology, Galileo's observations and brilliant writings convinced most of his scientific contemporaries of the reality of the Copernican theory.

REVIEW QUESTIONS

1. From where on Earth could you observe all of the stars during the course of a year? What fraction of the sky can be seen from the North Pole?

2. Describe a practical way to determine the constellation the Sun is in at any time of the year.

3. Explain retrograde motion, from both the geocentric and the heliocentric perspectives.

4. Draw a picture that explains why Venus goes through phases like the Moon, according to the heliocentric cosmol-ogy. Does Jupiter also go through phases as seen from the Earth?

5. Can you think of any forces or other influences that depend on the positions of the planets at your birth? Can any of these be used to justify the supposed predictive power of astrology?

6. In what ways did the work of Copernicus and Galileo differ from the traditional views of the ancient Greeks and of the Catholic Church?

THOUGHT QUESTIONS

7. Where on the Earth **(a)** are all stars at some time visible above the horizon? **(b)** is only half the sky ever above the horizon?

8. Show with a simple diagram how the lower parts of a ship disappear first as it sails away from you on a spherical Earth. Use the same diagram to show why lookouts on old sailing ships could see farther from the masthead than from the deck. Would there be any advantage to posting lookouts on the mast if the Earth were flat? (Note that these nautical arguments for a spherical Earth were quite familiar to Columbus and other mariners of his time).

9. Parallaxes of stars were not observed by ancient astronomers. How can this fact be reconciled with the heliocentric hypothesis of the solar system proposed by Copernicus?

10. From our modern perspective, what are the strongest arguments you could use to convince a skeptical neighbor that the Earth is not flat and **(b)** that the Earth is not stationary but goes around the Sun?

11. Copernicus accepted the traditional view that celestial motions are circular and uniform. What problems did this assumption cause for him? Do you think he should have questioned this assumption, given the information available to him?

12. Design an experiment that would allow you to test whether or not the planets and their motions influence human behavior.

13. Consider three cosmological perspectives: (1) the geocentric perspective; (2) the heliocentric perspective; and (3) the modern perspective in which the Sun is a minor star on the outskirts of one galaxy among billions. Discuss some of the cultural and philosophical implications of each point of view.

PROBLEMS

14. The Moon moves relative to the background stars. Go outside at night and note the position of the Moon relative to nearby stars. Repeat the observation a few hours later. How far has the Moon moved relative to the stars? (For reference, the diameter of the Moon is about $1/2°$.) Based on your estimate of the Moon's motion, how long will it take for the Moon to return to the position relative to the stars in which you first observed it?

15. The north celestial pole appears at an altitude above the horizon that is equal to the observer's latitude. Identify Polaris, the North Star, which lies very close to the north celestial pole. Measure its altitude. (This can be done with a protractor. Alternatively, your fist, extended at arm's length, spans a distance approximately equal to $10°$.) Compare this estimate with your latitude. The next time you travel several hundred miles north or south, determine the altitude of Polaris again. Can you detect a difference? (This experiment cannot easily be performed in the Southern Hemisphere because Polaris itself is, of course, not visible, and there is no bright star near the south celestial pole.)

16. Suppose Eratosthenes had found that at Alexandria at noon on the first day of summer the line to the Sun makes an angle of $30°$ with the vertical. What then would he have found for the Earth's circumference?

17. Suppose Eratosthenes' results for the Earth's circumference were quite accurate. If the diameter of the Earth is 12,740 km, evaluate the length of his stadium in kilometers.

18. Suppose you are on a strange planet and observe, at night, that the stars do not rise and set but circle parallel to the horizon. Now you walk in a constant direction for 8000 miles, and at your new location on the planet you find that all stars rise straight up in the east and set straight down in the west, perpendicular to the horizon.

a. How could you determine the circumference of the planet without any further observations?

b. What evidence is there that the Greeks could have done what you suggest?

c. What is the circumference of that planet, in miles?

Astronaut in space, circling the Earth like any other satellite and subject to the same laws of motion. *(NASA)*

Johannes Kepler (1571–1630),
German mathematician and
astronomer, was a contemporary of
Galileo, living during the tumultuous
period of the Counter-Reformation
and the Thirty Years' War. His
discovery of the basic quantitative
laws that describe planetary motion
placed the heliocentric cosmology
of Copernicus on a firm
mathematical basis and made
possible Newton's later formulation
of the laws of motion and of
universal gravitation.

ORBITS AND GRAVITATION

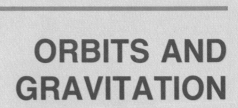

We all know that the planets "circle" the Sun, but what does this mean? Do planets (and other members of the solar system) literally follow circular paths around the Sun? No, nature is not that simple. If planetary orbits were really circular, understanding celestial motions would not have presented so formidable a problem to early astronomers, and the history of science might have been quite different.

If we could look down on the solar system from somewhere out in space, far above the plane of the planets' orbits, the problem of interpreting planetary motions would be much simpler. But the fact is that we must observe the positions of all the other planets from our own moving planet, and scientists of the Renaissance did not know the nature of the Earth's orbit any better than that of the other planets. Their problem, therefore, was to deduce the nature of the Earth's motion as well as that of each of the other planets, using only observations of the apparent positions of the other planets as seen from our own shifting vantage point.

Kepler discovered the descriptive laws that govern planetary motion, but it was Newton who provided the conceptual framework for understanding these motions. His development of the concept of universal gravitation gave a physical basis to Kepler's laws and ultimately provided the means to calculate interplanetary trajectories for spacecraft as well as to study the orbits of planets with remarkable accuracy.

2.1 The Laws of Planetary Motion

At about the time that Galileo was beginning his experiments with falling bodies, the efforts of two other scientists dramatically advanced our understanding of the motions of the planets. These two astronomers were the observer Tycho Brahe and the mathematician Johannes Kepler. Together, they placed the speculations of Copernicus on a sound mathematical basis and paved the way for the work of Newton in the next century.

Tycho Brahe

Three years after the publication of Copernicus' *De Revolutionibus*, Tycho Brahe (1546–1601) was born to a family of Danish nobility. He developed an early interest in astronomy and as a young man made significant astronomical observations. The reputation of the young Tycho Brahe gained him the patronage of the Danish King Frederick II, and at the age of 30, Tycho was able to establish a fine astronomical observatory on the North Sea island of Hveen. Tycho was the last and greatest of the pretelescopic observers in Europe.

Tycho's observations at Hveen included a continuous record of the positions of the Sun, Moon, and planets. His extensive and precise observations enabled him to note that the positions of the planets varied from those given in published tables, which

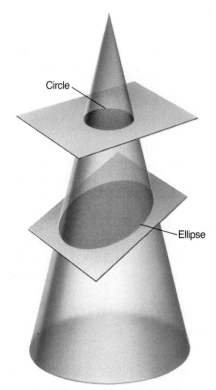

FIGURE 2.1 The circle and the ellipse are formed by the intersection of a plane with a cone. This is why both curves are called conic sections.

were based on the work of Ptolemy. But Tycho was an extravagant and cantankerous fellow, and he accumulated enemies among government officials. When his patron, Frederick II, died in 1597, Tycho was forced to leave Denmark. He took up residence near Prague, where he became Court Astronomer to the Emperor Rudolf of Bohemia. There, in the year before his death, Tycho secured the assistance of a most able young mathematician, Johannes Kepler.

Kepler

Johannes Kepler (1571–1630) was born in the German city of Württemberg and lived much of his life amid the turmoil of the Thirty Years' War. He attended college at Tübingen and studied for a theological career. There he learned the principles of the Copernican system and became converted to the heliocentric hypothesis. Kepler, a Protestant refugee from his Catholic homeland, went to Prague to serve as an assistant to Tycho, who set him to work trying to find a satisfactory theory of planetary motion—one that was compatible with the long series of observations made at Hveen.

But Tycho, being jealous of the young Kepler's brilliance, was reluctant to provide him with much material at any one time for fear Kepler would discover the secrets of the universal motions by himself,

thereby robbing Tycho of some of the glory. It was not until after Tycho's death in 1601 that Kepler obtained possession of the majority of the priceless records. Their study occupied most of Kepler's time for more than 20 years.

Kepler's most detailed study was of Mars, for which the observational data were the most extensive. He published the first results of his work in 1609 in *The New Astronomy*; it is there that we find his first two laws of planetary motion. Their discovery was a profound step in the development of modern science.

The Orbit of Mars

Kepler began his research with the idea that the orbits of planets were circles, but the observations contradicted this assumption. Working with data for Mars, he eventually discovered that the orbit of that planet had the shape of a flattened circle, or **ellipse.** Next to the circle, the ellipse is the simplest kind of closed curve, belonging to a family of curves known as conic sections (Figure 2.1).

An ellipse is a curve for which the sum of the distances from any point on the ellipse to two points inside the ellipse is the same. These two points inside the ellipse are called its **foci** (singular: **focus**). This property suggests a simple way to draw an ellipse. The ends of a length of string are tied to two tacks pushed through a sheet of paper into a drawing board, so that the string is slack. If a pencil is pushed against the string, so that the string is held taut, and then slid against the string around the tacks (Figure 2.2), the curve that results is an ellipse. At any point where the pencil may be, the sum of the distances from the pencil to the two tacks is a constant length—the length of the string. The tacks are at the two foci of the ellipse.

The maximum diameter of the ellipse is called the major axis. Half this distance, that is, the distance from the center of the ellipse to one end, is the **semimajor axis,** which is the quantity usually given to specify the size of the ellipse. For example, the semimajor axis of the orbit of Mars, which is also the

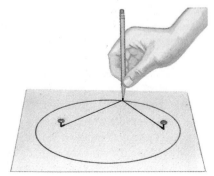

FIGURE 2.2 Drawing an ellipse with two tacks and a string.

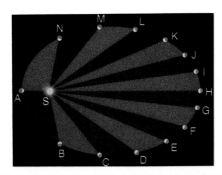

FIGURE 2.3 Kepler's second law: the law of equal areas. A planet moves most rapidly on its elliptical orbit when it is at position *A*, nearest the Sun (*S*) at one focus of the ellipse. The orbital speed of the planet varies in such a way that in equal intervals of time it moves distances *AB*, *BC*, *CD*, and so on, so that the regions swept out by the line connecting the planet to the Sun (alternating shaded and clear zones) are always the same area. The eccentricities of the planets are substantially less than is shown here.

planet's average distance from the Sun, is 228 million km. The shape (roundness) of an ellipse depends on how close together the two foci are, compared with the major axis. The ratio of the distance between the foci to the major axis is called the **eccentricity** of the ellipse. If an ellipse is drawn as just described, the length of the major axis is the length of the string, and the eccentricity is the distance between the tacks divided by the length of the string.

If the foci (or tacks) coincide, the eccentricity is zero and the ellipse is a circle; thus a circle is an ellipse of zero eccentricity. Ellipses of various shapes are obtained by varying the spacing of the tacks (as long as they are not farther apart than the length of the string). The greater the eccentricity, the more elongated is the ellipse, up to a maximum eccentricity of 1.0.

The size and shape of an ellipse are completely specified by its semimajor axis and its eccentricity. Kepler found that Mars has an orbit that is an ellipse and that the Sun is at one focus (the other focus is empty). The eccentricity of the orbit of Mars is only about 0.1; the orbit, drawn to scale, would be practically indistinguishable from a circle. Yet the difference is critical for understanding planetary motions.

Kepler's first law describes the elliptical shape of planetary orbits. The second law deals with the speed with which each planet moves along the ellipse. Working with Tycho's observations of Mars, Kepler discovered that the planet speeds up as it comes closer to the Sun and slows down as it pulls away from the Sun. He expressed this relation by imagining that the Sun and Mars are connected by a straight, elastic line. As Mars travels in its elliptical orbit around the Sun, in equal intervals of time the areas swept out in space by this imaginary line are always

equal (Figure 2.3). This is also a general property of the orbits of planets. A planet in a circular orbit always moves at the same speed, but in an eccentric orbit the planet's speed varies considerably.

Laws of Planetary Motion

Kepler's first two laws of planetary motion describe the shape of a planet's orbit and allow us to calculate the speed of its motion at any point in the orbit. Kepler was pleased to have discovered such fundamental rules, but they did not satisfy his quest to understand planetary motions. He wanted to know why the orbits of the planets were spaced as they are, and to find a pattern in their orbital periods—a "harmony of the spheres." For many years he worked to discover mathematical relationships governing planetary spacings and periods of revolution.

In 1619, Kepler succeeded in finding the simple algebraic relation that links the semimajor axes of the planets' orbits and their periods of revolution. The relation is now known as Kepler's third law. It applies to all of the planets, including the Earth, and it provides a means for calculating the relative distances of the planets from the Sun, given their revolution periods.

Kepler's third law takes its simplest form when the period is expressed in years (the revolution period of the Earth) and the semimajor axis of the orbit in terms of the Earth's average distance from the Sun, called the **astronomical unit (AU)**. One astronomical unit is equal to 1.5×10^8 km/s. Kepler's third law is, then,

$$(\text{distance})^3 = (\text{period})^2$$

For example, this relationship tells us how to calculate the average distance from the Sun of Mars (the semimajor axis of its orbit) from its period of 1.88 years. The period squared is $1.88 \times 1.88 = 3.53$, and the cube root of 3.53 is 1.52, which is equal to the semimajor axis in astronomical units.

Kepler's three laws of planetary motion can be summarized as follows:

Kepler's First Law Each planet moves about the Sun in an orbit that is an ellipse, with the Sun at one focus of the ellipse.

Kepler's Second Law The straight line joining a planet and the Sun sweeps out equal areas in space in equal intervals of time.

Kepler's Third Law The squares of the periods of revolution of the planets are in direct proportion to the cubes of the semimajor axes of their orbits.

These three laws provided a precise geometric description of planetary motion within the framework of

ESSAY A Scale Model of the Solar System

Astronomy often deals with dimensions and distances that far exceed our ordinary experience. What does a billion kilometers really mean to anyone? Sometimes it helps to visualize the system in terms of a scale model.

In our imagination, let us build a scale model of the solar system, adopting a scale factor of 1 billion (10^9)—that is, we reduce the actual solar system by dividing every dimension by a factor of 10^9. The Earth then has a diameter of 1.3 cm, or about the size of a grape. The Moon is a pea orbiting this grape at a distance of 40 cm, or a bit over 1 ft. The Earth-Moon system fits into an attaché case.

In this model the Sun is nearly 1.5 m in diameter, about the average height of an adult, at a distance of 150 m—about one city block. Jupiter is five blocks away, and its diameter is 15 cm, about the size of a very large grapefruit. Saturn is 10 blocks away, Uranus 20, and Neptune and Pluto each 30 (Pluto is now closer to the Sun than is Neptune). Most of the satellites of the outer solar system are the size of seeds of various kinds orbiting the grapefruit, oranges, and lemons we use to represent the outer planets.

In our scale model, a human is reduced to the dimensions of a single atom, and autos (and spacecraft) to the size of molecules. Sending the Voyager spacecraft to Neptune involves navigating a single molecule from the Earth-grape toward a lemon 5 km distant with an accuracy equivalent to the width of a thread in a spider's web. In this fruity model, the nearest stars are tens of thousands of kilometers away—on the other side of the Earth or beyond.

the Copernican system. With these tools, it is possible to calculate planetary positions with undreamed-of precision. But Kepler's laws are purely descriptive; they do not help us understand what forces of nature constrain the planets to follow this particular set of rules. That step was left to Newton.

2.2   Newton's Great Synthesis

It was the genius of Isaac Newton (1643–1727) that unified the natural laws discovered by Galileo, Kepler, and others. Newton was born in Lincolnshire, England, in the year after the death of Galileo (Figure 2.4). He entered Trinity College at Cambridge in 1661 and eight years later was appointed Professor of Mathematics there. Among Newton's contemporaries in England were architect Christopher Wren, authors Samuel Pepys and Daniel Defoe, and composer G.F. Handel.

Newton's Laws of Motion

As a young man in college, Newton became interested in natural philosophy, as science was then called. He worked out many of his ideas on mechanics and optics during the plague years of 1665 and 1666. Eventually, his friend Edmund Halley prevailed on him to collect and publish the results of his investigations in mechanics and gravitation. The result was *Philosophiae Naturalis Principia Mathematica.* The *Principia,* as the book is generally known, was published at Halley's expense in 1687.

At the very beginning of the *Principia,* Newton states three laws that he presumes to govern the motions of all objects:

Newton's First Law   Every body continues in a state of rest, or of uniform motion in a straight line, unless it is compelled to change that state by forces impressed upon it.

FIGURE 2.4   Newton's birthplace in Lincolnshire, England. *(G.O. Abell)*

Newton's Second Law The change of motion is proportional to the force impressed, and is made in the direction of the straight line in which that force is impressed.

Newton's Third Law To every action there is always an equal and opposite reaction; or, the mutual actions of two bodies upon other are always equal and act in opposite directions.

In the original Latin, the three laws contain only 59 words, but those few words set the stage for modern science. We shall have to examine them carefully.

Interpretation of Newton's Laws

Newton's first law is a restatement of one of Galileo's discoveries, which Newton called the conservation of **momentum,** a measure of the motion of a body. The law states that in the absence of any outside influence, a body's momentum remains unchanged. In other words, a stationary object stays put and a moving object keeps moving. Momentum depends on three factors. The first is speed—how fast a body moves (zero if it is stationary). The second is direction. The term velocity describes both speed and direction. For example, 20 km/h due south is velocity, while 20 km/h is speed. The third factor in momentum is what Newton called mass. Mass is a measure of the amount of matter in a body, as we will discuss further below.

The momentum of a body can change only under the action of an outside influence. Newton's second law defines *force* in terms of its ability to *change momentum.* Force has both size and direction, and when force is applied to a body, the momentum changes in the direction of the applied force. This means that a force is required to change either the speed or the direction of a body or both—that is, to start it moving, to speed it up, to slow it down, to stop it, to change its direction, or to do all these things. Any such change in an object's state of motion is called *acceleration.*

Newton's third law is the most profound. Basically, it is a generalization of the first law, but it also gives us a way to define mass. If we consider a system of two or more objects, isolated from outside influences, Newton's first law suggests that the total momentum of the system of objects should remain constant. Therefore, any change of momentum within the system must be balanced by another change that is equal and opposite to it, so that the momentum of the entire system is not changed. Thus, all forces occur as pairs of forces that are mutually equal to and opposite each other. If a force is exerted on an object, it must be exerted by something else, and the object will exert an equal and opposite force on that something.

Suppose a boy jumps off a table down to the ground. The force pulling him down is a mutual gravitational force between him and the Earth. Both he and the Earth suffer the same total change of momentum because of the influence of this mutual force. Of course, the boy does most of the moving; because of the greater mass of the Earth, it can experience the same change of momentum by accelerating only a very small amount.

A more obvious manifestation of the mutual nature of forces between objects is familiar to all who have batted a baseball. The recoil of the bat shows that the ball exerts a force on the bat during the impact, just as the bat does on the ball. The momentum imparted to the bat by the ball is transmitted through the batter to the Earth, so the acceleration produced is far less than that suffered by the ball. Similarly, when a rifle is discharged, the force pushing the bullet out of the muzzle is equal to that pushing backward upon the gun and the person shooting it. Here, in fact, is the principle of jet engines and rockets—the force that discharges the exhaust gases from the rear of the rocket is balanced by a force that shoves the rocket forward. The exhaust gases need not push against air or the Earth; a rocket operates best in a vacuum (Figure 2.5).

FIGURE 2.5 The U.S. Space Shuttle at launch, powered by three liquid fuel engines burning liquid oxygen and liquid hydrogen, with two solid fuel boosters. *(NASA)*

Mass, Volume, and Density

It is important not to confuse mass, volume, and density. Mass is a measure of the amount of material in an object. Volume, in contrast, is a measure of the physical space occupied by a body, say in cubic centimeters or liters. In short, the volume is the "size" of an object—it has nothing to do with its mass. A penny and an inflated balloon may both have the same mass, but they have very different volumes.

The penny and balloon are also very different in **density,** which is a measure of how much mass is contained within a given volume. Specifically, density is the ratio of mass to volume:

$$\text{density} = \text{mass/volume}$$

Note that often in everyday language we use "heavy" and "light" as indications of density (rather than weight), as, for instance, when we say that iron is heavy or that a puff pastry is light.

The units of density that will be used in this book are grams per cubic centimeter (g/cm³) or, alternatively, metric tons per cubic meter.* Familiar materials span a considerable range in density, from gold (19 g/cm³) to artificial materials such as plastic insulating foam (less than 0.1 g/cm³) (Table 2.1). In the astronomical universe, much more remarkable densities can be found, all the way from a comet's tail (10^{-16} g/cm³) to a neutron star (10^{15} g/cm³).

To sum up, then, mass is "how much," volume is "how big," and density is "how tightly packed."

Angular Momentum

The concept of **angular momentum** is a bit more complex, but it is important for understanding many astronomical objects. Angular momentum is a measure of the momentum of an object as it rotates or revolves about some fixed point. Whenever we deal with the evolution of spinning objects, from planets to galaxies, we shall have to consider angular momentum. Just as ordinary momentum is defined as the product of two quantities—mass and velocity—the angular momentum of an object is defined as the product of three quantities: mass, velocity, and the distance from the fixed point around which the object turns.

If these three quantities remain constant, that is, if the motion takes place at a constant speed and at a fixed distance from the point of origin, then the angular momentum is also a constant. More generally, angular momentum is constant, or is conserved,

TABLE 2.1 Densities of Materials

Material	Density (g/cm³)
Gold	19.3
Lead	11.4
Iron	7.9
Earth (bulk)	5.6
Rock (typical)	2.5
Water	1.0
Wood (typical)	0.8
Insulating foam	0.1
Silica gel	0.02

in any rotating system in which no external forces act or in which the only forces are directed toward or away from the point of origin. An example of such a system is a planet orbiting the Sun. Kepler's second law is an example of the conservation of angular momentum. When a planet approaches the Sun on its elliptical orbit, it speeds up; when it recedes, it moves more slowly.

Just as a planet speeds up when it approaches the Sun, a shrinking cloud of dust or a stream of matter falling into a black hole increases its spin rate as it contracts. The concept can also be illustrated by figure skaters, who bring their arms and legs in to spin more rapidly and extend their arms and legs to slow down. A diver does the same thing by taking a tuck to spin and then stretching out for a smooth entry into the water.

2.3 Universal Gravitation

The Law of Gravitation

Newton's laws of motion showed that, left to themselves, objects at rest stay at rest and those in motion continue in uniform motion in a straight line. Thus it is the *straight line*, not the circle, that defines the most natural state of motion. In the case of the planets, some *force* must act to bend their paths from straight lines into ellipses. That force is gravitation.

It is obvious that the Earth exerts a gravitational force upon all objects at its surface. Newton reasoned that the Earth's gravity might extend as far as the Moon and produce the acceleration required to curve the Moon from a straight path and keep it in its orbit. He further speculated that there is a general force of attraction between all material bodies. If so, the attractive force between the Sun and each of the planets could keep each in its respective orbit.

Thus Newton hypothesized that there is a universal attraction between all bodies everywhere in space.

---

* Generally we use the standard metric (or SI) units in this book. The proper metric unit of density is kg/m³. But to most people g/cm³ provides a more meaningful unit because the density of water is exactly 1 g/cm³. Density expressed in g/cm³ is sometimes called specific density or specific weight.

Next he had to determine the mathematical nature of the attraction. The precise mathematical description of that gravitational force must dictate that the planets move exactly as Kepler had observed them to (as expressed in Kepler's three laws). Moreover, at the same time, the law of gravitation must predict the correct behavior of falling bodies on the Earth, as observed by Galileo. How, then, must the gravitational force depend on distance for these conditions to be met?

The answer to this question involved mathematics not yet developed. But this did not deter Isaac Newton, who invented what we today call calculus to deal with this problem. Eventually he concluded that the force should drop off with increasing distance between the Sun and a planet in proportion to the inverse square of their separation. Newton also concluded that the gravitational attraction between two bodies must be proportional to their masses. Expressed as a formula, the gravitational attraction between any two objects is given by

$$\text{force} = \frac{GM_1M_2}{R^2}$$

where M_1 and M_2 are the masses of the two objects and R is their separation. The number represented by G is called the constant of gravitation. With such a force and the laws of motion, Newton showed mathematically that the only orbits permitted were exactly those described by Kepler's laws.

Falling Apples and the Orbit of the Moon

Newton's law of gravitation was observed to hold for the planets. The gravitational theory should also predict the observed acceleration of the Moon toward the Earth, falling about the Earth at a distance of 60

Earth radii, as well as of an object (say an apple) dropped near the Earth's surface. The gravitational attraction of the Earth for an object at (or near) its surface is the object's weight. Does Newton's gravitational law predict the correct weights for objects at the Earth's surface?

The weight of an object at the surface of the Earth is the result of the simultaneous attractions of the many parts of the Earth pulling from many different directions. Exactly what is the resultant gravitational effect of the many parts of a sphere, each pulling independently on a mass outside the surface of the sphere? Using the new mathematical tools he had developed, Newton found a beautifully simple answer. A spherical body acts gravitationally as though all its mass were concentrated at a point at its center (Figure 2.6). This means that we can consider the Earth, the Moon, the Sun, and the planets as geometrical points as far as their gravitational influences are concerned.

Newton's theory of gravitation predicted that the acceleration of an object toward the Earth should be proportional to the Earth's mass and inversely proportional to the square of its distance from the center of the Earth. This can be written as follows:

$$\text{acceleration} = \frac{GM}{R^2}$$

Objects at the surface of the Earth ($R = 1$ Earth radius from its center) are observed to accelerate downward at 9.8 meters per second per second (9.8 m/s^2). Therefore, if the law holds, the Moon, 60 Earth radii from its center, should experience an acceleration toward the Earth that is 60^2, or 3600, times less, that is, about 0.00272 m/s^2. This is precisely the observed acceleration of the Moon in its orbit. What a triumph! Can you imagine the thrill—the glorious

(a)

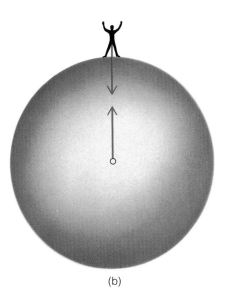

(b)

FIGURE 2.6 The gravitational attraction of a sphere is as though all its mass were concentrated at its center, called the center of mass.

exhilaration—Newton must have felt to realize he had discovered and verified a law that holds for the Earth, apples, the Moon, and, as far as we know, everything in the universe!

Orbital Motion

Kepler's laws are descriptions of the orbits of objects moving according to Newton's laws of motion and the law of gravitation. Knowing that gravity is the force that attracts planets toward the Sun, however, allowed Newton to introduce an important modification of Kepler's third law. As before, we express distances in astronomical units and periods in years, but we introduce as well the masses of the Sun and planet, both expressed in units of the Sun's mass. If D is the distance, M is the mass, and P is the period, Newton's law of gravitation can be used to show mathematically that

$$D^3 = (M_1 + M_2)P^2$$

In the solar system, where most of the mass is in the Sun itself, the sum of the masses $(M_1 + M_2)$ is very nearly equal to M_1, the Sun's mass. This is the reason that Kepler did not realize that both masses had to be included in the calculation. There are many cases in which we *do* need to include the two mass terms, however, and for this very reason observations of motions of objects acting under their mutual gravitation permit the astronomer to deduce their masses.

Newton's reformulation of Kepler's third law is one of the most powerful concepts in astronomy. Our ability to deduce the masses of objects from their motions is key to understanding the nature and evolution of many astronomical bodies. We will use this law repeatedly throughout this text in calculations that range from the orbits of comets to the interactions of galaxies.

2.4   Orbits in the Solar System

Description of an Orbit

Celestial mechanics is the study of the motions of astronomical objects, using gravitational theory. The path of an object through space is called its orbit, whether that object is a planet, star, or galaxy. An orbit, once determined, allows the future positions of the object to be calculated. In the case in which only two objects are involved (such as a planet in orbit about the Sun), three quantities are required to describe the orbit. These are the *size* (the semimajor axis), the *shape* (the eccentricity), and the *period* of revolution.

Two points in any orbit have been given special names. The place where the planet is closest to the Sun is called the **perihelion** of its orbit, and the place where it is farthest away and moves the most slowly is the **aphelion.** For a satellite orbiting the Earth, the corresponding terms are **perigee** and **apogee.**

Orbits of the Planets

There are nine planets, beginning with Mercury closest to the Sun and extending outward to Pluto. Pluto has the largest semimajor axis of any planet (almost 40 AU, or 5 billion km). However, it is not at present the most distant planet. Because Pluto is currently near the perihelion of its orbit, it is closer to the Sun than is Neptune. During the final 20 years of the 20th century, Neptune has the distinction of being the most distant planet, at 30 AU from the Sun. The orbital data for the planets are summarized in Table 2.2.

According to Kepler's laws, Mercury must have the shortest period of revolution (88 Earth days) and the highest orbital speed (averaging 48 km/s). At the opposite extreme, Pluto has a period of 249 years and an average orbital speed of just 5 km/s.

All of the planets have orbits of rather low eccentricity. The most eccentric orbits are those of Mercury (0.21) and Pluto (0.25); all of the rest have eccentricities of less than 0.1. It is fortunate for the development of science that Mars has an eccentricity greater than that of most planets, for otherwise the pretelescopic observations of Tycho would not have been sufficient for Kepler to deduce that its orbit had the shape of an ellipse and not a circle.

The planetary orbits are also confined close to a common plane, which is near the plane of the Earth's orbit (the ecliptic). The orbit of Pluto is inclined about 17° to the average, but all of the other planets lie within 10° of the common plane of the solar system.

TABLE 2.2   Orbital Data for the Planets

Planet	Semimajor Axis (AU)	Period (yr)	Eccentricity
Mercury	0.39	0.24	0.21
Venus	0.72	0.62	0.01
Earth	1.00	1.00	0.02
Mars	1.52	1.88	0.09
(Ceres)	2.77	4.60	0.08
Jupiter	5.20	11.86	0.05
Saturn	9.54	29.46	0.06
Uranus	19.19	84.07	0.05
Neptune	30.06	164.80	0.01
Pluto	39.60	248.60	0.25

For a long time scientists looked for evidence of regularity in the spacing of the orbits of the planets. Look at Table 2.2. You will see that the four inner planets (called the **terrestrial planets**) have orbits that are almost evenly spaced, with semimajor axes of 0.4, 0.7, 1.0, and 1.5 AU. In contrast, the distances between the outer planets, or **jovian planets,** are much greater. There is a very large spacing between the orbits of Mars and Jupiter. Is there some law of nature reflected in these spacings, or some grand "harmony of the spheres," as sought by Kepler?

In the 19th century, astronomers thought they had discovered just such a relationship, called the Titius-Bode law. The semimajor axes of the orbits of the planets from Mercury through Uranus are matched quite well by the series of numbers formed by first writing down the numbers 0, 3, 6, 12, . . . , with each number in the sequence (after the first) being double the previous one. If 4 is added to each number and the result divided by 10, you will get a good approximation of the spacing of the planets. Note that there is no planet, however, at the distance of the fifth number in the sequence, 2.8 AU. Also note that this "law" breaks down completely for the outer two planets, Neptune and Pluto. The Titius-Bode law is an interesting piece of numerology, but it really doesn't tell us anything fundamental about why the planets are spaced as they are.

Orbits of Asteroids and Comets

In addition to the nine planets, there are many smaller objects in the solar system. Some of these are natural satellites that orbit the planets (all except Mercury and Venus, which have no satellites). In addition, there are two classes of objects in heliocentric orbits: the asteroids and the comets. Their properties will be discussed in some detail in Chapter 11.

Asteroids and comets differ from each other in composition and orbits. In general, the asteroids have orbits with smaller semimajor axes than do the comets. The great majority of them lie between 2.2 and 3.3 AU, in the region known as the **asteroid belt.** The asteroid belt is thus in the middle of the gap between the orbits of Mars and Jupiter. It is because these two planets are so far apart that stable orbits of small bodies can exist in the region between them. In Table 2.2, the orbit is shown for the largest, Ceres.

Comets generally have orbits of larger size and greater eccentricity than those of the asteroids. Typically, the eccentricities of their orbits are 0.8 or higher. According to Kepler's second law, therefore, they spend most of their time far from the Sun, moving very slowly. As they approach perihelion, the comets speed up and whip through the inner parts of their orbits rapidly.

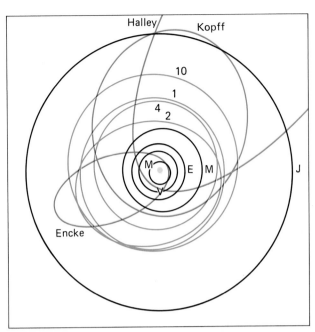

FIGURE 2.7 Orbits of typical comets and asteroids compared with those of the planets Mercury, Venus, Earth, Mars, and Jupiter. Shown in red are three comets: Halley, Kopff, and Encke. In blue are the four largest asteroids: 1 Ceres, 2 Pallas, 4 Vesta, and 10 Hygeia.

2.5 Motions of Satellites and Spacecraft

Space flight and Satellite Orbits

Gravitation and Kepler's laws describe the motions of Earth satellites and interplanetary spacecraft as well as the planets. Sputnik, the first artificial Earth satellite, was launched by the U.S.S.R. on October 4, 1957. Since that time, thousands of satellites have been placed into orbit around the Earth, and spacecraft have also orbited the Moon, Venus, and Mars.

Once an artificial satellite is in orbit, its behavior is no different from that of a natural satellite, such as our Moon. If the satellite is high enough to be free of atmospheric friction, it will remain in orbit forever, following Kepler's laws in a perfectly respectable way. However, although there is no difficulty in maintaining a satellite once it is in orbit, a great deal of energy is required to lift the spacecraft off the Earth and accelerate it to orbital speed.

To illustrate how a satellite is launched, imagine a gun on top of a high mountain, firing a bullet horizontally (Figure 2.8a—adapted from a similar diagram by Newton, shown in Figure 2.8b). Imagine, further, that the friction of the air could be removed and that all hindering objects, such as other mountains, buildings, and so on, are absent. Then the only force that

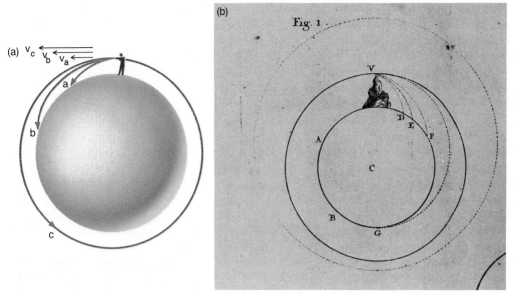

FIGURE 2.8 (a) Firing a bullet into a satellite orbit. (b) A diagram by Newton in his *De Mundi Systematic*, 1731 edition, illustrating the same concept shown in panel a. *(Crawford Collection, Royal Observatory, Edinburgh)*

acts on the bullet after it leaves the muzzle is the gravitational force between the bullet and Earth.

If the bullet is fired with velocity v_a, it continues to have that forward speed, but meanwhile the gravitational force acting upon it pulls it downward, so that it strikes the ground at a. However, if it is given a higher muzzle velocity v_b, its higher forward speed carries it farther before it hits the ground, for, regardless of its forward speed, the downward gravitational force is the same. Thus, this faster moving bullet strikes the ground at b. If the bullet is given a high enough muzzle velocity v_c, as it falls toward the

ground, the curved surface of the Earth causes the ground to tip out from under it so that it remains the same distance above the ground and "falls around" the Earth in a complete circle. This speed—called the **circular satellite velocity**—is about 8 km/s.

Suppose that a rocket is fired to an altitude of a few hundred miles, then turned so that it is moving horizontally, and finally given a forward horizontal thrust. It proceeds in an orbit the size and shape of which depend critically on the exact direction and speed of the rocket at the instant of its "burnout," that is, the instant when the thrust supplied by its fuel

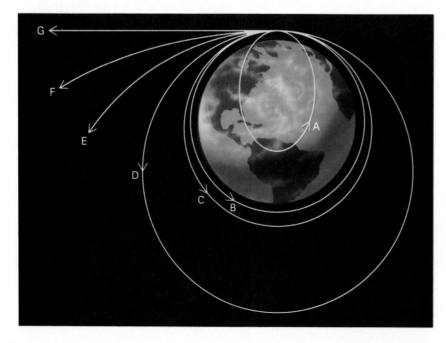

FIGURE 2.9 Various satellite orbits that result from different initial velocities parallel to the Earth's surface. *A* is an orbit that is intercepted by the solid Earth (like that of a military ballistic missile); *B* is a circular orbit; and *C*, *D*, *E*, and so forth are orbits of increasing energy but all with the same perigee at the point of injection.

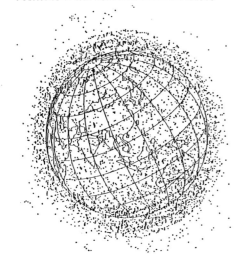

FIGURE 2.10 Plot of all known satellites and satellite debris in Earth orbit in March 1986.

is shut off. For simplicity, suppose that it is moving exactly horizontally, or parallel to the ground, at burnout. The possible kinds of orbits it can enter are shown in Figure 2.9.

If the rocket's burnout speed is less than 8 km/s, its orbit is an ellipse, with the center of the Earth at one focus of the ellipse. The apogee point of the orbit, that point farthest from the center of the Earth, is the point of burnout; the perigee, the point of closest approach to the center of the Earth, is halfway around the orbit from burnout. If the burnout speed is substantially below the circular satellite velocity, most of

the rocket's elliptical orbit lies beneath the surface of the Earth (orbit *A*), where, of course, the satellite cannot travel. Consequently, it will traverse only a small section of its orbit before colliding with the surface of the Earth. This is the kind of trajectory followed by a ballistic missile. If the burnout speed is just slightly below the circular satellite velocity, the rocket may clear the surface of the Earth (orbit *B*), although its orbit will probably lie too low in the atmosphere for the satellite to be long-lived.

If the burnout speed were exactly the circular satellite velocity, a circular orbit centered on the center of the Earth would result (orbit *C*). A slightly greater burnout speed produces an elliptical orbit with perigee at burnout point and apogee halfway around the orbit (orbit *D*).

Each year about a hundred new satellites are launched into orbit, most of them by the U.S.S.R. Other satellite-launching nations, in addition to the U.S.S.R. and the United States, include China, Japan, India, Israel, and the European Space Agency (ESA), a consortium of European nations (Figure 2.10). Most satellites are launched into low Earth orbit, since this requires the minimum launch energy. At the orbital speed of 8 km/s, they circle the planet in about 90 min. These orbits are not stable indefinitely, since the drag generated by friction with the thin upper atmosphere eventually leads to a loss of energy and "decay" of the orbit. Upon re-entering the denser parts of the atmosphere, most satellites are burned up by atmospheric friction, although some solid parts may reach the surface. Of course, the U.S. and Soviet shuttles (Figure 2.11) and other recoverable payloads are designed to survive re-entry intact.

FIGURE 2.11 The U.S. Space Shuttle in flight. *(NASA)*

Interplanetary Spacecraft

The exploration of the solar system has been carried out largely by robot spacecraft sent to the other planets. To escape Earth, these craft must achieve **escape velocity,** which is about 11 km/s, after which they coast to their targets, subject only to minor trajectory adjustments provided by small thruster rockets on board. Note that the escape velocity from a planet is equal to its circular satellite velocity multiplied by $\sqrt{2}$ (approximately 1.4). In interplanetary flight, these spacecraft follow Keplerian orbits around the Sun, modified only when they pass near one of the planets.

While close to its target, a spacecraft is deflected by the planet's gravitational force into a modified orbit, either gaining or losing energy in the process. By carefully choosing the aim point in a planetary encounter, controllers have been able to redirect a flyby spacecraft to a second target. Voyager 2 (Figure 2.12) used a series of gravity-assisted encounters to yield successive flybys of Jupiter (1979), Saturn (1980), Uranus (1986), and Neptune (1989). The Galileo spacecraft to Jupiter, launched in 1989, is flying past Venus once and Earth twice to gain the energy required for it to reach its ultimate goal at Jupiter.

If we wish to orbit a planet, we must slow the spacecraft with a rocket when the spacecraft is near its destination, allowing it to be captured into an elliptical orbit. Additional rocket thrust is required to bring a vehicle down from orbit for a landing on the surface. Finally, if a return trip to Earth is planned, the landed payload must include enough propulsive power to repeat the entire process in reverse; it must take off from the other planet, achieve escape velocity for the return trip to Earth, and then slow down in successive steps to reach Earth orbit and ultimately descend to its landing on the Earth's surface.

2.6 Gravitation With More Than Two Bodies

Until now we have considered the Sun and a planet, or a planet and one of its satellites, as a pair of mutually revolving bodies. Actually, the planets (and different satellites of a planet) exert gravitational forces upon one another as well. These interplanetary attractions cause slight variations from the orbits that would be expected if the gravitational forces between planets were neglected. Unfortunately, the problem of treating the motion of a body that is under the gravitational influence of two or more other bodies is very complicated and can be handled properly only with large computers.

If the exact position of each other body is specified at any given instant, we can calculate the combined gravitational effect of the entire ensemble on any one member of the group. Knowing the force on the body

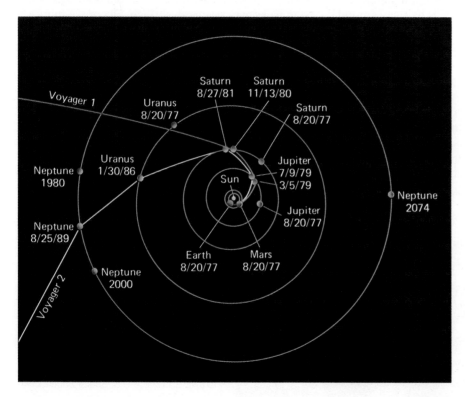

FIGURE 2.12 The flight paths of the two Voyager spacecraft through the outer solar system, taking advantage of the gravitation of each planet to adjust the trajectory toward the next target. *(NASA/JPL)*

in question, we can find how it will accelerate; a knowledge of its initial velocity, therefore, is enough to calculate how it will move in the next instant of time, thus tracking its motion. However, the problem is complicated by the fact that the gravitational acceleration of one body depends on the positions of all the other bodies in the system. Since they, in turn, are accelerated by all the members of the system, we must simultaneously calculate the acceleration of each body produced by the combination of the gravitational attractions of all the others to track the motions of all of them, and hence of any one. Such complex calculations have been carried out, with modern computers, to track the evolution of hypothetical clusters of stars with up to a million members (Figure 2.13).

Gravitational Perturbations

Within the solar system, the problem of computing the orbits of planets and spacecraft is somewhat simpler. We have seen that Kepler's laws, which do not take into account the gravitational effects on an orbit caused by the other planets, really work quite well. This is because these additional influences are very small in comparison to the dominant gravitational attraction of the Sun. Under such circumstances, it is possible to treat the effects of other bodies as small **perturbations** on the force exerted by the Sun. During the 18th and 19th centuries, mathematicians developed many elegant techniques for calculating perturbations, permitting them to predict very precisely the positions of the planets. Such calculations also led to the discovery of a new planet, Neptune, in 1846.

Discovery of Neptune

The discovery of the planet Neptune provides an example of the use of perturbation theory, as well as representing one of the high points in the development of gravitational theory. In 1781, William Herschel discovered, by accident, the seventh planet: Uranus. It happens that Uranus had been observed even a century earlier, but in none of those several earlier observations was it recognized as a planet; rather it was simply recorded as a star.

By 1790, an orbit had been calculated for Uranus on the basis of observations of its motion in the decade following its discovery. Even after allowance was made for the perturbing effects of Jupiter and Saturn, however, it was found that Uranus did not move on an orbit that fitted exactly the earlier observations of it made since 1690. By 1840, the discrepancy between the positions observed for Uranus and those predicted from its computed orbit amounted to about

FIGURE 2.13 Supercomputers at NASA Ames Research Center are capable of tracking the motions of more than a million objects under their mutual gravitation. *(NASA/ARC)*

0.03°— an angle barely discernible to the unaided eye but still larger than the probable errors in the orbital calculations. In other words, Uranus did not seem to move on an orbit that would have been predicted from Newtonian theory.

In 1843, John Couch Adams, a young Englishman who had just completed his studies at Cambridge, began an analysis of the irregularities in the motion of Uranus to see whether they could be produced by an unknown planet. His calculations indicated the existence of a planet more distant than Uranus from the Sun. In October 1845, Adams delivered his results to George Airy, the Astronomer Royal, informing him where in the sky to find the new planet. Adams' predicted position for the unknown body was correct to within 2°.

Meanwhile, Urbain Jean Joseph Leverrier, a French mathematician, unaware of Adams or his work, attacked the same problem and published its solution in June 1846. Airy, noting that Leverrier's predicted position for the unknown planet agreed to within 1° with that of Adams, suggested to James Challis, director of the Cambridge Observatory, that he begin a search for the new object. The Cambridge astronomer, having no up-to-date star charts of the region of the sky in Aquarius where the planet was predicted to be, proceeded by recording the positions of all the faint stars he could observe with his telescope in that location. It was Challis' plan to repeat such plots at intervals of several days, in the hope

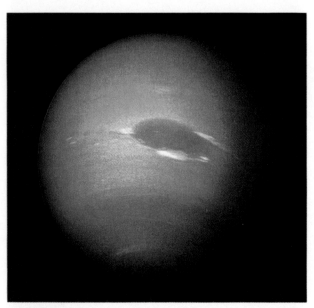

FIGURE 2.14 The planet Neptune as photographed by Voyager in 1989. *(NASA/JPL)*

that the planet would reveal its presence and distinguish itself from a star by its motion. Unfortunately, he was negligent in examining his observations; al-

though he had actually seen the planet, he did not recognize it.

About 1 month later, Leverrier suggested to Johann Galle, an astronomer at the Berlin Observatory, that he look for the planet. Galle received Leverrier's letter on September 23, 1846, and, possessing new charts of the Aquarius region, he found and identified the planet that very night. It was less than a degree from the position Leverrier predicted (Figure 2.14).

The discovery of the eighth planet, now known as Neptune (Latin for the Greek Poseidon, god of the sea), was a major triumph for gravitational theory, for it dramatically confirmed the generality of Newton's laws. The honor for the discovery is properly shared by the two mathematicians, Adams and Leverrier.

The discovery of Neptune was not a complete surprise to astronomers, who had long suspected the existence of the planet. On September 10, 1846, John Herschel, son of the discoverer of Uranus, remarked in a speech before the British Association, ''We see it as Columbus saw America from the shores of Spain. Its movements have been felt trembling along the far-reaching line of our analysis with a certainty hardly inferior to ocular demonstration.''

SUMMARY

2.1 The Dane Tycho Brahe (1546–1601) was the most skillful of the pre-telescopic astronomical observers. His accurate observations of planetary positions provided the data used by the German Johannes Kepler (1571–1630) to derive the three fundamental laws of planetary motion that bear his name. Kepler's laws are as follows (1) Planetary orbits are **ellipses** (described by the **semimajor axis** and **eccentricity**) with the Sun at one **focus.** (2) In equal intervals, a planet sweeps out equal areas. (3) If times are expressed in years and distances in **astronomical units,** the relationship between period (P) and semimajor axis (A) of an orbit is given by ($P^2 = D^3$).

2.2 In his *Principia* Isaac Newton (1643–1727) established the three laws that govern the motion of objects: (1) the law of inertia; (2) the law that relates force and acceleration; and (3) the law that relates each action to an equal and opposite reaction. **Momentum** is proportional to mass multiplied by speed, and **angular momentum** is a measure of motion of a spinning or revolving object. **Density** is mass divided by volume.

2.3 Gravitation is the force that keeps the planets in orbit. Newton's law of gravitation relates gravitational force to mass and distance ($F = GM_1M_2/R^2$). Since a spherical object exerts the same gravitational force as a point mass, Newton was able to show the equivalence of weight on

Earth to the gravitational force between objects in space. When Kepler's laws are re-examined in the light of gravitational theory, it becomes clear that the masses of both Sun and planet are important for the third law, which becomes ($(M_1 + M_2)P^2 = D^3$). Mutual gravitational effects permit us to calculate the masses of astronomical objects, from planets to galaxies.

2.4 The lowest point in a satellite orbit is its **perigee,** and the highest point is its **apogee** (corresponding to **perihelion** and **aphelion** for an orbit about the Sun). The planets all follow orbits about the Sun that are nearly circular and in the same plane. Mercury, Venus, Earth, and Mars are the inner or **terrestrial planets.** The **jovian planets,** much more widely spaced, are Jupiter, Saturn, Uranus, and Neptune. Most asteroids are found between Mars and Jupiter in the **asteroid belt,** while comets generally follow orbits of high eccentricity.

2.5 The orbit of an artificial satellite depends on the circumstances of its launch. The **circular satellite velocity** at the Earth's surface is 8 km/s, and the **escape velocity** of 11 km/s is greater by a factor of $\sqrt{2}$. There are many possible interplanetary trajectories, including those that use gravity-assisted flybys of one object to redirect the spacecraft toward its next target.

2.6 Gravitational problems that involve more than two interacting bodies are much more difficult to deal with than two-body problems. They require large computers for accurate solutions. If one object dominates gravitationally, it is possible to calculate the effects of a second object in terms of small **perturbations**. This approach was used by Adams and Leverrier to predict successfully the position of Neptune from its perturbations on the orbit of Uranus.

REVIEW QUESTIONS

1. State Kepler's three laws in your own words.

2. Look up the revolution periods and distances from the Sun for Venus, Earth, Mars, and Jupiter and verify that they obey Kepler's third law.

3. Explain how a quantity of lead could be less massive than a quantity of feathers. Could a certain mass of lead also be less dense than an equal mass of feathers?

4. Explain how Kepler was able to find a relationship (his third law) between the periods and distances of the planets that did not depend on the masses of the planets or the Sun.

5. Is it possible to escape the force of gravity by going into orbit about the Earth? How does the force of gravity in the Soviet Space Station Mir compare with that on the ground?

6. Earth satellites in low orbits (like the U.S. Space Shuttle) require 90 min. for each orbit. Calculate their speed, and note that this circular satellite velocity is equal to the escape velocity (11 km/s) divided by $\sqrt{2}$.

THOUGHT QUESTIONS

7. What is the momentum of a body whose velocity is zero? Does the first law of motion include the case of a body at rest?

8. A body moves in a perfectly circular path at constant speed. What can be said about the presence or absence of forces in such a system?

9. As air friction causes a satellite to spiral inward closer to the Earth, its orbital speed increases. Why?

10. Describe how a space vehicle must be launched if it is to fall into the Sun. Is it harder to reach the Sun or the planet Jupiter from Earth?

11. Use a history book or encyclopedia to find out what was happening in England during Newton's lifetime, and discuss what trends of the time might have contributed to his accomplishments and the rapid acceptance of his work.

PROBLEMS

12. What is the semimajor axis of a circle of diameter 24 cm? What is its eccentricity?

13. If 24 g of material fills a cube 2 cm on a side, what is the density of the material?

14. Draw an ellipse by the procedure described in the text, using a string and two tacks. Arrange the tacks so that they are separated by one-tenth the length of the string. Comment on the appearance of your ellipse. This (if you have been careful in your construction) is approximately the shape of the orbit of Mars.

15. The Earth's distance from the Sun varies from 147.2 million to 152.1 million km/s. What is the eccentricity of its orbit?

Answer: 0.016

16. What would be the period of a planet whose orbit has a semimajor axis of 4 AU? Of an asteroid with a semimajor axis of 10 AU?

17. What is the distance of an asteroid from the Sun (in astronomical units) if it has a period of revolution of eight years? What would be the distance of a planet whose period is 45.66 days?

18. Newton showed that the periods and distances in Kepler's third law depend on the masses of the objects. What would be the period of revolution of the Earth (at 1 AU from the Sun) if the Sun had twice its present mass? What would be the period if the Sun kept its present mass but expanded to three times its current diameter?

19. By what factor would a person's weight at the surface of the Earth be reduced if **(a)** the Earth had its present mass but eight times its present volume? **(b)** the Earth had its present size but only one-third its present mass?

20. If the Sun had eight times its present mass and the Earth's orbit were twice its present size, what would then be the Earth's orbital period?

The Jantar Mantar, a pre-telescopic observatory in Jaipur, India, built in 1734 for the determination of celestial positions and times. The observing techniques were similar to those developed by Tycho. *(David Morrison)*

3

EARTH, MOON, AND SKY

Isaac Newton (1643–1727) had the insight to realize that the force that makes planets fall around the Sun and the force that makes apples fall to the ground are different manifestations of the same thing: gravitation. Newton's work on the laws of motion, gravitation, optics, and mathematics laid the foundation for almost all physical science up to the 20th century.

It has been written that Galileo, following his retraction (forced by the Roman Inquisition) of the doctrine that the Earth rotates and revolves about the Sun, said under his breath, "But nevertheless it moves." The story is perhaps apocryphal, but certainly Galileo knew that the Earth was in motion, whatever the Catholic Church said. It is the motions of the Earth that produce the seasons and give us our concepts of time and date. The most important motions are its *rotation* about an axis that runs through the Earth and its *revolution* about the Sun.

3.1 Earth and Sky

Locating Places on the Earth

We denote positions of places on the Earth by a system of coordinates on the Earth's surface. The Earth's axis of rotation, which fixes the locations of its North and South Poles and of the equator, is the basis for such a system. Two other directions are also defined: East is the direction toward which the Earth rotates, and west is its opposite. At almost any point on the Earth, the four directions—north, south, east, and west—are well defined, in spite of the fact that our planet is round and not flat. The only exceptions are exactly at the North and South Poles, where the directions east and west are ambiguous.

Coordinates on a sphere, such as latitude and longitude, are a little more complicated than on a flat surface. We must define lines on the sphere that are equivalent to the rectangular grid that specifies position on a plane. A **great circle** is any circle on the surface of a sphere whose center is at the center of the sphere. The Earth's equator is a great circle on the Earth's surface halfway between the North and South Poles. We can also imagine a series of great circles that pass through the North and South Poles. These circles are called **meridians;** they intersect the equator at right angles.

A meridian can be imagined passing through any point on the surface of the Earth (Figure 3.1). This meridian specifies the east-west location of that place. Its *longitude* is defined by international agreement as the number of degrees of arc along the equator between the meridian passing through the place and the one passing through Greenwich, England, the site of the old Royal Observatory (Figure 3.2). Longitudes are measured either to the east or to the west of the Greenwich meridian from 0° to 180°. As an example, the longitude of the benchmark in the clock house of the U.S. Naval Observatory in Washington, D.C., is 77.066° W. Note in Figure 3.1 that the number of degrees along the equator between the meridians of Greenwich and Washington is also the angle at which the planes of those two meridians intersect at the Earth's axis.

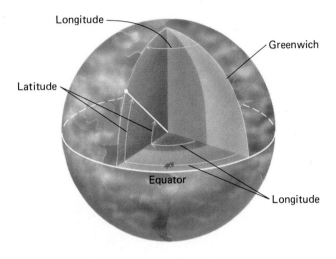

The *latitude* of a place is the number of degrees of arc measured north or south along its meridian from the equator to the place. Latitudes are measured either to the north or to the south of the equator from 0° to 90°. As an example, the latitude of the previously mentioned Naval Observatory benchmark is 38.921° N. Note that the latitude of Washington, D.C., is also the angular distance between it and the equator as seen from the center of the Earth.

Locating Places in the Sky

Positions on the sky are measured in a manner very similar to those on the surface of the Earth, except that instead of latitude and longitude, astronomers use coordinates called declination and right ascension (RA). In denoting positions of objects in the sky, it is often convenient to make use of the fictitious celestial

FIGURE 3.2 The Royal Greenwich Observatory, England— internationally agreed upon zero point of longitude on the Earth. *(David Morrison)*

sphere, a concept, we recall, that many early peoples accepted literally. We saw in Chapter 1 that the sky appears to rotate about points in line with the North and South Poles of the Earth—the north celestial pole and the south celestial pole. Halfway between the celestial poles, and thus 90° from each, is the celestial equator, a great circle on the celestial sphere that is in the same plane as the Earth's equator.

Declination on the celestial sphere is measured the same way that latitude is measured on the sphere of the Earth, from the equator toward the north (positive) or south (negative). **Right ascension (RA)** is like longitude, except that instead of Greenwich, its point of origin is the vernal equinox, the point on the sky where the ecliptic crosses the celestial equator. RA can be expressed in units of angle (degrees) or in units of time, recognizing that the celestial sphere seems to turn around the Earth once in 24 h. Each 15° of arc is equal to 1 h of time. Thus the celestial coordinates of the bright star Vega (Appendix 14) are RA 18h 36.2m (=279.05°) and declination +38.77°.

It helps to visualize these circles in the sky if we imagine the Earth as a hollow, transparent, spherical shell with the terrestrial coordinates (latitude and longitude) painted on it. Then we imagine ourselves at the center of the Earth, looking out through its transparent surface to the sky. The terrestrial poles, equator, and meridians will be superimposed upon the celestial ones.

An alternative approach indicates the position of objects in the sky simply from the perspective of the observer standing on the Earth. Here the positions are measured with respect to the horizon and the point directly overhead, the zenith. Angles measured around the horizon are called **azimuth,** and those measured up from the horizon toward the zenith are called **altitude.** The altitude-azimuth system works very well for specifying, for example, the position of a mountain peak or the flight of a plane. It is less useful for astronomical objects, since the altitude and azimuth of a star constantly change as the celestial sphere appears to rotate overhead.

The Orientation of the Celestial Sphere

Try to visualize the orientation of the celestial sphere with respect to the zenith and horizon of a particular observer on the Earth. At the North (or South) Pole, the problem is simple. The north celestial pole, directly over the Earth's North Pole, appears at the zenith. The celestial equator, 90° from the celestial poles, lies along the horizon. As the Earth rotates, the sky turns about the zenith, and the stars neither rise nor set but circle parallel to the horizon. Only that

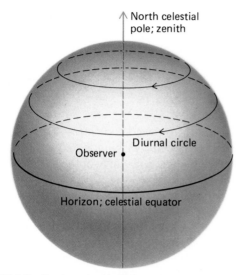

FIGURE 3.3 Sky from the North Pole; the apparent paths of all the stars are parallel to the horizon.

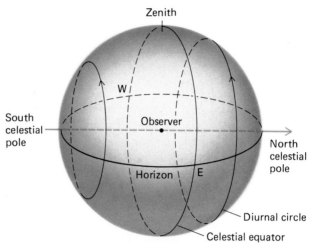

FIGURE 3.4 Sky from the equator; here, stars rise and set perpendicular to the horizon.

half of the sky that is north of the celestial equator is ever visible to an observer at the North Pole (Figure 3.3). Similarly, an observer at the South Pole would see only the southern half of the sky.

At the equator, the problem is almost as simple. The celestial equator, in the same plane as the Earth's equator, passes through the zenith. It runs east and west, intersecting the horizon at the east and west points. The celestial poles, being 90° from the celestial equator, must be at the north and south points on the horizon. As the sky turns, all stars rise and set; they move straight up from the east side of the horizon and set straight down on the west side. During a 24-h period, all stars are above the horizon exactly half the time (Figure 3.4).

Evidently, at points on the Earth between the equator and poles, one of the celestial poles must be a certain distance above the horizon. For an observer between the equator and North Pole, say, at 34° north latitude, the situation is as depicted in Figure 3.5. Here the north celestial pole is 34° above the observer's northern horizon. The south celestial pole is 34° below the southern horizon. As the Earth turns, the whole sky seems to pivot about the north celestial pole. For this observer, stars within 34° of the North Pole can never set. They are always above the horizon, day and night. This part of the sky is called the north **circumpolar zone.** To observers in the continental United States, the Big and Little Dippers and Cassiopeia are examples of star groups that are in the north circumpolar zone. On the other hand, stars within 34° of the south celestial pole never rise. That part of the sky is the south circumpolar zone. To most U.S. observers, the Southern Cross is in that zone. At

the North and South Poles, half the sky is circumpolar above the horizon and half below.

Stars north of the celestial equator, but outside the north circumpolar zone, lie above the horizon in the greater parts of their daily paths; hence they are up more than half the time. Stars on the celestial equator are up exactly half the time, while stars south of the celestial equator, but outside the south circumpolar zone, are up less than half the time. You are probably most conscious of this fact in the case of the Sun; when it is north of the equator (spring and summer), it is up more than 12 h each day, and when it is south of the equator (fall and winter), there are fewer than 12 h

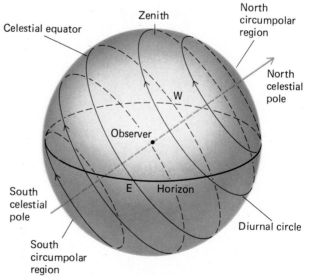

FIGURE 3.5 Sky from latitude 34° N. Stars in the north circumpolar region never set, while those in the south circumpolar region never rise at this latitude.

of daylight. In the Southern Hemisphere, these directions are reversed, and summer takes place when the Sun is south of the equator.

The Foucault Pendulum

We have seen that the apparent rotation of the celestial sphere could be accounted for either by a daily rotation of the sky around a stationary Earth or by the rotation of the Earth itself. Since the 17th century, it has been generally accepted that the Earth turns, but not until the 19th century did the French physicist Jean Foucault provide a direct and unambiguous demonstration of this rotation. In 1851, he suspended a 60-m pendulum weighing about 25 kg from the domed ceiling of the Pantheon in Paris and started the pendulum swinging evenly. If the Earth were not turning, there would be no force to alter the plane of oscillation of the pendulum, and it would continue tracing the same path. Yet after a few minutes it was apparent that the pendulum's plane of motion was turning. Foucault knew as we do that it was not the pendulum that was shifting, but rather the Earth that was turning underneath it.

A full understanding of the principle of the Foucault pendulum is easiest if we imagine the experiment being performed at the North Pole. There the Earth turns under the pendulum, rotating the building and the observers once each day. At other latitudes the effects are somewhat more complicated, with the

apparent rotation taking longer than 24 h, but the idea is the same. This experiment was one of the first direct proofs that the Earth does indeed rotate on its axis.

3.2 The Seasons

The Earth's orbit around the Sun is an ellipse, its distance from the Sun varying by about 3 percent. However, the changing distance of the Earth from the Sun is *not* the cause of the seasons. The seasons result because the plane in which the Earth revolves is not the same as the plane of the Earth's equator. The planes of the equator and ecliptic are inclined to each other by about 23°, the tilt of the rotational axis of the Earth. The Northern Hemisphere is inclined toward the Sun in June and away from it in December.

The Seasons and Sunshine

Figure 3.6 shows the Earth's path around the Sun. The line *EE* is in the plane of the celestial equator. We see in the figure that on about June 22 (the date we who live in the Northern Hemisphere call the *summer solstice*), the Sun shines down most directly upon the Northern Hemisphere of the Earth. It appears 23° north of the equator and thus on that date passes through the zenith of places on the Earth that are at 23° north latitude. The situation is shown in detail in

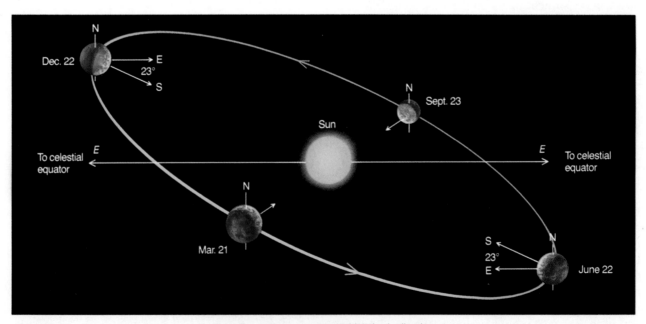

FIGURE 3.6 The Earth's path around the Sun. The seasons are caused by the inclination of the plane of the Earth's orbit to the plane of the equator.

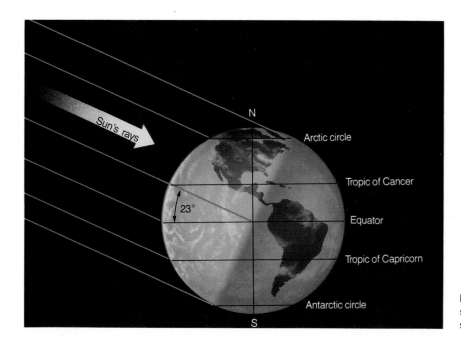

FIGURE 3.7 The Earth on June 22, the summer solstice in the Northern Hemisphere.

Figure 3.7. To a person at latitude 23° N (in Hawaii, for example), the Sun is overhead at noon. This latitude on the Earth, at which the Sun can appear at the zenith at noon on the first day of summer, is called the Tropic of Cancer. We see also in Figure 3.7 that the Sun's rays shine down past the North Pole; all places within 23° of the pole have sunshine for 24 h on the first day of summer. The Sun is as far north on this date as it can get; thus, 67° N is the southernmost latitude where the Sun can ever be seen for a full 24-h period (the midnight Sun). That circle of latitude is called the Arctic Circle. At the same time, all places within 23° of the South Pole—that is, south of the Antarctic Circle—do not see the Sun at all.

The situation is reversed six months later, about December 22 (the date of the *winter solstice*), as shown in Figure 3.8. Now it is the Arctic Circle that has a 24-h night and the Antarctic Circle that has the midnight Sun. At latitude 23° S, the Tropic of Capricorn, the Sun passes through the zenith at noon. It is winter in the Northern Hemisphere and summer in the Southern Hemisphere.

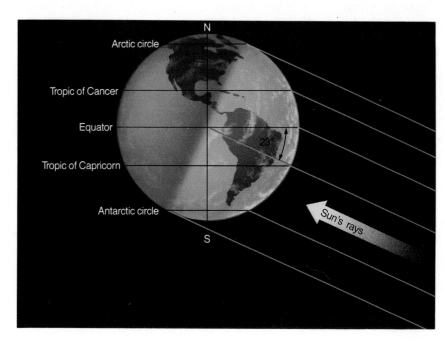

FIGURE 3.8 The Earth on December 22, the winter solstice in the Northern Hemisphere.

Finally, we see in Figure 3.6 that on about March 21 and September 23 the Sun appears to be in the direction of the celestial equator. Every place on the Earth then receives exactly 12 h of sunshine and 12 h of night. As we have seen, the points where the Sun crosses the celestial equator are called the *vernal* (spring) and *autumnal* (fall) *equinoxes*.

The Seasons at Different Latitudes

Figure 3.9 shows the aspect of the sky at a typical latitude in the United States or Europe. During spring and summer, the Sun is north of the equator and is thus up more than half the time. A typical spot in the United States, on the first day of summer (about June 22), receives about 15 h of sunshine. Also, notice that the Sun appears high in the sky, and sunlight is more effective in heating than in the fall and winter, when the Sun appears at a lower altitude in the sky.

In the fall and winter, the Sun is south of the equator, so it is up less than half the time. On about December 22, a typical city at, say, 35° north latitude, receives only 9 or 10 h of sunshine. Also, the Sun is low in the sky; a bundle of its rays is thus spread out over a larger area on the ground than in summer. Because the energy is spread out over a larger area, there is less for each square meter, and so the Sun at low altitudes is less effective in heating the ground.

The seasonal effects are different at different latitudes. At the equator, for instance, all seasons are much the same. Every day of the year, the Sun is up half the time, so there are always 12 h of sunshine and 12 h of night. As one travels north or south, the seasons become more pronounced, until we reach extreme cases in the Arctic and Antarctic. At the

North Pole, all celestial objects that are north of the celestial equator are always above the horizon and, as the Earth turns, circle around parallel to it. The Sun is north of the celestial equator from about March 21 to September 23, and so at the North Pole, the Sun rises when it reaches the vernal equinox and sets when it reaches the autumnal equinox. Each year there are six months of sunshine at each pole, followed by six months of darkness.

A Clarification

In the above paragraphs we have been describing the rising and setting of the Sun and stars as they would appear if the Earth had little or no atmosphere. In reality, however, the atmosphere has the curious effect of allowing us to see a little way "over the horizon." This effect is a result of refraction, an optical phenomenon we will discuss in Chapter 6. Because of this atmospheric refraction, the Sun appears to rise earlier, and to set later, than it would if the atmosphere were not present. In addition, the atmosphere scatters light and provides some twilight illumination even when the Sun is below the horizon. Astronomers define morning twilight as beginning when the Sun is still 18° below the horizon, while evening twilight extends until the Sun sinks more than 18° below the horizon.

These atmospheric effects require small corrections in many of the statements we have made about the seasons. At the equinoxes, for example, the Sun appears to be above the horizon for a few min longer than 12 h and below the horizon for less than 12 h. These effects are most dramatic at the Earth's poles, where the Sun actually rises more than a week before it reaches the equator. Also, as a consequence of twilight, the period of real darkness at each pole lasts for only about three months of each year, rather than six.

3.3 Keeping Time

The measurement of time is based on the rotation of the Earth. Throughout most of human history, time has been reckoned by the positions of the Sun and stars in the sky. Only recently have mechanical and electronic clocks taken over this function in regulating our lives.

The Length of the Day

The most fundamental astronomical unit of time is the day, measured in terms of the rotation of the Earth.

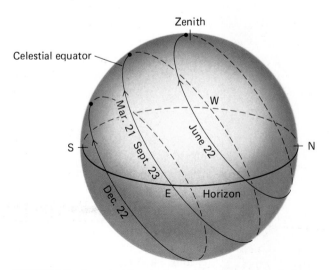

FIGURE 3.9 Aspect of the sky for various dates at a typical place in the temperate Northern Hemisphere.

There is, however, more than one way to define the day. Usually, the day is the rotation period of the Earth with respect to the Sun, called the **solar day.** After all, for most people the time of sunrise is more important than the time Arcturus or some other star rises, so we set our clocks to some version of Sun time. However, astronomers also use a **sidereal day,** which is defined in terms of the rotation period of the Earth with respect to the stars.

A solar day is slightly longer than a sidereal day, as a study of Figure 3.10 will show. Suppose we start the day when the Earth's orbital position is at *A*, with the Sun on the meridian of an observer at point *O* on the Earth. When the Earth has completed one rotation with respect to the distant stars (*C*), the Sun will not yet have advanced to the meridian, as a result of the movement of the Earth along its orbit from *A* to *B*. To complete a solar day, the Earth must rotate an additional amount, equal to 1/365 of a full turn. The time required for this extra rotation is 1/365 of a day, or about 4 min, so the solar day is about 4 min longer than the sidereal day.

Apparent Solar Time

Apparent solar time is reckoned by the position of the Sun in the sky (or, during the night, its position below the horizon). This is the kind of time that would be indicated by a sundial, and it probably represents the earliest measure of time used by ancient civilizations. Today we adopt the middle of the night as the starting point and measure time in hours elapsed since midnight.

During the first half of the day, the Sun has not yet reached the meridian. We designate those hours as before midday (*ante meridiem,* or A.M.). We customarily start numbering the hours after noon over again, and designate them by P.M. (*post meridiem*) to distinguish them from the morning hours. On the other hand, it is often useful to number the hours from 0 to 24, starting from the beginning of the day at midnight. For example, in various conventions, 7:46 P.M. may be written as 19h 46m, 19:46, or simply 1946.

Although apparent solar time seems simple, it is not really very convenient to use. The exact length of an apparent solar day varies slightly during the year. The eastward progress of the Sun in its annual journey around the sky is not uniform because of the slightly varying speed of the Earth in its elliptical orbit and the fact that the Earth's axis of rotation is not perpendicular to the plane of its revolution. Thus apparent solar time does not advance at a uniform rate. After the invention of clocks that could run at a uniform rate, it became necessary to abandon the

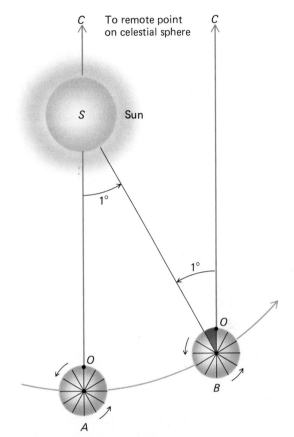

FIGURE 3.10 Sidereal and solar days.

apparent solar day as the fundamental unit of time. Otherwise, all clocks would have to be adjusted to run at a different rate each day.

Mean Solar Time and Standard Time

Mean solar time is based on the mean solar day, which is defined as having a duration equal to the average length of an apparent solar day. Although mean solar time has the advantage of progressing at a uniform rate, it is still somewhat inconvenient for practical use, because it is referred to the apparent position of a fictitious "average" Sun. Naturally, this position will differ, depending on the longitude of the observer. Thus observers on different north-south lines on the Earth keep different mean solar times. If mean solar time were strictly observed, people traveling east or west would have to reset their watches continually as the longitude changed, if they were always to read the local mean time correctly. For instance, a commuter traveling from Oyster Bay to New York City would have to adjust the time on the trip through the East River tunnel, because Oyster Bay time is actually about 1.6 min more advanced than that of Manhattan.

Until near the end of the last century, every city and town in the United States kept its own local mean time. With the development of railroads and the telegraph, however, the need for some kind of standardization became evident. In 1883, the nation was divided into four standard time zones (now five, including Hawaii and Alaska). Within each zone, all places keep the same standard time, the local mean solar time of a standard meridian running more or less through the middle of each zone. Now travelers reset their watches only when the time change has amounted to a full hour. For local convenience, the boundaries between the U.S. time zones are chosen to correspond, as much as possible, to divisions between states. Since 1884, standard time has been in use around the world by international agreement. Almost all countries have adopted one or more standard time zones, although one of the largest nations, India, has settled on a half-zone, being $5\frac{1}{2}$ hours from Greenwich standard.

Daylight saving time is simply the local standard time of the place plus 1 h. It has been adopted for spring and summer use in most states in the United States and in many countries to prolong the sunlight into evening hours, on the apparent theory that it is easier to change the time by government action than it would be for individuals or businesses to adjust their own schedules to produce the same effect.

The International Date Line

The fact that time is always more advanced to the east presents a problem. Suppose you travel eastward around the world. You pass into a new time zone, on the average, for about every 15° of longitude you travel, and each time you dutifully set your watch ahead an hour. By the time you have completed your trip, you have set your watch ahead through a full 24 h, and thus gained a day over those who stayed at home.

The solution to this dilemma is the **international date line,** set by international agreement to run approximately along the 180° meridian of longitude. The date line runs about down the middle of the Pacific Ocean, although it jogs a bit in a few places to avoid cutting through groups of islands and through Alaska. By convention, at the date line, the date of the calendar is changed by one day. Crossing the date line from west to east, thus advancing your time, you compensate by decreasing the date; crossing from east to west, you increase the date by one day. We simply must accept that the date will differ in different cities at the same time. A well-known example is the date when the Imperial Japanese Navy bombed Pearl Harbor in Hawaii, known in the United States as Sunday, December 7, 1941, but taught to Japanese students as taking place on Monday, December 8.

3.4 The Calendar

The Challenge of the Calendar

There are two traditional functions of any calendar. First, it must keep track of the passage of time, allowing people to anticipate the cycle of the seasons and to honor special religious anniversaries. Second, in order to be useful to a large number of people, it must address the problem of the relationships between the basic natural time intervals defined by the motions of the Earth, Moon, and sometimes even the planets. The natural units of the calendar are the *day,* based on the period of rotation of the Earth; the *month,* based on the period of revolution of the Moon about the Earth; and the *year,* based on the period of revolution of the Earth about the Sun. Difficulties have resulted from the fact that these three periods are not commensurable—that is, one does not divide evenly into any of the others.

The rotation period of the Earth is, by definition, 1.0000 day. The period required by the Moon to complete its cycle of phases, called the lunar month, is 29.5306 days. The basic period of revolution of the Earth, called the tropical year, is 365.2422 days. The ratios of these numbers (1.0000/29.5306/365.2422) are not very convenient for calculations. Our natural clocks run on different time. This is the historic challenge of the calendar, dealt with in various ways by different cultures.

Early Calendars

Even the earliest cultures were concerned with the keeping of time and the calendar. Particularly interesting are monuments left by Bronze Age people in northwestern Europe, especially in the British Isles. The best preserved of the monuments is Stonehenge (Figure 3.11), about 13 km from Salisbury in southwest England. It is a complex array of stones, ditches, and holes arranged in concentric circles. Carbon dating and other studies show that Stonehenge was built during three periods ranging from about 2500 B.C. to about 1700 B.C. Some of the stones are aligned with the directions of the Sun and Moon during their risings and settings at critical times of the year (such as the beginnings of summer and winter), and it is generally believed that at least one function of the monument was connected with the keeping of a calendar.

FIGURE 3.11 Stonehenge, a megalithic monument, possibly an observatory or calendar-keeping device, located in the Salisbury Plain of England. *(David Morrison)*

The Maya in Central America were also concerned with the keeping of time. The Mayan calendar was as sophisticated, and perhaps more complex, than contemporary calendars in use in Europe. The Maya did not attempt to correlate their calendar accurately with the length of the year or lunar month. Rather, their calendar was a system for keeping track of the passage of days and for counting time far into the past or future. Among other purposes, their calendar was useful for predicting astronomical events—for example, the positions of Venus in the sky (Figure 3.12).

The ancient Chinese developed an especially complex calendar, largely limited to a few privileged hereditary court astronomer-astrologers. In addition to the motions of the Earth and Moon, they were able to fit in the approximately 12-year cycle of Jupiter, which was central to their system of astrology. The Chinese still preserve some aspects of this system in their cycle of 12 "years"—the Year of the Dragon, the Year of the Pig, and so on—that are defined by the position of Jupiter in the zodiac.

Our calendar derives from Greek calendars dating from at least the eighth century B.C. They led, eventually, to the calendar introduced by Julius Caesar, which approximated the year by 365.25 days, fairly close to the actual value of 365.2422. The Romans achieved this approximation by declaring years to have 365 days each, with the exception of every fourth year. The *leap year*, which was to have one extra day, bringing its length to 366 days, thus brought the average length of the year in the Julian calendar to 365.25 days.

By this time, the Romans had dropped the almost impossible task of trying to base their calendar on the Moon as well as the Sun, although a vestige of older lunar systems can be seen in the fact that our months have an average length of about 30 days. However, lunar calendars remained in use in other cultures, and Islamic calendars are still primarily lunar rather than solar.

FIGURE 3.12 Ruins of the Caracol, a Mayan Observatory at Chichin Itza in the Yucatan, Mexico. *(David Morrison)*

The Gregorian Calendar

Although the Julian calendar (which was adopted by the early Christian Church) represented a great advance, its average year still differed from the true year by about 11 min, an amount that accumulates over the centuries to an appreciable error. By 1582, that 11 min per year had added up to the point where the first day of spring was occurring on March 11 instead of March 21. If the trend were allowed to continue, eventually Easter would be occurring in early winter. Pope Gregory XIII, a contemporary of Galileo, felt it necessary to institute a further calendar reform.

The Gregorian calendar reform consisted of two steps. First, ten days had to be dropped out of the calendar to bring the vernal equinox back to March 21; by proclamation, the day following October 4, 1582, became October 15. The second feature of the new Gregorian calendar was that the rule for leap year was changed so that the average length of the year would more closely approximate the tropical year. Gregory decreed that three of every four century years, all leap years under the Julian calendar, would be common years henceforth. The rule was that only century years divisible by 400 should be leap years. Thus, 1700, 1800, and 1900—all divisible by 4 but not by 400—were not leap years in the Gregorian

calendar. On the other hand, the years 1600 and 2000, both divisible by 400, are leap years under both systems. The average length of this Gregorian year, 365.2425 mean solar days, is correct to about one day in 3300 years.

The Catholic countries immediately put the Gregorian reform into effect, but countries under control of the Eastern Church and most Protestant countries did not adopt it until much later. It was 1752 when England and the American colonies finally made the change. By parliamentary decree, September 2, 1752, was followed by September 14. Although special laws were passed to prevent such abuses as landlords' collection of a full month's rent for September, there were still riots, and people demanded their 12 days back. Russia did not abandon the Julian calendar until the time of the Bolshevik revolution. The Russians then had to omit 13 days to come into step with the rest of the world.

3.5 Phases and Motions of the Moon

After the Sun, the Moon is the brightest and most obvious object in the sky. Equally obvious is its monthly cycle of phases, which results from the changing angle of its illumination by the Sun.

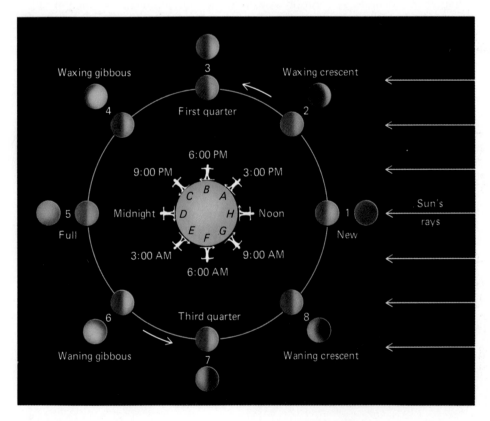

FIGURE 3.13 Phases of the Moon and the time of day. (The other series of figures shows the Moon at various phases as seen in the sky from the Earth's surface.)

Lunar Phases

The Moon is said to be *new* when it is in the same general direction from Earth as the Sun (position *A* in Figure 3.13), with its daylight side turned away from us and its night side turned toward us. In this phase it is invisible. A day or two after the new phase, the thin *crescent* first appears (position *B*) as we see a small part of the Moon's daylight hemisphere. The illuminated crescent increases in size on successive days as the Moon moves farther and farther around the sky away from the direction of the Sun.

After about one week, the Moon is one-quarter of the way around its orbit (position *C*) and is at the *first quarter* phase. Here half of the Moon's daylight side is visible. During the week after the first quarter phase, we see more and more of the Moon's illuminated hemisphere (position *D*), until the Moon (at *E*) and the Sun are opposite each other in the sky. The side of the Moon turned toward the Sun is then also turned toward the Earth, and we have *full phase*. During the next two weeks, the Moon goes through the same phases again in reverse order, returning to new phase after about 29½ days.

If you have difficulty picturing the phases of the Moon from this verbal account, try this simple experiment: Stand about 6 ft in front of a bright electric light outdoors at night and hold in your hand a small round object such as a tennis ball or an orange. If the object is then viewed from various sides, the portions of its illuminated hemisphere that are visible represent the analogous phases of the Moon.

The best way to become fully acquainted with these lunar phases is to watch the real Moon in the sky. Observe its shape, its direction from the Sun, and its time of setting and rising. During most of the month, you will have no trouble seeing the Moon in broad daylight.

The Moon's Revolution and Rotation

The Moon's sidereal period, that is, the period of its revolution about the Earth measured with respect to the stars, is a little over 27 days—27.3217 days, to be exact. The time interval in which the phases repeat — say from full to full—is 29.5306 days. The Moon changes its position on the celestial sphere rather rapidly, moving about 13° to the east each day. Even during a single evening, the Moon creeps visibly eastward among the stars, traveling its own width in a little less than 1 h. The delay in moonrise from one day to the next, caused by this eastward motion, averages about 50 min.

The Moon *rotates* on its axis with exactly the same period at which it *revolves* about the Earth. As a consequence, the Moon keeps the same face turned toward the Earth; the differences in its appearance from one night to the next are due to changing illumination by the Sun, not to its own rotation (Figure 3.14).

We sometimes hear the back side of the Moon (the side we never see) called the "dark side." Of course, the back side is dark no more frequently than the

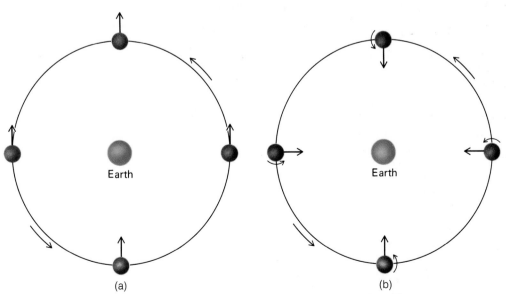

(a)   (b)

FIGURE 3.14 (a) If the Moon did not rotate, it would turn all its sides to our view. (b) Actually, it does rotate in the same period as it revolves, so we always see the same side. (Thus there is a "back side" or "far side," but certainly not a "dark side.")

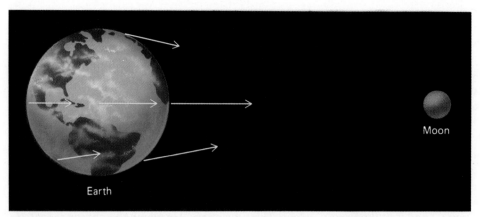

FIGURE 3.15 The Moon's differential attraction of different parts of the Earth.

front side. Since the Moon rotates, the Sun rises and sets on all sides of the Moon.

3.6   Ocean Tides

Anyone living near the sea is familiar with the twice-daily rising and falling of the tides. Early in history it was realized that tides are related to the Moon, because the daily delay in high tide is the same as the daily delay in successive times of moonrise. A satisfactory explanation of the tides, however, awaited the theory of gravitation, supplied by Newton.

The Pull of the Moon on the Earth

The gravitational forces exerted by the Moon at several arbitrarily selected places on the Earth are illustrated in Figure 3.15. These forces differ slightly from one another because of the Earth's finite size; all parts are not equally distant from the Moon, nor are they all in exactly the same direction from the Moon. If the Earth retained a perfectly spherical shape, the combination of all these forces would be that of the force on a point mass, equal to the mass of the Earth and located at the Earth's center. This is approximately true, because the Earth is nearly spherical.

The Earth, however, is not perfectly rigid. Consequently, the differences between the forces of the Moon's attraction on different parts of the Earth (called differential forces) cause the Earth to distort slightly. The side of the Earth nearest the Moon is attracted toward the Moon more strongly than is the center of the Earth, which, in turn, is attracted more strongly than is the side of the Earth opposite the Moon. Thus, the differential forces tend to "stretch" the Earth slightly into a prolate spheroid (a football), with its long diameter pointed toward the Moon.

If the Earth were fluid, like water, it would distort

until the Moon's differential forces over different parts of its surface came into equilibrium with the Earth's own gravitational forces pulling it together. Calculations show that in this case the Earth would distort from a sphere by amounts ranging up to nearly 1 m. Measurements have been made to investigate the actual deformation of the Earth, and it is found that the solid Earth does distort, as would a liquid, but only about one-third as much because of the great rigidity of the Earth's interior.

Since the tidal distortion of the solid Earth amounts at its greatest to only about 20 cm, the Earth does not distort enough to achieve an equilibrium shape, and the Moon's differential gravitational forces are not completely balanced by the Earth's own gravitation. Hence objects at its surface experience tiny horizontal tugs, tending to make them slide about. These so-called tide-raising forces are too insignificant to notice or to affect solid objects or crustal rocks, but they do affect the waters in the oceans.

The Formation of Tides

The tide-raising forces, acting over a number of hours, produce motions of the water that result in measurable tidal bulges in the oceans. Water on the lunar side of the Earth is drawn toward the sublunar point (the point on the Earth where the Moon appears in the zenith), piling up water to greater depths on that side of the Earth, with the greatest depths at the sublunar point. On the opposite side of the Earth, water flows in the opposite direction, producing a tidal bulge on the side of the Earth opposite the Moon (Figure 3.16).

Note that the tidal bulges in the oceans do not result from the Moon's compressing or expanding the water, nor from the Moon's lifting the water "away from the Earth." Rather, they result from an actual flow of water over the Earth's surface, toward the

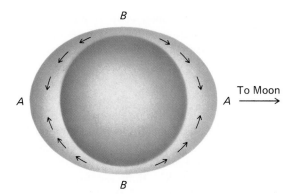

FIGURE 3.16 Tidal bulges in the "ideal" oceans.

regions below and opposite the Moon, causing the water to pile up to greater depths at those places.

In the idealized picture just described, not actually realized even in the largest oceans, the tides would cause the depths of the ocean to range through only a few feet. The rotation of the Earth would carry an observer at any given place alternately into regions of deeper and shallower water. An observer being carried toward the regions under or opposite the Moon where the water was deepest, would say, "The tide is coming in"; when carried away from those regions, the observer would say, "The tide is going out." During a day, the observer would be carried through two tidal bulges (one on each side of the Earth) and so would experience two high tides and two low tides.

The Sun also produces tides on the Earth, although the Sun is less than half as effective a tide-raising agent as the Moon. Actually, the gravitational attraction between the Sun and the Earth is about 150 times as great as that between the Earth and the Moon.

However, the Sun is so distant that it attracts all parts of the Earth with almost equal strength. The Moon, on the other hand, is close enough for its attraction on the near side of the Earth to be substantially greater than its attraction on the far side. In other words, its differential gravitational pull on the Earth is greater than the Sun's, even though its total gravitational attraction is less.

If there were no Moon, the tides produced by the Sun would be all we would experience, and the tides would be less than half as great as those we now have. The Moon's tides, therefore, dominate. On the other hand, when the Sun and Moon are lined up, that is, at new or full phase, the tides produced by the Sun and Moon reinforce each other and are greater than normal. These are called spring tides. *Spring tides* (which have nothing to do with spring) are approximately the same, whether at new Moon or full Moon, because tidal bulges occur on both sides of the Earth—the side toward the Moon (or Sun) and the side away from the Moon (or Sun). In contrast, when the Moon is at first quarter or last quarter (at right angles to the Sun's direction), the tides produced by the Sun partially cancel the tides of the Moon, and the tides are lower than usual. These are *neap tides* (Figure 3.17).

The "simple" theory of tides, described in the preceding paragraphs, would be sufficient if the Earth were completely surrounded by very deep oceans, and if it rotated very slowly. However, the presence of land masses stopping the flow of water, the friction in the oceans and between oceans and the ocean floors, the rotation of the Earth, the variable depth of the ocean, winds, and so on all complicate the picture.

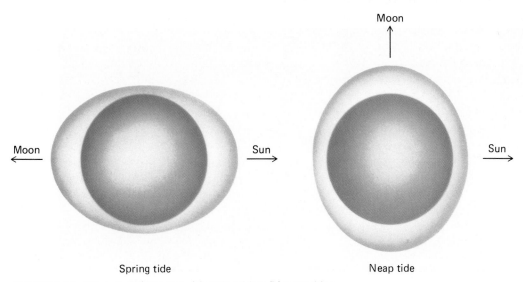

Spring tide

Neap tide

FIGURE 3.17 Tides of different size: (a) spring tides; (b) neap tides.

The Earth's rapid rotation causes the tide-raising forces within a given mass of water to vary too rapidly for the water to adjust completely to them. These forces, however, recurring periodically, set up oscillations, making the oceans slosh back and forth in their basins, so that the water over a large area rises and lowers in step. Consequently, the highest water does not necessarily occur when the Moon is highest in the sky (or lowest below the horizon) but rather when the oscillations of the ocean, produced by the tidal forces acting upon it, pile up the water to its greatest depth at that location. The magnitude of the tide thus depends critically upon the shape and depth of the adjacent ocean basin.

3.7   Eclipses of the Sun and Moon

One of the fortunate coincidences of nature is that the two most prominent astronomical objects, the Sun and the Moon, have so nearly the same apparent size in the sky. Although the Sun is about 400 times as large in diameter as the Moon is, it is also about 400 times as far away, so both the Sun and the Moon have the same angular size—about 1/2°. Consequently, the Moon, as seen from Earth, can appear to barely cover the Sun, producing one of the most impressive events in nature.

Any solid object in space casts a shadow by blocking the light of the Sun from a region behind it; this shadow becomes apparent whenever another object moves into it. In general, an eclipse occurs whenever any part of either the Earth or the Moon enters the shadow of the other. When the Moon's shadow strikes the Earth, people on Earth within that shadow see the Sun covered at least partially by the Moon;

that is, they witness a **solar eclipse.** When the Moon passes into the shadow of the Earth, people on the night side of the Earth see the Moon darken—a **lunar eclipse.**

Eclipses of the Sun

The apparent or angular sizes of both Sun and Moon vary slightly from time to time, as their respective distances from the Earth vary. Much of the time, the Moon is slightly smaller than the Sun and cannot cover it completely, even if the two are perfectly aligned. However, if an eclipse of the Sun occurs when the Moon is somewhat nearer than its average distance, the Moon can completely hide the Sun, producing a total solar eclipse. In other words, a total eclipse of the Sun occurs whenever the dark cone of the Moon's shadow reaches the surface of the Earth.

The geometry of a total solar eclipse is illustrated in Figure 3.18. If Sun and Moon are properly aligned, then the Moon's shadow intersects the ground at a small point on the Earth's surface. Anyone on the Earth within this small area covered by the tip of the Moon's shadow will not see the Sun and will witness a total eclipse. On the other hand, within a larger area of the Earth's surface, one will see part, but not all, of the Sun eclipsed by the Moon—a partial solar eclipse.

As the Moon moves eastward in its orbit, the tip of its shadow sweeps eastward at about 1500 km/h along a thin band across the surface of the Earth, and the total solar eclipse is observed successively along this band. This path across the Earth within which a total solar eclipse is visible (weather permitting) is called the eclipse path. Within a zone about 3000 km on either side of the eclipse path, a partial solar eclipse is visible.

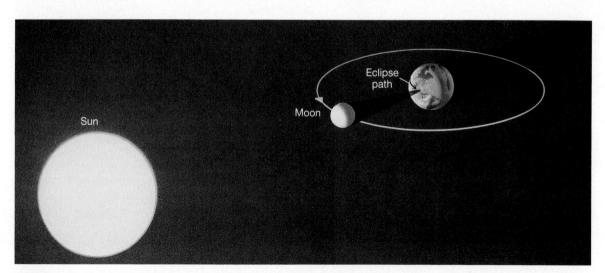

FIGURE 3.18   Geometry of a total solar eclipse (not to scale).

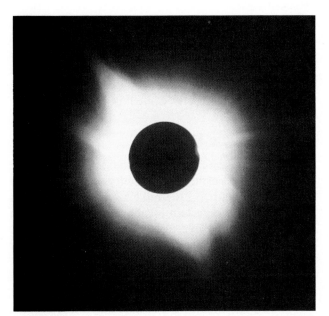

FIGURE 3.19 The corona visible during the July 11, 1991 total solar eclipse photographed from Lomas de Palmira, La Paz, Baja California. Note the two long streamers. *(Stephen J. Edberg, ©1991)*

Because the Moon's conical shadow just barely reaches the Earth, the width of the eclipse path, within which a total eclipse can be seen, is small. Under the most favorable conditions, the path is only 269 km wide in regions near the Earth's equator. At far northern or southern latitudes, because the Moon's shadow falls obliquely on the ground, it can cover a path somewhat more than 269 km wide. It does not take long for the Moon's shadow to sweep past a given point on Earth. The duration of totality may be only a brief instant; it can never exceed about 7 min.

Appearance of a Total Eclipse

Almost anyone who has witnessed a total solar eclipse will advise you to make the effort to see one if you are anywhere near the path of totality; it is a rare and impressive event!

A solar eclipse begins when the Moon just begins to silhouette itself against the edge of the Sun's disk. The partial phase follows, during which more and more of the Sun is covered by the Moon. About an hour after the eclipse begins, the Sun becomes completely hidden behind the Moon. In the few minutes immediately before this period of totality begins, the sky noticeably darkens; some flowers close up, and chickens may go to roost. In the last instant before totality, the only parts of the Sun that are visible are those that shine through the lower valleys in the Moon's irregular profile and line up along the periphery of the advancing edge of the Moon—a phenome-

non called Bailey's beads. A final flash of sunlight through a lunar valley produces a brilliant flare on the disappearing crescent of the Sun—the diamond ring. During totality, the sky is dark enough that planets are visible, and usually the brighter stars as well.

As Bailey's beads disappear and the bright disk of the Sun becomes entirely hidden behind the Moon, the corona flashes into view (Figure 3.19). The **corona** is the Sun's outer tenuous part, consisting of sparse gases that extend for millions of miles in all directions from the apparent surface of the Sun. It is ordinarily not visible because the light of the corona is feeble compared with that from the underlying layers of the Sun that radiate most of the solar energy into space. Only when the brilliant glare from the Sun's visible disk is blotted out by the Moon during a total eclipse is the pearly white corona visible.

The total phase of the eclipse ends, as abruptly as it began, when the Moon begins to uncover the Sun. Gradually, the partial phases of the eclipse repeat themselves, in reverse order, until the Moon has completely uncovered the Sun.

How to Observe Solar Eclipses

The progress of an eclipse can be observed safely by holding a card with a small (1-mm) hole punched in it several feet above a white surface, such as a concrete sidewalk. The hole in the cardboard produces a pinhole camera image of the Sun (Figure 3.20).

FIGURE 3.20 How to watch a solar eclipse safely during its partial phases.

Although there are safe filters through which one can look at the Sun directly, people have suffered eye damage by looking at the Sun through improper filters (or no filter at all). In particular, neutral-density photographic filters are not safe, for they transmit infrared radiation that can cause severe damage to the retina.

Common sense (and pain) prevents most of us from looking at the Sun directly on an ordinary day for more than a brief glance. There is nothing about an eclipse that makes sunlight more dangerous than it is any other time; on the contrary, we receive less light from the Sun when it is partly hidden by the Moon. It is never safe, however, to look at the Sun directly when it is still in partial eclipse; even the thin crescent visible a few minutes before totality has a surface brightness great enough to burn the retina. Unless you have a filter prepared especially for viewing the Sun, it is best to watch the partial phases with a pinhole camera device, as described previously.

It is perfectly safe, however, to look at the Sun directly when it is totally eclipsed, even through binoculars or telescopes. Unfortunately, unnecessary panic has often been created by uninformed public officials acting with the best intentions. One of us witnessed two marvelous total eclipses in Australia, during which townspeople held newspapers over their heads for protection, and schoolchildren cowered indoors, with their heads under their desks. What a cheat to those people to have missed what would have been one of the most memorable experiences of their lifetimes! During totality, by all means look at the Sun.

Nor should you be terrified of accidentally catching a glimpse of the Sun outside totality. How many times have you glanced at the Sun on ordinary days while driving a car or playing ball or tennis? Common sense made you look away at once. Do the same if you glimpse the Sun directly while it is partially eclipsed.

Eclipses of the Moon

A lunar eclipse occurs when the Moon enters the shadow of the Earth. The geometry of a lunar eclipse is shown in Figure 3.21. Unlike a solar eclipse, which is visible only in certain local areas on the Earth, a lunar eclipse is visible to everyone who can see the Moon. Since a lunar eclipse can be seen (weather permitting) from the entire night side of the Earth, lunar eclipses are observed far more frequently from a given place on Earth than are solar eclipses.

An eclipse of the Moon is total only if the Moon's path carries it through the dark central shadow cast by the Earth, which has a diameter about five times the apparent diameter of the Moon. A lunar eclipse, of course, can take place only when the Moon is full. About 20 min or so before the Moon reaches the dark shadow, it dims somewhat as the Earth partly blocks the sunlight. As the Moon begins to dip into the shadow, the curved shape of the Earth's shadow upon it is soon apparent. Aristotle listed the round shape of the Earth's shadow as one of the earliest proofs of the fact that the Earth is spherical.

Even when totally eclipsed, the Moon is still faintly visible, usually appearing a dull coppery red. The illumination on the eclipsed Moon is sunlight that has

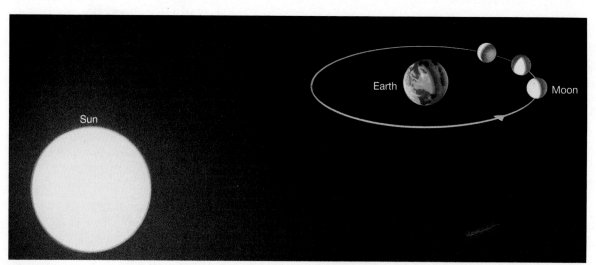

FIGURE 3.21   Geometry of a lunar eclipse (not to scale).

ESSAY Observing The Earth's Shadow

You don't have to wait for an eclipse of the Moon to see the Earth's shadow. Go outside and look at the eastern horizon just after sunset. If it is clear to the west and if there is some haze in the atmosphere to the east, you may see a dark band of bluish-gray extending upward a few degrees above the eastern horizon. This bluish-gray band is the Earth's shadow. If you have a long, clear horizon you should even be able to see that the shadow is curved, as it must be, since the Earth itself is curved. Just above

the dark band, the sky may be tinged with pink. This coloration is produced by the red light of the setting Sun, which passes through the Earth's atmosphere from west to east, with a small part of it being scattered back to the observer.

The shadow of the Earth can be seen more easily from a high altitude. Look for it the next time you are on a mountain or flying in a jet plane at sunset.

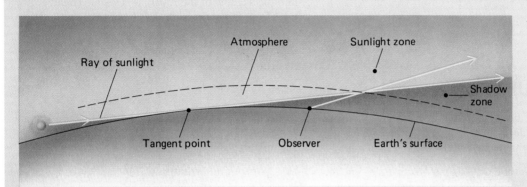

This diagram shows why it is possible to see the shadow of the Earth shortly after sunset. For the observer shown, the Sun is below the horizon, and the Earth blocks the light that would otherwise reach the eastern horizon, thereby casting a shadow.

passed through the Earth's atmosphere and has been bent by the air into the Earth's shadow.

After totality, the Moon moves out of the shadow and the sequence of events is reversed. The total duration of the eclipse depends on how closely the Moon's path approaches the axis of the shadow dur-

ing the eclipse. For a central eclipse, each partial phase consumes at least 1 h, while totality can last as long as 1 h and 40 min. Total eclipses of the Moon occur, on the average, about once every two or three years.

SUMMARY

3.1 The terrestrial system of latitude and longitude makes use of the **great circles** called **meridians.** An analogous celestial coordinate system is based upon **right ascension** and **declination,** with the vernal equinox serving as reference point (like the prime meridian on the Earth). A second system for locating objects in the sky is based on **azimuth** and **altitude.** These coordinate systems help us to visualize the rising and setting of the stars as the Earth rotates on its axis. Stars in the **circumpolar zone** (which depends on the observer's latitude) neither rise nor set.

3.2 The familiar cycle of the seasons results from the 23° tilt of the Earth's axis of rotation. Once we understand the apparent motions of the Sun as seen from different latitudes on the Earth, we can see how changes in the amount of sunlight produce the seasons.

3.3 The basic unit of astronomical time is the day (either the **solar day** or the **sidereal day**). **Apparent solar time** is based on the position of the Sun in the sky, while **mean solar time** relates to a fictitious mean Sun. Variations on mean solar time include standard time and daylight saving time. The convention of the **international date line** is necessary to reconcile times in different parts of the Earth.

3.4 The fundamental problem of the calendar is to reconcile the incommensurable lengths of the day, the month, and the year. Most modern calendars, beginning with the Roman (Julian) calendar of the first century B.C., neglect the problem of the month and concentrate on determining the correct number of days in a year, using such conventions as the leap year. Today, most of the world has adopted the Gregorian calendar established in 1582 by Pope Gregory XIII.

3.5 The Moon's monthly cycle of phases results from the changing angle of its illumination by the Sun. Since its period of revolution is the same as its period of rotation, the Moon always keeps the same face toward the Earth.

3.6 The twice-daily ocean tides are the result primarily of the differential gravitational force of the Moon on the material of the Earth's crust and ocean. These tidal forces cause ocean water to flow into two tidal bulges on opposite sides of the Earth. Actual ocean tides are complicated by the additional effects of the Sun and by the shape of the coasts and ocean basins.

3.7 The Sun and Moon have nearly the same angular size (about 0.5°). A **solar eclipse** occurs when the Moon moves between the Sun and the Earth, casting its shadow on a part of the Earth's surface. If the eclipse is total, the light from the bright disk of the Sun is completely blocked, and the solar atmosphere (the **corona**) comes into view. Solar eclipses take place rarely in any one location, but they are among the most spectacular sights in nature, and you should not miss an opportunity to see one if you have the chance. A **lunar eclipse** takes place when the Moon moves into the shadow of the Earth; it is visible (weather permitting) from the entire night hemisphere of the Earth.

REVIEW QUESTIONS

1. Compare the three coordinate systems: (1) latitude and longitude, (2) declination and right ascension, and (3) altitude and azimuth. What are the advantages of each?

2. What is the latitude of (**a**) the North Pole? (**b**) the South Pole? Why has longitude no meaning at the North or South Pole?

3. When and where on Earth is it possible for the ecliptic to lie on the horizon? Where is it possible for the Sun to be at the north point on the horizon at midnight? If a star rises in the northeast, in what direction does it set?

4. What are the advantages and disadvantages of apparent solar time? How is the situation improved by introducing mean solar time and standard time?

5. Why is it difficult to construct a calendar based on the Moon's cycle of phases?

6. Explain why there are two high tides and two low tides every day. Strictly speaking, should it be a 24-h period during which there are two high tides? If not, what should the interval be?

7. What is the phase of the Moon at the time of a total solar eclipse? At the time of a total lunar eclipse?

THOUGHT QUESTIONS

8. Tell where you are on the Earth from the following descriptions:
   a. The stars rise and set perpendicular to the horizon.
   b. The stars circle the sky parallel to the horizon.
   c. The celestial equator passes through the zenith.
   d. In the course of a year, all stars are visible.
   e. The Sun rises on September 23 and does not set until March 21.

9. In far northern countries, the winter months tend to be so cloudy that astronomical observations are nearly impossible. Why cannot good stellar observations be made at those places during the summer months?

10. About what time does the Moon rise when it is at each of the following phases:

   a. New?
   b. Full?
   c. Third quarter?
   d. Two days past new?
   e. Three days past full?
   f. First quarter?

11. What is the phase of the Moon if (**a**) it rises at 3:00 P.M.? (**b**) it is on the meridian at 7:00 A.M.? (**c**) it sets at 10:00 A.M.?

12. Suppose you lived in the crater Copernicus on the Moon. (**a**) How often would the Sun rise? (**b**) How often would the Earth set? (**c**) Over what fraction of the time would you be able to see the stars?

13. Does the Moon enter the shadow of the Earth from the east or west side? Explain why.

14. Describe what an observer at the crater Copernicus would observe while the Moon is eclipsed. What would the same observer see during what would be a total solar eclipse as viewed from the Earth?

PROBLEMS

15. From the Earth we always see the same hemisphere of the Moon. Verify this fact by making a map of the dark markings on the Moon a few nights before the Moon is full. Now observe the Moon for several more nights. Do the dark markings appear to move?

16. (a) If a star rises at 8:30 P.M. tonight, approximately what time will it rise two months from now? (b) What is the altitude of the Sun at noon on December 22 as seen from a place on the Tropic of Cancer?

17. Suppose the tilt of the Earth's axis were only 16°. What, then, would be the difference in latitude between the Arctic Circle and the Tropic of Cancer? If the tilt were only 16°, what would be the effect on the seasons compared with that produced by the actual tilt of 23°?

18. Consider a calendar that is based entirely on the day and the month (the Moon's period from full phase to full phase). How many days are there in a month? Can you figure out a scheme analogous to leap year to make this calendar work? Can you also incorporate the idea of a week in your lunar calendar?

19. Show that the Gregorian calendar will be in error by one day in about 3300 years.

Molten lava fountaining from a volcanic vent in Hawaii—testament to the dynamic nature of the Earth's crust. *(G. Briggs, USGS)*

Alfred Wegener (1880–1930), German astronomer and meteorologist, suggested the idea of continental drift in 1912 from a study of geological similarities between the two sides of the Atlantic Ocean. His arguments, although not accepted then, are now regarded as precursors of the theory of plate tectonics—the most important development in Earth sciences of the 20th century. *(Historical Pictures Service, Chicago)*

4

EARTH AS A PLANET

The Copernican revolution established that the Earth is a planet, orbiting the star we call the Sun. At about the same time, the first circumnavigation of the Earth by the Portuguese explorer Fernando de Magellan proved that the Earth is round, and in subsequent centuries adventurers and scientists roamed its surface and explored its varied landscapes.

In previous chapters we discussed the size and orbit of the Earth, and we described terrestrial phenomena, such as seasons and tides, that are related to the motions of the Earth. In this chapter we will examine the composition and structure of our planet with its envelope of ocean and atmosphere. We will ask how the surface of our planet came to be the way we see it, touching on the subjects of geology and geophysics and noting the crucial role played by life in determining the nature of the oceans and atmosphere. This direct knowledge helps us use the Earth as a reference point for understanding other worlds where we are limited to remote observations.

4.1 The Global Perspective

Basic Properties

The Earth is a medium-sized planet, with a diameter of approximately 13,000 km (Figure 4.1). It is composed primarily of a few heavy elements, such as iron, silicon, and oxygen—very different from the composition of the Sun and stars, which are dominated by the light elements hydrogen and helium. The Earth's orbit is nearly circular, and it is close enough to the Sun to support liquid water on its surface—the only planet that is neither too hot nor too cold, but "just right" for the development of life. Some of the basic properties of the Earth are summarized in Table 4.1.

Earth's Interior

The interior of a planet is difficult to study—even that of our own Earth—and its composition and structure must be determined indirectly. Our only direct experience is with the outermost skin of the Earth's crust, a layer no more than a few kilometers in depth. It is important to remember that in many ways we know less about our own planet a few kilometers beneath our feet than we do about the surfaces of Venus or Mars.

The Earth is composed largely of metal and of **silicate** rock—rocks composed of various minerals containing silicon and oxygen. Some of this material is in a solid state, and some is molten. The structure of the material in the interior has been probed in considerable detail from measurements of the transmission of **seismic waves** through the Earth. Such waves are produced by natural earthquakes or by

FIGURE 4.1 The Earth from space. *(NASA)*

artificial impacts or explosions. The seismic waves travel through a planet rather like sound waves through a bell, and the response of the planet to various frequencies is characteristic of interior structure, just as the sound of a bell reveals its size and construction. Some of these vibrations travel along the surface; others pass directly through the interior.

Seismic studies have shown that most of the interior of the Earth is solid and that it consists of several distinct layers of different composition (Figure 4.2). The top layer is the **crust,** the part of the Earth we know best. The oceanic crust, which covers 55 percent of the surface, is typically about 6 km thick and is composed of volcanic rocks called **basalts.** The continental crust, which covers 45 percent of the surface, is 20 to 70 km thick and is predominantly

made of a different volcanic class of silicates called **granites.** These crustal rocks typically have densities of about 3 g/cm³. The crust makes up only about 0.3 percent of the mass of the Earth.

The largest part of the solid Earth, which is called the **mantle,** stretches from the base of the crust down to a depth of 2900 km. The mantle is more or less solid, but at the temperatures and pressures found there, the mantle rock can deform and flow slowly. The density in the mantle increases downward from about 3.5 g/cm³ to more than 5 g/cm³, as a result of the compression produced by the weight of overlying material. Samples of upper mantle material are occasionally ejected from volcanoes, permitting a detailed analysis of its chemistry.

Beginning at a depth of 2900 km, we encounter the dense metallic **core** of the Earth. The core has a diameter of 7000 km (substantially larger than the planet Mercury). The outer core is liquid. The innermost part of the core (about 2400 km in diameter) is extremely dense and probably solid. The core is composed primarily of iron, but probably also contains substantial quantities of nickel and sulfur.

Differentiation

Why should the Earth have a metal core? And more generally, why should any planet be separated into layers like the crust, mantle, and core of the Earth? The answer goes back to the early history of the solar system, when the larger bodies in the planetary sys-

TABLE 4.1 Some Properties of the Earth	
Semimajor axis	1.00 AU
Period	1.00 yr
Mass	5.98×10^{24} kg
Diameter	12,756 km
Escape velocity	11.2 km/s
Rotation period	23h 56m 4s
Surface area	5.1×10^{8} km²
Atmospheric pressure	1.00 bar

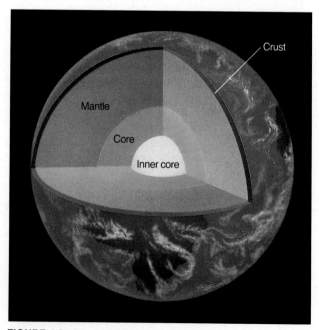

FIGURE 4.2 Interior structure of the Earth.

TABLE 4.2 Structure of the Earth

Region	Radius (km)	Composition	Mass (%)
Core	3500	Iron, nickel	33
Mantle	6370	Silicates	67
Crust	6378	Granite, basalt	0.3

tem differentiated. **Differentiation** is the name given to the process by which a planet organizes its interior into layers of different composition and density.

Differentiation results from heating of the planet during its formation. Once the planet becomes molten, the heavier metal tends to sink to form a core, while the lightest minerals float to the surface to form a crust. Later, when the planet cools, this layered structure is preserved. In order for a rocky planet to differentiate, it must be heated to the melting point of rocks, typically above 1300 K. But an object composed in large part of water ice will differentiate as soon as its temperature rises above the melting point of water, at 273 K.

Additional clues concerning the Earth's interior are provided by its magnetic field. This field is similar to that produced by a bar magnet, approximately aligned with the rotational poles of the Earth. The field is generated by moving material in the Earth's liquid metallic core. Energy to drive these motions is obtained from the slow escape of interior heat and from the rotation of the planet.

The Magnetosphere

The Earth's magnetic field extends into surrounding space. Above the atmosphere, this field is able to trap small quantities of electrons and other atomic particles. This volume, which is called the Earth's **magnetosphere,** is defined as the region within which Earth's magnetic field dominates the weak interplanetary field that extends outward from the Sun (Figure 4.3). Typically, the Earth's magnetosphere extends about 60,000 km, or ten Earth radii, in the direction of the Sun. It is shaped like a wind sock pointing away from the Sun, where it can reach as far as the orbit of the Moon.

The existence of the magnetosphere was discovered in 1958 by instruments on the first U.S. Earth satellite, Explorer 1, which recorded the ions trapped in its inner part. The regions of high-energy ions in the magnetosphere are often called the Van Allen belts, in recognition of the University of Iowa professor who built the scientific instrumentation for Explorer 1 and correctly interpreted the satellite measurements. Since 1958, hundreds of spacecraft have explored various regions of the magnetosphere.

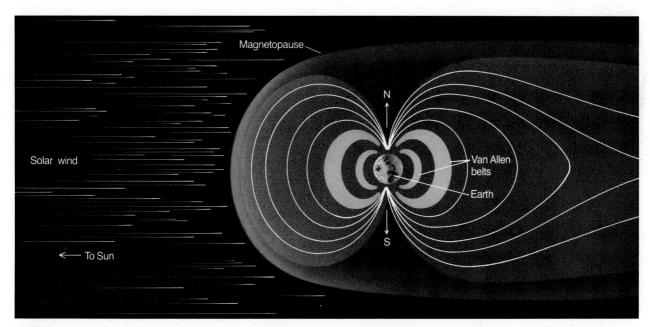

FIGURE 4.3 Cross-section of the Earth's magnetosphere as revealed by numerous spacecraft.

4.2   The Crust of the Earth

The Earth's crust is a dynamic place. Volcanic eruptions, erosion, and large-scale movements of the continents constantly rework the surface of our planet. Geologically, ours is the most active planet. Many of the geological processes described in this section have taken place on other planets as well but were usually confined to periods in the planet's distant past.

Composition of the Crust

The crust of the Earth is made up in large part of the oceanic basalts and the continental granites. These are both examples of **igneous** rock, which is the term used for any rock that has cooled from a molten state. All volcanically produced rock is igneous (Figure 4.4).

There are two other kinds of rock with which we are familiar on the Earth, although it turns out that neither is common on other planets. These are the sedimentary rocks and the metamorphic rocks. **Sedimentary** rocks are made of fragments of igneous rocks or of living organisms, deposited by wind or water and cemented together without melting. On Earth, these include the common sandstones, shales, and limestones. **Metamorphic** rocks are produced by the chemical and physical alteration of igneous or sedimentary rocks at high temperature and pressure. Metamorphic rocks are produced on Earth because

geological activity carries surface rocks to considerable depths and then brings them back up to the surface.

There is a fourth, very important category of rock that can tell us much about the early history of the planetary system. This is **primitive** rock, which is what we call rock that has largely escaped chemical modification by heating. Primitive rock represents the original material out of which the planetary system was made. There is no primitive material left on the Earth because it was strongly heated early in its history. To find primitive rocks, we must look to smaller objects: comets, asteroids, and small planetary satellites.

A block of marble on Earth is composed of materials that have gone through all four kinds of rock: Beginning as primitive material before the Earth was born, it was heated in the early Earth to form igneous rock, subsequently eroded and redeposited (perhaps many times) to form sedimentary rock, and finally transformed several kilometers below the Earth's surface into the hard white metamorphic stone we see today.

Plate Tectonics

Geology is the study of the crust of the Earth, and particularly of the processes that have shaped the surface throughout history. Heat escaping from the interior provides the energy for the formation of mountains, valleys, volcanoes, and even the conti-

FIGURE 4.4   The formation of igneous rock as liquid lava cools and freezes. This is a lava flow from a basaltic eruption in Hawaii. *(David Morrison)*

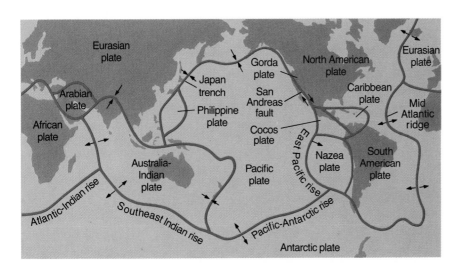

FIGURE 4.5 The major plates into which the crust of the Earth is divided. Arrows indicate the motion of the plates.

nents and ocean basins themselves. But it was not until the middle of the 20th century that geologists succeeded in understanding how these landforms are created through the process of plate tectonics.

Plate tectonics is a theory that explains how slow motions within the mantle of the Earth move large segments of the crust, resulting in a gradual drifting of the continents as well as the formation of mountains and other large-scale geologic features. It is a concept as basic to geology as evolution by natural selection is to biology or gravitation is to understanding the orbits of planets.

The Earth's crust and upper mantle (to a depth of about 60 km) are divided into about a dozen major plates that fit together like the pieces of a jigsaw puzzle (Figure 4.5). In some places, such as the Atlantic Ocean, these plates are moving apart, while in others they are forced together. The motive power for plate tectonics is provided by slow **convection** of the mantle, a process by which heat escapes from the interior through the upward flow of warmer material and the slow sinking of cooler material. Convection is a process we encounter often in astronomy where energy is being transported from a warm region like the interior of the Earth to a cooler region.

Four basic kinds of interactions between crustal plates are possible at their boundaries: (1) They can pull apart; (2) one plate can burrow under another; (3) they can slide alongside each other; or (4) they can jam together. Each of these activities is important in determining the geology of the Earth.

Rift and Subduction Zones

Plates pull apart from each other along **rift zones,** such as the Mid-Atlantic Ridge, driven by upwelling currents in the mantle (Figure 4.6a). A few rift zones are found on land, the best known being the central African rift, an area in which the African continent is slowly breaking apart. Most rift zones, however, are in the oceans. The new material that rises to fill the space between the receding plates is basaltic lava, the kind of igneous rock that forms most of the ocean basins.

From a knowledge of sea-floor spreading, we can calculate the average age of the oceanic crust. About 60,000 km of active rifts have been identified, with average separation rates of about 4 cm per year. The new area added to the Earth each year is about 2 km^2, enough to renew the entire oceanic crust in a little more than 100 million years. This is a very short interval in geological time, less than 3 percent of the age of the Earth. The present oceans are among the youngest features on the planet.

When two plates come together, one plate often dives down beneath another in what is called a **subduction zone** (Figure 4.6b). Generally, continental masses cannot be subducted, but the thin oceanic plates can be rather readily forced down into the upper mantle. Often a subduction zone is marked by an oceanic trench, a fine example being the deep Japan Trench along the coast of Asia. The subducted plate is forced down into regions of high pressure and temperature, eventually melting several hundred kilometers below the surface. Its material is recycled back into a downward-flowing convection current, ultimately balancing the material that rises along rift zones.

A part of the subducted material reaches the surface more directly through volcanic eruptions. All along the subduction zone, earthquakes and volcanoes mark the death throes of the plate. Some of the most destructive earthquakes in history have taken place along subduction zones. These include

(a) RIFT ZONE

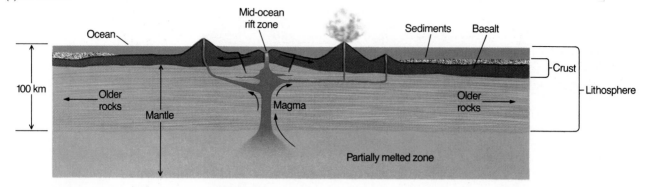

(b) SUBDUCTION ZONE

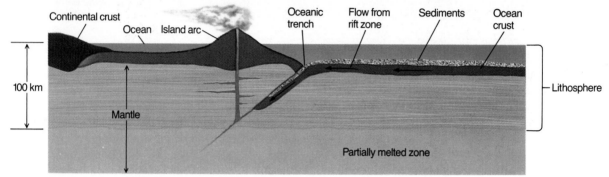

FIGURE 4.6   Rift zones and subduction zones in the Earth's crust.

the 1923 Yokohama earthquake and fire, which killed 100,000 people, and the 1976 earthquake that leveled the Chinese city of Tangshan and resulted in more than half a million deaths.

Fault Zones and Mountain Building

Along much of their lengths, the crustal plates slide parallel to each other. These plate boundaries are marked by cracks, or **faults.** Along active fault zones, the motion of one plate with respect to the other is several centimeters per year, about the same as the spreading rates along rifts.

One of the most famous faults is the San Andreas Fault, lying on the boundary between the Pacific Plate and the North American Plate. This fault runs from the Gulf of California in the south to the Pacific Ocean just west of San Francisco in the north. The Pacific Plate, to the west, is moving northward, carrying Los Angeles, San Diego, and parts of the southern California coast with it. In several million years, Los Angeles will be an island off the coast of San Francisco.

Unfortunately for us, the motion along most fault zones does not take place smoothly. The creeping motion of the plates against each other builds up

stresses in the crust that are released in sudden, violent slippages, generating earthquakes. Since the average motion of the plates is constant, the longer the interval between earthquakes, the greater the stress and the larger the energy released when the surface finally moves. For example, the part of the San Andreas Fault near the central California town of Parkfield has slipped about every 22 years during the past century, moving an average of about 1 m each time. In contrast, the average interval between major earthquakes in the Los Angeles region is about 140 years, and the average motion is about 7 m. The last time the San Andreas Fault slipped in this area was in 1857; tension has been building ever since, and sometime soon it is bound to be released.

When two continental masses are brought together by the motion of the crustal plates, they are forced against each other under great pressure. The Earth buckles and folds, forcing some rock deep below the surface and raising other folds to heights of many kilometers. This is the way most of the mountain ranges on Earth were formed; as we will see, however, quite different processes produced the mountains on other planets.

At the same time that a mountain range is being formed by upthrusting of the crust, its rocks are sub-

FIGURE 4.7 The Alps, a young region of the Earth's crust where sharp mountain peaks are being sculpted by glaciers. *(David Morrison)*

ject to the erosional force of water and ice. The sharp peaks and serrated edges that are characteristic of our most beautiful mountains (Figure 4.7) have little to do with the forces that make mountains. Instead, they result from the processes that tear them down. Ice is an especially effective sculptor of rock. In a planet without moving ice or running water, mountains remain smooth and dull.

Volcanoes

Volcanoes mark locations where molten rock (called **magma**) rises to the surface. One example is the mid-ocean ridges, which are volcanic features formed by mantle convection currents at plate boundaries. A second, major kind of volcanic activity is associated with subduction zones, and volcanoes sometimes also appear in regions where continental plates are colliding.

Another location for volcanic activity on our planet is found above so-called mantle hot spots, areas far from plate boundaries where heat is nevertheless rising from the interior of the Earth. Perhaps the best known such hot spot is under the island of Hawaii, where it supplies the energy to maintain three currently active volcanoes, two on land and one under the ocean (Figure 4.8). The tallest Hawaiian volcanoes are among the largest individual mountains on

Earth, up to 100 km in diameter and rising 9 km above the ocean floor.

Not all volcanic eruptions produce mountains. If the lava is very fluid and flows rapidly from long

FIGURE 4.8 Mauna Loa, a large volcano in Hawaii, in eruption. The stars of the Southern Cross are visible to the left of the volcanic plume. *(Dale P. Cruikshank)*

FIGURE 4.9 Sand dunes in Death Valley National Monument are sculpted by wind. *(David Morrison)*

4.3 The Earth's Atmosphere

We live at the bottom of the ocean of air that envelops our planet. The atmosphere, weighing down upon the surface of the Earth under the force of gravitation, exerts a pressure at sea level of 1 **bar,** which equals the weight of 1.03 kg over each square centimeter. The total mass of the atmosphere is about 5×10^{18} kg, or about a millionth of the total mass of the Earth. Thus the atmosphere represents a smaller fraction of the Earth than the share of your mass represented by the hair on your head.

Structure of the Atmosphere

The structure of the atmosphere is illustrated in Figure 4.10. Most of the atmosphere is concentrated near the surface of the Earth, within about the bottom 10 km. That is where clouds form and airplanes fly. Within this region, called the **troposphere,** warm air, heated by the surface, rises and is replaced by descending currents of cooler air—another example of convection. This circulation generates clouds and

cracks, it spreads out to form lava plains. The largest known terrestrial eruptions, such as those that produced the Snake River Basalts in the U.S. Northwest or the Deccan Plains in India, are of this type. As we will see later, similar lava plains are found on the Moon and the other terrestrial planets.

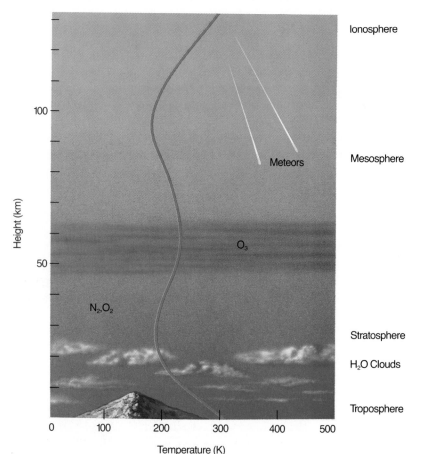

FIGURE 4.10 The structure of the Earth's atmosphere. The curving line shows the temperature (scale at bottom of figure).

other manifestations of weather. Within the troposphere, the temperature drops rapidly with increasing elevation to values near 50°C below freezing at its upper boundary, where the **stratosphere** begins. Most of the stratosphere, which extends to about 80 km, is cold and free of clouds.

Near the top of the stratosphere is a layer of **ozone** (O_3), a heavy form of oxygen having three atoms per molecule instead of the usual two. Since ozone is a good absorber of ultraviolet light, it protects the surface from some of the Sun's dangerous ultraviolet radiation, making it possible for life to exist on land. Because ozone is essential to our survival, we react with justifiable concern to indications that human civilization—particularly the industrial chemicals called CFCs (chlorofluorocarbons)—are depleting atmospheric ozone to a significant degree.

At heights above 100 km, the atmosphere is so thin that orbiting satellites can pass right through it without being dragged down. At these elevations, individual atoms can occasionally escape completely from the gravitational field of the Earth. Thus there is a continuous slow leaking of atmosphere from the planet, especially of lightweight atoms, which move faster than heavy atoms. The Earth's atmosphere cannot, for example, hold on to hydrogen or helium, which escape into space.

Atmospheric Composition and Origin

At the Earth's surface, the atmosphere consists of 78 percent nitrogen (N_2), 21 percent oxygen (O_2), and 1 percent argon (A), with traces of water vapor (H_2O), carbon dioxide (CO_2), and other gases. Variable amounts of dust particles and water droplets are also found suspended in the air.

A complete census of the Earth's atmosphere, however, should take into account more than the gas now present. Suppose, for example, that our planet were heated to above the boiling point of water (100°C, or 373 K); the oceans would boil, and this water vapor would become a part of the atmosphere. To estimate how much water vapor would be released, we note that there is enough water to cover the entire Earth to a depth of about 3000 m. Since the pressure exerted by 10 m of water is about equal to 1 bar, the average pressure at the ocean floor is about 300 bars. Since water weighs the same whether it is in liquid or vapor form, the atmospheric pressure of water if the oceans boiled away would also be 300 bars. Water would therefore greatly dominate the Earth's atmosphere, with N_2 and O_2 reduced to the status of trace constituents.

On a warmer Earth another source of additional atmosphere would be found in the sedimentary carbonate rocks of the crust. These minerals contain abundant CO_2, which, if released by heating, would generate about 70 bars of CO_2, far more than the current CO_2 pressure of only 0.0005 bar. Thus the atmosphere of a warm Earth would be dominated by water vapor and CO_2, with a surface pressure close to 400 bars.

We do not know the origin of the Earth's original atmosphere. Today we see that CO_2, H_2O, sulfur dioxide (SO_2), and other gases are released from volcanoes. Much of this apparently new gas, however, is probably recycled material that has been subducted through plate tectonics. With regard to the original source of the atmosphere and oceans, there are three possibilities: (1) The atmosphere could have been formed with the rest of the Earth as it accumulated from the debris left over from the formation of the Sun; (2) it could have been released from the interior through volcanic activity, subsequent to the formation of the Earth; or (3) it may have been derived from impacts by comets or other icy materials from the outer parts of the solar system. Current opinion favors the cometary hypothesis, but all three mechanisms may have contributed.

Weather and Climate

All planets with atmospheres have *weather*, which is just a name we give to the circulation of the atmosphere. The energy that powers the weather is derived primarily from the sunlight that heats the surface. As the planet rotates and also as slower seasonal changes vary the deposition of sunlight, the atmosphere and oceans try to redistribute the heat from warmer to cooler areas. Weather on any planet represents the response of its atmosphere to changing inputs of energy from the Sun (Figure 4.11).

Climate is a term used to refer to the effects of the atmosphere on a time scale of decades and centuries.

TABLE 4.3 Composition of the Dry Atmosphere	
Element or Compound	Amount (%)
Nitrogen (N_2)	78.1
Oxygen (O_2)	21.0
Argon (Ar)	0.93
Carbon dioxide (CO_2)	0.03
Neon (Ne)	0.002

FIGURE 4.11 A large tropical storm, marked by clouds swirling in a cyclonic direction around a low-pressure region. *(NASA)*

Changes in climate (as opposed to the random variations in weather from one year to the next) are often difficult to detect over short periods, but as they accumulate, their effect can be devastating. Modern farming is especially sensitive to temperature and rainfall, and calculations indicate that a drop of only 2°C throughout the growing season would cut the wheat production by half in Canada and the United States.

The best documented changes in the climate of the Earth are the great ice ages, which have periodically lowered the temperature of the Northern Hemisphere over the past million years or so. Today we are in a relatively warm period, interpreted by many scientists as a fairly short-lived interglacial interval between major ice ages.

It is generally believed that the ice ages are primarily the result of changes in the tilt of the Earth's rotational axis, produced by the gravitational effects of the other planets. This idea of an astronomical cause of climate changes was first proposed in 1920 by the Serbian scientist Milutin Milankovich. As we will see in Chapter 8, there is also evidence of periodic climate changes on Mars, and modern calculations suggest that these also have an origin in slow changes in the orbit and rotational axis of that planet.

4.4 Life and Chemical Evolution

Earth is the only inhabited planet in the solar system, as we know from the investigation of other planets by spacecraft. Terrestrial life forms an important part of the story of our planet. Life arose early in Earth's history, and living organisms have been interacting with their environment for billions of years. We all recognize that lifeforms have evolved to adapt themselves to the environment on Earth, and now we are beginning to realize that the Earth itself has been changed in important ways by the presence of living matter.

Origin of Life

The record of the birth of life on Earth has been lost in the restless motions of the crust. By the time the oldest surviving rocks were formed, about 3.8 billion years ago, life already existed. At 3.5 billion years ago, life had achieved the sophistication to build large colonies called stromatolites (Figure 4.12), a form so successful that stromatolites survive on Earth today. But there is little surviving crust from these ancient times, and abundant fossils have been produced only during the past 600 million years—less than 15 percent of the planet's history.

Any theory of the origin of life must therefore be partly speculative, since there is little direct evidence to go on. All we really know is that the atmosphere of the early Earth, unlike that today, contained abundant CO_2 but no O_2. In the absence of O_2, many complex chemical reactions that lead to the production of amino acids, proteins, and other chemical building blocks of life are possible. It now seems relatively certain that life arose from these building blocks very early in our history.

For tens of millions of years after its formation, life (perhaps little more than large molecules like the viruses of today) probably existed in warm, nutrient-rich seas, living off accumulated organic chemicals. Eventually, however, as this easily accessible food became depleted, life began the long evolutionary road that led to the proliferation of different organisms on Earth today. As it did so, life influenced the chemical evolution of the atmosphere.

Evolution of the Atmosphere

Studies of the chemistry of ancient rocks show that the atmosphere of the Earth lacked O_2 until about 2 billion years ago, in spite of the presence of plants releasing O_2 by photosynthesis. Apparently, chemical reactions with the crust removed the O_2 as quickly as it formed. Slowly, however, the increasing evolutionary sophistication of life led to a growth in plant population, finally reaching the point where O_2 was

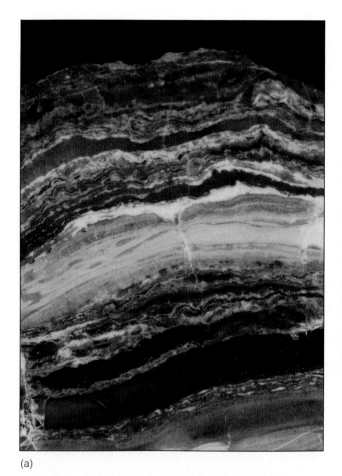

(a)

(b)

FIGURE 4.12 Cross-section of stromatolites, both fossil and contemporary. These colonies of microorganisms date back more than 3 billion years. *(NASA/ARC, courtesy of David DesMarais)*

produced faster than it could be removed, and the atmosphere became more and more oxidizing.

The appearance of O_2 between 1 and 2 billion years ago eventually led to the formation of the Earth's ozone layer, which protects the surface from lethal solar ultraviolet light. Before that, it was unthinkable for life to venture outside the protective oceans, and the land masses of Earth were barren. The presence of O_2 and hence ozone thus allowed the colonization of the land. It also made possible a tremendous proliferation of animals, who lived by oxidizing the organic materials produced by plants. When the animals moved to the land, they developed techniques for breathing O_2 directly from the atmosphere.

On a planetary scale, one of the most important consequences of life has been a decrease in atmospheric CO_2. In the absence of life, Earth would probably have a much more massive atmosphere dominated by CO_2. But life has effectively stripped us of most of this gas.

Most of the CO_2 is locked up in sediments composed of carbonates, derived from the shells of marine creatures, which have evolved techniques for extracting CO_2 from the water. When they die, their shells sink to the ocean floor. The CO_2 remains trapped on the ocean floor until it is subducted, when much of it returns to the atmosphere in volcanic eruptions. But evidently the marine organisms quickly cycle it back into sediments, so that at any time only a minute fraction of the Earth's CO_2 is present in the atmosphere.

Another way that life removes CO_2 is by producing deposits of the fossil fuels coal and oil. These substances are primarily carbon, mostly extracted from atmospheric CO_2 hundreds of millions of years ago, when the first great forests populated the land.

The Greenhouse Effect and Global Warming

We have a special interest in the CO_2 content of the atmosphere because of the special role this gas plays in retaining heat from the Sun, through a process called the **greenhouse effect.** To understand how the greenhouse effect works, imagine the fate of the sunlight that strikes the surface of the Earth. It is absorbed, heats the surface, and is re-emitted as infrared or heat radiation (Figure 4.13). However, CO_2, which is a colorless, transparent gas to visible light, is

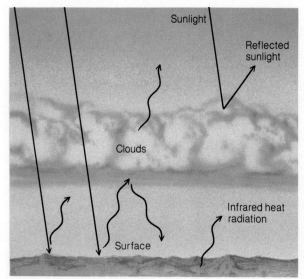

FIGURE 4.13 The operation of the greenhouse effect. Sunlight that penetrates the lower atmosphere and surface is reradiated as infrared or heat radiation, which the atmosphere restrains from escaping. The result is an elevated surface temperature.

largely opaque to infrared energy. As a result, it acts as a blanket, impeding the flow of heat back to space. The surface heats up and establishes an equilibrium temperature higher than it would have had without the CO_2. The more CO_2, the higher this temperature will be.

The greenhouse effect in a planetary atmosphere is similar to the heating of a gardener's greenhouse or the inside of a car left out in the Sun with the windows rolled up. In these examples, the window glass plays the role of CO_2, letting sunlight in but impeding the outward flow of heat radiation. You all know the result: a greenhouse or car interior much hotter than would otherwise be expected from solar heating. On

Earth, the current greenhouse effect elevates the surface temperature by about 23°C. Without this greenhouse effect, the average surface temperature on Earth would be well below the freezing point, and we would be locked in a global ice age.

Modern industrial society depends on energy extracted from burning fossil fuels. As these ancient coal and oil deposits are oxidized, additional CO_2 is released into the atmosphere. The situation is exacerbated by the widespread destruction of tropical forests, which we depend upon to extract CO_2 from the atmosphere and replenish our supply of O_2. So far in this century, the amount of CO_2 in the atmosphere has increased by about 25 percent, and it is continuing to rise at 0.5 percent per year. By early in the 21st century, the CO_2 level is predicted to reach twice its preindustrial value (Figure 4.14). The consequences of such an increase are being studied with elaborate computer models, but the fact is that we are not sure of the effects, which will vary from one location to another.

Already some global warming is apparent. In the United States, Europe, and the U.S.S.R., summer temperatures throughout the 1980s reached record highs. The sea level has also risen measurably, owing to expansion of the water as its average temperature climbs. These effects, of course, are superimposed on the usual year-to-year fluctuations in weather, but most scientists are convinced that global warming due to an enhanced greenhouse effect is a reality. These effects of an increased greenhouse effect come at a particularly difficult time, when the planet is already in an unusually warm interglacial period. We are rapidly entering unknown territory, where human activities are contributing to the highest temperatures on Earth in more than 50 million years.

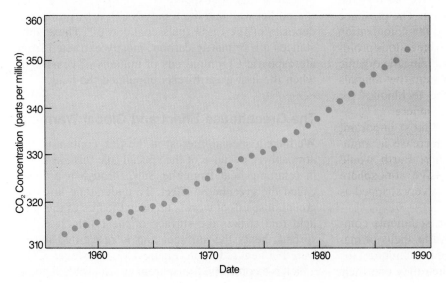

FIGURE 4.14 Increase with time of atmospheric CO_2, which is expected to double by the middle of the 21st century. *(Adapted from data obtained at the Mauna Loa Observatory of NOAA)*

4.5 Cosmic Influences on the Evolution of Earth

Where Are the Craters on Earth?

In discussing the geology of the Earth in Section 4.2, we dealt only with the effects of internal forces, expressed through the processes of plate tectonics and volcanism. In contrast, when we look at the Moon, we see primarily craters, produced from the impacts of interplanetary debris. Why do we not see more evidence here of the kinds of craters that are so prominent on the Moon and other planets?

Is it possible that the Earth has escaped being struck by the interplanetary debris that has pockmarked the Moon? Our atmosphere provides no shield against the large impacts that form craters a kilometer or more in diameter, nor is there any other known way to avoid these cosmic events. Certainly the Earth has been cratered as heavily as the Moon. The difference is that on the Earth these craters are destroyed by our active geology before they can accumulate. As plate tectonics constantly renews our crust, evidence of past cratering events is destroyed. Only in the past few decades have geologists succeeded in identifying the eroded remnants of many old impact craters. Even more recent is our realization that these impacts may have had an important influence on the evolution of life on Earth.

Recent Impacts

The collision of interplanetary debris with the Earth is not a hypothetical idea. The best studied such collision took place on June 30, 1908, near the Tunguska River in Siberia. In this desolate region, a remarkable explosion took place in the atmosphere about 8 km above the surface. The shock wave flattened more than a thousand square kilometers of forest; herds of reindeer and other animals were killed; and a man at a trading post 80 km from the blast was thrown from his chair and knocked unconscious. The blast wave spread around the world, recorded by instruments designed to measure changes in atmospheric pressure (Figure 4.15).

Despite this violence, no craters were formed by the explosion. While we do not know exactly what caused the Tunguska event, it certainly represented the disintegration of an impacting body weighing approximately 100,000 tons. The force of the blast was equivalent to a 10-megaton nuclear bomb. Apparently, the projectile did not have the strength to survive its plunge to the surface but rather gave up its energy of motion in the atmosphere, creating the

FIGURE 4.15 Aftermath of the Tunguska explosion. This photograph, taken 21 years after the blast, shows a part of the forest that was devastated by the 10-megaton explosion. *(Novosty)*

equivalent of an "air burst" in nuclear weapons jargon.

The most recent impact that produced a substantial crater took place 50,000 years ago in Arizona. The projectile in this case was a lump of iron about 100 m across. The crater, called Meteor Crater and now a major tourist attraction, is about a mile across and has all of the features associated with similar sized lunar impact craters. Meteor Crater is one of the few impact feature on the Earth that remains relatively intact; other, older craters are so eroded that only a trained eye can distinguish them (Figure 4.16).

FIGURE 4.16 Meteor Crater in Arizona, a 50,000-year-old impact scar. While impact craters are common on less active bodies like the Moon, this is one of very few well-preserved craters on the Earth. *(Meteor Crater, Northern Arizona)*

FIGURE 4.17   Artist's impression of the impact of a 10-km asteroid on the Earth. This is the approximate magnitude of the event that ended the Cretaceous period of geological history. *(Painting by Don Davis)*

Extinction of the Dinosaurs

The impact that produced Meteor Crater would have been dramatic indeed to any humans who witnessed it, since the energy release was equivalent to a 15-megaton nuclear bomb. But such explosions have no global consequences. Much larger (and rarer) impacts, however, can disturb the ecological balance of the planet and have a major influence on the course of evolution.

The best documented such impact took place 65 million years ago, at the boundary between the Cretaceous and Tertiary periods of geological history. This break in the Earth's history is marked by a **mass extinction,** when more than half of the species on our planet became extinct. While there are a dozen or more mass extinctions in the geological record, this particular event has always intrigued paleontologists because it marks the end of the age of the dinosaurs. For tens of millions of years, these great warm-blooded reptiles had ruled the world. Then, suddenly, they disappeared, and thereafter the mammals began their development and diversification.

The body that collided with the Earth at the end of the Cretaceous period had a mass of more than a

trillion tons and a diameter of at least 10 km (Figure 4.17). We know this because of a worldwide layer of sediment deposited from the dust cloud that enveloped the planet after the impact. First identified in 1980, this sediment layer is enriched in the rare metal iridium and other elements that are relatively abundant in asteroids and comets but very rare in the crust of the Earth. Even diluted by the terrestrial material excavated from the crater, this cosmic component is easily identified. In addition, the sediment contains many minerals that are characteristic of the temperatures and pressures of a gigantic explosion.

The end-Cretaceous impact released energy equivalent to 5 billion Hiroshima-sized nuclear bombs, excavating a crater almost 200 km across and deep enough to penetrate through the Earth's crust. The explosion lifted about a hundred trillion tons of dust into the atmosphere, as can be determined by measuring the thickness of the sediment layer formed when this dust settled to the surface. Such a quantity of material would have blocked sunlight completely from reaching the surface, plunging the Earth into a period of cold and darkness that lasted several months. Other effects include global acid rain and large-scale fires that probably destroyed most of the planet's forests and grasslands. Presumably, these environmental effects, rather than the explosion itself, were responsible for the mass extinction, including the death of the dinosaurs.

Impacts and the Evolution of Life

Several other mass extinctions in the geological record have been tentatively identified with large impacts, although none is so dramatic as the event that destroyed the dinosaurs. But even without such specific documentation, it is clear that impacts of this size do occur and that their effects can be catastrophic for life. What is a catastrophe for one group of living things, however, may create opportunities for another group. Following each mass extinction, there is a sudden evolutionary burst as new species develop to fill the ecological niches opened by the event.

Impacts by comets and asteroids represent the only mechanisms we know of that could cause global catastrophes and seriously influence the evolution of life all over the planet. According to some estimates, the *majority* of all extinctions of species may be due to such impacts. As noted by Stephen Jay Gould of Harvard, such a perspective fundamentally changes our view of biological evolution. A central issue for the survival of a species is not just its success in competing with other species and adapting to slowly changing environments, as envisioned by Darwinian natural selection. Of at least equal importance is its ability to survive random global ecological catastrophes due to impacts.

Still earlier in its history, the Earth was subject to even larger impacts from the leftover debris of the formation of the planets. We know the Moon was repeatedly struck by objects more than 100 km in diameter, a thousand times more massive than the object that wiped out most terrestrial life 65 million years ago. The Earth must have experienced similar large impacts during its first 700 million years of existence. Some of them were probably violent enough to strip the planet of most of its atmosphere and to boil away its oceans. Such events would sterilize the planet, utterly destroying any life that had begun. Life may have formed and been wiped out several times before our own ancestors took hold sometime about 4 billion years ago.

The Earth is a target in a cosmic shooting gallery, subject to random violent events that were unsuspected a few decades ago. We owe our existence to such events. Had the impact of 65 million years ago not redirected the course of evolution, mammals might never have become the dominant, large animals they are today. But even more fundamentally, if impacts with comets and asteroids had not occurred throughout our planet's history, could biological evolution in more stable conditions have produced the wondrous diversity of life that populates the Earth today?

SUMMARY

4.1 The Earth is the prototype terrestrial planet. It's interior composition and structure are probed using **seismic waves.** Such studies reveal the metal **core** and **silicate mantle.** The **crust** consists primarily of oceanic **basalt** and continental **granite.** This structure is the result of **differentiation,** a process that takes place when a planet is heated. A global magnetic field, generated in the core, produces the Earth's **magnetosphere** of charged atomic particles.

4.2 Terrestrial rocks can be classified as **igneous, sedimentary,** or **metamorphic.** A fourth type, **primitive** rock, is not found on the Earth. Our planet's geology is dominated by **plate tectonics,** in which crustal plates move slowly in response to mantle **convection.** The surface expression of plate tectonics includes continental drift, recycling of the ocean floor, mountain building, **rift zones, subduction, faults,** earthquakes, and volcanic eruptions of **magma** from the interior.

4.3 The atmosphere has a surface pressure of 1 **bar** and is composed primarily of N_2 and O_2 and includes such important trace gases as H_2O, CO_2, and **ozone** (O_3). Its structure consists of **troposphere, stratosphere,** and tenuous higher regions. Atmospheric O_2 is the product of life. Atmospheric circulation (weather) is driven by the seasonally changing deposition of sunlight. Longer term climatic variations, such as the ice ages, are probably due to changes in the planet's orbit and axial tilt.

4.4 Life originated on Earth at a time when the atmosphere consisted mostly of CO_2 and there was no O_2. Later, photosynthesis gave rise to O_2 and ozone, and most of the CO_2 was trapped in oceanic sediments. CO_2 in the atmosphere heats the surface through the **greenhouse effect,** and increasing atmospheric CO_2 is leading to global warming.

4.5 Earth, like the Moon and other planets, has been influenced by the impacts of cosmic debris, including such small recent examples as Meteor Crater and the Tunguska explosion. Larger past impacts are responsible for at least some **mass extinctions,** including the large extinction 65 million years ago that ended the Cretaceous period, and have probably played an important role in the evolution of life.

REVIEW QUESTIONS

1. What are the compositions of the core and mantle of the Earth? Explain how we know, and indicate what the uncertainties in your conclusions might be.

2. Describe the differences between primitive, igneous, sedimentary, and metamorphic rock and relate these differences to their origin.

3. Consider several familiar landforms on the Earth, such as mountains, volcanoes, or canyons, and briefly describe how they were formed. Try to relate your answers to the theory of plate tectonics, which provides the basis for understanding much of terrestrial geology.

4. What is the origin of each of the main gases in the Earth's atmosphere? Explain the relationship between the composition of the atmosphere and the evolution of life.

5. What is the role of impacts by comets and asteroids in influencing the Earth's geology, its atmosphere, and the evolution of life?

THOUGHT QUESTIONS

6. If you wanted to live where the chances of a destructive earthquake are small, would you pick a location near a fault zone, near a midocean ridge, near a subduction zone, or on a volcanic island like Hawaii? What are the relative risks of earthquakes at each of these locations?

7. If all life were destroyed on Earth, would new life eventually form to take its place? Explain how conditions would have to change for life to start again on our planet.

8. Why will a decrease in the Earth's ozone be harmful? Why will an increase in our CO_2 also be harmful?

9. Suppose that society should decide to give high priority to reducing the emission of CO_2 and other greenhouse gases. How could this be accomplished? What would be the costs of such a program (both monetary and in terms of impact on our life style)?

10. Is there evidence of changes in climate in your area over the past century? How would you distinguish a true climate change from the random variations in weather that take place from one year to the next?

11. Why does the Earth have so few impact craters, relative to the Moon and many other planets and satellites?

12. How might the history of the Earth have been different if it were closer to the Sun? Farther from the Sun?

PROBLEMS

13. What fractions of the volume of the Earth are occupied by the inner core, the entire core, the mantle, and the crust?

14. Suppose that the next slippage along the San Andreas Fault in southern California takes place in the year 2000, and that it completely relieves the accumulated strain in this region. How much slippage will take place?

15. Measurements using Earth satellites have shown that Europe and North America are moving apart by 4 m per century owing to plate tectonics. As the continents separate, new ocean floor is created along the Mid-Atlantic rift. If the rift is 5000 km long, what is the total area of new ocean floor created in the Atlantic Ocean each century? How much new area per year?

16. Over the entire Earth, there are 60,000 km of active rift zones, with average separation rates of 4 m per century. How much area of new ocean crust is created each year over the entire planet? This is also approximately equal to the amount of ocean crust that is subducted, since the total area of the oceans remains about the same.

17. With the information from Problem 16, you can calculate the average age of the ocean floor. First find the total area of ocean floor (equal to about 60 percent of the surface area of the Earth). Then compare this with the area created (or destroyed) each year. The average lifetime is just the ratio of these numbers: the total area of ocean crust compared with the amount created (or destroyed) each year.

18. What is the volume of new oceanic basalt that is added to the Earth's crust each year? Assume that the thickness of the new crust is 5 km, that there are 60,000 km of active rifts, and that the average speed of plate motion is 4 cm per year. What fraction of the entire volume of the Earth does this annual addition of new material represent?

19. The sea-level pressure of the atmosphere (1 bar) corresponds to 10^4 kg of mass above each square meter of the Earth's surface. Calculate the total mass of the atmosphere in kilograms and in tons (1 ton equals 1000 kg). Then compute the total mass of ocean, given that the oceans would be 3000 m deep if water covered the globe uniformly. (*Note:* 1 m^3 of water has a mass of one ton.) Compare the mass of atmosphere and the mass of the oceans with the total mass of the Earth to determine what percentage of our planet is represented by the atmosphere and oceans.

20. Suppose that a major impact that produces a mass extinction takes place on the Earth once every 5 million years. Further suppose that if such an event occurred to-day, you and most other humans would be killed (this is true even if the human species as a whole survived). Such impact events are random, and one could take place any time. Calculate the probability that such an impact will occur within the next 50 years (within your lifetime). This is equal to the probability that you will be killed by this means, rather than dying from an auto accident or from heart disease or some other "natural" cause. How do the risks of dying from an asteroidal or cometary impact compare with other risks we are concerned about?

The McMath solar telescope at Kitt Peak National Observatory in Arizona. *(National Optical Astronomy Observatories)*

5

ELECTROMAGNETIC RADIATION

Niels Bohr (1885–1962), a Danish physicist, was awarded the Nobel prize in 1922 for his investigation of the structure of atoms and of the unique patterns of spectral lines that they produce. *(AIP Niels Bohr Library, Margarethe Bohr Collection)*

The Earth, as described in the preceding chapter, is accessible to direct investigation. But in astronomy, most of the objects that we study are completely out of reach. The temperature of the Sun is so high that a spacecraft would be destroyed by its heat long before reaching the solar surface, and the stars are too far away to visit in our lifetimes with the technology now available. Even light, which travels at a speed of 300,000 km/s, takes more than four years to reach us from the *nearest* star. If we want to learn about the Sun and stars, or even about most members of our solar system, we must devise techniques that will allow us to analyze them from a distance.

To study the astronomical universe, we must rely on the information contained in the light, x rays, radio waves, and other radiation emitted by objects in the universe. In this chapter we shall explore the nature of the radiation that is approaching us with the speed of light from all directions, ready to be sampled by our telescopes. What are the secrets it holds? What are the revelations it will give us about those objects it left years, centuries, even billions of years ago?

5.1 The Nature of Light

With one simple equation, Newton's theory of gravitation accounts for the motions of both the planets and objects on the Earth. Application of this theory to

a variety of problems dominated the work of scientists for nearly two centuries. In the 19th century, many physicists turned to the study of electricity and magnetism. The scientist who played a role analogous to that of Newton was physicist James Clerk Maxwell (1831–1879), who was born and educated in Scotland. Maxwell developed a single theory that describes in a small number of elegant equations both electricity and magnetism. It is this theory that allows us to understand the behavior of light.

Maxwell's Theory of Electromagnetism

Maxwell's theory deals with electric charges and their effects, especially when they are moving. In the vicinity of an electric charge, another charge feels a force of attraction or repulsion, depending on whether the two charges have the opposite or the same sign, respectively. It was known experimentally in the 19th century that changing magnetic fields could produce electrical currents. Maxwell showed through theoretical calculations that changing electrical currents could also produce magnetic fields, a result that was subsequently confirmed in laboratory experiments. The word *field* is a technical term used in physics to describe the consequences of forces that act on distant objects. For example, the Sun produces a *gravitational field* that controls the Earth's orbit, even though the Sun and the Earth do not come

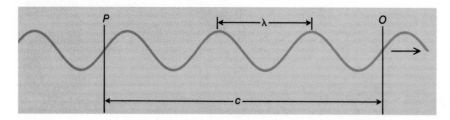

FIGURE 5.1 Electromagnetic radiation has wave-like characteristics. The relationship between the length of the wave (λ), the frequency of the wave, and the speed (c) with which it moves is shown.

directly into contact. Stationary electric charges produce *electric fields*, and as Maxwell showed, moving electric charges produce *magnetic fields*.

Maxwell found that electric and magnetic fields propagate through space. Maxwell was able to calculate the speed at which an electromagnetic disturbance moves through space and found that it was equal to the speed of light, which had been measured experimentally. On that basis he speculated that light was one form of **electromagnetic radiation,** a conclusion that was again confirmed in the laboratory. When light enters a human eye, its changing electric and magnetic fields stimulate nerve endings, which then transmit the information contained in these changing fields to the brain.

Since the word *radiation* will be used frequently in this book, it is important to understand what it means. In the modern world, "radiation" is commonly used to describe certain kinds of dangerous subatomic particles released by radioactive materials in our environment. But this is *not* what we mean when we speak of radiation in an astronomy text. Radiation, as used in this book, is a general term for light, x rays, and other forms of electromagnetic waves. This radiation provides almost our only link with the universe beyond our own solar system.

The Wave-like Characteristics of Light

Most of the characteristics of electromagnetic radiation can be described adequately if this radiation is represented as *energy propagated in waves*, although no material is required to transmit the electromagnetic waves. Any wave motion can be characterized by a **wavelength.** Ocean waves provide an analogy. The wavelength is simply the distance separating successive wave crests. Various forms of electromagnetic energy differ from one another only in their wavelengths. Those with the longest waves, ranging up to many kilometers in length, are called *radio waves*. Forms of electromagnetic energy of successively shorter wavelengths are called, respectively, *infrared radiation, light, ultraviolet radiation, x rays*, and *gamma rays*. All these forms (which will be

described in greater detail in Section 5.2) are the same basic kind of energy and could be thought of as different kinds of light.

The energy carried by electromagnetic radiation is directly related to the frequency of the wave motion. The **frequency** associated with any wave motion is the rate at which wave crests pass a given point, that is, the number of wave crests that pass per second. Imagine a long train of waves moving to the right, past point O (Figure 5.1), at a speed c. If we measure the distance of c centimeters to the left of O, we arrive at the point P along the wave train that will just reach point O after a period of 1 s. The frequency f of the wave train—that is, the number of waves between P and O—times the length of each, λ*, is equal to the distance c. The same idea can be expressed by the simple formula

$$c = \lambda f$$

Thus, we see that for any wave motion the speed of propagation equals the frequency times the wavelength.

Propagation of Light

As electromagnetic radiation moves away from its source, it spreads out and covers an ever-widening area. The increase in area is proportional to the square of the distance that the radiation has traveled (Figure 5.2). For example, when light from the Sun reaches the Earth, it is spread out over a sphere 1 AU in radius. Because the surface area of a sphere is proportional to the square of its radius, when it has gone twice as far, to 2 AU from the Sun, that same light is spread over an area four times as great. When the Sun's radiation reaches Saturn, 10 AU from the Sun, it is spread over an area 100 times that at the Earth's distance.

The **apparent brightness** of a light source depends on how much of its energy (that is, light) enters the

---

*The Greek letter for "l"—lambda or λ—is almost always used to denote wavelength.

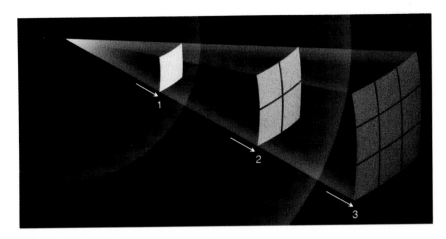

FIGURE 5.2 The inverse-square law for light. As light energy radiates away from its source, it spreads out, so that the energy passing through a unit area decreases as the square of the distance from its source.

pupil of our eye or telescope. Since the collecting area of the eye or telescope is constant, the larger the area over which light is spread, the smaller is the fraction of the total light that is observed. We see, then, that the amount of energy from a source that passes through a unit area *decreases* with the *square of the distance from the source*.

The formula for the area of a sphere of radius d is

$$A = 4\pi d^2$$

Then at distances d_1 and d_2 from a light source, the amounts of energy received by a telescope (or other detecting device), ℓ_1 and ℓ_2 are in the following proportion:

$$\frac{\ell_1}{\ell_2} = \frac{4\pi d_2^2}{4\pi d_1^2} = \left(\frac{d_2}{d_1}\right)^2$$

The above relation is known as the **inverse-square law** of light propagation. In this respect, the propagation of radiation is similar to the effects of gravity. Remember that the force of gravitation between two attracting masses is also inversely proportional to the square of their separation.

The inverse-square law for light explains why the stars appear so faint relative to the Sun. A typical star emits about the same total energy as does the Sun, but even the nearest star is about 270,000 times farther away, and so it appears about 73 billion times fainter ($73 \times 10^9 = 270,000 \times 270,000$).

5.2 The Electromagnetic Spectrum

Types of Electromagnetic Radiation

The types of electromagnetic radiation are shown in Figure 5.3. Electromagnetic radiation with the shortest wavelengths, not larger than 0.01 nm, is called

gamma radiation. (One nanometer [nm] is 10^{-9}m; see Appendix 4.) Gamma rays are often emitted in the course of nuclear reactions and by radioactive atoms. Gamma radiation is generated in the deep interior of stars.

Electromagnetic radiation with wavelengths between 0.01 nm and 20 nm is referred to as **x rays,**

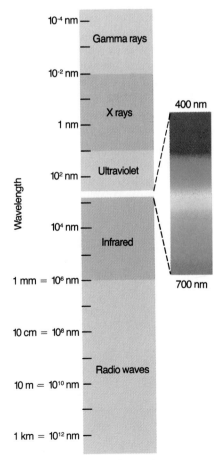

FIGURE 5.3 The electromagnetic spectrum.

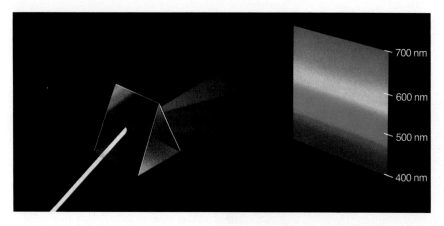

FIGURE 5.4 When Newton passed a beam of white sunlight through a prism, he saw a rainbow-colored band of light that we now call a continuous spectrum.

while radiation intermediate between x rays and visible light is **ultraviolet** (meaning higher energy than violet). Between visible light and radio waves are the wavelengths of **infrared** or heat radiation. The **microwaves** used in shortwave communication and in television are radio waves with wavelengths ranging from a few centimeters to a few meters. Other radio radiation can have wavelengths as long as several kilometers.

In 1672, in the first paper that he submitted to the Royal Society, Newton described an experiment in which he permitted sunlight to pass through a small hole and then through a prism. Newton found that sunlight, which gives the impression of being white, is actually made up of a mixture of all the colors of the rainbow (Figure 5.4). The scientific term for the array of colors produced by visible light is **spectrum.** The array of radiation of all wavelengths, from gamma rays to radio waves, is called the **electromagnetic spectrum.** Table 5.1 summarizes the types of electromagnetic radiation and indicates the temperatures and types of astronomical objects that emit specific types of electromagnetic radiation.

Radiation Laws

Some astronomical objects emit mostly infrared radiation, others mostly visible light, and still others mostly ultraviolet radiation. What determines the type of electromagnetic radiation emitted by the Sun, stars, and other astronomical objects? The answer is *temperature*.

A *solid* is composed of molecules and atoms that are in continuous vibration. A *gas* consists of molecules that are flying about freely at high speed, continually bumping into one another, and bombarding the surrounding matter. That energy of motion is called *heat*. The hotter the solid or gas, the more rapid is the motion of the molecules; temperature is just a measure of the average energy of those particles.

To understand in more quantitative detail the relationship between temperature and electromagnetic

TABLE 5.1   Electromagnetic Radiation

Type of Radiation	Wavelength Range (nm)	Radiated by Objects at This Temperature	Typical Sources
Gamma rays	Less than 0.01	More than 10^8 K	No astronomical sources this hot; some gamma rays produced in nuclear reactions
X rays	0.01–20	10^6–10^8 K	Gas in clusters of galaxies; supernova remnants; solar corona
Ultraviolet	20–400	10^4–10^6 K	Supernova remnants; very hot stars
Visible	400–700	10^3–10^4 K	Stars
Infrared	10^3–10^6	10–10^3 K	Cool clouds of dust and gas; planets; satellites
Radio	More than 10^6	Less than 10 K	No astronomical objects this cold; radio emission produced by electrons moving in magnetic fields (synchrotron radiation)

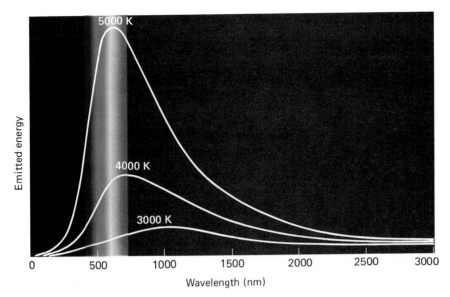

FIGURE 5.5 Energy emitted at different wavelengths for blackbodies at three different temperatures. At hotter temperatures, more energy is emitted at all wavelengths. The peak amount of energy is radiated at shorter wavelengths for higher temperatures (Wien's law).

radiation, it is useful to consider an idealized object that absorbs all the electromagnetic energy that impinges on it. Such an object is called a perfect radiator or a **blackbody.** A blackbody absorbs all of the energy incident upon it and heats up until it is emitting energy at the same rate that energy is being absorbed.

The radiation from a blackbody has several characteristics, which are illustrated in Figure 5.5. First, a blackbody with a temperature higher than absolute zero emits some energy at *all* wavelengths. Second, a blackbody at higher temperature emits *more* energy at all wavelengths than does a cooler one. Third, the higher the temperature, the shorter the wavelength at which the maximum energy is emitted.

This third characteristic is one that we have all observed in everyday life. For example, when a burner on an electric stove is turned on low, it emits heat, which is infrared radiation. If the burner is set to a higher temperature, it will glow a dull red. At a still higher setting, it will glow a brighter orange-red. At still higher temperatures, which cannot be reached with ordinary stoves, metal can appear brilliant yellow or even blue-white.

The Sun and stars emit energy that approximates that from a blackbody, so it is possible to estimate temperatures by measuring the energy that they emit as a function of wavelength—that is, by measuring their colors. The temperature at the surface of the Sun, which is where the radiation that we see is emitted, turns out to be 5800 K. (Throughout this text we use the Kelvin or absolute temperature scale. On this scale, water freezes at 273 K and boils at 373 K. All molecular motion ceases at 0 K. The various temperature scales are described in Appendix 5.)

The wavelength at which a blackbody emits its maximum energy can be calculated according to the equation

$$\lambda_{max} = \frac{3 \times 10^6}{T}$$

where the wavelength is in nanometers and the temperature is in Kelvins. This relationship is called **Wien's law.** For the Sun, the wavelength at which the maximum energy is emitted is 520 nm, which is near the middle of that portion of the electromagnetic spectrum that is called visible light. It is surely no coincidence, but rather a consequence of evolutionary adaptation, that human eyes are most sensitive to electromagnetic radiation at those wavelengths at which the Sun puts out the most energy. Characteristic temperatures of other astronomical objects, and the wavelengths at which they emit most of their energy, are listed in Table 5.1.

If we sum up the contributions from all parts of the electromagnetic spectrum, we obtain the total energy emitted by a blackbody over all wavelengths. That total energy, emitted per second per square meter by a blackbody at a temperature T, is proportional to the fourth power of its absolute temperature. This relationship is known as the **Stefan-Boltzmann law,** which can be written in the form of an equation as

$$E = \sigma T^4,$$

where E stands for the luminosity and σ is a constant number. If the Sun, for example, were twice as hot—that is, if it had a temperature of 11,600 K—it would radiate 2^4, or 16, times more energy than it does now.

5.3 Spectroscopy in Astronomy

The most powerful tool available to an astronomer is the **spectrometer**—a device that breaks light up into a spectrum for analysis. Through the study of spectra, we can learn a remarkable amount about the nature of stars and other astronomical sources.

Optical Properties of Light

Light and other forms of electromagnetic energy obey certain laws that are important to the design of telescopes and other optical instruments. For example, light is *reflected* from a surface. If the surface is smooth and shiny, the direction of the reflected light beam can be accurately calculated from a knowledge of the shape of the reflecting surface. Light is also bent, or *refracted*, when it passes from one kind of transparent medium into another, say, from the air into a glass lens.

Reflection and refraction of light are the basic properties that make optical instruments possible—from eyeglasses to giant astronomical telescopes. Such instruments are generally combinations of glass lenses, which bend light according to the principles of refraction, and curved mirrors, which depend on the properties of reflection. Small optical devices, such as eyeglasses or binoculars, generally use lenses, while large telescopes depend almost entirely on mirrors for their main optical elements. In Chapter 6, we will discuss a number of astronomical instruments and their uses.

When light passes from one transparent medium to another, an interesting effect occurs in addition to simple refraction. Because the bending of the beam depends on the wavelength of the light as well as the properties of the medium, different wavelengths or colors are bent by different amounts. This phenomenon is called **dispersion.**

Figure 5.4 shows how light can be separated into different colors with a prism—a piece of glass with a triangular cross-section. Upon entering one face of the prism, light is refracted once, the violet light more than the red, and upon leaving the opposite face, the

light is bent again and further dispersed. If the light leaving the prism is focused on a screen, the different wavelengths or colors that compose white light are lined up side by side (Figure 5.6), as in a rainbow, which is formed by the dispersion of light through raindrops (see Essay).

The Value of Stellar Spectra

If the spectrum of the white light from the Sun and stars were simply a continuous rainbow of colors, astronomers would have little interest in the study of stellar spectra. To Newton, who first described the laws of refraction and dispersion in optics, the solar spectrum did appear as just a continuous band of colors. However, in 1802, William Wollaston built an improved spectrometer that included a lens to focus the spectrum of the Sun on a screen. With this device, Wollaston saw that the colors were not spread out uniformly but that some ranges of color were missing, appearing as dark bands in the solar spectrum. He attributed these lines to natural boundaries between the colors. Later, in 1815, the German physicist Joseph Fraunhofer, upon a more careful examination of the solar spectrum, found about 600 such dark lines (Figure 5.7).

These physicists called what they saw *dark lines* because the rainbow-colored spectrum as they viewed it was interrupted at various wavelengths, and in their spectrometers such interruptions looked like dark lines. There are two possible explanations for dark lines in the solar spectrum. These lines could be narrow ranges of wavelengths (or frequencies) in which the Sun does not emit radiation. Alternatively, the Sun might be emitting radiation at these colors, too, but something between the Sun and the spectrometer absorbs these particular wavelengths.

Subsequently, researchers found that similar dark spectral lines can be produced in the spectra of artificial light sources by passing their light through various apparently transparent substances (usually gases). These gases are not transparent at all colors but are quite opaque at a few sharply defined wavelengths. On the other hand, the spectra emitted by

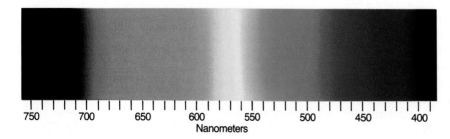

FIGURE 5.6 When white light is passed through a prism, the effect of dispersion is to form a continuous spectrum of visible light.

ESSAY The Rainbow

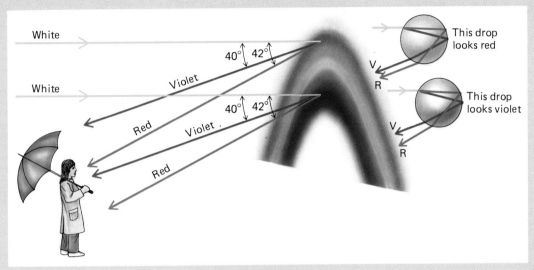

FIGURE 5A This diagram shows how light from the Sun, which is located behind the observer, can be refracted by raindrops to produce a rainbow.

Rainbows are an excellent illustration of refraction. You have a good chance of seeing a rainbow anytime you are between the Sun and a rain shower, and this situation is illustrated in Figure 5A. Remember that white sunlight is composed of all the colors from violet to red. Now suppose that a ray of sunlight encounters a raindrop and passes into it. The light changes direction, or to use the proper technical term, the light is refracted (Figure 5B), when it passes from air to water, and the blue light is refracted more than the red. The light is then reflected at the backside of the drop and re-emerges from the front, where it is again refracted. Now the white light has been spread out into a rainbow of colors.

Note that in Figure 5B, violet light lies above the red light after it emerges from the raindrop. When you look at a rainbow, however, it is the red light that is higher in the sky. Why? Look again at Figure 5A. If the observer looks at a raindrop that is high in the sky, the violet light passes over her head, while the red light enters the eye. Similarly, if the observer looks at a raindrop that is low in the sky, the violet light reaches her eye and the drop appears violet, while the red light from that same drop strikes the ground and is not seen. Colors of intermediate wavelengths are refracted to the eye by drops that are intermediate in altitude between the drops that appear violet and the ones that appear red. Thus, a single rainbow always has red on the outside and violet on the inside.

For an even simpler example of refraction, put a pencil at a slanted angle in a glass of water. What do you see? Can you offer an explanation?

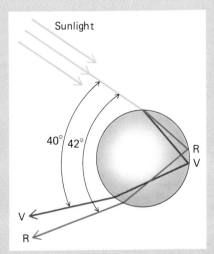

FIGURE 5B A diagram showing the path of light passing through a raindrop. Refraction separates white light into its component colors.

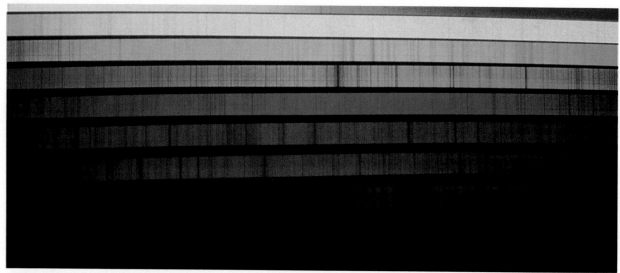

FIGURE 5.7   The visible spectrum of the Sun. The spectrum is crossed by dark lines produced by atoms in the solar atmosphere that absorb light at certain wavelengths. *(National Solar Observatory/National Optical Astronomy Observatories)*

certain glowing gases are observed to consist of several separate *bright lines*; that is, these gases emit light only at discrete wavelengths or colors. Thus we can begin to see that different substances show distinctive *spectral signatures*, by which their presence can be detected (Figure 5.8).

We distinguish, then, among three types of spectra. A **continuous spectrum** is an array of all wavelengths or colors of the rainbow. A bright line, or **emission line spectrum,** appears as a pattern or series of bright lines; it is formed from light in which only certain discrete wavelengths are present. A dark line, or **absorption line spectrum,** consists of a series or pattern of dark lines—missing colors—superimposed

upon the continuous spectrum of a source of white light.

Each particular chemical element or compound, when in the gaseous form, produces its own characteristic pattern of spectral lines: its spectral signature. In other words, each particular gas can absorb or emit only certain wavelengths of light peculiar to that gas. The temperature and other conditions determine whether the lines that we see are bright or dark, but the wavelengths are the same in either case, and it is the precise pattern of wavelengths that makes the signature of each element unique. The presence of a particular pattern of dark (or bright) lines characteristic of a certain element is evidence of the presence of

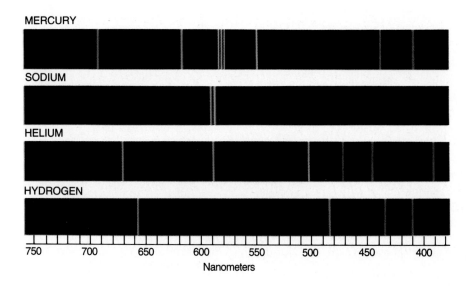

MERCURY

SODIUM

HELIUM

HYDROGEN

750   700   650   600   550   500   450   400

Nanometers

FIGURE 5.8   The line spectra produced by several different kinds of hot gas. Each gas produces its own unique pattern of lines, so the composition of a gas can be identified from observations of its spectrum.

that element somewhere along the path of the light whose spectrum has been analyzed. Liquids and solids also can generate spectral lines or bands, but they are broader and less well defined, and are more difficult to interpret.

The dark lines in the solar spectrum thus give evidence of certain chemical elements between us and the Sun, absorbing those wavelengths of light. Similarly, we can use the presence of absorption and emission lines to analyze the composition of stars and gas in space. The analysis of spectra is the key to modern astrophysics. In 1835, the French philosopher Auguste Comte speculated that it would eventually be possible to measure the sizes, distances, and motions of stars, but that it would never be possible by any means to learn their chemical composition. Only 25 years later, the German Gustav Kirchhoff proved him resoundingly wrong by using spectroscopy to identify the element sodium in the Sun. In the years that followed, astronomers found many other chemical elements in the Sun and stars. But it was only in the 20th century, with the development of a model for the atom, that scientists learned how spectral lines are formed.

5.4 The Structure of the Atom

The idea that matter is composed of tiny particles called atoms is at least 25 centuries old. It was not until the 20th century, however, that scientists invented instruments that permitted them to probe inside an atom and discover that it was not, as had been thought, hard and indivisible. Instead, the atom is a complex structure composed of still smaller particles.

Probing the Atom

The first of these smaller particles was discovered by British physicist J.J. Thomson in 1897. Named the electron, this particle is negatively charged. Since an atom in its normal state is electrically neutral, each electron in an atom must be balanced by the same amount of positive charge.

The next problem was to determine where in the atom the positive and negative charges are located. In 1911 British physicist Ernest Rutherford devised an experiment that provided part of the answer to this question. What he did was to bombard a piece of gold foil, which was about 400 atoms thick, with a beam of alpha particles emitted from a radioactive material (Figure 5.9). We now know that alpha particles are helium atoms that have lost all of their electrons. Most of the alpha particles passed through the gold foil just as if it and the atoms composing it were nearly empty space. About 1 in 8000 of the alpha particles, however, completely reversed direction and bounced backward from the foil. Rutherford wrote, "It was quite the most incredible event that has ever happened to me in my life. It was almost as incredible as if you fired a 15-inch shell at a piece of tissue paper and it came back and hit you."

The only way to account for the alpha particles that reversed direction when they hit the gold foil is to assume that nearly all of the mass, as well as all of one type of charge, either positive or negative, in each individual gold atom is concentrated in a tiny nucleus. We now know that it is the positive charge that is located in the nucleus. When an alpha particle strikes a nucleus, it reverses direction, much as a cue ball reverses direction when it hits another billiard ball.

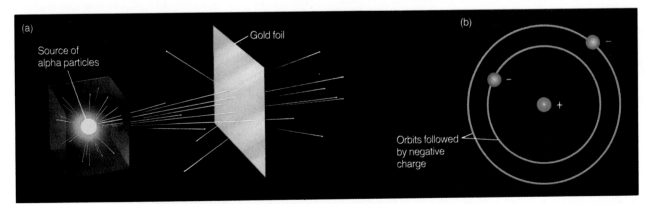

FIGURE 5.9 (a) When Rutherford allowed alpha particles from a radioactive source to strike a target of gold foil, he found that some of the alpha particles rebounded back in the direction from which they came. (b) From this experiment, he concluded that the atom must be constructed like a miniature solar system, with the positive charge concentrated in the nucleus. The negative charge was assumed to orbit in the large volume around the nucleus.

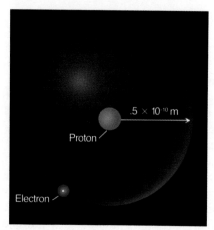

FIGURE 5.10 Schematic diagram of a hydrogen atom in its lowest energy state, which is also called the ground state. The proton and electron have equal but opposite charges, which exert an electromagnetic force that binds the hydrogen atom together.

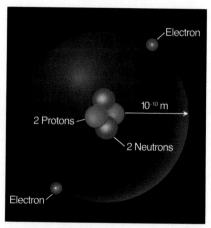

FIGURE 5.11 Schematic diagram of a helium atom in its lowest energy state. Two protons are present in the nucleus of all helium atoms. In the most common variety of helium, the nucleus also contains two neutrons, which have nearly the same mass as the proton but carry no charge.

Rutherford's model placed the other type of charge—the electrons as we now know—in orbit around this nucleus.

Rutherford's model requires that the electrons be in motion. Since positive and negative charges attract each other, stationary electrons would fall into the nucleus. Because most of the atom is empty, nearly all of Rutherford's alpha particles were able to pass right through the gold foil without colliding with anything.

The Atomic Nucleus

The simplest atom is hydrogen, and the nucleus of ordinary hydrogen contains a single positively charged particle called a proton. Moving around this proton is a single electron. The mass of an electron is nearly 2000 times smaller than the mass of a proton, but the electron carries an amount of charge that is exactly equal in magnitude but opposite in sign to that of the proton. The electron's charge is negative instead of positive (Figure 5.10). Opposite charges attract each other, so it is the electromagnetic force that binds the proton and electron together, just as gravity is the force that keeps the planets in orbit around the Sun.

There are, of course, other types of atoms. Helium, for example, is the second most abundant element in the Sun. Helium has two protons in its nucleus, instead of the single proton that characterizes hydrogen. In addition, the helium nucleus contains two neutrons, particles with a mass slightly larger than that of the proton but with no electric charge. Moving around this nucleus are two electrons, so that the

total net charge of the helium atom is zero (Figure 5.11).

From this description of hydrogen and helium, perhaps you have guessed the pattern for building up all of the elements that we find in the universe. The specific element is determined by the *number of protons* in the nucleus. Carbon has 6 protons, oxygen has 8, iron has 26, and uranium has 92. In its normal state, each atom has the same number of electrons as protons, and these electrons follow complex orbital patterns around the nucleus.

Although the number of neutrons in the nucleus is usually approximately equal to the number of protons, the number of neutrons is not necessarily the same for all atoms of a given element. For example, most atoms of hydrogen contain no neutron at all. There are, however, hydrogen atoms that contain one proton and one neutron, and others that contain one proton and two neutrons. The various types of nuclei of hydrogen are called **isotopes** of hydrogen (Figure 5.12), and other elements have isotopes as well.

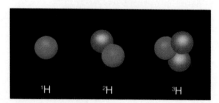

FIGURE 5.12 Schematic diagram of the nuclei of the isotopes of hydrogen. A single proton in the nucleus defines the atom to be hydrogen, but there may be zero, one, or two neutrons. By far the most common isotope is the one with only a single proton. A hydrogen nucleus with one neutron is called deuterium; one with two neutrons is called tritium.

The Bohr Atom

There is one serious problem with Rutherford's model for atoms. As we have already seen, Maxwell's theory says that when electrons change either their speed or direction of motion, they emit energy in the form of electromagnetic radiation. Since orbiting electrons constantly change their direction of motion, they should emit a constant stream of energy. Earth-orbiting satellites spiral back toward Earth as they lose energy through friction with the Earth's atmosphere. So, too, should electrons spiral into the nucleus of the atom as they radiate electromagnetic energy.

It was the Danish physicist Niels Bohr (1885–1962) who solved the mystery of the electrons. He suggested that the spectrum of hydrogen can be understood if it is assumed that only orbits of certain sizes are possible for the electron. Bohr further assumed that so long as the electron moves only in one of these allowed orbits, it radiates no energy. If the electron moves from one orbit to another closer to the atomic nucleus, then it must give up some energy in the form of electromagnetic radiation, just as a satellite gives up energy when it spirals back into the Earth's atmosphere. Conversely, energy is required to boost the electron from a smaller orbit to one farther from the nucleus, and one way to obtain the necessary energy is to absorb electromagnetic radiation if some is streaming past the atom from an outside source.

According to Bohr's model, one fundamental difference between an electron moving about an atomic nucleus and a satellite orbiting the Earth is that only certain orbits are allowed for the electron. The amount of energy that the electron must either absorb or emit to move from one orbit to another is therefore fixed and definite.

When an electron moves from a larger to a smaller orbit, it emits a discrete packet of electromagnetic energy, which is called a **photon.** Photons must not be thought of as particles, for, as we have said, electromagnetic energy travels with a wave-like motion. Photons can be regarded as waves propagating through space, each spreading in all directions from its source.

Each photon carries a certain amount of energy that depends only on the frequency of the radiation. Specifically, the energy of a photon is proportional to the frequency; the constant of proportionality, h, called Planck's constant, is named for Max Planck, the German physicist who was one of the originators of the quantum theory. If metric units are used (that is, if energy is measured in joules and frequency in cycles or waves per second), Planck's constant has the value $h = 6.626 \times 10^{-20}$ joule/s. Since the frequency times the wavelength is equal to the speed of light, the energy of a photon is also inversely proportional to the wavelength. Photons of violet and blue light are thus of higher energy than those of red light. The highest energy photons of all are gamma rays; those of lowest energy are radio waves.

5.5 Formation of Spectral Lines

The Hydrogen Spectrum

In order to understand how spectral lines are formed, suppose a beam of white light (which consists of photons of all wavelengths) shines through a gas of atomic hydrogen. Since a photon of wavelength 656 nm has the right energy to raise an electron in a hydrogen atom from the second to the third orbit, it can be absorbed by those hydrogen atoms that are in their second to lowest energy states. Other photons will have the right energies to raise electrons from the second to the fourth orbit, or the first to the fifth orbit, and so on. Only photons that have exactly these correct energies can be absorbed. All of the other photons will stream past the atom untouched. Thus, the hydrogen atoms absorb light only at certain wavelengths and produce spectral absorption lines. Conversely, hydrogen atoms in which electrons move from larger to smaller orbits emit light—but, again, only light of those energies or wavelengths that correspond to the energy difference between permissible orbits. The changes in orbits of electrons giving rise to spectral lines are shown in Figure 5.13.

In the case of hydrogen, the spectral lines are produced at wavelengths that can be calculated by a simple formula. For the lines in the visible, which are called the *Balmer lines* after the Swiss mathematician and schoolteacher who first discovered the formula, the wavelengths are give by the following equation:

$$\frac{1}{\lambda} = \frac{1}{911.8}\left(\frac{1}{2^2} - \frac{1}{n^2}\right)$$

where λ is the wavelength in nanometers and n is an integer that can take any value from 3 on. If $n = 3$, the wavelength of the first line in the Balmer series of hydrogen is obtained (at 656 nm in the red). For $n = 4, 5$, and so on, the wavelengths of the second, third, and higher Balmer lines are obtained. As n approaches larger and larger values, the wavelengths of the successive Balmer lines become more and more nearly equal. The lines of hydrogen in stellar spectra are observed to do just this; they approach a limit at

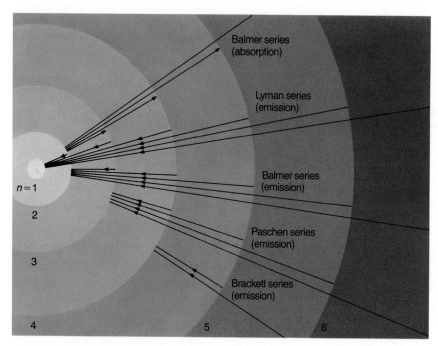

FIGURE 5.13 Emission and absorption of light by a hydrogen atom according to the Bohr model. Several different series of spectral lines are shown, corresponding to transitions of electrons from or to certain allowed energy levels. Each series of lines that terminates on a specific inner orbit is named for the physicist who studied it.

about 365 nm (Figure 5.14), corresponding to a value of $n = \infty$.

After Balmer's work, other series of hydrogen lines were found. The Lyman series, in the ultraviolet, approaches a limit at about 91.2 nm. The Paschen series, in the infrared, approaches a limit at about 820 nm. Still farther in the infrared are found the Brackett series, the Pfund series, and so on. All these series (including the Balmer series) can be predicted by the more general formula, known as the Rydberg formula:

$$\frac{1}{\lambda} = R \left(\frac{1}{m^2} - \frac{1}{n^2} \right)$$

where m is an integer and n is any integer greater than m. The Rydberg constant, R, has the value 1.097×10^7 if λ is measured in meters. For the Lyman series, $m = 1$. For the Balmer series $m = 2$, for Paschen $m = 3$, and so on.

Similar pictures can be drawn for atoms other than hydrogen. However, since these other atoms ordinar-ily have more than one electron each, the orbits of their electrons are much more complicated, and the spectra are more complex as well. Each type of atom has its own unique pattern of orbits, and that is why it also has its own unique set of spectral lines.

Energy Levels of Atoms and Excitation

Bohr's model of the hydrogen atom was a great step forward in our understanding of the atom. However, we know today that atoms cannot be represented by quite so simple a picture as the Bohr model. Even the concept of sharply defined orbits of electrons is not really correct. Since an electron has wave-like char-acteristics, we can only estimate the *probability* that it will follow a particular orbit. Nevertheless, since the most likely orbits are fairly narrow with respect to the size of an atom, we still retain the concept that only certain discrete energies are allowable for an atom. These energies, called **energy levels,** can be thought of as representing certain average distances

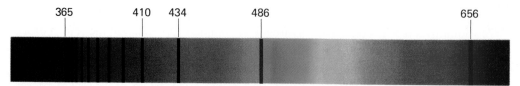

FIGURE 5.14 Balmer lines in the spectrum of hydrogen. Note how the lines come more closely together as the number n of the lower energy level becomes very large. (For the purposes of illustration, colors have been extended to wavelengths somewhat shorter than those that can be seen by the human eye.)

of the most likely of the electron's possible orbits around the atomic nucleus.

Differences in energy of the electron orbits in atoms are usually expressed in electron volts. An electron volt (abbreviated eV) is the small amount of energy acquired by an electron after being accelerated through a potential difference of 1 volt. The energy needed to raise the hydrogen atom from its lowest energy level to the next lowest one is about 10.2 eV. One eV = 1.602×10^{-19} joules. One joule of energy expended in 1 s is a *watt* of power. Typical electric light bulbs use about 100 watts.

Ordinarily, an atom is in the state of lowest possible energy, its **ground state.** In the Bohr model, the ground state corresponds to the electron's being in the innermost orbit. However, an atom can absorb energy, which raises it to a higher energy level (corresponding, in the Bohr picture, to the movement of an electron to a larger orbit). The atom is then said to be in an **excited state.** Generally, an atom remains excited for only a very brief time. After a short interval, typically a hundred-millionth of a second or so, it drops back down to its ground state, with the simultaneous emission of light. The atom may return to its lowest state in one jump, or it may make the transition in steps of two or more jumps, stopping at intermediate levels on the way down. With each jump, it emits a photon of the wavelength that corresponds to the energy difference between the levels at the beginning and end of that jump.

An energy-level diagram for a hydrogen atom and several possible atomic transitions are shown in Figure 5.15; compare this figure with the Bohr model, shown in Figure 5.13.

Because atoms that have absorbed light and have thus become excited generally de-excite themselves and emit that light again, we might wonder why *dark* spectral lines are ever produced. In other words, why doesn't this re-emitted light "fill in" the absorption lines? Some of the re-emitted light actually is received by us, but this light fills in the absorption lines only to a slight extent. The reason is that the atoms re-emit light in mostly random directions, and only a small fraction of the re-emitted light is directed toward the observer. We can observe the re-emitted light as emission lines if and only if we can view the absorbing atoms from a direction from which little or no background light is coming—as we do, for example, when we look at clouds of hot gas located in the space between the stars. Figure 5.16 illustrates the situation.

Atoms in a gas are moving at high speeds and continually collide with one another and with electrons. They can be excited and de-excited by these

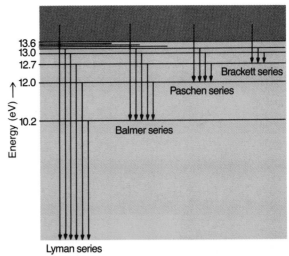

FIGURE 5.15 *Energy-level diagram for hydrogen. The shaded region represents energies at which the atom is ionized.*

collisions as well as by absorbing and emitting light. The velocity of atoms in a gas depends on the temperature of the gas. When the temperature is higher, so are the velocity and the energy of the collisions. The hotter the gas, therefore, the more likely it is that electrons will occupy the outermost orbits, which correspond to the highest energy levels.

Ionization

We have described how certain discrete amounts of energy can be absorbed by an atom, raising the atom to an excited state and moving one of its electrons farther from its nucleus. If enough energy is absorbed, the electron can be removed completely from the atom. The atom is then said to be **ionized.** The minimum amount of energy required to ionize an atom from its ground state is called its ionization energy or ionization potential. The ionization potential for hydrogen is about 13.6 eV.

Still greater amounts of energy must be absorbed by the ionized atom (called an **ion**) to remove an additional electron. Successively greater energies are needed to remove the third, fourth, and fifth electrons from the atom, and so on. If enough energy is available (in the form of very short wavelength photons or in the form of a collision with a very fast-moving electron or another atom), an atom can become completely ionized, losing all of its electrons. A hydrogen atom, having only one electron to lose, can be ionized only once; a helium atom can be ionized twice, and an oxygen atom, up to eight times.

An atom that has become ionized has lost a negative charge—that carried away by the electron—and

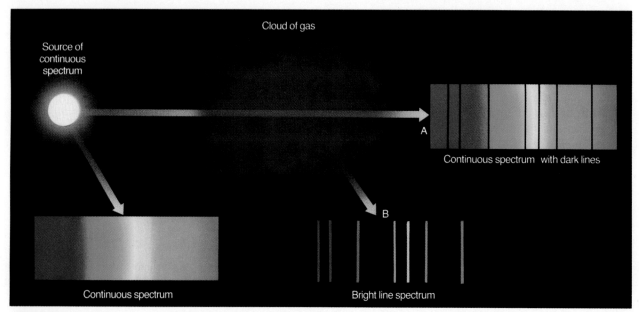

FIGURE 5.16   Production of bright and dark spectral lines. The atoms in the gas cloud produce absorption lines in the continuous spectrum of the white light source when viewed from direction *A*, but they produce emission lines (of the light they re-emit) when viewed from direction *B*.

thus is left with a net positive charge. It has, therefore, a strong affinity for a free electron. Eventually, an electron will be captured, and the atom will become neutral (or ionized to one less degree) again. During the capture process, the atom emits one or more photons, depending on whether the electron is captured at once to the state corresponding to the lowest energy level of the atom or whether it stops at one or more intermediate levels on its way to the ground state.

Just as the excitation of an atom can result from a collision with another atom, ion, or electron (collisions with electrons are usually most important), so also can ionization result from collisions. The rate at which such collisional ionizations occur depends on the atomic velocities and hence on the temperature of the gas. The rate of recombination of ions and electrons also depends on their relative velocities, that is, on the temperature. In addition, it depends on the density of the gas; the higher the density, the greater the chance for recapture, because the different kinds of particles are crowded more closely together. From a knowledge of the temperature and density of a gas, it is possible to calculate the fraction of atoms that have been ionized once, ionized twice, and so on. In the Sun, for example, we find that most of the hydrogen and helium atoms in its atmosphere are neutral, whereas most of the atoms of calcium, as well as many other metals, are once ionized.

The energy levels of an ionized atom are entirely different from those of the same atom when it is neutral. In each degree of ionization, the energy levels of the ion, and thus the wavelengths of the spectral lines it can produce, have their own characteristic values. In the Sun, therefore, we find lines of neutral hydrogen and helium, but lines of ionized calcium. Ionized hydrogen, having no electron, can produce no absorption lines.

Summary of Emission and Absorption Processes

Atoms are characterized by energy levels that correspond to various distances of their electrons from their nuclei. By absorbing or emitting radiant energy, an atom can move from one to another of these levels, thus raising or lowering its energy. Since only certain discrete energy levels exist for each kind of atom in a gas, the absorbed or emitted radiation occurs only at certain energies or wavelengths, producing dark or bright spectral lines.

An atom is said to be excited if it is in any but its lowest allowable energy level. It is said to be ionized if, by the absorption of energy, it has lost one or more of its electrons. It can be excited or de-excited by collisions as well as by the absorption and emission of radiation; it can be ionized by collision or by absorbing radiation.

5.6 The Doppler Effect

In 1842 Christian Doppler pointed out that if a light source is approaching or receding from the observer, the light waves will be, respectively, crowded more closely together or spread out. The principle, known as the **Doppler effect,** is illustrated in Figure 5.17. In part a, the light source is stationary with respect to the observer. As successive wave crests 1, 2, 3, and 4 are emitted, they spread out evenly in all directions, like the ripples from a splash in a pond. They approach the observer at a distance λ behind one another, where λ is the wavelength of the ripple. On the other hand, if the source is moving with respect to the observer, as in part b, the successive wave crests are emitted with the source at different positions, S_1, S_2, S_3, and S_4, respectively. Thus, to observer A, the waves seem to follow one another more closely, at a decreased wavelength and increased frequency, whereas to observer C they are spread out and arrive at an increased wavelength and decreased frequency. To observer B, in a direction at right angles to the motion of the source, no effect is observed.

The effect is produced only by a motion toward or away from the observer, a motion called **radial velocity.** Observers between A and B and between B and C would observe some shortening or lengthening of the light waves (increase or decrease in frequency), respectively, for a component of the motion of the source is in their line of sight.

The Doppler effect is also observed in sound. We have all heard the higher than normal pitch of an approaching siren and the lower than normal pitch of a receding one. Standing by the highway, you can hear the characteristic "whoosh" or declining pitch of each car or truck as it speeds past. The wavelengths are shortened if the distance between the source and observer is decreasing, and they are lengthened if the distance is increasing.

If the relative motion is entirely in the line of sight, the formula for the Doppler shift of light is

$$\frac{\Delta\lambda}{\lambda} = \frac{v}{c}$$

where λ is the wavelength emitted by the source, $\Delta\lambda$ is the difference between λ and the wavelength measured by the observer, c is the speed of light, and v is the relative line of sight velocity of the observer and source. The variable v is counted as positive if the velocity is one of recession and negative if it is one of approach. Solving this equation for the velocity, we find

$$v = c\,\frac{\Delta\lambda}{\lambda}$$

If a star approaches or recedes from us, the wavelengths of light in its continuous spectrum appear shortened or lengthened, respectively, as well as those of the dark lines. However, unless its speed is tens of thousands of kilometers per second, the star does not appear noticeably bluer or redder than normal. The Doppler shift is thus not easily detected in a continuous spectrum and cannot be measured accurately in such a spectrum. On the other hand, the wavelengths of the absorption lines can be measured accurately, and their Doppler shift is relatively simple to detect.

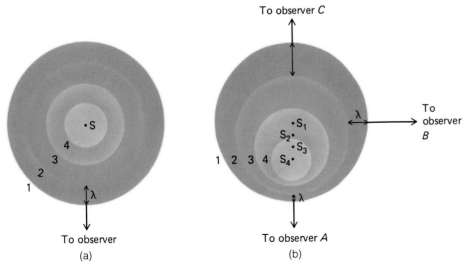

FIGURE 5.17 The Doppler effect. See text for explanation.

SUMMARY

5.1. Light is one form of **electromagnetic radiation.** The **wavelength** of light determines the color of visible radiation. Wavelength (λ) is related to **frequency** (f) and the speed of light (c) by the equation $c = \lambda f$. The **apparent brightness** of a source of electromagnetic energy decreases with increasing distance from that source in proportion to the square of the distance. The mathematical equation describing this relationship is known as the **inverse-square law.**

5.2. The **electromagnetic spectrum** consists of **gamma rays, x rays,** and **ultraviolet radiation** (all forms of electromagnetic radiation with wavelengths shorter than that of visible light) and **infrared, microwave,** and longer wave radio radiation (with wavelengths longer than that of light). The higher the temperature of a **blackbody,** the shorter the wavelength at which the maximum amount of electromagnetic radiation is emitted. The mathematical equation describing this relationship ($\lambda_{max} = 3 \times 10^6/T$) is known as **Wien's law.** The total energy emitted per square meter increases with increasing temperature. The relationship between emitted energy and temperature ($E = \sigma T^4$) is known as the **Stefan-Boltzmann law.**

5.3 A **spectrometer** is a device that forms a spectrum, often utilizing the optical phenomenon of **dispersion.** The light from an astronomical source can consist of a **continuous spectrum,** a bright line or **emission line spectrum,** or a dark line or **absorption line spectrum.** Spectral analyses reveal the composition of the Sun and stars.

5.4. Atoms consist of a nucleus containing one or more positively charged protons. All atoms except hydrogen also contain one or more neutrons in the nucleus. Negatively charged electrons orbit the nucleus, and the number of electrons normally equals the number of protons. The number of protons defines the element (hydrogen, helium, and so on) of the atom. Nuclei with the same number of protons but different numbers of neutrons are different **isotopes** of the same element.

5.5. When an electron moves from one orbit to another closer to the atomic nucleus, a **photon** is emitted, and a spectral emission line is formed. Absorption lines are formed when an electron moves to an orbit farther from the nucleus. Since each atom has its own characteristic set of orbits, each atom has associated with it a unique pattern of spectral lines. An atom in its lowest **energy level** is said to be in the **ground state.** If an electron is in an orbit other than the least energetic one possible, an atom is said to be **excited.** If an atom contains fewer electrons than protons, it is called an **ion** and is said to be **ionized.**

5.6. If an atom is moving toward an observer when an electron changes orbits and produces a spectral line, that line will be shifted slightly to the blue of its normal wavelength. If the atom is moving away, the line will be shifted to the red. This shift is known as the **Doppler effect** and can be used to measure the **radial velocities** of distant objects by the formula $v = c(\Delta\lambda/\lambda)$.

REVIEW QUESTIONS

1. What distinguishes one type of electromagnetic radiation from another? Identify types of electromagnetic radiation that you make use of, or are exposed to, in everyday life.

2. What is a wave? Use the terms "wavelength" and "frequency" in your definition.

3. What is a blackbody? How does the energy emitted by a blackbody depend on its temperature?

4. Where in an atom would you expect to find electrons? Protons? Neutrons?

5. Explain how emission lines and absorption lines are formed.

6. Explain how the Doppler effect works for sound waves, and give some familiar examples.

THOUGHT QUESTIONS

7. Suppose the Sun radiates like a blackbody. Explain how you would calculate the total amount of energy radiated into space by the Sun each second. What solar data would you need to make this calculation?

8. What type of electromagnetic radiation is best suited to observing a star with a temperature of 5800 K? A gas heated to a temperature of 1 million K? A human being on a dark night?

9. Why is it dangerous to be exposed to x rays but not (or at least very much less) dangerous to be exposed to radio waves? Are gamma rays likely to be more dangerous than x rays?

10. Go outside at night and look carefully at the brightest stars. Some should look red and others blue. The primary factor determining the color of a star is its temperature. Which is hotter—a blue star or a red one? Explain.

11. Water faucets are often labeled with a red dot for hot water and a blue dot for cold water. Given Wien's law, does this labeling make sense?

12. The planet Jupiter appears yellowish, while Mars is red. Does this observation mean that Mars is cooler than Jupiter?

13. Suppose that you are standing at the exact center of a park surrounded by a circular road. Suppose an ambulance drives completely around this circular road. How will the pitch of the ambulance's siren change as the ambulance circles around you?

14. How could you measure the Earth's orbital speed by photographing the spectrum of a star at various times throughout the year? (*Hint:* Suppose the star lies in the plane of the Earth's orbit.)

PROBLEMS

15. "Tidal waves," or tsunamis, are waves of seismic origin that travel rapidly through the ocean. If tsunamis travel at the speed of 600 km/h and approach a shore at the rate of one wave crest every 15 min, what would be the distance between those wave crests at sea?

16. How many times brighter or fainter would a star appear if it were moved to (**a**) twice its present distance? (**b**) ten times its present distance? (**c**) half its present distance?

17. Two stars are at the same distance. The two stars have identical diameters. One has a temperature of 5800 K; the other has a temperature of 2900 K. Which is brighter? How much brighter is it?

18. If the emitted infrared radiation from Pluto has a wavelength of maximum intensity at 50,000 nm (50 μm), what is the temperature of Pluto?

19. What is the temperature of a star with a wavelength of maximum light of 290 nm?

20. Suppose that a spectral line of some element, normally at 500 nm, is observed in the spectrum of a star to be at 500.1 nm. How fast is the star moving toward or away from the Earth?

Answer: About 60 km/s away from the Earth

The 3.6-m Canada-France-Hawaii telescope on Mauna Kea, Hawaii, at an altitude of 4200 m
(Charles Kaminski, University of Hawaii)

George Ellery Hale (1868–1939) was one of the founders of the new science of astrophysics at the beginning of the 20th century. He also had the vision and leadership to initiate the construction of the world's largest telescope no less than four times! The 5-m telescope on Palomar Mountain is named in his honor. *(Caltech)*

6

ASTRONOMICAL OBSERVATIONS

Electromagnetic radiation from space comes at all wavelengths, from gamma rays to radio waves. Unfortunately for astronomical observations (but fortunately for biological organisms), much of the electromagnetic spectrum is filtered out by the terrestrial atmosphere. There are two *spectral windows* in the atmospheric filter through which we can observe (Figure 6.1). One of these is the optical window, which includes the near-ultraviolet (wavelengths longer than about 300 nm or 0.3 μm) and portions of the infrared (with wavelengths up to about 30 μm). The other is the radio window, which includes radio waves that range in length from about a millimeter to about 20 m. We call these regions "windows" because the atmosphere is transparent at these wavelengths. Equally important, but less easily studied, is radiation reaching the Earth from parts of the spectrum where the atmosphere is opaque; to observe in these regions, astronomers must have telescopes in orbit.

6.1 Telescopes

Archeoastronomy, which is the study of the practice of astronomy in the ancient world, has shown that the construction of astronomical observatories has been a continuing concern of civilizations of diverse cultural values and levels of sophistication (Figure 6.2). For the most part, ancient observatories were not used to study the heavens with any thought of trying to understand the processes that control the celestial phenomena that we observe. Their prime purpose was the measurement of the positions of objects and determinations of time and calendar. Telescopes were a relatively late addition to the complement of instruments housed within observatories, becoming important only in the past 300 years.

The Astronomical Telescope

Galileo built the first telescopes for astronomical observations in 1610. As we have seen, these were simple tubes that could be held in your hand. However, astronomical telescopes have come a long way since Galileo's time. Now they tend to be huge devices, constructed at costs of tens of millions of dollars.

In order to study astronomical objects, it is useful to form an *image* of a source of radiation. The image can then be detected, recorded, measured, reproduced, and analyzed in a host of ways. At first, astronomers simply viewed the image with their eyes, but today one rarely actually looks through an astronomical telescope. The image is recorded electronically or photographically, and this permanent record becomes the object of subsequent detailed study.

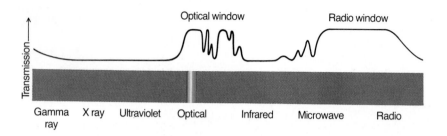

FIGURE 6.1 A portion of the electromagnetic spectrum, showing those regions (windows) in which the Earth's atmosphere is transparent.

The most important functions of a telescope are to *collect* the faint light from an astronomical source and to *focus* it into an image. The measure of the size and capability of an astronomical telescope is its light-gathering ability, which is determined by the diameter, or **aperture,** of its primary lens or mirror. The telescope is like a large light-gathering bucket; the bigger the cross-sectional area of the bucket, the more photons are collected.

Formation of an Image by a Lens or a Mirror

A telescope, just like a camera lens, forms an image of a distant object. Figure 6.3 shows how a simple lens forms an image. If the curvatures of the surfaces of the lens are just right, light passing through the lens is bent, or refracted, in such a way that it converges toward a point, called the **focus** of the lens. At the focus, an image of the light source appears. The distance of the focus, or image, behind the lens is called the **focal length** of the lens.

Rays of light can also be focused to form an image with a concave mirror—one curved like the inner surface of a sphere (Figure 6.4) and coated with silver or aluminum to make it highly reflecting. If the mirror has the correct concave shape, all parallel rays are reflected back through the same point, the focus of the mirror. Thus images are produced by a mirror exactly as they are by a lens.

Most people, when they think of a telescope, picture a long tube with a large glass lens at one end. This design is called a **refracting telescope.** Galileo's telescopes were refractors, as are binoculars or opera

(a)

(b)

FIGURE 6.2 Two surviving pretelescopic observatories. (a) The Jantar Mantar, built in 1724 by Maharaja Jai Singh in Delhi, India. (b) Seventeenth-century bronze instruments from the old Chinese imperial observatory, Beijing. *(David Morrison)*

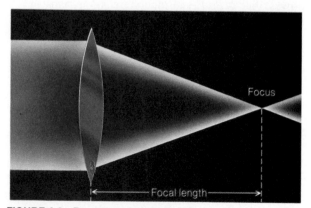

FIGURE 6.3 Formation of an image by a simple convex lens.

FIGURE 6.4 Formation of an image by a concave mirror.

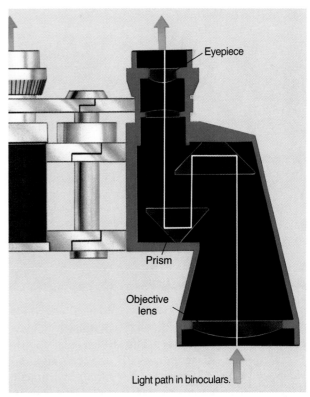

FIGURE 6.5 Binoculars are common examples of refracting telescopes.

glasses today (Figure 6.5). But refractors are not very good for most astronomical applications. In large sizes they are bulky and expensive to manufacture. Most astronomical telescopes (both amateur and professional) use a mirror rather than a lens as their primary optical part; these are called **reflecting telescopes.** Since reflecting telescopes are standard in astronomy, we will refer to mirrors in most of what follows. However, similar statements can also be made about telescopes that use a lens as their primary optical element.

Images have certain properties that depend on the diameter (aperture) and focal length of the primary mirror. One property is the *size* of an image; the size is proportional to the angular size of the object in the sky and also to the focal length of the mirror. If the angular diameter is measured in degrees and the focal length (f) in meters, the diameter of the image (in meters) is given by the following equation:

$$D_{\text{image}} = 0.018 D_{\text{object}} \times f$$

For example, the Moon has an angular diameter of $1/2°$ in the sky. A small telescope with a focal length of 1 m produces an image of the Moon just under 1 cm across. The 5-m (200-in) mirror of the Hale telescope on Palomar Mountain has a focal length of 17 m and produces a lunar image about 15 cm (6 in) across.

The *brightness* of an image is a measure of the amount of light energy that is concentrated into a unit area—say, a square millimeter—of the image. This brightness determines how much time is required to record the image electronically or photographically. The brightness of the image is greater the greater the amount of light focused into it by the primary mirror and is less when the area of the image over which the light must be spread is greater.

The image of a star, however, is so small that it is nearly a point. The total amount of light energy in this point depends only on the area of the mirror. Dou-

bling the aperture of the mirror, therefore, results in images of stars that are four times brighter. This is the gain produced, for example, in going from the 5-m Hale telescope at Palomar to the 10-m Keck telescope on Mauna Kea in Hawaii.

Resolution refers to the fineness of detail present in the image. The larger the telescope aperture, the sharper the image will be. However, additional limits on resolution are imposed by fluctuations in the Earth's atmosphere, which always introduce a certain amount of blurring into images formed with ground-based telescopes. It is important to locate observatories at locations where this atmospheric distortion is minimal.

The resolution of an astronomical image is measured by the angular size of a point source, such as a star. This size is expressed in seconds of arc, or arcsec, where 1 **arcsec** is 1/3600 degree. One arcsec is the apparent or angular diameter of a quarter at a distance of 50 km. During the past century, a resolution of about 1 arcsec has been considered the standard in astronomy, only achieved at most observatory sites when the atmosphere is unusually steady.

In the 1990s, however, astronomers are raising their standards. At the best observing locations, such

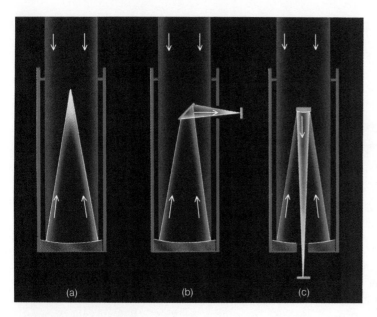

FIGURE 6.6 Various focus arrangements for reflecting telescopes: (a) prime focus, (b) Newtonian focus, (c) Cassegrain focus.

as Mauna Kea in Hawaii, it is possible to achieve resolutions of 0.3 arcsec. The design goal for the Hubble Space Telescope was 0.1 arcsec. And as we shall see, radio astronomers have built instruments with resolution better than 0.001 arcsec—the width of a quarter at 50,000 km.

The Complete Telescope

Now that we have seen how an image is produced, we are able to understand the operation of a telescope. Since its main purpose is to collect light from faint sources, astronomers usually want a telescope with as large an aperture as possible. Thus they must obtain a large disk of glass, which is laboriously ground and polished to produce a concave mirror of high optical quality. The mirror must be mounted so that it can point toward and track various astronomical sources, and a number of sophisticated instruments are built to analyze and record the light that is collected at the telescope's focus.

The reflecting telescope was first conceived by James Gregory in 1663, and the first successful model was built by Newton in 1668. The concave mirror is placed at the bottom of a tube or open framework. The mirror reflects the light back up the tube to form an image near the front end at a location called the *prime focus*. The image may be observed at the prime focus, or, alternatively, various systems of auxiliary mirrors can be used to intercept the light and bring it a focus at more convenient locations. Two such arrangements are shown in Figure 6.6. The most popular design for optical telescopes today is the *Cassegrain* system, in which a small convex mirror

intercepts the light from the primary mirror and sends it back down through a small hole in the primary.

The telescope must be mounted to point toward any direction in the sky and follow the apparent motion of the source under study. Until very recently, almost all astronomical optical telescopes had **equatorial mounts** (Figure 6.7). An equatorial mount

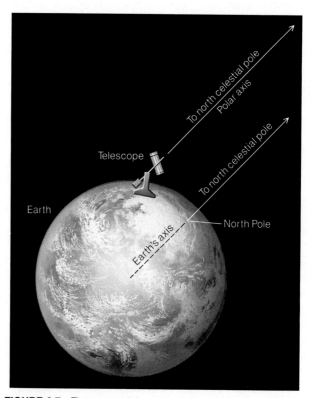

FIGURE 6.7 The equatorial mount, until recently the standard design for astronomical telescopes.

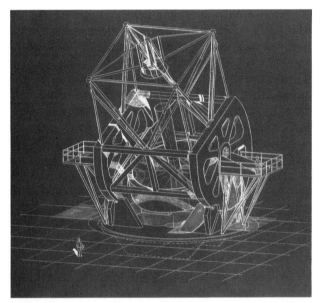

FIGURE 6.8 Model of the 8-m altitude-azimuth telescope under development by the U.S. National Optical Astronomy Observatories. *(National Optical Astronomy Observatories)*

allows the telescope to turn to the north and south about one axis and to the east and west about another. Since the axis for the east-west motion of the telescope is parallel to the axis of the Earth's rotation, a slow and steady rotation of the telescope about this axis compensates for the rotation of the Earth, making it very simple to track celestial objects.

An equatorial mount, however, is a disadvantage for a very large telescope because it is bulky and therefore expensive to construct. Consequently, the world's largest optical telescopes, the 6-m reflector in the Soviet Union and the 10-m Keck telescope in Hawaii, have simpler **altitude-azimuth mounts,** in which the telescope rotates about one vertical and one horizontal axis (Figure 6.8). Other large telescopes now under design will be similarly mounted, and so are large radio telescopes. It is more complicated for an altitude-azimuth mount to compensate for the Earth's rotation and track the stars, but today that problem can be handled with computer-controlled driving mechanisms.

6.2 Optical Detectors and Instruments

The popular view of the astronomer is of a person in a cold observatory peering through a telescope all night. Most astronomers, however, do not live at observatories but near the universities or laboratories where they work. Many astronomers work only with

radio telescopes or with space experiments. Still others work at purely theoretical problems and never observe at a telescope of any kind.

Even optical astronomers seldom inspect telescopic images visually except to center the telescope on a desired region of the sky or to make adjustments. On the contrary, electronic detectors are used to record the data permanently for detailed analysis after the observations are completed. A typical astronomer might spend a total of only a week or so each year observing at the telescope, and the rest of the time measuring or analyzing data.

The instrumentation on a large telescope is of three basic types. First, there is *imaging*—photographing or otherwise recording the appearance of a small portion of the sky. Second is the accurate measurement of the *brightness* and *color* of objects. Finally, there is *spectroscopy*—the measurement of the spectra of astronomical sources. All three require similar detectors to record and measure light, as we now describe.

Photographic and Electronic Detectors

Throughout most of the 20th century, photographic plates and film served as the prime astronomical detectors, whether for direct imaging or for photographing spectra. To photograph an astronomical object, the image of that object is allowed to fall on a light-sensitive coating that, when developed, provides a permanent record of the image—one that can be measured, studied, enlarged, published, and inspected by many individuals. When used for imaging photography, a telescope serves as nothing more than a large camera; the lens or mirror of the telescope serves as the camera lens.

The photographic plate is a superb device for collecting a large amount of information, and plates have been used by astronomers for more than a century. Photographic plates do, however, have serious limitations. Perhaps the most obvious is that they are inefficient—only about 1 percent of the light actually falling on the plate contributes to the image, with the rest wasted. For this and other reasons, astronomy has moved from the photographic to the electronic detection of light.

The most important electronic detectors are called **charge-coupled devices (CCDs).** The CCD offers several advantages over photographic plates. For example, CCDs are more sensitive and record as much as 60 to 70 percent of all the photons that strike them. CCDs also provide more accurate measurements of the brightness of astronomical objects than do photographic plates.

Infrared Observations

The infrared part of the spectrum extends from wavelengths near 1 μm, which is about the longwave sensitivity limit of both CCDs and photographic plates, out to 100 μm or longer. Throughout this region of the spectrum, special detectors are required that are different from those developed for visible light. However, the main challenge to infrared astronomers is a consequence of the emission of heat by the telescopes and atmosphere.

Typical temperatures on the Earth's surface are near 300 K, and the atmosphere through which observations are made is only a little cooler. According to Wien's law, the telescope, the observatory, and even the sky are radiating infrared energy with a peak wavelength of about 10 μm. To infrared eyes, everything is brightly aglow. The problem is to detect faint cosmic sources against this sea of light. The infrared astronomer must always contend with the sort of interference that would face a visible-light observer working in broad daylight with a telescope and optics lined with bright fluorescent lights.

The first problem is to protect the infrared detector from this radiation, just as you would shield photographic film from bright daylight. Since anything that is warm radiates infrared energy, the detector must be isolated in very cold surroundings—often held near absolute zero (1 to 3 K) by surrounding it with a bath of liquid helium. The second problem is radiation from the telescope structure and optics. To reduce the emission from the optics, they are kept very clean, since every bit of dust is an infrared source. The telescope is itself designed so that the thermal emission from the telescope does not reach the detector.

Spectroscopy

We have discussed detectors as if they were always used to record an image of a portion of the sky, but they can also be used to record a spectrum. Spectroscopy is one of the astronomer's most powerful tools, and more than half the time on most large telescopes is used for spectroscopy.

We saw in Chapter 5 that the light of the Sun or stars is a mixture of all wavelengths and that these can be separated by passing the light through a prism, forming a spectrum.* A spectrometer is an instrument designed to record the spectrum of a light source.

The design of a simple spectrometer is illustrated in Figure 6.10. Light from the source (actually, the image of a source produced by the telescope) enters the instrument through a small hole or a narrow slit and is then collimated (made into a beam of parallel rays) by a lens. The light then passes through a prism, producing a spectrum; different wavelengths leave the prism in different directions because of dispersion. A second lens placed behind the prism focuses the multiplicity of different images of the slit or entrance hole on the CCD or other detecting device. This is the desired spectrum.

6.3 Optical and Infrared Observatories

The Astronomical Observatory

An observatory is a collection of astronomical telescopes (Figure 6.11). Since the end of the 19th century, it has been obvious that the best observatory sites are on mountains far from the lights and pollution of cities. Although a number of urban observatories remain, especially in the large cities of Europe, they have by now become purely administrative centers or museums. The real action takes place far away, often on desert mountains or on isolated peaks in the Atlantic and Pacific Oceans.

Typically, the astronomer travels for hours by plane or car in order to reach the observatory. Sometimes the trip is halfway around the world, as for the

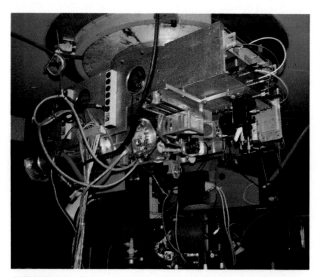

FIGURE 6.9 Modern telescope instrumentation is complex. This is a photograph of an infrared spectrometer at the Cassegrain focus of the 3-m NASA Infrared Telescope Facility on Mauna Kea. The spectrometer is operated remotely by the astronomer, who sits in a heated control room and records the data on a computer disk. *(University of Hawaii photo by Alan Tokunaga)*

---

*In practice, it is more common to use a different dispersing piece of optics called a *grating* to form the spectrum in an astronomical spectrometer, but the principle is exactly the same as with a prism spectrometer.

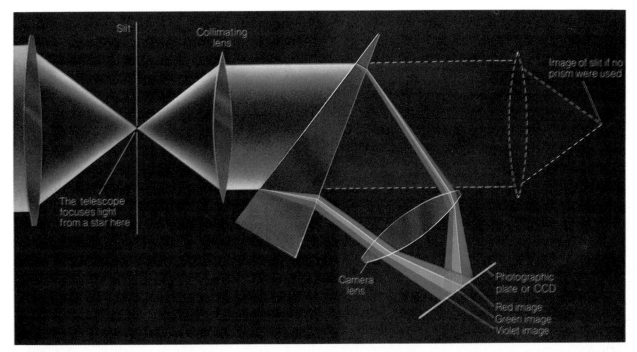

FIGURE 6.10 Design of a simple prism spectrometer for astronomy.

British and French astronomers who operate large telescopes in Hawaii. At the observatory are living quarters, computers, electronic and machine shops, and of course the telescopes themselves. A large observatory today requires a supporting staff of 20 to 100 persons, in addition to the astronomers.

The performance of an optical telescope is determined not only by the size of its mirror but also by where it is located. Today, major new telescopes are being built in Chile, in the Canary Islands, and on Mauna Kea in Hawaii, a mountain that is 13,700 ft (4200 m) high. What has led astronomers to seek such remote locations for observatories?

The Earth's atmosphere, so vital to life, is the biggest headache to the observational astronomer. In at least four ways the air imposes limitations upon the usefulness of telescopes. (1) The most obvious limitation is weather—clouds, wind, rain, and the like. At the best sites the weather may be clear as much as 75 percent of the time. (2) Even on a clear night, the

FIGURE 6.11 Kitt Peak National Observatory in Arizona with the dome of the 4-m Mayall telescope in the foreground. *(National Optical Astronomy Observatories)*

FIGURE 6.12   Several telescope domes are shown in this view of Mauna Kea Observatory, located at an altitude of 4200 m (14,000 ft) on an extinct volcano in Hawaii. *(University of Hawaii photo by David Morrison)*

atmosphere filters out a certain amount of starlight, especially in the infrared, where the absorption is due primarily to water vapor. Astronomers therefore prefer dry sites, generally found at the highest altitudes. (3) The sky should also be dark. Near cities the air scatters the glare from lights, producing an illumination that hides the faintest stars and limits the distances that can be probed by telescopes. It is best if an observatory is located at least a hundred miles from the nearest large city. (4) Finally, the air is often unsteady; light passing through this turbulent air is disturbed, with the result that star images are blurred. The astronomers call these effects "bad **seeing.**" When seeing is bad, images of celestial objects are distorted by the constant twisting and bending of light rays by turbulent air.

The best observatory sites are located far from cities and in locations where the seeing has been found to be good most of the time. Seeing of 1 arcsec or better is considered the standard for major observatories. For infrared work, sites should also be high and dry, usually above 3000 m in elevation. And, of course, observatories must be in locations with many clear nights, which usually means in desert regions at latitudes between about 20° and 35°. Sites that meet these criteria are found along the western mountains that border the North and South American continents from California to Chile, in the desert mountains of Spain and the U.S. states of Arizona and New Mexico, and on island mountaintops in Hawaii and the Canaries (Figure 6.12).

Optical Telescopes in the 20th Century

All of the major research telescopes in the world today were built within this century. The giant among early telescope builders was surely George Ellery Hale (1868–1939). Not once but four times he initiated projects that led to construction of what was at the time the world's largest telescope.

Hale's training and early research were in solar physics. In 1892, at age 24, he was named Associate Professor of Astral Physics and Director of the astronomical observatory at the newly founded University of Chicago. At that time, the largest telescope in the world was the 36-in refractor at the Lick Observatory near San Jose, California. Taking advantage of an existing glass blank for a 40-in telescope, Hale set out to raise the money for a larger telescope than the one at Lick. One prospective donor was Charles T.

Yerkes, who, among other things, ran the trolley system in Chicago. Hale wrote a letter to Yerkes, trying to persuade him to support construction of the giant telescope by saying, ". . . and the donor could have no more enduring monument. It is certain that Mr. Lick's name would not have been nearly so widely known today were it not for the famous observatory established as a result of his munificence." Yerkes agreed, and the new telescope was completed in May 1897 (Figure 6.13).

Even before the completion of the Yerkes refractor, Hale was not only dreaming of building a still larger telescope but also taking concrete steps to achieve that goal. In the 1890s, there was a major controversy about the relative quality of refracting and reflecting telescopes. Hale realized that 40 in was close to the maximum feasible aperture for refracting telescopes. If telescopes a factor of 2 or more larger in aperture were to be built, then they would have to be reflecting telescopes.

Using funds borrowed from his own family, Hale set out to construct a 60-in reflector. For a site, he left the Midwest for the much better conditions on Mount Wilson, then a wilderness peak above the small city of Los Angeles. In 1904, at the age of 36, Hale re-

FIGURE 6.13 The 40-in. (1-m) refracting telescope of Yerkes Observatory was the largest telescope in the world when completed in the 1890s by George Ellery Hale. *(Yerkes Observatory)*

ceived funds from the Carnegie Foundation to establish the Mount Wilson Observatory. The 60-in mirror was placed in its mount in December 1908.

Two years earlier, in 1906, Hale had already approached John D. Hooker, who had made his fortune in hardware and steel pipe, with a proposal to build a 100-in telescope. The technological risks were substantial. The 60-in telescope was not yet complete, and the utility of large reflectors for astronomy had not yet been demonstrated. George Ellery Hale's brother called him "the greatest gambler in the world." Once again, Hale was successful in obtaining funds, and the 100-in telescope was completed in November 1917.

Hale was not through dreaming. In 1926 he wrote an article in *Harper's* magazine about the scientific value of a still larger telescope. This article came to the attention of the Rockefeller Foundation, which granted 6 million dollars for the construction of a 200-in telescope. Hale died in 1938, and the 200-in (5-m) Hale telescope on Palomar Mountain was dedicated ten years later.

Following the success of the 200-in telescope and the construction of a 120-in (3-m) telescope by the University of California, astronomers from universities in the East and Midwest began a campaign for a national observatory to provide comparable facilities for the rest of the astronomical community. The National Science Foundation eventually agreed, and Kitt Peak National Observatory in Arizona began operations in 1960. A few years later, a second U.S. national observatory was established at Cerro Tololo in the Chilean Andes to provide access to southern skies. Meanwhile, a consortium of European nations founded their own European Southern Observatory in the Andes, while the British and Australians undertook the construction of major facilities in Australia. The largest telescopes at these observatories are about 4 m in aperture. Still larger is the Soviet 6-m telescope, but this instrument has been much less productive, as a result of a poorer site and a long history of technical problems.

Telescopes of the Future

Today, construction of ground-based telescopes is proceeding at an unprecedented pace. Worldwide, there are now about 20 telescopes with apertures exceeding 2.5 m (Table 6.1). While most of these have followed engineering designs similar to those developed for the 200-in Hale telescope, new technologies are available to construct instruments with twice the aperture of the 200-in telescope and four times its light-gathering power. These instruments have thin-

TABLE 6.1   Large Optical Telescopes of the World

Aperture (m)	Telescope Name	Location
10.0	Keck	Mauna Kea, Hawaii, U.S.A.
6.0	Bolshoi Alt-Azimuth	Mount Pastukhov, Russia, U.S.S.R.
5.0	Hale	Palomar Mountain, California, U.S.A.
4.5	Multi-Mirror	Mount Hopkins, Arizona, U.S.A.
4.2	William Herschel	Canary Islands, Spain
4.0	NOAO Cerro Tololo	Cerro Tololo, Chile*
3.9	Anglo-Australian	Siding Spring, Australia
3.8	NOAO Mayall	Kitt Peak, Arizona, U.S.A.*
3.8	United Kingdom Infrared	Mauna Kea, Hawaii, U.S.A.
3.6	Canada-France-Hawaii	Mauna Kea, Hawaii, U.S.A.
3.6	ESO	Cerro La Silla, Chile†
3.6	ESO New Technology	Cerro La Silla, Chile†
3.5	Max Planck Institut	Calar Alto, Spain
3.0	Shane (Lick Observatory)	Mount Hamilton, California, U.S.A.
3.0	NASA IRTF	Mauna Kea, Hawaii, U.S.A.
2.7	McDonald	Mount Locke, Texas, U.S.A.
2.6	Haute Province	St. Michele, France
2.6	Crimean Astrophysical	Simferopol, Crimea, U.S.S.R.
2.6	Byurakan Astrophysical	Mount Argatz, Armenia, U.S.S.R.
2.5	Las Compañas	Las Compañas, Chile‡
2.5	Hooker	Mount Wilson, California, USA‡
2.5	Isaac Newton	Canary Islands, Spain

*U.S. National Optical Astronomy Observatories

†European Southern Observatory

‡Mount Wilson and Las Compañas Observatories, Carnegie Institution of Washington

ner, lighter mirrors, altitude-azimuth mountings, and more compact mechanical designs.

The 10-m Keck telescope on Mauna Kea, Hawaii (Figure 6.14), is both the first of these new technology telescopes and the largest optical telescope in the world. Operated by the University of California and the California Institute of Technology, the Keck telescope was built with private rather than government funds. Instead of constructing a single primary mirror of 10-m diameter, the Keck achieves its large aperture by combining the light from 36 separate hexagonal mirrors, each about 1 m in width (Figure 6.15). Computer-controlled actuators constantly adjust these 36 mirrors so that they collect and focus the light with high accuracy. The first Keck telescope began operation in 1991, and plans are now being made to construct a twin 10-m instrument (sometimes called "Side-Keck") on an adjacent site.

Several additional large telescopes are under design or construction by various governments or consortia of universities around the world. These all will use single mirrors with apertures of 7.5 to 8.1 m, assuming that such giant glass disks can be successfully constructed and polished. By the late 1990s, we can expect to see in operation on Mauna Kea both a Japanese 7.5-m telescope and an 8-m telescope to be built by the U.S. National Observatories in partnership with Britain.

Several European nations, through their European Southern Observatory (ESO), are planning the world's largest optical telescope for a site on Cerro Paranal, an isolated peak in Chile's Atacama Desert.

FIGURE 6.14 The W. M. Keck Observatory is located at an altitude of 13,600 feet on Hawaii's Mauna Kea—acknowledged as the world's best site for astronomical observations. The Observatory houses the Keck Telescope, which has a primary mirror that is 10-m in diameter. *(California Association for Research in Astronomy)*

FIGURE 6.15 This picture shows nine of the 36 hexagonal mirror segments that make up the primary mirror of the Keck Telescope. This picture was taken in late 1990, and the remaining segments were installed in the winter of 1991-92. The Keck Telescope is being operated by the California Association for Research in Astronomy (CARA), a partnership of the California Institute of Technology and the University of California. *(California Association for Research in Astronomy)*

This instrument, called the Very Large Telescope (VLT), will use four 8-m mirrors to achieve an aperture equivalent to a single 16-m mirror. The first of these four elements is scheduled to be completed in 1995.

6.4   Radio Telescopes

Origins of Radio Astronomy

In 1931, Karl G. Jansky of the Bell Telephone Laboratories was experimenting with antennas for long-range radio communication when he encountered interference in the form of radio radiation coming from an unknown source (Figure 6.16). He discovered that this radiation came in strongest about 4 min earlier on each successive day and correctly concluded that since the Earth's sidereal rotation period is 4 min shorter than a solar day, the radiation must be originating from some region of the celestial sphere. Subsequent investigation showed that the source of the radiation was part of the Milky Way.

In 1936, Grote Reber, an amateur astronomer and radio ham, built from galvanized iron and wood the first antenna specifically designed to receive these cosmic radio waves. Over the years, Reber built several such antennas and used them to carry out pioneering surveys of the sky for celestial radio sources; he remained active in radio astronomy for more than 30 years. During the first decade, he worked practically alone, for professional astronomers had not yet recognized the vast potential of radio astronomy.

FIGURE 6.16 The rotating radio antenna used by Jansky in his serendipitous discovery of radio radiation from the Milky Way. *(Bell Laboratories)*

Many of the objects that Reber discovered became subjects of intensive investigation years later. His original radio telescope is on display at the National Radio Astronomy Observatory (Figure 6.17).

Detection of Radio Energy from Space

It is important to understand that radio waves are not "heard"; they are not the same as sound waves. Although in commercial radio broadcasting, radio

FIGURE 6.17 Grote Reber's original radio telescope, now reconstructed at the National Radio Astronomy Observatory in Green Bank, West Virginia. *(National Radio Astronomy Observatory)*

waves are modulated or coded to carry sound information, the sound itself is not transmitted. The radio waves merely carry the information that a radio receiver must decode and convert into sound by means of a loudspeaker or earphones. Sound is a physical vibration of matter; radio waves, like light, are a form of electromagnetic radiation.

Most astronomical objects emit all forms of electromagnetic radiation—radio waves as well as light, infrared and ultraviolet radiation, and so on. The radio waves we can receive from space are those that can penetrate the ionized layers of the Earth's atmosphere, that is, those with wavelengths in the range from a few millimeters to about 20 m. The human eye and CCDs are not sensitive to radio waves; we must detect this form of radiation by different means.

Radio waves induce a current in conductors of electricity. An *antenna* is such a conductor; it intercepts radio waves, which induce a feeble current in it. The current is then amplified in a radio receiver until it is strong enough to measure or record. Receivers can be tuned to select a single frequency, but today it is more common to use sophisticated data processing techniques to allow thousands of separate frequency bands to be detected simultaneously. Thus the astronomical radio receiver operates much like a spectrometer on an optical telescope. The signals, after computer processing, are recorded on magnetic disk or tape.

Radio Telescopes

Radio waves are reflected by conducting surfaces just as light is reflected from an optically shiny surface,

FIGURE 6.18 The 100-m radio telescope near Bonn, West Germany. *(Max-Planck Institut für Radioastronomie)*

and according to the same laws of optics. A radio reflecting telescope consists of a concave metal reflector (called a *dish*), analogous to a telescope mirror. The radio waves collected by the dish are reflected to a primary or Cassegrain focus, where they form a radio image.

Radio astronomy is a young field compared with optical astronomy, but it has prospered in the 1970s and 1980s. Several European nations that do not possess high-quality sites for optical telescopes have instead chosen to concentrate their astronomical efforts in the radio part of the spectrum; these include the Netherlands, Britain, and Germany. The world's largest radio reflectors that can be pointed to any direction have an aperture of 100 m. One of these is at the Max-Planck Institute for Radio Astronomy, located near Bonn, Germany (Figure 6.18); the other is under construction at the U.S. National Radio Astronomy Observatory in West Virginia. Table 6.2 lists some of the major radio telescopes of the world.

Radio Interferometry

Because the wavelengths of radio radiation are far greater than those of visible light, the resolving power of a radio telescope of a given size is correspondingly less than that of an optical telescope. One solution to this limitation is provided by the use of the radio **interferometer.** Here two radio dishes are placed far apart. The radio waves from a source strike one antenna a brief instant before the other, so that the two antennas receive the same waves at slightly different times and thus become "out of phase" with each other. The difference in phase between the waves detected at the two antennas can be measured electronically, and the direction to the source can thus be calculated. The farther apart the components of an interferometer are placed (the longer the *baseline*), the more accurately we can pinpoint the direction of the source.

The use of pairs of radio telescopes as interferometers was pioneered in England. If one of the telescopes is mounted on rails and can be moved from time to time, interferometers with different baselines can be established. The next step is to combine a large number of dishes into an **interferometer array.** In effect, such an array works as a large number of two-dish interferometers, all observing the same part of the sky together. Computer processing of the results permits the reconstruction of a radio *image*, with its resolution determined by the largest spacing in the array, just as the resolution of an optical telescope is determined (in the absence of seeing) by the diameter of the primary mirror.

The most extensive radio array is the National Radio Astronomy Observatory's Very Large Array (VLA) near Socorro, New Mexico. It consists of 27 movable radio telescopes, each having an aperture of

TABLE 6.2 Major Radio Observatories of the World

Observatory	Location	Description (Main Instruments)
National Astronomy and Ionospheric Center	Arecibo, Puerto Rico	305-m fixed dish
National Radio Astronomy Observatory	Green Bank, West Virginia	100-m steerable dish* and 43-m steerable dish
Max-Planck Institut für Radioastronomie	Bonn, Germany	100-m steerable dish
Jodrell Bank Radio Observatory	Manchester, England	76-m steerable dish and 66-m steerable dish
Goldstone Tracking Station (NASA/JPL)	Barstow, California	70-m steerable dish
Australia Tracking Station (NASA/JPL)	Tidbinbilla, Australia	70-m steerable dish
Spain Tracking Station (NASA/JPL)	Madrid, Spain	70-m steerable dish
Parks National Radio Observatory	Parks, Australia	64-m steerable dish
Owens Valley Radio Observatory (Caltech)	Big Pine, California	2 27-m dishes
Very Large Array (NRAO)	Socorro, New Mexico	27-element array of 25-m dishes (36 km in length)
Westerbork Radio Observatory	Westerbork, the Netherlands	12-element array of 25-m dishes (1.6 km in length)
Very Long Baseline Array (NRAO)	Ten U.S. sites	10-element array of 25-m dishes (9000 km in length)*
Nobeyama Cosmic Radio Observatory	Minamimaki-Mura, Japan	25-m steerable millimeter-wave dish
James Clerk Maxwell Telescope	Mauna Kea, Hawaii	15-m steerable millimeter-wave dish
Hat Creek Radio Observatory (U. California)	Cassel, California	9-element array of 6-m millimeter-wave dishes*

*Still under construction (1992)

25 m, spread over a total span of about 36 km (Figure 6.19). The VLA normally operates at four wavelengths: 1.3 cm, 2 cm, 6 cm, and 20 cm. By electronically combining the signals from all of its individual telescopes, this array permits the radio astronomer for the first time to make "pictures" of the sky at radio wavelengths that are comparable to those obtained with an optical telescope, with resolution of about 1 arcsec.

Very Long Baseline Interferometers

Ordinary interferometers and interferometer arrays are limited in size by the requirement that all of the dishes be accurately wired together. Their maximum dimensions are thus only a few tens of kilometers. However, larger interferometer baselines can be achieved if the telescopes do not require a physical connection. Using high-precision clocks at each site, astronomers have learned to build **very long baseline interferometers** with baselines as long as from California to Parkes (in Australia) and from Greenbank, West Virginia, to the Crimea (in the U.S.S.R.). The resulting angular resolution far surpasses the angular resolution of optical telescopes.

The next step is the construction of an array of very long baseline interferometers. U.S. astronomers are completing a telescope called the Very Long Baseline

FIGURE 6.19   Part of the **Y**-shaped Very Large Array (VLA) near Socorro, New Mexico. The individual antennas are 25 m in aperture. *(National Radio Astronomy Observatory)*

Array (VLBA) made up of ten individual telescopes stretching from the Virgin Islands to Hawaii (Figure 6.20) and linked together into one gigantic instrument of nearly planetary dimensions. The VLBA can form astronomical images with a resolution of 0.0001 arcsec, permitting features as small as 10 AU to be distinguished at the center of our Galaxy.

The addition of one or more telescopes in space is planned as a means to increase further the capability of the ground-based VLBA beyond the limits currently set by the dimensions of the Earth. The U.S.S.R. has announced plans to orbit such a receiver in the early 1990s (called Radioastron) and has invited the participation of astronomers from other

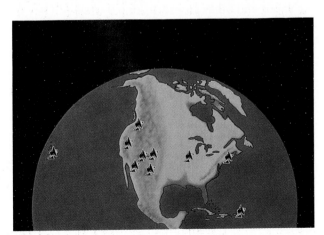

FIGURE 6.20   Distribution of the ten antennas that constitute the U.S. VLBA (Very Long Baseline Array). *(National Radio Astronomy Observatory)*

nations in this project. This telescope will be, in a sense, larger than the Earth itself!

Radar Astronomy

Radar is the technique of transmitting radio waves to an object and then detecting the radio radiation that the object reflects back. The time for the radio waves to make the round trip can be measured electronically, and because they travel with the known speed of light, the distance to the object is determined.

Radar observations of the Moon and of the planets have yielded our best knowledge of the distances of these worlds and played an important role in navigating spacecraft throughout the solar system. In addition, as will be discussed in later chapters, radar observations have determined the rotation periods of Venus and Mercury, probed the tiny Earth-approaching asteroids and the nuclei of comets, analyzed the rings of Saturn, and investigated the surfaces of Mercury, Venus, Mars, and the large satellites of Jupiter.

Any radio dish can be used as a radar telescope if it is equipped with a powerful transmitter as well as a receiver. The most spectacular facility in the world for radar astronomy is the 1000-ft (305-m) telescope at Arecibo in Puerto Rico (Figure 6.21). The Arecibo telescope is too large to be pointed directly toward different parts of the sky. Instead, the huge "bowl" (more than a mere dish!) is carved out of the ground and lined with a reflecting surface. The Arecibo bowl has a volume roughly equal to the world's annual beer consumption. A limited ability to track astronomical

FIGURE 6.21 The 305-m (1000-ft) dish at the National Astronomy and Ionosphere Center, Arecibo, Puerto Rico, operated by Cornell University and sponsored by the National Science Foundation. *(Cornell University)*

sources is achieved by moving the receiver system, which is suspended on cables 100 m above the surface of the dish.

6.5   Observations Outside the Earth's Atmosphere

The Earth's atmosphere blocks most radiation at wavelengths shorter than visible light. Ultraviolet, x-ray, and gamma-ray observations can be made only from space. It is also advantageous for many observations at visible and infrared wavelengths to get above the distorting effects of the atmosphere.

Airborne and Space Infrared Telescopes

Water vapor, the main source of atmospheric interference throughout the infrared region, is concentrated in the lower part of the Earth's atmosphere. That is why a gain of even a few hundred meters in elevation can make an important difference to the quality of an observatory site. Given the limitations of high mountains, most of which attract clouds and violent storms, it is natural to investigate the possibility of observing from airplanes and ultimately from space.

The first airborne infrared observations were made in the 1960s by Frank J. Low of the University of Arizona. On the basis of the success of these efforts, NASA decided to construct and operate a larger airborne observatory, flying regularly out of the Ames Research Center south of San Francisco. The Gerard P. Kuiper Airborne Observatory—named for one of the pioneers of infrared astronomy, who first

applied modern detectors to astronomy after World War II—consists of a 0.9-m Cassegrain reflector mounted in a Lockheed C-141. The telescope views the sky through a large hole in the side of the airplane. The KAO (as it is called) flies at elevations of 12 to 13 km, where it is above 99 percent of the atmospheric water vapor. NASA plans to build a second infrared telescope, with a 2.5-m aperture, to fly in a modified Boeing 747SP.

Taking the next step, to observations from space itself, has important advantages for infrared astronomy. First, of course, is the elimination of all interference from the atmosphere. But equally important, it provides the opportunity to cool the entire optical system of the telescope so as to reduce, and nearly eliminate, infrared radiation from this source as well. If we tried to cool a telescope within the atmosphere, it would quickly become coated with condensing water vapor and other gases, making it useless. Only in the vacuum of space can optical elements be cooled to hundreds of degrees below freezing and still remain operational. The first orbiting infrared observatory, launched in 1983, was the Infrared Astronomy Satellite (IRAS), built as a joint project by the United States, the Netherlands, and Britain (Figure 6.22).

IRAS was equipped with a 0.6-m telescope cooled to a temperature of less than 10 K, with a series of detectors sensitive to wavelengths of 12, 25, 60, and 100 μm. For the first time, the infrared sky could be seen at night, as it were, rather than through a bright foreground of atmospheric and telescope emission. IRAS carried out a rapid but comprehensive survey of the entire infrared sky over a ten-month period, until its liquid helium coolant was exhausted. A second small infrared telescope called the Cosmic Background Explorer (COBE) was launched in 1990 to

FIGURE 6.22 The Infrared Astronomical Satellite (IRAS), a joint project of NASA, the Netherlands, and the United Kingdom. Here the spacecraft is being tested before its 1983 launch. *(NASA/JPL)*

carry out a very high-precision survey of the sky at low resolution.

The next step in infrared observations from space will require larger cooled telescopes that can make detailed observations of individual objects. The European Space Agency plans to launch such a facility, called the Infrared Space Observatory (ISO), in 1994. Toward the end of the decade, NASA hopes to complete the Space Infrared Telescope Facility (SIRTF), a 1-m infrared telescope with a planned lifetime of 5 years.

Ultraviolet, X-Ray, and Gamma-Ray Astronomy

The atmosphere is opaque to electromagnetic radiation of wavelength less than about 300 nm. Consequently, ultraviolet, x-ray, and gamma-ray observations must be made from space. Such observations first became possible in 1946 with V2 rockets captured from the Germans. The U.S. Naval Research Laboratory instrumented these rockets for a series of pioneering flights, used initially to detect far-ultraviolet radiation from the Sun. Subsequently, many rockets have been launched to make x-ray and ultraviolet observations of the Sun, and later of other celestial objects as well.

Since the 1960s, Earth satellites have been launched to carry out astronomical observations. These have included the orbiting solar observatories (OSOs), the orbiting astronomical observatories (OAOs), and the high-energy astronomy observatories (HEAOs). The U.S.-built x-ray satellite Uhuru (Swahili for "freedom"), launched in 1970 from Kenya, discovered about 100 x-ray sources, some of which were found to emit higher energy gamma rays as well. Another highly successful space observatory has been the International Ultraviolet Explorer, or IUE. This telescope, operated for the past decade from the Goddard Space Flight Center outside Washington, D.C., is a joint U.S.-European effort.

X-ray astronomy received a major boost in 1978 with the launching of HEAO-2, called Einstein. Einstein was the first x-ray telescope capable of forming an image of a source, and it had a sensitivity a thousand times as great as anything that preceded it. This advance is equivalent to changing from a small amateur telescope to the 200-in Hale reflector on Palomar.

6.6 Space Observatories of the Future

Most of the orbiting observatories we have described were launched in the 1970s and operated for a few years at most. During the 1980s, there was a gap in this sequence of space observations. (A similar gap exists in planetary exploration, with only two NASA spacecraft launches during the entire decade of the 1980s, compared with an average of almost one per year to the Moon and planets during the previous two decades.) This interruption can be attributed in large part to the development of the Space Shuttle, which absorbed a disproportionate share of NASA funding during the 1970s. The problem became worse after the destruction of the Challenger in 1986, just when the Shuttle was becoming operational and scientific flights were about to resume.

For the decade of the 1990s, scientists and NASA managers expect to reverse this decline and establish a series of space observatories. First came the COBE infrared telescope (1989) and a joint U.S.-German x-ray telescope called Rosat (1990), followed shortly by two larger astronomical observatories, the Hubble Space Telescope (HST) and the Gamma Ray Observatory (GRO). These instruments have larger apertures and are designed for longer operational lifetimes than most previous astronomical satellites.

The Hubble Space Telescope

The first of the large observatories is the Hubble Space Telescope (HST), a Cassegrain reflector with an aperture of 2.4 m, launched in 1990 and operated from the NASA-funded Space Telescope Science Institute in Baltimore. The HST is initially instrumented for imaging and spectroscopy at visible and ultraviolet wavelengths, although the addition of other instruments is expected in the late 1990s (Figure 6.23).

The primary objective of the HST is to take advantage of the absence of atmospheric seeing to achieve a resolution to 0.1 arcsec and thereby to probe deeper into space than is possible with any ground-based telescope. Unfortunately, astronomers discovered after launch that a serious error had been made in shaping and polishing the mirror in the early 1980s. Although the mirror was highly precise, it was figured to the wrong shape—"perfectly wrong" as one NASA scientist termed it. Because of this error, HST cannot achieve a resolution of 0.1 arcsec. On bright sources it can yield images with 0.2 arcsec resolution, but it offers no improvement over ground-based telescopes for faint stars and galaxies.

NASA plans to correct the problem with the HST optics in 1993, when astronauts are scheduled to visit the telescope to install a series of correcting lenses and mirrors that should restore its full capability. Until then, astronomers must wait for the promised capability to see farther into space than had been possible from the ground.

Other Orbiting Observatories

The second of the U.S. space observatories is a more specialized facility called the Gamma Ray Observatory (GRO), launched in April 1991. Its objectives are to conduct an all-sky survey for gamma-ray sources as well as to study individual objects in detail at a variety of energies. The potential for new discoveries from this instrument is great, since it will provide us with the first in-depth look at the gamma-ray universe.

FIGURE 6.23 The Hubble Space Telescope at the time of its launch in 1990. *(NASA)*

The third of the observatories, the Advanced X-Ray Astrophysics Facility (AXAF), is planned for launch about 1996. AXAF is the logical successor to the highly successful Einstein x-ray satellite, with much larger imaging optics and advanced detector systems.

The fourth and final spectral region that needs to be covered from space is the infrared. The proposed U.S. infrared observatory is the Space Infrared Telescope Facility (SIRTF), mentioned in Section 6.5. This 1-m cooled telescope could be in operation by the late 1990s (Figure 6.24).

NASA calls these four orbiting telescopes—HST, GRO, AXAF, and SIRTF—its "great observatories in space." These are all long-lived telescopes with multiple detectors, available for the use of astronomers from all over the world to carry out a wide variety of research projects. These space observatories, like the VLBA, the Keck telescope, and the new 8-m optical telescopes under construction, hold the promise of an unprecedented era of astronomical discovery during the next two decades.

FIGURE 6.24   Artist's impression of the Space Infrared Telescope Facility (SIRTF) in orbit 100,000 km above the Earth. *(NASA/ARC)*

SUMMARY

6.1 The astronomical telescope collects light and forms an image, using a convex lens in a **refractor** or a concave mirror in a **reflector** to bring it to a **focus.** The distance from the lens or mirror to the focus is called the **focal length.** The diameter, or **aperture,** of the telescope determines the size, brightness, and **resolution** of the image. For optical telescopes, the resolution is typically about 1 **arcsec.** The traditional **equatorial mounts** for astronomical telescopes are now being supplanted by **altitude-azimuth** mounts, which are less expensive for very large instruments.

6.2 Optical detectors include the eye, photographic film or plates, and the **charge-coupled device (CCD)** detector. Detectors sensitive to infrared radiation must be cooled to very low temperatures. A spectrometer can be built using either a prism or grating to disperse the light into a spectrum.

6.3 Telescopes are housed in domes and controlled by computers. Observatory sites must be carefully chosen for clear weather, dark skies, low water vapor, and excellent atmospheric **seeing** (low atmospheric turbulence). Hale pioneered in construction of telescopes, including the Yerkes 40-in refractor and the large reflectors at Mount Wilson Observatory. The 200-in (5-m) Hale telescope at Palomar Mountain was completed in 1948. In the 1990s, a new generation of instruments is being constructed, including the 10-m Keck telescope at Mauna Kea and the four 8-m instruments that constitute the European Large Optical Telescope in Chile.

6.4 In the 1930s, radio astronomy was pioneered by Jansky and Reber. A radio telescope is basically a radio antenna (often a large parabolic dish) connected to a receiver. Much enhanced resolution is obtained by **interferometry,** including the development of **interferometer arrays** like the 27-element VLA. Expanding to **very long baseline interferometers,** radio astronomers can achieve resolutions as good as 0.001 arcsec. **Radar** involves transmitting as well as receiving. The largest radar telescope is the 305-m dish at Arecibo.

6.5 Infrared observations are made from the Kuiper Airborne Observatory and from space, as well as from ground-based facilities. IRAS and COBE provided the first all-sky surveys, and the larger ISO and SIRTF space infrared telescopes are under development. Ultraviolet, x-ray, and gamma-ray observations must be made from above the atmosphere. Many orbiting observatories have been flown, notably the Uhuru and Einstein x-ray observatories and the IUE for ultraviolet observations.

6.6 Few space observatories were launched during the 1980s, but many are promised in the 1990s. These include the NASA Great Observatories in Space: the Hubble Space Telescope (HST), the Gamma Ray Observatory (GRO), the Advanced X-Ray Astrophysics Facility (AXAF), and the Space Infrared Telescope Facility (SIRTF).

REVIEW QUESTIONS

1. Prepare a table, based on Figure 6.1, listing the wavelengths of radiation that can reach the Earth's surface.

2. For astronomical telescopes, bigger is better. Why?

3. Compare the advantages of reflecting and refracting telescopes for (a) professional astronomy, (b) amateur astronomy, and (c) birdwatching.

4. Compare the eye, photographic film, and CCDs as detectors for light. What are the advantages and disadvantages of each?

5. Radio and radar observations are often made with the same antenna, but otherwise they are very different techniques. Compare and contrast radio and radar astronomy, in terms of the equipment needed, the methods used, and the kind of results that are obtained.

6. Why do astronomers place telescopes in Earth orbit? What are the advantages for different spectral regions?

7. Describe the four NASA Great Observatories in Space and summarize their capabilities.

THOUGHT QUESTIONS

8. What happens to the image produced by a lens if the lens is "stopped down" with an iris diaphragm—a device that covers its periphery?

9. What would be the properties of an ideal astronomical detector? How closely do the actual properties of a CCD approach this ideal?

10. Fifty years ago, the astronomers on the staff of Mount Wilson and Palomar Observatories each received about 60 nights per year for their observing programs. Today an astronomer feels fortunate to receive 10 nights per year on a large telescope. Can you suggest some reasons for this change?

11. The largest observatory complex in the world is on Mauna Kea in Hawaii, at an altitude of 4.2 km. Consider other potential sites on high mountains and discuss their advantages and disadvantages. Should astronomers, for example, consider building an observatory on Mount McKinley (Denali) or Mount Everest?

12. Another potential site sometimes considered for astronomy is the Antarctic plateau. Discuss its advantages and disadvantages.

13. Suppose you are looking for a site for an optical observatory, an infrared observatory, and a radio observatory. What are the main criteria of excellence for each of these cases? What sites on Earth are actually thought to be the best today?

14. Radio astronomy involves wavelengths of radiation much longer than those of visible light, while many orbiting observatories have probed the universe in radiation of very short wavelengths. What sorts of objects and physical conditions would you expect to be associated with emission of radiation of very long and very short wavelengths?

15. Astronomers who hope to see the HST, GRO, AXAF, and SIRTF all operating by the end of the century refer to an era of the "great observatories in space." If all four of these instruments were in operation at the same time, together with ground-based instruments, what would be the coverage of the electromagnetic spectrum? Are any important wavelength regions left out?

PROBLEMS

16. The 3-m reflector at the Lick Observatory can be operated as an $f/5$ prime-focus telescope or as an $f/17.2$ Cassegrain telescope. (a) In which of these modes can an extended source be exposed fastest, and by what factor? (b) In which of these modes can a point source be exposed fastest and by what factor?

17. Suppose you are observing with a 1024 × 1024 element CCD. How many individual picture elements (pixels) are there? If the CCD can distinguish 2^{20} (about 1 million)

different brightness levels, how many binary data bits are there in one image? Suppose you obtain one image every 6 min for a 10-h night; how many digital data bits must you store for later processing?

18. The resolution of a radio interferometer is proportional to the maximum spacing of the antennas. The VLA, with a maximum antenna separation of 36 km, achieves a resolution of 1 arcsec at a wavelength of 6 cm. (a) What is the resolution of a very long baseline interferometer at this wavelength if the antennas are separated by 3600 km? **(b)** What would be the resolution of the VLBA, with antennas on opposite sides of the Earth?

19. A typical large telescope today requires about 4 million dollars per year to operate. Suppose you wish to observe with that telescope. What is the cost per night for the telescope time? The cost per hour? The cost per minute? Can you think of any other activities that have such a high associated cost, in dollars per hour?

20. The HST cost about 1.7 billion dollars for construction, 300 million dollars for its Shuttle launch, and 100 million dollars per year for operations. If the telescope lasts a total of 10 years, what is the cost per year? The cost per day? If the telescope can be used just 30 percent of the time for actual observations, what is the cost per hour and per minute for the astronomer's observing time on this instrument?

INTERVIEW

JOHN JEFFERIES

John Jefferies is best known for his leadership in the development of Mauna Kea, which is quite simply the finest observing site on Earth.

Jefferies was born in a small country town in western Australia. He left school at age 15 to work in a bank; it was during World War II and what he really wanted to do was join the Air Force. Banking, he soon concluded, was not the career for him, although it took him the next two years of daytime work and nighttime school to escape by qualifying for entrance to the University of Western Australia. After completing his schooling in Australia and Cambridge, England, and working in solar physics with Ron Giovanelli, who was one of the finest solar physicists in this century, Jefferies moved to Boulder, Colorado. There he was part of a small group that started the Joint Institute for Laboratory Astrophysics under the leadership of Lewis Branscomb. After 19 years in Hawaii, Jefferies was appointed as the first Director of the National Optical Astronomy Observatories, an organization formed to unite Kitt Peak, Cerro Tololo, and the National Solar Observatories under a single management. Recently, he resigned that position to return to the work of trying again to understand the atmosphere of the Sun and the many strange and curious phenomena that are found within it.

When did you first get interested in astronomy?

I got interested in astronomy purely by accident. I was over at Cambridge [England], working toward a

degree in physics, under a scholarship supported by the Australian government research agency (CSIRO). One day in 1949, Ron Giovanelli came into my rooms at Cambridge and asked if I would like to go back to Australia and work for him. He told me what he was doing, and it sounded interesting so I accepted his offer. A week later, I got a letter from a friend who was working in solid state physics. He asked me to work for him because he thought there was a vacancy coming up. Had I received that letter a week earlier, I probably would have become a solid state physicist.

John Jefferies is a solar physicist who played the leading role in developing Mauna Kea into the finest observing site in the world.

Can you describe a problem you're working on now?

The problem has to do with the Sun and the state of the Sun's atmosphere. How do you determine the magnetic field of the Sun? The magnetic field underlies so much of the Sun's activity: the flares, the energy output. The Earth's weather is responsive to the state of solar magnetism, and so is the Earth's environment, including the magnetosphere around the Earth. All of these things make an understanding of the magnetic state of the Sun's atmosphere important.

How do you decide what problem to work on?

The things that challenge me are the things that seem to be of the greatest contemporary interest, the things that are going to tell me what I want to know most about the Sun's atmosphere. Sometimes it's nice to select one little problem of your own that not too many other people are interested in—but one you think is important—and work away on that. It has to be something that interests you or you're not going to do well. It's also got to be of substantial importance; otherwise you're not going to feel that you're using your time in the best way.

Your biggest contribution to astronomy is founding what turned out to be the finest observatory in the world. Can you describe a little bit about how that came about?

Like much of my life, it was more or less by accident. One factor was an inherent restlessness, I suppose. It's part of my genetic makeup, and it led me to consider leaving a good job in a first-rate scientific organization in Boulder to take a plunge in the middle of the Pacific with a few colleagues. I rationalized going out there because in Hawaii there were clearly excellent conditions for observing the Sun and the solar corona. That opportunity combined with the inherent appeal of the tropics and Polynesia to attract me.

There was lots of interest in building facilities in Hawaii for observing the planets and the Moon because of the longitude and the latitude, and because of the clarity of the skies and evident good properties for observing nighttime as well as daytime skies. It was inevitable that Mauna Kea happened.

Would you explain why Mauna Kea is such a special place?

Well, the quality of Mauna Kea—the quality of any observatory site—is measured by a number of more or less quantifiable factors. Some of them are obvious and some of them are less obvious. The obvious ones are cloud-free and dust-free atmosphere, and accessibility. One of the more subtle ones is the dryness of the site, which leads to its higher transmission in the infrared and so to better conditions for observing in the infrared. Observing in this part of the spectrum was just coming to be a real possibility about the time that I went to Hawaii, and this was one of the particular factors that had persuaded the federal government to consider building a telescope on Mauna Kea. Dryness increases with height, and in tropical locations you get very dry conditions above the inversion layer. It may seem surprising that a site surrounded by thousands of miles of ocean is exceedingly dry, but the fact is that the tropical trade wind inversion produces a natural barrier between the lower wet maritime air and the more or less free atmosphere above. Stability of the atmosphere, which leads to excellent quality images, is another quantifiable factor. Even from the relatively crude measurements that one was able to take at that time, it was clear that Mauna Kea was a superb site in all of those respects. Certainly it had drawbacks. It didn't have a road; it didn't have any power; it was remote from all other astronomical centers. And most particularly, it was very high—far higher than any other site that had been proposed for a major astronomical telescope.

And what was the reaction at the time to the proposal that an observatory should be built at that altitude?

Some people thought it was a crazy idea. Of course, Gerard Kuiper had been out there about a year or so

Mauna Kea is now home to national observatories sponsored by the United States, Canada, France, the United Kingdom, and the Netherlands. The Keck 10-m telescope, which is the world's largest, is also located on Mauna Kea. Several additional telescopes are planned for construction between now and the year 2000.

before I got there. He had been to Mauna Kea, and he was loudly proclaiming its virtues. He was not particularly concerned with the fact that the altitude was as great as it was, but a lot of other people were. It was certainly a reasonable concern. Mauna Kea *was* nearly 14,000 feet in altitude, and the working conditions were unknown.

I became convinced that the altitude was not a serious problem. We already had people going up there on a nightly basis, measuring the seeing, measuring the humidity, measuring the wind speed, temperature, and various other factors. The staff did that for about a year or so, but even after a few months it became clear that they weren't dropping like flies. They were perfectly healthy; they enjoyed it, in fact.

Suppose you were setting out to do that kind of development again with the kind of divided opinion that existed in the astronomy community at the time, do you think you could do it now, thirty years later?

I doubt it. It's hard to put yourself back into that situation from the current situation when Mauna Kea is so intimately established in the mind of every astronomer as a place where you can do astronomy of the very highest quality. I think the attitudes that made it possible have dis-

appeared. Perhaps there were more opportunities around for everyone then. We were able to get support from the planetary office in NASA to build the telescope. NASA was such an optimistic agency, carrying forward this enormously exciting program, and this was a component part of it. The whole approach was that anything was possible, and it's not quite the same way anymore.

Why did you decide to become a scientist?

My first recollection of wanting to be a scientist was during World War II. I dropped out of high school and then finally decided that working in a bank was not what I wanted to do the rest of my life. I remember I was on a train, and I was going to the university for the first time. People were asking me, "Why do you want to go into science?" Science seemed pretty far out at that time in this little town of western Australia, and I said it seemed to me that science was going to be the key to the world for the next 50 years at least and that I wanted to position myself to be a part of it. I wouldn't take any great credit for my prediction because it was a very easy one to make, but I think it was right and at the heart of why I wanted to be a scientist. I wanted to be involved in something that was moving the world.

Apollo astronaut on the surface of the Moon. *(NASA)*

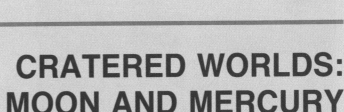

7

CRATERED WORLDS: MOON AND MERCURY

Grove K. Gilbert (1843–1918), a founding member of the U.S. Geological Survey, was among the first scientists to recognize the importance of impact cratering on the Earth and Moon. In the 1890s, his careful arguments for the impact origin of the lunar craters, although decades ahead of their time, laid the foundations for the modern science of lunar geology and the development of a chronological sequence for the evolution of the terrestrial planets. *(USGS)*

We begin our discussion of the planets as worlds with two relatively simple objects: the Moon and Mercury. Unlike our planet, the Moon is geologically dead, a world that has exhausted its internal energy sources. As such, it provides us a window on the earlier eras of solar system history. The planet Mercury is in many ways similar to the Moon, which is why the two are discussed together. Both are relatively small, lacking in atmospheres, deficient in geological activity, and dominated by the effects of impact cratering from outside. Most of the processes that have molded their surfaces acted on all the members of the planetary system.

7.1 General Properties of the Moon

Some Basic Facts

The Moon has only 1/80 the mass of the Earth and a surface gravity too low to retain an atmosphere. If, early in its history, the Moon outgassed an atmosphere from its interior or collected a temporary envelope of gases from impacting comets, such an atmosphere was lost before it could leave any recognizable evidence of its short existence. All sign of water is similarly absent. Indeed, the Moon is dramatically deficient in a wide range of **volatiles,** those elements and compounds that evaporate at moder-

ately low temperatures. Examples of volatiles include such common ices as H_2O and CO_2 (dry ice).

Some of the properties of the Moon are summarized in Table 7.1, along with comparative values for Mercury.

Exploration of the Moon

Most of what we know about the Moon today derives from the U.S. Apollo program, which sent nine manned spacecraft to our satellite between 1968 and 1972 and landed 12 astronauts on its surface. Before the era of spacecraft studies, astronomers had mapped the side of the Moon that faces the Earth to a best telescopic resolution of about 1 km, but lunar geology hardly existed as a scientific subject. All that changed beginning in the early 1960s.

Initially, the U.S.S.R. took the lead in lunar exploration with Luna 3, which returned the first photographs of the lunar far side in 1959, and Luna 9, which landed on the surface in 1966 and transmitted pictures and other data to Earth. However, these efforts were overshadowed once the American astronauts set foot on the Moon, on July 20, 1969 (Figure 7.2).

Table 7.2 summarizes the nine Apollo flights—six landings and three other missions that circled the Moon but did not land. The initial landings were on flat plains selected out of considerations of safety, but with increasing experience and confidence, NASA

FIGURE 7.1 The Moon as photographed from a spacecraft near the Earth. The hemisphere shown consists primarily of heavily cratered highlands. The resolution of such an image is several kilometers, similar to that of high-powered binoculars or a small telescope. *(NASA)*

TABLE 7.1   Properties of the Moon and Mercury

	Moon	Mercury
Mass (Earth = 1)	0.012	0.055
Diameter (km)	3476	4878
Density (g/cm³)	3.3	5.4
Surface gravity (Earth = 1)	0.2	0.4
Escape velocity (km/s)	2.4	4.3
Rotation period (days)	27.3	58.6
Surface area (Earth = 1)	0.07	0.15

orate equipment. Finally, on the last Apollo landing, NASA included one scientist-astronaut, geologist Harrison Schmitt.

In addition to landing on the lunar surface and studying it at close hand, the Apollo missions accomplished three objectives of major importance for lunar science. First, the astronauts collected nearly 400 kg of samples to return to Earth for detailed laboratory analysis (Figure 7.3). These samples, which are still being extensively studied, have probably revealed more about the Moon and its history than all other lunar studies combined. Second, each Apollo landing after the first deployed an Apollo Lunar Surface Experiment Package (ALSEP), which continued to operate for years after the astronauts departed. (The ALSEPs were turned off by NASA in 1978 as a cost-cutting measure.) Third, the orbiting Apollo Com-

targeted the last three missions to more geologically interesting locales. The level of scientific exploration also increased with each mission, as the astronauts spent longer time on the Moon and carried more elab-

FIGURE 7.2 Geologist (and later U.S. Senator) Jack Schmitt in front of a large boulder in the Litrow Valley at the edge of the lunar highlands. *(NASA)*

TABLE 7.2 Apollo Flights to the Moon

Flight	Date	Landing Site	Accomplishments
8	Dec. 68	Orbiter	First human circumlunar flight
10	May 69	Orbiter	First lunar orbit rendezvous
11	Jul. 69	Mare Tranquillitatis	First human landing; 22 kg of samples returned
12	Nov. 69	Oceanus Procellarum	First ALSEP. Visit to Surveyor 3
13	Apr. 70	Flyby	Landing aborted owing to explosion in Command Module
14	Jan. 71	Mare Nubium	First "rickshaw"
15	Jul. 71	Imbrium/Hadley	First "rover." Visit to Hadley Rille; 24-km traverse
16	Apr. 72	Descartes	Highland landing site; 95 kg of samples returned
17	Dec. 72	Taurus Mts.	Geologist present; 111 kg of samples returned

mand Modules carried a wide range of instrumentation to photograph and analyze the lunar surface from above.

The last human left the Moon in December 1972, just a little more than three years after Neil Armstrong took his "giant leap for mankind." The program of lunar exploration was cut off in midstride as the result of political and economic pressures. It had cost just about $100 per American, spread over ten years—the equivalent of one pizza and a six-pack per year. The giant Apollo rockets built to travel to the Moon have been left to rust on the lawns of NASA

FIGURE 7.3 Lunar samples collected in the Apollo project are analyzed and stored in NASA facilities at the Johnson Space Center in Houston, Texas. *(NASA)*

centers at Cape Canaveral and Houston (Figure 7.4). Today, no nation on Earth has the capability of returning to the Moon; NASA estimates indicate it would require more than a decade to mount such an effort. Having reached our nearest neighbor in space, we humans have retreated to our own planet. How long before we will venture out again into the solar system?

The scientific legacy of Apollo remains, however, as we will see in the following sections of this chapter.

Composition and Structure of the Moon

The composition of the Moon is not the same as that of the Earth. With its density of only 3.3 g/cm³, the Moon must be made almost entirely of silicate rock. Relative to the Earth it is *depleted* in iron and other metals. We also know from the study of lunar samples that water and other volatiles are absent. It is as if the Moon were composed of the same basic silicates as the Earth's mantle and crust, with the core metals and the volatiles selectively removed. The differences in composition between Earth and Moon provide important clues concerning the origin of the Moon, a topic we shall return to later in this chapter.

Probes of the interior carried out with seismometers taken to the Moon as part of the Apollo program confirm the absence of a large metallic core. The Moon also lacks a global magnetic field like that of the Earth, a result consistent with the theory that such a field is generated by motions in a liquid metal core. Not only is the metal lacking, but, in addition, the Moon is cold and has a solid interior. The level of seismic activity on the Moon is correspondingly less than that on the Earth; no moonquake measured by any of the Apollo seismometers could have been felt by a person standing on the Moon, and the total energy released by moonquakes is a hundred billion times less than that of earthquakes on our planet.

FIGURE 7.4 One of the unused Saturn 5 rockets built to go to the Moon, but now a tourist attraction at NASA's Johnson Space Center in Houston. *(NASA)*

7.2 The Lunar Surface

General Appearance

Through the telescope (or photographed from a spacecraft), the Moon is seen to be covered by impact craters of all sizes. However, none of these craters or other topographic features are large enough to be seen without optical aid. The most conspicuous of the Moon's surface features—those that can be seen with the unaided eye and are familiarly called "the man in the Moon"—are splotches of darker material of volcanic origin. When Galileo first looked at the Moon through a telescope, he saw mountains, craters, and valleys, in addition to these dark regions that he thought were seas. The idea developed that the Moon might be a world, perhaps not unlike our own.

The early names given to the lunar "seas" are still in use today: Mare Nubium (Sea of Clouds), Mare Tranquillitatis (Tranquil Sea), and so on. In contrast, the "land" areas between the seas do not have individual names. Thousands of craters have been named, however, mostly for great scientists and philosophers. Among the most prominent are Plato, Copernicus, Tycho, and Kepler.

The early lunar observers regarded the Moon as having continents and oceans and as being a possible abode of life. We know today, however, that the resemblance of lunar features to terrestrial ones is superficial. Even when these features look somewhat similar, the origins of lunar features such as craters and mountains may be very different from the origins of their terrestrial counterparts. The Moon's relative lack of internal activity, together with the absence of air and water, makes most of its geological history unlike anything we know on Earth. Much of the lunar surface is also older than the rocks of the Earth's crust, as scientists learned when the first lunar samples were analyzed in the laboratory.

Ages of Lunar Rocks

In order to trace the detailed history of the Moon or of any planet, we must have a way to determine the ages of individual rocks. Once lunar samples were brought back by the Apollo astronauts, the techniques that had been developed to date rocks on Earth were applied to establish a chronology for our satellite.

The ages of rocks are measured using the properties of natural radioactivity. Radioactive nuclei of atoms are unstable and spontaneously convert to other nuclei, with the emission of particles such as electrons or of radiation in the form of gamma rays. For any given nucleus the decay process is random, and it might happen at any time. But for a very large collection of identical radioactive atoms, there is a specific period, called its **half-life,** during which the chances are 50-50 that decay will occur. A particular nucleus may last a shorter or longer time than its half-life, but in a large sample almost exactly half will have decayed after a time equal to one half-life, and half of those remaining (three-quarters of the sample) will have decayed in two half-lives. After three half-lives,

ESSAY Observing the Moon

The Moon is one of the most beautiful sights in the sky, and it is the only object that reveals its topography without requiring a visit from a spacecraft. A fairly small amateur telescope easily shows craters or mountains on the Moon as small as a few kilometers across. An angular resolution of 1 arcsec, which is typical for a professional telescope on a good night, corresponds to about 1 km on the Moon. Since topographic features such as mountains and craters are typically a few kilometers to a few tens of kilometers in size, they are readily detected.

Even as seen through a good pair of binoculars, the appearance of the Moon's surface changes dramatically with its phase. At full phase it shows almost no topographic detail, and you must look closely to see more than a few craters. This is because the sunlight illuminates the surface straight-on, and in this flat lighting no shadows are cast. Much more revealing is the view near first or last quarter, when the sunlight streams in from the side and topographic features cast sharp shadows. At a given resolution, it is almost always more rewarding to study a

planetary surface under such *oblique lighting*, when the maximum information about the surface topography can be obtained.

The flat lighting at full phase does accentuate brightness contrasts on the Moon, however, such as the boundaries between the maria and highlands. Notice also that several of the large mare craters seem to be surrounded by aprons of white material, and from them extend light streaks or *rays* that can stretch for hundreds of kilometers across the surface. These lighter features are ejecta, splashed out from the crater-forming impact.

One lunar phenomenon easily seen without optical aid is called "the new Moon in the old Moon's arms." Look at the Moon when it is a thin crescent, and you can often make out the faint circle of the entire lunar disk, even though the sunlight shines on only a narrow crescent. The disk is illuminated not by sunlight but by earthlight: reflected light from the Earth. The light of the full Earth on the Moon is about 50 times brighter than that of the full Moon shining on the Earth.

The appearance of the Moon at different phases. Illumination from the side brings craters and other topographic features into sharp relief. At full phase, there are no shadows and it is difficult to see topographic features. However, the flat lighting at full phase brings out some classes of surface features, such as the bright rays of ejecta that stretch out from a few large, young craters. *(Lick Observatory)*

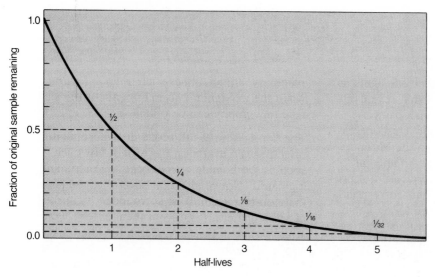

FIGURE 7.5 Radioactive decay and half-life.

only one-eighth of the original sample remains, and so on (Figure 7.5).

Thus, radioactive elements provide accurate nuclear clocks; by comparing the relative abundance of a radioactive element with that of the element it decays into, we can learn how long the process has been going on and hence the age of the sample. Table 7.3 summarizes the decay reactions used most often to date lunar and terrestrial rocks. Typically, more than one element is used to date a rock, and the process is carried out for several different mineral grains in the same rock to eliminate uncertainties. The final age, checked by the agreement of the different methods, is usually accurate to within a few percent, that is, to a few tens of millions of years in a rock several billion years old.

In order to interpret such measurements, it is important to recognize what the age represents. A radioactive age is the time during which the rock has remained undisturbed, so that both the radioactive element and its decay product are still present in each mineral grain. In most cases it can be thought of as the **solidification age** of the rock: the time since it cooled from the molten state. It is such solidification ages that are quoted in the discussion of the Moon's geological features. As we will see, for the Moon these ages range from about 3.3 to 4.4 billion years—substantially older than most of the rocks on the Earth.

Radioactive dating techniques can be used to determine the date of the formation of a planet as well as the solidification age of individual rocks. To obtain such dates, the total quantities of the radioactive isotopes (such as uranium-238) and their decay products (lead-206, in this example) are used. The results of this technique agree for both Earth and Moon; each was formed 4.5 billion years ago, with an uncertainty of less than 100 million years.

Geological Features

Most of the surface of the Moon (83 percent) is heavily cratered and consists of relatively light-colored silicate rocks called anorthosites. These regions are known as the lunar **highlands.** With ages of more than 4.0 billion years, the highlands are the oldest surviving part of the lunar crust. They represent material that solidified on the crust of the cooling Moon like slag floating on the top of a smelter. Because they formed so early in lunar history, the highlands are also extremely heavily cratered, bearing the scars of billions of years of impacts by interplanetary debris (Figure 7.6).

TABLE 7.3   Radioactive Decay Reactions Used to Date Rocks*

Parent	Daughter	Half-Life (billion yr)
Samarium-147	Neodymium-143	106
Rubidium-87	Strontium-87	48.8
Thorium-232	Lead-208	14.0
Uranium-238	Lead-206	4.47
Potassium-40	Argon-40	1.31

*The number given following each element name is the atomic weight, equal to the number of protons and neutrons in the nucleus. Since isotopes of a given element differ from one another in the number of neutrons in the nucleus, we need to specify the atomic weight to identify the particular isotope being discussed.

FIGURE 7.6   The old, heavily cratered lunar highlands make up 83 percent of the Moon's surface. The width of the frame is 250 km. *(NASA)*

The most prominent lunar features are the so-called seas, still called **maria** (Latin for "seas"). These dark plains (Figure 7.7), which are much less cratered than the highlands, cover just 17 percent of the lunar surface, mostly on the side facing the Earth. The maria are, of course, dry land. They are volcanic plains, laid down in eruptions billions of years ago and partly filling huge depressions called *impact basins*, which were produced by collisions of asteroids with the Moon (Figure 7.8).

FIGURE 7.7   About 17 percent of the Moon's surface consists of the maria—flat plains of basaltic lava. This view of Mare Imbrium includes numerous secondary craters and other ejecta from the large crater Copernicus, on the upper horizon. Copernicus is almost 100 km in diameter. *(NASA)*

FIGURE 7.8 The youngest of the large lunar impact basins is Orientale, formed 3.8 billion years ago. Unlike most of the other basins, this scar, 1000 km in diameter, has not been filled in with lava flows, so it retains its striking "bull's eye" appearance. *(NASA)*

The lunar maria are all composed of basalt, very similar in composition to the oceanic crust of the Earth or to the lavas erupted by many terrestrial volcanoes. A series of large eruptions between 3.3 and 3.8 billion years ago (dated from laboratory measurements of returned samples) formed smooth flows typically a few meters thick but extending over distances of hundreds of kilometers. Eventually, these flows filled in the lowest parts of the basins to form the mare surfaces we see today.

Volcanic activity may have begun very early in the Moon's history, although most evidence of the first half-billion years is lost. What we do know is that the major mare volcanism, which involved the release of highly fluid magmas from hundreds of kilometers below the surface, ended about 3.3 billion years ago (Figure 7.9). After that, the Moon's interior cooled, and by 3.0 billion years ago all volcanic activity ceased. Since then, our satellite has been a geologically dead world, changing only slowly as the result of random impacts.

FIGURE 7.9 Sample of basalt from the mare surface. The gas bubbles are characteristic of a lava rock. *(NASA)*

FIGURE 7.10 Mount Hadley on the edge of Mare Imbrium, photographed by the Apollo 15 astronauts. Note the smooth contours of the lunar mountains, which have not been sculpted by water or ice. *(NASA)*

FIGURE 7.11 Apollo photograph of bootprint on the Moon. *(NASA)*

There are several mountain ranges on the Moon, generally along the edges of the maria. Most of them bear the names of terrestrial ranges—the Alps, Apennines, Carpathians, and so on—but their mode of origin is entirely different from that of their terrestrial namesakes.

The major lunar mountains are all the result of the Moon's history of impacts. The long, arc-shaped ranges that border the maria are debris ejected from the impacts that formed these giant basins. These mountains have low, rounded profiles that resemble old, eroded mountains on Earth (Figure 7.10). But appearances can be deceiving. The mountains of the Moon have not been eroded, except for the effects of meteoritic impacts. They are rounded because that is the way they formed, and there has been no water or ice to carve them into cliffs and sharp peaks.

On the Lunar Surface

The surface of the Moon is everywhere covered with a fine-grained soil of tiny, shattered rock fragments. The dark basaltic dust of the lunar maria was kicked up by every astronaut footstep, and this dust eventually worked its way into all of the astronauts' equipment. The upper layers of the surface are porous, consisting of loosely packed dust into which their boots sank several centimeters (Figure 7.11).

This lunar dust, like so much else on the Moon, is the product of impacts. Each cratering event, large or small, breaks up the rock of the lunar surface and scatters the fragments. The impacts also melt tiny droplets of rock, producing spherules of *impact glass* that are found throughout the soil. Especially important are the multitudes of very small impacts by grains of interplanetary dust that never strike the Earth's surface because they are filtered out by our atmosphere (Figure 7.12).

In the absence of any air, the lunar surface experiences much greater temperature extremes than does the surface of the Earth, even though the Earth is virtually the same distance from the Sun. Near local noon, the temperature of the dark lunar soil rises to just above the boiling point of water, while during the long lunar night (two Earth weeks) it drops to about 100 K ($-173°C$). The extreme cooling is a result not only of the absence of air but also of the porous nature of the dusty soil, which cools more rapidly than would solid rock.

7.3 Impact Craters

The Moon provides an important benchmark for understanding the history of the planetary system. Most solid bodies show the effects of impacts, often extending back to the era when a great deal of debris from the formation process was still present. On the Earth, this long history has been erased by our active geology. On the Moon, in contrast, most of the impact history is preserved. If we can understand what has happened on the Moon, we may be able to extrapolate this knowledge to the other cratered planets and satellites.

Volcanic Versus Impact Origin of Craters

Until the middle of the 20th century, the impact origin of the craters of the Moon was not widely recognized. Since impact craters are extremely rare on Earth, geologists did not consider them to be the major feature of lunar geology. They reasoned (perhaps unconsciously) that since the craters we have on Earth are volcanic, the lunar craters must have a similar origin.

(a)

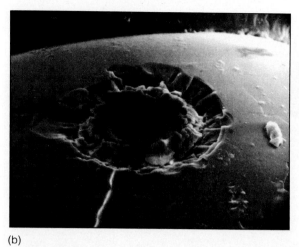
(b)

FIGURE 7.12 The lunar soil, which consists of impact fragments and tiny spheres of dark glass (impact glass). (a) Soil collected from the mare. (b) Photomicrograph of microscopic impact pits on a tiny sphere of impact melted glass. *(NASA)*

One of the first geologists to argue for an impact origin of lunar craters was G.K. Gilbert (1843–1918), a scientist with the U.S. Geological Survey in the 1890s. Gilbert pointed out that the large lunar craters, which are mountain-rimmed, circular features with floors generally below the level of the surrounding plains, are different in form from terrestrial volcanic craters. Volcanic craters are smaller and deeper and almost always occur at the tops of volcanic mountains (Figure 7.13). Therefore, he concluded that the lunar craters were not volcanic.

Gilbert believed that the lunar craters were of impact origin, but he still had difficulty explaining why all of them are circular. Impacts of the sort we are most familiar with, such as those made by throwing a stone into a sandbox, make circular features only when falling straight down; otherwise, the outline of the crater is more or less elliptical.

The solution to this problem is readily seen, however, when we note the speed with which projectiles

(a)   Terrestrial volcano

(b)   Lunar impact crater

FIGURE 7.13 Profiles of (a) terrestrial volcanic craters and (b) lunar impact craters are quite different from each other.

approach the Earth or Moon. Attracted by the gravity of the larger body, the projectile strikes with at least escape velocity, which is 11 km/s for the Earth and 2.4 km/s for the Moon. To this escape velocity is added whatever speed the projectile already had with respect to the Earth or Moon, typically 10 km/s or more. The corresponding energy of impact leads to an *explosion*, and we know from experience with bomb and shell craters on Earth that explosion craters are always essentially circular. Thus, recognition of the similarity between *impact craters* and *explosion craters* removed the last important objection to the impact theory for the origin of lunar craters.

The Cratering Process

Let us see how an impact at these high speeds produces a crater. When an impacting projectile strikes a planet at a speed of several kilometers per second, it penetrates two or three times its own diameter before stopping. During these few seconds, its energy of motion is transferred into a shock wave, which spreads through the target body, and into heat, which vaporizes most of the projectile and some of the surrounding target. The shock wave fractures the rock of the target, while the hot silicate vapor generates an explosion not too different from that of a nuclear bomb detonated at ground level (Figure 7.14). The size of the crater that is excavated depends primarily on the speed of impact, but generally it is about ten times the diameter of the projectile.

An impact explosion of the sort described above leads to a characteristic kind of crater, as illustrated in Figure 7.15. The central cavity is initially bowl-

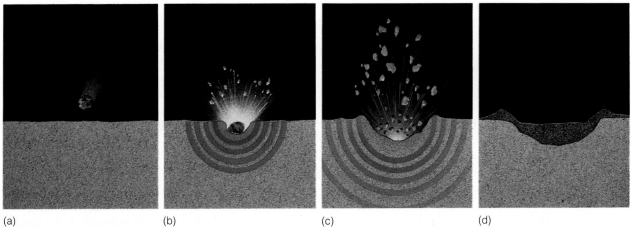

(a)   (b)   (c)   (d)

FIGURE 7.14 Stages in the formation of an impact crater: (a) the impact; (b) the projectile vaporizes, and a shock wave spreads through the lunar rock; (c) ejecta are thrown out of the crater; (d) most of the ejected material falls back to fill the crater and form an ejecta blanket.

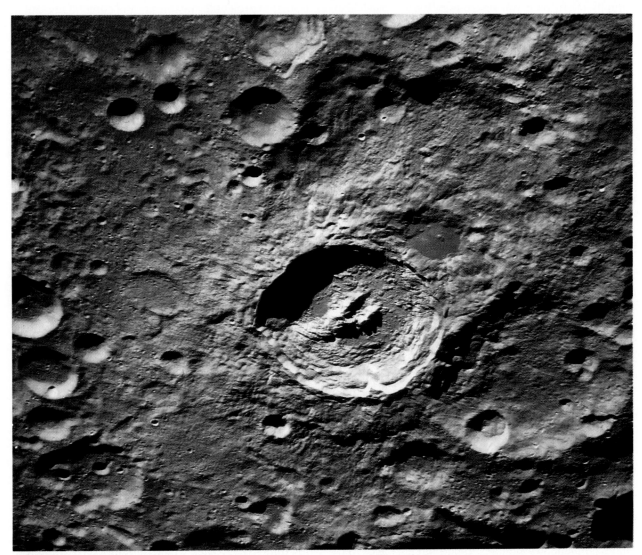

FIGURE 7.15 King Crater on the far side of the Moon, a fairly recent lunar crater 75 km in diameter, clearly showing most of the features associated with large lunar impact craters. *(NASA)*

shaped ("crater" comes from the Greek word for "cup"), but the gravitational rebound of the crust partially fills it in, producing a flat floor and sometimes creating a central peak. Around the rim, landslides create a series of terraces.

The rim of the crater is turned up by the force of the explosion, so it rises above both the floor and the adjacent terrain. Surrounding the rim is an *ejecta blanket*, consisting of material thrown out by the explosion that falls back to create a rough, hilly apron, typically about one crater diameter wide. Additional, higher speed ejecta fall at greater distances from the crater, often digging small *secondary craters* where they strike the surface. Some of these streams of ejecta can extend for hundreds or even thousands of kilometers from the crater, creating on the Moon the bright *crater rays* that are so prominent in photographs taken near full phase (see Essay, page 129). The brightest lunar crater rays are associated with large, young craters, such as Kepler and Tycho.

Using Crater Counts to Date Planetary Surfaces

If a planet has had little erosion or internal activity, like the Moon during the past 3 billion years, it is possible to use the number of impact craters counted on its surface to estimate the age of that surface. By "age" we mean the time since there was a major disturbance, such as the volcanic eruptions that produced the lunar maria.

This technique works because the rate at which impacts have occurred has been roughly constant for several billion years. Thus, in the absence of forces to eliminate craters, the number of craters is simply proportional to the length of time the surface has been exposed.

Estimating ages from crater counts is a little like an experience you might have walking along a sidewalk in a snowstorm, when the snow has been falling steadily for a day or more. You may notice that in front of some houses the snow is deep, while next door the sidewalk may be almost clear. Do you conclude that less snow has fallen in front of Mr. Jones' house than Ms. Smith's? Of course not. Instead, you conclude that Jones has recently swept the walk clean, while Smith has not. Similarly, the numbers of craters indicate how long since a planetary surface was last "swept clean" by lava flows or the ejecta from a nearby large impact.

Except for those from the Moon and Earth, we have no planetary samples to permit exact ages to be calculated from radioactive decay. Without samples, the planetary geologist cannot tell if a surface is a million years old or a billion. But the number of craters provides an important clue. On a given planet, the more heavily cratered terrain will always be the older (as we have defined age above). And if we can calibrate the relationship between crater numbers and ages, using what we know about the Moon, we may be able to estimate the ages of surfaces on other cratered planets, such as Mars or Mercury.

Cratering Rates

The rate at which craters are being formed in the vicinity of the Earth and Moon cannot be measured directly, since the interval between large crater-forming impacts is longer than the span of human history. Remember that Meteor Crater in Arizona is 50,000 years old. However, the cratering rate can be estimated from the number of craters on the lunar maria, or it can be calculated from the numbers of potential projectiles (asteroids and comets) present in the solar system today, as will be further discussed in Chapter 11. Fortunately, both lines of reasoning lead to about the same answer.

For the entire land area of the Earth, these calculations indicate that a crater 1 km in diameter should be produced about every 10,000 years, several 10-km craters every million years, and one or two 100-km craters every 50 million years. Comets and asteroids appear to contribute about equally to these statistics. For the Moon, the numbers are about 1/20 as great, primarily as a consequence of the Moon's smaller total area.

If these cratering rates are extrapolated back in time, they lead to the conclusion that the craters on the lunar maria should accumulate to currently observed values in several billion years. Since the maria are 3.3 to 3.8 billion years old, this agreement suggests that comets and asteroids in approximately their present numbers have been impacting planetary surfaces for at least 3.8 billion years. Calculations carried out for other planetary bodies indicate that they also have been subject to about the same number of interplanetary impacts.

Earlier than 3.8 billion years ago, however, the impact rates must have been a great deal higher. This conclusion becomes immediately evident when we compare the numbers of craters on the lunar maria with those on the highlands. Typically, there are ten times as many craters on the highlands as on a similar area of maria, yet the highlands are only a little older than the maria, typically 4.2 billion years rather than 3.8 billion years. If the rate of impacts had been constant, the highlands would have had to be at least ten times older. They would therefore have formed 38 billion years ago—long before our universe began.

The Moon, therefore, experienced a period of *heavy bombardment* previous to 3.8 billion years ago, as illustrated in Figure 7.16. This heavy bom-

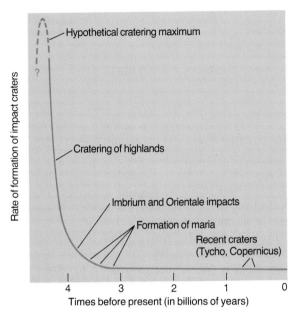

FIGURE 7.16 Schematic diagram of cratering flux on the Moon, covering the past 4.3 billion years.

bardment produced most of the craters we see today in the highlands. Was this a local event, or did similar high impact rates apply throughout the solar system? A partial answer is provided by Voyager pictures of the satellites of Jupiter and Saturn, which reveal some surfaces to be as heavily cratered as the lunar highlands. Since it is impossible to accumulate this many craters *at the present rate* within the lifetime of the solar system, there must have been a period of high bombardment in the outer solar system as well.

Since we do not have any samples of other planets to date in the laboratory, we do not know for sure that the heavy bombardments elsewhere coincided exactly with events on the Moon. However, for most purposes it does not matter whether the heavy bombardments were really simultaneous. The main point is that they all occurred a long time ago, so that any heavily cratered surface can be assigned an age going back to the first billion years of solar system history. Thus, the measurement of craters on planets and satellites throughout the solar system allows scientists to establish a rough chronology for planetary evolution, even without samples. We will make use of these ideas in the following chapters as we discuss the interpretation of spacecraft photographs of other planets.

7.4 Mercury

The planet Mercury is similar to the Moon in many ways. Like the Moon, it has no atmosphere, and its surface is heavily cratered. As we will see later in this chapter, it also shares with the Moon a violent birth history.

Mercury's Orbit

Mercury is the nearest to the Sun of the nine planets and, in accordance with Kepler's third law, has the shortest period of revolution about the Sun (88 of our days) and the highest average orbital speed (48 km/s). It is appropriately named for the fleet-footed messenger god of the Greeks and Romans.

The semimajor axis of the orbit of Mercury, that is, the planet's average distance from the Sun, is 58 million km or 0.39 AU. However, because Mercury's orbit has the high eccentricity of 0.206, its actual distance from the Sun varies from 46 million km at perihelion to 70 million km at aphelion. Pluto is the only planet with a more eccentric orbit. Furthermore, the 7° inclination of Mercury's orbit to the plane of the ecliptic is also greater than that of any other planet except Pluto.

Composition and Structure

Table 7.1 compared some basic properties of Mercury with those of the Moon. Mercury's mass is 1/18 that of the Earth; Pluto is the only planet with a smaller mass. Mercury is also the second smallest of the planets, having a diameter of only about 4878 km, less than half that of the Earth. Its density is 5.4 g/cm³, much larger than the density of the Moon, indicating that the composition of these two objects differs substantially.

One of the most interesting things about Mercury is its composition, which is unique compared with that of the other planets. Mercury's high density tells us that it is composed largely of metals. Apparently, this planet formed without a full complement of silicates of the sort that make up the mantle of the Earth and the bulk of the Moon. The most likely models for the interior of Mercury suggest a metallic iron-nickel core with a mass amounting to 60 percent of the total, with the rest of the planet made up primarily of silicates. The core has a diameter of 3500 km and extends to within 700 km of the surface (Figure 7.17). We could think of Mercury as a metal ball the size of the Moon surrounded by a rocky crust 700 km thick. Unlike the Moon, Mercury also has a weak magnetic field. The existence of this field is consistent with a large metal core, and it suggests that at least a part of the core must be liquid in order to generate the observed field.

The escape velocity is too low and the surface temperature too high for Mercury to retain any substantial atmosphere. In 1985, however, an extremely thin atmosphere of sodium was detected spectroscopically. Apparently, atoms of this metal are ejected

FIGURE 7.17   The interior of Mercury is dominated by a metallic core about the same size as our Moon. *(University of Arizona, courtesy of Robert Strom).*

from the surface through bombardment by the solar wind.

Mercury's Strange Rotation

Visual studies of Mercury's indistinct surface markings were once thought to indicate that the planet kept one face to the Sun, and for many years it was widely believed that Mercury's rotation period equaled its period of revolution of 88 days.

Radar observations of Mercury in the mid-1960s, however, showed conclusively that Mercury does rotate with respect to the Sun. Recall (Section 6.4) that the frequency of a transmitted radar pulse can be controlled precisely. Any motion of the target then introduces a measurable change in this frequency from the Doppler effect (Section 5.6). Such target motions can include rotation; if a planet is turning, one side seems to be approaching the Earth while the other is moving away from it. The result is to spread or broaden the precise transmitted frequency into a

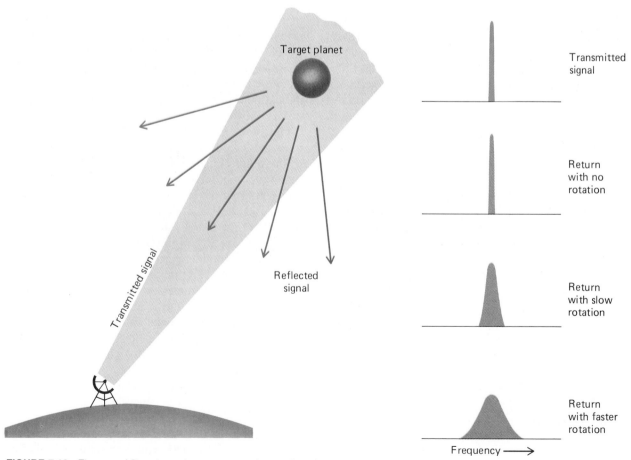

FIGURE 7.18   The use of Doppler radar to measure the rotation of a planet.

range of frequencies in the reflected signal (Figure 7.18). The degree of broadening provides an exact measurement of the rotation rate of the planet.

Mercury's sidereal period of rotation (that is, with respect to the distant stars) was found to be about 59 days. Shortly after this period was discovered by radar astronomers, the Italian dynamicist Giuseppe Colombo pointed out that this is very nearly two-thirds of the planet's period of revolution, and subsequently it was found that there are theoretical reasons for expecting that Mercury can rotate stably with a period of exactly two-thirds that of its revolution— 58.65 days. In this configuration, alternate ends of the tidal bulge face the Sun at successive perihelions, when the tidal forces are largest, so that the gravitational force of the Sun maintains this exact period.

Mercury is very hot on its daylight side, but because it has no appreciable atmosphere, the planet gets surprisingly cold during its long nights. The temperature on the surface climbs to 700 K at noontime. After sunset, however, the temperature drops, reaching 100 K just before dawn. The range in temperature on Mercury is thus 600 K, more than on any other planet.

The Surface of Mercury

The first closeup look at Mercury came in 1974, when the U.S. spacecraft Mariner 10 passed 9500 km from the surface of the planet and televised more than 2000 photographs to Earth, revealing details with a resolution down to 150 m. Mercury strongly resembles the Moon in appearance (Figure 7.19 and Figure 7.20). It is covered with thousands of craters up to hundreds of kilometers across, and larger basins up to 1300 km

FIGURE 7.19 The planet Mercury as photographed by Mariner 10 in 1974. *(NASA/JPL)*

FIGURE 7.20 The Caloris basin on Mercury, 1300 km in diameter. This partially flooded impact basin is the largest structural feature on Mercury seen by Mariner 10. Compare this photograph with that of the Orientale basin on the Moon (Figure 7.8). *(NASA/JPL)*

in diameter. There are also scarps (cliffs) over a kilometer high and hundreds of kilometers long, as well as ridges and plains. Some of the brighter craters are rayed, and many have central peaks. Most of the mercurian features have been named in commemoration of artists, writers, composers, and other contributors to the arts and humanities, in contrast with the scientists commemorated on the Moon. Among the most prominent craters are Bach, Shakespeare, Tolstoy, Mozart, and Goethe.

The larger basins resemble the lunar maria, in both size and appearance. They show evidence of flooding from lava, although the flows do not have the distinctive dark color that characterizes the lunar maria. Since Mercury is so difficult to study telescopically and has been visited by only one spacecraft, we actually know very little about the chemistry of its surface or the details of its geological history.

There is no evidence of plate tectonics on Mercury. Mercury's distinctive long scarps, however, which sometimes cut across craters, seem to be due to a limited degree of crustal compression. Apparently this planet shrank, wrinkling the crust, and it did so after most of the craters on its surface were formed. If the standard cratering chronology applies to Mercury, this shrinkage must have taken place during the past 4.0 billion years. If we understood better the interior structure and composition of this planet, we could probably calculate how internal changes in temperature might have led to this global compression, which has no counterpart on the Moon or the other terrestrial planets.

7.5   Origin of the Moon and Mercury

Understanding the origin of the Moon has proved to be extremely difficult for planetary scientists. Part of the problem is simply that we know so much about our satellite. There is a great wealth of data, particularly on the details of the elemental and isotopic composition of the Moon, which present a challenge to any simplified theory of lunar origins.

Theories for the Origin of the Moon

Most of the various theories for the Moon's origin follow one of three general ideas: (1) that the Moon was once part of the Earth but separated from it early in their history (the fission theory); (2) that the Moon formed together with (but independent of) the Earth, as we believe many satellites of the outer planets formed (the sister theory); and (3) that the Moon formed elsewhere in the solar system and was captured by the Earth (the capture theory).

Unfortunately, there are fundamental problems with each of these ideas. Perhaps the easiest theory to reject is the capture theory. The difficulty is primarily that no one knows of any way that the early Earth could have captured a large satellite from elsewhere. One body approaching another cannot go into orbit around it without a substantial loss of energy—this is the reason that spacecraft destined to orbit other planets are equipped with retrorockets. Further, if such a capture did take place, it would be into a very eccentric orbit, rather than the nearly circular orbit the Moon occupies today. And finally, there are too many compositional similarities between the Earth and the Moon, particularly an identical fraction of the major isotopes of oxygen, to justify seeking a completely independent origin.

The first possibility, that the Moon separated from the Earth, was suggested in the late 19th century by George Darwin, the astronomer son of naturalist Charles Darwin. It is often called the fission theory. More modern dynamical calculations have shown that the sort of spontaneous fission or splitting imagined by Darwin is impossible. Further, it is difficult to understand how a Moon made out of terrestrial material in this way could have developed the many distinctive chemical differences that are now known to characterize our satellite.

Scientists are therefore left with the sister theory, that the Moon formed alongside the Earth, or with some modification of the fission theory that finds a more acceptable way for the lunar material to be separated from the Earth.

The Giant Impact Theory

The giant impact theory for the origin of the Moon is in part a response to the problems of lunar chemistry posed by the sister theory and in part a reflection of interest among planetary scientists in the role of impacts in the late stages of planet growth. There is increasing evidence that large projectiles—objects of essentially planetary mass—were orbiting in the inner solar system at the time of formation of the terrestrial planets. What if one of these very large projectiles struck the Earth?

The *giant impact theory* envisions the Earth's being struck obliquely by an object of about one-tenth the Earth's mass, or about the size of Mars. This is very nearly the largest impact the Earth could experience without being shattered. Computer calculations (Figure 7.21) show that such an impact would disrupt much of the Earth, penetrating the core, ejecting a vast amount of material into space, and releasing almost enough energy to break the planet apart.

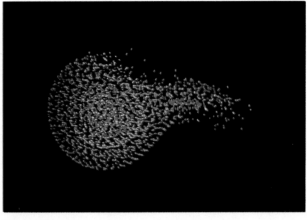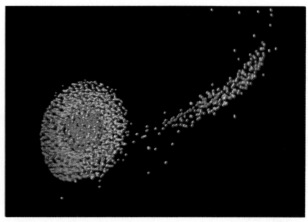

(a) (b)

FIGURE 7.21 Two stages in a computer simulation of an oblique giant impact on the Earth. The metal cores of the Earth and the Mars-sized projectile are shown in blue, the mantles and crusts in orange tones. (a) Shortly after the impact, a plume of material from both the projectile and the Earth's mantle is blasted into space. (b) Later, part of the ejecta begin to orbit the Earth. Note that the fastest moving ejecta, which may condense to form the Moon, consist entirely of mantle material; the core metal of the projectile is moving more slowly and will eventually collapse back onto the Earth. *(Courtesy of A.G.W. Cameron, Center for Astrophysics)*

Material totaling several percent of the mass of the Earth could be ejected in such an impact. Most of the material would be from the mantles of the Earth and the impacting body, and it would initially be ejected at high speed in the form of hot vapor. Perhaps—and the calculations are not really clear on this point— much of this hot vapor could condense into a ring of material in orbit around the Earth. Alternatively, it might form a sort of superheated silicate atmosphere in rapid rotation around the planet. Either way, it is suggested that this ejected material ultimately condensed to form the Moon, as described in the sister theory.

The giant impact hypothesis offers potential solutions to most of the major problems raised by the chemistry of the Moon. First, since the raw material for the Moon is derived from the mantles of the Earth and the projectile, the absence of metals is easily understood. Second, most of the volatile elements could have been lost during the high-temperature phase following the impact. Yet by making the Moon

primarily of terrestrial mantle material, it is also possible to understand similarities such as identical oxygen isotopic abundances.

The Origin of Mercury

The problem with Mercury is the reverse of that posed by the composition of the Moon. Unlike the Moon, Mercury is composed mostly of metal. However, Mercury should have formed with about the same ratio of metal to silicate as the Earth or Venus. How did it lose so much of its rocky material?

The most probable explanation for Mercury's loss of silicates may be similar to the explanation for the Moon's lack of a metal core. Mercury is likely to have experienced several giant impacts very early in its youth, and one or more of these may have torn away a fraction of its mantle and crust, leaving a body dominated by its iron core. This planet, like the Moon, bears testimony to the violence that must have characterized the solar system during its youth.

SUMMARY

7.1 Most of what we know about the Moon derives from the Apollo program, including 400 kg of lunar samples still being intensively studied. The Moon has 1/80 the mass of the Earth and is severely depleted in both metals and **volatiles.** It is made almost entirely of silicates like those in the Earth's mantle and crust.

7.2 Lunar rocks are dated by their radioactivity, using atoms with different **half-lives.** Like the Earth, the Moon was formed 4.5 billion years ago. The heavily cratered **highlands** are made of rocks up to 4.4 billion years old. The darker volcanic plains of the **maria** were erupted between 3.3 and 3.8 billion years ago. Generally, the surface is dominated by impacts, including continuing small impacts that produce its fine-grained soil.

7.3 A century ago Gilbert suggested that the lunar craters were of impact origin, but the cratering process was not well understood until more recently. High-speed impacts produce explosions and excavate craters with raised rims, ejecta blankets, and often central peaks. Cratering rates have been roughly constant for the past 3 billion years but earlier were much greater. Crater counts can be used to derive approximate ages for geologic features on the Moon and other planets.

7.4 Mercury is the nearest planet to the Sun and the fastest moving. Mercury is similar to the Moon in having no

atmosphere and a heavily cratered surface, but it differs in having a very large metal core. Early in its evolution, it appears to have lost part of its silicate mantle.

7.5 There are three standard theories for the origin of the Moon: the fission theory, the sister theory, and the capture theory. All have problems, and recently they have been supplanted by the giant impact theory, which ascribes the origin of the Moon to the impact 4.5 billion years ago of a Mars-sized projectile on the Earth. A similar giant impact early in Mercury's history may have been responsible for the loss of much of its mantle.

REVIEW QUESTIONS

1. What is the composition of the Moon, and how does it relate to the composition of the Earth and the planet Mercury?

2. Outline the chronology of the Moon's geological history.

3. Explain how high-speed impacts form craters, and indicate the characteristic features of impact craters.

4. Explain the evidence for a period of heavy bombardment on the Moon about 4 billion years ago. What

might have been the source of this high flux of impacting debris?

5. Compare and contrast Mercury with the Moon.

6. Summarize the four main theories for the origin of the Moon.

7. What do current ideas about the origins of the Moon and Mercury have in common?

THOUGHT QUESTIONS

8. One of the primary scientific objectives of the Apollo program was the return of samples of lunar material. Why was this so important? What can be learned from samples, and are they still of value now?

9. Apollo astronaut David Scott dropped a hammer and a feather together on the Moon, and both reached the ground at the same time. There are two reasons why this experiment on the Moon had distinct advantages over the same experiment as performed by Galileo on the Earth. What are these advantages?

10. Galileo thought the lunar maria were seas of water. If you had no better telescope than Galileo's, could you prove that the maria are not composed of water?

11. Why did it take so long for geologists to recognize that the lunar craters had an impact origin rather than being volcanic?

12. How would a crater made by the impact of a comet with the Moon differ from a crater made by the impact of an asteroid?

13. Why are the lunar mountains smoothly rounded rather than having sharp, pointed peaks (the way they had almost always been depicted by artists before the first lunar landings)?

14. The lunar highlands have about ten times more craters on a given area than do the maria. Does this mean that the highlands are ten times older? Explain your reasoning.

15. Give several reasons why Mercury would be a particularly unpleasant place to live.

16. Summarize the main weakness of each of the traditional theories for the origin of the Moon: the fission theory, the capture theory, and the sister theory.

17. The Moon has too little iron, Mercury too much. How can both these anomalies be the result of giant impacts? Explain how the same process can yield such apparently contradictory results.

PROBLEMS

18. The Moon was once closer to the Earth than it is now. When it was at half its present distance, how long was its period of revolution?

19. In any one mare there are a variety of rock ages, spanning typically about a hundred million years. The individual lava flows as seen in Hadley Rill by the Apollo 15 astronauts were about 4 m thick. Estimate the average interval between lava flows if the total depth of the lava in the mare is 2 km.

20. The Moon requires about one month (0.08 year) to orbit the Earth. Its distance is about 400,000 km (0.0027 AU). Use Kepler's third law, as modified by Newton, to calculate the mass of the Earth relative to the Sun.

Heavily eroded canyonlands on Mars. This Viking photograph of the Valles Marineris shows
an area about 60 km across. *(NASA/USGS, courtesy of Alfred McEwen)*

Carl Sagan (b. 1934) of Cornell University is one of the best known scientists in the United States, as a consequence of his writings and television appearances. Sagan was the first to develop greenhouse models to explain the high surface temperature on Venus, and later he provided leadership to the Viking search for life on Mars. As a founder of The Planetary Society, Sagan remains a leading advocate of the scientific exploration of the planets. *(Brent Peterson, Parade)*

8

EARTH-LIKE PLANETS: VENUS AND MARS

The Moon and Mercury are geologically dead. In contrast, the larger terrestrial planets—the Earth, Venus, and Mars—are more active and interesting worlds. We discussed the Earth in Chapters 3 and 4, and now we turn to Venus and Mars. These are the nearest planets and the most accessible to spacecraft. Not surprisingly, the greater part of the efforts of the United States and the U.S.S.R. in planetary exploration has been devoted to these fascinating worlds. In this chapter, we will discuss some of the results of more than two decades of scientific exploration of Mars and Venus. In Chapter 12, we will further compare Mars and Venus with each other and with the Earth, Moon, and Mercury.

8.1 The Nearest Planets

Appearance and Orbits

Venus and Mars are among the brightest objects in the night sky. The average distance of Mars from the Sun is 227 million km, but its orbit is somewhat eccentric (0.093), and its actual distance varies by 42 million km. Venus approaches the Earth more closely than any other planet—at its nearest it is only 40 million km away. Its orbit is very nearly circular at a distance of 108 million km from the Sun. Like Mercury, Venus sometimes appears as an "evening star" and sometimes as a "morning star." Under favorable

circumstances, Venus is so bright that it even casts a visible shadow.

Seen through a telescope, Mars is both tantalizing and disappointing. The planet has a distinctly red color, due (as we now know) to the presence of iron oxides in its soil. At its nearest, Mars has an apparent diameter of 25 arcsec, and the best telescopic resolution obtainable is about 100 km, or about the same as the Moon seen with the unaided eye. At this resolution, however, no hint of topographic structure can be detected: no mountains, no valleys, not even impact craters (Figure 8.1). On the other hand, the bright polar caps of the red planet can easily be seen, together with dusky surface markings that gradually change in outline and intensity from season to season. Of all the planets, only Mars has a surface that can clearly be made out from Earth, and this surface exhibits changes that bespeak a dynamic atmosphere.

For a few decades around the turn of the 20th century, some astronomers believed that they saw evidence on Mars of a civilization of intelligent beings. The controversy began in 1877, when the Italian astronomer Giovanni Schiaparelli announced the discovery of long, faint, straight lines on Mars that he called *canale*, or channels. In English-speaking countries, the term was translated as canals, implying an artificial origin.

Until his death in 1916, the most effective proponent of intelligent life on Mars was Percival Lowell,

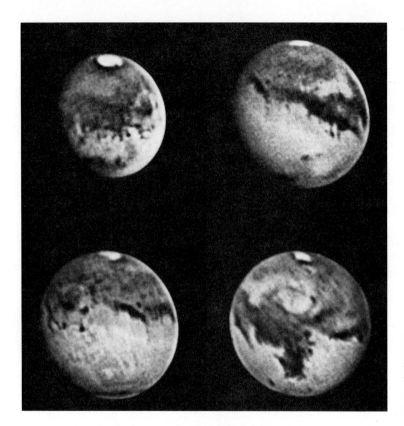

FIGURE 8.1   These are among the best Earth-based photographs of Mars, taken in 1988 when the planet was exceptionally close to the Earth. The polar caps and dark surface markings are evident, but no topographic features can be seen. *(Steve Larson, University of Arizona)*

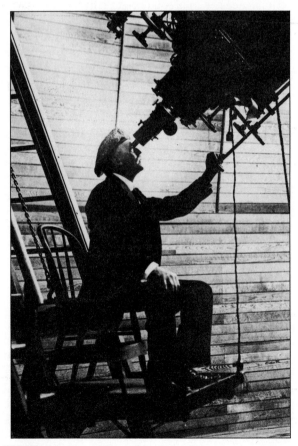

FIGURE 8.2   Percival Lowell in about 1910 observing with his 24-in. telescope at Flagstaff. *(Lowell Observatory)*

self-made American astronomer and member of the wealthy Lowell family of Boston (Figure 8.2). In order to pursue his study of the martian canals, he established the Lowell Observatory in Flagstaff, in the Territory of Arizona. An effective author and public speaker, Lowell made a convincing case for intelligent Martians, who, he believed, had constructed huge canals to carry water from the polar caps in an effort to preserve their existence in the face of a deteriorating climate.

The argument, however, hinged on the reality of the canals, a matter that remained in dispute among astronomers. The canals were always difficult to study, glimpsed only occasionally as atmospheric conditions caused the tiny image of Mars to shimmer in the telescope. Lowell saw them everywhere, but many other observers were skeptical. When larger telescopes failed to confirm the existence of canals, the skeptics seemed to be vindicated. Astronomers had given up the idea by the 1930s, although it persisted in the public consciousness until the first spacecraft photographs clearly showed that there were no martian canals. Now it is generally accepted that the canals were an optical illusion, the result of the human mind's tendency to see order in random features glimpsed dimly at the limits of the eye's resolution.

Venus, in contrast to Mars, exhibits little of interest through even the largest telescope. As first discovered by Galileo, Venus displays a full range of phases as it revolves about the Sun. But the planet's actual surface is not visible because it is shrouded by dense clouds. These clouds reflect about 70 percent of the incident sunlight, a circumstance that contributes greatly to the planet's brightness but frustrates efforts to study the underlying surface (Figure 8.3).

Rotation of Mars and Venus

The presence of permanent surface markings enables us to determine the rotation period of Mars with great accuracy; its sidereal day is $24^h 37^m 23^s$, just a little greater than the rotation period of the Earth. This high precision is not obtained by watching Mars for a single rotation, but by noting how many turns it makes in a long period. Good observations of Mars date back for more than 200 years, a period during which tens of thousands of martian days have passed. The rotation period is known now to within a few hundredths of a second. The rotational pole of Mars has a tilt of about 25°, very similar to that of the Earth's pole. Thus Mars experiences seasons very much like those on Earth. Because of the longer martian year, however, seasons there each last about six of our months.

Since no surface detail can be seen on Venus, its rotation period can be found only by using radar. The first radar observations of the planet's rotation were made in the early 1960s. Surprisingly, they showed Venus to rotate from east to west—in the reverse direction from the rotation of most other planets—in a period of about 250 days. Subsequently, topographical surface features were identified on the planet that showed up in the reflected radar signals. The rotation period of Venus, determined from the motion of such features across its disk, is 243.08 days retrograde (east to west). The rotation period of Venus is about 19 days longer than its period of revolution about the Sun. The length of a day on Venus—the time between successive noons—is 116.67 Earth days.

Basic Properties of Venus and Mars

Before discussing these planets individually, it is appropriate to compare them with each other and with the Earth, in terms of basic properties. Venus is the Earth's twin, with a mass 0.82 times the mass of the Earth and an almost identical density. The level of geological activity is also relatively high, creating perhaps the most geologically complex and diverse surface in the solar system. In addition, Venus has a much more massive atmosphere than the Earth's, with a surface pressure nearly 100 times greater than ours. Its surface is also remarkably hot, with a temperature of 730 K. One of the major challenges presented by Venus is to understand why its atmosphere and surface environment have diverged so sharply from those of our own planet.

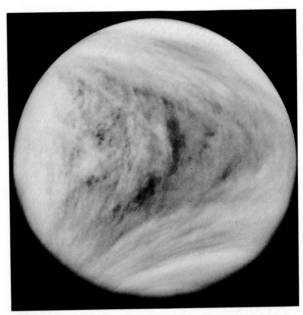

FIGURE 8.3 Venus, as photographed by the Pioneer Venus orbiter. This ultraviolet image shows upper atmosphere cloud structure that would be invisible at visible wavelengths. *(NASA/ARC)*

As a planet, Mars is rather small, with a mass only 0.11 times the mass of the Earth. It is larger than either the Moon or Mercury, however, and unlike them it retains a thin atmosphere. Mars is also large enough to have supported considerable geological activity, some apparently persisting to the present day. But the most fascinating thing about Mars is that it probably once had a thick atmosphere and seas of liquid water, and there is a high probability that an indigenous life flourished there in the distant past.

Table 8.1 summarizes some of the basic data for Venus and Mars.

8.2 Geology of Venus

Since Venus has about the same size and composition as the Earth, we might expect its geology to be similar. This is partly true, but Venus does not exhibit the same kind of plate tectonics as does the Earth, and its lack of erosion results in a very different surface appearance.

Spacecraft Exploration of Venus

More spacecraft have been sent to Venus than to any other planet. Although the U.S. Mariner 2 flyby was the first (in 1962), the U.S.S.R. has flown most of the

TABLE 8.1   Properties of Earth, Venus and Mars

	Earth	Venus	Mars
Semimajor axis (AU)	1.00	0.72	1.52
Period (yr)	1.00	0.61	1.88
Mass (Earth = 1)	1.00	0.82	0.11
Diameter (km)	12756	12102	6790
Density (g/cm³)	5.5	5.3	3.9
Surface gravity (Earth = 1)	1.00	0.91	0.38
Escape velocity (km/s)	11.2	10.4	5.0
Rotation period	23.9 hr	−243 days	24.6 hr
Surface area (Earth = 1)	1.00	0.94	0.28
Atmospheric pressure (bar)	1.00	90	0.007

missions to our sister planet. The early Soviet Venera entry probes were crushed by the high pressure of the atmosphere before they could reach the surface, but in 1970 Venera 7 successfully landed and broadcast data from the surface for 23 min before succumbing to the high surface temperature. Additional Venera probes and landers followed, photographing the surface and analyzing the atmosphere and soil. In 1985, two instrumented balloons were deployed into the planet's atmosphere by the Soviet VEGA flyby missions, which then continued to intercept Comet Halley.

Recently, the primary objective of spacecraft missions to Venus has turned from the atmosphere to the surface. Since thick clouds always obscure the surface from view, radar must be used to penetrate the clouds and image the underlying topography. The first crude global radar map was made by the U.S. Pioneer 12 orbiter in the late 1970s, followed by the Soviet Venera 15 and 16 radar orbiters in the early 1980s. However, most of our information on the geology of Venus is derived from the U.S. Magellan spacecraft, which went into orbit about Venus in 1990 (Figure 8.4). With its powerful imaging radar, Magellan was able to study the surface at a resolution of 100 m, much higher than that of previous missions, yielding our first detailed look at the surface of our sister planet.

FIGURE 8.4 The Magellan radar orbiter spacecraft being launched from the Space Shuttle in 1989. *(NASA)*

Probing Through the Clouds

The radar maps of Venus reveal a planet that looks much the way the Earth might look if our planet's surface were not constantly being modified by erosion and deposition of sediment. Since there is no water or ice on Venus and the surface wind speeds are very low, almost nothing obscures or erases the complex geological features produced by widespread crustal forces and volcanic eruptions. Having finally penetrated below the clouds of Venus, we find its surface to be naked, revealing the history of hundreds of millions of years of geological activity. Venus is a geologist's dream planet.

About 75 percent of the surface of Venus consists of lowland lava plains. Superficially, these resemble the basaltic ocean basins of the Earth, but they do not have the same origin. There is no evidence of subduction zones on Venus, indicating that this planet never experienced plate tectonics. Although internal mantle convection has generated great stresses in the crust of Venus, it was never able to initiate plate motion. The formation of the lava plains of Venus resembled that of the lunar maria, which were also the result of widespread eruptions without the crustal spreading associated with plate tectonics. The venerian lava plains are much younger than the lunar plains, however, and

they have been distorted by stresses in the crust, as we shall describe below. In contrast, the lunar plains have languished almost unchanged since their formation more than 3 billion years ago.

Rising above the lowland lava plains are individual mountains and mountain ranges, as well as two full-scale continents. The largest continental area on Venus, called Aphrodite, is about the size of Africa. Aphrodite stretches along the equator for about one third of the way around the planet. Next in size is the northern highland region called Ishtar, which is about the size of Australia. Ishtar contains the highest region on the planet, the Maxwell Mountains, which rise about 11 km above the surrounding lowlands.

Craters and the Age of the Surface

One of the first questions to be addressed with the high-resolution Magellan images is the age of the surface of Venus. A young age implies an active geology. As described in the previous chapter, the age can be derived from counting impact craters (Figure 8.5a); the more densely cratered the surface, the greater its age. The largest crater on Venus (called Mead) is 275 km in diameter, slightly larger than the largest known terrestrial crater but much smaller than the lunar impact basins.

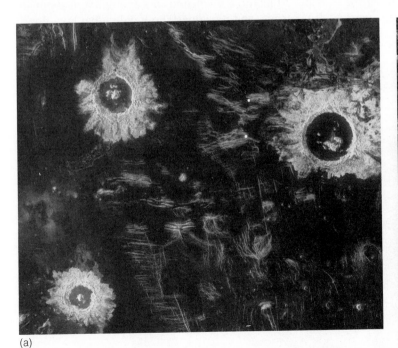
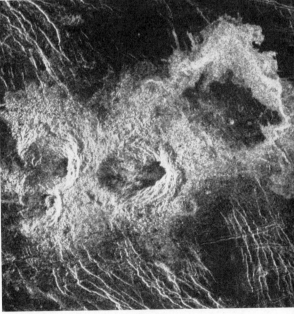

(a)   (b)

FIGURE 8.5   (a) Large impact craters in the Lavinia region of Venus. In this Magellan radar image, the rough crater rims and ejecta are excellent radar reflectors and hence appear brighter than the surrounding smooth volcanic plains. The largest of these craters has a diameter of 50 km. (b) Crater Stein, a triple-impact that resulted when the incoming asteroid broke apart in the atmosphere of Venus. The projectile had an initial diameter of between 1 and 2 km. Most projectiles smaller than about 1 km do not survive intact their passage through the thick atmosphere of Venus. *(NASA/JPL)*

You might think that the thick atmosphere of Venus would protect the surface from impacts, but this is not true for large projectiles. The effect of the atmosphere is readily seen in the crater statistics, which show very few craters less than 10 km in diameter. In addition, those craters with diameters from 10 to 30 km are frequently distorted, apparently because the incoming projectile broke apart and exploded in the atmosphere before it could strike the ground. There are also examples of multiple craters resulting when the projectile broke into several pieces before striking the surface (Figure 8.5b). But for impacts that would produce craters with diameters of 30 km or greater, crater counts are as useful here as on airless bodies like the Moon.

The numbers of craters on the plains of Venus are typically only about 15 percent of the lunar mare values, indicating a surface age only about 15 percent as great, or about 500 to 600 million years. These results indicate that Venus is a planet with persistent geological activity, intermediate between that of the Earth ocean basins (which are younger and more active) and that of its continents (which are older and less active).

Almost all of the craters look fresh, with little degradation or filling in by either lava or windblown dust. This is one way we know that the rates of erosion or deposition of sediment are very low. We have the impression that most of the venerian plains were resurfaced by large-scale volcanic activity a few hundred million years ago, and that relatively little has happened since. We see a similar situation in the lunar maria, which were all formed within a relatively short interval of time (geologically speaking), but on Venus this period of widespread volcanic activity was much more recent. It is not possible to determine if there was a global surge in volcanism; more likely, the lowland flooding by lava took place at different times in different parts of the planet.

Volcanoes on Venus

Like the Earth, Venus is a planet that has experienced widespread volcanism. In the lowland plains, volcanic eruptions are the principal way the surface is renewed, with large flows of highly fluid lava destroying old craters and generating a fresh surface every few hundred million years. In addition, there are numerous volcanic mountains and other structures associated with surface hot spots—places where convection in the planet's mantle transports the interior heat to the surface.

One of the largest individual volcanoes, called Sif Mons, is about 500 km across and 3 km high—broader but lower than the Hawaiian volcano Mauna Loa. At its top is a volcanic crater or caldera about 40 km across, and its slopes show individual lava flows

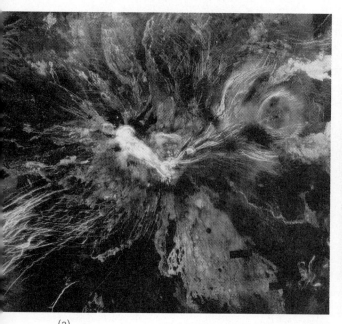

(a)

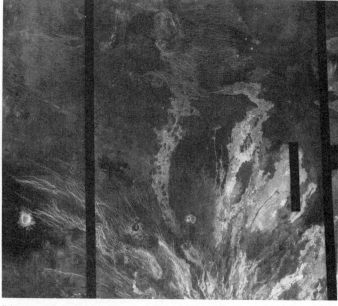

(b)

FIGURE 8.6 Large volcanic complex on Venus. (a) Two broad volcanic mountains, Sif and Gula, each rise about 4 km above the surrounding plains. (b) Complex lava flows extend down the flanks of Sif for hundreds of kilometers. Smooth flows are dark and rough flows are light in this Magellan radar image.

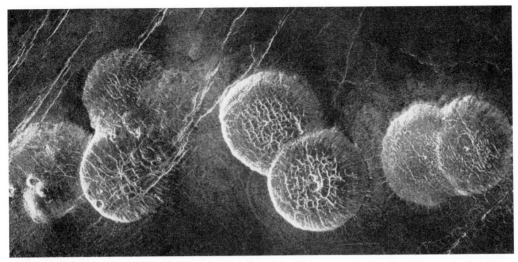

FIGURE 8.7 Pancake-shaped volcanoes on Venus. These remarkable circular domes, each about 25 km across, are the result of eruptions of highly viscous lava. *(NASA/JPL)*

up to 500 km in length (Figure 8.6). Thousands of smaller volcanoes dot the surface, down to the limit of visibility of the Magellan images, which corresponds to cones or domes about the size of a shopping mall parking lot. Most of these seem to be similar to terrestrial volcanoes.

Much more striking are circular, flat-topped volcanoes informally known as "pancake domes" (Figure 8.7). The larger examples are about 25 km in diameter and 2 km high, with steep ramparts around their rims. These volcanoes, which have no terrestrial counterpart, appear to be the product of very thick, viscous lava.

All of the volcanism described above is the result of eruption of lava onto the surface of the planet. But the hot magma rising from the interior of a planet does not always make it to the surface. On both the Earth and Venus, this magma can collect to produce bulges in the crust. Many of the granite mountain ranges on Earth, such as the Sierra Nevada in California, involve subsurface collections of magma. Such features are common on Venus, and in the absence of plate tectonics or surface erosion, they are much more visible than on the Earth. Their characteristic visible expression is a large circular or oval feature called a *corona* (Figure 8.8). Coronae are typically several

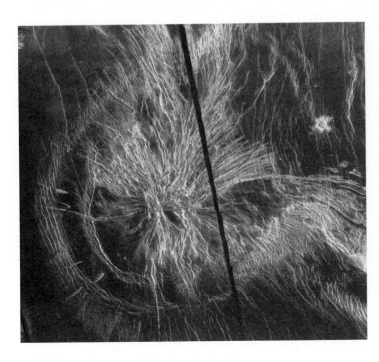

FIGURE 8.8 Pandora Corona, a circular feature 350 km in diameter produced when magma pressure built up below the surface of Venus. This Magellan radar image illustrates the extensive radial and concentric cracks that resulted from this subsurface pressure. The black streak is due to missing radar data. *(NASA/JPL)*

hundred kilometers across, with a slightly raised interior surrounded by a depressed ring or moat. They are a unique feature of venerian geology, not seen on any other planet or satellite in the solar system.

Tectonic Activity

The mantle convection currents on Venus do more than bring magma to the surface. As on the Earth, these convection currents exert forces on the crust. Although full-scale plate motion has not occurred on Venus, its crust is constantly subjected to pushing and stretching. These forces are called **tectonic,** and the geological features that result from such forces are called tectonic features. The geology of Venus is dominated by stationary tectonics associated with mantle convection, and this unique geology is sometimes called ''blob tectonics'' to distinguish it from terrestrial plate tectonics.

On the lowland plains, tectonic forces have broken the lava surface to create remarkable patterns of ridges and cracks (Figure 8.9). In a few places, the crust has been torn apart to generate great rift valleys. The circular features associated with coronae are tectonic ridges and cracks, and most of the mountains of Venus also owe their existence to tectonic forces.

The Ishtar continent, which has the highest elevations on Venus, is the product of tectonic forces. In many ways, Ishtar and its high Maxwell Mountains resemble the Tibetan Plateau and Himalayan Mountains on the Earth. Both are the product of compres-

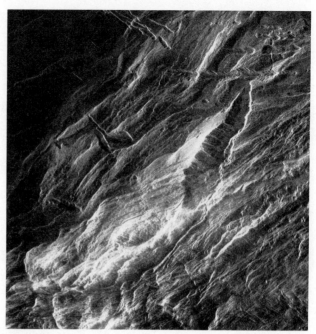

FIGURE 8.10 Folded mountains with individual peaks rising about 2 km high. This region of the Danu Mountains in the Ishtar Continent may be similar in origin to the Himalayan Mountains on the Earth. *(NASA/JPL)*

sion of the crust, and both are maintained by the continuing forces of mantle convection. Figure 8.10 illustrates the steep flanks of the Ishtar plateau, where folded mountains rise from the lowland plains to the south. These features look much more like the mountains of the Earth than anything seen on the Moon, Mars, or other planets—except, of course, for the lack of erosion and the absence of overlying soil or vegetation.

On the Surface

The successful Soviet Venera landers have found themselves on an extraordinarily inhospitable planet, with a surface pressure of 90 bars and a temperature of 730 K. The surface of Venus is hot enough to melt lead and zinc.

Despite these unpleasant conditions, the Soviet spacecraft have photographed their surroundings and collected surface samples for chemical analysis. The rock in the landing areas is igneous, primarily basalts. Examples of the Venera photographs are shown in Figure 8.11. Each picture shows a desolate, flat landscape with a variety of rocks, some of which may be ejecta from impacts. Other areas show flat, layered lava flows. Unfortunately, there are no chemical studies of sites that have been photographed and no photographs of sites studied chemically, so direct comparisons are not possible.

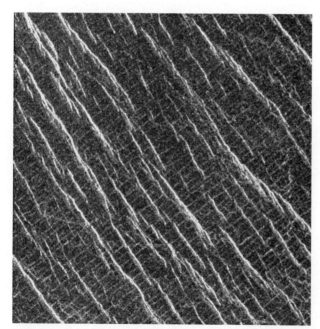

FIGURE 8.9 Tectonic features on Venus. This region of the Lakshmi Plains has been fractured to produce a grid of cracks and ridges. This width of the radar image is 40 km. *(NASA/JPL)*

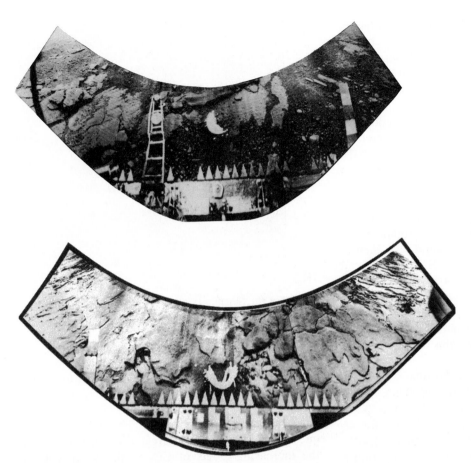

FIGURE 8.11 Views of the surface of Venus from two of the Venera landers. *(U.S.S.R. Academy of Science and Brown University, courtesy of James Head)*

The Sun cannot shine directly through the heavy, opaque clouds, but the surface is fairly well lit by diffused light. The illumination is about the same as that on Earth under a very heavy overcast, but with a strong red tint, since the massive atmosphere blocks blue light. The weather is unchanging: hot and dry with calm winds. Because of the heavy blanket of clouds and atmosphere, one spot on the surface of Venus is similar to any other as far as weather is concerned. The explanation for the uniform conditions and blistering temperatures is to be found in the atmosphere.

8.3   The Massive Atmosphere of Venus

Composition and Structure

The most abundant gas in the atmosphere of Venus is carbon dioxide (CO_2), which accounts for 96 percent of the atmosphere. The second most abundant gas is nitrogen (N_2). The predominance of CO_2 over N_2 is not surprising when you remember (Section 3.3) that the Earth's atmosphere would also be mostly CO_2 if the CO_2 were not locked up in marine sediments.

Table 8.2 compares the compositions of the atmospheres of Venus, Mars, and the Earth. Expressed in this way, as percentages, the proportions of major gases are very similar for Venus and Mars, but in other respects their atmospheres are dramatically different. With its surface pressure of 90 bars, the venerian atmosphere is more than 10,000 times more massive than its martian counterpart.

In addition to these gases, there are measurements of sulfur dioxide (SO_2) in the middle atmosphere of Venus. The atmosphere is very dry; the absence of

TABLE 8.2   Atmospheric Compositions of Earth, Venus, and Mars (in Percent)

Gas	Earth	Venus	Mars
Carbon dioxide (CO_2)	0.03	96.5	95.3
Nitrogen (N_2)	78.1	3.5	2.7
Argon (Ar)	0.93	0.006	1.6
Oxygen (O_2)	21.0	0.003	0.15
Neon (Ne)	0.002	0.001	0.0003

water (H₂O) on Venus is one of the important characteristics that distinguish Venus from Earth.

The venerian atmosphere (Figure 8.12) has a huge troposphere that extends up to at least 50 km above the surface. Within the troposphere, the gas is heated from below and circulates slowly, rising near the equator and descending over the poles. With no rapid rotation to break up this flow, the atmospheric circulation is highly stable. In addition, the very size of the atmosphere maintains stability. Being at the base of the atmosphere of Venus is something like being a kilometer below the ocean surface on the Earth. There, also, the mass of water evens out temperature variations and results in a uniform environment.

The thick clouds of Venus are composed primarily of sulfuric acid (H₂SO₄) droplets. H₂SO₄ is formed from the chemical combination of SO₂ and H₂O. In the atmosphere of the Earth, SO₂ is one of the primary gases emitted by volcanoes, but it is quickly diluted and washed out of the atmosphere by rainfall. In the dry atmosphere of Venus, this unpleasant substance is apparently stable.

The clouds lie in the upper troposphere, between 30 and 60 km above the surface. Below 30 km, the air is clear. In the middle of the cloud layer, at the 53-km altitude where the VEGA balloons floated for 46 hours each in 1985, the conditions are almost Earth-like. The pressure here is 0.5 bar and the temperature a comfortable 305 K, just a little warmer than the room in which you are reading this book. Were it not for the absence of oxygen and the nasty sulfuric acid clouds, this would not be a bad place to visit. Certainly it beats conditions on the *surface* of our "sister" planet.

Surface Temperature

The high surface temperature of Venus was discovered by radio astronomers in the late 1950s and confirmed by Mariner 2 observations and by the early Venera probes. It was not easy to understand, however, how this planet could be so much hotter than would be calculated from solar heating. A major question therefore arose: What is heating the surface of Venus to a temperature above 700 K? The answer is to be found in the *greenhouse effect*.

The greenhouse effect works on Venus just as it does for the Earth (Section 4.3). But since Venus has so much more CO₂—almost a million times more—the effect is much stronger. Sunlight that diffuses through the atmosphere of Venus heats the surface, but the CO₂ acts as a blanket, making it very difficult for the infrared radiation to leak back to space. In consequence, the surface heats up until eventually it is emitting so much heat that an energy balance is reached with incoming sunlight.

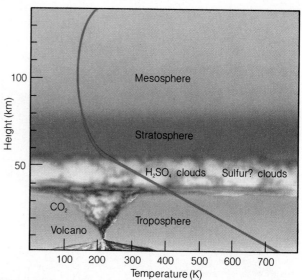

FIGURE 8.12 Structure of the massive atmosphere of Venus, based on data from the Pioneer and Venera entry probes.

Has Venus always had such a massive atmosphere and high surface temperature, or might it have evolved to such conditions from a climate that was once more nearly Earth-like? The answer to this question is of particular interest to us when we look at the increasing levels of CO₂ in the Earth's atmosphere, as discussed in Section 4.4. Are we in any danger of transforming our own planet into a hellish place like Venus?

Let us try to reconstruct the possible evolution of Venus from an Earth-like beginning to its present state. Imagine that it began with moderate temperatures, with water oceans, and with much of its CO₂ dissolved in the ocean or chemically combined with the surface rocks, as is the case on Earth today. In our thought experiment, we then allow for modest additional heating—for example, by a small rise in the energy output of the Sun or by an increase in atmospheric CO₂. One consequence is a further increase in atmospheric CO₂ and H₂O, as a result of increased evaporation from the oceans and release of gas from surface rocks. These two gases would in turn produce a stronger greenhouse effect, further raising the temperature and leading to still more CO₂ and H₂O in the atmosphere. Unless some other processes intervene, the temperature will continue to rise. Such a situation is called the **runaway greenhouse effect.**

The runaway greenhouse is not just a larger greenhouse effect; it is a process whereby an atmosphere evolves from a state in which the greenhouse effect is small, such as on the Earth, to one with a much larger effect, such as we see today on Venus. Once the larger greenhouse conditions develop, the planet es-

ESSAY   How a Greenhouse Works

The warming of a planet that occurs when atmospheric gases trap infrared radiation is called the greenhouse effect. But do greenhouses actually work by trapping infrared radiation? You may be surprised to learn that scientists have actively debated this issue. One school of thought says that greenhouses do indeed work because glass is transparent to solar radiation but opaque to infrared radiation. The solar radiation heats the ground, which reradiates in the infrared. The glass then absorbs this infrared radiation, warms up and thereby keeps the air and soil in the greenhouse warmer than they would otherwise be.

The other school of thought says that greenhouses work because they provide shelter from the wind. If wind could blow freely across soil warmed by the Sun, it would carry the heat away efficiently. Some meterologists feel so strongly that this idea is right that they argue that the greenhouse effect should actually be called the *atmospheric effect*.

In fact, greenhouses work for both of these reasons, and which one is more important depends on the greenhouse and its environment. If you build a thin-walled greenhouse in a very windy site, then it will work primarily because it provides protection from the wind. After all, you know that you seek shelter to keep warm during a blizzard to keep out the wind—not because the shelter traps infrared radiation. On the other hand, in a place where the winds are calm, a greenhouse will trap infrared radiation and also be warmer than its surroundings.

You can judge the heating ability of the greenhouse effect yourself if you have a car with a trunk. Let the car sit in the Sun all day with the windows rolled up. Then compare the temperature inside with the temperature in the trunk. The inside and the trunk were equally exposed to the Sun and protected from the wind. Any difference in temperature must be due to the trapping of infrared radiation by the glass windows.

---

tablishes equilibrium, and reversing the situation is difficult, if not impossible.

If large bodies of water are available, the runaway greenhouse leads to their evaporation, creating an atmosphere of hot water vapor, which itself is a major contributor to the greenhouse effect. Water vapor is not stable in the presence of solar ultraviolet light, however, which tends to dissociate, or break apart, the molecules of H_2O into their constituent parts, oxygen and hydrogen. As we have seen, hydrogen can escape from the atmospheres of the terrestrial planets, leaving the oxygen to combine chemically with surface rock. The loss of H_2O is therefore an *irreversible* process; once the H_2O is gone, it cannot be restored. There is evidence that this is exactly what happened to the H_2O once present on Venus.

8.4 Geology of Mars

Spacecraft Exploration of Mars

The U.S. Mariner 4 spacecraft flew by Mars in 1965 and radioed 22 photographs to Earth. These pictures showed an apparently bleak planet with abundant impact craters. In those days craters were unexpected,

and perhaps people who were romantically inclined still expected to see canals; in any case, the Mariner 4 results represented something of a shock to the public. In 1971, Mariner 9 became the first spacecraft to orbit another planet, mapping the entire surface of Mars at a resolution of about 1 km and discovering a great variety of geological features missed by the previous flybys, including volcanoes, huge canyons, intricate layers on the polar caps, and channels that appeared to have been cut by running water.

Mariner 9 set the stage for the Viking landers, one of the most ambitious and successful of all U.S. planetary missions. Two Viking orbiters surveyed the planet and served to relay communications for the two landers. Viking 1 touched down on the surface of Chryse Planitia (the Plains of Gold) on July 20, 1976, exactly seven years after Neil Armstrong's historic first step on the Moon. Two months later, Viking 2 landed with equal success in another plain farther north, called Utopia. Most of the information about Mars in this chapter is derived from the four Viking spacecraft.

The Viking mission continued until November 5, 1982, when Viking 1 sent its last message after more than six Earth years on the surface. Even then, the

FIGURE 8.13   Mars as photographed from the Viking spacecraft in 1976. The red color is due to a pervasive dust of clay containing iron oxides (rust). Darker areas have similar composition but different texture. *(NASA/USGS)*

mission was terminated by a human programming error on Earth, not by failure of the robot spacecraft on Mars. In 1981, NASA transferred ownership of the Viking 1 lander to the National Air and Space Museum. It is the only museum exhibit located on another planet!

For more than a decade after Viking, no new missions were launched toward Mars, but in the 1990s there is renewed interest in the exploration of the red

planet. In 1988 the Soviet Phobos 2 spacecraft orbited Mars, to be followed in 1992 by the U.S. Mars Observer orbiter. However, these are small missions compared with those on the drawing boards in both the United States and the U.S.S.R. Both nations have indicated plans for detailed robotic exploration of the planet as a prelude to human flights in the second or third decade of the next century. These robotic missions will include surface rovers and the return of samples of martian material to Earth for detailed analysis (Figure 8.14).

Global Properties

Mars has a diameter of 6790 km, just over half that of the Earth, resulting in a surface area very nearly equal to the continental area of our planet. Its density of 3.9 g/cm^3 suggests a composition consisting primarily of silicates but with the possibility of a substantial metal core. The planet has no detectable magnetic field, suggesting that its core is not liquid.

Like the surface of the Earth, Moon, and Venus, that of Mars divides into continental or highland terrain and lower volcanic basins or plains. Approximately half the planet, lying primarily in the southern hemisphere, consists of upland cratered terrain. The other half, which is mainly in the north, contains younger, lightly cratered volcanic plains at an average elevation about 4 km lower than the uplands. This kind of division into older uplands and younger lowland plains seems to be a characteristic of all of the terrestrial planets except Mercury. The uplands are shown in blue and the lowlands in yellow in Figure 8.15, which is a map of the major geological

FIGURE 8.14   Full-scale model of a Soviet automatic rover vehicle being tested in martian-analog terrain in the volcanic fields of Kamchatka Peninsula. *(U.S.S.R. Space Research Institute, courtesy of Lev Muhkin)*

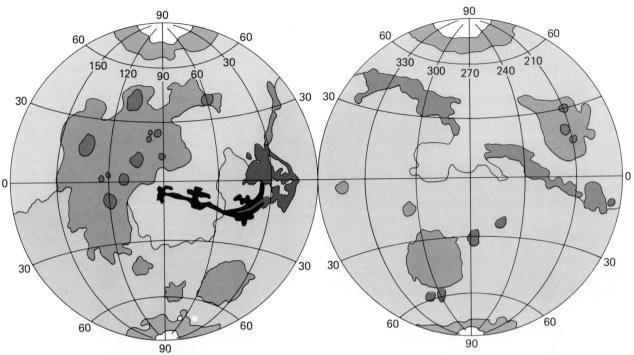

FIGURE 8.15 Global geological map of Mars. The color code is as follows: polar caps (white); polar deposits (light green); cratered uplands (light blue); impact basins (dark blue); lowland plains (yellow); volcanic terrain (orange); volcanic mountains (red); heavily eroded terrain (light brown); channels and chaotic terrain (dark brown); Valles Marineris (black).

areas of Mars, many of which are described in more detail below.

Mars has a few large impact basins in the old southern uplands, shown in dark blue in Figure 8.15. The largest basin, called Hellas, is about 1800 km in diameter and 6 km deep, larger than the largest basin on the Moon.

In addition to the division of the planet into old uplands and younger volcanic plains, Mars displays an impressive uplift or bulge the size of North America. This is the 10-km-high Tharsis bulge, a volcanically active region crowned by four great volcanoes that rise another 15 km into the martian sky (orange and red in Figure 8.15).

Volcanoes on Mars

The lowland volcanic plains of Mars look very much like the lunar maria, and they have about the same numbers of impact craters. Like the lunar maria, they probably formed between 3 and 4 billion years ago. Apparently, Mars experienced extensive volcanic activity at about the same time the Moon did, producing similar basaltic lavas.

The largest volcanic mountains of Mars are found in the Tharsis area, although many smaller volcanoes dot much of the surface of the younger northern half of the planet. The most dramatic is Olympus Mons

(Mount Olympus), which is more than 500 km in diameter, with a summit 25 km above the surrounding plains (Figure 8.16). The volume of this immense volcano is nearly 100 times greater than that of the largest terrestrial volcano, Mauna Loa in Hawaii.

The Viking imagery permits a detailed examination of the slopes of these volcanoes in the search for impact craters. Many of the volcanoes show fair numbers of such craters, suggesting that they ceased activity a billion years or more ago. However, Olympus Mons has very, very few impact craters. Its present surface cannot be more than about a hundred million years old, and it could even be much younger. Some of the fresh-looking lava flows we see might have been formed a hundred years ago, or a thousand, or a million, but geologically speaking they are young. It is probable that Olympus Mons remains intermittently active today.

The volcanoes of Mars look more like terrestrial counterparts than do the volcanoes of Venus. Apparently, martian volcanism has been characterized by the same sort of basaltic lava that is erupted on the Earth.

Cracks and Canyons

The Tharsis region consists of more than a collection of huge volcanoes. In this part of the planet, the

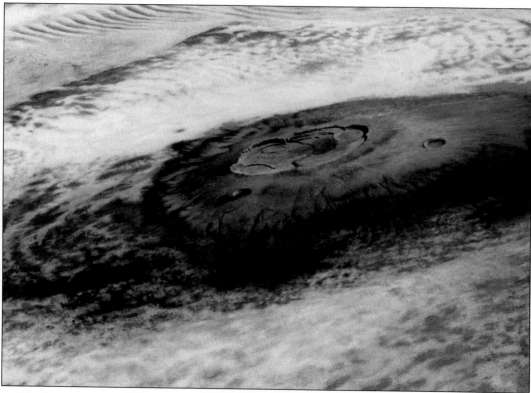

FIGURE 8.16   The largest volcano on Mars, and probably the largest in the solar system, is Olympus Mons, illustrated in this Viking orbiter photograph. Also note the extensive clouds over the lower slopes of the volcano. *(NASA/JPL)*

surface itself has bulged upward, forced by great pressures from below, resulting in extensive tectonic cracking of the crust. None of these cracks show any evidence of sliding motion, such as that associated with most faults on Earth. Like Venus, Mars never reached the stage of plate tectonics. Nor has it experienced the widespread "blob tectonics" of Venus.

Instead, there is just the single feature of Tharsis, forced upward but not induced to shift sideways. It is as if martian tectonic forces began to act but then subsided at least a billion years ago (Figure 8.17).

Among the most spectacular tectonic features on Mars are the great canyons called the Valles Marineris, which extend for about 5000 km (nearly a quar-

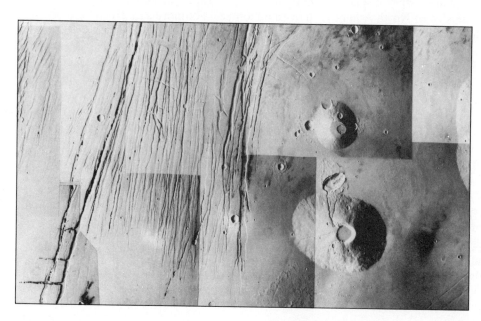

FIGURE 8.17   Tectonic features in the Tharsis region of Mars, produced by tension in the crust. Two large volcanoes are also visible. The width of this image is about 400 km. *(NASA/JPL)*

ter of the way around Mars) along the slopes of the Tharsis bulge (see photograph on page 144). The main canyon is about 7 km deep and up to 100 km wide. It is so large that the Grand Canyon of the Colorado River would fit comfortably into one of its side canyons.

The term canyon is somewhat misleading, because the Valles Marineris canyons were not cut by running water. They have no outlets. They are basically tectonic cracks, produced by the same crustal tensions that caused the Tharsis uplift. However, water is believed to have played a later role in shaping the canyons, primarily through undercutting of the cliffs by seepage from deep springs. This undercutting led to landslides, gradually widening the original cracks into the great valleys we see today.

Why is there less pervasive tectonic structure on Mars than on Venus? In part, this may reflect a lower general level of geological activity, as would be ex-

pected for a smaller planet. But it is also possible that evidence of widespread faulting has been hidden by wind-deposited sediment over much of Mars. Like the Earth, Mars may have hidden part of its geological history under a cloak of soil.

The View from the Surface

Viking 1 landed at a latitude of 22° N, on a 3-billion-year-old windswept plain near the lowest point of a broad basin. Its desolate but strangely beautiful surroundings included numerous angular rocks, some more than a meter across, interspersed with dune-like deposits of fine-grained soil. On the horizon, the low profiles of several distant impact craters could be seen (Figure 8.18a). At the Viking 2 site in Utopia, at latitude 48° N, the surface was somewhat similar, but with substantially greater numbers of rocks (Figure 8.18b). At both sites, it seems that winds, blowing

(a)

(b)

FIGURE 8.18 The martian surface as photographed by Viking. (a) The Viking 1 landing site in Chryse. (b) The Viking 2 landing site in Utopia. *(NASA/JPL)*

sometimes up to 100 km/h or more, have stripped the surface of loose, fine material to leave the rocks exposed.

Each lander peered at its surroundings through color stereo cameras, sniffed the atmosphere with a variety of analytical instruments, and poked at nearby rocks and soil with its mechanical arm. As part of its mission of searching for martian life, each lander collected soil samples and brought them on board for analysis, as described in more detail below. The soil was found to consist of clays and iron oxides, as had long been expected from the red color of the planet.

Each lander also carried a weather station to measure temperature, pressure, and wind. As expected, temperatures vary much more on Mars than on Earth, owing to the absence of moderating oceans and clouds. Typically, the summer maximum was 240 K ($-33°C$), dropping to 190 K ($-83°C$) at the same location just before dawn. The lowest air temperatures, measured farther north by Viking 2, were about 173 K ($-100°C$). During the winter, Viking 2 also photographed water frost deposits on the ground (Figure 8.19).

Most of the winds measured at the Viking sites were low to moderate, only a few kilometers per hour. However, Mars is capable of great windstorms, which can shroud the entire planet in dust. At such times the Sun was greatly dimmed at the Viking sites, and the sky turned a dark red color.

Martian Samples

Much of what we know of the Moon, including the circumstances of its origin, comes from studies of lunar samples, but spacecraft have not yet returned martian samples to Earth for laboratory analysis. It is with great interest, therefore, that scientists have recently concluded that samples of martian material are already available for study on Earth.

There are eight of these martian rocks, all members of a rare class of meteorites called *SNC meteorites* (Figure 8.20). The most obvious special characteristic of these meteorites is that they are volcanic *basalts*, and they are relatively *young*, about 1.3 billion years. We know from details of their composition that they are not from the Moon, and in any case there was no lunar volcanic activity as recently as 1.3 billion years ago. Theory suggests that it would be impossible for ejecta from impacts to escape from Venus, with its thick atmosphere. By process of elimination, the only reasonable origin seems to be Mars, where the Tharsis volcanoes were certainly active at that time.

The martian origin of the SNC meteorites has been confirmed by the analysis of tiny bubbles of gas trapped in several of these meteorites; the features of the trapped gas were found to match the atmospheric properties of Mars as measured directly by Viking.

FIGURE 8.19 Surface frost photographed at the Viking 2 landing site during late winter. *(NASA/JPL)*

FIGURE 8.20 One of the SNC meteorites, believed to be fragments of basalt ejected from Mars. *(NASA/JSC)*

Apparently the atmospheric gas was trapped in the rock by the shock of the impact that ejected it from Mars and started it on its way toward Earth. As for helping us understand Mars, the work on these meteorites has just begun, and we do not know how much they will tell us. This is an intriguing chapter in martian science that is only beginning to be written.

8.5 Martian Polar Caps, Atmosphere, and Climate

The Martian Atmosphere

The atmosphere of Mars has an average surface pressure of only 0.007 bar, less than 1 percent that of the Earth. It is composed primarily of CO_2 (95 percent), with about 3 percent N_2 and 2 percent argon (Ar). The proportions of different gases are similar to those in the atmosphere of Venus (Table 8.2).

Several types of clouds can form in the martian atmosphere. First, there are dust clouds, raised by winds, which can sometimes grow to cover a large fraction of the surface. Second are water ice clouds similar to those on Earth. These often form around mountains, just as happens on our planet (see Figure 8.16). Finally, the CO_2 of the atmosphere can itself condense at high altitudes to form hazes of dry ice crystals. The CO_2 clouds have no counterpart on Earth, since on our planet temperatures never drop low enough (about 150 K) for this gas to condense.

Although there is water vapor in the atmosphere and occasional clouds of water ice can form, *liquid* H_2O is not stable under present conditions on Mars. Part of the problem is the low temperatures on the planet. But even if the temperature on a sunny summer day rises above the freezing point, liquid H_2O cannot exist. At pressures of less than 0.006 bar, only the solid and vapor forms are possible. In effect, the boiling point is as low or lower than the freezing point, and water changes directly from solid to vapor without an intermediate liquid state.

The Polar Caps

Through a telescope the most prominent surface features on Mars are the bright polar caps, which change with the seasons. These seasonal caps are similar to the seasonal snow cover on Earth. We do not usually think of the winter snow as a part of our polar caps. But seen from space, the thin snow would blend with the thick permanent ice caps to create a situation much like that seen on Mars.

The seasonal caps on Mars are composed not of ordinary snow, but of frozen CO_2 (dry ice). These deposits condense directly from the atmosphere

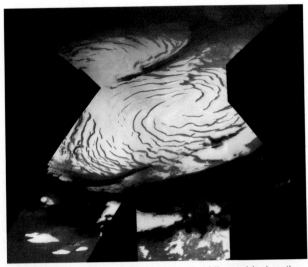

FIGURE 8.21 Viking orbiter photograph of the residual north polar cap of Mars, which is about 1000 km across and is composed of water ice. *(NASA/USGS)*

when the surface temperature drops below about 150 K. The caps develop during the cold martian winters, extending down to about latitude 50° by the start of spring.

Quite distinct from these thin seasonal caps of CO_2 are the permanent, or residual, caps that are always present near the poles (Figure 8.21). As the seasonal cap retreats during spring and early summer, it reveals a brighter, thicker cap beneath. The southern permanent cap has a diameter of 350 km and is composed of deposits of frozen CO_2 together with an unknown thickness of water ice. Throughout the southern summer, it remains at the freezing point of CO_2, 150 K, and this cold reservoir is thick enough to survive the summer heat intact.

The northern permanent cap is different. It is much larger, never shrinking below a diameter of 1000 km, and it is composed of ordinary H_2O. Summer temperatures are too high for the frozen CO_2 to be retained. We do not know the thickness of the water ice in either cap, but it may be as much as several kilometers. In any case, the polar caps represent a huge reservoir of H_2O, in comparison with the very small amounts of water vapor in the atmosphere. The two caps are different because the seasons are complicated by the substantial variation in distance from the Sun that Mars experiences during its year.

The terrain surrounding the permanent polar caps and seen in ice-free areas within the caps is remarkable (Figure 8.22). At latitudes above 80° in both hemispheres, the surface consists of recent, layered sedimentary deposits that entirely cover the older cratered ground below. Individual layers are typically

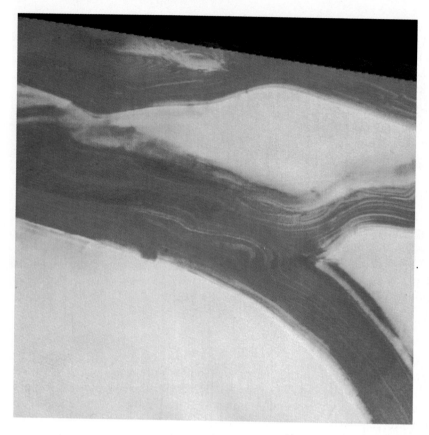

FIGURE 8.22 Detail of the southern polar cap showing terracing on exposed slopes. These terraces preserve a record of past variations in climate. The width of this frame is about 50 km. *(NASA/JPL)*

a few tens of meters in thickness, marked by alternating light and dark bands of sediment. Probably the material in the polar deposits is dust carried by wind from equatorial regions.

The time scales represented by the polar layers are tens of thousands of years. Apparently the martian climate experiences periodic changes with this frequency, which is similar to the intervals between ice ages on the Earth. Calculations indicate that the causes are probably also similar: variations in the orbit and tilt of the planet induced by gravitational perturbations from other planets.

Channels and Floods

Although no liquid H_2O is present today on Mars, there is fascinating evidence that rain once fell and rivers flowed. Two kinds of geological features appear to be the remnants of ancient water courses: the runoff channels of the old uplands and the larger outflow channels that lead from the uplands down to the great northern basins.

In the upland equatorial plains are multitudes of small, sinuous (twisting) channels typically a few meters deep, some tens of meters wide, and perhaps 10 or 20 km long (Figure 8.23). They are called *runoff*

channels because they appear to have carried the surface runoff of ancient rainstorms. These runoff channels seem to be telling us that the planet had a very different climate long ago. How can we tell when

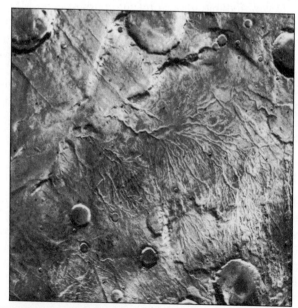

FIGURE 8.23 Runoff channels in the old martian highlands, interpreted as the valleys of ancient rain-fed rivers. The width of this image is about 250 km. *(NASA/JPL)*

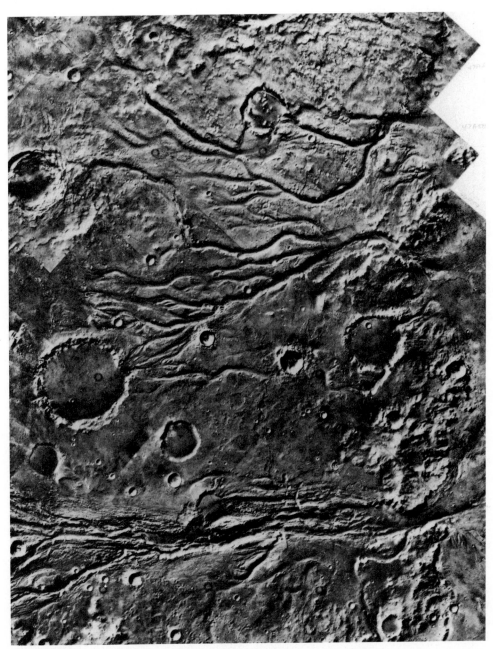

FIGURE 8.24 Large outflow channels photographed by Viking. These features appear to have been formed in the distant past from massive floods of water. The width of this image is about 400 km. *(NASA/JPL)*

rain might have last fallen on Mars? Crater counts show that this part of Mars is more cratered than the lunar maria but less so than the lunar highlands. Thus the runoff channels are older than the maria, presumably at least 3.9 billion years.

The *outflow channels* (Figure 8.24) are much larger than the older runoff channels. The largest of these, which drain into the Chryse basin where Viking 1 landed, are 10 km or more in width and hundreds of kilometers long. Many features of these outflow chan-nels have convinced geologists that they were carved by huge volumes of running water, far too great to be produced by ordinary rainfall. Where did this flood-water come from? As far as we can tell, the source regions contained abundant water frozen in the soil as permafrost. Some local source of heating must have released this water, leading to catastrophic flooding. Perhaps this heating was associated with the forma-tion of the volcanic plains, which are con-temporaneous with the channels.

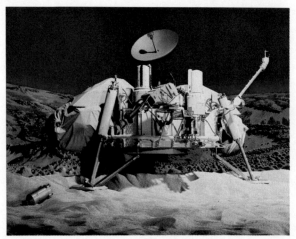

FIGURE 8.25 Engineering model of the Viking lander in a Mars simulation laboratory. The boom picks up soil and rocks for analysis. The two cylinders at the top of the lander are survey cameras. Below the right-hand camera is one of three rocket engines used during the final soft landing on Mars. *(NASA/JPL)*

Climate Change

The evidence cited above suggests that the climate of Mars has varied on at least two different time scales. Billions of years ago, temperatures were warmer, rain fell, and the atmosphere must have been much more substantial than it is today. And much more recently, there has been a cyclical variation, as recorded in the layered deposits of the martian polar regions. It is the long-term changes that are most interesting, since they are closely related to the possibility of life on Mars in the distant past.

The long-term cooling of Mars and loss of its atmosphere are a result of its small size (and low escape velocity) relative to the Earth and Venus. Presumably, Mars formed with a much thicker atmosphere, and the atmosphere maintained a higher surface temperature owing to the greenhouse effect. Escape of the atmosphere to space, however, gradually lowered the temperature. Eventually it became so cold that the water froze out of the atmosphere, further reducing its ability to retain heat—a sort of runaway refrigerator effect, just the opposite of the runaway greenhouse effect that occurred on Venus. The result is the cold, dry Mars we see today. Probably this loss of atmosphere took place within a few hundred million years; from the absence of runoff channels in the northern plains, it seems that rain has not fallen on Mars for at least 3 billion years.

The Search for Life on Mars

If there was running water on Mars in the past, perhaps there was life as well; perhaps life, in some form,

remains in the martian soil. Testing this possibility, however unlikely, was one of the primary objectives of the Viking landers (Figure 8.25).

The Viking landers, in addition to the instruments we have already discussed, carried miniature biological laboratories to test for microorganisms in the martian soil. They looked for evidence of *respiration* by living animals, *absorption of nutrients* offered to organisms that might be present, and *exchange of gases* between the soil and its surrounding for any reason whatsoever. In various tests the martian soil was scooped up by the spacecraft's long arm and provided to the experimental chambers, where it was isolated and incubated in contact with a variety of gases, radioactive isotopes, and nutrients to see what would happen. A fourth instrument pulverized the soil and analyzed it carefully to see what organic (carbon-bearing) material it contained.

The Viking experiments were sensitive enough that had one of the spacecraft landed anywhere on Earth, with the possible exception of Antarctica, it would easily have detected life. Those experiments that tested martian soil for absorption of nutrients and gas exchange did show activity, but this could have been caused by inorganic chemical reactions. In fact, these experiments showed that the martian soil seems to be much more chemically active than terrestrial soils, as a result of its exposure to solar ultraviolet radiation. The organic chemistry experiment showed no trace whatever of organic material, which is apparently destroyed by the sterilizing effect of this ultraviolet light. While the possibility of martian life has not been eliminated, most experts consider the chance of life on that planet to be negligible. Although Mars has the most Earth-like environment of any planet, nobody seems to be home.

8.6 Divergent Planetary Evolution

Venus, Mars, and our own planet Earth form a remarkably diverse triad of worlds. Although all three orbit at about the same distance from the Sun, and all apparently started with about the same chemical mix of silicates and metals, they have diverged in their evolutionary paths. As a result, Venus became hot and dry, Mars became cold and dry, and only the Earth ended with what we would consider a hospitable climate.

We have discussed the runaway greenhouse effect on Venus and the runaway refrigerator effect on Mars. But we do not understand exactly what started these two planets down these particular evolutionary paths. Was the Earth ever in danger of a similar fate? Or might it still be diverted into one of these paths,

ESSAY Why Explore Mars?

The following essay is taken from a 1991 NASA document discussing robotic science missions to Mars, but the arguments apply also to possible future human exploration of the Red Planet. (Compliments of Scott Hubbard, NASA Ames Research Center, and Steven Squyres, Cornell University)

Of all the nine planets and their dozens of moons, Mars still stands out as the one other potentially habitable body in the solar system. It is the only other planet that we can hope one day to understand as well as our own, and perhaps to live on as on our own. As a result of the highly visible and successful Viking missions in the late 1970s, the general nature of Mars has become familiar to most people. The Red Planet is intermediate in size between Earth and our Moon. Part of its surface resembles the Moon and shows massive impact basins, cratered highland regions, and extensive flooding by lavas. Much more exciting are the regions that resemble Earth: mountains, volcanoes, dried-up river beds, desert sand dunes, an atmosphere, variable cloud patterns, and seasonal polar caps. Obviously, Mars has evolved to an advanced stage, and the path that it has taken in its evolution has been both similar to yet intriguingly different from that taken by the Earth.

The fact that Mars has a measurable atmosphere (mostly carbon dioxide, with small amounts of nitrogen, argon, and water vapor) and a planetary surface whose temperature may locally rise above the freezing point of water inevitably raises the question of whether Mars could have developed indigenous life. In 1972, the Mariner 9 images provided the first evidence that the martian surface had been extensively eroded by flowing water. This discovery suggested that Mars might have been even warmer—and its atmosphere thicker—in the past. Under those gentler conditions, could life have developed?

Four years later, two Viking landers, thousands of kilometers apart, sought the answer to this question by collecting and analyzing the martian soil itself. The results of these investigations were either negative or ambiguous. No evidence was found for the presence of martian life. Moreover, it was determined that virtually no organic material was present in the martian soil analyzed. These results make it hard to sustain the notion that life exists on Mars at the present, even though the basic ingredients (water, carbon compounds, energy) are there. However, the prospects for life earlier in martian history may be much better. We know from the Mariner and Viking Orbiter results that early Mars and early Earth were more similar than present Mars and present Earth, perhaps significantly so. On Earth, life had evolved to a fair degree of sophistication within 0.8 billion years after the planet's formation. The period of time required for its origin may have been significantly less than this, but the absence of a suitable fossil record precludes that determination. On Mars, however, there is a much better record of conditions in the first billion years preserved in the geologic record—record waiting to be read.

Whether or not Mars has ever been an oasis for life, the nature of the planet—and its direct relevance to understanding Earth and other planets—makes it a compelling target for in-depth exploration. A range of geologic processes has operated on Mars to produce landscapes that are alien and yet familiar: individual shield volcanoes that would stretch from Boston to Washington, DC; a canyon that would extend from New York to Los Angeles; and seas of sand dunes that girdle the entire north polar region. Mars is thus a natural laboratory for studying various geologic forces that shape a planet.

Mars is also a natural laboratory for the study of planetary weather and climate. The diurnal and seasonal cycles on Mars are remarkably similar to those of Earth, but the extremely thin atmosphere, the rapid heating and cooling of the surface, the lack of oceans, and the enormous vertical scale of the landscape also create critical differences. The condensation cycle of atmospheric carbon dioxide and the exaggerated role of dust in heating the martian atmosphere also provide fascinating contrasts to Earth. Despite these differences, martian meteorology is exciting, complicated, and capable of being understood. A better understanding of the weather of Mars could teach us much about the weather of our own world as well.

Martian climatology, which is in essence the long-term history of the planet's weather, is a science still in its infancy, but it holds great promise to illuminate not only the history of Mars, but also that of Earth. There is ample evidence of massive climate changes on both planets—ice ages, changing shorelines, and species extinctions on Earth; regional flooding, glaciation, and periodic polar sedimentary layering on Mars. Common mechanisms could have been at work on both planets—solar luminosity changes, periodic orbital variations, episodes of volcanic eruption, and asteroidal impacts. With comparative study of both Mars and Earth, we may make important progress toward understanding matters of both intellectual and practical importance.

perhaps as a result of stress on the atmosphere generated by human pollutants? One of the reasons for studying Venus and Mars is to try to gain insight into these questions.

Some people have even suggested that if we understood the evolution of Mars and Venus better, there is a possibility that we could reverse their evolution and restore a more Earth-like environment. This process, which is highly speculative, is called *terraforming*.

While it seems unlikely that humans could ever make either Mars or Venus into a replica of the Earth, considering such possibilities is a useful part of our more general quest to understand the delicate environmental balance that distinguishes our planet from its two neighbors.

In Chapter 12 we will return to the comparative study of the terrestrial planets and their divergent evolutionary histories.

SUMMARY

8.1 Venus, the nearest planet, is a great disappointment as seen through the telescope, because of its impenetrable cloud cover. Mars is more tantalizing, with dark markings and polar caps. Early in the 20th century, it was widely believed that the "canals" of Mars indicated intelligent life there. Mars has only 11 percent the mass of the Earth, but Venus is nearly our twin in size and mass. Mars rotates in 24 h and has seasons like the Earth; Venus has a retrograde rotation period of 243 days. Both planets have been extensively explored by spacecraft.

8.2 Venus has been mapped by radar, from Earth and from orbit. Its crust consists of lowland lava plains (75 percent); numerous volcanic features, including strange pancake domes; and many large coronae, which are the expression of subsurface volcanism. The planet has been modified by widespread **tectonics** driven by mantle convection, forming complex patterns of ridges and cracks and building high continental regions such as Ishtar. The surface is extraordinarily inhospitable, with a pressure of 90 bars and a temperature of 730 K, but several U.S.S.R. Venera landers have investigated it successfully.

8.3 The atmosphere of Venus is primarily CO_2 (97 percent). Thick clouds at altitudes of 30 to 60 km are made of H_2SO_4, and a CO_2 greenhouse maintains the high surface temperature. Venus presumably reached its current state from initially more Earth-like conditions as a result of a **runaway greenhouse effect,** which included the loss of large quantities of H_2O.

8.4 Mars has heavily cratered uplands in its southern hemisphere but younger, lower, volcanic plains over much of its northern half. The Tharsis bulge, as big as North America, includes several huge volcanoes; Olympus Mons is 25 km high and 500 km in diameter. The Valles Marineris canyons are tectonic features widened by erosion. The Viking landers revealed barren, windswept plains at Chryse and Utopia. Currently, there is great interest in the SNC meteorites, which appear to be samples of martian basalts.

8.5 The martian atmosphere has less than 0.01 bar surface pressure and is made up primarily (95 percent) of CO_2. There are dust clouds, water clouds, and CO_2 (dry ice) clouds. Liquid H_2O is not possible, but there may be subsurface permafrost. Seasonal polar caps are made of dry ice, but the residual caps contain water ice. Evidence of a very different climate in the past is found in water erosion features, both runoff channels and outflow channels, the latter carved by catastrophic floods. The Viking landers searched for martian life in 1976, with negative results, but life might have flourished long ago.

8.6 Earth, Venus, and Mars have diverged in their evolution. We need to understand why if we are to protect the environment of the Earth.

REVIEW QUESTIONS

1. Compare the basic data on orbits, mass, size, density, and rotation for Venus, Earth, and Mars.

2. Describe the main geological features on Venus, and compare with those of the Earth.

3. Describe the current state of the venerian atmosphere, and indicate how it might have evolved to this condition through a runaway greenhouse effect.

4. Describe the main geological features of Mars, and compare with those of the Earth, Venus, and the Moon.

5. List the main differences between the martian canals, the runoff channels, and the outflow channels.

6. Describe the current state of the martian atmosphere, and indicate how it must have been different in the past, judging from the evidence that there was once abundant water on the surface.

7. Is it likely that there was ever life on either Venus or Mars? Justify your answer.

THOUGHT QUESTIONS

8. What are the advantages of using radar imaging rather than ordinary cameras to study the topography of Venus? What are the relative advantages of these two approaches to mapping the Earth or Mars?

9. Venus and Earth are nearly the same size and distance from the Sun. What are the main differences in the geology of the two planets, and what might be some of the reasons for these differences?

10. Why is there so much more carbon dioxide in the atmosphere of Venus than in that of the Earth? Why so much more than in the martian atmosphere?

11. Compare Mars with Mercury and the Moon in terms of its bulk properties. What are the main similarities and differences?

12. Why is Mars red? Why aren't the Moon or Earth the same color?

13. Explain the major division of the surface of Mars into two distinct kinds of topography. Compare with the distinction on the Moon between the highlands and the maria.

14. How do the mountains on Mars and Venus compare with those on the Earth and Moon?

15. Explain how the theory that all of the terrestrial planets had similar impact histories can be used to date the formation of the martian uplands, the martian basins, and the Tharsis volcanoes. How certain are the ages derived for these features?

PROBLEMS

16. At its nearest, Venus comes within about 40 million km of the Earth. How distant is it at its farthest?

17. Calculate the relative *land* areas of Earth, Moon, Venus, and Mars. (*Note:* 70 percent of the Earth is covered with water.)

18. Where is the water on Mars? Try to estimate how much water might be present in various forms, such as in the polar caps (using the dimensions given in the text) or in subsurface permafrost (assuming various thicknesses for the permafrost, from 1 to 10 km, and a concentration of ice in the permafrost of 10 percent by volume).

19. Mariner 2 to Venus was the first successful interplanetary spacecraft. It traveled to Venus from the Earth on an elliptical orbit in which the Earth was at aphelion and Venus at perihelion. Calculate the total period of revolution about the Sun for a spacecraft on such an orbit. Then calculate the one-way travel time from Earth to Venus on this same trajectory.

20. A similar trajectory from Earth to Mars requires a longer trip than that to Venus. Calculate the one-way trip time to Mars using the same approach as in Problem 19. Explain in general why it takes longer to go to Mars than to Venus. (*Hint:* The reason is *not* just that Mars is slightly farther away.)

The four jovian planets to scale, in a composite Voyager image. *(NASA/JPL)*

9

THE GIANT PLANETS

James Van Allen (b. 1914), American physicist from the University of Iowa, was one of the originators of the discipline of space physics. Van Allen has extended his studies from the Earth to the larger and more complex magnetospheres of the outer planets as an experimenter on the Pioneer spacecraft, and his strong advocacy was critical in the approval of the Galileo mission to Jupiter. He is also a frequent critic of NASA's policy of emphasis on manned space flight at the expense of robotic missions.

The members of the outer solar system are very different from the inner planets. The four giant planets are much larger, distances between them are vastly increased, and they are accompanied by extensive systems of satellites and rings. From many perspectives, the outer solar system is where the action is, and the giant planets are the most important members of the Sun's family. In contrast, the little cinders of rock and metal that orbit closer to the Sun seem like an insignificant afterthought. In this chapter we describe the giant planets, and in Chapter 10 we shall investigate the smaller members of the outer solar system.

9.1 Outer Planet Overview

There are five planets in the outer solar system: Jupiter, Saturn, Uranus, Neptune, and Pluto. The first four of these are often called the giant or jovian planets, while Pluto is a small world, by far the smallest in the planetary system. Pluto is physically similar to the satellites of the giant planets and therefore is discussed with them in Chapter 10.

Mass Distribution

Most of the mass in the planetary system is found in the outer solar system. Jupiter alone accounts for 1/1000 the mass of the Sun and exceeds the mass of all the other planets combined. Table 9.1 summarizes the masses of members of the planetary system.

Chemistry in the Outer Solar System

Many of the properties of the solar system depend on distance from the Sun. The temperatures of planets and satellites are determined by this distance; the farther we are from the source of the Sun's energy, the colder we will be. But in addition, the temperatures in the original solar nebula out of which the planets formed 4.5 billion years ago were critical in deciding the composition of the planets. The **solar nebula** is the name we give to the contracting, spinning cloud of gas and dust that was the nursery of the planets. As we describe in some detail in Chapter 12, the chemistry of the solar system today is directly related to the temperatures in the solar nebula.

Thus, it should come as no surprise that in moving outward beyond Mars and the asteroid belt, we enter a region of different planetary composition. More distant than about 4 AU from the Sun, temperatures are always below the freezing point of H_2O (273 K). Here water ice was available in the solar nebula as a raw material, in addition to the silicates and metals present in the inner solar system. Since the atoms that constitute H_2O—hydrogen and oxygen—are among the most abundant in the universe, a great deal of

TABLE 9.1   Mass in the Solar System

Object	Percentage of Mass
Sun	99.80
Jupiter	0.10
Comets	0.05
All other planets	0.04
Satellites and rings	0.00005
Asteroids	0.000002
Dust and debris	0.0000001

water ice formed. Beyond 10 AU, additional ices are stable, but none of these is nearly as plentiful as water ice. Table 9.2 lists the main materials available to form a planet, based on the solar abundances.

With so much hydrogen available, the chemistry of the outer solar system is *reducing*. Most of the oxygen present is chemically combined with hydrogen to make H_2O, and it is therefore unavailable to form many oxidized compounds with other elements. The compounds detected in the atmospheres of the giant planets are thus hydrogen-based gases, such as methane (CH_4) and ammonia (NH_3), or more complex hydrocarbons, such as ethane (C_2H_6) and acetylene (C_2H_2).

Exploration of the Outer Solar System

Four spacecraft, all from the United States, have penetrated beyond the asteroid belt to initiate the exploration of the outer solar system. The challenges of probing so far from Earth are considerable. Flight times to the outer planets are measured in years to decades, rather than the few months required to reach Venus or Mars. Spacecraft must be highly reliable, and they must also be capable of a fair degree of

TABLE 9.2   Abundances in the Outer Solar System

Material	Percent (by mass)
Hydrogen (H_2)	77
Helium (He)	22
Water (H_2O)	0.6
Methane (CH_4)	0.4
Ammonia (NH_3)	0.1
Rock (including metal)	0.3

independence and autonomy, since the light-travel time between Earth and the spacecraft is several hours. If a problem develops near Saturn, for example, the spacecraft computer must deal with it directly. To wait hours for the alarm to reach Earth and instructions to be routed back to the spacecraft could spell disaster. These spacecraft also must carry their own electrical energy sources, since sunlight is too weak to supply energy through solar cells. Heaters are required to keep instruments at proper operating temperatures, and spacecraft must have powerful radio transmitters and large antennas if their precious data are to be transmitted to receivers on Earth a billion kilometers or more distant.

The first spacecraft to the outer solar system were Pioneers 10 and 11, launched in 1972 and 1973 as pathfinders to Jupiter. Their main objectives were to determine whether a spacecraft could navigate through the asteroid belt without collision with small particles and to measure the radiation hazards in the magnetosphere of Jupiter. Both spacecraft passed through the asteroid belt without incident, but the energetic ions associated with Jupiter nearly wiped out their electronics, providing information necessary for the design of subsequent missions. Pioneer 10 flew past Jupiter in 1974, after which it sped outward toward the limits of the solar system. Pioneer 11 undertook a more ambitious program, using the gravity of Jupiter during its 1975 encounter to divert it toward Saturn, which it reached in 1979.

The primary scientific missions to the outer solar system were Voyagers 1 and 2, launched in 1977 (Figure 9.1). The Voyagers each carried 11 scientific instruments, including cameras and spectrometers as well as devices to measure the magnetic field and characteristics of planetary magnetospheres. Voyager 1 reached Jupiter in 1979 and used a gravity assist from that planet to take it on to Saturn in 1980 (see Figure 2.12). The second Voyager, arriving four months later at Jupiter, followed a different path to accomplish a full grand tour of the outer planets: Saturn in 1981, Uranus in 1986, and Neptune in 1989. Most of the information in this chapter and in Chapter 10 is derived from the Voyager missions.

Voyager followed a trajectory made possible by the alignment of the four giant planets on the same side of the Sun. About once every 175 years, these planets are in such a position that a single spacecraft can visit them all, using gravity-assisted flybys to adjust its course for the next encounter. We are fortunate that this opportunity was seized. Because of this alignment, every planet in the outer solar system except Pluto has been visited by spacecraft; otherwise, it would probably have been well into the next century before this basic reconnaissance of the planetary system was accomplished.

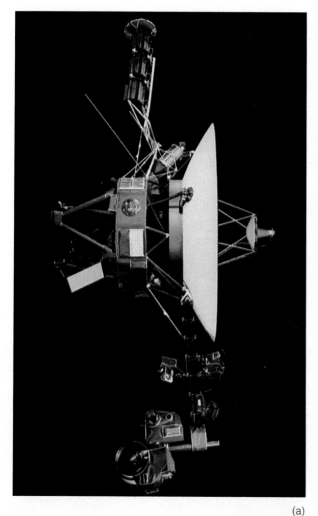

(a)

(b)

FIGURE 9.1 The Voyager missions. (a) These 1-ton robot explorers revealed much of the true nature of the outer solar system for the first time. (b) Scientists Brad Smith and Larry Soderblom admiring images returned during the Voyager 2 encounter with Saturn. *(NASA/JPL)*

The next steps in the exploration of the outer solar system involve extended study of Jupiter, Saturn, and their satellites. The Galileo mission to Jupiter was launched in 1989 and will arrive in 1995. It will deploy an entry probe into the planet for direct studies of the atmosphere, before beginning a three-year orbital tour, during which there will be repeated close flybys of the four large Galilean satellites. The similar Cassini mission to Saturn is under development as a cooperative venture between NASA and the European Space Agency. Cassini is planned for launch in 1996 and arrival at Saturn in 2002.

Table 9.3 summarizes the encounter dates for the spacecraft missions to the outer solar system.

9.2 The Jovian Planets

We now look at the four giant or jovian planets in some detail. Our approach is to compare them with one another, noting their similarities and differences and attempting to relate these properties to their differing masses and distances from the Sun.

Basic Properties

The median distance of Jupiter from the Sun is 778 million km, 5.2 times that of the Earth; its period of revolution is just under 12 years. Saturn is about

TABLE 9.3   Missions to the Outer Solar System

Planet	Spacecraft	Encounter Date
Jupiter	Pioneer 10	Dec. 73
	Pioneer 11	Dec. 74
	Voyager 1	Mar. 79
	Voyager 2	Jul. 79
	Galileo	Dec. 95
Saturn	Pioneer 11	Sept. 79
	Voyager 1	Nov. 80
	Voyager 2	Aug. 81
	Cassini	Dec. 02
Uranus	Voyager 2	Jan. 86
Neptune	Voyager 2	Aug. 89

TABLE 9.4 Basic Properties of the Jovian Planets

Planet	Distance (AU)	Period (yr)	Diameter (km)	Mass (Earth = 1)	Density (g/cm³)	Rotation (hr)
Jupiter	5.2	11.9	142,800	318	1.3	9.9
Saturn	9.5	29.5	120,540	95	0.7	10.7
Uranus	19.2	84.1	51,200	14	1.2	17.2
Neptune	30.1	164.8	49,500	17	1.6	16.1

twice as far away as Jupiter, at an average distance from the Sun of 1427 million km, or 9.5 AU. Saturn completes one revolution in very nearly the standard human generation of 30 years. Its slow cycle around the sky provided the longest natural time interval available to ancient peoples. Uranus and Neptune are more distant yet. Uranus orbits at 19 AU with a period of 84 years, and Neptune at 30 AU requires 165 years for each circuit of the Sun. Not until 2010 will a full Neptune "year" have passed since its discovery in 1845.

Jupiter and Saturn have many similarities in composition and internal structure, although Jupiter is nearly four times more massive. In contrast, Uranus and Neptune are smaller worlds with different composition and structure. Some of the main properties of these four planets are summarized in Table 9.4.

Jupiter is 318 times as massive as the Earth, a value that is very close to 1/1000 the mass of the Sun. Its diameter is about 11 times the Earth's diameter and about 1/10 the diameter of the Sun. Jupiter's density is 1.3 g/cm³, much lower than that of any of the terrestrial planets. The mass of Saturn is 95 times that of the Earth, and its density is only 0.7 g/cm³—the lowest of any planet. In fact, Saturn would be light enough to float if an ocean large enough to contain it existed.

Both Uranus and Neptune have a mass about 15 times that of the Earth and hence only 5 percent as great as that of Jupiter. Their densities of 1.2 g/cm³ and 1.6 g/cm³, respectively, are much higher than that of Saturn, in spite of their smaller mass and weaker gravity. This must be because their composition is fundamentally different, consisting, for the most part, of heavier materials than the hydrogen and helium that are the primary constituents of Jupiter and Saturn. We will discuss the details of their composition below.

Appearance and Rotation

When we look at the giant planets, we see only their atmospheres, composed primarily of hydrogen and helium gas (Figure 9.2). If any solid surface existed, it would be invisible to us, hidden by opaque clouds. The uppermost cloud deck of Jupiter and Saturn, and therefore the part of the planet we see when looking down from above, is composed of NH_3 crystals. On Neptune, the upper cloud deck is CH_4. On Uranus, we detect no obvious cloud deck at all but see only a deep and featureless haze.

Seen through a telescope, Jupiter is a colorful and dynamic planet. Distinct details in its cloud patterns allow us to determine the rotation rate of the atmosphere at the cloud level, although such an apparent rotation of the atmosphere may have little to do with the spin of the underlying planet. Much more fundamental is the rotation of the mantle and core, which can be determined by periodic variations in the magnetic field. Since the magnetic field originates deep inside the planet, it shares the rotation of the interior. This period of $9^h 56^m$ gives Jupiter the shortest "day" of any planet.

The rotation period of Saturn is $10^h 40^m$, as derived from variations in its radiation at radio wavelengths, which are in turn linked to its magnetic field. Uranus and Neptune have slightly longer rotation periods of about 17^h, also determined from the rotation of their magnetic fields.

The axis of rotation of Jupiter is tilted by only 3°, so there are no seasons to speak of. However, Saturn does have seasons, since its axis of rotation is inclined at 27° to the perpendicular to its orbit. Neptune has about the same tilt as Saturn (29°); therefore, it experiences similar seasons. However, Uranus has an axis of rotation that is tilted by 98° with respect to the north direction. This unusual tilt creates very strange seasons, with each pole alternately facing toward the Sun for about 40 years at a time.

Composition and Structure

Astronomers are confident that the interiors of Jupiter and Saturn are composed primarily of hydrogen and helium. Of course, these gases have been measured only in their atmospheres, but calculations first

FIGURE 9.2 Jupiter, as photographed by Voyager. The banded structure on the clouds represents strong east-west winds in the atmosphere. *(NASA/JPL)*

carried out 50 years ago by Yale University astronomer Rupert Wildt (1905–1976) have shown that these two gases are the only possible materials out of which a planet with the observed masses and densities of Jupiter and Saturn could be constructed. There remain some uncertainties in calculating models, however, primarily because of our incomplete knowledge of the compressibility of liquid hydrogen and helium at the temperatures and pressures that exist inside Jupiter and Saturn. The best models of Jupiter predict a central pressure of over 100 million bars and a central density of about 31 g/cm^3.

The internal structures of the giant planets are very different from those of the rocky inner planets, and the materials of which they are composed can take on strange forms. At depths of only a few thousand kilometers below their visible clouds, pressures become so large that hydrogen changes from a gaseous to a liquid state. Still deeper, this liquid hydrogen can act like a metal, if the pressure is great enough. On Jupiter, the greater part of the interior is liquid metallic hydrogen. Because Saturn is less massive, it has only a small volume of metallic hydrogen, but most of its interior is liquid. Uranus and Neptune are probably too small to reach internal pressures sufficient to liquefy hydrogen.

Each of these planets has a core composed of heavier materials, as demonstrated by detailed analyses of their gravitational fields. Presumably these cores are the original rock-and-ice bodies that formed before the capture of gas from the surrounding nebula. The cores exist at pressures of tens of millions of bars, compared with a pressure of 4 million bars at the center of the Earth. While scientists speak of the giant planet cores being composed of rock and ice, we can be sure that neither rock nor ice assumes any familiar forms at such pressures and the accompanying high temperature. What is really meant by "rock" is any materials made up primarily of iron, silicon, and oxygen. By "ice" is meant materials composed primarily of the elements carbon, nitrogen, and oxygen in combination with hydrogen.

Figure 9.3 illustrates the interior structures of the four jovian planets. It appears that all four have similar cores of "rock" and "ice." On Jupiter and Saturn, the cores constitute only a few percent of the total mass, consistent with the initial composition of raw materials shown in Table 9.2. However, most of the mass of Uranus and Neptune resides in these cores, demonstrating that these two planets were unable to attract massive quantities of hydrogen and helium.

It is interesting that Jupiter has very nearly the maximum possible size for a body of "cold" hydrogen, that is, one that is not generating energy as does a star. Less massive bodies than Jupiter would occupy a smaller volume (like Saturn). More massive

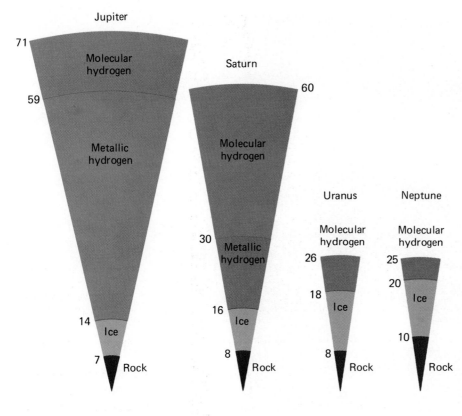

FIGURE 9.3 Internal structures of the four jovian planets, drawn to scale. Jupiter and Saturn are composed primarily of hydrogen and helium, but the mass of Uranus and Neptune consists in large part of compounds of carbon, nitrogen, and oxygen.

bodies, by virtue of their greater gravitation, would also be compressed to a smaller volume than Jupiter's. Such an object, with a mass larger than Jupiter's but not large enough to maintain nuclear reactions, is called a brown dwarf or infrared dwarf.

Internal Heat Sources

Because of its large size, each of the giant planets was strongly heated during its formation by the collapse of surrounding nebular gas onto its core. Jupiter, being the largest, was by far the hottest. In addition, it is possible for giant, largely gaseous planets to generate heat after formation by slow contraction. The effect of these internal energy sources is to raise the temperatures in the interiors and atmospheres to values that are higher than would be possible by solar heating alone.

Jupiter has the largest internal energy source, amounting to 4×10^{17} watts. It is glowing with the equivalent of 4 million billion hundred-watt light bulbs. This is about the same as the total solar energy absorbed by Jupiter. The atmosphere of Jupiter is therefore somewhat of a cross between a normal planetary atmosphere, which obtains most of its energy from the Sun, and the atmosphere of a star, which is entirely heated by an internal energy source. Most of the internal energy of Jupiter is primordial heat, left over from the formation of the planet 4.5 billion years ago.

Saturn has an internal energy source about half as large as that of Jupiter, which means (since its mass is only about one-quarter as great) that it is producing twice as much energy per kilogram of material as does Jupiter. Since Saturn is expected to have much less primordial heat, there must be another source at work generating most of this 2×10^{17} watts of power. This source is believed to be the separation of helium from hydrogen in the interior. In the liquid hydrogen mantle, the heavier helium forms drops that sink toward the core, releasing gravitational energy. In effect, Saturn is still differentiating. This precipitation of helium is possible in Saturn because it is cooler than Jupiter; at the temperatures in Jupiter's interior, hydrogen and helium remain well mixed.

Uranus and Neptune are different. Neptune has a small internal energy source, while Uranus does not emit enough internal heat for it to have been measured. As a result, these two planets have almost the same temperatures, in spite of the greater distance of Neptune from the Sun. No one knows why these two planets differ in their internal heat.

9.3   Atmospheres of the Jovian Planets

The atmospheres of the jovian planets are the parts that we can observe directly. Since these planets have no solid surfaces, their atmospheres are also

FIGURE 9.4 Jupiter's complex and colorful clouds. Also shown in this Voyager photograph are two of Jupiter's satellites: Io *(left)* and Europa *(right)*. *(NASA/JPL)*

more representative of the bulk composition of the planets than was the case with the terrestrial planets.

Atmospheric Composition

Spectroscopic observations of the jovian planets began in the 19th century, but for a long time the observed spectra could not be interpreted. As late as the 1930s, the most prominent absorption bands photographed in these spectra remained unidentified. Then methane (CH_4) was identified in the atmospheres of Jupiter and Saturn, followed by ammonia (NH_3).

At first it was thought that CH_4 and NH_3 might be the primary constituents of these atmospheres, but now we know that hydrogen and helium are actually the dominant gases. But neither hydrogen nor helium has easily detected spectral features, and it was not until the Voyager spacecraft measured the far-infrared spectra of Jupiter and Saturn that an abundance of the elusive helium could reliably be found on

either planet. Table 9.5 summarizes the compositions of these two atmospheres.

The compositions of the two atmospheres are generally similar, except that on Saturn there is only about half as much helium—the result of the precipitation of helium that contributes to Saturn's internal energy source. The measurement by Voyager of this depletion of atmospheric helium represents an impressive confirmation of the theory that helium can precipitate in Saturn but not in Jupiter. The atmospheres of Uranus and Neptune have about the same abundance of helium relative to hydrogen as is found on Jupiter.

Clouds and Atmospheric Structure

The clouds of Jupiter are among the most spectacular sights in the solar system, much beloved by makers of science fiction films. They range in color from white to orange to red to brown, swirling and twisting in a constantly changing kaleidoscope of patterns (Figure 9.4). Saturn shows similar but very much subdued cloud activity; instead of vivid colors, its clouds are a nearly uniform butterscotch hue (Figure 9.5).

At the temperatures and pressures of the upper atmospheres of Jupiter and Saturn, CH_4 remains a gas, but NH_3 can condense, just as water vapor condenses in the Earth's atmosphere, to produce clouds. The primary clouds that we see when we look at these planets, whether from a spacecraft or through a telescope, are composed of crystals of frozen NH_3. The NH_3 cloud deck marks the upper edge of the convective troposphere; above it is the cold stratosphere.

On both planets the temperature near the cloud tops is about 140 K (only a little cooler than the polar caps of Mars). On Jupiter this cloud level is at a

TABLE 9.5	Atmospheric Compositions of Jupiter and Saturn (Number of Atoms Relative to Hydrogen)	
Gas	Jupiter	Saturn
H_2	1	1
He	0.12	0.06
CH_4	2×10^{-3}	2×10^{-3}
NH_3	2×10^{-4}	2×10^{-5}
C_2H_2	8×10^{-7}	1×10^{-7}
C_2H_6	4×10^{-5}	8×10^{-6}
PH_3	4×10^{-7}	3×10^{-6}

FIGURE 9.5 Saturn and its rings, photographed by Voyager. The clouds of Saturn are less colorful than those of Jupiter, but the structure and dynamics of the atmosphere are similar. *(NASA/JPL)*

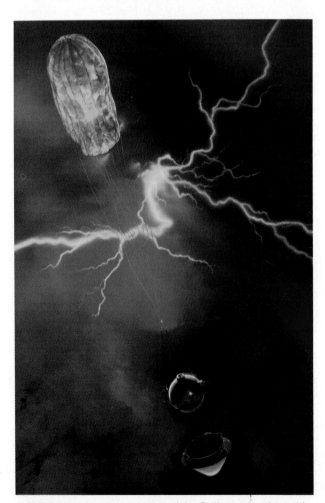

FIGURE 9.6 Artist's impression of the Galileo probe descending through the jovian clouds in 1995. *(NASA/ARC)*

pressure of about 0.1 bar, but on Saturn it occurs at about 1 bar of pressure. Because the NH_3 clouds lie so much deeper on Saturn, they are more difficult to see, and the overall appearance of the planet is much more bland than that of Jupiter.

Within the tropospheres of these planets, the temperature and pressure both increase with depth. Through breaks in the NH_3 clouds, we can see other layers of cloud that exist in these deeper regions of the atmosphere—regions that will be sampled directly by the Galileo probe. In 1995 this probe (Figure 9.6) will enter the atmosphere of Jupiter and descend to a pressure level between 10 and 20 bars before its battery power is exhausted. Below the thin NH_3 clouds it should pass through a clear region, but at a pressure of about 3 bars we expect the probe to enter another thick deck of condensation clouds, composed of ammonium hydrosulfide (NH_4SH). The NH_4SH clouds probably also contain some sulfur particles, which color them a darker yellow or brown.

As it descends to a pressure of 10 bars and ever-higher temperatures, the Galileo probe should pass next into a region of frozen H_2O clouds, then below that into clouds of liquid H_2O droplets perhaps similar to the common clouds of the terrestrial troposphere. This region corresponds almost to a "shirt sleeve" environment, in which astronauts could exist quite comfortably if they carried scuba gear for breathing. But with no solid surface to stop it, the probe will continue to descend, penetrating to dark regions of higher and higher pressure and temperature. No matter how strongly it was built, eventually the probe will

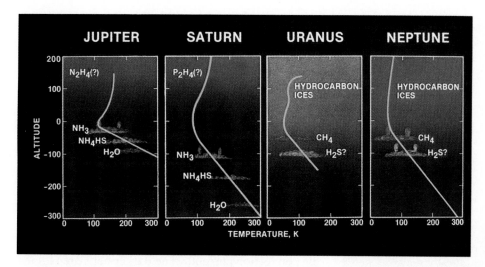

FIGURE 9.7 The structure of the atmosphere and clouds of the jovian planets. *(NASA/JPL)*

be crushed and swallowed in the black depths, where the great pressures finally transform the atmospheric hydrogen into a hot, dense liquid.

Above the visible NH_3 clouds, the atmosphere of Jupiter is clear and cold, reaching a minimum temperature near 120 K. At still higher altitudes, temperatures rise again, just as they do in the upper atmosphere of the Earth, owing to the absorption of solar ultraviolet light. We call chemical reactions induced by ultraviolet light **photochemistry**. In this region, photochemical reactions create a variety of fairly complex hydrocarbon compounds that form a thin layer of photochemical smog far above the visible clouds.

There is one other mystery of the jovian clouds that should be mentioned, and that concerns their colors. The NH_3 condensation clouds identified on the planet should be white, like water clouds on Earth, yet we see beautiful and complex patterns of red, orange, and brown. Some additional chemical or chemicals must be present to lend the clouds such colors, but we do not know what they are. Various photochemically produced organic compounds have been suggested, as well as sulfur and red phosphorus. But there are no firm identifications, nor any immediate prospects of solving this mystery.

The atmospheric structure of Saturn is similar to that of Jupiter. Temperatures are somewhat colder, and the atmosphere is more extended as the result of Saturn's lower surface gravity, but qualitatively the same atmospheric regions, and the same condensation clouds and photochemical reactions, should be present. Figure 9.7 compares the atmospheric structures of the four jovian planets.

Unlike Jupiter and Saturn, Uranus is almost entirely featureless as seen at wavelengths that range from the ultraviolet to the infrared (Figure 9.8). Calculations indicate that the basic atmospheric structure of this planet should resemble that of Jupiter and Saturn, although now the upper condensation clouds (at the 1-bar pressure level) are composed of CH_4, rather than NH_3. However, the absence of an internal heat source suppresses convection and leads to a very stable atmosphere with little visible structure. In addition, the troposphere is hidden from our view by a deep, cold, hazy stratosphere.

Neptune differs dramatically from Uranus in its appearance (Figure 9.9), although the basic atmospheric temperatures are almost identical. The upper clouds are composed of CH_4, which forms a thin cloud near the top of the troposphere at a temperature of 70 K and a pressure of 1.5 bars. Most of the atmosphere above this level is clear and transparent, with less haze than on Uranus. Scattering of sunlight lends Neptune a deep blue color similar to that of the Earth's atmosphere. Another cloud layer exists at a pressure of 3 bars, perhaps composed of hydrogen sulfide ice particles.

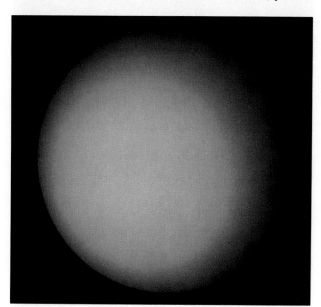

FIGURE 9.8 The planet Uranus as photographed by Voyager in 1986, when its rotation pole was tipped toward the Sun. *(NASA/JPL)*

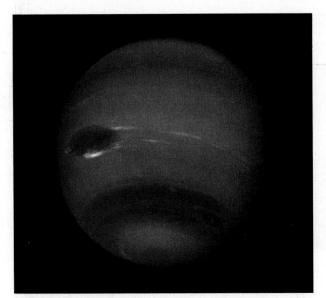

FIGURE 9.9 The planet Neptune as photographed in 1989 by Voyager. *(NASA/JPL)*

The primary difference between Uranus and Neptune is the presence on Neptune of convection currents from the interior, powered by the planet's internal heat source. These currents carry warm gas above the 1.5-bar cloud level, forming additional clouds at elevations about 75 km higher. The high-altitude clouds form bright white patterns against the blue planet beneath. They can even cast distinct shadows on the CH_4 cloud tops, permitting their altitudes to be calculated (Figure 9.10).

Winds and Weather

Observations of the changing cloud patterns in the atmospheres of the jovian planets permit us to measure wind speeds and track the circulation of the atmosphere. The atmospheric dynamics observed on these planets differ fundamentally from those of the terrestrial planets. There are three primary reasons for these differences: (1) These planets have much deeper atmospheres, with no solid lower boundary; (2) they spin faster than the terrestrial planets, suppressing north-south circulation patterns and accentuating east-west airflow; (3) on all except Uranus, internal heat sources contribute about as much energy as sunlight, forcing the atmospheres into deep convection to carry the internal heat outward.

The main features of the visible clouds of Jupiter are alternating dark and light bands that stretch around the planet parallel to the equator. These bands are semipermanent features, although they shift in intensity and position from year to year. Consistent with the small tilt of Jupiter, there are no seasonal effects detectable.

More fundamental than these bands are the underlying east-west wind patterns in the atmosphere, which do not appear to change at all, even over many decades (Figure 9.11). The main such feature on Jupiter is an eastward-flowing equatorial jet stream with a speed of 300 km/h, similar to the speed of jet streams in the Earth's upper atmosphere. At higher latitudes there are alternating east- and west-moving streams, with each hemisphere an almost perfect mirror image of the other. Saturn shows a similar pattern, but with a much stronger equatorial flow at a speed of 1300 km/h (almost 400 m/s).

Generally, the light zones on Jupiter are regions of upwelling air, capped by white NH_3 cirrus clouds. They apparently represent the tops of upward-moving convection currents. The darker belts are regions where the cooler atmosphere moves downward, completing the convection cycle; they are darker because there are fewer NH_3 clouds and it is possible to see deeper in the atmosphere, perhaps down to the NH_4SH clouds.

In spite of the strange seasons induced by the 98° tilt of its axis, Uranus' basic circulation is parallel with its equator, just as it is on Jupiter and Saturn.

FIGURE 9.10 The atmosphere of Neptune. The bright narrow cirrus clouds are composed of crystals of methane ice injected into the lower stratosphere. These clouds cast their shadows on the solid cloud layer about 75 km beneath. *(NASA/JPL)*

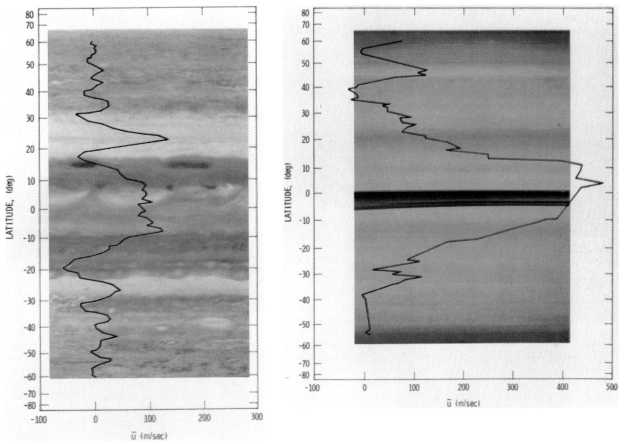

FIGURE 9.11 Zonal (east-west) winds on Jupiter and Saturn as measured by Voyager. The wind speeds are referred to the rotation of the core as determined from magnetic field and radio measurements. *(NASA/JPL)*

The mass of the atmosphere and its capacity to store heat are so great that the alternating 40-year periods of sunlight and darkness have little effect; in fact, Voyager measurements show that the atmospheric temperatures are a few degrees higher on the dark, winter side than on the hemisphere facing the Sun. The dynamics of all the jovian atmospheres are complex, and we do not understand their seasonal effects in detail.

Neptune's weather is characterized by strong east-west winds generally similar to those observed on Jupiter and Saturn. The highest wind speeds near the equator reach 2100 km/h (600 m/s), nearly twice as fast as the peak winds on Saturn. The Neptune equatorial jet stream actually approaches supersonic speeds.

Storms

Superimposed on the regular atmospheric circulation patterns described above are many local disturbances—weather systems or storms, to borrow terrestrial terminology. The most prominent of these are large oval high-pressure regions on both Jupiter and Neptune.

The largest and most famous "storm" on Jupiter is the Great Red Spot, or GRS, a reddish oval in the southern hemisphere that is almost 30,000 km long—big enough to hold two Earths side by side (Figure 9.12). First seen 300 years ago, the GRS is clearly much longer lived than storms in our own atmosphere. The GRS also differs from terrestrial storms in being a high-pressure region; its counterclockwise rotation has a period of six days. Three similar but smaller disturbances on Jupiter formed about 1940, called the "white ovals"; these are only about 10,000 km across.

We don't know what causes the GRS or the white ovals, but it is possible to understand how they can last so long once they do form. On Earth, a large oceanic hurricane or typhoon typically has a lifetime of a few weeks, or even less when it moves over the continents and encounters friction with the land. On Jupiter, there is no solid surface to slow down an

FIGURE 9.12 The Great Red Spot of Jupiter. Below it and to the right is one of the white ovals, which are similar smaller high-pressure features. *(NASA/JPL)*

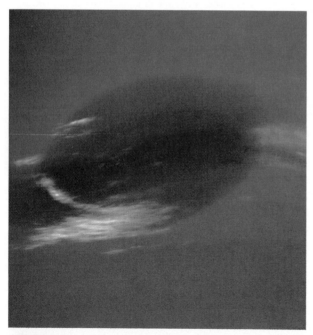

FIGURE 9.13 The Great Dark Spot of Neptune. The spot is accompanied by streamers of bright methane cirrus clouds that form around it but at a higher elevation. *(NASA/JPL)*

atmospheric disturbance, and furthermore the sheer size of these features lends them stability. It is possible to calculate that on a planet with no solid surface, the lifetime of anything as large as the GRS should be measured in centuries, while lifetimes for the white ovals should be measured in decades. These time scales are consistent with the observed lifetimes of jovian storms.

In spite of its smaller size and different cloud composition, Neptune has an atmospheric feature surprisingly similar to the jovian GRS. Neptune's Great Dark Spot (Figure 9.13) is nearly 10,000 km long. Like Jupiter's GRS, it is found at latitude 20° S, and its size and shape are similar relative to the size of the planet. This Great Dark Spot rotates with a period of 17 days. Finally, just as with the GRS, we do not yet understand the origin of the Great Dark Spot.

Large storms on Saturn are much rarer and appear to be seasonal in their genesis, controlled by the large tilt of Saturn's axis. Approximately once every 30 years, observers have seen outbreaks of spots in the equatorial regions of Saturn, with the most recent such outbreak occurring in 1990. As photographed by

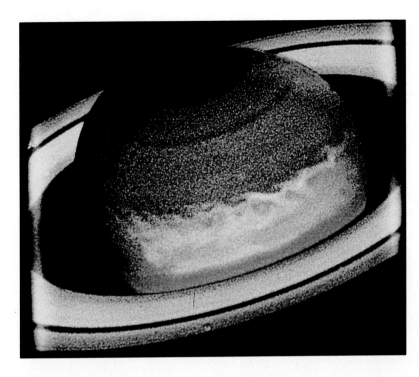

FIGURE 9.14 Storm on Saturn as photographed by the Hubble Space Telescope in November 1990. Note similarities to the structure visible on Jupiter in Figure 9.12. *(NASA, courtesy Space Telescope Science Institute)*

the Hubble Space Telescope about a month after its appearance (Figure 9.14), this storm had spread most of the way around the planet and resembled some of the turbulent cloud structure that accompanies the jovian GRS. These seasonal storms on Saturn appear to represent large bubbles or plumes of warmer air that rise into the stratosphere and are then gradually dissipated by the strong equatorial jet streams.

9.4 Magnetospheres

The largest features of the giant planets are their magnetospheres. Like the magnetosphere of the Earth, these regions are defined as the cavities within which the planet's magnetic field dominates the interplanetary magnetic field. Inside the magnetosphere, ions and electrons can be accelerated to high energies. These physical processes are similar to those dealt with by astrophysicists in many distant objects, from pulsars to quasars. The magnetospheres of the giant planets provide nearby analogs of these cosmic processes.

Planetary Magnetic Fields

In the late 1950s, radio energy was observed from Jupiter that is more intense at longer than at shorter wavelengths—just the reverse of what is expected from thermal radiation. It is typical, however, of the radiation emitted by electrons accelerated by a magnetic field, called **synchrotron radiation.** Later obser-

vations showed that the radio energy originated from a region surrounding the planet whose diameter is several times that of Jupiter itself. The evidence suggested, therefore, that there are a vast number of charged atomic particles circulating around Jupiter, spiraling through the lines of force of a magnetic field associated with the planet. This phenomenon is like the Van Allen belts around the Earth.

The Pioneer and Voyager spacecraft supplemented these indirect measurements with direct studies of the magnetic field and magnetosphere of Jupiter (Figure 9.15). They found Jupiter's surface magnetic field to be from 20 to 30 times as strong as the Earth's field. Because of Jupiter's great size, moreover, its total magnetic energy is enormous compared with the Earth's.

The jovian magnetic axis, like that of the Earth, is not aligned exactly with the axis of rotation of the planet but is tipped at some 10°. The magnetic axis also does not pass exactly through the planet's center but is offset by about 18,000 km. In addition, the jovian field has the opposite polarity from the Earth's current value. However, the Earth's field is known to reverse polarity from time to time, and the same may be true of Jupiter's field.

Saturn does not emit strong synchrotron radiation, because its magnetosphere is depleted in electrons by collisions between electrons and its rings. It does have a substantial magnetic field, however, as discovered by Pioneer 11 and the Voyagers. Unlike the fields of Earth and Jupiter, Saturn's field is almost perfectly aligned with its rotation axis.

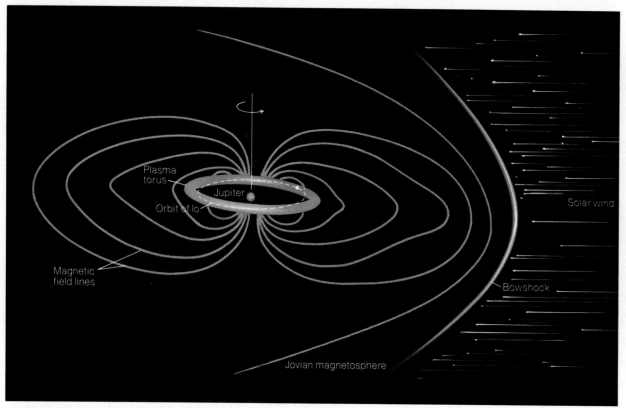

FIGURE 9.15 The magnetosphere of Jupiter as mapped out by the Pioneer and Voyager missions.

The magnetic field of Uranus was not discovered until the Voyager flyby in 1986. The strength of the field is comparable to that of Saturn, about what would be expected from the size of the planet. However, the orientation of the magnetic field of Uranus is very different. Like Jupiter's field, it is offset from the center of the planet, but to a greater degree (by about one-third of the planet's radius). In addition, the magnetic field of Uranus is tilted by 60° with respect to the axis of rotation—the extreme opposite case from that of Saturn.

Neptune's magnetic field was discovered in 1989, when the Voyager 2 spacecraft reached this distant world. Its configuration is similar to that of Uranus, with a magnetic axis tilted by 55° from the rotational axis (Figure 9.16). The offset of the neptunian field is the greatest of any planet, amounting to nearly half the planet's radius. The magnetic fields of the four giant planets are compared with that of the Earth in Table 9.6.

Presumably, the magnetic fields of the outer planets are generated in much the same way as the field of the Earth. All of these planets spin rapidly, so there is a ready source of energy to power their internal magnetic generators. Jupiter and Saturn have large interior regions of metallic liquid hydrogen that act like the liquid iron core of the Earth. In the case of Uranus and Neptune, however, the metallic region may be in the hydrogen-water mantle, possibly accounting for the large offset of the field from the center of the planet. Although the detailed mechanisms may not be well understood, these planets seem to meet the conditions required for the generation of a planetary magnetic field in a spinning metallic core.

Magnetosphere of Jupiter

The jovian magnetosphere is one of the largest features in the solar system. It is actually much larger than the Sun and completely envelops the innermost satellites of Jupiter. If we could see the magnetosphere, it would appear the size of our Moon. The total mass of the ions and electrons in the magnetosphere, however, is less than the mass of the Great Pyramid of Giza in Egypt.

On its upstream side (facing toward the solar wind), the magnetosphere is bounded by a pressure balance between the ions inside and the solar wind streaming toward it at about 400 km/s. At the outer planets, the magnetic fields are stronger than Earth's, and the

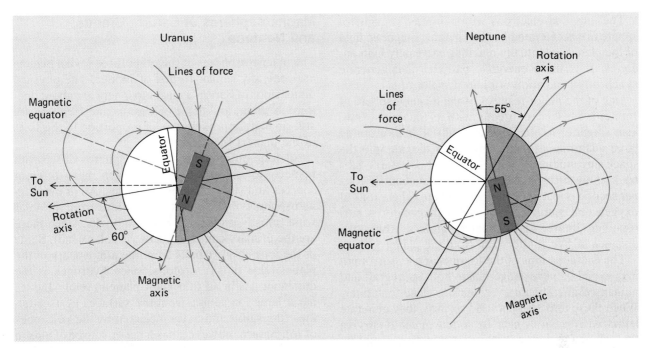

FIGURE 9.16 The magnetic fields of Uranus and Neptune as revealed by Voyager; note the large offset from the center of the planet and the tilt with respect to the planet's axis of rotation.

solar wind is weaker, contributing to the large size of their magnetospheres. The actual borders vary, however, with the changing pressure of the solar wind.

The jovian magnetosphere is characterized as much by the particles trapped within it as by the planetary magnetic field. The two primary sources of ions in the Earth's magnetosphere are the solar wind (mostly protons and electrons) and atmospheric atoms (mostly nitrogen) that escape upward from the planet. On Jupiter both of these sources also apply, but they are supplemented by a much stronger source on the large Galilean satellites. Unlike our Moon, these satellites are enveloped by the magnetosphere. Some atoms are ejected, or *sputtered*, from their surfaces by the impact of energetic magnetospheric

ions. An even larger source of ions is provided by oxygen and sulfur from the active volcanoes of Io. In the inner magnetosphere of Jupiter, the dominant ions are sulfur and oxygen.

The ultimate fate of ions in the jovian magnetosphere is similar to that of the ions in the terrestrial magnetosphere. For both planets, some magnetospheric particles escape and some are lost by collision with the planet's atmosphere (where they generate auroral discharges). In addition, Jupiter loses some by collision with its satellites, but in general more new ions are released by sputtering than old ones destroyed. Since the Earth's Moon is far outside our magnetosphere, it does not contribute as either a source or a sink of magnetospheric ions.

TABLE 9.6   Planetary Magnetic Fields

Planet	Average Surface Field (gauss)	Dipole Moment (Weber-m)	Tilt	Offset (planet radii)
Jupiter	4	1×10^{19}	10°	0.1
Saturn	0.2	3×10^{17}	1°	0.0
Uranus	0.3	3×10^{16}	60°	0.3
Neptune	0.2	2×10^{16}	55°	0.5
Earth	0.3	6×10^{14}	11°	0.0

The ions and electrons within Jupiter's magnetosphere are accelerated by the spinning magnetic field of the planet, eventually reaching extremely high energies. It is these energetic particles that generate synchrotron radiation at radio wavelengths.

One of the major features of the magnetosphere of Jupiter is associated directly with its innermost Galilean satellite, Io. Io is volcanically active, erupting large quantities of sulfur and sulfur dioxide into the space surrounding it. While most of this material falls back to the surface, it is estimated that about 10 tons per second is lost to the magnetosphere. These ions of oxygen and sulfur form a donut-shaped torus surrounding Jupiter approximately at Io's orbit, at a distance of five Jupiter radii from the planet.

The energetic ions of the Io torus and its surroundings would be very dangerous to both spacecraft and humans if any should ever venture close to Jupiter. When these ions strike a solid surface, they generate lethal x rays. On or near Io, a human could survive for only a few minutes. Special shielding would be required for spacecraft electronics, and no spacecraft has been built that could last more than a few hours in this environment. Thus an Io lander or orbiter is far beyond our present capabilities, and it is probably safe to predict that human exploration of the inner jovian system will never be possible.

Magnetospheres of Saturn, Uranus, and Neptune

The magnetospheres of the other three jovian planets are generally similar to that of Jupiter. The physical dimensions of Saturn's magnetosphere are about one-third as great, and those of Uranus and Neptune are still smaller, approximately in proportion to the sizes of the planets themselves.

All four outer-planet magnetospheres differ with respect to composition, and each has different sources and sinks of atomic particles. All of the magnetospheres obtain a part of their ions from the solar wind (mostly protons and helium nuclei, since these are the primary constituents of the solar wind). Some of the ions also originate from the atmosphere of the planet (also mostly protons, since hydrogen is the dominant gas in all four giant planets). Only Jupiter has a large ion source from the surfaces of its satellites (the sulfur and oxygen ejected by the volcanoes of Io). Saturn, however, has a major source of ions in the atmosphere of its large satellite Titan, which is constantly losing nitrogen to the saturnian magnetosphere. Both Uranus and Neptune derive some oxygen ions from sputtering of ice on the surfaces of their icy satellites, but their magnetospheres consist primarily of protons and electrons, derived from the planet's atmosphere and the solar wind.

SUMMARY

9.1   The outer solar system contains the four jovian planets and Pluto. Because temperatures were lower in the outer parts of the **solar nebula,** the composition of the outer planets and their satellites is different from that of the terrestrial planets. The chemistry is generally reducing, and Jupiter and Saturn have an overall composition similar to that of the Sun. Exploration has been carried out by Pioneers 10 and 11 and by the two Voyager spacecraft. Voyager 2, perhaps the most successful of all space science missions, explored Jupiter (1979), Saturn (1981), Uranus (1986), and Neptune (1989)—a grand tour of the jovian planets.

9.2   Jupiter is 318 times more massive than the Earth. Saturn is about 25 percent and Uranus and Neptune only 5 percent as massive as Jupiter. All four have deep atmospheres and opaque clouds, and all rotate quickly (periods from 10 to 17 h). Jupiter and Saturn have extensive mantles of liquid hydrogen. Uranus and Neptune are depleted in hydrogen and helium relative to Jupiter and Saturn (and the Sun). Each jovian planet has a core of "ice" and "rock" of about 10 Earth masses. Jupiter, Saturn, and Neptune have major internal heat sources, obtaining as much (or more) energy by convection from their interiors as by radiation from the Sun. Uranus has no measurable internal heat.

9.3   The four jovian planets have generally similar atmospheres, composed mostly of hydrogen and helium. The atmospheres of the jovian planets contain small quantities of methane (CH_4) and ammonia (NH_3) gas, both of which also condense to form clouds. Deeper (invisible) cloud layers consist of H_2O and possibly ammonium hydrosulfide (NH_4SH) (Jupiter and Saturn) and hydrogen sulfide (H_2S) (Neptune). In the upper atmospheres, hydrocarbons and other trace compounds are produced by **photochemistry.** We do not know what colors the clouds of Jupiter. Atmospheric dynamics are dominated by east-west circulation. Jupiter displays the most active cloud patterns, with Neptune second. Saturn is generally bland, and Uranus is featureless (perhaps owing to its lack of an internal heat source). The two major storms (the Great Red Spot on Jupiter and the Great Dark Spot on Neptune) are similar oval-shaped, high-pressure systems.

9.4   The jovian planets have substantial magnetic fields, approximately in proportion to their sizes. Within their large magnetospheres, trapped ions and electrons are accelerated to high energies and emit **synchrotron radiation.** Jupiter has the most active magnetosphere, partly because of the ions provided by the volcanoes of Io.

REVIEW QUESTIONS

1. Describe the differences in the chemistry of the inner and outer parts of the solar system. What is the relationship between chemistry and temperature?

2. What is the solar nebula?

3. What are the main differences between terrestrial and jovian planets?

4. What are the visible clouds composed of on the four jovian planets, and why are they different from one another?

5. Describe the seasons of Uranus.

6. Compare the atmospheric circulation (weather) for the four jovian planets.

7. Compare the magnetic fields and magnetospheres of the four jovian planets.

THOUGHT QUESTIONS

8. Jupiter is denser than water, yet it is composed for the most part of two light gases, hydrogen and helium. How can it be so dense?

9. Would you expect to find oxygen gas in the atmospheres of the giant planets? Why or why not?

10. The water clouds believed to be present on Jupiter and Saturn exist at temperatures and pressures similar to those at locations in the clouds in the terrestrial atmosphere. What would it be like to visit such a location on Jupiter or Saturn? In what ways would the environment differ from that in the clouds of Earth?

11. Describe the different processes that lead to substantial internal heat sources for Jupiter and Saturn. Since these two objects generate much of their energy internally, should they be called stars instead of planets? Justify your answer.

12. Give several reasons why the magnetosphere of Jupiter is bigger than the magnetosphere of the Earth.

PROBLEMS

13. As the Voyager spacecraft penetrated the outer solar system, the illumination from the Sun declined. Relative to the situation on Earth, how bright is the sunlight at each of the jovian planets?

14. Jupiter's Great Red Spot rotates in six days and has a circumference equivalent to a circle with radius 10,000 km. Neptune's Great Dark Spot is one-quarter as large and rotates in 17 days. For each, calculate the wind speeds at the outer edges of the spots. How do these compare with the winds in terrestrial hurricanes?

15. The ions in the inner parts of the jovian magnetosphere rotate with the same period as Jupiter. Calculate how fast they are moving at the orbit of Io. Will these ions strike Io from behind or in front as it moves about Jupiter?

16. Use Kepler's laws to calculate the flight times to Saturn, Uranus, and Neptune for simple spacecraft orbits in which the Earth is at perihelion and the target planet is at aphelion. Compare these with the flight times for the Voyager 2 mission. Why are they so different?

17. Estimate how frequently all four giant planets are approximately in alignment (for example, with longitudes differing by no more than 60°), permitting a "grand tour" trajectory like that of Voyager 2.

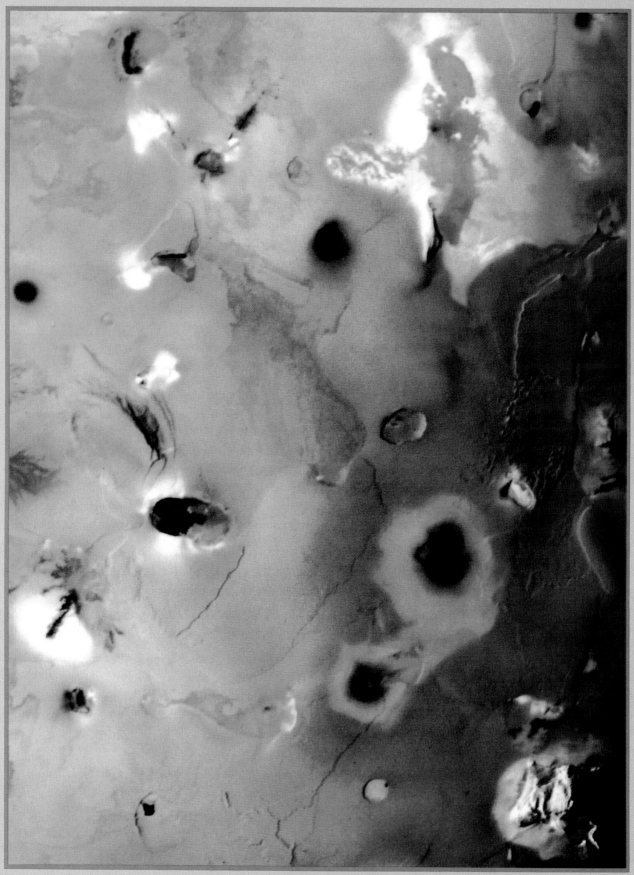

The multi-hued surface of Io, Jupiter's volcanically active satellite. *(NASA/JPL)*

Gerard P. Kuiper (1905–1973), Dutch-born American astronomer, contributed to many aspects of planetary studies. He discovered one satellite of Neptune, carried out pioneering studies of the rings of Saturn, discovered the atmosphere of Titan, founded several observatories, and was an influential architect of the early NASA program of lunar and planetary exploration. *(University of Arizona)*

10

RINGS, MOONS, AND PLUTO

The giant planets are accompanied by satellites, which orbit about them like planets in a miniature solar system. There are 59 known satellites in the outer solar system, too many to discuss individually in any detail (see Appendix 11). In this chapter we make no effort to cover them all, but instead focus attention on a few of the more interesting satellites and on Pluto. Equally interesting in the outer solar system are the ring systems, which provide fascinating examples of physical processes with applications ranging from the formation of planetary systems to the spiral structure of galaxies.

10.1 Ring and Satellite Systems

General Properties

The rings and satellites of the outer solar system are chemically distinct from objects in the inner solar system, as is to be expected from the fact that they formed in regions of lower temperature. The primary difference results from the availability of large quantities of water ice as building materials for bodies beyond the asteroid belt. Notable, in addition, is the presence of dark, organic compounds. Mixed with the ice that is present in these objects, this dark primitive material often results in low reflectivities.

Paradoxically, therefore, the ring and satellite systems contain many objects that are both icy and black.

Most of the satellites in the outer solar system are in direct or regular orbits; that is, they revolve about their parent planet in an east-to-west direction and very nearly in the plane of the planet's equator. Ring systems are also in direct revolution in the planet's equatorial plane. Such objects probably formed at about the same time as the planet by processes similar to those that formed the planets in orbit around the Sun.

In addition to the regular satellites, there are irregular satellites that orbit in a retrograde (west-to-east) direction or else have orbits of high eccentricity or inclination. These are usually smaller satellites, located relatively far from their planet, and they were probably formed between the planets and subsequently captured into orbit.

The Jupiter System

Jupiter has 16 satellites and a faint ring. The 16 satellites include the four large Galilean satellites (Figure 10.1) discovered in 1610 by Galileo: Callisto, Ganymede, Europa, and Io. The smallest of these, Europa and Io, are about the size of the Moon. The largest, Ganymede and Callisto, are larger than Mer-

FIGURE 10.1   The four large Galilean satellites of Jupiter shown with the planet in a montage of Voyager images. The satellite in the foreground is Callisto. *(NASA/JPL)*

cury. We shall discuss the Galilean satellites individually below.

The other 12 jovian satellites are much smaller. They divide themselves conveniently into three groups of four each. The inner four all circle the planet inside the orbit of Io; one of these, Amalthea, has been known for about a century, but the other three were discovered by Voyager. The outer satellites consist of four in direct but highly inclined orbits and four farther out in retrograde orbits. These eight are believed to be captured objects. The two groupings may indicate two parent bodies that were broken up in a collision early in the history of the jovian system. These eight outer satellites are dark, apparently primitive objects.

The Saturn System

Saturn has 19 known satellites in addition to its magnificent rings (Figure 10.2). The largest of the satel-

lites, Titan, is almost as big as Ganymede in the jovian system, and it is the only satellite with a substantial atmosphere. We shall discuss it in detail below. The composition of six other regular satellites, with diameters between 400 and 1600 km, is about half water ice. Saturn also has two distant irregular satellites, one of which is in a retrograde orbit.

The rings of Saturn are broad and flat, with only a few gaps. Individual ring particles are composed of water ice and are typically the size of tennis balls. Gravitational interactions between various small inner satellites and the rings are responsible for much of the detailed ring structure observed by Voyager.

The Uranus System

The regular ring and satellite system of Uranus shares the 98° tilt of the planet. It consists of 11 rings and 15 regular satellites. The five largest satellites are similar in size to the regular satellites of Saturn, with diameters from 500 to 1600 km, while the ten smaller satellites and the ring particles are very dark, reflecting only a few percent of the sunlight that strikes them.

The rings of Uranus, discovered in 1977, are narrow ribbons of material with broad gaps between— fundamentally different from the broad rings of Saturn. Presumably the ring particles are confined to these narrow paths by the gravitational effects of small satellites.

The Neptune System

Neptune has eight satellites: six regular satellites close to the planet and two irregular satellites. The most interesting of these is Triton, a relatively large satellite in a retrograde orbit. Triton has an atmosphere, and active volcanic eruptions were discovered there by Voyager in its 1989 flyby. We shall compare it in detail with the planet Pluto later in this chapter.

The rings of Neptune are narrow and faint. Like those of Uranus, they are composed of dark materials. One ring is distinguished by the presence of three bright regions that represent unexplained concentrations of ring material.

10.2   The Galilean Satellites and Titan

In this section we discuss the five largest satellites of the outer solar system. Table 10.1 summarizes their properties, with the Moon also listed for comparative purposes.

FIGURE 10.2 Montage of Voyager images of Saturn and several of its satellites. The satellite in the foreground is Enceladus. *(NASA/JPL)*

The Three Largest Satellites

The three largest satellites are Ganymede and Callisto in the jovian system and Titan in the saturnian system. All three of these have about the same diameter (from 5270 km for Ganymede down to 4820 km for Callisto) and nearly identical density (1.9 g/cm³). Each therefore appears to have the same composition, and we would expect them to have experienced parallel, and perhaps nearly identical, evolution. It is thus with considerable interest that we note that the three are different in several fundamental ways.

Since they have the same size and composition, these three largest satellites probably have the same general interior structure. They are almost surely dif-

ferentiated into a central, Moon-sized core of rock and mud, surrounded by a thick mantle of ice or, possibly, liquid water (H_2O). The crust is hard, brittle ice at the temperature prevalent on these satellites.

Geology of Ganymede and Callisto

Callisto and Ganymede provide an excellent introduction to the geology of icy worlds. We begin with Callisto, the simpler of the two. The entire surface of Callisto is covered with impact craters, like the lunar highlands (Figure 10.3). The existence of this heavily cratered surface tells us three important facts not known before Voyager: (1) An icy planet retains impact craters in its surface if its temperature is low

TABLE 10.1 The Largest Satellites

Name	Diameter (km)	Mass (Moon = 1)	Density (g/cm³)	Reflectivity (percent)
Moon	3476	1.0	3.3	12
Callisto	4820	1.5	1.8	20
Ganymede	5270	2.0	1.9	40
Europa	3130	0.7	3.0	70
Io	3640	1.2	3.5	60
Titan	5150	1.9	1.9	20

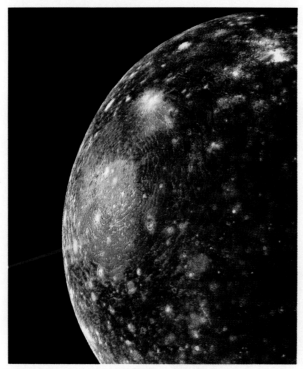

FIGURE 10.3   The heavily cratered surface of Jupiter's outermost Galilean satellite, Callisto. *(NASA/JPL)*

FIGURE 10.4   Jupiter's largest satellite, Ganymede. *(NASA/JPL)*

enough; (2) there was a heavy bombardment by debris in the outer solar system as well as nearer the Sun; (3) Callisto has experienced little, if any, geological activity other than impacts for a long time—probably billions of years.

The craters of Callisto do not look exactly like their counterparts in the inner solar system. They tend to be flatter, as if the surface did not have the strength to support much vertical relief. Such subdued topography is to be expected for an ice crust at the temperatures of 130 to 140 K measured near local noon on Callisto, since ice loses some of its strength as it is warmed. Farther from the Sun, in the Saturn system, temperatures are so low that ice is as strong as rock.

Ganymede, the largest satellite in the solar system, is also cratered, but less so than Callisto (Figure 10.4). About one-third of its surface seems to be contemporary with Callisto; the rest formed later, after the end of the heavy bombardment period. This younger terrain on Ganymede is probably 3 to 4 billion years old, like the lunar maria or the martian volcanic plains, judging from crater counts (Figure 10.5). It was produced when the crust cracked, flooding many of the craters with water from the interior and forming extensive parallel mountain ridges. This mountainous terrain, with its ridges evenly spaced about 15 km apart, covers more than one-quarter of the surface.

Ganymede experienced expansion and consequent resurfacing during the first billion years after its for-

mation, while Callisto did not. Apparently the small difference in size and internal heating between the two led to this difference in their evolution.

The Atmosphere of Titan

Titan, found in 1655 by the Dutch astronomer Christian Huygens, is the largest satellite of Saturn. It was the first satellite discovered since Galileo had seen the four moons of Jupiter that bear his name. Titan's

FIGURE 10.5   Detail of the younger, mountainous terrain on Ganymede. Width of the image is about 300 km, and the individual ridges are spaced about 15 km apart. *(NASA/JPL)*

atmosphere was discovered in 1944 by another Dutch astronomer, Gerard P. Kuiper; his spectra, obtained at the McDonald Observatory in Texas, showed absorptions due to methane gas (CH_4). Subsequent observations established the presence of dense clouds, obscuring the surface from our view.

The Voyager 1 flyby of Saturn was designed to yield as much information as possible about Titan and its atmosphere. Voyager passed within 4000 km, and it also flew behind the satellite as seen from the Earth, producing an occultation. Astronomers use the term **occultation** to describe any situation when one object passes in front of another, blocking its light. In this example, Titan passed in front of the spacecraft (or, equivalently, the spacecraft passed behind Titan). As seen from the Earth, the Voyager radio signal traversed successive paths through Titan's atmosphere, generating data from which scientists could reconstruct the atmospheric profile all the way down to the invisible surface. The measured surface pressure was 1.6 bars, higher than that on any of the terrestrial planets except Venus.

The composition of Titan's atmosphere is primarily nitrogen (N_2), another respect in which Titan resembles the Earth. CH_4 and argon amount at most to a few percent each. Additional compounds detected in Titan's upper atmosphere include carbon monoxide (CO), various hydrocarbons, and nitrogen compounds such as hydrogen cyanide (HCN). The discovery of HCN was particularly interesting, since this molecule is the starting point for formation of some of the components of deoxyribonucleic acid (DNA), the fundamental genetic molecule essential to life on Earth.

There are multiple cloud layers on Titan (Figure 10.6). The lowest clouds are in the troposphere, within the bottom 10 km of the atmosphere; these are condensation clouds composed of CH_4. This gas plays the same role in Titan's atmosphere as H_2O does on Earth; the gas is only a minor constituent of the atmosphere, but it condenses to form the major clouds in the troposphere. Much higher, photochemical reactions have produced a dark reddish haze or smog consisting of complex organic chemicals. Formed at an altitude of several hundred kilometers, this aerosol slowly settles downward, where it presumably has built up a deep layer of tar-like organic chemicals on the surface of Titan.

Titan's surface temperature is about 90 K, held uniform by the blanketing atmosphere. At such a low temperature, there may be seas of liquid methane and ethane. Organic compounds are chemically stable at Titan's temperatures, unlike the situation on the warmer, oxidizing Earth. Therefore Titan's surface probably records a chemical history that goes back

FIGURE 10.6 Enhanced color photograph of the upper atmosphere of Titan, showing multiple haze layers. *(NASA/JPL)*

billions of years. Many people believe that this satellite will provide more insights into the early history of Earth's atmosphere, and even into the origin of life, than any other object in the solar system.

Why does Titan have an atmosphere, while Ganymede and Callisto do not? Part of the answer lies in Titan's greater distance from the Sun, producing cooler temperatures that slow down the molecules in the atmosphere and decrease their rate of escape. But the primary reason must be that Titan outgassed from its interior more gas than was ever present on the two jovian satellites. Where Titan formed, small but significant amounts of methane and ammonia were present, while Ganymede and Callisto apparently had none. Subsequently, photochemical reactions dissociated most of the ammonia (NH_3) to release nitrogen (N_2), while the hydrogen escaped from Titan's atmosphere to produce the conditions seen today.

Europa and Io

Europa and Io, the inner two Galilean satellites, are not icy worlds like most of the satellites of the outer planets. Similar in density and size to our Moon, they appear to be predominantly rocky objects. How did they fail to acquire the ice that must have been plentiful at the time of their formation? The most probable cause is Jupiter itself, which became hot and radiated a great deal of infrared energy during the first few million years after its formation. Temperatures there-

FIGURE 10.7 Europa has a surface of water ice, crossed by complex cracks and low ridges. The width of this enhanced-color image is about 1000 km. *(NASA/USGS)*

In spite of its mainly rocky composition, Europa (Figure 10.7) has an ice-covered surface. In this way it is like the Earth, which also has global oceans of water, except that most of Europa's ocean may be frozen. There are very few impact craters, indicating that the surface of Europa has been capable of some degree of self-renewal. Additional indications of continuing internal activity are provided by an extensive network of cracks in its icy crust.

Io, the innermost of Jupiter's large satellites, might have been expected to be a twin of Europa. Instead, it displays a high level of volcanic activity, setting it off from the other objects in the planetary system.

Volcanoes of Io

The discovery of active volcanism on Io was the most dramatic event of the Voyager flybys of Jupiter. Eight volcanoes were seen erupting when Voyager 1 passed in March 1979, and six of these were still active four months later when Voyager 2 passed. These eruptions consisted of graceful plumes that extended hundreds of kilometers into space (Figure 10.8). The material erupted is not lava or steam or carbon dioxide (CO_2), all of which are vented by terrestrial volcanoes, but sulfur and sulfur dioxide (SO_2). Both of these can build up to high pressure in the crust of Io and then be ejected to tremendous heights. As the

fore rose in the disk of material near the planet, and the ice evaporated, leaving Europa and Io with compositions more appropriate to bodies in the inner solar system.

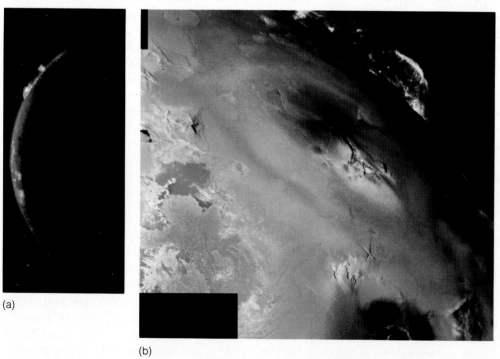

(a)

(b)

FIGURE 10.8 Two views of erupting volcanoes on Io. (a) Crescent view with two eruptions near the edge of the image. (b) A large plume rising above the volcano called Pele. *(NASA/JPL)*

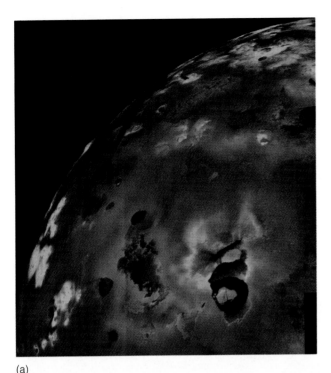

(a)

(b)

FIGURE 10.9 The Loki hot spot on Io. (a) The black, horseshoe-shaped "lava lake" about 200 km across. *(NASA/JPL)* (b) Image of the Loki hot spot obtained at 3.5-μm wavelength on Christmas night, 1989, with the NASA 3-m IRTF telescope in Hawaii. At this resolution (about 0.3 arcsec), the thermal glow of the volcano is easily visible against the sunlit surface of the satellite. *(University of Hawaii, courtesy of John Spencer)*

rising plume cools, the sulfur and SO_2 recondense as solid particles, which fall back to the surface in gentle "snowfalls" that extend as much as a thousand kilometers from the vent. The SO_2 snow is white, while sulfur forms red and orange deposits. Another sulfur compound detected on Io is hydrogen sulfide (H_2S). The surface of Io is slowly buried in these deposits, which accumulate at an average rate of a millimeter or so per year. Over millions of years, this is sufficient to cover any impact craters, so it is no surprise that such craters have not been seen on Io's surface.

Io displays other types of volcanic activity in addition to the spectacular plume eruptions. Images of its surface show numerous volcanoes and lava flows. From their bright colors, these lava flows are thought to be sulfur. Further volcanic activity is indicated by hot spots, surface areas that are hundreds of degrees warmer than their frigid surroundings. (Note that on Io, where the average daytime temperature is only 130 K, even a 300-K area, the surface temperature of the Earth, would classify as a hot spot.) The largest of these hot spots is a type of "lava lake" 200 km in diameter near the Loki eruption (Figure 10.9).

The SO_2 and other gases belched out by Io's volcanoes form a tenuous atmosphere. Io, however, orbits deep within the jovian magnetosphere, and its

surface is subject to a tremendous bombardment by energetic ions of sulfur and oxygen. The molecules in Io's thin atmosphere are broken apart and ionized by these charged particles. Once ionized, they are swept up in Jupiter's magnetic field and have a major influence on the huge magnetosphere of Jupiter.

How can Io maintain this remarkable level of volcanism, which exceeds that of much larger planets such as Earth and Venus? The answer lies in tidal heating of the satellite by Jupiter. Io is about the same distance from Jupiter as is our Moon from the Earth, yet Jupiter is more than 300 times as massive as the Earth, causing tremendous tides on Io. These tides pull the satellite into an elongated shape, with a bulge several kilometers high extending toward Jupiter. Now if Io always kept exactly the same face turned toward Jupiter, this tidal bulge would not generate heat. However, Io's orbit is not exactly circular, because of gravitational perturbations from Europa and Ganymede. In its slightly eccentric orbit, Io twists back and forth with respect to Jupiter, at the same time moving nearer and farther from the planet on each revolution. The twisting and flexing of the tidal bulge heats Io, much as repeated flexing of a wire coathanger heats the wire. In this way, the complex interaction of orbit and tides pumps energy into Io, melting its interior and providing power to drive its volcanic eruptions.

After billions of years, this tidal heating has taken its toll on Io, driving away H_2O and CO_2 and other gases, until now sulfur and sulfur compounds are the most volatile materials remaining. The inside is entirely melted, and the crust itself is constantly recycled by volcanic activity. Although Io was well mapped by Voyager, we expect that when re-imaged by the Galileo spacecraft in 1995, its surface will wear a partly unfamiliar face.

10.3 Triton and Pluto

In this section we discuss two apparently similar objects: Neptune's satellite Triton and the planet Pluto. Pluto is the only planet not visited by a spacecraft, and therefore it is especially useful to compare it with the better studied Triton.

Triton and Its Volcanoes

One of the most remarkable objects in the outer solar system is Neptune's retrograde-orbiting satellite Triton (not to be confused with Saturn's Titan). Triton was discovered in 1846 by the English amateur astronomer William Lassel, and it was investigated in some detail by Voyager 2 in 1989—the last target in Voyager's remarkable grand tour. Triton has a diameter of 2720 km and a density of 2.1 g/cm^3, indicating that it is probably composed of a mixture of 75 percent rock and 25 percent water ice.

The surface material of Triton is fresh ice or frost, with a very high average reflectivity of about 80 percent. This frost may include mixtures of H_2O, CH_4, and N_2, all of which are frozen at Triton's temperature. Because its reflectivity is so high, Triton reflects most of the incident solar energy, and its surface temperature is proportionately low—between 35 and 40 K. Most potential atmospheric gas is frozen at

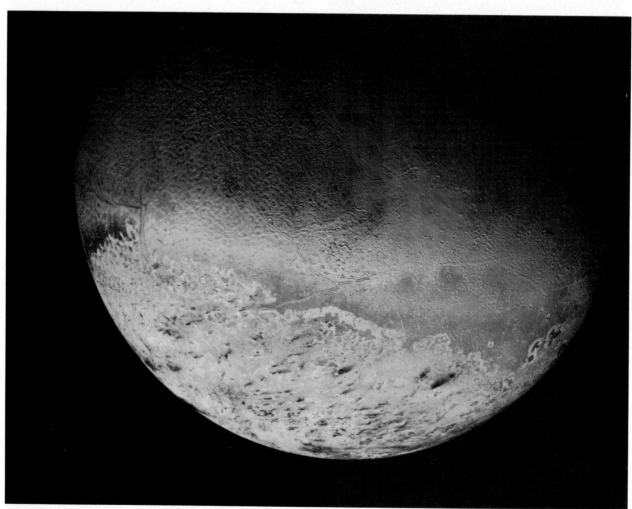

FIGURE 10.10 Global mosaic of Voyager images of Triton, showing a wide variety of surface features. The large polar cap dominates the lower part of this image. *(NASA/JPL)*

these temperatures, but a small quantity of N_2 vapor persists to form an atmosphere. The surface pressure of this atmosphere is only 16 millionths of a bar, yet this is sufficient to maintain a substantial ionosphere and to support haze or cloud layers.

Triton's surface, like that of many other satellites in the outer solar system, reveals a long history of geological evolution (Figure 10.10). While there are some impact craters, there are also many regions that have been flooded by "lava" (perhaps H_2O or H_2O/NH_3 mixtures). The evidence for such low-temperature volcanism includes a number of frozen "lava lakes" more than 100 km across (Figure 10.11). There are also mysterious regions of jumbled or mountainous terrain that resemble the mountainous regions of Ganymede.

The Voyager flyby of Triton took place at a time when the satellite's southern pole was tipped toward the Sun, and this part of the surface was enjoying a period of relative warmth. A polar cap covers much of the southern hemisphere, apparently evaporating along its northern edge. This polar cap may consist of frozen N_2, deposited during the previous winter. Remarkably, the evaporation of this polar cap generates geysers or volcanic plumes of N_2 that fountain to altitudes of about 10 km above the surface. These plumes differ from the volcanic plumes of Io in their composition, and they differ also in that they derive their energy from sunlight warming the surface rather than from internal heat.

Discovery of Pluto

Now we turn to Pluto, a planet that may be similar to Triton in many ways. Pluto was discovered through a careful, systematic search, not (like Neptune) as the result of a position calculated from gravitational theory. Nevertheless, the history of the search for Pluto began with indications of departures of Uranus and Neptune from their predicted orbits. Early in the 20th century, several astronomers became interested in this problem, most notably Percival Lowell, then at the peak of his fame as an advocate of intelligent life on Mars.

At the time Lowell made his calculations, Neptune had moved such a short distance since its discovery that it could not be used effectively to search for perturbations by an unknown ninth planet. Therefore Lowell and his contemporaries based their calculations primarily on the minute remaining irregularities of the motion of Neptune. Lowell's computations indicated two places where a perturbing planet could be, the more likely of the two being in the constellation of Gemini. He predicted a mass for the planet

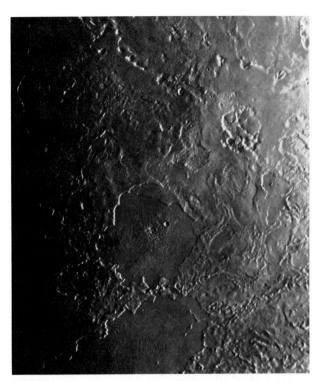

FIGURE 10.11 Old flooded "lava lakes" on Triton. These features, from 100 to 200 km in diameter, date from an earlier period of water volcanism on Triton. *(NASA/JPL)*

intermediate between that of the Earth and that of Neptune (his calculations gave about 6.6 Earth masses). Other astronomers, however, obtained other solutions, including one that indicated *two* exterior planets. Lowell searched for the unknown planet at his Arizona observatory from 1906 until his death in 1916, without success. Subsequently, Lowell's brother donated to the observatory a 33-cm photographic telescope that could record a 12° by 14° area of the sky on a single photograph. The new camera went into operation in 1929, and the search was continued for the ninth planet.

In February 1930, Clyde Tombaugh (Figure 10.12), comparing photographs made on January 23 and 29 of that year, found an object whose motion appeared to be about right for a planet far beyond the orbit of Neptune. It was within 6° of the position Lowell predicted for the unknown planet. The new planet was named for Pluto, the god of the underworld.

Although at the time the discovery of Pluto appeared to be a vindication of gravitational theory similar to the triumph of Adams and Leverrier in predicting the position of Neptune, we now know that Lowell's calculations were wrong. When the mass of Pluto was finally measured, it was found to be much less than that of the Moon. It could not possibly have exerted any measurable pull on either Uranus or Nep-

FIGURE 10.12 Clyde Tombaugh, at the time of his discovery of the planet Pluto in 1930. At this writing (in 1991), Tombaugh is still active in his study of the planets. *(Lowell Observatory, courtesy of Robert Millis)*

tune. To the degree that the discrepancies in the orbits of these two planets are real, they must be due to some other cause. A survey of the entire sky in the infrared region carried out in 1983 by the Infrared Astronomical Satellite (IRAS), however, has revealed no hidden "Planet X," and today it is generally accepted that the supposed perturbations of Uranus and Neptune are not, and never were, real.

Pluto's orbit has the highest inclination to the ecliptic (17°) of any planet and also the largest eccentricity (0.248). Its average distance from the Sun is 40 AU, or 5.9 billion km, but its perihelion distance is under 4.5 billion km, within the orbit of Neptune. Pluto is now closer to us than Neptune, and it will remain so until 1999. Even though the orbits of these two planets cross, there is no danger of collision because of the high inclination of Pluto's orbit. Pluto completes its orbital revolution in a period of 248.6 years; since its discovery in 1930, the planet has traversed less than one-quarter of its long path around the Sun.

Pluto's satellite Charon was discovered in 1978. As might be expected, Charon is very faint and difficult to see or photograph (Figure 10.13). The exact nature of its orbit was not confirmed until 1985, when the system had turned to the point at which the satellite and the planet began to occult each other on each satellite orbit. These observations confirmed that the satellite was in a retrograde orbit and indicated that it had a diameter of about 1200 km, nearly half the size of Pluto itself.

The Nature of Pluto

Pluto has not been visited by spacecraft, and it is so faint that studies require the use of the largest telescopes in the world. The diameter of Pluto is 2200 km, only 60 percent as large as the Moon. From the diameter and mass, we find a density of 2.1 g/cm³, suggesting the presence of water ice in about the same proportions as in Triton.

Pluto's surface is highly reflective, and its spectrum demonstrates the presence of methane ice. The surface temperature ranges from about 50 K near aphelion to 60 K near perihelion, resulting in partial evaporation of this methane ice to generate an atmosphere. Pluto is now in its warmest period, near perihelion, and its atmosphere is accordingly near maximum size and density. Observations of a stellar occultation suggest that the surface pressure is near 10^{-4} bar and that the atmosphere consists of a mixture of CH_4 and N_2 gases. Because Pluto is a few degrees warmer than Triton, its atmospheric pressure is about ten times greater.

Charon's surface shows the spectral signature of water ice rather than CH_4. Perhaps Charon once contained CH_4, but with its smaller surface gravity Charon was not able to retain this gas. Water ice, in contrast, remains tightly frozen at this distance from the Sun, and Charon therefore does not develop an atmosphere near perihelion, as Pluto does. The probable appearance of the Pluto-Charon system is illustrated in Figure 10.14.

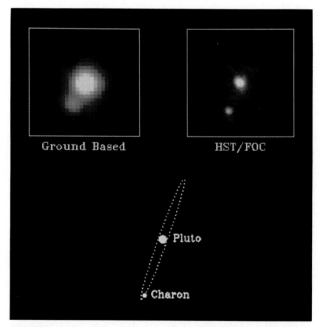

FIGURE 10.13 Pluto and its satellite Charon in an image from the Hubble Space Telescope. The separation of the two images is approximately 1 arcsec. *(Space Telescope Science Institute)*

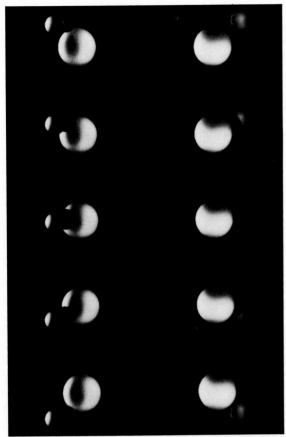

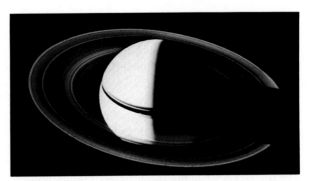

FIGURE 10.15 Voyager image of the rings of Saturn, photographed from below. Note that except for the B Ring, these rings transmit enough sunlight to be clearly visible. *(NASA/JPL)*

FIGURE 10.14 The appearance of Pluto and Charon as modeled in the computer by Marc Buie and David Tholen, based on observations of mutual occultations. On the left, Charon moves in front of Pluto; on the right, it moves behind the planet. *(Space Telescope Science Institute, courtesy of Marc Buie)*

10.4 Planetary Rings

All four of the jovian planets have ring systems, consisting of billions of small particles or moonlets orbiting close to their planet. Each ring system displays complex structure apparently related to interactions between the ring particles and the larger satellites. However, these rings are also very different from one another. Saturn's system, which is by far the largest, is made up primarily of small icy particles spread out into a vast flat ring with a great deal of fine structure (Figure 10.15). The uranian rings (with much smaller mass) are nearly the reverse, consisting of very dark particles confined to a few narrow rings, with broad gaps between. The neptunian rings are more tenuous but otherwise similar to those of Uranus. Finally, the jovian ring is merely a faint and transient dust band, constantly renewed by erosion of dust grains from its inner satellites. The main properties of these ring systems are summarized in Table 10.2.

Ring Origin and Dynamics

A ring is a collection of vast numbers of particles, each obeying Kepler's laws as it follows its own orbit around the planet. Thus the inner particles orbit faster than those farther out, and the ring as a whole does not rotate as a solid body. In fact, it is better not to think of a ring *rotating* at all, but rather to consider the *revolution* of its individual moonlets.

If the particles were widely spaced, they would move independently, like separate small satellites. However, in the rings of Saturn and Uranus the particles are close enough to one another to exert mutual

TABLE 10.2 Properties of Ring Systems				
	Outer Radius		**Mass**	**Reflectivity**
Planet	**(km)**	**(R_{planet})**	**(kg)**	**%**
Jupiter	128,000	1.8	10^{10} (?)	?
Saturn	140,000	2.3	10^{19}	60
Uranus	51,000	2.1	10^{14}	5
Neptune	63,000	2.5	10^{12}	5

gravitational influence, and occasionally even to rub together or bounce off of one another in low-speed collisions. Because of these interactions, phenomena such as waves can be produced that move across the rings, like water waves moving over the surface of the ocean.

There are two basic theories of ring origin. First is the breakup theory, which suggests that the rings are the remains of a shattered satellite. The second theory, which takes the reverse perspective, suggests that the rings are made of particles that were unable to come together to form a satellite in the first place.

In either theory, an important role is played by tidal forces. Around each planet there exists a **tidal stability limit,** which is the distance within which a satellite with no internal strength (like a pile of gravel) would be disrupted by tides. Alternatively, we may think of it as the distance within which the individual particles in a disk cannot attract one another to form a satellite. For most objects in the outer solar system, the tidal stability limit is at about 2.5 planetary radii from the center of a planet. All four ring systems lie within the tidal stability limits for their respective planets (Figure 10.16).

This stability limit applies only to a satellite with no intrinsic strength. A solid object held together by its own strength will not necessarily break up inside the limit. This is why we find small satellites (up to 100 km in diameter) orbiting within all four ring systems.

If the satellite is large enough, however, its intrinsic strength becomes less important in comparison to the tidal forces, and breakup is more likely. Also, a satellite within the limit will break up if it is fractured by a large impact, while if it is outside the limit, it is likely to fall back together under its own gravitation after such an impact.

In the breakup theory of ring formation, we can imagine a satellite or even a passing comet coming too close and being torn apart by tidal forces. A more likely variant of this idea suggests that a small satellite near the stability limit might be broken apart in a collision, with the fragments then dispersing into a disk. The third possibility is that the rings represent primitive material left over from the time of formation of the planet and its satellite system.

The Rings of Saturn

The rings of Saturn circle the planet in its equatorial plane, which is tilted by 27° to the planet's orbit plane. As Saturn revolves about the Sun, we see one side of the rings for about 15 years, followed by the other side for the same period. The three brightest rings of Saturn, visible from Earth, are labeled (from outer to inner) the A, B, and C Rings. The outer radius of the A Ring is 136,780 km, while the inner edge of the C ring is just 12,900 km above the cloud tops of Saturn.

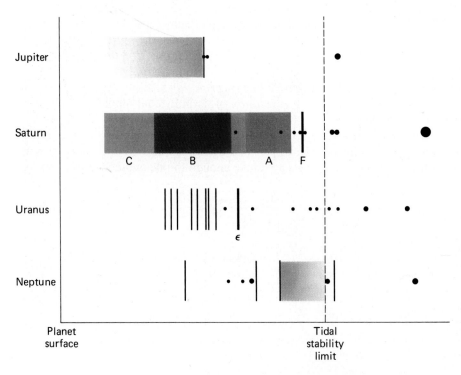

FIGURE 10.16 The ring systems of Jupiter, Saturn, Uranus, and Neptune, compared with the location of the tidal stability limit for each planet. All four ring drawings are scaled to the diameters of their respective planets. The black dots are the inner satellites of each planet on the same scale.

FIGURE 10.17 Saturn's B Ring in detail. The structure seen here has scales of tens to hundreds of kilometers. Most of this structure has not been explained. *(NASA/JPL)*

snowflakes and hailstones, including a number of snowballs and larger objects, many of which are loose aggregates of smaller particles.

As revealed by Voyager, the rings of Saturn have a great deal of complex structure, including about a dozen gaps, each tens to hundreds of kilometers wide (Figure 10.17). Some of these gaps contain peculiar eccentric ringlets, that is, ribbons of particles that do not share the circular orbits of the other ring particles.

The Pioneer and Voyager spacecraft revealed additional rings not visible from Earth. A faint D Ring lies inside the C Ring, and a very narrow F Ring, of radius 140,180 km, lies outside the A Ring. The F Ring is one of the most interesting features of the saturnian system, and it is the one ring of Saturn that is similar in many ways to the rings of Uranus and Neptune. This ring (Figure 10.18) has a mass equivalent to an icy satellite a few kilometers in diameter. Within its 100-km width there are many ringlets, including a double bright ring with two components just a few hundred meters wide. In some places, the F Ring breaks up into two or three parallel strands, which sometimes show bends or kinks. Further, the F Ring as a whole is eccentric.

The major B Ring has no gaps, but it contains intricate structure, partly in the form of waves. Each

The B Ring is the brightest and has the most closely packed particles, while the A and C Rings are translucent. The total mass of the B Ring is about equal to that of an icy satellite 300 km in diameter. The B and A Rings are separated by a gap easily visible from the Earth, discovered in 1675 by the Italian-French astronomer J.D. Cassini and called the Cassini Division. Although it looks empty from the Earth, the Cassini Division actually contains many ring particles with considerable structure. The Cassini Division looks like a gap only by contrast with the denser A and B Rings on either side of it.

The rings of Saturn are very broad but very thin. The width of the main rings is 70,000 km, yet their thickness is only about 20 m. If we made a scale model of the rings out of paper the thickness of the sheets in this book, we would have to make the rings a kilometer across—about eight city blocks. On this scale, Saturn itself would loom as high as an 80-story building.

The ring particles are composed primarily of water ice, and they range in size from grains of sand to house-sized boulders. An insider's view of the rings would probably resemble a bright cloud of floating

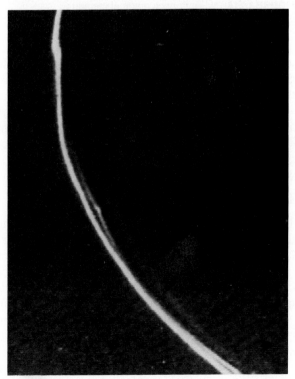

FIGURE 10.18 Voyager photograph of the narrow but complex F Ring of Saturn. *(NASA/JPL)*

FIGURE 10.19 Voyager photograph of the narrow, dark rings of Uranus. *(NASA/JPL)*

pattern of ring occultations was repeated later, as the opposite side of each arc passed in front of the star. Additional occultations led to the discovery of a total of nine narrow rings, and two more were added by Voyager in 1986.

Since the 1977 discovery, many more occultations have been observed to map out the uranian rings in detail, and in 1986 the Voyager 2 spacecraft was able to study them at close range. Despite their low reflectivity—typically about 5 percent—they could be photographed by the spacecraft cameras (Figure 10.19).

The broadest and outermost of the rings of Uranus is called the Epsilon Ring. The main Epsilon Ring is about the width of the Saturn F Ring and has an eccentricity of 0.008, although it also has a much wider component of lower density. Its thickness is probably no more than 100 m, and from probes with the spacecraft radio system it appears that most of the particles are relatively large—several meters or more in diameter. The Epsilon Ring circles Uranus at a distance of 2.1 Uranus radii—near the position of the tidal stability limit; it probably contains as much mass as all of the other ten rings combined.

The rings of Neptune are also invisible from the Earth, and they too were discovered from their occultation of starlight. Beginning in 1985, several ring occultations were observed, but their meaning remained in dispute. Unlike the symmetrical occultations observed at Uranus, the obscuration of a star by the "rings" on one side of Neptune was not repeated on the other side. At some occultation opportunities, the stars did not dim at all. It was inferred that if there really were rings at Neptune, they must be discontinuous or clumpy.

As photographed by Voyager 2, the rings of Neptune revealed themselves as real, but much fainter than the rings of Uranus (Figure 10.20). They are composed of dark particles and appear to contain a larger proportion of fine material (dust) than the uranian rings. With one exception, they are too tenuous to block starlight and thus generate an observable occultation. That exception applies to three arcs or condensations in the main ring (N63), each about 10° in length, which are the features that had been detected previously from the Earth.

wave corresponds to alternating ringlets in which the ring particles are bunched together or spread more thinly. The A Ring has even more of this wave-like structure. However, the bulk of the structure in the A and B Rings is not wave-like, but apparently random and irregular. This structure has not been satisfactorily explained.

The Rings of Uranus and Neptune

The rings of Uranus are narrow and black, making them almost invisible from the Earth. They were discovered accidentally in 1977 during observations of the occultation of a bright star by Uranus. A team of Cornell University astronomers observed the occultation from above the middle of the Indian Ocean using the NASA Kuiper Airborne Observatory, while others operated telescopes in Australia, China, India, and South Africa.

About 20 min before its predicted occultation by the planet, the star briefly dimmed several times as it disappeared behind successive narrow rings. This

10.5 Satellite-Ring Interactions

Much of the current fascination with planetary rings is a consequence of the intricate structures discovered by the Voyager spacecraft. There is a remark-

FIGURE 10.20 The rings of Neptune as photographed by Voyager in 1989. Note the three denser regions of the outer (N63) ring. The crescent image of Neptune is greatly overexposed in this time exposure. *(NASA/JPL)*

able amount of fine detail in the rings, as revealed by both photographs and occultations. The challenge is to understand the origin of this complex structure.

Ring Gaps

Most of the structure in the rings of Uranus and Saturn owes its existence to the gravitational effects of satellites. If there were no satellites, the rings would be flat and featureless. Indeed, if there were no satellites, there would probably be no rings at all, since left to themselves thin disks of matter gradually spread out and dissipate. The sharp edges as well as the fine structure of rings are due to the satellites.

For the most part, the rings lie closer to the planets than do the satellites. All of the larger satellites of Saturn, Uranus, and Neptune orbit well beyond the outer edge of the main rings. However, a few very

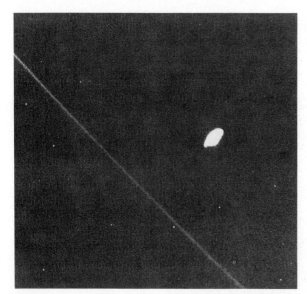

FIGURE 10.21 Voyager photograph of one of the shepherd satellites (Pandora) with Saturn's narrow F Ring. *(NASA/JPL)*

small satellites are more intimately connected with the rings, and these play an especially important role in governing the structure within the rings, as we shall see below.

Most of the gaps in Saturn's rings, and the location of the outer edge of the A Ring, apparently result from gravitational **resonances** with the innermost large satellites. A resonance takes place when two objects have orbital periods that are exact ratios of each other, such as 1:2, 1:3, and so on. For example, any particle in the gap at the inner side of the Cassini Division would have a period exactly equal to one-half that of the satellite Mimas. Such a particle would be nearest Mimas in the same part of its orbit every second revolution, and the repeated gravitational tugs of Mimas, acting always in the same direction, would perturb it, forcing it into a new orbit that does not represent a resonance with a satellite.

Resonances can form gaps by ejecting material with periods that are an exact multiple of satellite periods, but they can also lead to circumstances in which the boundary of a ring is stabilized by these gravitational effects. Such is the case for the sharp outer edge of the A Ring of Saturn, which is in a 7:6 resonance with the two satellites Janus and Epimetheus.

Narrow Rings and Shepherd Moons

One of the most interesting rings of Saturn is the F Ring, which is only 100 km wide. All but two of the rings of Uranus and Neptune are narrow ribbons like the F Ring of Saturn. What defines these narrow rings and keeps their particles from spreading out?

The best theory is that the rings are controlled gravitationally by small satellites orbiting very close to them. This certainly seems to be the case for Saturn's F Ring, which is bounded by the orbits of the satellites Pandora and Prometheus (Figure 10.21). These two small objects are referred to as shepherd satellites, since their gravitation serves to "shepherd" the ring particles and keep them confined to a narrow ribbon. A similar situation applies to the Epsilon Ring of Uranus, which is shepherded by Desdemona and Cordelia. These two shepherd satellites, each about 50 km in diameter, orbit about 2000 km inside and outside the ring, respectively.

Theoretical calculations suggest that the other narrow rings in the uranian system should also be controlled by shepherd satellites, but none have been located. The calculated diameter for such shepherds—about 10 km—was just at the limit of detectability for the Voyager cameras, so it is impossible to say if they are present or not. Satellites orbiting within rings can also clear narrow gaps, and such "embedded satellites" cause several of the gaps in the Saturn rings.

Making Waves in Rings

Saturn's major B Ring has no gaps. The Voyager data, however, indicate an intricate structure, partly

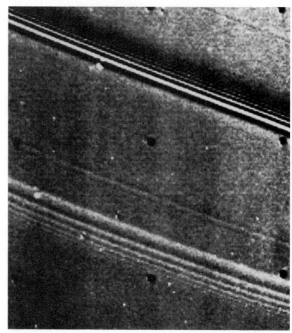

FIGURE 10.22 Spiral density waves and bending waves in the A Ring of Saturn as photographed by Voyager. *(NASA/JPL)*

in the form of waves. Photographed from the spacecraft, these waves, which are typically separated by 100 km or so, look like the grooves in a phonograph record (Figure 10.22). The A Ring has even more of this wave-like structure, and smaller waves appear to be present even in narrow rings, such as Saturn's F Ring and Uranus' Epsilon Ring.

Many of the observed waves are in the form of tightly wound spirals. These are **spiral density waves,** produced at distances from Saturn corresponding to resonances, mostly with the inner satellites. Astronomers are especially interested in these waves, since the effect is very similar to that which is thought to generate the spiral arms of galaxies, with the role of individual stars being played here by the ring particles.

Enceladus and the E Ring of Saturn

There is one other peculiar association of a satellite and a ring. Saturn has a very tenuous ring of fine ice particles called the E Ring, which envelops the orbit of its satellite Enceladus. Enceladus itself is a fascinating icy object (Figure 10.23); although its diameter is only about 500 km, about half of its surface is nearly free of craters. This evidence of powerful internal activity in the geologically recent past—within the past few hundred million years. In addition, the surface of Enceladus is among the most highly reflective of any planet or satellite, suggesting that it is covered with fine particles of fresh crystalline ice, like the glass beads on a projection screen. Has a recent volcanic eruption or impact sprayed out water droplets to freeze and form the E Ring, and subsequently to

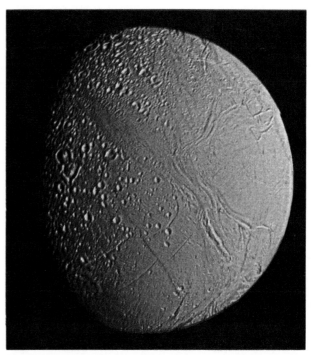

FIGURE 10.23 Enceladus, Saturn's most geologically interesting satellite, showing both heavily cratered and smooth regions of its surface. Enceladus orbits within the tenuous E Ring of Saturn, but the relationship between the satellite and the ring is not clear. *(NASA/JPL)*

coat the surface of Enceladus with bright material? Is such an event related to the crater-free areas on the surface? And most puzzling, what could maintain internal activity on a body as small as Enceladus, which should have cooled down very quickly after its formation? No one knows the answers to these questions.

SUMMARY

10.1 The four jovian planets are accompanied by systems of satellites and rings. Jupiter has 16 satellites, Saturn 19, Uranus 15, and Neptune 8, many discovered by Voyager. Most of these satellites are composed in part of water ice (up to 60 percent by mass). Saturn has the largest ring system, Uranus and Neptune have narrow rings of dark material, and Jupiter has a tenuous ring of dust.

10.2 The largest satellites are Ganymede, Callisto, and Titan (diameters about 5000 km, masses about twice that of the Moon). They are all differentiated objects, half-composed of H_2O. Callisto has an ancient cratered surface; Ganymede once experienced tectonic and volcanic activity; and Titan has a cloudy atmosphere (mostly N_2) with a surface pressure of 1.6 bars and interesting organic chemistry, as revealed by **occultation** observations. Smaller and

denser are Europa and Io, each about the size of the Moon. Io is the most volcanically active object in the solar system; its eruptions of sulfur and SO_2 are powered by tidal interactions with Jupiter.

10.3 Triton is the retrograde satellite of Neptune, smaller than Europa and made of about 75 percent rock, and 25 percent ice. It has a thin atmosphere and eruptions of N_2 from within its evaporating polar cap. The planet Pluto, discovered in 1930 by Tombaugh, has never been visited by a spacecraft, but we know it resembles Triton in size, composition, and temperature. Pluto has a relatively large satellite, Charon.

10.4 Rings are composed of vast numbers of individual particles orbiting a planet within its **tidal stability limit**

(about 2.5 planetary radii). They may have formed from the breakup of one or more satellites or from material that never formed a satellite in the first place. The saturnian rings are broad, flat, and nearly continuous except for a handful of gaps. The particles are water ice and are typically a few centimeters in dimension. The uranian rings are 100 times less massive and are narrow ribbons separated by wide gaps. The neptunian rings are similar but 100 times smaller yet. Both are made of dark particles.

10.5 Much of the complex structure of the rings is due to waves and **resonances** induced by the gravitational effects of inner satellites or by the effects of shepherd satellites embedded within the rings. Part of the structure is due to **spiral density waves** generated by satellites. The strange E Ring of Saturn may be connected with the geological history of its satellite Enceladus.

REVIEW QUESTIONS

1. What are the satellites of the outer planets made of, and why is their composition different from that of the Moon?

2. Compare the geology of Callisto, Ganymede, and Titan. Why does Titan have an atmosphere, while the other two large satellites do not?

3. Explain the energy source that powers the volcanoes of Io.

4. How are Triton and Pluto similar, and in what important ways might they be different?

5. Describe and compare the rings of Saturn and Uranus.

6. Three possibilities were suggested in the text for the origin of the rings of Saturn. List them and briefly summarize the arguments in favor of each.

THOUGHT QUESTIONS

7. Why do you think the outer planets have such extensive systems of rings and satellites, while the inner planets do not?

8. Which would have the greater period, a satellite 1 million km from the center of Jupiter or a satellite 1 million km from the center of Earth? Why?

9. Ganymede and Callisto were the first icy objects to be studied from a geological point of view. Summarize the main differences between their geology and that of the rocky terrestrial planets.

10. Compare the properties of the atmosphere of Titan with those of the Earth's atmosphere.

11. Compare the properties of the volcanoes of Io with those of terrestrial volcanoes.

12. Would you expect to find more impact craters on Io or Callisto? Why?

13. Where did the nitrogen in Titan's atmosphere come from? Compare with the origin of the nitrogen in our atmosphere.

14. Do you think there are many impact craters on the surface of Titan? Why or why not?

15. Explain why a large satellite is more likely to break up inside the tidal stability limit than a small satellite.

16. Why do you suppose the rings of Saturn are made of bright particles, whereas the particles in the rings of Uranus and Neptune are black?

17. Suppose you miraculously removed all of the satellites of Saturn. What would happen to its rings?

PROBLEMS

18. Saturn's A, B, and C Rings extend from a distance of about 75,000 to 137,000 km from the center of the planet. What is the approximate variation factor for the periods for various parts of the rings to revolve about the planet?

19. Occultations of stars by the rings of Uranus have yielded resolutions of 10 km in determining ring structure. What would be the angular resolution (in arcsec) that a space telescope would have to achieve to obtain equal resolution from Earth orbit? How close to Uranus would a spacecraft have to come to obtain equal resolution with a camera having angular resolution of 2 arcsec?

20. The main ring of Neptune (N63) has a radius of 63,000 km and a width of 15 km. Calculate the periods of revolution about Neptune for the inner and outer edges of the rings, and hence the relative velocities of the inner and outer edges. Now suppose a clump of material were placed at one location in the rings (like the three observed clumps in this ring). How long would it take for the differential speeds at the inner and outer edges to result in smearing the clump all the way around the planet?

Comet Halley photographed from Australia in 1985. *(Anglo-Australian Observatory)*

11

COMETS AND ASTEROIDS: DEBRIS OF THE SOLAR SYSTEM

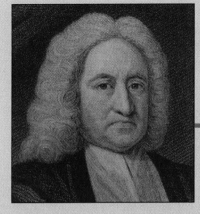

Edmund Halley (1656–1742), British astronomer, was a close friend of Isaac Newton. Halley was active in many areas of astronomy, but he is best known for his pioneering study of comets and for the realization that the comets of 1531, 1607, and 1682 were the same object. He predicted the 1758 return of this comet, which now bears his name. *(Pulkova Observatory, courtesy of Don Yeomans)*

Where did we come from? This is one of the most basic questions asked by astronomers (or anyone else). In previous chapters we have traced much of the history of the Earth and the other planets since they were formed. But we have not addressed the more basic problem of the origin of the solar system itself.

On questions of origins, the planets themselves are largely mute. Melted, battered by giant impacts, repeatedly resurfaced, they retain little evidence of their births. Rather, we must turn to clues provided by the surviving remnants of the creation process.

Asteroids are small objects, differing from the planets primarily in size. Sometimes they are called minor planets. A **comet** is also small, but it is defined as an object with a visible transient atmosphere and an extended tail of gas and dust. This definition reflects primarily a difference in composition between comets and asteroids: Comets contain water ice and other volatiles, whereas asteroids are rocky objects with little volatile material. In this chapter we discuss the asteroids and then the comets. In Chapter 12 we continue this story with the meteorites, those samples of cosmic material that survive their fall to the surface of the Earth.

11.1 Asteroids

Discovery of the Asteroids

Most of the asteroids have orbits between those of Mars and Jupiter, in a region called the **asteroid belt.** The asteroids are too small to be seen without a telescope, and the first of them was not discovered until the beginning of the 19th century, at a time when many astronomers thought there should be another planet in the large gap between the orbits of Mars and Jupiter.

In January 1801, the Sicilian astronomer Giovanni Piazzi thought he had found this missing planet when he discovered the first asteroid, which he named Ceres, orbiting at 2.8 AU from the Sun. However, this discovery was followed the next year by the discovery of another little planet in a similar orbit, and two more were found in 1804 and 1807. There was not a single "missing planet" between Mars and Jupiter, but rather a whole group of objects, each no more than 1000 km across. By 1890 more than 300 were known. In that year, Max Wolf of Heidelberg introduced astronomical photography to the search for asteroids, greatly accelerating the discovery of addi-

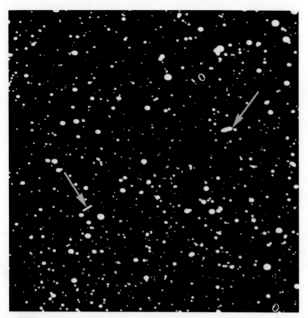

FIGURE 11.1   Time exposure showing trails left by asteroids (*arrows*). *(Yerkes Observatory)*

tional objects (Figure 11.1). More than 5000 asteroids now have well-determined orbits.

Originally, the names of asteroids were chosen from goddesses in Greek and Roman mythology. These, however, were soon used up. After exhausting other female names (including those of wives, friends, flowers, cities, colleges, pets, and the like), astronomers have recently turned to the names of astronomers who have studied the asteroids; asteroid number 2410 is named Morrison, for one of the authors of this text.

It would be a formidable task to discover, determine orbits for, and catalogue all the asteroids bright enough to be photographed with modern telescopes. Nevertheless, the total number of such objects can be estimated by systematically sampling regions of the sky. These studies indicate that there are approximately 100,000 (10^5) asteroids with a diameter greater than 1 km.

The largest asteroid is Ceres, with a diameter just under 1000 km. Two (Pallas and Vesta) have diame-

TABLE 11.1   The Twenty Largest Asteroids

Name	Discovery	Semimajor Axis (AU)	Diameter (km)	Class
Ceres	1801	2.77	940	C
Pallas	1802	2.77	540	C
Vesta	1807	2.36	510	*
Hygeia	1849	3.14	410	C
Interamnia	1910	3.06	310	C
Davida	1903	3.18	310	C
Cybele	1861	3.43	280	C
Europa	1868	3.10	280	C
Sylvia	1866	3.48	275	C
Juno	1804	2.67	265	S
Psyche	1852	2.92	265	M
Patientia	1899	3.07	260	C
Euphrosyne	1854	3.15	250	C
Eunomia	1851	2.64	245	S
Bamberga	1892	2.68	235	C
Camilla	1868	3.49	230	C
Herculina	1904	2.77	230	S
Doris	1857	3.11	225	C
Amphitrite	1854	2.55	225	S
Fortuna	1852	2.44	220	C

*Vesta has a very unusual (once thought unique) basaltic surface.

ters near 500 km, and about 15 are larger than 250 km (Table 11.1). The number of asteroids increases rapidly with decreasing size; there are about 100 more objects 10 km across than 100 km across. The total mass, which probably represents just a tiny fraction of the original asteroid population, is less than that of the Moon. The orbits of the four largest asteroids are illustrated in Figure 2.7.

The asteroids all revolve about the Sun in the same direction as the planets (from west to east), and most of them have orbits that lie near the plane of the Earth's orbit. The asteroid belt is defined to contain all asteroids with semimajor axes in the range 2.2 to 3.3 AU, with corresponding periods of orbital revolution about the Sun from 3.3 to 6 years (Figure 11.2). Although more than 75 percent of the asteroids are in

the main belt, they are not closely spaced. The volume of the belt is actually very large, and the typical spacing between objects (down to 1 km in size) is several million kilometers.

In 1917, the Japanese astronomer Kiyotsuga Hirayama found that a number of the asteroids fall into *families*, or groups with similar orbital characteristics. He hypothesized that each family may have resulted from an explosion of a larger body or from the collision of two bodies. Slight differences in the initial velocities of the fragments account for the small spread in orbital characteristics now observed for the different asteroids in a given family. There are several dozen such families, and for the larger ones observations have shown that their individual members are physically similar, as if they were fragments

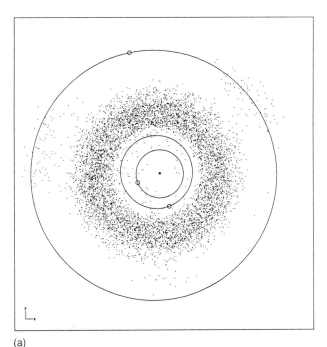

(a)

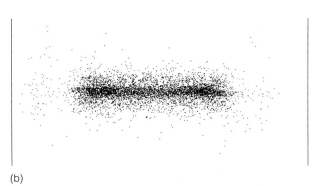

(b)

FIGURE 11.2 Three views of the asteroid belt. (a) The positions of more than 6000 asteroids in February 1990, seen from above. Also shown are the planets Earth, Mars, and Jupiter. (b) The same distribution viewed from the plane of the solar system. (c) A plot of the number of asteroids with various semimajor axes (in astronomical units). Some of the resonances are indicated at gaps where the period of an asteroid would be a simple fraction of the period of Jupiter.

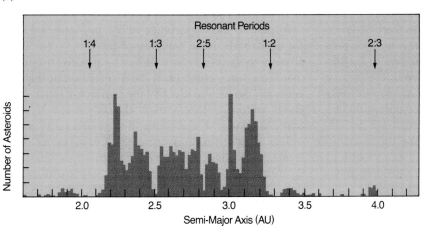

(c)

of a common parent. The existence of these families testifies to the frequency of asteroid collisions in the past.

Composition and Classification

Asteroids are as different as black and white. The majority are very dark, with reflectivities of only 3 or 4 percent, like a lump of coal. However, there is another large group with typical reflectivities of about 16 percent, a little greater than that of the Moon, and still others have reflectivities as high as 60 percent.

The dark asteroids are revealed from spectral studies to be primitive (chemically unchanged) bodies, composed of silicates mixed with dark organic carbon compounds. Two of the largest asteroids, Ceres and Pallas, are primitive, as are almost all of the objects in the outer third of the belt. Most of the primitive belt asteroids are classed as C asteroids, where C stands for carbonaceous or carbon-rich, but recently several other classes of primitive objects with different silicate minerals have also been identified.

The second most populous asteroid group comprises the S asteroids, where S stands for stony. In these asteroids the dark carbon compounds are missing, resulting in higher reflectivities and clearer spectral signatures of silicate minerals. The minerals present are common ones, similar to those that make up many meteorites, but the exact composition of the S asteroids remains in dispute. We are not even sure whether the S asteroids are primitive or differentiated.

Asteroids of a third class, much less numerous than those of the first two, are composed primarily of metal. These are called the M asteroids. Spectroscopically, the identification of metal is difficult, but for at least the largest M asteroid, called Psyche, this identification has been confirmed by radar. A metal asteroid, like an airplane or ship, is a much better reflector of radar than is a stony object. There is enough metal in even a 1-km M asteroid to supply the world's needs of iron and most other industrial metals for the foreseeable future, if we could only bring one to Earth.

No more than 5 percent of the asteroids are clearly differentiated objects. In addition to the M asteroids (presumably the cores of differentiated parent bodies that were shattered in collisions), there are a few asteroids that have basaltic surfaces like the volcanic plains of the Moon and Mars. The large asteroid Vesta is in the latter category. Apparently some of the asteroids were heated early in the history of the solar system, but why these, and why only a small percentage of the total number, we do not know.

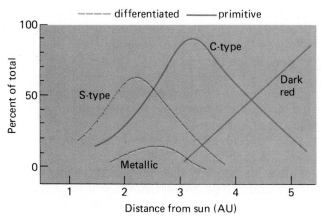

FIGURE 11.3 The distribution of asteroid compositional types with distance from the Sun. *(Adapted from work carried out by D. Tholen and J. Gradie [University of Hawaii] and E. Tedesco [JPL].)*

One of the most interesting products of compositional studies of asteroids has been the discovery that different classes of asteroids are grouped together at different distances from the Sun (Figure 11.3). Apparently the asteroids are still located near their birthplaces, and by tracing how their composition varies with distance from the Sun, we can reconstruct some of the properties of the solar nebula from which they originally formed.

Vesta: A Volcanic Asteroid

Vesta is one of the most interesting of the asteroids. It orbits the Sun with a semimajor axis of 2.4 AU, and its relatively high reflectivity of almost 30 percent makes it the brightest of the main belt objects, visible to the unaided eye if you know just where to look. But its real claim to fame is the fact that its surface is covered with basalt, indicating that Vesta was once volcanically active in spite of its small size (a diameter of about 500 km).

Adding to the importance of Vesta is the fact that we apparently have samples of its surface to study directly in the laboratory. It has long been suspected that meteorites come from the asteroids, but there is generally no way to identify the particular source of a given meteorite that strikes the Earth. In Vesta's case, however, this identification seems fairly firm.

The meteorites that are believed to come from Vesta are called the eucrites, a group of about 30 basaltic meteorites of very similar composition (Figure 11.4). Chemical analysis of the eucrites has shown that they cannot have come from the Earth, Moon, or Mars. On the other hand, their spectra

FIGURE 11.4 Photograph of one of the eucrite meteorites, believed to be fragments from the crust of asteroid Vesta. *(NASA)*

(measured in the laboratory) match perfectly the spectra of Vesta obtained telescopically. The age of the lava flows from which the eucrites derived has been measured at 4.4 to 4.5 billion years, very shortly after the formation of the solar system. This age is consistent with what we might expect for Vesta; whatever process heated such a small object was probably intense and short-lived.

Phobos and Deimos

The two satellites of Mars are probably captured asteroids. Since they have been investigated in some detail from several different spacecraft, we will discuss them here as possible analogs for "real" asteroids. However, there is no guarantee that Phobos and Deimos resemble other asteroids in detail. If they were captured, it must have been very early in solar system history. For more than 4 billion years they have been in orbit about Mars, and not out among the other asteroids.

Phobos and Deimos were first studied at close range by the Viking orbiters in 1977. In 1989, Phobos was the target of the Soviet spacecraft named "Phobos." Unfortunately, the Phobos spacecraft failed before it could land experiments on the surface, but this mission did add to our knowledge of this little world.

Both Phobos and Deimos (Figure 11.5) are rather irregular, somewhat elongated, and heavily cratered. The largest diameters of Phobos and Deimos are about 25 km and 13 km, respectively. Each is a dark brownish-gray in color, and spectral analysis suggests that each is composed of dark materials similar to those out of which most asteroids are made. Apparently these two satellites are chemically primitive.

Some additional clues to the compositions of Phobos and Deimos can be derived from their densities, which have been calculated from the masses and volumes measured by the Viking and Phobos spacecraft. Each has a density of only 2.0 g/cm³, a remarkably low value for a rocky object—

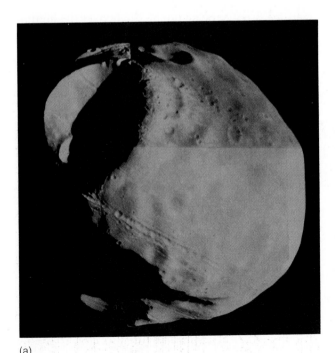

(a)

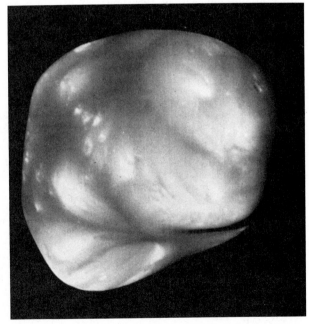

(b)

FIGURE 11.5 The satellites of Mars as photographed by Viking. (a) Phobos. (b) Deimos.
(NASA/JPL, courtesy of Peter Thomas, Cornell University)

FIGURE 11.6 Long "grooves" in the surface of Phobos, presumably features related to the ancient impact that produced the large crater Stickney. The width of the frame is 15 km. *(NASA/JPL, courtesy of Peter Thomas, Cornell University)*

substantially lower than the densities of Ceres or Pallas, for example, which are about 2.5 g/cm³. In fact, each has about the same density as do Pluto and Triton, which are thought to be made in part of water ice. At the temperatures of Phobos and Deimos, how-ever, ice is not expected to be stable over the lifetime of the solar system. So their composition remains something of a mystery.

When studied in detail, Phobos is found to have some quite remarkable surface features. It is laced by

FIGURE 11.7 A portion of the surface of Phobos that is unaffected by grooves. This is probably what the surface of an asteroid would look like imaged at a resolution of tens of meters. The width of the frame is 11 km. *(NASA/JPL, courtesy of Peter Thomas, Cornell University)*

long grooves or troughs associated with the large crater Stickney (Figure 11.6). Apparently the impact that produced Stickney very nearly ruptured the satellite. If we want to imagine what a more "normal" asteroid might look like imaged up close, Figure 11.7 is a better representation. Note the dark, rough surface, the frequent, small impact craters, and the irregular contours—all probably typical of the appearance of small asteroids.

11.2 Asteroids Far and Near

Not all of the asteroids are in the main asteroid belt. In this section, we consider some of the special groups of asteroids that stray outside the belt's boundaries.

The Trojans

The Trojan asteroids are objects located far beyond the main belt, orbiting the Sun at about the same distance as Jupiter, 5.2 AU. Calculations first carried out in the 18th century show that there should be two points in the orbit of Jupiter near which an asteroid can remain almost indefinitely. These are the two points that, with Jupiter and the Sun, make equilateral triangles (Figure 11.8). Between 1906 and 1908, four such asteroids were found; the number has now increased to several hundred. These asteroids are named for the Homeric heroes from the *Iliad* and are collectively called the Trojans.

Measurements of the reflectivities and spectra of the Trojans show that they are dark, presumably primitive objects like those in the outer part of the

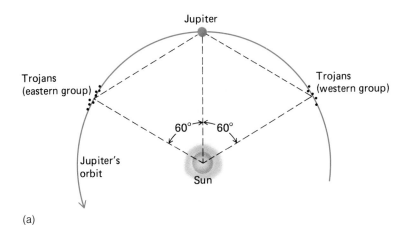

(a)

(b)

FIGURE 11.8 (a) Locations of the Trojan points of the orbit of Jupiter. (b) Plot of the actual positions of the 132 known Trojan asteroids in February 1990. *(Part b courtesy of Edward Bowell, Lowell Observatory)*

asteroid belt. They appear faint because they are so dark and far away, but actually the larger Trojans are quite sizable. Four of them—Hektor, Diomedes, Agamemnon, and Patroclus—have diameters between 150 and 200 km. Hektor is about twice as long as it is wide, leading to the suggestion that it is a double asteroid, with two similar objects orbiting in contact with each other.

In 1990 the first asteroids in Trojan-type orbits were discovered in association with Mars. Like Mars, these asteroids have semimajor axes of 1.5 AU. Other planets, including the Earth, may have their own cloud of small Trojan asteroids.

Hidalgo and Chiron

There are two known asteroids with orbits that carry them far beyond Jupiter. Hidalgo, with its semimajor axis of 5.9 AU and a very large eccentricity of 0.66, has an aphelion outside the orbit of Saturn. Still more distant is Chiron, which has a semimajor axis of 13.7 AU. Its orbit carries it from just inside the orbit of Saturn at perihelion out almost to the orbit of Uranus. Its diameter is estimated to be about 200 km.

Ever since the discovery of Hidalgo and Chiron, astronomers have wondered about the relationship between such distant asteroids and the comets. If these two objects are composed of volatiles (like water or dry ice), they would probably develop atmospheres if they were heated by the Sun. In that case we would call them comets, not asteroids.

In 1988 this speculation was settled for Chiron, when astronomers found that it had brightened by about a factor of 2. Presumably the additional light was being reflected from gas or dust ejected from the surface. One year later, this new atmosphere was photographed. Chiron has been slowly approaching its perihelion for several years, and the gradual increase in its temperature was sufficient by 1988 to initiate cometary activity.

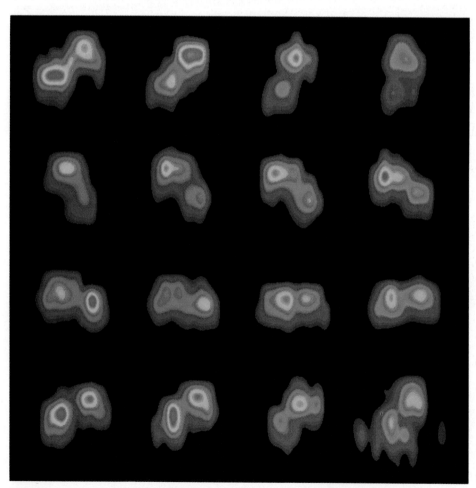

FIGURE 11.9   Radar images of the Earth-approaching asteroid 1989 PB over a period of 3 h, showing the elongated or "dumbbell" shape of the asteroid and its rotation. The resolution of these images is about 50 m. *(Arecibo Observatory; courtesy of Steven Ostro, JPL)*

Chiron and Hidalgo are much larger than the nuclei of any known comet. What a spectacular show they would put on if either were diverted into the inner solar system! Chiron has been shown to be in an unstable orbit, so sometime in the future this is exactly what may happen.

Earth-Approaching Asteroids

Of great interest are the asteroids with orbits that come close to or cross the orbit of the Earth. Some of these are the nearest approaching celestial objects, excluding the Moon and meteorites. In 1989 a 200-m object passed within 800,000 km, and in 1991 a tiny 10-m object was picked up passing just 170,000 km from our planet. Some of these objects have collided with the Earth in the past, and others will continue to do so, as we saw in Section 4.5. These are known collectively as **Earth-approaching asteroids.**

At present, fewer than 150 Earth-approaching asteroids have been located, although the total population of such objects is calculated at approximately 1000 for those larger than 1 km in diameter, with many more that are smaller yet. The largest is about 40 km across. Figure 11.9 illustrates radar images of an Earth-approaching asteroid that is very elongated and perhaps even dumbbell-shaped. Searches for additional Earth-approaching asteroids result in the discovery of several dozen new objects each year. Information on some of these is found in Table 11.2.

The orbits of Earth-approaching asteroids are unstable. These objects will meet one of two fates: Either they will impact one of the terrestrial planets, or they will be ejected gravitationally from the inner solar system as the result of a near-encounter with a planet. The probabilities of these two outcomes are about the same. The time scale for impact or ejection is only about 100 million (10^8) years, very short in comparison with the age of the solar system.

If the current population of Earth-approaching asteroids will be removed by impact or ejection in 10^8 years, there must be a continuing source of new objects. Some of these apparently come from the asteroid belt, where collisions between asteroids can eject fragments into Earth-crossing orbits. Others may be dead comets that have exhausted their volatiles. Possibly as many as half of the Earth approachers are the solid remnants of former comets.

TABLE 11.2 Some Earth-Approaching Asteroids*

Name	Orbit Type	Discovery	Semimajor axis (AU)	Eccentricity	Diameter (km)	Class
Eros	Amor	1898	1.46	0.22	22.	S
Ganymed[1]	Amor	1924	2.66	0.54	37.	S
Amor	Amor	1932	1.92	0.44	—	—
Icarus[2]	Apollo	1949	1.08	0.83	0.9	—
Geographos	Apollo	1951	1.24	0.34	2.0	S
Apollo	Apollo	1932	1.47	0.56	1.5	†
Sisyphus[3]	Apollo	1972	1.89	0.54	8.2	—
Aten	Aten	1976	0.97	0.18	0.9	S
Ra-Shalom	Aten	1978	0.97	0.18	2.4	C
Phaethon[2]	Apollo	1983	1.27	0.89	6.9	†
1989FC[4]	Apollo	1989	1.02	0.36	0.2	—
1989PB[5]	Apollo	1989	1.06	0.48	0.6	—

*Data in part from L.A. McFadden, in *Asteroids II*, University of Arizona Press, 1989
†Unusual spectral class
[1]Largest Earth-approaching asteroid
[2]High eccentricity; possibly a dead comet
[3]Largest currently Earth-crossing asteroid
[4]Passed within 700,000 km of Earth on 22 March 1989
[5]Dumbbell-shaped (see Figure 11.9)

11.3 The Long-Haired Comets

Appearance of Comets

Comets have been observed from the earliest times; accounts of spectacular comets are found in the histories of virtually all ancient civilizations. A typical comet has the appearance of a rather faint, diffuse spot of light, somewhat smaller than the Moon and many times less brilliant (Figure 11.10). There may be a very faint, nebulous **tail,** extending several degrees away from the main body of the comet.

Like the Moon and planets, comets slowly shift their positions in the sky from night to night, remaining visible for periods that range from a few days to a few months. Unlike the planets, however, most comets appear at unpredictable times, perhaps explaining why they have frequently inspired fear and superstition.

Today we recognize comets as the best preserved, most primitive material available in the solar system. Stored in the deep freeze of space, these icy objects are messengers from the distant past, providing us unique access to the initial material from which the planets formed 4.5 billion years ago.

FIGURE 11.10   Comet Halley in the spring of 1986 had the appearance typical of a moderately bright comet with a tail a few degrees long. It is shown here rising above Mauna Kea Observatory in Hawaii. *(William Golisch, University of Hawaii)*

Comet Orbits

The study of comets as members of the solar system dates from the time of Newton, who first suggested that their orbits were extremely elongated ellipses. Newton's colleague Edmund Halley developed these ideas, and in 1705 he published calculations of 24 cometary orbits. In particular, he noted that the orbits of the bright comets of 1531, 1607, and 1682 were so similar that the three could well be the same comet, returning to perihelion at average intervals of 76 years. If so, he predicted that the object should return about 1758. When the comet did return as predicted, it was given the name Comet Halley, in honor of the man who first recognized it to be a permanent member of the solar system.

Comet Halley has been observed and recorded on every passage near the Sun at intervals from 74 to 79 years since 239 B.C. The period varies somewhat because of changes in its orbit produced by the jovian planets. In 1910 the Earth was brushed by the comet's tail, and Comet Halley last appeared in 1986.

A substantial number of comets have periods near six years and aphelia near the orbit of Jupiter. These are the Jupiter family of comets, which have been deflected by the gravity of Jupiter into short-period orbits. Because they are easier to reach than the long-period comets, several of these comets in the Jupiter

family have been considered as targets for spacecraft missions.

Observational records exist for about a thousand comets. Today, new comets are discovered at an average rate of five to ten per year. Most never become conspicuous and are visible only on photographs made with large telescopes. Every few years, however, a comet may appear that is bright enough to be seen easily with the unaided eye. The brightest comet of recent years was Comet West in 1976. Comet Halley, in 1986, was less bright as a result of its poor placement relative to the Earth, although it put on a good show for a few weeks when seen from the Southern Hemisphere. Table 11.3 lists some well-known comets.

The Comet's Nucleus

When we look at a comet, we see its transient atmosphere of gas and dust, illuminated by sunlight. Since the escape velocity from such small bodies is very low, the atmosphere we see is rapidly escaping; therefore it must be coming from somewhere. The source is in the heart of the comet's head, usually hidden in the glow of its atmosphere; there is found the small, solid **nucleus.** The nucleus is the *real* comet, the fragment of primitive material preserved since the beginning of the solar system (Figure 11.11).

TABLE 11.3 Some Well-Known Comets

Name	Period	Special Interest
Great Comet of 1577	Long	Found by Tycho to be beyond Moon
Great Comet of 1811	Long	Largest head (>2 million km)
Great Comet of 1843	Long	Brightest ever; visible in daylight
Donati's Comet of 1858	Long	Multiple tails; very beautiful
Daylight Comet of 1910	Long	Brightest comet of 20th century
Kohoutek (1973)	Long	Widespread public interest
West (1976)	Long	Best recent comet; nucleus broke up
Halley	76 yr	First periodic; 1986 spacecraft flybys
Schwassmann-Wachmann 1	16 yr	Outbursts far from Sun
Biela	6.7 yr	Broke up in 1846; disappeared
Giacobini-Zinner	6.5 yr	First spacecraft encounter, 1985
Tempel 2	5.3 yr	Target of NASA rendezvous mission
Encke	3.3 yr	Shortest period

Until about the past decade, no comet nucleus had been seen or measured; the very existence of the nucleus as a single solid object was speculative. Now most of the studies being made of comets are directed toward a better understanding of the nucleus.

The modern theory of the physical and chemical nature of comets was first proposed by Harvard astronomer Fred L. Whipple in 1950 (Figure 11.12). Before Whipple's work, many astronomers had thought that the nucleus of a comet might be a loose aggregation of solids of meteoritic nature—a sort of orbiting "gravel bank." Whipple proposed instead that the nucleus is a solid object a few kilometers across, composed in substantial part of water ice, mixed with silicate grains and dust. This proposal became known as the "dirty snowball" model for the nucleus of a comet.

The water vapor and other volatiles that escape from the nucleus when it is heated can be detected telescopically in the comet's head and tail. We are somewhat less certain of the non-icy component of the nucleus, however. No large fragments of solid matter from a comet have ever survived passage through the Earth's atmosphere to be studied as me-

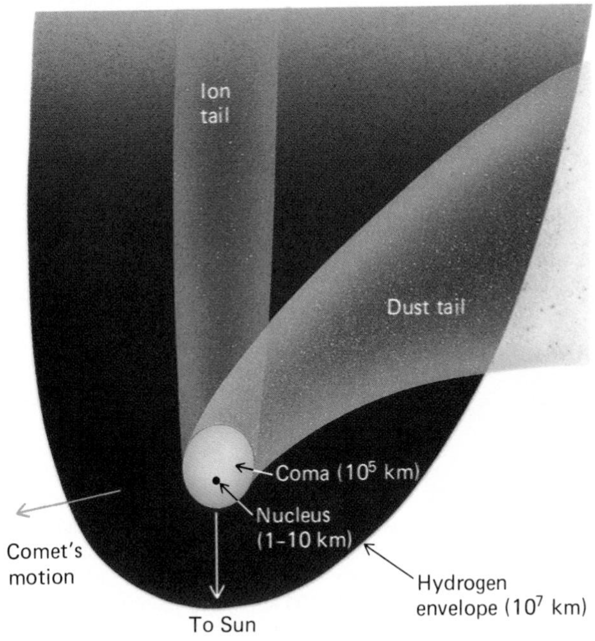

Ion tail

Dust tail

Coma (10^5 km)

Nucleus (1–10 km)

Comet's motion

To Sun

Hydrogen envelope (10^7 km)

FIGURE 11.11 Schematic illustration of the parts of a comet.

FIGURE 11.12 Fred Whipple, the father of modern comet studies, conferring with Roald Sagdeev (*right*), leader of the team that sent the VEGA spacecraft into the head of Comet Halley. This photograph was taken in Moscow on the date of the VEGA 1 encounter with the comet. *(David Morrison)*

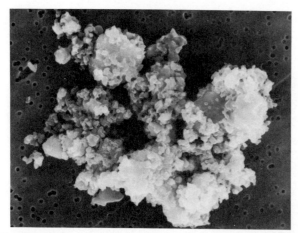

FIGURE 11.13   A particle that is believed to be a tiny fragment of cometary dust, collected in the upper atmosphere of the Earth. *(Donald Brownlee, University of Washington)*

teorites. Some very fine, microscopic grains of comet dust have been collected in the Earth's upper atmosphere, however, and have been studied in the laboratory (Figure 11.13). The spacecraft that encountered Comet Halley in March 1986 also carried dust detectors. From these various investigations it seems that much of the "dirt" in the dirty snowball is in the form of tiny bits of dark, primitive hydrocarbons and silicates, rather like the material thought to be present on the dark, primitive asteroids.

Since the comet nucleus is small and dark, it is a difficult object for astronomers to study. Even measuring its diameter has been a problem. However, recent radar observations of several faint comets have indicated diameters for the nucleus of 5 to 10 km. Comet Halley, as measured in 1986, has dimensions of about 8 by 12 km.

Activity on the Nucleus

The spectacular activity of comets that gives rise to their atmospheres and tails results from the evaporation of cometary ices when they are heated by sunlight. Beyond the asteroid belt, where comets spend most of their time, the ices are solidly frozen. But as a comet approaches the Sun, it begins to warm up. If water (H_2O) is the dominant ice, significant quantities vaporize as temperatures rise toward 200 K, somewhat beyond the orbit of Mars. The evaporating H_2O in turn releases the dust that was mixed with the ice. Since the comet nucleus is so small, its gravity cannot hold back either the gas or the dust, both of which flow away into space at speeds of about 1 km/s.

The comet continues to absorb energy as it approaches the Sun. A great deal of this energy goes into the evaporation of its ice, as well as into heating the surface. However, observations of many comets indicate that the evaporation is not uniform, and that most of the gas is released in sudden spurts, perhaps confined to a few areas of the surface. Such jets were observed directly on the surface of Comet Halley by the spacecraft that photographed it in 1986, which showed that the jets resembled volcanic plumes or geysers (Figure 11.14). Most of the surface is apparently inactive, with the ice buried under a layer of black silicates and carbon compounds.

The Comet's Atmosphere

The atmosphere of a comet is composed of the gas released from the nucleus, together with the dust and other solid material being carried along with it. Expanding at a speed of about 1 km/s, the atmosphere can reach an enormous size. The diameter of the

FIGURE 11.14   Composite photograph of the black, irregularly shaped nucleus of Comet Halley. This image, obtained by the Giotto spacecraft at a distance of about 1000 km, has a resolution of better than 1 km. *(Max Planck Institut für Aeronomie and Ball Aerospace Corporation, courtesy of Harold Reitsema)*

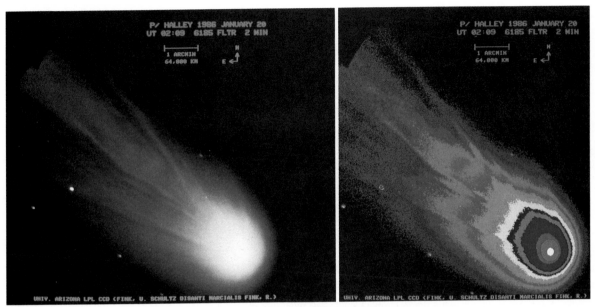

FIGURE 11.15 Two versions of the same photograph of the head of Comet Halley taken January 20, 1986. False color is frequently used in modern image processing to display the wide range in brightness that is recorded in images obtained with electronic charge-coupled device (CCD) cameras. *(University of Arizona, courtesy of Uwe Fink)*

comet's head is usually as large as Jupiter, and it often approaches 1 million km (Figure 11.15). The composition of the gas is primarily H_2O (about 80 percent in the case of Comet Halley), a few percent of carbon dioxide (CO_2), and small quantities of many additional gases, including hydrocarbons. Hydrogen is produced when solar ultraviolet light breaks up the molecules of H_2O. Huge hydrogen clouds, up to tens of millions of kilometers across, are formed around comets.

Many comets develop tails as they approach the Sun. The tail of a comet is an extension of its atmosphere, consisting of the same gas and dust that make up the head. As early as the 16th century, observers realized that comet tails point away from the Sun (Figure 11.16), not back along the comet's orbit. Newton attempted to account for comet tails by a repulsive force of sunlight driving particles away from the comet's head, an idea that is close to our modern view. In addition to sunlight, however, we

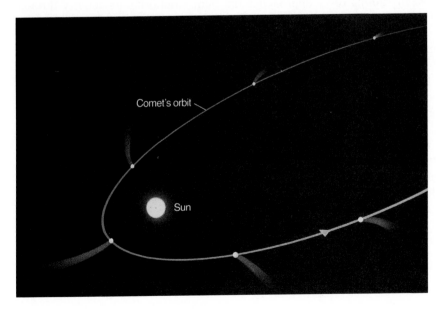

Comet's orbit

Sun

FIGURE 11.16 Orientation of a typical comet tail as the comet passes perihelion.

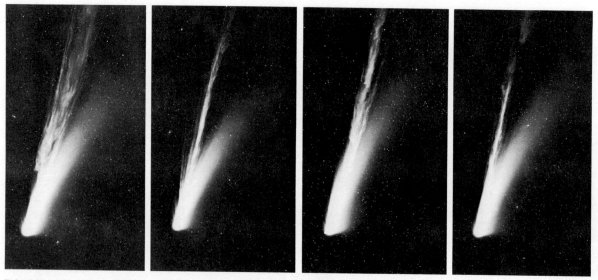

FIGURE 11.17   Comet Mrkos, photographed with the Schmidt telescope at Palomar Observatory. Note the features in the tail that are swept back by the solar wind. *(CalTech/Palomar Observatory)*

now know that cometary gas is also repulsed by streams of ions that are emitted by the Sun (Figure 11.17).

Spacecraft Encounters with Comets

In 1985 the first spacecraft encounter with the tail of a comet took place. A U.S. Earth satellite called the International Sun-Earth Explorer was diverted in a complicated set of maneuvers so that it left our planet and intercepted Comet Giacobini-Zinner. This spacecraft, renamed the International Comet Explorer (ICE), was not instrumented, however, to measure the nucleus; that was left to the flotilla of spacecraft that encountered Comet Halley the following year.

Six spacecraft made measurements of Comet Halley in March of 1986. Three of these—the U.S. ICE spacecraft and the Japanese craft named Sakagaki and Suisei—served to monitor the comet from a distance. The primary exploration tasks were undertaken by three craft targeted for the nucleus itself. The Soviet VEGA 1 and VEGA 2 were the first to arrive, on March 6 and 9, 1986. Each plunged deeply into the inner atmosphere and dust cloud of the comet, passing within about 8000 km of the nucleus. Both VEGA craft were severely damaged by dust impacts, losing most of their solar cells and suffering the loss of several instruments at the time of closest approach.

The trajectory data for the VEGA craft were provided to the European Space Agency to allow them to target their Giotto spacecraft for an even closer encounter on March 14, 1986, just 605 km from the comet's nucleus. Giotto obtained a beautiful photograph of the nucleus (see Figure 11.14) and also carried out many measurements of its near environment, confirming and extending the Soviet results.

Halley is the only bright comet with a predictable orbit that can be targeted for spacecraft investigation. To take the next step, mission planners are turning to the smaller short-period comets. In 1996, NASA expects to launch a spacecraft that will match the orbit of Comet Tempel 2 near aphelion, when the comet is inactive. The spacecraft will stay with the comet as it approaches the Sun, monitoring its activity through perihelion. By thus extending the observation period from a few minutes to several months, this mission should provide much more information on cometary activity than did the Halley flybys.

11.4   Origin and Evolution of Comets

The Oort Comet Cloud

Although comets are part of the solar system, observations show that they come initially from very great distances. Observationally, the aphelia of new comets typically have values near 50,000 AU. This clustering of aphelion distances was first noted by Dutch astronomer Jan Oort, who in 1950 proposed a scheme for the origin of the comets that is still accepted today.

It is possible to calculate that the *gravitational sphere of influence* of a star—the distance within which it can exert sufficient gravitation to hold onto orbiting objects—is about one-third of the distance to the nearest other stars. In the vicinity of the Sun, stars are spaced such that the Sun's sphere of influence extends only a little beyond 50,000 AU, or about 1 LY. At such distances, objects in orbit about the Sun are perturbed by the gravitation of passing stars. Oort suggested, therefore, that the new comets were objects orbiting the Sun with aphelia near the edge of its sphere of influence, and that the perturbing effects of other nearby stars modified their orbits to bring them close to the Sun where we can see them. This region of space from which the new comets are derived is called the **Oort comet cloud.**

There are thought to be about a trillion (10^{12}) comets in the Oort cloud. In addition, it is hypothesized that about ten times this number of potential comets are orbiting the Sun in the volume of space between the planets and the Oort cloud at 50,000 AU. These objects remain undiscovered because their orbits are too stable to permit any of them to be deflected inward close to the Sun. The total number of cometary objects is thus on the order of 10^{13}.

What is the mass represented by 10^{13} comets? We can make an estimate if we assume something about typical comet sizes and masses. Let us suppose that the nucleus of Comet Halley is typical. Its observed volume is about 600 km^3. If the primary constituent is water ice with a density of about 1 g/cm^3, the total mass of the nucleus must be about 6×10^{14} kg, or 10^{-10} Earth masses. The corresponding mass for all the comets is about 1000 Earth masses—greater than the mass of all the planets put together. Therefore, cometary material is probably the most important constituent of the solar system after the Sun itself.

The Fate of Comets

Any comet we see today will have spent nearly its entire existence in the Oort cloud, at a temperature near absolute zero. But once a comet enters the inner solar system, its previously uneventful life history begins to accelerate. It may, of course, survive its initial passage near the Sun and return to the cold reaches of space where it spent the previous 4.5 billion years. At the other extreme, it may impact the Sun or pass so close that it is destroyed on its first perihelion passage. Observations from space indicate that at least one comet collides with the Sun every year. Frequently, however, the new comet does not come that close to the Sun, but instead it interacts with one or more of the planets.

A comet coming within the gravitational influence of a planet has three possible fates: (1) It can impact the planet, ending the story at once; (2) it can be speeded up and ejected, leaving the solar system forever; or (3) it can be perturbed into a shorter period. In the last case, its fate is sealed. Each time it approaches the Sun, it will lose part of its material, and it also has a significant chance of collision with a planet. Once the comet is in a short-period orbit, the comet's lifetime is measured in thousands, not billions, of years.

Measurements of the amount of gas and dust in the atmosphere of a comet permit an estimate of the total losses during one orbit. Typical loss rates are up to a million tons per day from an active comet near the Sun, adding up to some tens of millions of tons per orbit. At that rate, the comet will be gone after a few thousand orbits.

Whether the comet evaporates completely is not known. If the gas and dust are well mixed, we would expect the nucleus to shrink each time around the Sun until it has entirely disappeared. However, there remains the suggestion that many of the Earth-approaching asteroids are extinct comets. If there is a silicate core in the comet, or if the dirty snowball includes large blocks of nonvolatile material that are held gravitationally to the surface, then there could be a substantial solid residue after the ices are gone. We simply do not know which of these alternatives is correct.

11.5 Comet Dust and Meteors

Whatever the fate of the remnants of the nucleus after the volatiles are exhausted, we do know what happens to the dust that is carried away from a comet by the evaporation of the nucleus. This dust fills the inner part of the solar system. The Earth is surrounded by it. And each of the larger dust particles that reach the Earth creates a shooting star, or **meteor.**

The Phenomenon of a Meteor

Although the layperson often confuses comets and meteors, these two phenomena are very different. Comets can be seen when they are many millions of miles away from the Earth and may be visible in the sky for weeks or even months, slowly shifting their positions from day to day. They rise and set with the stars, and during a single night they appear motionless to the casual glance. Meteors, on the other hand, are small, solid particles that enter the Earth's atmosphere from interplanetary space. Since they move at speeds of many kilometers per second, the high friction they encounter in the air vaporizes them. The light caused by the luminous vapors formed in such an encounter appears like a star moving rapidly across the sky, fading out within a few seconds.

On a dark, moonless night, an alert observer can see half a dozen meteors per hour. To be visible, a meteor must be within 200 km of the observer. Over the entire Earth, the total number of meteors bright enough to be visible must total about 25 million per day.

The typical bright meteor is produced by a particle with a mass less than 1 g—no larger than a pea. Of course, the light you see comes from the much larger region of glowing gas surrounding this little grain of interplanetary material. A particle the size of a golf ball produces a bright fireball when it strikes the atmosphere, and one as big as a bowling ball has a fair chance of surviving its fiery entry to become a meteorite if its approach speed is not too high. The total mass of meteoritic material entering the Earth's atmosphere is estimated to be about 100 tons per day.

Meteor Showers

Many—perhaps most—of the meteors that strike the Earth can be associated with specific comets. These interplanetary dust particles retain approximately the orbit of their parent comet, and the particles travel together through space. When the Earth crosses such a dust stream, we see a sudden burst of meteor activity, usually lasting several hours. These events are called **meteor showers.**

The dust particles that produce shower meteors are moving together in space before they encounter the Earth. From our perspective, as we look up at the atmosphere, their parallel paths seem to diverge from a place in the sky called the radiant, which is the direction in space from which the meteor stream is moving (Figure 11.18). Thus we distinguish a meteor shower by the common radiant from which the meteors move, as well as by their frequency.

The meteoric dust is not always evenly distributed along the orbit of the comet, so that sometimes more meteors are seen when the Earth intersects the dust stream and sometimes fewer. A very clumpy distribution is associated with the Leonid meteors, which in 1833 and again in 1866 (after an interval of 33 years—the period of the comet) yielded the most spectacular showers ever recorded. The last good Leonid shower was on November 17, 1966, when in some southwestern states up to 100 meteors could be observed per second.

The best meteor shower that can be depended on at present is the Perseid shower, which appears for about three nights near August 11 each year. In the

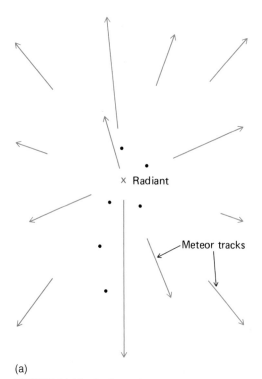

(a)

(b)

FIGURE 11.18 Radiant of a meteor shower.

TABLE 11.4   Major Annual Meteor Showers

Shower Name	Date of Maximum	Associated Comet	Comet's Period Yr
Quadrantid	January 3	—	—
Lyrid	April 21	1861 I	415
Eta Aquarid	May 4	Halley	76
Delta Aquarid	July 30	—	—
Perseid	August 11	1862 III	105
Draconid	October 9	Giacobini-Zinner	7
Orionid	October 20	Halley	76
Taurid	October 31	Encke	3
Leonid	November 16	1866 I	33
Geminid	December 13	Phaethon*	1.4

*An Earth-approaching asteroid, not a comet.

absence of bright moonlight, meteors can be seen with a frequency of about one per minute during a typical Perseid shower. It is estimated that the total combined mass of the particles in the Perseid swarm is nearly 5×10^{11} kg; this gives at least a lower limit for the original mass of its associated comet, called Comet 1862 III. The mass of Comet Halley, however, is nearly a thousand times greater, suggesting that only a very small fraction of the original material survives in the meteor stream.

The characteristics of some of the more famous meteor showers are summarized in Table 11.4. Other spectacular meteor showers can occur, however, at almost any time, just as some bright comets appear unexpectedly.

No shower meteor has ever survived its flight through the atmosphere and been recovered for laboratory analysis. However, there are other ways to investigate the nature of these particles and thereby to gain additional insight into the comets from which they are derived. Analysis of the photographic tracks of meteors shows that most of them are very light or porous, with densities typically less than 1.0 g/cm^3. Apparently a fist-sized lump, if you placed it on a table, would fall apart under its own weight. Such particles break up very easily in the atmosphere, accounting for the failure of even relatively large shower meteors to produce meteorites. Comet dust, apparently, is fluffy, rather inconsequential stuff. This fluff, by its very nature, does not reach the Earth's surface intact. However, the more substantial fragments from asteroids do make it into our laboratories, as we will see in the next chapter.

SUMMARY

11.1   Ceres is the largest **asteroid** (diameter of 940 km); about 15 are larger than 250 km, and 10^5 are larger than 1 km. Most are in the **asteroid belt**, between 2.2 and 3.3 AU from the Sun. Resonances with Jupiter introduce structure into the belt, and asteroid families represent many objects that are the remnants of asteroid collisions and fragmentation. The asteroids include both primitive and differentiated objects. Most asteroids are classed as C type, meaning they are composed of carbonaceous materials. Dominating the inner belt are S-type stony asteroids, with a few M-type (metallic) ones. Vesta is rare in having a vol-

canic (basaltic) surface; it is the parent body of the eucrite meteorites.

11.2   The Trojan asteroids are dark, primitive objects orbiting at 5.2 AU. The most distant asteroids known are Hidalgo and Chiron, and recently Chiron has developed a tenuous atmosphere, like that of a comet. Phobos and Deimos, the satellites of Mars, are probably captured primitive asteroids. Of great interest are the **Earth-approaching asteroids,** estimated to number about 1000 (down to 1 km in diameter). These are on unstable orbits, and on time scales of 10^8 years, they will either impact one of the terrestrial planets or be ejected. Most probably come from the asteroid belt, but some may be dead comets.

11.3   Halley first showed that some **comets** are on closed orbits and return periodically to the Sun. The heart of a comet is its **nucleus,** a few kilometers in diameter and composed of volatiles (primarily H_2O) and solids (including both silicates and carbonaceous materials). Whipple first suggested this "dirty snowball" model in 1950, and it has been confirmed by spacecraft studies of Comet Halley (by the Soviet VEGA and European Giotto probes). As the nucleus approaches the Sun, its volatiles evaporate (perhaps in localized jets or explosions) to form the comet's head or atmosphere, which escapes at about 1 km/s. The atmosphere streams away from the Sun to form a long **tail.**

11.4   Oort proposed in 1950 that comets are derived from the **Oort comet cloud,** which surrounds the Sun out to about 50,000 AU (near the limit of the Sun's gravitational sphere of influence) and contains between 10^{12} and 10^{13} comets. Comets are primitive bodies left over from the formation of the outer solar system. Once a comet is diverted into the inner solar system, it survives no more than about 1000 perihelion passages before it loses all its volatiles.

11.5   When a fragment of interplanetary dust strikes the Earth's atmosphere, it burns up to create a **meteor.** Streams of dust particles traveling through space together produce **meteor showers,** in which we see the meteors diverging from a spot in the sky called the radiant of the shower. Many meteor showers recur each year and are associated with particular comets.

REVIEW QUESTIONS

1.   Why are asteroids and comets important to our understanding of the solar system?

2.   Describe the main differences between C asteroids and S asteroids.

3.   Compare asteroids of the asteroid belt with the Earth-approaching asteroids. What are the main differences between the two groups?

4.   Describe the nucleus of a typical comet, and compare it with an asteroid of similar size.

5.   Describe the origin and eventual fate of comets.

6.   What is a meteor, and what does it have to do with comets?

7.   Comets and asteroids are considered to be relics from the origin of the solar system. Why do astronomers think they are older and more primitive than the planets and their satellites?

THOUGHT QUESTIONS

8.   There is a great deal of interest today in the discovery of additional Earth-approaching asteroids. Can you think of several reasons for this high level of interest?

9.   If Vesta is not the parent body of the eucrite meteorites, what might be an alternative source for these objects?

10.   Comets are considered to be the "most primitive" solid bodies in the solar system. What does this statement mean?

11.   Suppose a comet is discovered approaching the Sun, and it is found to be on an orbit that will cause it to collide

with the Earth 20 months later, after perihelion passage. (This is approximately the situation described in the science fiction novel *Lucifer's Hammer*, by Larry Niven and Jerry Pournelle.) What could we do? Is there any way to protect ourselves from a catastrophe?

12.   From the descriptions of comets and asteroids given in this chapter, discuss how an impact of a comet with the Earth might differ from an impact by an asteroid. Which kind of impact would do more damage?

PROBLEMS

13.   What is the period of revolution about the Sun for an asteroid in the middle of the asteroid belt (semimajor axis of 3.0 AU)?

14.   What is the period of revolution for a comet with aphelion at 5 AU and perihelion at the orbit of the Earth?

15.   Suppose that the Oort comet cloud contains 10^{12} comets, with an average comet diameter of 10 km. Calculate the mass of a comet 10 km in diameter, assuming that is it composed mostly of water ice (density of 1 g/cm³). Next calculate the total mass of the comet cloud. Finally, compare this mass with that of the Earth and Jupiter.

16. The calculation in Problem 15 refers to the known Oort cloud, the source for the comets we see. If, as some astronomers suspect, there are ten times this many cometary objects in the solar system, how does the total mass of cometary matter compare with the total mass of the planets?

17. If the Oort comet cloud contains 10^{12} comets and 10 new comets are discovered each year, what percentage of the comets have been used up since the beginning of the solar system?

INTERVIEW

GENE SHOEMAKER

Eugene Shoemaker began his career as a field geologist, studying volcanic features in northern Arizona. His Ph.D. thesis from Caltech was devoted to a careful study of Meteor Crater, one of the best-preserved impact structures on the Earth. Extending his research to the Moon and planets, Shoemaker was an active participant in the Apollo program and in the Voyager investigations of the satellites of the outer planets. He also initiated a highly successful program to search for Earth-approaching asteroids and comets using the 18-inch telescope on Palomar Mountain, and he has become the world's foremost expert on the risks they pose to the Earth.

How did you first recognize that impacts were important in the history of the Earth?

I got interested in Meteor Crater in the mid-1950s when I was doing field geology in Northern Arizona. Someone had the idea of recovering plutonium, which was in short supply, from controlled explosions of nuclear devices. I went out to the Nevada test site and looked at the two small craters that had been produced by subsurface bursts of nuclear bombs. I had to think of how to scale from these little craters, which were only roughly 100 m across, to the energy range that would be of interest to the experiment. So I thought, "Well, I'll go back out and look at Meteor Crater." This was also a good way to really learn how craters of this kind are made. And, to my surprise, Meteor Crater turned out to be a pretty good, scaled-up version of one of the nuclear craters. Having

done this, one of the questions that comes immediately to mind is, "How often do you make a crater like this on the Earth?" And so I undertook to try to understand what was then known, which wasn't much, about Earth-crossing asteroids.

Did you have a special interest in the Moon?

When I first went to work for the Geological Survey in 1948, I had just graduated fresh out with a Masters degree from Caltech at the tender age of 20. I was aware of the V-2 rockets and others that were being developed and flown out of White Sands. And it came to me in a flash, "By golly, I'll bet they're going to build a rocket big enough to take

Gene Shoemaker leading a geological field trip at Meteor Crater. *(David Morrison)*

people to the Moon in my lifetime." And I thought, "What could be more exciting than to be the first geologist on the Moon?" So I had a mental image of going to the Moon. That ambition influenced the direction I took, with the notion that I'd try to be prepared to stand at the head of the line when applications were taken for scientists to go to the Moon. I even pictured how that would happen . . . that the National Academy of Sciences would be called upon to constitute a committee to review the scientific qualifications of the applicants. The irony of it is that I was excluded on health grounds from being considered for the astronaut corps, and when the time came to select scientists to go to the Moon, I ended up chairing that very selection committee!

You were part of the Apollo support team on the ground. Was this a vicarious substitute?

Well, sure, I had founded the Branch of Astrogeology of the U.S. Geological Survey and had a whole team of people working on the geology of the Moon. When the Apollo program came along, we planned and practiced and worked intensively with the astronauts. But science had been mostly thrown off of Apollo 11, the first flight, so we were participating in that mission only to a limited extent. It wasn't really anywhere close to the kind of scientific mission that we had envisioned. There was not much interaction between the astronauts and the science team in the back room.

By the time of Apollo were you convinced that both the Earth and Moon had been heavily cratered from impacts?

In the late 1950s, once having understood Meteor Crater, I went back and I looked at what were being called cryptovolcanic structures in the United States, and I recognized that these were also ancient impact scars. Before Apollo, we used this terrestrial impact record to try to estimate the ages of features on the Moon. I made a very rough estimate of the rate of crater formation on the Earth, and from this derived a timescale for the geology of the Moon. Even with our limited data, we got the answer approximately right. We concluded that the lunar maria must be several billion years old, perhaps nearly as old as the Moon, and the formation of the highlands therefore must be crowded into very early lunar history.

Where did this flux of impacts come from?

I had been interested for some time in these Earth-crossing asteroids. By 1959, a total of nine had been discovered, but only three were well-enough observed to yield good orbits. In 1969, I hired a young woman, Eleanor Helin, and set her on the problem of tracking down every scrap of information about Earth-crossing asteroids. After probing the problem for a couple of years, Helin and I decided that the best way to get information is go out and try to discover more asteroids. From our discovery rate, we ought to be able to estimate the flux much more closely. This was a feasible thing to do because at Palomar there was a grand old Schmidt telescope built in 1936. It was being used occasionally for student theses, but there was a lot of time available on it. At the time I estimated that we ought to be able to find about four Earth-approaching asteroids per year. Sure enough, six months into this effort, she found the first Earth-crossing asteroid, and I thought, "Hot damn! The old back-of-the-envelope calculation is right!" But then it was more than two years till we got the next one, so the pickings were a little slim. I'd overestimated the population by a factor of about two.

To maintain a program like this year after year, you must have a stronger motive than just the fun of discovering objects. What are you learning from these surveys?

There have been enough discoveries to estimate what's really out there. There are a little bit more than 1000 objects bigger than a kilometer in diameter. And those are all objects capable of hitting the Earth. Whether

they actually hit the Earth depends upon a fate that can only be determined on a statistical or probabilistic basis. Of that batch, about one third will hit the Earth, about one third will hit Venus. A small fraction will hit some other planet or the Moon, and the remainder are ejected from the solar system. We can calculate what the rate of collision is, and we find that the number of craters 10 km or more in diameter that would be formed on the whole Earth would be about 1 every 100,000 years.

What are the consequences for the Earth of the kind of impacts we're talking about?

A crater diameter of somewhere between 10 and 30 kilometers is probably the threshold size at which there may be global effects. For a 30-km crater, the amount of dust that's thrown up if the impact is on land, or the amount of water vapor that may be injected into the high atmosphere if the impact is in the ocean, will begin to produce global climatic effects. With still bigger craters, at the 50 km to 100 km diameter level, we're into the regime where the global climatic effects become truly severe.

What if you discovered a smaller object and found that its orbit would intersect that of the Earth in a decade or so?

Then I think it would be time to pursue, on an international level, a program to change the orbit. Technically it could be done. I don't know whether one could really get an international agreement to carry out the engineering to change the orbit. I think the political problem is much more difficult than the technical one. I think one could change the orbit with a series of stand-off explosions in very small steps. We'd send out a flotilla of ten or a dozen spacecraft, do one explosion, see what happened, then track the asteroid and do the next explosion. And, do it *very carefully*.

You have simultaneously maintained a variety of careers with very different styles: the field geologist, the

Gene and Carolyn Shoemaker at Palomar Observatory, scanning sky photographs for new comets and Earth-approaching asteroids.

spacecraft experimenter, the observational astronomer. How do you feel about those different modes of doing science?

Well, they are rather different. I think one of the most enjoyable aspects of having been a participant in planetary exploration has been rubbing shoulders with scientists and engineers in a very broad range of disciplines. This was a great eye-opener, because I really did start out my career as a straight-out field geologist. It's very stimulating, very broadening, to see how different the styles are in different branches of science. I was lucky enough to participate in all this. I absorb new ideas and teach myself new ways of doing things as I go along.

I think scientists in general ought to hitch up their britches and go off and do something different about every 10 years. You probably exhaust your originality in any one field after a decade's time. You make the most significant contributions you're going to make in that period of time. If you continue pursuing the same topic, you're just digging a deeper trench and probably not getting very far. So that's a general recommendation. Young man and young woman, don't stay in the same field; go out and plow new territory about every 10 years.

Collision between planetesimals during the formative stage of the solar system. *(Painting by Don Dixon)*

Harold C. Urey (1893–1981) won the Nobel prize for chemistry in 1934 for the discovery of deuterium. Later in his career he became interested in the origin of the Earth and planets, and during the 1950s and 1960s his recognition of the role of primitive meteorites as remnants from the birth of the solar system laid the foundation for much of the modern interest in both the meteorites and the broader study of cosmochemistry.

12

METEORITES AND THE HISTORY OF THE SOLAR SYSTEM

We conclude our survey of the solar system with a discussion of its origin and evolution. We begin with an examination of the oldest material available for our study: the meteorites. Much of what we know today about the events 4.5 billion years ago when the planets were born has been derived from the study of meteorites.

12.1 Meteorites: Stones From Heaven

Any fragment of interplanetary debris that survives its fiery plunge through the Earth's atmosphere is called a **meteorite.** Meteorites fall only very rarely in any one locality, but over the entire Earth hundreds of meteorites fall each year. These rocks from the sky carry a remarkable record of the formation and early history of the solar system.

Extraterrestrial Origin of Meteorites

While occasional meteorites have been recovered throughout history, their extraterrestrial origin was not accepted by scientists until the beginning of the 19th century. Before that, these strange stones were either ignored or treated with supernatural respect.

The earliest recovered meteorites are lost in the fog of mythology. A number of religious texts speak of stones from heaven, which sometimes arrive at opportune moments to smite the enemies of the authors of the texts. At least one sacred meteorite has survived in the form of the Ka'aba, the holy black stone in Mecca that is revered by Islam as a relic from the time of the Patriarchs.

The modern scientific history of the meteorites begins in the late 18th century, when a few scientists suggested that some of the strange stones that had been found around the world were of such peculiar composition and structure that they were probably not of terrestrial origin. The general acceptance that indeed "stones fall from the sky" occurred after the French physicist Jean Baptiste Biot described the circumstances of a well-observed fall in 1803, in which many meteoritic stones were found, reportedly still warm, on the ground.

Meteorites sometimes fall in groups or showers. Such a fall may result when a group of particles were moving together in space before they collided with the Earth, but more likely the different stones are fragments of a single particle that broke up during its violent passage through the atmosphere. It is important to remember that such a *shower of meteorites* has nothing to do with a *meteor shower*. No meteorites have ever been recovered in association with meteor showers. Whatever the ultimate source of the meteorites, they do not appear to come from the comets or their associated particle streams.

FIGURE 12.1 An iron meteorite lying on the Antarctic ice just before it was added to our collections. *(NASA)*

Meteorite Falls and Finds

Meteorites are found in two ways. First, sometimes bright meteors (fireballs) are observed to penetrate the atmosphere to very low altitudes. A search of the area beneath the point where the fireball was observed to burn out may reveal one or more remnants of the particle. Observed **meteorite falls,** in other words, may lead to the recovery of fallen meteorites.

Second, unusual-looking rocks that turn out to be meteoritic are occasionally discovered. These are termed **meteorite finds.** Now that the public has become meteorite-conscious, many suspected meteorites are sent to experts each year. Some scientists refer to these objects as "meteorites" and "meteorwrongs." Outside Antarctica (see below), genuine meteorites are turned up at an average rate of 25 or so per year. Most of these end up in natural history museums or specialized meteoritical laboratories throughout the world.

Recently a new source of meteorites from the Antarctic has dramatically increased the rate of discovery of meteorites and is greatly enriching our knowledge of these objects. Thousands of meteorites have been recovered from the Antarctic ice, as a result of the low precipitation and peculiar motion of the ice in some parts of that continent (Figure 12.1).

Meteorites that fall in regions where ice accumulates are buried and then carried slowly, with the motion of the ice, to other areas where the ice is gradually worn away. After thousands of years, the rock again finds itself on the surface, along with other meteorites carried to these same locations. The ice thus concentrates the meteorites that have fallen both in a large area and during a long period of time.

Meteorite Classification

The meteorites in our collections include a wide range of compositions and histories, but traditionally they have been placed into three broad classes. First, there are the **irons,** which are composed of nearly pure metallic nickel-iron. Second are the **stones,** which is the term used for any silicate or rocky meteorite. Third are the much rarer **stony-irons,** which are (as the name implies) made of mixtures of stony and metallic iron materials.

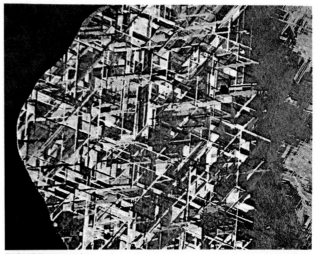

FIGURE 12.2 Slice of the Kamkas iron meteorite, polished and etched to show the crystal pattern in the metal. *(Ivan Dryer)*

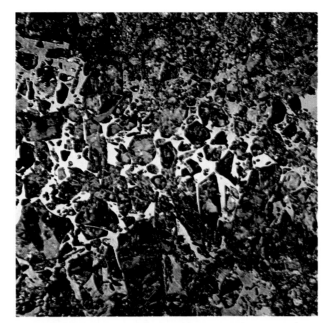

FIGURE 12.3 Polished slice of the Albin stony-iron meteorite. This type of meteorite, called a pallasite, consists of nickel-iron mixed with crystals of the green mineral olivine. *(Ivan Dryer)*

TABLE 12.1 Frequency of Occurrence of Different Meteorite Classes

	Falls (%)	Finds (%)	Antarctic (%)
Primitive stones	87	51	85
Differentiated stones	8	1	12
Irons	3	42	2
Stony irons	1	5	1

particular sample is really of extraterrestrial origin, especially if it has lain on the ground for some time and been subject to weathering. The most scientifically valuable stones are those that are collected immediately after they fall or the Antarctic samples that have been preserved in a nearly pristine state by the ice.

Table 12.1 summarizes the frequencies of ocurrence of the different classes of meteorites among falls, finds, and the Antarctic meteorites.

Ages and Compositions of Meteorites

It was not until the ages of meteorites were measured and techniques developed for the detailed analysis of their compositions that their true significance, as the oldest and most primitive materials available for direct study in the laboratory, was appreciated. The ages of stony meteorites can be determined from the careful measurement of radioactive isotopes and their decay products, as described for the lunar samples in Section 7.2. Almost all meteorites have radiometric ages between 4.48 and 4.55 billion years. The few exceptions are igneous rocks that are ejecta from cratering events on the Moon or Mars.

The average age for all of the old meteorites, calculated using the best data and the most accurate values now available for the radioactive half-lives, is 4.53 billion years, with an uncertainty of less than 0.1 billion years. This value is taken to represent the *age of the solar system*—the time since the first solids condensed and began to form into larger bodies.

The traditional classification of meteorites into irons, stones, and stony-irons is easy to use because it is obvious from inspection which category a meteorite falls into (although it may be much more difficult to distinguish a meteoritic stone from a terrestrial rock). Much more significant, however, is the distinction between *primitive* and *differentiated* meteorites.

The differentiated meteorites are, as the name implies, fragments of differentiated parent bodies. But

Of these three types, the irons (Figure 12.2) and stony-irons (Figure 12.3) are the most obviously extraterrestrial in origin because of their metallic content. Native, or unoxidized, iron almost never occurs naturally on Earth. This metal is always found here as an oxide or other mineral ore. Therefore, if you ever come across a chunk of metallic iron, it is sure to be either man-made or a meteorite.

The stones (Figure 12.4) are much more common than the irons but more difficult to recognize. Often a laboratory analysis is required to demonstrate that a

FIGURE 12.4 Stony meteorite of the type called ordinary chondrites. To the layperson, such a meteorite looks very much like a terrestrial rock. *(NASA/JSC)*

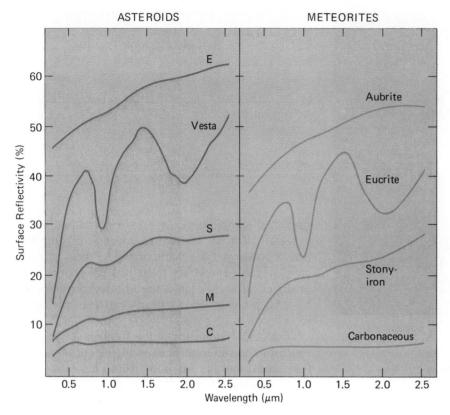

FIGURE 12.5 Examples of visible-infrared spectra of several asteroids of diverse compositional types, compared with similar spectra of meteorites as measured in the laboratory. *(Data courtesy of Clark R. Chapman, Planetary Science Institute)*

for information on the earliest history of the solar system, we turn to the primitive meteorites. A primitive meteorite is one that is made of materials that have never been subject to great heat or pressure since their formation. The fiery passage of the meteorite through the air takes place so rapidly that the interior (below a burned crust a few millimeters thick) never even becomes hot, so a fragment of primitive interplanetary debris is still primitive (in the sense in which we use the word) after it lands on the Earth. What are the parent bodies of these primitive meteorites?

Almost certainly the parent bodies are asteroids, as we can tell from a comparison of asteroid and meteorite spectra. Figure 12.5 illustrates the reflectivity in the visible and infrared part of the spectrum for several asteroid classes, as observed telescopically. Also shown are laboratory measurements of several meteorite classes. The similarities of the two curves demonstrate that the meteorites contain at least approximately the same minerals as the asteroids, and studies of this kind produce tentative associations between groups of meteorites and their parent bodies.

The great majority of the meteorites that reach the Earth are primitive stones. Many of them are composed of light-colored gray silicates with some metallic grains mixed in, but there is also an important group of darker stones called **carbonaceous meteor-**

ites. As their name suggests, these meteorites contain carbon, various complex organic compounds, and often chemically bound water; they are also depleted in metallic iron. The carbonaceous meteorites probably originate among the dark, carbonaceous asteroids, which are concentrated in the outer part of the asteroid belt.

The Allende and Murchison Primitive Meteorites

The carbonaceous meteorites are the most primitive materials available for laboratory study, excepting the tiny dust particles from comets. Two large carbonaceous meteorites that fell within a few months of each other have proved particularly valuable in probing the birth of the solar system. The Allende meteorite fell in Mexico and the Murchison meteorite in Canada, both in 1969. Like other meteorites, Allende and Murchison are named for the towns near which they fell.

Murchison is best known for the variety of organic, or carbon-bearing, chemicals that it has yielded. Most of the carbon compounds in these meteorites are complex, tar-like substances that defy exact analysis. However, Murchison also contains 16 amino acids, 11 of which are rare on Earth. Unlike terrestrial amino acids, which are formed by living things, the

FIGURE 12.6 The Allende carbonaceous meteorite that fell in Mexico in 1969. *(NASA/JSC)*

Murchison chemicals include equal numbers with right-handed and left-handed molecular symmetry. The presence of these naturally occurring amino acids and other complex organic compounds in Murchison demonstrates that a great deal of interesting chemistry must have taken place when the solar system formed. Perhaps some of the molecular building blocks of life on Earth were derived directly from the primitive meteorites and comets.

The Allende meteorite (Figure 12.6) is a rich source of information on the formation of the solar system because it contains many individual grains with varied chemical histories. As much as 10 percent of the material in Allende has been estimated to be of pre–solar system origin—interstellar dust grains that were not destroyed in the processes that gave rise to our own system.

Meteorites from the Moon and Mars

In this chapter we are interested primarily in the primitive meteorites, since they tell us the most about the origin of the solar system. However, we note that recently a number of surprising discoveries have been made among the differentiated meteorites, including objects from the Moon and Mars.

The first meteorite to yield a definitive identification was ALHA 81005, which was found at the Allan Hills Antarctic site in 1981. This meteorite is clearly lunar, similar in many ways to the samples returned in the Apollo program. Half a dozen additional lunar samples have been identified subsequently among the U.S. and Japanese collections of Antarctic meteorites. Another group of basaltic meteorites is the SNC meteorites, discussed in Section 8.4, which have solidification ages of 1.3 billion years. These eight

stones, including four from the Antarctic, are now generally believed to represent samples of the martian surface.

12.2 Formation of the Solar System

The comets, asteroids, and meteorites are surviving remnants from the origin of the solar system. The planets and the Sun, of course, also are the products of the formation process. We are now ready to put together the information from the past five chapters in order to discuss what is known of the origin of the solar system.

Observational Constraints

There are certain basic properties of the planetary system that any theory of formation should explain. These may be summarized under three categories: dynamical constraints, chemical constraints, and age constraints.

Dynamical Constraints   The planets all move around the Sun in the same direction and approximately in the plane of the Sun's own rotation. In addition, most of the planets share this same sense of rotation, and most of the satellites also move in counterclockwise orbits. With the exception of the comets, the members of the system define a disk shape. On the other hand, exceptions are possible in the form of retrograde rotation, like that of Venus.

Chemical Constraints   The planets Jupiter and Saturn have approximately the same composition as the Sun and stars, dominated by hydrogen and helium. Each of the other members is, to some degree, lacking in the light elements. A careful examination of the composition of solar system objects shows a striking progression from the metal-rich inner planets through those made predominantly of rocky materials out to objects with ice-dominated composition in the outer solar system. The comets are also icy objects, whereas the asteroids represent a transitional rocky composition with abundant dark, carbon-rich material. This general chemical pattern can be interpreted as a temperature sequence, with the inner parts of the system strongly depleted in materials that could not condense at the high temperatures found near the Sun. Again, however, there are important exceptions to the general pattern. In particular, it is difficult to explain the presence of water on Earth and Mars if these planets had formed in a region where the temperature was too hot for ice to condense, unless the ice or water was brought in later from cooler regions. There are also problems with the composition of the Moon and Mercury, as we discussed in Chapter 7.

Age Constraints Radioactive dating demonstrates that there are rocks on the surface of the Earth that have been present for at least 3.8 billion years and lunar samples that are 4.4 billion years old. In addition, the primitive meteorites all have radioactive ages near 4.5 billion years. The age of these unaltered building blocks is considered the age of the planetary system. The similarity of the measured ages tells us that planets formed and their crusts cooled within a few hundred million years, at most, of the beginning of the solar system. Further, detailed examination of primitive meteorites indicates that they are made primarily from material that condensed or coagulated out of a hot gas; few identifiable fragments or grains survived from before this hot vapor stage 4.5 billion years ago.

The Solar Nebula

All of the above constraints lead to the conclusion that the solar system formed 4.5 billion years ago out of a rotating cloud of hot vapor called the *solar nebula*. The initial composition of the solar nebula was similar to that of the Sun today. As the cloud cooled, grains of solid material coagulated from it, like raindrops condensing from moist air as it rises over a mountain. The meteorites and comet dust are remnants of this original condensate.

Initially, the solar nebula was large—much larger than the present dimensions of the solar system—and cool—perhaps only a few tens of degrees Kelvin. Slowly, the pull of gravity caused this cool, diffuse mass of gas and dust to collapse. As the solar nebula shrank, it was heated by its own gravitational energy, and its rotation speed increased. To understand why its rotation speeded up, we need to introduce the idea of **angular momentum.** Angular momentum is a property of any object that rotates or revolves about some fixed point. It depends on the mass of the moving object, its speed, and its size. Whenever we study spinning or rotating objects, from planets to galaxies, we must consider angular momentum.

Angular momentum is a property that is *conserved*, or stays constant, for any rotating system in which no external forces act. This means that if one part of the system changes—for example, if its size becomes smaller—then in order for the total angular momentum to remain the same, some other property must also adjust itself. In this example, the rotation speed must increase to compensate for the smaller size. Thus a shrinking cloud of dust or a stream of matter

falling into a black hole increases its spin rate as it contracts.

Increasing temperatures in the shrinking nebula vaporized most of the solid material that was originally present. As shown schematically in Figure 12.7, the nebula eventually collapsed into a disk shape, with most of the material confined to a thin, spinning

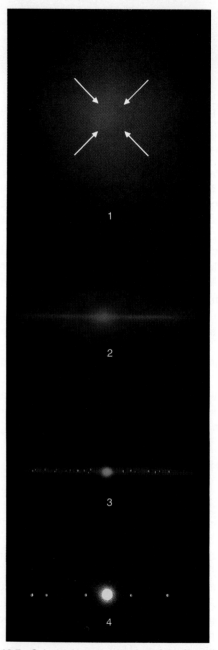

FIGURE 12.7 Schematic representation of the formation of the solar system. (1) The solar nebula contracts. (2) As the nebula shrinks, its rotation causes it to flatten until (3) the nebula is a disk of matter with a concentration near the center, which (4) becomes the protosun. Meanwhile, solid particles condense as the nebula cools, giving rise to the planetesimals, which are the building blocks of the planets.

sheet. At the center, continuing collapse ultimately led to the birth of the Sun. The existence of this disk-shaped rotating nebula explains the primary dynamical properties of the solar system as described above.

Condensation in the Solar Nebula

Picture the solar nebula at the end of the collapse phase, when it was at its hottest. With no more gravitational energy to heat it, most of the nebula began to cool. In the center, however, the newly formed Sun maintained high temperatures. The temperature within the nebula therefore decreased with increasing distance from the Sun, much as the planetary temperatures vary with position today. As the nebula cooled, the gases interacted chemically to produce compounds, and eventually these compounds condensed into liquid droplets or solid grains.

The *chemical condensation sequence* (Figure 12.8) in the cooling nebula was calculated by geochemists in the 1970s. The first materials to form grains were the metals and various rock-forming silicates. As the temperature dropped, these were joined throughout much of the solar nebula by sulfur compounds and by carbon- and water-rich silicates, such as those now found abundantly among the asteroids. However, in the inner parts of the nebula, the temperature never dropped low enough for such materials as ice or car-

bonaceous organic compounds to condense, so they are lacking on the innermost planets. Far from the Sun, where temperatures continued to decline, the oxygen combined with hydrogen and condensed in the form of water (H_2O) ice. Beyond the orbit of Saturn, carbon and nitrogen combined with hydrogen to condense as additional ices such as methane (CH_4) and ammonia (NH_3). This chemical condensation sequence explains the basic chemistry of the solar system, and it also tells us why the oldest materials (the primitive meteorites) all have about the same age—the age corresponding to the time when all of these solid grains formed.

It is thought that the grains that condensed in the solar nebula rather quickly formed into larger and larger aggregates, until most of the solid material was in the form of **planetesimals** a few kilometers to a few tens of kilometers in diameter. These planetesimals then became the building blocks for the planets. Some of them survive today as the comets and asteroids.

12.3 Birth of the Planets

The planetesimals are thought to have been no more than 100 km in diameter, like the majority of the comets and asteroids we see today. A substantial step in size is required, however, to go from planetesimals to planets.

Formation of the Terrestrial Planets

Some planetesimals were large enough to attract their neighbors gravitationally and thus to grow by the process called **accretion.** While the intermediate steps are not well understood, ultimately several dozen centers of accretion seem to have developed in the inner solar system. Each of these attracted surrounding planetesimals until it had aquired a mass similar to that of Mercury or Mars. At this stage we may think of these objects as *protoplanets*.

Each of these protoplanets continued to grow by the accretion of planetesimals. Now, however, every incoming planetesimal was accelerated by the gravity of the protoplanet, so that it struck with great energy—sufficient energy to melt both the projectile and a part of the impact area. Soon the entire protoplanet was heated above the melting temperature of rocks. The result was planetary differentiation, with metals sinking toward the core and lighter silicates rising toward the surface. At the same time, the protoplanets lost some of their more volatile constituents through outgassing and atmospheric escape.

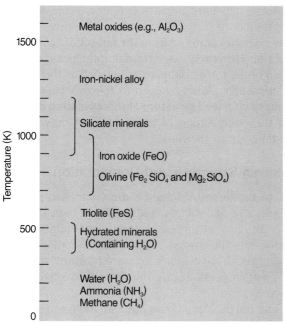

FIGURE 12.8 The chemical condensation sequence in the solar nebula, showing the primary chemical species that would be expected to form in a cooling gas cloud of solar composition under equilibrium conditions. *(Adapted from diagrams published by John Lewis, University of Arizona)*

Formation of the Giant Planets

In the outer solar system, where the building blocks included ices as well as silicates, much larger protoplanets grew, with masses 10 to 20 times the mass of the Earth. These protoplanets of the outer solar system became so large that they were able to attract and hold the surrounding gas. As the hydrogen and helium rapidly collapsed onto their cores, the giant planets were heated by the energy of contraction, just as the contraction of the solar nebula had ignited the nuclear fires of the Sun. But these giant planets were far too small to achieve the central temperatures and pressures necessary to initiate self-sustaining nuclear reactions. After glowing dull red for a few thousand years, they gradually cooled to their present state.

The collapse of nebular gas onto the cores of the giant planets explains how these objects came to have about the same hydrogen-rich composition as the Sun itself. The process was most efficient for Jupiter and Saturn, so that their composition is most nearly "cosmic." Much less gas was captured by Uranus and Neptune, which is why these two planets have compositions dominated by the icy and rocky building blocks that made up their large cores, rather than by hydrogen and helium.

Dynamical Evolution of the System

All of the processes described above, from the collapse of the solar nebula to the formation of protoplanets, took place within at most a few million years, and possibly even less time. However, the story of the formation of the solar system is not complete at this stage—there remains the fate of the planetesimals and other debris that did not initially accumulate to form the planets.

The Oort comet cloud was probably formed from icy planetesimals in the outer solar system. The gravitational influence of the giant planets is thought to have ejected the comets from their initial orbits in the disk, probably near the present orbits of Uranus and Neptune. If this idea is correct, then the comets are leftovers from the building blocks of the outer planets, preserved in the Oort cloud.

In the inner parts of the system, remnant planetesimals and perhaps several dozen protoplanets continued to whiz about and to interact gravitationally with the prototerrestrial planets. Collisions between these objects were inevitable. Giant impacts at this stage probably stripped Mercury of part of its mantle and crust, reversed the rotation of Venus, and broke the Earth apart to create the Moon.

Smaller-scale impacts also added mass to the inner protoplanets. This impacting material could have come from almost anywhere within the solar system.

In contrast to the previous stage of accretion, therefore, this new material did not represent just a narrow range of compositions as specified by the initial temperatures in the solar nebula. Much of the debris striking the inner planets, for example, was ice-rich material that had condensed in the outer part of the solar nebula. As this comet-like bombardment progressed, the Earth accumulated the water and various organic compounds that would later be critical to the formation of life. Mars and Venus should also have acquired water and organic materials from the same source.

Gradually, as the planets swept up the remaining debris, most of the leftover planetesimals disappeared. In the region between Mars and Jupiter, however, there exist stable orbits where small bodies can avoid impacting the planets or being ejected from the system. The remaining objects that survive in this special location are what we call the asteroids.

12.4 Planetary Evolution: A Comparison of the Terrestrial Planets

The era of giant impacts was probably confined to the first 100 million years of solar system history, ending by 4.4 billion years ago. Shortly thereafter, the planets cooled and began to assume their present aspects. Up until about 4.0 billion years ago, they continued to acquire volatiles, and their surfaces were heavily cratered from the tail-off of accretionary debris. However, as external influences declined, each of the terrestrial planets, as well as the satellites of the outer planets, began to follow its own evolutionary course. The nature of this evolution depended on the composition of the object, its mass, and its distance from the Sun.

Internal Structure and Composition

All of the terrestrial planets are composed primarily of silicates and metals. Judging from their densities, Mercury has the highest proportion of metals, and the Moon has the lowest. As we have seen, these two objects both have anomalous compositions as a result of giant impacts during the formative stages of the solar system.

The Earth, Venus, and Mars all have roughly similar bulk compositions, consisting of about one-third iron-nickel metal or iron-sulfur combinations and two-thirds silicates (mass fractions). All are thought to be differentiated, although the Earth and Moon are the only objects that have been probed by seismic studies.

FIGURE 12.9 The ancient lunar highlands, which record impacts dating back 4.4 billion years ago. *(NASA)*

Where there is a combination of a liquid metal core and rapid rotation, an internal magnetic field is generated, as we see on the Earth. Little or no magnetic field is present on the Moon or Mars, presumably because they lack a liquid metal core. Venus probably has such a core, but it rotates so slowly that the absence of a magnetic field can also be understood. However, the weak magnetic field of Mercury remains something of a mystery, since we would not expect such a small planet to have retained enough internal heat to permit a liquid core. The theory seem-

ingly breaks down here, suggesting that the process of generation of a planetary magnetic field is not very well understood after all.

In the outer solar system, most of the satellites acquired a large quantity of water ice. For Ganymede and Callisto and the satellites of Saturn, the mass fraction of water is about 50 percent. These objects differentiated easily, allowing the rock and metal to sink to the center and producing a mantle and crust of water ice.

Geological Activity

We have seen a wide range in the level of geological activity on the terrestrial planets and icy satellites. Internal sources of geological activity require energy, either in the form of primordial heat left over from the formation of a planet or from decay of radioactive elements in the interior. The larger the planet or satellite, the more likely it is to retain its internal heat, and therefore the more we expect to see evidence on the surface of continuing geological activity (Figure 12.10).

The amount of radioactive heat generated in a planet is proportional to its mass. Doubling the amount of material doubles the energy released. To escape, however, this heat must work its way to the surface and be radiated to space. Doubling the mass of a planet increases the thickness of blanketing material, and instead of doubling the surface area it increases it only by a factor of 1.6; thus, larger objects cannot rid themselves easily of their internal heat, while smaller objects of the same composition tend to cool down more rapidly. In a similar way, a large baked potato retains its heat, but if you cut it into smaller pieces, they cool quickly.

Accretion, heating,
    differentiation

Formation of solid
crust, heavy cratering

Widespread mare-like
volcanism

Reduced volcanism,
possible plate tectonics

Mantle solidification,
end of tectonic activity

Cool interior, no
activity

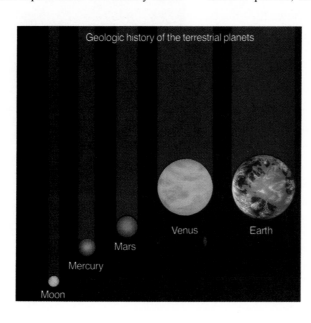

FIGURE 12.10 Stages in the geological history of a terrestrial planet. The smaller the planet, the more quickly it passes through these stages and reaches "old age."

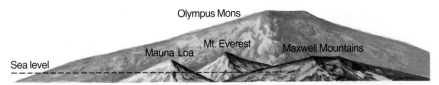

FIGURE 12.11 Comparison of the highest mountains on Mars, Venus, and Earth. Mountains can rise taller on Mars because the surface gravity is less. The vertical scale is exaggerated 3:1 in this sketch.

For the most part, the history of volcanic activity seen on the terrestrial planets conforms to the expectation of this simple theory. The Moon, the smallest of these objects, was internally active until about 3.3 billion years ago, when the major mare volcanism ceased. Since that time the mantle has cooled and become solid, and today even internal seismic activity has declined almost to zero. The Moon is a geologically dead world. Although we know much less about Mercury, it seems likely that this planet, too, ceased most volcanic activity about the same time that the Moon did.

Mars represents an intermediate case, and it has been much more active than the Moon. The southern hemisphere crust had formed by 4 billion years ago, and the northern hemisphere volcanic plains seem to be contemporary with the lunar maria. However, the Tharsis bulge formed somewhat later, and activity in the large Tharsis volcanoes has apparently continued intermittently to the present.

The Earth is the largest and the most active terrestrial planet, with its global plate tectonics driven by mantle convection. As a result, our surface is continually reworked, and over most of the planet the age of the surface material is less than 200 million years. Venus, with its similar size, has generally similar levels of volcanic activity, but the radar data suggest that it has not experienced plate tectonics, as has the Earth. Instead, much of its surface shows the effects of "blob tectonics" due to mantle hot spots. A better understanding of the geological differences between Venus and Earth is a high priority for planetary geologists, now that we have a detailed radar map of our sister planet.

The geological evolution of the icy satellites is distinct from that of the terrestrial planets. The energy source is the same, but the materials are different. Here we see evidence of low-temperature volcanism, with the silicate lava replaced by sulfur and sulfur compounds on Io and by water and other ices on the colder and more distant moons.

Elevation Differences

The mountains on the planets have very different origins. On the Moon and Mercury, the major mountains are just ejecta thrown up by the large basin-forming impacts that took place billions of years ago. On Mars, most large mountains are volcanoes, produced by repeated eruptions of lava from the same vents. There are also similar volcanoes up to 10 km high on Earth and Venus. However, the highest mountains on Earth and Venus are the result of compression and uplift of the surface. On the Earth, this crustal compression results from collisions of one continental plate with another.

It is interesting to compare the maximum heights of the volcanoes on Earth, Venus, and Mars (Figure 12.11). On both of the larger planets, the maximum elevation differences between these mountains and their surroundings are about 10 km. Olympus Mons, in contrast, towers 26 km above its surroundings, and nearly 30 km above the lowest elevation areas on Mars (Figure 12.12).

One reason that Olympus Mons is so much larger than its terrestrial counterparts is that the crustal plates on Earth never stop long enough to let a really

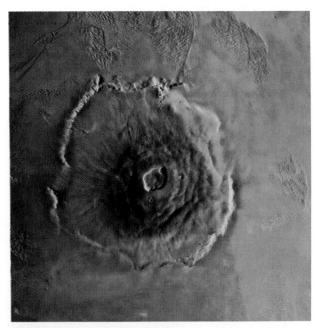

FIGURE 12.12 The martian volcano Olympus Mons is 26 km high and nearly 400 km across. It is able to support its large bulk in part because the surface gravity on Mars is low. This Viking view from above is a composite of many individual orbiter images. *(NASA/USGS, courtesy of Alfred McEwen)*

large volcano grow. Instead, the moving plate creates a long row of volcanoes, like the Hawaiian Islands. On Mars (and perhaps Venus), the crust remains stationary with respect to the underlying hot spot, and the volcano can continue to grow for hundreds of millions of years.

A second difference relates to the force of gravity on the three planets. The surface gravity on Venus is nearly the same as that on Earth, but on Mars it is only about one-third as great. In order for a mountain to survive, its internal strength must be great enough to support its weight against the force of gravity. Volcanic rocks have known strengths, and it is apparent that on Earth 10 km is about the limit. For instance, when new lava is added to the top of Mauna Loa in Hawaii, the mountain eventually slumps downward under its own weight. The same height limit applies on Venus, where the force of gravity is the same. On Mars, however, with its lesser surface gravity, much greater elevation differences can be supported, and it should be no surprise that Olympus Mons is more than twice as high as the mountains of Venus or Earth.

The same kind of calculation that determines the limiting height of a mountain can be used to ascertain the largest body that can have an irregular shape. All of the planets and larger satellites are nearly spherical, as a result of the force of their own gravity. But the smaller the object, the greater the departures from spherical shape that can be supported by the strength of its rocks. For silicate bodies, the limiting diameter is about 400 km; larger objects will always be approximately spherical, while smaller ones can have almost any shape.

Atmospheres

Planetary atmospheres were formed by a combination of outgassing from their interiors and the impacts of volatile-rich debris from the outer solar system. Each of the terrestrial planets must have originally had similar atmospheres, but Mercury was too small and too hot to retain its gas. The Moon probably never had an atmosphere, since the material out of which it is made was depleted in volatiles.

The predominant atmospheric gas on the terrestrial planets is now carbon dioxide (CO_2), but initially there were probably also hydrogen-containing gases. In this more chemically reduced environment, there should have been large amounts of carbon monoxide (CO) and perhaps also traces of NH_3 and CH_4. Solar ultraviolet light destroyed the reducing gases, however, and most of the hydrogen escaped to leave behind the oxidized atmospheres seen today on Earth, Venus, and Mars.

The fate of water was different for each of these three planets, depending on its size and distance from the Sun. Early in its history, Mars apparently had a thick atmosphere with abundant liquid water, but it could not retain these conditions. The CO_2 necessary for a substantial greenhouse effect was lost, the temperature dropped, and eventually the water froze. On Venus the reverse process took place, with a runaway greenhouse effect leading to the permanent loss of water. Only on Earth was the delicate balance maintained that permits liquid water to persist.

With the water gone, Venus and Mars each ended up with an atmosphere that is about 96 percent carbon dioxide and a few percent nitrogen. On Earth, the presence first of water and then of life led to a very different kind of atmosphere. The CO_2 was removed to be deposited in marine sediment, while proliferation of photosynthetic life eventually led to the release of enough oxygen to overcome natural chemical reactions that eliminate this gas from the atmosphere. As a result, we on Earth find ourselves with a great deficiency of CO_2 and with the only planetary atmosphere containing free oxygen.

In the outer solar system, Titan is the only satellite with a substantial atmosphere. Apparently this object contained sufficient volatiles—such as ammonia, methane, and nitrogen—to form an atmosphere. It also was large enough and cold enough to retain the heavier gases. Thus, as we saw in Chapter 10, Titan has developed an atmosphere consisting today primarily of nitrogen. In comparison with the inner planets, we can see that temperatures on Titan are too low for either carbon dioxide or water to be in the vapor form. With these two common gases frozen solid, it is perhaps not too surprising that nitrogen should end up as the primary atmospheric constituent.

Conclusion

We now come to the end of our study of the planetary system. Although we have learned a great deal about the other planets during the past few years of spacecraft exploration, much remains unknown. Now that we have learned to use the comparative approach to study these different worlds and the processes that form them and their environments, we see how limited our knowledge really is. We do not understand even our own planet well enough to assess the consequences of the changes humans are inadvertently inflicting on the atmosphere and oceans. The exploration of the solar system is one of the greatest human adventures, and it will continue, in part so that we can learn to appreciate and protect our own planet.

SUMMARY

12.1   **Meteorites** are the debris (mostly fragments from asteroids) that survives to reach the surface of the Earth. Meteorites are called **finds** or **falls** according to how they are found; the most productive source today is the Antarctic ice cap. The compositional classification of meteorites is into **irons, stony-irons,** and **stones.** Most stones are primitive objects, dated to the origin of the solar system 4.5 billion years ago. The most primitive are the **carbonaceous meteorites,** such as Murchison and Allende.

12.2   A theory of solar system formation must take into account dynamical, chemical, and age constraints. Meteorites, comets, and asteroids are survivors from the formation of the solar system out of the solar nebula. The solar nebula formed by the collapse of an interstellar cloud of gas and dust, which contracted (conserving its **angular momentum**) to form the protosun surrounded by a thin disk of dust and hot vapor. Condensation in the disk led to **planetesimals,** which were the building blocks of the planets.

12.3   **Accretion** of planetesimals led to the formation of the terrestrial planets, which were heated by infalling material and differentiated. The giant planets were also able to attract and hold gas from the solar nebula. After a few million years of violent impacts, most of the debris was swept up or ejected, leaving only the asteroids and cometary remnants surviving to the present. All of these violent events terminated by about 4.4 billion years ago.

12.4   Understanding of the evolution of planets is illustrated by comparison of the terrestrial planets (Earth, Venus, Mars, Mercury, and Moon). All are rocky, differentiated objects. The level of geological activity is (as expected) proportional to mass: greatest for Earth and Venus, less for Mars, and absent for Moon and Mercury. Mountains can be the result of impacts, volcanism, or uplift. Whatever their origin, higher mountains can be supported on smaller planets, where the surface gravity is less. All the terrestrial planets may have acquired their atmospheric volatiles from comet impacts. The Moon and Mercury lost their atmospheres, most volatiles on Mars are frozen owing to its greater distance from the Sun, and Venus retained CO_2 but lost H_2O when it developed a massive greenhouse effect. Only Earth still has liquid H_2O and hence can support life.

REVIEW QUESTIONS

1.   What do we mean by primitive material? How can we tell if a meteorite is primitive?

2.   In what ways are meteorites different from meteors? What is the probable origin of each?

3.   How do we know when the solar system formed? Usually we say that the solar system is about 4.5 billion years old. To what does this age correspond?

4.   Describe the solar nebula, and outline the sequence of events within the nebula that gave rise to the planetesimals.

5.   Why do the giant planets and their satellites have compositions different from those of the terrestrial planets?

6.   Explain the role of impacts in planetary evolution, both giant impacts and more modest ones.

7.   Why are some planets more active geologically than others?

8.   Summarize the origin and evolution of the atmospheres of Venus, Earth, and Mars.

THOUGHT QUESTIONS

9.   Meteors apparently come primarily from comets, while the meteorites are thought to be fragments of asteroids. This may seem contradictory. Explain why we do not believe meteorites come from comets, or meteors from asteroids.

10.   Explain why iron meteorites represent a much higher percentage of finds than of falls.

11.   Why is it more useful to classify meteorites according to whether they are primitive or differentiated, rather than into stones, irons, and stony-irons?

12.   Which meteorites are the most useful for defining the age of the solar system? Why?

13.   Suppose a new primitive meteorite is discovered and analysis shows that it contains a trace of amino acids, all of which show the same rotational symmetry (unlike the Murchison meteorite). What might you conclude from this finding?

14.   Give some everyday examples of the conservation of angular momentum.

15. Describe the chemical building blocks that are thought to have been available in the grains that condensed from the solar nebula. If each planet formed in place from these grains, what would be the chemical composition of objects at 0.4 AU, 1.0 AU, 5.0 AU, and 25 AU from the Sun?

16. We have seen how Mars can support greater elevation differences than can Earth or Venus. According to the same arguments, the Moon should have higher mountains than any of the other terrestrial planets, yet we know it does not. What is wrong with this line of reasoning?

PROBLEMS

17. Consider the differentiated meteorites. We think the irons are from the cores, the stony-irons from the interface between mantle and core, and the stones from the mantles of their differentiated parent bodies. If these parent bodies were like the Earth, what fraction of these meteorites would you expect to consist of irons, stony-irons, and stones? Is this consistent with the observed numbers of each?

18. Estimate the maximum height of the mountains on a hypothetical planet similar to the Earth but with twice the surface gravity of our planet.

19. The angular momentum of an object is proportional to the square of its size divided by the period of rotation (D^2/P). If angular momentum is conserved, then any change in size must be compensated for by a proportional change in period, so as to keep D^2 divided by P a constant. Suppose that the solar nebula began with a diameter of 10,000 AU and a rotation period of 1 million years. What would be its rotation period when it had shrunk to the size of the orbit of Pluto? To the orbit of Jupiter? To the orbit of the Earth?

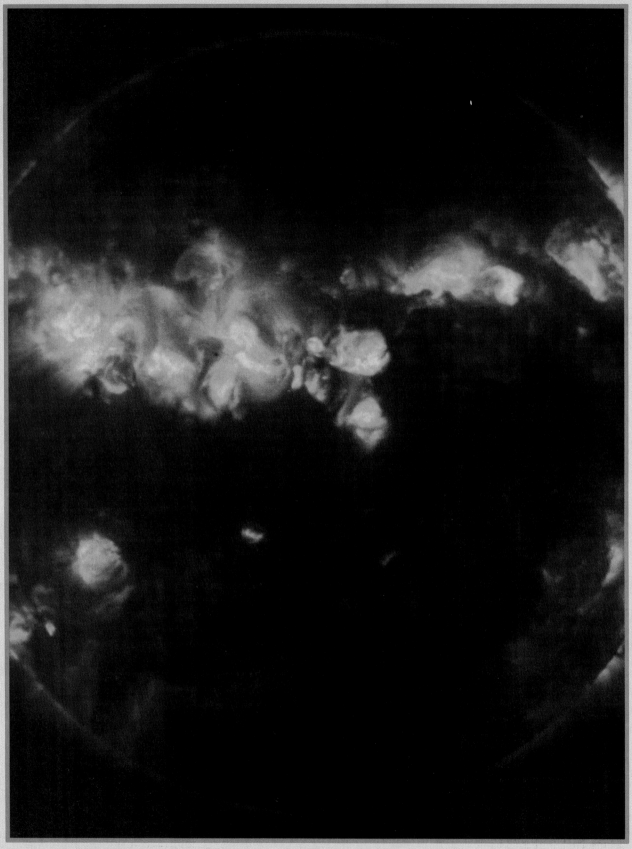

The sharpest x-ray image to date of the Sun's corona. Many active regions (bright areas) with temperatures of 2 to 3 million K are shown. The image was taken from a NASA sounding rocket launched to an altitude of about 150 miles. *(Leon Golub/Smithsonian Astrophysical Observatory and IBM)*

Arthur Stanley Eddington (1882–1944), British mathematician and astrophysicist, organized two expeditions to observe the total solar eclipse of 1919 in order to test a prediction of Einstein's general theory of relativity. Eddington is best known among astronomers for his development of theoretical methods of investigating the internal structure of the Sun and stars.

13

THE SUN: A GARDEN-VARIETY STAR

The Sun is a star.

Grade-school children memorize this fact, but what does it mean? It is far from obvious that this "fact" is even correct. In appearance and in its influence on the Earth, the Sun is very different from a star. The Sun appears much brighter. It also seems to be much larger than stars, which look like mere points of light. The Sun is the source of heat that sustains life on Earth and controls its climate; there is evidence that tiny decreases in the output of energy from the Sun can cause ice ages. The Sun warms some parts of the Earth more than others and so sets up winds that determine weather patterns. It bombards the Earth with charged atomic particles that can disrupt radio communications. Given its influence, it is little wonder that the Sun was worshipped as a god by many ancient civilizations.

The idea that the stars might also be suns, perhaps surrounded by their own families of planets and faint only because of their large distances from us, is relatively modern. The Italian philosopher Giordano Bruno (1548–1600) was one of the first to speculate that the universe is infinite, with an uncountable number of suns. For his modern views about astronomy, as well as for his attacks on the Catholic Church, Bruno was burned at the stake in 1600.

Modern astronomers have found that the Sun is a rather ordinary star—not unusually hot or cold, old or young, large or small. It is fortunate that we have such a garden-variety star so near at hand (astronomically speaking!), since by observing the Sun closely, we can learn a great deal about the physical processes that must determine the structure and evolution of other stars. The Sun plays a role in stellar astronomy similar to that of the Earth in planetary studies. It is the one well-studied example that serves as a benchmark for all of the much fainter stars for which we cannot obtain the same enormously detailed information.

Some basic data for the Sun are summarized in Table 13.1.

13.1   Outer Layers of the Sun

The only parts of the Sun that can be observed directly are its outer layers, collectively known as the Sun's atmosphere. There are three general regions in the solar atmosphere, each having substantially different properties, but with a gradual transition from one region to the next. These regions are the photosphere, the chromosphere, and the corona.

The Solar Photosphere

The Sun, although composed of the same chemical elements found on Earth, is so hot that all of those elements are in the gaseous state. There is no solid

TABLE 13.1   Solar Data

Datum	How Found	Value
Mean distance	Radar reflection from planets	1 AU
		149,597,892 km
Maximum distance from Earth		1.521×10^8 km
Minimum distance from Earth		1.471×10^8 km
Mass	Orbit of Earth	333,400 Earth masses
		1.99×10^{30} kg
Mean angular diameter	Direct measure	31'59".3
Diameter of photosphere	Angular size and distance	109.3 times Earth diameter
		1.39×10^6 km
Mean density	Mass/volume	1.41 g/cm^3
Gravitational acceleration at photosphere (surface gravity)	$\dfrac{GM}{R^2}$	27.9 times Earth surface gravity
		273 m/s^2
Solar constant	Measure with instrument sensitive to radiation at all wavelengths	1370 watts/m^2
Luminosity	Solar constant times area of spherical surface 1 AU in radius	3.8×10^{26} watts
Spectral class	Spectrum	G2 V
Effective temperature	Derived from luminosity and radius of Sun	5800 K
Visual magnitude		
Apparent	Visible flux	− 26.7
Absolute	Apparent magnitude and distance	+ 4.8
Rotation period at equator	Sunspots and Doppler shift in spectra taken at the edge of the Sun	24^{d}16^h
Inclination of equator to ecliptic	Motions of sunspots	7°10'.5

surface. The gas decreases in density outward in the Sun until, in the outer corona, it disperses into interplanetary space. The outer parts of the solar atmosphere are quite transparent to visible light, and we can look through them to deeper layers of the Sun. The **photosphere** (Figure 13.1) is that depth in the Sun past which we can see no deeper. Beneath the photosphere, the gases are opaque, absorbing and re-emitting the radiation they receive from lower levels. Thus the photosphere, like a cloud seen from an airplane window, looks solid even though it really is not.

The solar atmosphere changes from being almost perfectly transparent to almost completely opaque in a distance of only just over 400 km, a distance that is very thin indeed compared with the size of the Sun. In effect, therefore, the photosphere defines a sharp boundary. When we speak of the "size" of the Sun, we generally mean the size of the region surrounded by the photosphere.

From analysis of the light received from the Sun, astronomers can calculate how the temperature, density, and pressure of the gases vary through the photosphere. The solar photosphere is not the same everywhere, but its average characteristics are listed in Table 13.2. This table shows that over a distance of a little more than 300 km the pressure and density increase by a factor of 10, while the temperature climbs from 4500 to 6000 K. At a typical point in the photosphere, the pressure is only a few hundredths of sea-level pressure on the Earth, and the density is about one ten-thousandth of the Earth's atmospheric density at sea level. At a depth of 400 km, only 4 percent of the light emitted escapes from the Sun. The remainder is absorbed and re-emitted again at higher levels before it finally escapes.

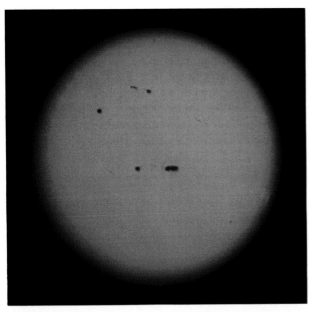

FIGURE 13.1 The visible surface of the Sun is called the photosphere. This photograph of the photosphere shows several sunspots. *(National Solar Observatory/National Optical Astronomy Observatories)*

Composition of the Sun

More than 60 of the elements known on the Earth have now been identified in the solar spectrum. Those that have not been identified in the Sun either do not produce lines in the observable part of the spectrum or are so rare in the universe that they are not expected to produce detectable lines. Most of the elements found in the Sun are in the atomic form, but more than 18 types of molecules have been identified. Most of the molecular spectra are observed only in the light from the cooler regions of the Sun, such as the sunspots (Section 13.2).

Nearly three-fourths of the Sun (by mass) is hydrogen and another 25 percent is helium. The remaining 2 percent is made up of all the other chemical elements. The ten most abundant gases in the solar photosphere are listed in Table 13.3. As we shall see when we turn to the study of stars, the relative abundances of the chemical elements in the Sun are similar to the relative abundances found in other stars.

The Chromosphere

The Sun's outer gases extend far beyond the photosphere, but they are transparent to most visible radiation. The region of the Sun's atmosphere that lies immediately above the photosphere is the **chromosphere.** Until this century, the chromosphere was observed only when the photosphere was concealed by the Moon during a total solar eclipse (Section 3.7). In the 17th century, several observers described what appeared to them as a narrow red "streak" or "fringe" around the edge of the Moon during a brief instant after the Sun's photosphere had been covered. The name chromosphere, which means "colored sphere," was given to this red streak.

Observations made during eclipses show that the chromospheric spectrum consists of bright lines, indicating that the chromosphere is composed of hot, transparent gases emitting light at discrete wavelengths. The reddish color of the chromosphere arises from one of the strongest emission lines in the visible part of its spectrum, the bright red line due to hydro-

TABLE 13.2 Properties of the Solar Photosphere

Depth in Atmosphere (km)	Percent of Light Emerging from That Depth	Temperature (K)	Pressure (bars)	Density (g/cm³)
0	99.5	4465	6.8×10^{-3}	2.3×10^{-8}
100	97	4780	1.7×10^{-2}	5.4×10^{-8}
200	89	5180	3.9×10^{-2}	1.2×10^{-7}
250	80	5455	5.8×10^{-2}	1.6×10^{-7}
300	64	5840	8.3×10^{-2}	2.1×10^{-7}
350	37	6420	1.2×10^{-1}	2.7×10^{-7}
375	18	6910	1.4×10^{-1}	3.0×10^{-7}
400	4	7610	1.6×10^{-1}	3.1×10^{-7}

TABLE 13.3   Abundance of Elements in the Sun

Element	Percentage by Number of Atoms	Percentage by Mass
Hydrogen	92.0	73.4
Helium	7.8	25.0
Carbon	0.02	0.20
Nitrogen	0.008	0.09
Oxygen	0.06	0.8
Neon	0.01	0.16
Magnesium	0.003	0.06
Silicon	0.004	0.09
Sulfur	0.002	0.05
Iron	0.003	0.14

gen (the Hα line—first line in the Balmer series—Section 5.5). One yellow line observed first in the chromospheric spectrum more than a century ago did not correspond to any previously known element on Earth. It was thought to be produced by a new element, which was named *helium* (from *helios*, the Greek word for "Sun"). In 1895, helium was discovered on Earth.

The chromosphere is about 2000 to 3000 km thick. The density of the chromospheric gases decreases upward, but the temperature *increases* from 4500 K at the photosphere to 10,000 K or so in the lower levels of the chromosphere. The temperature in the outermost layers of the Sun (the corona) exceeds 10^6 K. This temperature increase is unexpected, since throughout the rest of the Sun the temperature steadily decreases from the center outward.

The Transition Region

The region in the solar atmosphere where the temperature changes from 10,000 K, which is typical of the lower chromosphere, to nearly a million degrees, which is characteristic of the corona, is called the **transition region.**

The transition region is very thin. In a specific small region on the Sun, it may be only a few tens of kilometers thick. The transition region does not, however, form a smooth shell around the Sun at a fixed height above the photosphere. Figure 13.2 shows that the chromosphere contains many jet-like spikes of gas rising vertically through it. These features are called **spicules.** When they are viewed near the limb of the Sun, so many are seen in projection that they give the

effect of a forest. They consist of gas jets moving upward at about 30 km/s and rising to heights of 5000 to 20,000 km above the photosphere. Individual spicules last only 10 min or so. Through the spicules, matter continually flows into the corona. It is believed that the transition region may be wrapped around the spicules like a cloak.

Figure 13.3 shows the temperature structure of the solar atmosphere from the photosphere to the corona.

The Corona

The chromosphere merges into the outermost part of the Sun's atmosphere, the **corona.** Like the chromosphere, the corona was first observed during total

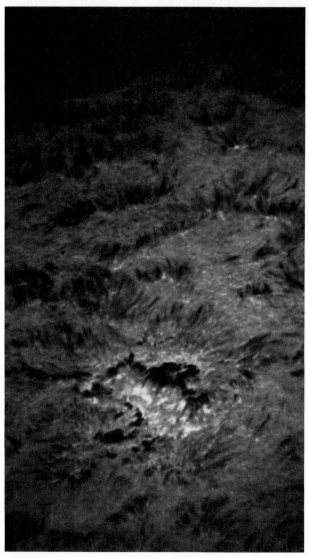

FIGURE 13.2   Solar spicules, photographed in the light of the red Balmer line of hydrogen, which is referred to as Hα. *(National Solar Observatory/National Optical Astronomy Observatories)*

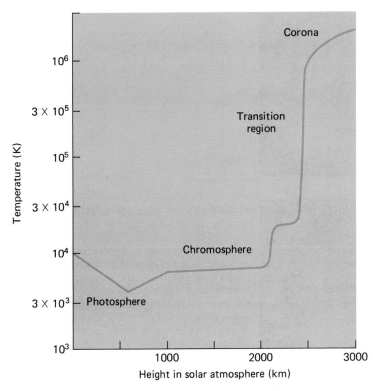

FIGURE 13.3 Temperatures in the solar atmosphere as a function of height above the photosphere. Note the very rapid increase in temperature over a very short distance in the transition region between the chromosphere and the corona.

eclipses (Figure 13.4). Unlike the chromosphere, the corona has been known for many centuries; it was referred to by Plutarch and was discussed in some detail by Kepler. The corona extends millions of kilometers above the photosphere and emits half as much light as the full moon. Under ordinary circumstances, we cannot see the corona because of the overpowering brilliance of the photosphere. While the best way to see the corona remains a total solar eclipse, the brighter parts of the corona can now be photographed with special instruments, even when there is no eclipse.

Studies of its spectrum show the corona to be very low in density. At the base of the corona, there are about 10^9 atoms per cubic centimeter, compared with about 10^{16} atoms per cubic centimeter in the upper photosphere and 10^{19} molecules per cubic centimeter at sea level in the Earth's atmosphere. The corona thins out very rapidly at greater heights, where it corresponds to a high vacuum by laboratory standards.

The corona is very hot. We know this because the spectral lines produced in the corona come from highly ionized atoms of iron, nickel, argon, calcium, and other elements. For example, lines of iron whose atoms have lost 16 electrons are observed. Such a high degree of ionization requires a temperature of millions of degrees Kelvin. The corona, in other words, is many times hotter than the photosphere. Because of this high temperature, the bulk of the spectral lines emitted by the corona lie in the far ultraviolet region (at wavelengths shorter than about 100 nm) and the corona is very bright at x-ray wavelengths (Figure 13.5).

The high temperature of the chromosphere and corona should seem surprising. The surface (photospheric) temperature of the Sun is only about 6000 K.

FIGURE 13.4 An image of the Sun taken at the time of the solar eclipse on February 16, 1980. Since the light from the brilliant surface (photosphere) of the Sun is blocked by the Moon, it is possible to see the tenuous outer atmosphere of the Sun, which is called the corona. *(High Altitude Observatory/NCAR)*

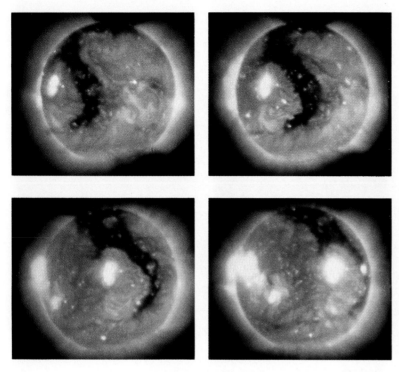

FIGURE 13.5 X-ray images of the Sun taken from the Skylab satellite show hot coronal gas. The corona is patchy, with bright spots indicating regions where hot gas is concentrated. The long dark area where there is no x-ray emission is called a coronal hole. In these regions, hot gas streams away from the solar surface out through the solar system. This stream of particles is called the solar wind. The four successive images of the Sun clearly show how the positions of solar features change as the Sun rotates. *(Harvard College Observatory/NASA)*

How, then, is it possible to heat the outer layers of the Sun's atmosphere to much higher temperatures? Observations indicate that magnetic fields play a major role. Magnetic fields apparently store energy and carry it to the chromosphere and corona, where it is converted to kinetic energy or electrical currents, which in turn heat the gases of the Sun's outer atmosphere. The precise way in which magnetic energy is converted to heat energy is not yet understood, and explaining this process in detail is one of the major challenges facing solar astronomers.

The Solar Wind

The **solar wind** is a stream of charged particles, mainly protons, flowing outward from the Sun at a speed of about 400 km/s. This solar wind exists because the gases in the corona are too hot to be confined by solar gravity. In just the same way, an atmosphere of light gas would quickly escape from the Moon.

In optical photographs, the solar corona appears to be fairly uniform and smooth. An x-ray picture, however, shows that the corona has loops, plumes, and both bright and dark regions (Figure 13.6). Sometimes large regions of the corona are relatively cool and quiet. These **coronal holes** are places of extremely low density and are usually (but not always) found in the polar regions of the Sun. They cause the empty spaces that can be seen on some of the eclipse photographs of the solar corona.

Measurements show that hot coronal gas is present mainly where magnetic fields have trapped and concentrated it. The coronal holes lie between these concentrations of gas. The solar wind comes predominantly from these coronal holes, streaming through them into space unhindered by magnetic fields.

The speed of the solar wind near the Earth's orbit averages about 400 km/s, and its density is usually 2 to 10 ions per cubic centimeter. Both the speed and the density of the solar wind, however, are highly variable. Some of the atomic particles streaming away from the Sun have energies in the low-energy cosmic-ray range. Thus the Sun can be an occasional source of weak cosmic rays. Most of the cosmic rays that reach the Earth, however, come from sources beyond the solar system.

At the surface of the Earth, we are protected from the solar wind by the atmosphere and by the Earth's magnetic field. The solar wind does, however, disrupt the ionized layers of gas in the ionosphere. This rain of particles is responsible for the aurora (Figure 13.7). As the particles in the solar wind strike atoms and molecules in the upper atmosphere, they excite them. When the electrons then rejoin the atoms and return to lower energy states, characteristic emission lines are produced. It is these emission lines that give rise to the aurora. The most spectacular auroras occur at altitudes of 75 to 150 km.

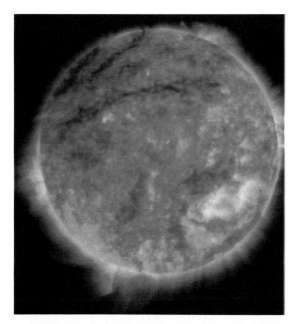

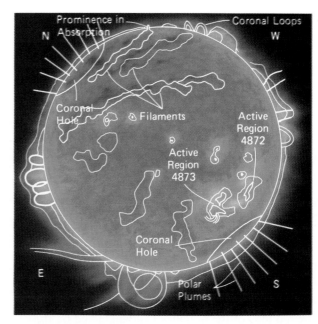

FIGURE 13.6 An x-ray/extreme ultraviolet image of the Sun taken from a sounding rocket 100 miles above White Sands Missile Range on October 23, 1987. The features shown in the photograph are identified in the drawing. *(Art Walker/Stanford University and NASA)*

13.2 The Active Sun

Overall, the Sun is quite stable, and we rely on that stability to maintain a life-sustaining environment on Earth. The surface of the Sun is not at all quiet, however. It is a seething, bubbling cauldron of hot gas. Sunspots come and go. Gas is ejected into the chromosphere and corona. Occasionally, there are giant explosions on the Sun that have major effects on the Earth.

Photospheric Granulation

Observations show that the photosphere has a mottled appearance resembling grains of rice spilled on a dark tablecloth. This structure of the photosphere is now generally called **granulation** (Figure 13.8). Typically, granules are 700 to 1000 km in diameter. They appear as bright areas surrounded by narrow darker regions, and individual granules persist for about 8 min.

FIGURE 13.7 An auroral arc in the Earth's atmosphere, as photographed from space. The frequency of occurrence of the aurora depends on the level of solar activity. *(NASA)*

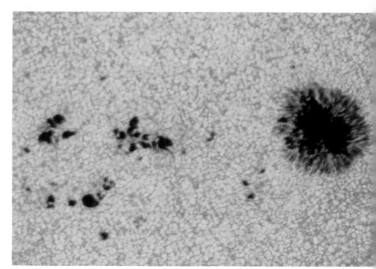

FIGURE 13.8 Solar granulation in the vicinity of sunspots. Each small, bright region is a rising column of hotter gas about 1000 km across. Cooler gas descends in the darker regions between the granules. *(National Solar Observatory/National Optical Astronomy Observatories)*

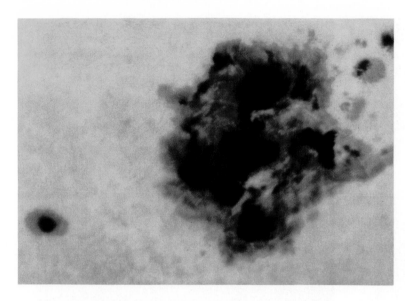

FIGURE 13.9   Photograph of a large sunspot group. Note the dark central regions (umbra) surrounded by less dark regions (penumbra). *(National Solar Observatory/National Optical Astronomy Observatories)*

The motions of the granules can be studied by the Doppler shifts in the spectra of gases just above them. It is found that the bright granules are columns of hotter gases rising from below the photosphere. As the rising gas reaches the photosphere, it spreads out and sinks down again in the darker regions between the granules. The centers of the granules are hotter than the intergranular regions by 50 to 100 K, and the vertical motions of gases in the granules have speeds of 2 or 3 km/s. The granules, then, are the tops of convection currents of gases rising through the photosphere.

Sunspots

The most conspicuous of the photospheric features are the **sunspots** (Figure 13.9). Occasionally, spots on the Sun are large enough to be visible to the naked eye.

Sunspots are darker than the photosphere in which they are embedded because the gases in sunspots are as much as 1500 K cooler than the surrounding gases. Sunspots are nevertheless hotter than the surfaces of many stars. If they could be removed from the Sun, they would be seen to shine brightly. They appear dark only by contrast with the hotter, brighter surrounding photosphere.

Individual sunspots have lifetimes that range from a few hours to a few months. If a spot lasts and develops, it is usually seen to consist of two parts: an inner darker core, the *umbra*, and a surrounding less dark region, the *penumbra*. Many spots become much larger than the Earth, and a few have reached diameters of 50,000 km. Frequently, spots occur in groups of 2 to 20 or more. If a group contains many spots, it is likely to include two large ones, one ap-

proximately east of the other, with many smaller spots clustered around the two principal ones. The largest groups are very complex and may have over a hundred spots. Like storms on the Earth, sunspots may move slowly on the surface of the Sun, but their individual motions are slow when compared with the solar rotation, which carries them across the disk of the Sun.

By recording the apparent motions of the sunspots as the turning Sun carried them across its disk, Galileo demonstrated that the Sun rotates on its axis (Figure 13.10). He found that the rotation period of the Sun is a little less than one month. Modern measurements show that the rotation period of the Sun is about 25 days at the equator, 28 days at latitude 40°, and 36 days at latitude 80°, in the direction west to east (like the orbital motions of the planets). The Sun, being a gas, need not rotate as a solid body.

13.3 The Sunspot Cycle

In 1851, a German apothecary and amateur astronomer, Heinrich Schwabe, published a paper in which he concluded that the number of sunspots visible, on the average, varied with a period of about ten years. Since Schwabe's work, the **sunspot cycle** has been clearly established. Although individual spots are short-lived, the total number of spots visible on the Sun at any one time is likely to be very much greater during certain periods, the periods of sunspot maximum, than at other times, the periods of sunspot minimum (Figure 13.11). Sunspot maxima have occurred at an average interval of 11.1 years, but the intervals between successive maxima have ranged from as lit-

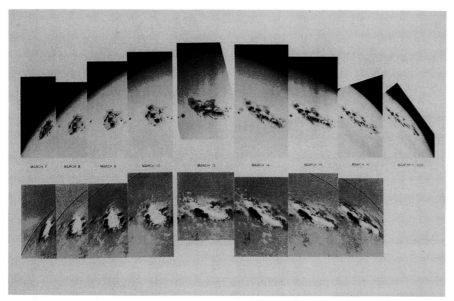

FIGURE 13.10 Photographs of the surface of the Sun showing a large group of sunspots. The series of exposures follows the rotation of sunspots across the visible hemisphere of the Sun. The top sequence shows the Sun in ordinary light; the bottom sequence is a filtergram that shows chromospheric emission. *(National Solar Observatory/National Optical Astronomy Observatories)*

tle as 8 years (from 1830 to 1838) to as long as 16 years (from 1888 to 1904). During sunspot maxima, more than 100 spots can often be seen on the Sun at once. During sunspot minima, the Sun sometimes has no visible spots. Activity was near maximum in 1990 and 1991.

Magnetism in the Solar Cycle

The solar cycle is closely related to magnetism in the Sun, and it is the changing magnetic field of the Sun that provides the driving force for many aspects of solar activity.

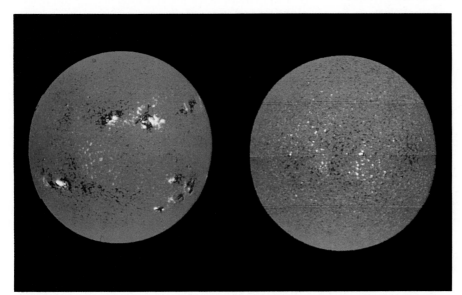

FIGURE 13.11 A comparison of the number of sunspots and magnetic activity on the active *(left-hand image)* and quiet Sun. The computer-generated images use yellow to indicate positive or north polarity, and blue for negative or south polarity. In the image of the active Sun, note that pairs of sunspots have opposite polarity. Note also that the polarity of the leading spot is different in the upper and lower hemispheres. At solar minimum *(right-hand image)*, there are no large sunspots and the magnetic fields are weak. *(National Solar Observatory/National Optical Astronomy Observatories).*

FIGURE 13.12   These photographs show how magnetic fields in sunspots are measured by means of the Zeeman effect. The vertical black line in the right-hand picture indicates the position of the spectrograph slit through which light passed in order to obtain the spectrum in the left-hand picture. Note that the strongest spectral line in the left-hand picture is split into three components. *(National Optical Astronomy Observatories)*

The solar magnetic field is measured using a property of atoms called the **Zeeman effect.** Recall (Chapter 5) that an atom has many energy levels and that spectral lines are formed by electrons that shift from one energy level to another. If each energy level is precisely defined, then only very specific transition energies between levels are permitted. The result is sharp, narrow spectral lines in either absorption or emission, depending on whether the electron increases or decreases its energy in the transition. In the presence of a strong magnetic field, however, each energy level is separated into several levels very close to one another. The separation of the levels is proportional to the strength of the field. As a result, spectral lines formed in the presence of a field are not single lines, but a series of very closely spaced lines corresponding to the subdivision of the atomic energy levels. This splitting of lines in the presence of a magnetic field is termed the Zeeman effect.

Measurements of the Zeeman effect in the spectra of the light from sunspot regions (Figure 13.12) show them to have strong magnetic fields. Whenever sunspots are observed in pairs or in groups containing two principal spots, one of the spots usually has the magnetic polarity of a north-seeking magnetic pole and the other has the opposite polarity. Moreover, during a given cycle, the leading spots of pairs (or leading principal spots of groups) in the northern hemisphere all tend to have the same polarity; those in the southern hemisphere all tend to have the opposite polarity.

During the next sunspot cycle, however, the polarity of the leading spots is reversed in each hemisphere. For example, if during one cycle the leading spots in the northern hemisphere all had the polarity of a north-seeking pole, the leading spots in the southern hemisphere would have the polarity of a south-seeking pole. During the next cycle, the leading spots in the northern hemisphere would have south-seeking polarity, and those of the southern hemisphere would have north-seeking polarity. We see, therefore, that the sunspot cycle does not repeat itself in regard to magnetic polarity until two maxima have passed. The solar activity cycle, which is fundamentally a magnetic cycle, is therefore, on average, 22 years in length, not 11.

Magnetic fields hold the key to explaining why sunspots are cooler and darker than the regions without strong magnetic fields. The forces produced by the magnetic field resist the motions of the bubbling columns of rising hot gases. Since these rising columns of hot gas carry most of the heat from inside the Sun to the surface by means of convection, there is less heating where there are strong magnetic fields. As a result, darker, cooler sunspots appear in regions where magnetic forces are strong.

13.4 Above the Photosphere

In order to see regions of the Sun that lie directly above the photosphere, we may observe at wavelengths where the photospheric gases are especially

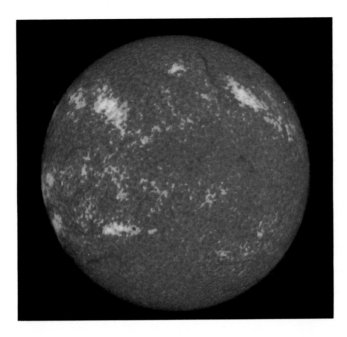

FIGURE 13.13 A filtergram showing the Sun in the light of singly ionized calcium. The picture was taken on March 18, 1990. *(National Solar Observatory/National Optical Astronomy Observatories)*

opaque—at the centers of strong absorption lines such as those of hydrogen and calcium. The parts of the solar atmosphere that give rise to these lines are thus at greater heights above the photosphere. Since many aspects of solar activity involve regions of gas in the chromosphere (above the photosphere), these strong spectral lines provide a means to monitor such activity.

There are filters that pass light only at these special wavelengths, and now astronomers routinely photograph the Sun through such monochromatic filters.

These photographs are called *filtergrams* (Figure 13.13).

Plages and Prominences

Filtergrams in the light of calcium and hydrogen show bright "clouds" in the chromosphere around sunspots; these bright regions are known as **plages** (Figure 13.14). The plages are not really clouds of any particular element but are regions of higher temperature and density.

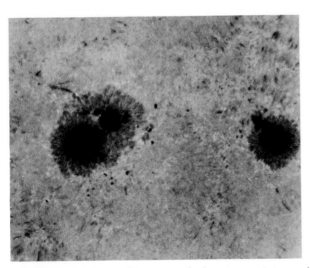

FIGURE 13.14 These pictures are of a large sunspot group photographed in 1972 at the Big Bear Solar Observatory. The left image was taken in normal (white) light and shows the photosphere. The right image, which was taken in the light of the red Balmer line of hydrogen, shows bright "clouds" or plages in the upper solar atmosphere (chromosphere) in the regions around sunspots. *(Caltech)*

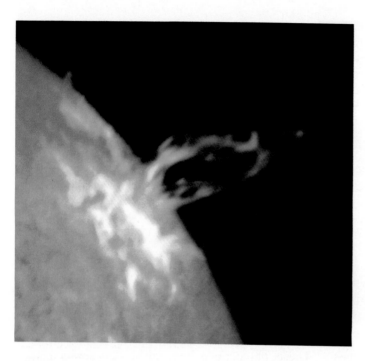

FIGURE 13.15 An eruptive prominence as seen in the light of the red line of the hydrogen Balmer series (Hα). This picture was taken on June 20, 1989. *(National Solar Observatory/National Optical Astronomy Observatories)*

Among the more spectacular of coronal phenomena are the **prominences** (Figure 13.15). Prominences have been viewed telescopically during solar eclipses for centuries, where they appear as red, flame-like protuberances rising high above the Sun. Some, the quiescent prominences, may remain nearly stable for many hours, or even days, and may extend to heights of tens of thousands of kilometers above the solar surface (Figure 13.16). Others, the more active prominences, move upward or have arches that surge slowly back and forth. The relatively rare eruptive prominences appear to send matter upward into the corona at speeds up to 700 km/s, and the most active surge prominences may move upward at speeds as high as 1300 km/s. Some eruptive prominences have reached heights of over 1 million km above the photosphere.

Prominences usually originate near regions of sunspot activity and lie on the boundary between regions of opposite magnetic polarity. Quiescent prominences are supported by coronal magnetic fields, and eruptive prominences evidently result

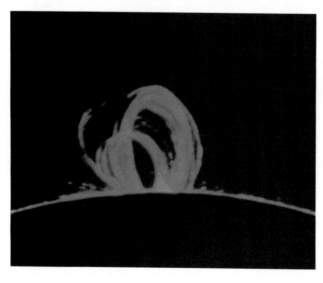

FIGURE 13.16 A loop prominence. The distinctive shape of the prominence results from strong magnetic fields in the region bending the hot, ionized gas into a loop. This picture was taken on September 29, 1989. *(National Solar Observatory/National Optical Astronomy Observatories)*

from sudden changes in the magnetic fields. Prominences seem to be further symptoms of the same general disturbances that produce spots and plages, that is, local magnetic fields.

Flares

The most awesome event on the surface of the Sun is a **solar flare.** A typical flare lasts for 5 to 10 min and releases a total amount of energy equivalent to that of perhaps a million hydrogen bombs. The largest flares last for several hours and emit enough energy to power the entire United States at its current rate of electrical consumption for 100,000 years.

The detailed process that leads to a solar flare is not well understood but apparently involves the liberation of energy stored in magnetic fields high in the solar corona. Near sunspot maximum, small flares occur several times per day, and major ones may occur every few weeks. Whatever the mechanism may be, the total amount of energy involved is astounding, and the effects on the Earth are profound.

Flares are often observed in the red light of hydrogen (Figure 13.17), but the visible emission is only a tiny fraction of what happens when a solar flare explodes. At the moment of the explosion, the matter associated with the flare is heated to temperatures as high as 10^7 K. At such high temperatures, a flood of x-ray and very-short-wavelength ultraviolet radiation is emitted, along with energetic particles, mainly protons and electrons, that stream outward into the solar system at speeds of 500 to 1000 km/s.

Effects of Flares on the Earth

The most obvious effect of solar flares on the Earth is the appearance of the aurora borealis. In March 1989, a gigantic flare, which occurred as the Sun approached maximum activity, produced an aurora visible as far south as Arizona (latitude 32°). Auroras occur preferentially near the magnetic poles of the Earth because the charged particles from the Sun tend to flow down into the Earth's atmosphere along the magnetic field and to penetrate to lowest altitudes near the poles. Only unusually bright auroras are seen in the southern United States.

Charged particles ejected by solar flares interact with the Earth's magnetic field and cause it to fluctuate. The changes in the Earth's magnetic field in turn generate changing electrical currents. The effects are most noticeable in long power lines, and solar flares can even cause components to burn out in power

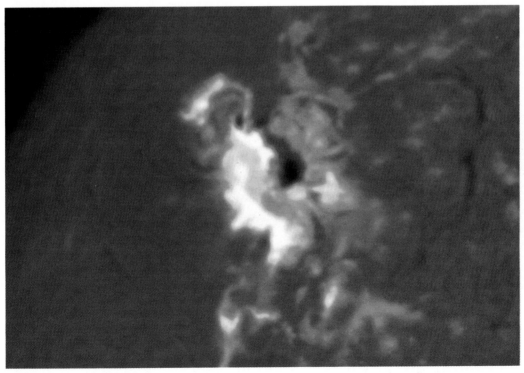

FIGURE 13.17 A giant solar flare is shown as seen in the red light of hydrogen. This flare occurred on March 10, 1989.

stations. As a result of the flare in March 1989, parts of Montreal and Quebec Province were without power for up to 9 h. Other effects occurred as well. For example, because of the electrical interference, people found their automatic garage doors opening and closing for no apparent reason. Flares can also overload telephone circuits.

The ultraviolet and x-ray emission from flares can affect the ability of the atmosphere (specifically the ionosphere) to reflect radio waves and can disrupt shortwave radio transmissions. The 1989 flare affected shortwave radio communications for 24 h.

The short-wavelength radiation produced during solar flares heats the outer atmosphere of the Earth.

September 11, 1989

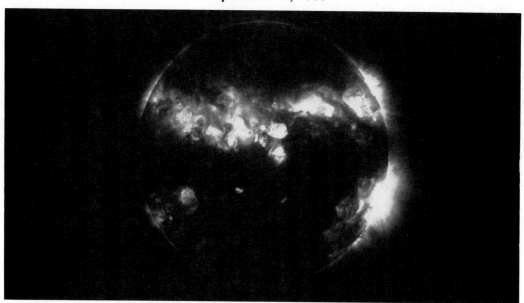

63.5 Å Normal Incidence X-Ray Telescope image (Leon Golub, SAO)

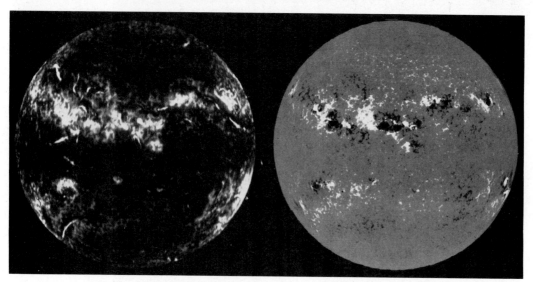

10830 Å spectroheliogram   Longitudinal magnetogram

FIGURE 13.18   Three images of the Sun taken at the same time. The top image shows where x-ray emission occurs. The lower left image, taken in the light of a strong line of neutral helium, shows where chromospheric emission is strong. In the lower right image, white and black show regions of strong magnetic field; white and black correspond to opposite magnetic polarities. Note that x-ray emission, chromospheric emission, and strong magnetic fields tend to occur in the same locations, which are called active regions. *(NASA/ National Solar Observatory/National Optical Astronomy Observatories)*

In 1981, a very large solar flare occurred while the space shuttle Columbia was in orbit. The astronauts aboard made measurements that showed that the flare, which lasted for 3 h, increased the temperature of the Earth's atmosphere at an altitude of 260 km from its normal value of 1200 K to 2200 K. When the outer atmosphere is heated, it also expands. As a consequence, friction between the atmosphere and spacecraft increases and drags satellites to lower altitude. At the time of the flare in March 1989, the system responsible for tracking some 19,000 objects orbiting the Earth temporarily lost track of 11,000 of them because their orbits were changed by the expansion of the Earth's atmosphere. During solar maximum, a great many satellites are brought to such a low altitude that they are destroyed by friction with the atmosphere.

The level of solar activity is a critical factor in calculating the lifetimes and orbits of the shuttle and satellites in near-Earth orbit. Flares could also be life-threatening to astronauts on a voyage to Mars. Obviously, it would be extremely valuable to be able to predict both the overall level of solar activity and the occurrence of individual flares. Solar astronomers are working very hard to try to learn how to make reliable predictions, but accurate forecasts of solar "weather" are proving to be as elusive a goal as reliable forecasts of the weather on Earth.

Active Regions

Sunspots, flares, and bright regions in the chromosphere and corona tend to occur together on the Sun. That is, they all tend to have similar longitudes and latitudes but, of course, to be located at different heights in the atmosphere. For example, the flare in March 1989 occurred in a region where there was a large and long-lived group of sunspots. A place on the Sun where these phenomena are seen is called an **active region** (Figure 13.18).

While we do not know what causes an active region to form, we do know that it characteristically possesses a strong magnetic field. Plages, prominences, solar flares, and strong magnetic fields all occur more frequently at times of sunspot maximum (Figure 13.19) and are related to the semiregular activity cycle of 22 years. The solar cycle is thus closely related to magnetism in the Sun, and it is the changing magnetic field of the Sun that provides the driving force for many aspects of solar activity. While astronomers have developed some understanding of the solar magnetic field, there remain many mysteries. Solar mag-

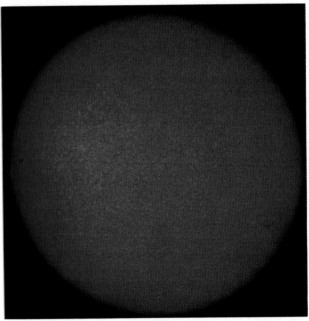

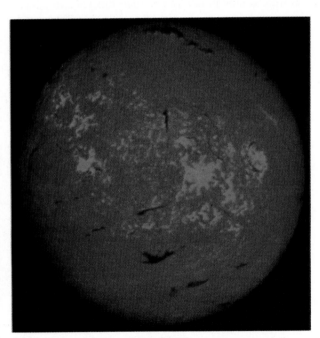

MINIMUM

MAXIMUM

FIGURE 13.19   These images show the appearance of the Sun in the red light of hydrogen at the times of a minimum *(left)* and a maximum *(right)* in solar activity. Hydrogen emission, the number of plages and prominences, and the number of flares all vary in the same cycle as the number of sunspots. *(National Solar Observatory/National Optical Astronomy Observatories)*

netism is an exceedingly complex subject, and our understanding of why the level of solar activity varies is really very limited.

13.5 Is the Sun a Variable Star?

The Sun is one of the few truly constant objects in our daily lives. It rises faithfully at a time that can be precisely calculated. Each day it deposits a constant amount of energy on the Earth, warming it and sustaining life. But is the Sun truly constant, day by day, year by year, millennium by millennium? Or does its energy output vary? Do these variations ever become large enough to affect the Earth or its climate?

We already know that variations in the total amount of energy emitted by the Sun, if any do exist, will be subtle. The existence of life on Earth demonstrates that there have been no major recent changes in the climate of the Earth. There is, however, growing evidence that long-term changes in the energy output of the Sun do have measurable effects on the Earth.

Variations in the Number of Sunspots

Astronomers have searched historical records to determine whether the number of sunspots has changed on time scales much longer than the 11- and 22-year intervals associated with the solar activity cycle. There is considerable evidence that the average number of sunspots was much lower from 1645 to 1715 than it is now. This interval of extremely low activity was first noted by Gustav Spörer in 1887 and by E.W. Maunder in 1890 and is now called the **Maunder Minimum.** The incidence of sunspots over the past four centuries is shown in Figure 13.20. According to the data presented in this figure, sunspot numbers were also somewhat lower than they are now during the

first part of the 19th century, and this period is called the Little Maunder Minimum.

When the number of sunspots is high, the Sun is active in a number of other ways as well, and this activity affects the Earth directly. As we have seen, auroras (see Figure 13.7) are caused by the impact of charged particles from the Sun on the Earth's magnetosphere. Energetic charged particles are much more likely to be ejected by the Sun when the Sun is active and when the sunspot number is high. There is a strong correlation between sunspot number and the frequency of auroral displays. Historical accounts indicate that auroral activity was abnormally low throughout the several decades of the Maunder Minimum.

The best quantitative evidence of long-term (several decades) variations in the level of solar activity comes from studies of the radioactive isotope carbon-14. The Earth is constantly bombarded by cosmic rays, which are high-energy charged particles, including protons and nuclei of heavier elements. The rate at which cosmic rays from sources outside the solar system reach the upper atmosphere depends on the level of solar activity. When the Sun is active, charged particles streaming away from the Sun out into the solar system carry the Sun's strong magnetic field with them. This magnetic field shields the Earth from incoming cosmic rays. At times of low activity, when the Sun's magnetic field is weak, cosmic rays reach the Earth in larger numbers.

When the energetic cosmic-ray particles impact the upper atmosphere, they produce several different radioactive isotopes. One such isotope is carbon-14, which is produced when nitrogen is struck by high-energy cosmic rays. The rate of production of carbon-14 is higher when the activity of the Sun is lower and the solar magnetic field does not shield the Earth from bombardment by cosmic rays.

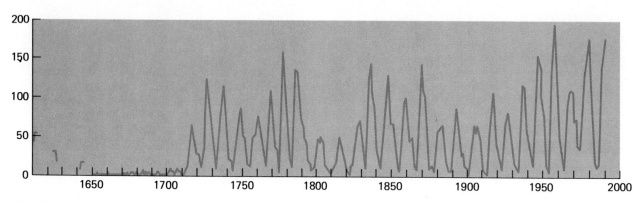

FIGURE 13.20   The relative numbers of sunspots as a function of time. Note the absence of sunspots from 1645 to 1715.

Some of the radioactive carbon is contained in carbon dioxide molecules, which are ultimately incorporated into trees through photosynthesis. By measuring the amount of radioactive carbon in tree rings, we can estimate the historical levels of solar activity. Correlations with visual estimates of sunspot numbers over the past 300 years indicate that the carbon-14 estimates of solar activity are indeed valid. Because it takes about 10 years, on the average, for a carbon dioxide molecule to be absorbed from the atmosphere or ocean into plants, this technique cannot provide data on the 11-year solar cycle. It can be used, however, to look for long-term (over several decades) changes in the level of solar activity.

Estimates of the amount of carbon-14 in tree rings now extend continuously back about 8000 years into the past. Variations in solar activity levels have occurred throughout this period, and the Sun has been at times both more and less active than it is now. The measurements confirm that the amount of carbon-14 was unusually high, and solar activity correspondingly low, during both the Maunder Minimum and the Little Maunder Minimum. During the past thousand years, activity was also low in the years 1410 to 1530 and 1280 to 1340. Between about 1100 and 1250, the level of solar activity may have been even higher than it is now.

Solar Variability and the Earth's Climate

Variations in the overall level of the Sun's activity seem to be well established. Did these variations have any direct impact on the Earth or its climate? It has long been known that the period of the Maunder Minimum was a time of exceptionally low temperatures in Europe—so low that this period is described as a Little Ice Age. The river Thames in London froze at least 11 times during the 17th century, ice appeared in the oceans off the coasts of southeast England, and low summer temperatures led to short growing seasons and poor harvests. The global climate also appears to have been unusually cool from 1400 to 1510, and this period was one of low solar activity as well.

The most obvious way in which the Sun and the Earth's climate might be linked is through variations in the luminosity of the Sun. If the Sun puts out less energy, then logically one might expect the Earth to become colder. Does the Sun become cooler at times of low activity, as would be required to account for the Little Ice Age? One might expect the opposite to occur. The most visible change at times of high activity is an increase in the number of sunspots, and sunspots are cooler than surrounding regions of the solar surface. There is, however, a second effect. At the time of sunspot maximum, large numbers of the bright, hot regions called plages (see Figure 13.14) also appear. Which effect is more important? Does the luminosity of the Sun go down when it is active because sunspots block some of the radiation? Or does it go up, because of extra radiation from the hot plages?

The relationship between solar luminosity and activity level was determined only recently from measurements made by a satellite orbiting the Earth. The changes in solar luminosity are too small to be measured reliably from the ground because of uncertainties in estimating how much of the Sun's energy is transmitted by the Earth's atmosphere. Precise measurements from space show that the luminosity of the Sun varies on time scales of weeks to months by 0.1 percent to, in extreme cases, 0.5 percent. Short-term variations in luminosity (associated with the rotation of the Sun) can be accurately estimated simply by knowing what fraction of the surface is covered by sunspots. The Sun is fainter when more sunspots are present.

Measurements also indicate, however, that there is a gradual overall decrease in luminosity that accompanies decreasing solar activity. In other words, over the cycle from sunspot maximum to minimum, the increased emission that results from the smaller numbers of sunspots is more than compensated for by decreased emission from plages and other bright regions. The total energy emitted from the Sun is least when it is least active.

These observations appear to support the idea that the Maunder Minimum was indeed associated with the Little Ice Age. The unusually cold temperatures at that time imply a drop in the solar luminosity of about 1 percent. It seems possible that such a drop in luminosity might have accompanied a long period of reduced solar activity.

There are, however, a wide variety of other phenomena that also affect the global climate, including variations in the shape of the Earth's orbit, the amount of carbon dioxide in the atmosphere, and changes in the transparency of the atmosphere because of injections of dust by volcanic explosions. Because of the complex circulation patterns of the Earth's atmosphere, local effects may differ from global effects. There can also be strong variations from one year to the next in the severity of either summers or winters that may mask long-term trends. We are still very far from having a quantitative model of the extent to which changes in the Sun affect the climate of the Earth.

SUMMARY

13.1   The outer layers of the solar atmosphere are, in order of increasing distance from the center of the Sun, the **photosphere,** with a temperature that ranges from 4500 K to about 6800 K; the **chromosphere,** with a typical temperature of 10^4 K; the **transition region**, a zone that may be only a few kilometers thick where the temperature increases rapidly from 10^4 K to 10^6 K; and the **corona,** with temperatures of a few million degrees K. **Solar wind** particles stream out into the solar system through **coronal holes.** Hydrogen and helium together make up 98 percent of the mass of the Sun. The Sun rotates more rapidly at its equator, where the rotation period is about 25 days, than near the poles, where the period is slightly greater than 36 days.

13.2   The Sun's surface is mottled with upwelling currents of hot, bright **granules. Sunspots** are dark regions where the temperature is up to 1500 K cooler than in the surrounding photosphere.

13.3   The number of visible sunspots varies on a time scale that averages 11 years in length. Spots frequently occur in pairs. During a given 11-year cycle, all leading spots in the northern hemisphere have the same magnetic polarity; all leading spots in the southern hemisphere have the opposite polarity. In the subsequent 11-year cycle, the polarity reverses. For this reason, the magnetic activity cycle of the Sun is often said to last for 22 years.

13.4   Sunspots, **solar flares, prominences,** and bright regions, including **plages,** tend to occur in **active regions**—that is, in places on the Sun with the same latitude and longitude but at different heights in the atmosphere.

13.5   There is evidence that long-term (100 years or more) variations in the level of solar activity and in the number of sunspots occur. For example, the number of sunspots was unusually low from 1645 to 1715, a period that is now called the **Maunder Minimum.** It appears that there is a tendency for the Earth to be cooler when the number of sunspots is unusually low for several decades.

REVIEW QUESTIONS

1.   Describe the main differences between the composition of the Earth and the composition of the Sun.

2.   Make a sketch of the Sun that shows the location of the photosphere, chromosphere, and corona. What is the approximate temperature of each of these regions?

3.   Why are sunspots dark?

4.   Describe some types of solar activity.

5.   Which aspects of the activity cycle have a period of about 11 years? Which vary on a time scale of about 22 years?

6.   Summarize the evidence that over a period of several decades or more there have been variations in the level of solar activity.

THOUGHT QUESTIONS

7.   The astronomer William Herschel (1738–1822) proposed that the Sun has a cool interior and is inhabited. Give at least one good argument against this idea.

8.   How might you convince an ignorant friend that the Sun is not hollow?

9.   Suppose you were to take two photographs of the Sun, one in light at a wavelength centered on a strong absorption line, and the other at a wavelength region in the continuum away from strong lines. In which photograph would you be observing deeper, hotter layers? Why?

10.   If the rotation period of the Sun is determined by observing the apparent motions of sunspots, must any correction be made for the orbital motion of the Earth? If so, explain what the correction is and how it arises. If not, explain why the Earth's orbital revolution does not affect the observations.

11.   Suppose an (extremely hypothetical) elongated sunspot formed that extended from a latitude of 30° to a latitude of 40° along a fixed line of longitude. How would the appearance of that sunspot change as the Sun rotates?

12.   Why is it difficult to determine whether or not small changes in the amount of energy radiated by the Sun have an effect on the Earth's climate?

PROBLEMS

13. Use the data in Table 13.1 to confirm the result that the density of the Sun is 1.4 g/cm³. What kinds of materials have similar densities? One such material is ice. How do you know that the Sun is not made of ice?

14. From the data in Section 13.1, find how long it takes solar wind particles, on the average, to reach the Earth from the Sun.

15. Suppose an eruptive prominence rises at 150 km/s. If it did not change speed, how far from the photosphere would it extend in 3 h?

16. From the Doppler shifts of the spectral lines in the light coming from the east and west edges of the Sun, it is found that the radial velocities of the two edges differ by about 4 km/s. Find the approximate period of rotation of the Sun.

INTERVIEW

ART WALKER

Art Walker is a solar astronomer who obtained his Ph.D. in Physics at the University of Illinois. His graduate training was supported by the Air Force reserve officer program, and after receiving his degree he began active service at the Air Force Weapons Laboratory, which was beginning a space research program in solar physics. Subsequently, he worked for the Aerospace Corporation. Today, Art Walker is a Professor at Stanford University, where he has also served as Dean of the Graduate College. A nationally-known astronomer, Walker was a member of the Presidential Commission that investigated the Challenger *accident, and he has chaired the Astronomy Advisory Committee for the National Science Foundation.*

You are one of the pioneers in space astronomy. What are some highlights of your research?

One of the first things I did when I began research in astronomy was to study the high-energy x-ray spectrum of the Sun, using a spectrometer developed for Air Force satellites. With this instrument I was able to obtain some of the first solar x-ray spectra, identifying several new emission lines and probing the temperature, structure, and composition of the corona. This was the first research I was involved in after I got my Ph.D., and I found it very exciting and a lot of fun.

Our early spectra of the Sun enabled us to identify some of the mechanisms by which x-rays are emitted and to understand the atomic processes that control the distributions and abundances in the corona. We discovered x-ray lines that were produced by peculiar recombination processes that occur

only in very low density plasmas, such as the corona. These had never been observed in the laboratory, so it was a thrill to find them in the Sun.

Is the x-ray corona that you were observing the same as the corona that is visible during a solar eclipse?

Yes and no. Most of the material you see during an eclipse is hydrogen, which scatters visible light from the photosphere. The x-ray observations see material that is at the same location, but most of the radiation arises not from hydrogen but from the heavier elements such as iron, oxygen, and neon.

What is it like doing space astronomy?

It's very challenging. In the laboratory, if you build an experiment and

it doesn't work quite right, you can always tinker with it until you get the instrumentation to operate. In the case of space observations, you don't have that opportunity. You have to build the instrument anticipating all of the things that might go wrong and eliminating each possibility of failure. Once the instrument is launched, it is out of your control, and you just have to hope that you have anticipated everything. But once you do have an instrument in space and it is working properly, it becomes highly productive. So space astronomy can be extremely exciting, but it is a very demanding way of doing experiments, and the penalties for failure are great.

Not all space experiments are alike. I've worked with rockets that provide just a few minutes observation above the atmosphere, and I've worked with long-lived satellite observatories. With rocket flights you have the opportunity to use new instrumentation and techniques very soon after they are discovered. For example, in the past several years, we have been using an optical technique that allows designs for x-ray telescopes that are similar to those in ground-based telescopes. These instruments yield very high resolution images and they are relatively simple to build. The use of rockets has allowed us to introduce these new instruments and to get results quickly and at relatively low cost. We could never have flown these instruments on satellites, because people are very hesitant to use new technology on satellites until it has been demonstrated and verified. Satellite experiments are very expensive, and they often require more than a decade from the inception of the project until launch. Rocket and satellite experiments each have their place in solar astronomy. Each technique has its advantages, and both approaches are required in an effective space research program.

The Space Shuttle was advertised as combining the best of both worlds, providing rapid access to space together with opportunities for long-term observations. Has it turned out that way?

The Shuttle has been a disappointment. The idea was that the Shuttle

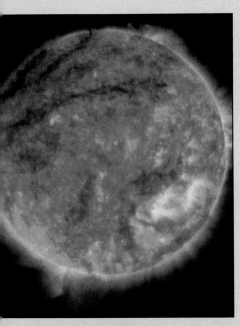

An x-ray/extreme ultraviolet image of the Sun taken from a sounding rocket 100 miles above White Sands Missile Range on October 23, 1987. *(Art Walker/Stanford University and NASA)*

would provide an inexpensive and cost-effective way of putting scientific instruments into space. However, it has turned out to be extraordinarily expensive, even more expensive than launching with the older, expendable rockets like Titan and Saturn. Instead of being able to use a relaxed and innovative approach as one does with rocket flights, the people running the Shuttle have imposed the same kinds of requirements for reliability and testing that they have on experiments put on free-flying space satellites. In fact, the requirements are even stricter, because the Shuttle is piloted and one must not inadvertently risk the crew's safety. The Shuttle is neither very fast nor very cost-effective as a launch vehicle for scientific instruments.

What is it like working on the planning and advocacy for a space experiment?

It can be very frustrating because of the long delay between the inception of an idea and the actual flight of that experiment. Since the planning is carried out before any commitment from the Congress is obtained, a great deal of effort is often

expended on experiments that will fail to be approved. The effort that went into the preliminary activity is thus wasted, and this is very disappointing for the scientists involved. NASA starts many more projects than it can complete. This difficulty is inherent in our system, in which programs are reevaluated and budget resources reallocated each year.

As a university teacher, what responses do you hear from students about possible careers in space science and technology?

Space exploration and space science are tremendously exciting to undergraduates, not only to those who are planning technical careers, but also to students who are studying scientific disciplines to round out their education and get a glimpse of how scientific research is carried out. I find that the space program is an effective way to excite and interest young people. It provides a window into scientific and technical careers for students at a time when they are making career choices.

I think a lot of students are motivated to study disciplines that they think will be financially rewarding. Law, business, and medicine are all lucrative fields. Students have a much better understanding of the rewards and challenge of those professions than they do for scientific careers because the media don't provide much information about science careers. Unfortunately, our society doesn't glamorize or romanticize scientific professions.

Women and minorities tend to be underrepresented in science careers. Why is this?

One of the reasons is that the early scientific and mathematical training required for a research career is expensive. Many minority children in this country attend school systems that are poorly funded and poorly run, so they do not have the opportunity to develop those skills. When they arrive at the university level, they are so far behind technically that it is difficult for them to catch up. As a result, they will turn to business or law or some other profession that does not require the

fundamental training in mathematics and science that is a prerequisite for a technical career.

In the case of women, things have changed a great deal. When I first came to Stanford University 20 years ago, we had very few women in our doctoral programs in physics or astronomy. Those few who did apply were often young women who had started out in some nontechnical area. My student, Sally Ride, is an example. She was an English major, and only midway in her undergraduate career did she discover the excitement of science. She changed her major to physics and went on to obtain her Ph.D. in astrophysics, and of course she was the first American woman to fly in space. Many of the young women who applied to our programs had similar backgrounds. Although often they were not well prepared from their previous education, they were very bright and highly motivated, and they were able to catch up and carry out their research. Today, however, we find that we are getting a large number of young women applying who have been science majors from the beginning of their university careers, and are as well trained as the young men who are applying for graduate work.

The American culture has changed so that young women can move into scientific careers without opposition. That is very exciting, but unfortunately we are not having as much success with regard to minorities. I think that economic and cultural factors still are impeding those students from obtaining the technical preparation at the intermediate and secondary level required for scientific and engineering careers.

One thing that it might be important for young people to realize is that a scientific career can provide varied experiences. Research can be exciting, but there are other challenges as well. I was asked to serve on the commission investigating the Shuttle *Challenger* accident, and I have also served on a number of scientific groups giving advice to the government. I have worked with the Air Force, in private industry, and in academia. A scientific career is varied and exciting with many different options available.

Atomic explosion. *(M. Meyer)*

James Clerk Maxwell (1831–1879),
Scottish physicist, unified electricity
and magnetism into a coherent
theory, much as Newton had unified
celestial and terrestrial mechanics.
Maxwell's theory was the
cornerstone on which Einstein built
his theory of special relativity.
*(American Institute of Physics, Niels Bohr
Library)*

14

THE SPECIAL THEORY OF RELATIVITY

In the previous chapter, we described what the Sun looks like and how such surface features as sunspots change with time. But did you notice that nowhere in that chapter could you find an answer to the most obvious question of all—what makes the Sun shine? Scientists throughout history have speculated about, and offered a variety of answers to, that question. It is only in this century that the real answer was found. To understand the answer, it is first necessary to make a detour and understand something about Einstein's special theory of relativity. That theory deals with the transmission of information via light and other forms of electromagnetic radiation and with what happens when objects travel at speeds close to that of light. It may seem surprising that a theory about motion at high speeds should have anything at all to do with how energy is generated in the Sun, but that is one of the wonderful things about scientific theories. Ideas and equations developed to explain one type of event may also turn out to make accurate predictions about what will happen in completely different situations.

14.1 The Speed of Light

Maxwell's theory of electromagnetism (Chapter 5) was published in 1873. Maxwell's equations predict that electromagnetic waves should move with a very definite speed. Maxwell recognized that this speed is similar to the speed that had been measured for light, and on this basis he suggested that light must be one form of electromagnetic radiation.

Roemer's Demonstration of the Finite Speed of Light

Long before Maxwell's time it was known that light travels with a finite (that is, not an infinitely great) speed, and this speed had been rather accurately measured. The speed of light is found by measuring the time required for it to travel an accurately known distance.

The first demonstration that light travels at a finite speed was provided by the Danish astronomer Olaus Roemer in 1675. Jupiter's inner satellites are regularly eclipsed when they pass into the shadow of the planet. Roemer was determining the period of revolution of Jupiter's satellites by measuring the time that elapsed between successive eclipses. He found that the eclipses were delayed by a few minutes, relative to the predictions of Kepler's laws, when Jupiter was far from the Earth. When Jupiter and the Earth were closer together, the eclipses were observed to take place earlier than predicted. Roemer correctly attributed the effect to the time it takes light to travel through space.

Observations like Roemer's indicate that light takes about 16 min to cross the orbit of the Earth. Because the Earth revolves about the circumference of its orbit in one year, it is easy to calculate that light must have a speed about 10,000 times that of the Earth. Later, when the distance from the Sun to the Earth, and hence the speed of the Earth, was well determined, the speed of light could be deduced in kilometers per second. The Earth's mean orbital speed is about 30 km/s, so the speed of light is about 300,000 km/s.

Measuring the Speed of Light

Even before Roemer's demonstration that light travels at a finite speed, Galileo suggested a way to measure the speed of light. His experiment assumes there are two people, separated by a mile or more and each equipped with a lantern that can be covered. The first opens his lantern, and the second, on seeing the light from the first, uncovers hers. The time that elapses between the time that the first experimenter opens his lantern and the time that he sees the light of his associate's lantern, after correction for the human reaction time, is how long light spends making the round trip. It is not clear whether Galileo actually conducted this experiment, but he correctly concluded that the speed of light is too great to be measured by so crude a technique.

Today, electronic timing can be used to perform a modified form of Galileo's experiment with high accuracy completely in the laboratory. The modern value for the speed of light in a vacuum (usually called c), is $c = 299,792.458$ km/s.

14.2 The Special Nature of the Speed of Light

The speed of light is different from other speeds that we normally encounter. First, of course, it is very great. More important, the speed of light is an absolute barrier. *Nothing* can go faster than light. In fact, no material body can ever even travel exactly at the speed of light. But light has an even more remarkable property—one that seems to violate "common sense." *The speed of light is always the same for all observers, no matter how fast they may be moving with respect to one another or to the source of the light.*

To see why it is so remarkable that the speed of light is the same for all observers, consider first the motion of a material body moving at velocities much less than that of light. Let's choose as an example a bullet fired from a pistol. The bullet has a certain velocity with respect to the gun, which we will call the muzzle velocity. The speed of the bullet with respect to a nearby observer, however, depends not only on the muzzle velocity but also on the speed of the gun at the time it is fired. If the gun is fired from a moving car, for example, the speed of the bullet would have the speed of the car added to its own muzzle velocity. It is possible, in principle, to move fast enough to catch up with a speeding bullet, or even outrun it. (Astronauts in orbit, for example, travel much faster than bullets.)

None of this is true for light! No matter how fast you approach or recede from a source of light, the speed of light, with respect to you, is always the same as if you and the source had no relative motion at all. Suppose you could race from the Earth in a spaceship at 99 percent the speed of light. If a colleague on Earth were to send you a light signal, then when that light caught up with your ship, its speed, with respect to the ship, would still be c. The speed of light depends in no way on your own speed. If you sent a light signal to your colleague back on Earth, then she, too, would measure the speed of that signal to be c. The speed of light does not depend on the speed of the source that emits it. You cannot catch up to, or outrun, a beam of light.

Light is therefore fundamentally different from other types of waves. You can outrun sound waves, as is done in a supersonic airplane, and you can swim into ocean waves, increasing their speed with respect to you. But nothing you can do will alter the speed of electromagnetic radiation. No matter how fast you move, or in what direction, light waves approach you with that same speed—c. You can race forward to meet the waves of light, like the swimmer in the ocean, and that light will reach you sooner than if you were stationary, but the speed of that light when it reaches you is nevertheless the same, with respect to you, as if you were not moving.

This constancy of the speed of light and of other electromagnetic radiation is predicted by Maxwell's theory of electromagnetism (Chapter 5). Light, according to Maxwell, does not act the way one would expect from considering only Newton's theories of motion. Does light point out a conflict between Newtonian mechanics and Maxwellian electrodynamics?

The existence of just such a contradiction was realized in 1895 by a 16-year-old schoolboy in the Luitpold Gymnasium in Munich. The boy was regarded as backward by his teachers and was advised to leave the school without a diploma, because he "would never amount to anything and his indifference was demoralizing." Ten years later this schoolboy, Albert Einstein, developed his ideas into the special theory of relativity.

Einstein reasoned that it should be possible to catch up with any uniformly moving object, after which the relative velocity of you and the object would be zero. But suppose you could catch up with a light beam—an electromagnetic wave. You would then find it still oscillating back and forth in time, and varying in intensity in space, but not moving! No such electromagnetic wave has ever been observed. Furthermore, theory says such an electromagnetic wave is impossible. According to Maxwell's equations, an electromagnetic wave must be moving with a speed c. Here is surely a contradiction: Either Maxwell's equations must be wrong, or our fundamental Newtonian concepts of motion must be wrong. Yet all the testable predictions of Maxwell's theory have turned out to be correct. The electronic technology available to us today certainly attests to the power and success of electromagnetic theory. The experimental evidence therefore seems to confirm Maxwell's ideas rather than Newton's.

To young Einstein there was also a strong philosophical reason for suspecting that Maxwell was right. That philosophical idea is the principle of relativity.

14.3 The Principle of Relativity

The **principle of relativity** states that all observers in uniform relative motion are equivalent. Uniform relative motion means that the observers are moving at constant velocity with respect to one another. Equivalent means that each observer finds that the *same* laws of motion are valid. Furthermore, there is *no* physical experiment by which an observer can detect his own state of uniform relative motion. What this means is that if two observers, moving at constant velocity with respect to each other, perform identical experiments in their own moving environment, they will obtain identical results. Neither can say, from anything the experiment tells him, that he is or is not moving, or how fast he is moving.

For example, two people standing in the aisle of an airliner going 600 mi/h can play catch exactly as they would on the ground. On that same airplane you can drop a heavy and a light object together, and they will hit the cabin floor at the same time. They fall at the same rate as they would if you had dropped them on the ground (provided that the airplane is moving uniformly—in a straight line at a constant speed). You can play table tennis quite normally on a moving ship on a calm sea. You can swing pendulums in an auto-

mobile (so long as it is not turning or accelerating in some other way), and they will swing in the same way, with the same periods, obeying the same pendulum laws as do pendulums in the laboratory.

When you are moving uniformly, you experience no physical sensation of speed or any other sensation that will tell you that you are in motion. You can, of course, look out the window and see the ground moving by, but if you were stationary and the ground were moving, you would feel the same and see the same thing. It is common to sit in a plane at the gate of the terminal and momentarily wonder whether it is your plane or the one at the next gate that starts to move. For that matter, none of us can feel the motion of the Earth carrying us about the Sun with its orbital speed of 30 km/s. Indeed, it is because we do not feel this motion that it took so long for people to accept the idea that the Earth *does* move.

But does this principle of relativity apply only to motion? Or does it apply to electromagnetic phenomena as well? All experiments indicate that it does apply to electromagnetic experiments and that the principle of relativity is quite general. Radios and tape recorders work the same on an airplane as in the house. You can pick up iron filings with a magnet just as easily in an automobile as in a classroom.

So far, this principle of relativity may seem rather simple and unimportant. But if we pursue its consequences just a bit more, then we will reach some astounding conclusions about time and distance. As we shall demonstrate, *if* the speed measured for light depended in any way on the velocity of the observer, then relatively simple experiments could be performed that *would* reveal the observer's motion. *All such experiments invariably fail.* Like Einstein, we will be forced to the conclusion that it is impossible to catch up with a light beam. Light will always have the same speed (in a vacuum)—c—and the prediction of Maxwell's electromagnetic theory is correct. The principle of relativity holds, and there is no experiment—whether mechanical or involving electricity and magnetism, or light, or anything else—by which we can detect our state of uniform motion.

The principle is profound. If it is impossible to detect uniform motion, the idea of *absolute motion*—motion with respect to absolute space, as envisioned by Newton—can have no meaning. There can be no absolute reference frame or coordinate system that is guaranteed to be at rest in the universe and with respect to which other motion can be measured. All we can define is *relative* motion with respect to something else. Thus the Earth moves 30 km/s with respect to the Sun; the Sun moves 20 km/s with respect to the average of its stellar neighbors; these nearby stars all move at about 220 km/s with respect to the center of our Galaxy; and so on. But there is no way to know how fast we are "really" moving, with respect to absolute space.

It is a bizarre phenomenon that two observers, one stationary with respect to a light source and one moving rapidly away from (or toward) it, will both measure the light from that source to be approaching them with the same speed. All the seemingly strange results of special relativity come about because of that one bizarre fact. Once we can swallow it, everything else in relativity makes perfect sense. Before proceeding, therefore, let's make sure we all understand that it is the real world we are talking about and not fantasy. We describe briefly in the next section a few (of very many) observations and experiments that demonstrate the absoluteness of the speed of light.

c Cannot Depend on the Motion of the Source

A simple experiment to show that the speed of light cannot depend on its source is provided by nature itself and was pointed out by the Dutch astronomer Willem de Sitter early in the 20th century. Many stars are found in double-star systems, in which the two stars revolve about their common center of mass. Take one star, say, the brighter, in such a system whose orbit lies roughly edge-on to our line of sight (Figure 14.1). Suppose the orbital speed of that star about the center of mass of the system is v. Now consider what would happen if the speed of the light it emits included the speed of the star itself, just as the speed of a bullet fired from a moving automobile has the speed of that car added to the bullet's muzzle velocity. Then when the star approaches us, at point A in Figure 14.1, light from it should be traveling toward the Earth at a speed of $c + v$. When the star is

moving away from us in its orbit, at B in the figure, its light should approach Earth with a speed of $c - v$. To be sure, the speeds of the stars in binary systems (v) are very small compared with the speed of light (c), but the stars are very far away. Over the many years it takes their light to reach us, the faster beam, traveling at $c + v$, can gain considerably over the slower beam, which is traveling only at $c - v$. If the distance to the binary system were just right, we could be receiving light from the star at position A at the same time as the light sent to us at a slower speed at an earlier time, when the star was at position B.

A little thought will show that under some circumstances we could be seeing the same star in a double-star system at many different places in its orbit at once, and analysis of the orbit would end in hopeless confusion. But we have actually analyzed the orbital motions of stars in thousands of double-star systems with distances ranging from a few light years to many hundreds of light years. In every case, the orbital motions are well behaved, with the stars moving in accordance with Newton's laws. The speed of light from them therefore cannot include the speeds of the stars themselves.

Further proof that the speed of light is independent of the speed of its source comes from the nuclear physics laboratory. In nuclear accelerators, subatomic particles moving at nearly the speed of light are often observed to change form (decay) and emit photons, but these photons are always observed to move with the normal speed of light, c, with respect to the laboratory.

c Cannot Depend on the Motion of the Observer

We have seen that the speed of light, c, cannot depend on the motion of the source of light. It cannot depend on the motion of the person who observes the light either. The most famous experiment demonstrating this result was performed in 1887 in Ohio by A. A. Michelson and E. W. Morley.

To illustrate the idea of the Michelson-Morley experiment, consider three hypothetical astronauts—Able, Baker, and Charley—as shown in Figure 14.2a, all stationary in space. Suppose (and the experiment, as we shall see, will prove this supposition wrong) that the speed of light is constant in every direction through absolute space. Baker and Charley are each 4 light years (LY) away from Able but in directions at right angles to each other. Able sends radio signals (which travel with the speed of light) to Baker and Charley at the same time, and those signals reach their destinations 4 years later. Immediately, Baker and Charley respond, and Able receives their an-

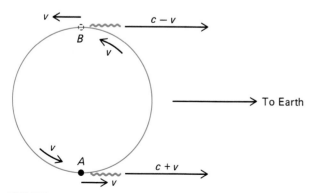

FIGURE 14.1 If the speed of light depended on the speed of the source, the light emitted by a star in a binary system would have a speed toward the Earth that depended on the location of the star in its orbit.

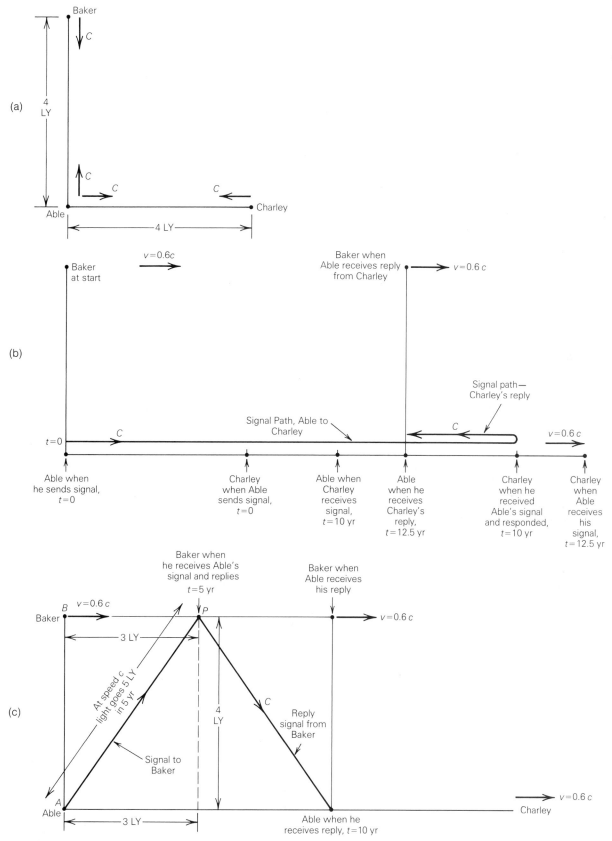

FIGURE 14.2   If the speed of light depended on the speed of the observer, a signal sent from Able to Baker and back would take less time than one sent from Able to Charley and back.

swers simultaneously, 8 years after his original transmission.

Now suppose the three astronauts maintain their relative positions, but all three are moving at 60 percent the speed of light and in the direction from Able toward Charley. As before, Able sends out the two messages, and, by our supposition, those signals move at the same speed, c, but with respect to stationary absolute space. Since we are assuming (erroneously, as it will turn out) that c is constant with respect to absolute space, the speed of light will depend on the motion of Able, Baker, and Charley. In this experiment Charley is moving away from the point where the signal was sent out by Able at $0.6c$ (Figure 14.2b). The waves from the signal, therefore, approach Charley at only $0.4c$ and take 2.5 times as long to reach him as before—that is, 10 years. On the other hand, Able is approaching the point where Charley sends back his response, at $0.6c$, so Able moves forward to meet Charley's transmission at a relative speed of $1.6c$. Thus those waves take only 2.5 years to span the 4 LY and reach Able 12.5 years after his original transmission.

But how about the message going from Able to Baker and back? The radio waves that reach Baker from Able must be directed *ahead* of Baker's position (B) as seen from Able (A) at transmission time, just as a hunter must "lead" his running prey (see Figure 14.2c). Thus the message that reaches Baker from Able travels an oblique path, and since Baker's speed is 60 percent that of the radio waves, Baker has traveled 0.6 times as far as the message has when it catches him at P. The three points—A, B, and P—make a right triangle. Since BP and AP are in the ratio 3:5, the theorem of Pythagoras for right triangles tells us that AB must be 4/5 of AP. But AB is 4 LY, so the radio message, traveling at the speed of light, took 5 years to reach Baker. The same geometry holds for the return message, so Able receives Baker's reply 10 years after his original transmission and 2.5 years ahead of hearing from Charley!

If Able actually does hear from Baker before he hears from Charley, then the principle of relativity is wrong. If Able had *not* known that his speed through space was 60 percent that of light, he could have figured it out from the difference in the time of arrival of the signals from Baker and Charley. Remember, however, that the principle of relativity says that there is no experiment we can perform that can tell us how fast we are moving through absolute space. In fact, if we were to perform this experiment, we would find that Able gets the signals from Baker and Charley at the same time and the principle of relativity is correct.

Michelson and Morley performed their experiment in a basement laboratory, but the principle is the same as in the Able-Baker-Charley experiment. Michelson and Morley hoped to determine the absolute speed of the Earth through space by measuring the difference in times required for light to travel across distances in the laboratory that were at right angles to each other. The two light paths were set up by multiple reflections between mirrors on the horizontal surface of a heavy stone slab. In advance, of course, Michelson and Morley had no way of knowing that the Earth, at the moment of the experiment, would be moving in absolute space in a direction parallel to one of the light paths (as our astronauts were moving along the line from Able to Charley). However, by rotating the granite slab through all possible directions, they reasoned that the difference between the light travel times would have to change.

It did not! Rotating the slab made absolutely no difference in the light travel time along the two beams at right angles to each other. Their experiment was accurate enough to detect velocities much smaller than the speed of the Earth in its orbit around the Sun, but there was no difference in light travel times. It was as if the Earth were absolutely stationary. But Copernicus could not be wrong, for gravitational theory shows that the Earth has to be moving. Michelson and Morley thought their experimental setup was at fault, so they repeated the experiment with even greater accuracy. There was still no difference in light travel times, and there has never been any in all the many, many times this and comparable experiments have been repeated. The only conclusion is that the speed of light does *not* depend on the motion of the observer. It is always c. Furthermore, we can never in the real world distinguish between a situation like the one shown in Figure 14.2a, with three "stationary" observers, and the situations shown in Figures 14.2b and 14.2c, in which the observers are "moving" at constant velocity.

14.4 Einstein's Special Theory of Relativity

How can we understand the bizarre properties of the propagation of light? Why didn't the Michelson-Morley experiment work? Why doesn't Able detect a difference in light travel time to Baker and Charley when all three are moving? Einstein gave the solution in his **special theory of relativity.** He showed that different observers in uniform relative motion (moving at constant velocity with respect to one another) perceive space and time differently. There are two assumptions on which the special theory is based: the principle of relativity and the absolute constancy of the speed of light. Let us now look at some of the implications of these two seemingly simple assumptions.

Time Dilation

Let us imagine the construction of an ideal clock. Of many possible designs, we shall choose a clock consisting of two parallel mirrors and a pulse of light reflecting back and forth perpendicularly between them. We shall count each time the pulse passes from one mirror to the other as a "tick" of the clock. Because light travels at an absolutely constant rate, by carefully standardizing the spacing of the mirrors, we can agree that all such clocks should keep identical time.

On the other hand, what if an observer is moving very rapidly to the right with respect to us, carrying her two-mirror clock with her? Further, suppose her direction of motion with respect to us is parallel to the surfaces of the mirrors (see Figure 14.3). As far as she is concerned, her clock, in her own system, is at rest, for there is no experiment by which she can detect her own motion. Consequently, as far as she is concerned, her clock is operating normally, with the light pulse reflecting perpendicularly back and forth between the mirrors. But as *we* see the situation, her clock is moving rapidly to the right. Therefore, the light pulse is not bouncing simply back and forth along a single line but is following a slanting path. In other words, we see the moving observer's light pulse traveling farther between ticks than she sees it traveling. But according to the principle of relativity, she and we agree on the *speed* of the light pulse, so *we* must conclude that the interval between *her* pulses is longer than it is between ours. Her seconds appear to us to be too long, and her clock is running slowly. On the other hand, she, aware of no motion on her part, argues that it is *we* who are moving to the left, that it is in *our* clock that light travels on a slanting path, and that it is *our* clock that runs slowly. Each of us insists that the other's clock is slow.

By isolating a triangle in Figure 14.3, with the most elementary algebra we can see by how much we disagree on the rate of passage of time. As far as our moving friend is concerned, she is stationary, and the light pulse has traveled vertically from A to B at a speed c, and the time it has taken to do so is t. Since distance equals rate times time, the distance from A to B must be ct. But we see the light taking the slanting path AC and requiring, at the same speed c, a longer time, t', to do so. Thus we say that the pulse

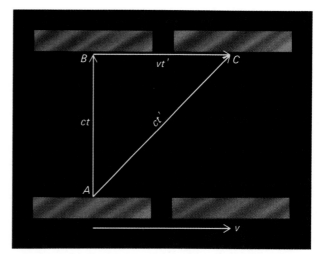

FIGURE 14.3   The light path in a moving observer's "ideal clock."

traveled a distance ct'. Meanwhile, our friend with her moving clock has gone from B to C. If her speed relative to us is v, and since we think it takes her a time t' to get to C, we calculate that the distance BC must be vt'. The theorem of Pythagoras for right triangles tells us that

$$c^2t^2 = c^2t'^2 - v^2t'^2,$$

from which we find, upon solving for t',

$$t' = \frac{t}{\sqrt{(1 - v^2/c^2)}}$$

Thus what the moving observer thinks is an interval t, we see to be a longer interval t', and it is longer by the factor $1/\sqrt{(1 - v^2/c^2)}$. She, of course, regards her time intervals as normal and *ours* as too long by the same factor.

Which of us is right? We both are. Time really *does* move at different rates in two different systems in uniform relative motion. We simply perceive time differently. Time is not absolute. Each of us has his or her own private time.

Reality of the Time Dilation

The stretching out of time between observers in uniform relative motion is called **time dilation.** It is not some artifact of the clock we choose to construct. It is a very real thing. All processes slow down in moving systems. Moving observers actually age more slowly than we do.

Nature provides a spectacular example of time dilation. The upper atmosphere of the Earth is continually bombarded by cosmic rays—atomic nuclei moving at very nearly the speed of light. When a cosmic-ray particle strikes a molecule in the upper air, it breaks the molecule into a number of subatomic particles. Common among these are particles called *muons*. A muon is rather like an electron but has about 200 times more mass. Muons spontaneously turn into electrons and emit certain other radiation, in an average time of 1.5 millionths of a second. The muons formed by collisions of cosmic rays high in the atmosphere are moving at speeds close to that of light. But even if they moved *at* the speed of light, they could travel, on the average, only about a half kilometer before decaying into electrons. The muons are formed at altitudes of 10 to 20 km, yet they rain down to the surface of the Earth in enormous numbers. In fact, the muon is the principal kind of cosmic-ray particle observed at ground level. If they are formed 15 km above the ground and decay before having time to travel as far as 1 km, how can we observe them at the bottom of the atmosphere?

As far as the muons are concerned, they *do* survive, on the average, for only 1.5 millionths of a second before turning into electrons. Because of their high speeds, however, it appears to us that their time has slowed down. As we see them, the muons have time to go very much farther than 0.5 km before decaying, and, in fact, most of them survive all the way to the Earth's surface. Muons are also observed to live very much longer when they are accelerated to high speeds in the nuclear physics laboratory. In 1976 at CERN, the international nuclear physics laboratory in Geneva, muons were accelerated to a speed of $0.9994c$. The formula for time dilation predicts that their time should slow down by a factor of 30 at that speed. Indeed, the average lifetime of those high-speed muons was 44 millionths of a second, 30 times the 1.5 millionths of a second they survive at rest. Note that we are not speaking here of light pulses bouncing between mirrors but of muons waiting to disintegrate. Time dilation is not just a strange property of our light clock but a fundamental property of time itself!

Would people live longer if they were rapidly moving? Not to their own way of thinking, of course, for they would have no sensation of moving. But relative to *us* they most certainly would age more slowly. In principle, long space trips could be made by astronauts if they were moving near enough to the speed of light. If we were to send a spaceship at a speed of $0.98c$ on a round trip to a star 100 LY away, the return journey would take just over 200 years of our time, but time for the astronauts aboard the spaceship would slow down by a factor of about 5. On their return they would be only 40 years older than when they left. (Although such relativistic space travel is theoretically possible, the virtually prohibitive energy requirements make it unfeasible in practice.)

Contraction of Length and Distance

Let us return to those astronauts traveling to a star 100 LY away at 98 percent the speed of light. If they make the trip in 20 years of their time (40 years round trip), does that mean that they have traveled at five times the speed of light? No, for lengths (and distances) as perceived by different observers in uniform relative motion are also different. If a system is in uniform motion with respect to us, we see all dimensions in that system that lie along the direction of relative motion to be *shorter* than as perceived by an observer in that moving system. The moving observer, on the other hand, sees lengths in *our* system (that lie parallel to the direction of relative motion) to be shorter than we see them. All objects in a moving system, in other words, appear foreshortened in the direction of motion.

As perceived by our astronauts moving $0.98c$ away from the Earth, the Earth and the star appear to be moving at that same speed in the opposite direction. The astronauts see the separation of the Earth and the star to be very much less than as perceived by Earthlings. In fact, they find the distance from the Earth to the star to be just under 20 LY.

We can see how this foreshortening must come about by reconsidering the astronauts Able, Baker, and Charley, who are lined up at right angles to one another. Suppose, as before, they are moving with respect to us along the direction from Able to Charley. We have already seen that according to relativity theory and the Michelson-Morley experiment, there is no way that they can detect their own motion. Thus if Able sends signals to Baker and Charley and receives simultaneous replies, he must conclude that they are equidistant from him.

But *we* don't see it that way. We see Charley moving away from Able's signal until it catches up with him. Then we see Able rushing forward to meet the return signal from Charley. And we see the signal from Able to Baker, and Baker's return signal, traveling on slanting paths, as shown in Figure 14.2. As we found before, if Baker and Charley were, according to our measurements, equidistant from Able, we should see Able receive Baker's reply first. What we actually see is that the two signals return to Able at the same time. Therefore, Baker and Charley cannot be, as we see them, at the same distance from Able. To explain what we see, we must conclude that Charley is *closer* to Able than Baker is. In the moving system, Baker and Charley are the same distance from Able, but in our system, the moving system of astronauts is foreshortened in the direction of motion.

Only a bit of algebra is needed to show that the factor of foreshortening is just the same factor by which time intervals in the moving system are too long. That is, a distance in the moving system, along the direction of motion, that the moving observer would say is D, we would say is only $D\sqrt{(1 - v^2/c^2)}$. Of course, the moving observer sees *our* distances as foreshortened, not his own. *Length* is just as private a matter as time is!

Increase in Mass

If different observers in uniform relative motion disagree on length and time, they must also disagree on velocity, which is distance covered in a given time. Thus they must, in turn, disagree on such things as momentum and energy, which depend on velocity. But they *do* agree on the laws of physics and the results of physical experiments—such as the conservation of momentum.

Suppose Jane and Mary are astronauts in space, moving together so that their relative velocity is zero (Figure 14.4a). At a given instant each fires an elastic missile, such as a billiard ball, toward the other. The two balls are identical and are fired at identical speeds. They meet halfway between the spaceships at C, rebound, and return to the ships from which they were launched. The balls had equal but opposite momentum before the impact (since they were moving in opposite directions), and since each was turned about, they had equal but opposite momentum after the collision. Total momentum must be conserved. (Remember that the momentum of an object is equal to its mass multiplied by its velocity.)

Now suppose that Jane and Mary are moving with equal speeds (with respect to us) but in opposite directions. As before, Jane and Mary discharge missiles toward each other, but because of their relative motion, they fire the balls at J and M, respectively, and in directions perpendicular to their relative velocity. Because the balls move forward with the spaceships, they follow the dashed paths shown in Figure 14.4b, meet at C, rebound, and return to their own ships at J' and M', respectively. Again, each is reversed in a symmetrical way, and momentum is conserved.

Let's hop on board Mary's spaceship and look at the last experiment from her point of view (Figure 14.4c). Mary is stationary (in her own system), so her missile moves straight out perpendicular to the path of Jane's ship. But Jane's missile is released when she was way back at J. As before, the two missiles meet

at C and rebound. Mary's missile returns to her ship, while Jane's returns to hers at J', as must happen, since the experiment is identical to the one we described in the last paragraph. Both Mary and Jane must agree that momentum is conserved (if not, one of them would be able to detect something about her own motion).

But now there is a problem. Jane is moving rapidly with respect to the stationary Mary, so Jane's time passes more slowly as seen by Mary. Similarly, all physical processes in Jane's system must slow down, including the component of velocity with which Jane's missile is fired toward Mary, perpendicular to the direction of their relative motion. But if Jane's missile is moving more slowly than Mary's, we would expect it to have less momentum as well, since the balls are of the same mass. But then how can each turn the other around, conserving momentum? We would expect Mary's missile, with the greater momentum, to suffer less change in the impact and not

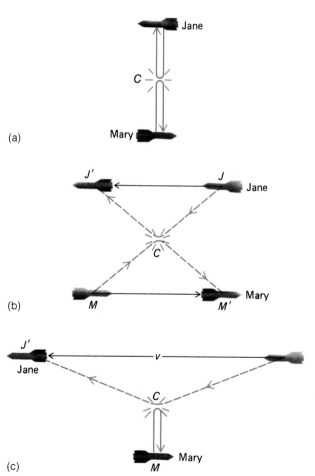

(a)

(b)

(c)

FIGURE 14.4 A hypothetical experiment involving collisions of elastic missiles, as seen from different perspectives. In all cases, momentum is conserved.

return to Mary's ship. But from our own vantage point (Figure 14.4b), we saw that it *did* return and that momentum *is* conserved. The only explanation is that Jane's missile, as observed by Mary, must have greater mass to compensate for its lower velocity.

If two observers are in uniform relative motion, each will say that the masses of objects in the other's system are greater than they would be if they were at rest. The factor by which mass is increased is exactly the same as the factor by which time is slowed. If an object has a mass m_0 when it is at rest, when it is moving with a speed v, its effective mass is $m_0/\sqrt{(1 - v^2/c^2)}$. The quantity m_0 is called the *rest mass* of the object.

The increase in mass of rapidly moving objects is not illusory. It is real. We observe it commonly in nuclear accelerators. As subatomic particles are sped up to nearly the speed of light, their masses increase manyfold, and enormously more power is required to provide them additional acceleration.

Note that if the speed of a body were equal to the speed of light, v/c would be 1, and $\sqrt{(1 - v^2/c^2)}$ would be zero. Anything divided by zero is infinite, so the mass of a particle moving with the speed of light would be infinite, which, of course, is impossible. Thus no material body (a body with nonzero rest mass) can ever travel at quite the speed of light. Here is the physical explanation of the fact that the speed of light is an absolute barrier that no body can cross. To accelerate a body of appreciable mass to a speed even very close to that of light would require absolutely tremendous amounts of energy. So far, we have succeeded in making only objects of the mass of subatomic particles reach speeds close to that of light.

Mass and Energy

All material bodies in motion possess energy of motion called **kinetic energy.** In Newtonian mechanics, the kinetic energy of a body of mass m and speed v is $(1/2)mv^2$. With a little algebra, it can be shown that the increase in mass of an object caused by its motion is its kinetic energy divided by the square of the speed of light. Equivalently, its kinetic energy equals its mass increase times c^2. Thus there is an equivalence between the mass and energy of a moving body. Einstein postulated that even when a body is at rest, there is an energy equivalence to its rest mass, so that its total energy is equal to its total mass times c^2, a concept made famous by the equation that is the hallmark of special relativity,

$$E = mc^2$$

The equivalence of mass and energy stated in the above equation suggests that matter can be converted

into energy and vice versa. Indeed, conversions in both directions are commonly observed in experiments with subatomic particles. For example, the electron has a twin called a positron, which has a charge that is equal but opposite to that of the electron but which has the same mass as the electron. When a positron and electron come into contact, they annihilate each other, turning into two photons of energy equal to the combined mass of the positron and electron times the square of the speed of light. Energetic photons can also combine to produce a positron and electron pair.

Because c^2 is a very large quantity, the conversion of even a small amount of mass results in a very great amount of energy. For example, the mutual annihilation of 1 g of electrons and 1 g of positrons (about 1/14 ounce in all) would produce as much energy as 30,000 barrels of oil. Here is the source of nuclear energy. Commercial nuclear power plants do not, however, involve the complete conversion of the nuclear fuel but only a small fraction of it. In the hoped-for hydrogen reactor of the future, hydrogen is converted to helium with the destruction of a little under half of 1 percent of the original hydrogen. Still, the conversion of only 15 kg of hydrogen into helium per hour annihilates enough matter to produce energy at the rate of the current U.S. oil consumption. We are still a long way from the technology to accomplish this, but the Sun and stars derive their energy by a similar process, as we shall describe in Chapter 15.

The fact that mass can be converted into energy and vice versa means that the old concepts of conservation of mass and conservation of energy are not strictly correct. However, if we multiply all mass by c^2 and add that number to the total energy, then the sum of the two quantities is conserved.

Faster Than Light?

According to the special theory of relativity, it is impossible for anything to travel faster than the speed of light. It is therefore impossible to transmit information of any kind at a speed larger than c. If we could communicate with infinite speed (instantaneously), there would be no special relativity at all. Michelson and Morley's experiment would have given a positive result, muons would not arrive at the ground, electrons and protons would not gain mass in accelerators, and all of the many thousands of extremely accurate tests of relativity would not have turned out the way they did. In particular, E would *not* equal mc^2, and we would not have nuclear bombs and reactors.

But this does not stop many people from feeling that somehow science and technology will find a way to "break the light barrier." Perhaps they read the wrong science fiction authors. Anyway, irrespective of Captain Kirk's taking the *Enterprise* to "warp II," it is impossible for a material body ever to reach the speed of light. It is not a technological problem but a fundamental principle of nature.

Nor is there any need to travel faster than light. At least in principle, a person can travel at a speed as close to that of light as he wishes (given enough energy). The closer his speed is to that of light, the smaller all distances around him become. Our hypothetical astronauts going only 98 percent the speed of light could reach a star 100 LY away in 20 years. By going even closer to the speed c, they could make the trip in a far shorter time. As you approach c, time slows and distances shrink, so that it is possible to go anywhere in as short a time as you like—from your own point of view but not in the eyes of your friends who remain back home.

14.5 CONCLUSION

Realm of the Universe is intended for the reader without mathematical training and is virtually without formal mathematics. Yet in this chapter, we have dipped into a little bit of algebra. The reason is that special relativity is an extremely important and fascinating subject—one that is mysterious to most people, yet one whose essence can be understood with only a tiny bit of mathematics. It would have seemed

a shame to cheat the reader out of such a wealth of knowledge that can be attained with so little extra effort.

Despite the little effort required in mathematics, the concepts are very difficult to grasp. The mathematics is easy enough, but the ideas are totally alien to our experience and present no easy conceptual hurdle. This is so because we have all grown up in a world where speeds around us are very small compared with that of light. All of the relativistic effects we have discussed depend on that factor $1/\sqrt{(1 - v^2/c^2)}$, a factor often denoted by the Greek letter gamma (γ). Values of gamma corresponding to several values of v/c are given in Table 14.1. Values in the table show by how much masses increase, lengths shrink, and clocks slow in moving systems. Until v/c is a pretty good-sized fraction, gamma is essentially equal to 1. In such low-velocity systems, Newton's laws of motion apply with admirable precision. Even the Earth's speed about the Sun—30 km/s—is only $0.0001c$, and gamma is equal to unity within one part in a hundred million. We have become used to the low-velocity world, and it has prejudiced our ideas of "common sense."

On the other hand, imagine a hypothetical civilization living on another world in an environment where speeds close to that of light are commonplace. Relativity would not seem strange to them. They, like us, given enough time, would discover the laws of physics, but not in the same order. As physicist Julian Schwinger has put it, "They would have their Maxwell and their Einstein, but alas, no Newton."

TABLE 14.1   Gamma ($1/\sqrt{(1 - v^2/c^2)}$) for Various Speeds

Moving Object	v	v/c	Gamma (γ)
Automobile	100 km/h	0.00000009	1.000000000
Concorde SST	2000 km/h	0.000002	1.000000000
Rifle bullet	1 km/s	0.000003	1.000000000
Earth escape speed	11 km/s	0.000037	1.000000001
Orbital speed of Earth	30 km/s	0.0001	1.000000005
10% light's speed	30,000 km/s	0.1	1.005
		0.5	1.155
		0.9	2.294
		0.98	5.025
		0.99	7.089
		0.999	22.37
		0.9994	28.87
		0.9999	70.71
		0.999999	707.1
		0.999999999	22360.7

SUMMARY

14.1 Roemer first showed experimentally that the speed of light is not infinite. The speed of light is measured by determining how long it takes light to travel a known distance. Today the speed of light, c, is known to be 299,792.458 km/s in a vacuum.

14.2 Maxwell's equations for electricity and magnetism show that light and all forms of electromagnetic radiation move through space at a constant speed, c, the speed of light. Unlike other speeds, this value is independent of the motion of either the light source or the observer.

14.3 Einstein based his special theory of relativity on the **principle of relativity,** namely, that there is no physical experiment by which one can detect a state of uniform relative motion. There is no absolute reference for motion, and correspondingly there is no reference place in the universe that can be guaranteed to be at rest. Michelson and Morley, in their famous 19th-century experiment, demonstrated that the speed of light does not depend on the motion of the observer (in their experiment, the motion involved was the orbital motion of the Earth).

14.4 There are several surprising consequences of the **special theory of relativity,** all of which can be derived from a few rather simple thought experiments. These are **time dilation,** foreshortening of distances, and increase in mass, all of which become evident at speeds near that of light. One consequence of this theory is that mass and energy are equivalent, and this equivalence is expressed by the famous equation $E = mc^2$. Because c^2 is a very large quantity, the conversion of even a small amount of mass releases a very great amount of energy.

14.5 The fact that nothing can exceed the speed of light is well established, despite the hopes of some science fiction authors. It is remarkable how little mathematics is required to understand special relativity. The results of those mathematical calculations, however, seem to violate common sense because we have become so used to our low-velocity world.

REVIEW QUESTIONS

1. Suppose two observers are moving at constant velocity with respect to each other. What is the one measurement about which they will always agree?

2. State the principle of relativity in your own words.

3. Suppose an astronaut is aboard a spacecraft moving at a velocity near the speed of light. How does your estimate of the rate at which time passes in her system differ from her own estimate?

4. One of the most famous equations in all of science is $E = mc^2$. Explain what this equation means.

THOUGHT QUESTIONS

5. Some distant galaxies are moving away from us at speeds of half the speed of light. What is the velocity of light from those galaxies when we detect it?

6. A spaceship that is spherical when it is on the ground moves past you at a velocity of $0.9c$. Does it still appear round?

7. Suppose a ball is thrown forward at 60 km/h from an automobile moving at 100 km/h. How fast is the ball moving with respect to an observer on the roadside? What if the ball is thrown toward the rear of the car with the same speed?

8. What is different about the way in which different observers compare the speed of the ball in the last exercise and the way in which they compare the speed of light?

PROBLEMS

9. Verify the statement in the text that the speed of light is 10,000 times greater than the orbital speed of the Earth. Assume that it takes 16 min for light to travel a distance equal to the diameter of the Earth's orbit.

10. Prepare a pendulum consisting of a small weight at the end of a string exactly 40 cm long. Start the pendulum swinging and time how long it takes to complete 10 oscillations (one oscillation is to and fro). What is your result?

Now take the pendulum into an automobile, try to arrange for the car to be driven as smoothly and at as constant a speed as possible, and repeat the experiment. Now what is your result?

11. Refer to Table 14.1. What would we measure for the mass of a 100-kg body moving past us with a speed of 90 percent that of light?

12. Isaac observes Albert to pass him, moving southward, at 99 percent the speed of light. Albert carries a 16-pound bowling ball, a meter stick (a ruler 1 m long) oriented north to south, and a poster, which he displays to Isaac (as he passes) for exactly 1 min of Albert's time.

    a. What is the value of gamma?

    b. What does Albert say is the mass of his bowling ball?

    c. What does Isaac say is the mass of Albert's bowling ball?

    d. What does Albert say is the length of his meter stick?

    e. What does Isaac say is the length of Albert's meter stick?

    f. How long does Isaac say Albert's poster was displayed?

13. According to Newton's laws, the ordinary kinetic energy of a body of mass m moving at a speed v is $(1/2)mv^2$. Calculate the kinetic energy of a body with a mass of 1 kg moving with a speed of 10^4 m/s (about one-third the orbital speed of the Earth). Now calculate the energy associated with the rest mass of the same body. How do the two energies compare?

14. By what factor does time slow for an astronaut moving 99.99 percent the speed of light? How long would it take an astronaut going that fast to make a round trip to a star 100 LY away: **(a)** according to people who stayed behind on Earth? **(b)** according to his own time?

15. By what factor is the mass of an object increased if it moves: **(a)** 0.6 times the speed of light? **(b)** 99.99 percent the speed of light?

The Sun, just rising above the horizon, with the silhouette of the mirror of the McMath Solar Telescope at Kitt Peak National Observatory. This mirror tracks the Sun across the sky and directs sunlight downward into the rest of the optics of the telescope. *(National Optical Astronomy Observatories)*

James Chadwick (1891–1974), British physicist. In addition to a knighthood in 1945, he received the Nobel prize in physics for his basic contributions to our understanding of the atomic nucleus. He proved the existence of the neutron in 1932 by bombarding beryllium nuclei with alpha particles (helium nuclei) and also worked on the generation of chain reactions and nuclear fission. (American Institute of Physics)

15

THE SUN: A NUCLEAR POWERHOUSE

What makes the Sun shine? The power output of the Sun has been, according to geological evidence, not very different since the formation of the Earth billions of years ago. Moreover, the amount of energy that the Sun has poured forth over these billions of years is enormous. The rate at which the Sun emits electromagnetic radiation into space, and thus the rate at which energy must be generated within it, is about 4×10^{26} watts. The challenge for scientists was to find what source of power can provide the gigantic amounts of energy required to keep stars like the Sun shining for so long.

15.1 Thermal and Gravitational Energy

Two large stores of energy in a star are its internal heat, or thermal energy, and its gravitational energy. The heat stored in a gas is simply the energy of motion (kinetic energy) of the particles that compose it. If the speeds of these particles decrease, the loss in kinetic energy is radiated away as heat and light. This is how a hot iron cools after it has been unplugged (except that the atoms in a solid vibrate within a crystalline structure, rather than moving freely, as in a gas).

The source of heat energy that is most familiar to us here on Earth is burning (the chemical term is oxidation) of wood, coal, gasoline, or other fuel. However, even if the immense mass of the Sun consisted of a burnable material like coal or wood, it could not produce energy at its present rate for more than a few thousand years. Geologists have found fossils in rocks that are 3.5 billion years old. We know, therefore, that the Sun must have been heating the Earth to nearly its current temperature for at least that long.

Conservation of Energy

In the 19th century, scientists used the *law of conservation of energy* to look for a source of energy for the Sun. The law of conservation of energy simply says that energy cannot be created or destroyed, but it can be transformed. The steam engine, which was the key to industrial development during the 19th century, relies on the transformation of heat energy to mechanical energy. The steam from a boiler drives the motion of a piston.

The reverse is also true. Mechanical motion can be transformed into heat. If you clap your hands vigorously, your palms will become hotter. If you rub ice on the surface of a table, the heat produced by friction will melt the ice. In the 19th century, scientists considered the possibility that the mechanical motion of meteorites falling into the Sun might provide an adequate source of heat. Calculations show, however, that in order to produce the total amount of energy—heat and light—emitted by the Sun, the mass in meteorites that would have to fall into it every 100 years would equal the mass of the Earth. The increase in the mass of the Sun would, according to Kepler's third law, change the period of the Earth's orbit by 2 s per year. Such a change would be easily measurable and has not been detected.

Gravitational Contraction As a Source of Energy

As an alternative, the German scientist Hermann von Helmholtz and the British physicist Lord Kelvin, in about the middle of the 19th century, proposed that the outer layers of the Sun might ''fall'' inward and thereby produce heat energy. The outer layer of the Sun is a gas made up of individual atoms, all moving about in random directions. Temperature is simply a measure of the speed of their motion. Now imagine that this outer layer starts to fall inward. The atoms acquire an additional velocity because of this falling motion. As the outer layer falls inward, it also contracts, and the atoms move closer together. Collisions become more likely. Some collisions serve to transfer the velocity associated with the falling motion to other atoms, increasing their velocities and so increasing the temperature of the Sun. Other collisions may actually excite electrons within the atoms to higher energy orbits. When these electrons return to their normal orbits, they emit photons, which can then escape from the Sun as heat or light.

Kelvin and Helmholtz calculated that a contraction of the Sun at a rate of only about 40 m per year would be enough to provide for its total energy output. Over the time span of human history, the decrease in the Sun's size from such a slow contraction would be undetectable. The amount of energy that has been released up until the present time by the contraction of the cloud of gas that forms the Sun is on the order of 10^{42} joules. This is the amount, according to the Helmholtz and Kelvin theory, that the Sun could have converted to thermal energy and luminosity. Since the present luminosity of the Sun is 4×10^{26} watts, or about 10^{34} joules/per year, its contraction can have kept it shining at its present rate for a period of about 100 million years.

In the 19th century, this length of time seemed adequate. But in the 20th century, geologists have shown that the Earth (and hence the Sun) has an age of several billion years. Contraction of the Sun therefore cannot account for the luminosity it has generated over its lifetime.

Even as geologists were ruling out one hypothesis about the source of the Sun's energy, physicists were developing a new one. The key lies in the nucleus of the atom and in Einstein's special theory of relativity.

15.2   Mass, Energy, and the Special Theory of Relativity

According to the law of conservation of energy, energy cannot be created or destroyed but only converted from one form to another. As we have seen in the previous chapter, one of the remarkable results of Einstein's special theory of relativity is that mass and energy are equivalent and are related by the following equation:

$$E = mc^2$$

Remember that in this equation E is the symbol for energy, m is the symbol for mass, and c, the constant that relates the two in a precise mathematical way, is the speed of light. What this equation says is that mass can be converted to energy, and energy can be converted to mass. Because c^2, the speed of light squared, is a very large quantity, the conversion of even a small amount of mass results in a very great amount of energy.

The application to the Sun is obvious. If we can find a set of interactions of atoms that lead to the destruction of some of the Sun's most abundant element (hydrogen) and the conversion of that lost mass into energy, then we will have identified a source of energy for the Sun that can last for billions of years. With Einstein's equation $E = mc^2$, it is possible to calculate that the amount of energy radiated by the Sun could be produced by the complete conversion of about 4 million tons of matter to energy each second. This sounds like a lot of matter, but in fact the Sun contains enough mass to continue shining at its present rate (given the efficiency of nuclear reactions as described below) for about 10 billion years before it exhausts its supply of fuel.

To understand how the conversion of mass to energy actually occurs, it is necessary to explore the structure of the atom.

Elementary Particles

The fundamental components of matter are called **elementary particles.** The most familiar of the elementary particles are the proton, neutron, and electron, which are the constituent particles of ordinary atoms (Section 5.4).

We have learned in the 20th century that protons, neutrons, and electrons are by no means all the particles that exist. First, for each kind of particle, there is a corresponding **antiparticle.** If the particle carries a charge, its anti has the opposite charge. The antielectron is the **positron,** of the same mass as the electron but positively charged. The antiproton has a negative charge. The antineutron, like the neutron, has no charge but interacts with other matter opposite to the way the neutron does. When a particle comes in contact with its antiparticle, the two are annihilated, turning into energy. Antimatter in our world of ordinary matter, therefore, is highly unstable, but individual antiparticles are found in cosmic rays and can be formed in the laboratory.

The existence of another type of particle, the **neutrino,** was originally postulated in 1933 by physicist Wolfgang Pauli to account for small amounts of energy that appeared to be missing in certain nuclear reactions. Neutrinos were presumed to be massless and to move with the speed of light. They interact very weakly with other matter and so are very difficult—but not impossible—to detect. Most of them pass completely through a star or a planet without being absorbed.

Experiments are not sufficiently precise to prove that neutrinos have *exactly* zero mass. If neutrinos turn out to have even a tiny mass, it could have interesting consequences for cosmology (Chapters 27 and 28) and for models of the interior of the Sun (Section 15.4).

The properties of the proton, electron, neutron, and neutrino are summarized in Table 15.1 .

Binding Energy

Just as gases give up gravitational energy when they fall together to form a star, so particles release energy in uniting to form an atomic nucleus. The energy given up is called the *binding energy* of the nucleus.

The binding energy is greatest for atoms with a mass near that of the iron nucleus, and it is less for both the lighter and the heavier atoms. In general, therefore, if light atomic nuclei come together to form a heavier one (up to iron), energy is released. This joining together of atomic nuclei is called nuclear **fusion.** On the other hand, if heavy atomic nuclei can be broken up into lighter ones (down to iron), energy is also released; this process is called nuclear **fission.** Nuclear fission sometimes occurs spontaneously, as in natural radioactivity.

According to Einstein's special theory of relativity, mass and energy are equivalent. Therefore, the energy produced by either fission or fusion must correspond to a loss of mass. Indeed, we find that the mass of every nucleus (other than the simple proton nucleus of hydrogen) is less than the sum of the masses of the nuclear particles that are required to build it. This slight deficiency in mass is always only a small fraction of the mass of a proton.

A nuclear transformation is a buildup of a heavier nucleus from lighter ones or a breakup of a heavier nucleus into lighter ones. In any such nuclear transformation, if the mass of the final nucleus is less than the mass of the nuclei that were present before the transformation, then the equivalent amount of energy is released. That energy, of course, is equal to the decrease in mass times the square of the speed of light. On the other hand, if, after the nuclear transformation, the mass of the final nucleus is larger, a corresponding amount of energy must be put into the system.

Nuclear Reactions in the Sun's Interior

The Sun taps the energy contained in the nuclei of atoms through nuclear fusion. Four hydrogen atoms combine or fuse to form a helium atom. The helium atom is slightly less massive than the four hydrogen atoms that combine to form it, and that lost mass is converted to energy.

The steps required to form one helium nucleus from four hydrogen nuclei are shown in Figure 15.1. First, two protons combine to make a deuterium nucleus, which by definition contains one proton and one neutron. In effect, one of the original protons has been converted to a neutron. Electric charge is conserved in nuclear reactions, so the positive charge originally associated with one of the protons is carried away by a positron.

This positron will instantly collide with an electron, and both will be annihilated, producing pure electromagnetic energy in the form of gamma rays. After about 10^7 years, this electromagnetic energy makes

TABLE 15.1	Properties of Some Elementary Particles	
Particle	Mass (kg)	Charge
Proton	1.67265×10^{-27}	+1
Neutron	1.67495×10^{-27}	0
Electron	9.11×10^{-31}	−1
Neutrino	0	0

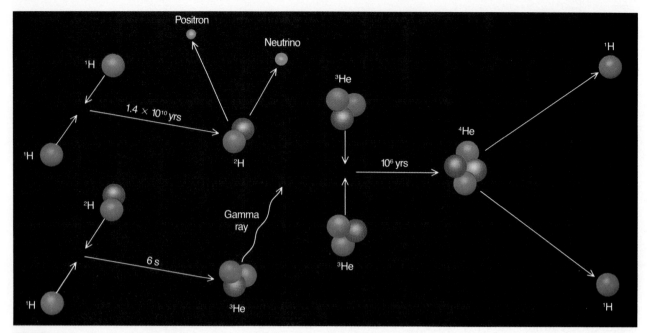

FIGURE 15.1   The Sun generates its energy by fusing four hydrogen nuclei to form helium. The steps involved in the process are shown.

its way to the surface of the Sun, being constantly absorbed and re-emitted by atoms along the way and converted to photons of longer wavelength and lower energy in the process. The photons that we observe directly are only those that are emitted so close to the surface of the Sun that they can escape without being absorbed again.

In addition to the positron, the fusion of two hydrogen atoms to form deuterium results in the emission of a neutrino. Neutrinos produced by fusion reactions near the center of the Sun travel directly to the Sun's surface and then on toward the Earth without interacting with other atoms along the way.

The next step in forming helium from hydrogen is to add a proton to the deuterium nucleus and form a helium nucleus that contains two protons and one neutron. In the process, more gamma radiation is emitted. Finally, this helium nucleus combines with another just like it to form normal helium, which has two protons and two neutrons in its nucleus. The two protons that are left over can participate in still more fusion reactions.

This series of reactions can be described succinctly through the following equations:

$$^1H + {}^1H \rightarrow {}^2H + e^+ + \nu$$

$$^2H + {}^1H \rightarrow {}^3He + \gamma$$

$$^3He + {}^3He \rightarrow {}^4He + 2\,{}^1H$$

where the superscripts indicate the total number of neutrons and protons in the nucleus, e^+ is the symbol for the positron, ν is the symbol for neutrino, and γ indicates that gamma rays are emitted.

How do these reactions take place? Protons are positively charged, and positive charges repel each other. These reactions can occur only in regions of very high temperature where the velocities of the protons are high enough to overcome the electrical forces that try to keep protons apart. In the Sun, hydrogen fusion takes place only in regions where the temperature is greater than about 10 million K and the velocities of the protons average 1000 km/s or more. Such extreme temperatures are reached only in the regions surrounding the center of the Sun, which has a temperature of 15 million K. Calculations show that nearly all of the Sun's energy is generated within about 150,000 km of its core, or within about one-quarter of its total radius.

Even at these high temperatures, it is exceedingly difficult to force two protons to combine. On average, a proton will rebound from other protons for about 14 billion years, at the rate of 100 million collisions per second, before it fuses with a second proton. Of course, some protons are lucky and take only a few collisions to achieve a fusion reaction. It is those protons that are responsible for producing the energy radiated by the Sun. Since the Sun is only about 4.5 billion years old, most of its protons have not yet been involved in fusion reactions. The low probability of the interaction of protons is fortunate for us, since it means that the Sun's fuel lasts for a long time—long enough to permit the slow biological processes on Earth to produce complex forms of life.

After the deuterium nucleus is formed, the remaining reactions happen very quickly. After about 6 s on average, the deuterium nucleus will be converted to ^{3}He. About a million years after that, the ^{3}He nucleus will combine with another to form ^{4}He.

We can compute the amount of energy generated by these reactions by calculating the difference in initial and final mass. The masses of hydrogen and helium atoms in the units normally used are 1.007825 u and 4.00268 u, respectively. (The unit of mass, u, is 1/12 the mass of an atom of carbon, or approximately the mass of a proton.) Here we include the mass of the entire atoms, not just the nuclei, because the electrons are involved as well. When hydrogen is converted to helium, two positrons are created, and these are annihilated with two free electrons, adding to the energy produced.

$$4 \times 1.007825 = 4.03130 \text{ u (mass of initial}$$
$$\text{hydrogen atoms)}$$
$$- \ 4.00268 \text{ u (mass of final}$$
$$\text{helium atom)}$$
$$0.02862 \text{ u (mass lost}$$
$$\text{in the}$$
$$\text{transformation)}$$

The mass lost, 0.02862 u, is 0.71 percent of the mass of the initial hydrogen. Thus if 1 kg of hydrogen is converted into helium, the mass of the helium is only 0.9929 kg, and 0.0071 kg of material is converted into energy. The velocity of light is 3×10^8 m/s, so the energy released by the conversion of 1 kg of hydrogen to helium is

$$E = 0.0071 \times (3 \times 10^8)^2$$
$$= 6.4 \times 10^{14} \text{ joules}$$

This amount of energy is more than ten times the Earth's annual consumption of electricity and fossil fuels.

To produce the Sun's luminosity of 4×10^{26} watts, some 600 million tons of hydrogen must be converted to helium each second, with the consequent conversion of about 4 million tons of matter into energy. As large as these numbers are, the store of nuclear energy in the Sun is still enormous. If half of the Sun's mass of 2×10^{30} kg is hydrogen that can ultimately be converted into helium, then the total store of nuclear energy would be 6×10^{44} joules. Even at the Sun's

current rate of energy expenditure, 10^{34} joules per year, the Sun could survive for more than 10^{10} years.

At temperatures that prevail in the Sun and in less massive stars, most of the energy is produced by the reactions that we have just described, and this set of reactions is called the **proton-proton cycle.** Protons collide directly with other protons to build into helium nuclei. In hotter stars, another set of reactions, called the **carbon-nitrogen-oxygen (CNO) cycle,** accomplishes the same net result. In the CNO cycle, carbon, nitrogen, and oxygen nuclei are involved in collisions with hydrogen nuclei (protons), eventually ending with carbon again and a new helium nucleus. The CNO cycle is important at temperatures above 15×10^6 K. The details of the CNO cycle are given in Appendix 8.

15.3 Interior of the Sun: Theory

Fusion of protons will occur in the center of the Sun only if the temperature exceeds 10^7 K. How do we know whether the Sun is actually this hot? To determine what the interior of the Sun is like, it is necessary to resort to mathematical calculations. In effect, astronomers teach a computer everything they know about the physical processes that are going on in the interior of the Sun. The computer then calculates the temperature and pressure at every point inside the Sun and determines what nuclear reactions, if any, are going on. The computer can also calculate how the Sun will change with time.

The Sun must change. In its center, the Sun is slowly depleting its supply of hydrogen and creating helium instead. Will this change in composition have measurable effects? Will the Sun get hotter? Cooler? Larger? Smaller? Brighter? Fainter? Ultimately, the changes must be catastrophic, since the hydrogen fuel will eventually be exhausted. Either a new source of energy must be found, or the Sun will cease to shine. What will happen to the Sun will be described in Chapters 21 and 22. For now, let's look at what we need to teach the computer about the Sun in order to carry out the calculations.

The Sun Is a Gas

The Sun is so hot that the material in it is gaseous throughout. The particles that constitute a gas are in rapid motion, frequently colliding with one another. This constant bombardment is the *pressure* of the gas (Figure 15.2). The greater the number of particles within a given volume of the gas, the greater the

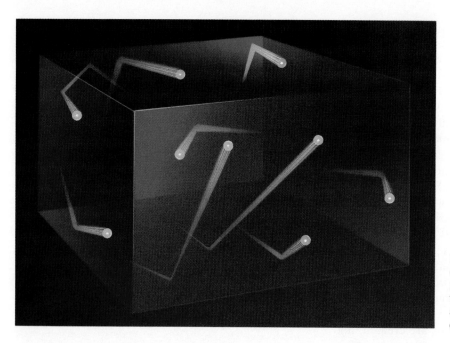

FIGURE 15.2 Gas pressure. The particles in a gas are in rapid motion and produce pressure through collisions with the surrounding material. Here particles are shown bombarding the sides of a container.

pressure, because the combined impact of the moving particles increases with their number. The pressure is also greater the faster the molecules or atoms are moving. Since their rate of motion is determined by the temperature of the gas, the pressure is greater the higher the temperature.

Most students have run across these concepts in high school, in the form of Boyle's law, which states that the pressure of a gas at constant temperature is proportional to its density, and Charles' law, which states that the pressure (at constant volume) is proportional to the temperature of the gas. These two ideas combine to give us the **perfect gas law,** which can be written in the form

$$P = nkT$$

where P is the pressure, n is the number of molecules per liter, and T is the temperature. The constant k is equal to 1.38×10^{-23} joules/K.

The perfect gas law thus provides a mathematical relation between the pressure, density, and temperature of a perfect, or ideal, gas (one in which intermolecular or interatomic forces can be ignored). The gases in most stars closely approximate an ideal gas; thus, they must obey this law. Exceptions are collapsed stars or the collapsed cores of stars, where the matter is degenerate (Chapter 22).

The Sun Is Stable

Apart from some very-low-amplitude pulsations (Section 15.4), the Sun, like the majority of other stars, is stable. It is neither expanding nor contrac-

ting. Such a star is said to be in a condition of *equilibrium.* All the forces within it are balanced, so that at each point within the star the temperature, pressure, density, and so on, are maintained at constant values. We shall see (Chapters 21 and 22) that even these stable stars, including the Sun, are changing as they evolve, but such evolutionary changes are so gradual that to all intents and purposes the stars are still in a state of equilibrium.

The mutual gravitational attraction between the masses of various regions within the Sun produces tremendous forces that tend to collapse the Sun toward its center. Yet the Sun has been emitting approximately the same amount of energy for billions of years and so has managed to resist collapse for a very long time. The gravitational forces must therefore be counterbalanced by some other force, and that force is the pressure of the gases within the Sun (Figure 15.3). To exert enough pressure to prevent the collapse of the Sun due to the force of gravity, the gases at the center of the Sun must be at a temperature of 15 million K. So we see that temperatures high enough to fuse protons are *required* by the fact that the Sun is not contracting.

If the internal pressure in a star were not great enough to balance the weight of its outer parts, the star would collapse somewhat, contracting and building up the pressure inside. If the pressure were greater than the weight of the overlying layers, the star would expand, thus decreasing the internal pressure. Expansion would stop, and equilibrium would be reached, when the pressure at every internal point again equaled the weight of the stellar layers above that point. An analogy is an inflated balloon, which

ESSAY   Hydrostatic Equilibrium

An everyday example of a gas that is in hydrostatic equilibrium is the Earth's own atmosphere. Air molecules have weight because they are attracted to the surface of the Earth by gravity. To keep the atmosphere from collapsing, there must be a force that acts to resist the Earth's gravitational attraction. That force is pressure. At any point in the Earth's atmosphere, the pressure must be exactly large enough to support the weight of all the air molecules above that point. As we move higher in the Earth's atmosphere, the number of molecules above us decreases, so the pressure also decreases with increasing altitude.

Figure 15A shows the change in pressure as a function of altitude. At the height of the world's highest major observatory on Mauna Kea, which has an altitude of nearly 14,000 feet, or about 4.2 km, atmospheric pressure is only about 60 percent of the pressure at sea level. Astronomers are above about 40 percent of the molecules in the Earth's atmosphere. At these low pressures, some chemical processes in the body, including the absorption of oxygen, occur less efficiently, and it is common for people to experience light-headedness or dizziness. If you have ever traveled to a mountain high in the Rockies or elsewhere, you may have had a similar experience.

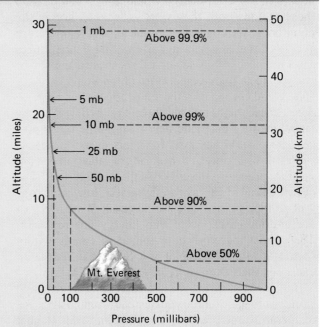

FIGURE 15A   Diagram showing that atmospheric pressure decreases rapidly with altitude. At the summit of Mt. Everest, which is about 9 km above sea level, atmospheric pressure is only about 300 millibars. A climber who reaches this height is above 70 percent of the molecules in the Earth's atmosphere.

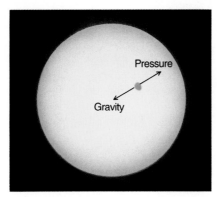

FIGURE 15.3   Hydrostatic equilibrium. In the interior of a star, the inward force of gravity is exactly balanced at each point by the outward force of gas pressure.

will expand or contract until an equilibrium is reached between the excess pressure of the air inside over that of the air outside and the tension of the rubber. This condition is called **hydrostatic equilibrium.** Stable stars are all in hydrostatic equilibrium; so are the oceans of the Earth, as well as the Earth's atmosphere. The pressure of the air keeps the air from falling to the ground.

The Temperature of the Sun Is Not Changing

From observations we know that electromagnetic energy flows from the surfaces of the Sun and stars. According to the second law of thermodynamics, heat always tries to flow from hotter to cooler regions. Therefore, as energy filters outward toward the surface of a star, it must be flowing from inner hotter regions. The temperature cannot ordinarily decrease inward in a star, or energy would flow in and heat up those regions until they were at least as hot as the outer ones. We conclude that the highest temperature occurs at the center of a star and that temperatures drop to successively lower values toward the stellar surface. (The high temperature of the Sun's chromosphere and corona may therefore appear to be a paradox. These high temperatures are believed to be maintained by magnetic heating or some other process that would not exist for a gas in which heat is simply flowing outward by means of radiation or convection, processes described in the next section.) The outward flow of energy through a star, however, robs it of its internal heat and would result in a cooling of the interior gases, were that energy not replaced. There must therefore be a source of energy within each star. In the case of the Sun, that source of energy is the fusion of hydrogen to form helium.

If a star is in a steady state (that is, in hydrostatic equilibrium and shining with a steady luminosity), the temperature and pressure at each point within it must remain approximately constant. If the temperature were to change suddenly at some point, the pressure would similarly change, causing the star to contract suddenly or to expand. Energy must be supplied, therefore, to each layer in the star at just the right rate to balance the loss of heat in that layer as it passes energy outward toward the surface. Moreover, the rate at which energy is supplied to the star as a whole must, at least on the average, exactly balance the rate at which the whole star loses energy by radiating it into space. That is, the rate of energy production in a star is equal to the total energy radiated by the star. We call this balance of heat gain and heat loss for the star as a whole and at each point within it the condition of **thermal equilibrium.**

Heat Transfer in a Star

Since the nuclear reactions that generate the Sun's energy occur deep within it, we must find a way to transport heat from the center of the Sun to its surface. There are three ways in which heat can be transported: by *conduction*, by *convection*, and by *radiation*. Conduction and convection are both important in planetary interiors. In stars, which are so much more transparent, radiation and convection are important, while conduction can be ignored unless the gas is degenerate (Chapter 22) or is very hot (as in solar flares, the solar corona, and the interstellar medium).

Stellar **convection** occurs as currents of gas flow in and out through the star (Figure 15.4). While these convection currents travel at moderate speeds and do not upset the condition of hydrostatic equilibrium or result in a net transfer of mass either inward or outward, they nevertheless carry heat very efficiently outward through a star. However, convection currents cannot be maintained unless the temperatures of successively deeper layers in a star increase rapidly in relation to the rate at which the pressures increase inward. In a similar way, convection in planetary atmospheres is important only in the troposphere, which also has a temperature that decreases rapidly with height.

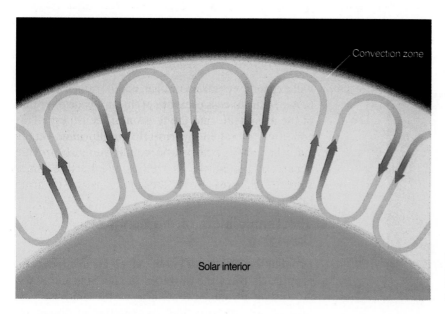

FIGURE 15.4 Rising convection currents carry heat from the interior of the Sun to its surface. Cooler material sinks downward.

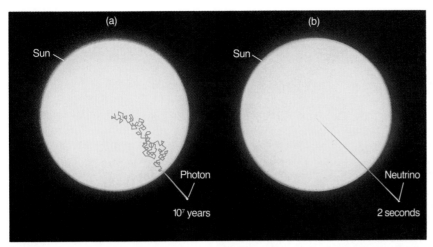

FIGURE 15.5 (a) Photons generated by fusion reactions in the solar interior travel only a short distance before they are absorbed. The re-emitted photons usually have lower energy and may travel in any direction. As a consequence, it takes about 10 million years for energy to make its way from the center of the Sun to its surface. (b) In contrast, neutrinos do not interact with matter but traverse the Sun at the speed of light, reaching the surface in only a little more than 2 s.

Unless convection occurs, the only significant mode of energy transport through a star is by electromagnetic **radiation,** which gradually filters outward as it is passed from atom to atom. However, radiative transfer is not an efficient means of energy transport, because gases in the interiors of stars are very opaque—that is, a photon does not go far before it is absorbed by an atom (typically, in the Sun, about 0.01 m). The energy absorbed by atoms is always re-emitted, but it can be re-emitted in *any* direction. A photon that is traveling outward in a star when it is absorbed has almost as good a chance of being re-radiated back toward the center of the star as toward its surface. A particular quantity of energy being passed from atom to atom, therefore, zigzags around in an almost random manner and takes a long time to work its way from the center of the star to the surface (Figure 15.5) . In the Sun, the time required is on the order of 10 million years. If the photons were not absorbed and re-emitted along the way, then they would travel at the speed of light and could reach the surface in a little over 2 s, just as the neutrinos do (Figure 15.5).

The measure of the ability of matter to absorb radiation is called its **opacity.** It should be no surprise that the gases in the Sun are opaque. If they were completely transparent, we would be able to see all the way through the Sun. We have discussed earlier (Section 5.5) the processes by which atoms and ions can interrupt the flow of energy—such as by becoming ionized. In addition, individual electrons can scatter radiation helter-skelter. For a given temperature, density, and composition of a gas, all of these processes can be taken into account, and the opacity can be calculated. The computations are very complicated and thus require powerful computers.

Model Stars

These ideas enable scientists to determine the internal structure of the Sun or of a star. They use the principles we have described: hydrostatic equilibrium, the perfect gas law, thermal equilibrium, energy transport, the opacity of gases, and the rate of energy generation from nuclear processes. These physical ideas are formulated into mathematical equations that are solved to determine the march of temperature, pressure, density, and other physical quantities throughout the stellar interior. The set of solutions so obtained, based upon a specific set of physical assumptions, is called a *theoretical model* for the interior of the star in question.

Figure 15.6 illustrates schematically what the interior of the Sun is like according to calculations. Energy is generated through fusion in the core of the Sun, which extends only about one-quarter of the way to the surface. This core contains about one-third of the total mass of the Sun. At the center, the temperature reaches a maximum of about 15 million K, and the density is nearly 150 times the density of water. The energy generated is transported toward the surface by radiation until it reaches a point about 70 percent of the distance from the center to the surface. At this point, convection begins, and energy is transported the rest of the way primarily by rising columns of hot gas.

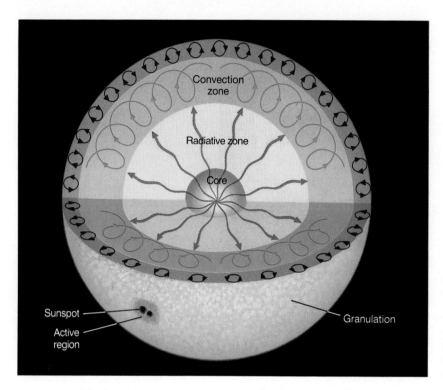

FIGURE 15.6   The interior structure of the Sun. Energy is generated in the core by the fusion of hydrogen to form helium. This energy is transmitted outward by radiation, that is, by the absorption and re-emission of photons. In the outermost layers, energy is transported mainly by convection.

Figure 15.7 shows how the temperature, density, composition, and rate of energy generation vary from the center of the Sun to its surface.

15.4   Interior of the Sun: Observations

Recently, astronomers have devised two types of observations that can be used to study the interior of the Sun directly and so check their calculations. One technique involves the study of tiny changes in the velocity of small regions at the surface of the Sun. The other relies on the measurement of neutrinos emitted by the Sun.

Solar Pulsations

The Sun pulsates—that is, it alternately expands and contracts—ever so slightly. This pulsation is detected by measuring the radial velocity of the surface of the Sun. The velocity is observed to change in a regular way, toward the Earth, away from the Earth, toward the Earth, and so on. Accurate measurements show that regions on the solar surface with diameters of 4000 to 15,000 km fluctuate back and forth every 2.5 to 11 min, with the dominant periods being about 5 min (Figure 15.8).

It is now known that these velocity fluctuations are produced by adding together millions of individual patterns of oscillation. Individual oscillations have velocities as small as 20 cm/s, and the combined sum of the velocities of all the patterns is only a few hundred meters per second. Since it takes only about 5 min to complete a full cycle from maximum to minimum velocity and back again, the change in the radius measured at any given point on the Sun is no more than a few kilometers. Since the total radius of the Sun is about 650,000 km, the percentage change in the size of the Sun is so small that it cannot be measured directly. It is only in the past 15 years or so that astronomers have had instruments that would permit them to detect the small velocity changes.

The discovery of the velocity variations was soon followed by the realization that they could be used to determine empirically what the interior of the Sun is like. The motion of the Sun's surface is caused by waves generated deep within it. Study of the amplitude and period of the velocity changes produced by this motion can yield information about the temperature, density, and composition beneath the Sun's surface. The situation is somewhat analogous to the use of earthquakes to infer the properties of the interior of the Earth. For this reason, studies of solar oscillations are referred to as **solar seismology.** It takes about an hour for these waves to traverse the Sun, so they,

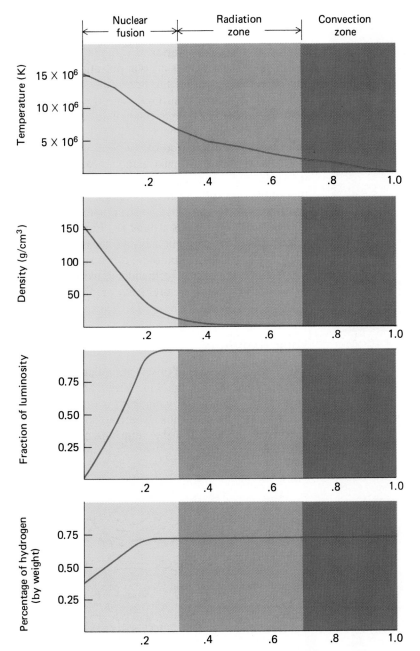

FIGURE 15.7   Diagrams showing how temperature, density, rate of energy generation, and the percentage (by mass) abundance of hydrogen vary inside the Sun.

like neutrinos, provide information about what the solar interior is like at the present time. As we have already noted, this statement is not true for visible light. Since photons take about 10 million years to travel from the center of the Sun to its surface, the luminosity that we now measure was generated millions of years ago.

Solar seismology has already yielded some important results. Calculations had suggested previously that the convection zone extends only 15 to 20 percent of the distance from the surface of the Sun to its center. Measurements of solar pulsation indicate that convection extends inward from the surface 30 percent of the way toward the center, as shown in Figure 15.6. The observed oscillations also indicate that the abundance of helium inside the Sun, except in the center where nuclear reactions have converted hydrogen to helium, is about the same as at its surface.

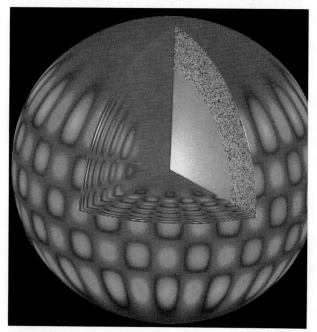

FIGURE 15.8 New observational techniques permit astronomers to measure small differences in velocity at the surface of the Sun to infer what the deep interior of the Sun is like. In this computer simulation, red denotes regions of the surface that are moving away from the observer; blue marks regions moving toward the observer. Note that the velocity changes deeply penetrate the interior of the Sun. *(National Optical Astronomy Observatories)*

That result means that it is correct to use the abundances measured in the solar atmosphere to construct models of the solar interior.

Recent measurements of solar pulsations show that the outer 30 percent of the Sun rotates at about the same rate as does the surface, with the rotation rate slower at the poles than at the equator (Figure 15.9). Deeper in the interior, the Sun appears to rotate like a solid body, with a period of about 27 days.

The next step is to make better measurements so that solar pulsations can be used to probe the structure of the Sun still closer to its center, where nuclear reactions are taking place. Astronomers are setting up a network of stations around the Earth to monitor solar oscillations continuously. As it is now, observations from any single station are interrupted by sunset. Completion of the worldwide network and analysis of the velocity measurements obtained will take perhaps 10 years.

Solar Neutrinos

There is another type of observation that can be used to probe the center of the Sun. About 3 percent of the total energy generated by nuclear fusion in the Sun is carried away by neutrinos. Neutrinos only rarely interact with matter, and the neutrinos created in the

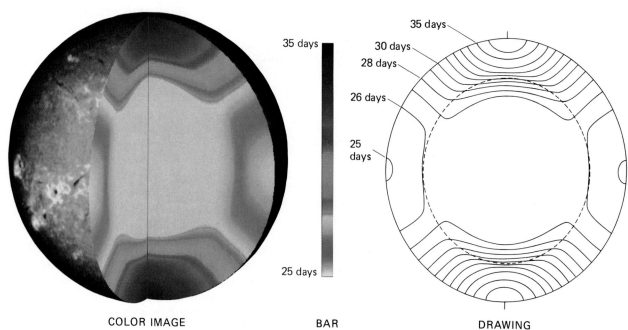

COLOR IMAGE   BAR   DRAWING

FIGURE 15.9 The period of solar rotation in days as a function of position within the Sun. Note that the rotation period increases from 25 days at the equator to 36 days at the pole. These same periods persist inward throughout the Sun's convective zone to a depth of 200,000 km. Deeper inside, the Sun appears to rotate as a solid body, with a period of 27 days. *(Kenneth Libbrecht/Caltech)*

center of the Sun make their way directly out of the Sun and to the Earth at the speed of light. As far as neutrinos are concerned, the Sun is transparent. If we can devise a way to detect some of the 300 billion billion (3×10^{20}) solar neutrinos that pass through each square meter of the Earth's surface every second, then we can obtain information directly about what is going on in the center of the Sun.

Unfortunately, the very property that makes neutrinos an interesting source of information about the interior of the Sun makes them very difficult to detect. Neutrinos pass through material on the Earth as readily as they escape from the Sun. On very, very rare occasions, however, a neutrino of the highest energy of those emitted from the Sun will react with the isotope chlorine-37 to produce argon-37 and an electron. To detect these rare events, Raymond Davis, Jr., and his colleagues at Brookhaven National Laboratory placed a tank containing nearly 400,000 liters of cleaning fluid (C_2Cl_4) 1.5 km beneath the surface of the Earth in a gold mine at Lead, South Dakota. Calculations show that solar neutrinos should produce about one atom of argon-37 daily in this tank.

Amazingly, it is possible to detect individual argon-37 atoms, and the results are that only about one-third as many neutrinos reach the Earth as are predicted by standard models of the solar interior. What is wrong? To answer that question, several other neutrino experiments have been started. In Japan, a neutrino detector has succeeded in measuring the direction from which the neutrinos are coming and has shown that they do originate in the Sun. This detector, like the one in South Dakota, measures fewer neutrinos than solar models predict.

Just beginning operation are two detectors that use reactions in gallium to detect neutrinos. The detectors in South Dakota and in Japan are sensitive only to very-high-energy neutrinos. The gallium detectors, in contrast, can detect the lower energy neutrinos produced when two protons collide in the first step of the proton-proton cycle. According to our understanding of both physics and the Sun, this step must occur at very nearly the calculated rate, or the Sun cannot shine. As of this writing only the first four months of data are available, and *no neutrinos were*

detected! If this shocking result is confirmed, and most scientists would guess that it will be, then we must be completely wrong either about the interior structure of the Sun or about the nature of neutrinos.

Right now the most likely possibility appears to be that we are wrong about neutrinos. It is known that there are not one but three types of neutrinos, and the cleaning fluid experiment can detect only the type that is generated by the nuclear reactions in the center of the Sun. If neutrinos have even a very small mass, and measurements are not yet precise enough to prove that neutrinos are completely massless, then they may change from one type to another. If the solar neutrinos can be converted to either of the other two types sometime during their journey from the center of the Sun to the Earth, then they would no longer be detectable by the cleaning fluid experiment.

Recent theoretical calculations suggest that this may be exactly what happens. The calculations may even be able to explain why we see some high-energy neutrinos from the Sun and (if the first measurements are confirmed) no or very few low-energy ones.

And so we are left with a puzzle. Something is fundamentally wrong with our understanding either of the Sun or of the basic physics of neutrinos. We can guess what might be wrong, but we have not yet proved it. While it may seem discouraging that there are questions for which definitive answers are not yet available, science very often works this way. Observations—in this case, of the mass, luminosity, composition, and other properties of the Sun—lead to the development of a model. This model then suggests a number of other measurements that can be made, in this case of neutrinos and solar pulsations, and predicts the outcome of those measurements. Frequently, the predictions are incorrect, and the models must be modified to take into account the new measurements. And so science moves forward by successive approximations, each step providing a better and more complete description of what is actually occurring. Rather than finding the lack of final answers discouraging or frustrating, scientists find this situation a source of never-ending challenge. The possibility of learning something never before known by *anyone* is what attracts many scientists to research.

SUMMARY

15.1 The Earth is 4.5 billion years old, so the Sun must have been shining for at least this long. Neither chemical burning nor gravitational contraction can account for the energy radiated by the Sun during this time.

15.2 The energy radiated by the Sun is produced by interactions of **elementary particles**—that is, protons, neutrons, electrons, and **neutrinos.** Specifically, the source of the Sun's energy is the **fusion** of hydrogen to form helium. A helium atom is about 0.71 percent less massive than the four hydrogen atoms that combine to form it, and that lost mass is converted to energy.

15.3 Even though we cannot see inside the Sun, it is possible to calculate what the solar interior must be like. As input for these calculations, we use what we know about the Sun. It is a gas whose behavior is described by the **perfect gas law.** Apart from some very tiny changes, the Sun is neither expanding nor contracting (it is in **hydrostatic equilibrium**) and puts out energy at a constant rate. Fusion of hydrogen occurs in the center of the Sun, and the energy generated is carried to the surface by **radiation** and **convection.** A stellar model describes the structure of the interior of a star. Specifically, it describes how the pres-

sure, temperature, mass, and luminosity depend on the distance from the center of the star.

15.4 Studies of solar oscillations and of neutrinos provide observational data about the interior of the Sun. The technique of **solar seismology** has so far shown that the composition of the interior of the Sun is much like that of the surface, except in the core, where hydrogen has been converted to helium; that the convection zone extends 30 percent of the way from the surface to the center of the Sun; and that the outer 30 percent of the Sun rotates at the same velocity as the solar surface. The standard solar model predicts that we should be able to detect more neutrinos than we do. It may be that this result indicates that the mass of the neutrino is not exactly equal to zero.

REVIEW QUESTIONS

1. How do we know that the Sun's energy is not supplied either by chemical burning, as in fires here on Earth, or by gravitational contraction?

2. What makes the Sun shine?

3. In what respect (or respects) is a neutrino very different from a neutron?

4. Describe in your own words what is meant by the statement that the Sun is in hydrostatic and thermal equilibrium.

5. Why do measurements of the number of neutrinos emitted by the Sun tell us about conditions deep in the solar interior?

THOUGHT QUESTIONS

6. Which of the following transformations is (are) fusion and which is (are) fission: the transformation of (**a**) helium to carbon; (**b**) carbon to iron; (**c**) uranium to lead; (**d**) boron to carbon; (**e**) oxygen to neon?

7. Stars that are hotter than the Sun also derive their energy by fusing hydrogen to form helium, but a different set of reactions is involved. This set of reactions is called the carbon-nitrogen-oxygen (CNO) cycle, and the individual steps in the process are given in Appendix 8. Draw a picture like Figure 15.1 that shows what happens in each step of the CNO cycle.

8. In the CNO cycle, carbon and nitrogen are referred to as *catalysts*, since the carbon and nitrogen nuclei are required to make the reactions proceed, but the total number of carbon and nitrogen atoms does not change when all of the steps are completed. Show from your drawing for the previous problem that this statement is true.

9. Why is a higher temperature required to fuse hydrogen to helium by means of the CNO cycle than by the process that occurs in the Sun, which involves only isotopes of hydrogen and helium?

10. After a star converts its hydrogen to helium, it must then use helium as a fuel. The specific reaction that is involved is the conversion of three helium nuclei to a car-

bon nucleus. The steps are given in Appendix 8. Again, draw a picture like Figure 15.1 that shows what happens in this two-step process. How do you think the interior structure of the star must change in order to make the conversion of helium to carbon possible?

11. The Earth's atmosphere is in hydrostatic equilibrium. Explain what this means. Would you expect the pressure in the Earth's atmosphere to increase or decrease as you climb from the bottom of a mountain to its summit. Why?

12. Give some everyday examples of the transport of heat by convection and by radiation.

13. Suppose the proton-proton cycle in the Sun were to slow down suddenly and generate energy at only 95 percent of its current rate. Would an observer on the Earth see an immediate decrease in the brightness of the Sun? Would she immediately see a decrease in the number of neutrinos emitted by the Sun?

14. Why do you suppose so great a fraction of the Sun's energy comes from its central regions? Within what fraction of the Sun's radius does practically all of the Sun's luminosity originate? (See Figure 15.7). Within what radius of the Sun has its original hydrogen been partially used up? Discuss what relation the answers to these questions bear to one another.

15. Why do we not expect nuclear fusion to occur in the surface layers of stars?

16. The Sun obtains its energy by fusing four hydrogen nuclei to form a helium nucleus. It is also possible to obtain energy by breaking up atomic nuclei of such heavy elements as uranium and plutonium to form lighter nuclei. This process is called nuclear *fission*. Why is fission not an important source of energy in the Sun?

PROBLEMS

17. Verify that some 600 million tons of hydrogen are converted to helium in the Sun each second.

18. If the atmospheric pressure were the same on two different days, but if one day were much hotter than the other, what could you say about the relative density of the air on the two days?

19. If an observed oscillation of the solar surface has a period of 10 min and the average radial velocity is 1 m/s in and out, calculate the total displacement of the surface that is involved in this particular oscillation mode. What percent is this of the total radius of the Sun?

The galaxy IC 5152 appears on the sky close to a star in our own Galaxy. The accurate determination of distances to individual objects like these was the first step in understanding what they are like. *(Copyright Anglo-Australian Telescope Board)*

Friedrich Wilhelm Bessel (1784–1846) made the first authenticated measurement of the distance to a star (61 Cygni) in 1838, a feat that had eluded many dedicated astronomers for almost a century. *(Yerkes Observatory)*

CELESTIAL SURVEYING: DISTANCES AND MOTIONS OF STARS

For centuries philosophers and scientists speculated about the nature of stars. Little by little, astronomers pieced together information about the stars—their distances, temperatures, energy output, composition, and mass. It was only in the 20th century, however, that theories of the structure of the atom and of relativity yielded true understanding of the life cycles of stars.

The determination of astronomical distances is central to understanding the nature of stars. As astronomy has developed over the centuries, both the precision of our measurements and the distances over which they are applied have increased. We have built up a sort of "cosmic distance ladder," stretching from the Earth to the stars to the farthest reaches of the universe. Each rung of that ladder has been carefully tested, although substantial uncertainties remain along the way. One of the properties of that ladder is that the largest distances—to faint galaxies and quasars—depend on the preceding steps, tracing back to the distance of the Earth from the Sun. Or, to use a different analogy, this entire chain is only as strong as its weakest link, and every increment is no more accurate than the sum of the steps already taken.

In this chapter, we begin with the fundamental definitions of distances on Earth and then extend our reach outward to the distant stars.

16.1 Fundamental Units of Distance

The first measures of distances were based on human dimensions: the inch as the distance between knuckles on the finger or the yard as the span from the extended index finger to the nose of the king. Later, the requirements of commerce led to some standardization of such units, but each nation tended to set up its own definitions. It was not until the middle of the 18th century that any real efforts were made to establish a uniform, international set of standards.

The Metric System

One of the enduring legacies of the Napoleonic era was the establishment of the *metric system* of units, officially adopted in France in 1799. The fundamental metric unit of length is the *meter*, originally defined as 1/10,000,000 of the distance along the Earth's surface from the equator to the pole. French astronomers of the 17th and 18th centuries had pioneered in the determination of the dimensions of the Earth, so it was logical to use this information as the foundation of the new system.

Practical problems exist with a definition expressed in terms of the size of the Earth, since anyone wishing to determine the distance from one place to another could hardly be expected to go out and remeasure the planet. Therefore, an intermediate standard meter was set up in Paris, consisting of a bar of platinum-iridium. In 1889, by international agreement this bar was defined to be exactly 1 m in length, and precise copies of the original meter bar were made to serve as standards for other nations. The U.S. standard meter is kept in a vault at the National Bureau of Standards in Washington.

Other units of length are derived from the meter. Thus 1 km equals 1000 m, 1 centimeter equals 1/100 m, and so on. Even the old American units, such as the inch and the mile, are now defined in terms of the metric system.

FIGURE 16.1   Radar telescope of the NASA Deep Space Net in the Mohave Desert of California. *(NASA/JPL)*

Modern Redefinitions of the Meter

In 1960 the definition of the meter was changed again. As a result of improved technology for generating spectral lines of precisely known length, the meter was redefined to be equal to 1,650,763.73 wavelengths of a particular atomic transition in krypton-86. The advantage of this redefinition is that anyone with a suitably equipped laboratory can reproduce a standard meter, without reference to any particular metal bar.

In 1983 the meter was again redefined, this time in terms of the velocity of light, which is perhaps the most fundamental constant of nature (Chapter 14). By this time, the length of the standard unit of time, the *second*, had been fixed by international agreement as 9,192,631,770 times the frequency of a cesium-133 atomic clock. The velocity of light was, in turn, defined to be 299,792,458.6 m/s. These two values thus served to define the meter as the distance light travels in a vacuum in a time interval of 1/299,792,458.6 s. Today, therefore, light travel time provides us our basic unit of length. Putting it another way, a distance of one *light second* is defined to be 299,792,458.6 m. We could just as well use the light second as the fundamental unit of length, but for practical reasons (and respecting tradition), we have defined the meter as a small fraction of the light second.

Distances Within the Solar System

Copernicus was able to determine the relative distances to the planets (in terms of the distance from the Earth to the Sun), but not their absolute distances in light seconds or meters or other standard units of length. Over the subsequent centuries, many improved measurements were made of planetary distances and hence of the fundamental measure of dis-

tance within the solar system, the *astronomical unit* *(AU)*, which is the average distance from the Earth to the Sun. But it was not until the past two decades that the length of the astronomical unit, and thus of planetary distances generally, could be measured with extremely high precision.

The key to our modern determination of solar system dimensions is *radar* (Figure 16.1). It is not possible to use radar to measure the distance to the Sun directly. The Sun does not reflect radar very efficiently, and the Sun itself is a source of radio waves that are much stronger than any radar signal we could send. Radar does tell us precisely the size of the orbits of the planets around the Sun, and from that information it is possible to calculate the distance from the Earth to the Sun.

In 1961, radar signals were bounced off Venus for the first time, providing a direct measurement of the distance from Earth to Venus in terms of light seconds (the round-trip travel time of the radar signal). Subsequently, radar was also used to determine the distances to Mercury, Mars, and the satellites of Jupiter. Years of painstaking analyses of such measurements have now led to a determination of the length of the astronomical unit to a precision of about one part in a billion.

The length of 1 AU can be expressed in light travel time as 499.004854 light seconds, or about 8.3 light minutes. If we use the definition of the meter given previously, this is equivalent to 1 AU = 149,597,892,000 m.

These distances are, of course, given here to a much higher level of precision than is normally needed. In this text, we are usually content to express numbers to a couple of significant places and leave it at that. For our following discussions it will be sufficient to round off these numbers:

speed of light: $c = 3.00 \times 10^8$ m/s
$= 3.00 \times 10^5$ km/s

length of light second: LS $= 3.00 \times 10^8$ m
$= 3.00 \times 10^5$ km

astronomical unit: AU $= 1.50 \times 10^{11}$ m
$= 1.50 \times 10^8$ km $= 500$ LS

But it is important to realize that we really do know the absolute distance scale within our own solar system with fantastic accuracy. This is the first step in the cosmic distance ladder, and we shall see that all of the subsequent measures of distances to stars and galaxies depend on our correct understanding of the length of the astronomical unit.

16.2 Surveying the Heavens

It is a huge step from the planets to the stars. The nearest star is tens of thousands of AU from the Earth. Yet in principle, we survey distances to the stars by the same technique that the civil engineer employs on Earth to survey the distance to an inaccessible mountain or tree. The method is that of *triangulation*.

Triangulation

Triangulation works as follows. Two observing stations are set up some distance apart. That distance (*AB* in Figure 16.2) is called the baseline. Now the direction to the remote object (*C* in the figure) in relation to the baseline is observed from each station. Note that *C* appears in different directions from the two stations. This apparent change in direction of the remote object due to a change in vantage point of the observer is called **parallax**. The parallax is also the angle at *C* between *A* and *B*—that is, the angle subtended by the baseline. A knowledge of the angles at

A and *B* and the length of the baseline, *AB*, allows the triangle *ABC* to be solved for any of its dimensions, say, the distance *AC* or *BC*. The solution could be accomplished by constructing a scale drawing or by numerical calculation with the technique of trigonometry.

Depth perception is an example of the same principle. Our eyes are separated by a baseline of a few inches, so our two eyes see an object in front of us from slightly different directions. The brain, like an electronic computer, solves the triangle and gives us an impression of the distance of the object. The greater the parallax, the nearer the object. Hold a pencil a few inches in front of your face and look at it first with one eye and then with the other. Note the large shift in direction against the more distant wall across the room. Now hold the pencil at arm's length and note how the parallax is less. If an object is fairly distant, the shift—the parallax—is too small to notice with the eyes. Thus depth perception fails for objects more than a few tens of meters away. It would take a larger baseline than the distance between the eyes to see the parallax of an object, say, 500 m distant.

Nearly all astronomical objects are very far away. To measure their distances requires either a very large baseline or highly precise angular measurements, or both. The Moon is the only object near enough that its distance can be found fairly accurately with measurements made without a telescope. Ptolemy determined the distance to the Moon correctly to within a few percent. He used the Earth itself as a baseline. Its rotation carries the observer from an observing station on one side of the Earth to one thousands of kilometers away in a few hours.

With the aid of telescopes, later astronomers were able to measure the distances to the nearer planets by using the Earth's diameter as a baseline. To reach for the stars, however, requires a much longer baseline for triangulation.

FIGURE 16.2 Triangulation of an inaccessible object.

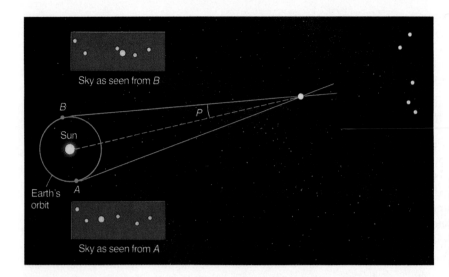

FIGURE 16.3 As the Earth revolves around the Sun, the direction in which we see a nearby star varies with respect to distant stars. The parallax of the nearby star is defined to be one-half of the total change in direction and is usually measured in arcseconds.

Surveying Distances to Stars

Aristotle argued that the Earth could not revolve about the Sun, or we would observe parallax of the nearer stars against the background of more distant objects as we view the sky from different parts of the Earth's orbit (Figure 16.3). Tycho Brahe advanced the same argument nearly two millennia later, when his careful measurements of stellar positions revealed no such shift. By the 18th century, when there was no longer serious doubt of the Earth's revolution, it was realized that the stars must be extremely distant and their parallaxes extremely tiny. Astronomers continued their efforts to measure stellar parallax, however, not as proof of the Earth's motion but as a means of determining the distances of the stars.

Before the first stellar parallax was actually determined, this search yielded another important discovery about the stars. Late in the 18th century, William Herschel carried out systematic telescopic surveys of the sky. He noted many examples of pairs of stars very close together, usually one brighter than the other. He thought the fainter star in each pair was more distant. If this were true, then the nearer, brighter star should appear to shift back and forth with respect to the fainter one as the Earth revolves about the Sun. Herschel catalogued many such pairs, thinking they would be good candidates for detecting stellar parallax.

Subsequently it was found that these close pairs of stars did, indeed, move with respect to each other, but not with a yearly period. Rather, each member of the pair was seen to be revolving about the other in an elliptical orbit. Herschel had discovered that double stars were common and that the stars in **binary-star** systems revolve about each other in accord with

Newton's and Kepler's laws. The discovery that these laws apply to stars far beyond the solar system was of immense importance. It offered evidence that the whole universe is governed by the same natural laws.

Eventually, the true parallaxes of some of the nearer stars were observed. The first successful detections were in the year 1838, when Friedrich Bessel (Germany), Thomas Henderson (Cape of Good Hope), and Friedrich Struve (Russia) measured the parallaxes of the stars 61 Cygni, Alpha Centauri, and Vega, respectively. However, even the nearest star, Alpha Centauri, showed a total displacement of only about 1.5 arcsec during the year. Small wonder that Tycho Brahe was unable to observe the stellar parallaxes and concluded that the Earth was stationary!

Even for the nearest stars, measured parallaxes are usually only a small fraction of an arcsecond. One minute of arc, written as 1 arcmin, is equal to 60 arcsec, and there are 60 arcmin in 1°. A coin the size of a quarter would appear to have a diameter of 1 arcsec if it were at a distance of about 3 miles, or 5 km.

Units of Stellar Distance

The baseline for the measurement of stellar distance is the diameter of the Earth's orbit, which equals 2 AU (or 1000 LS or 300 million km). Suppose that the total displacement in the position of a star, as seen from opposite sides of the Earth's orbit, is 1 arcsec (Figure 16.4). Imagine a circle, centered on the star, that passes through points A and B on opposite sides of the Earth's orbit. The diameter of the Earth's orbit must be in the same proportion to the circumference

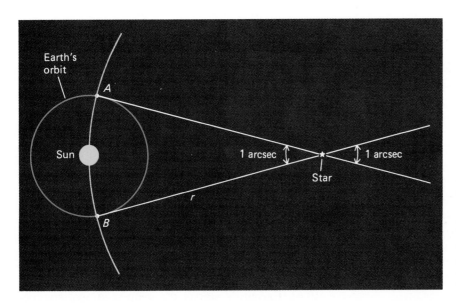

FIGURE 16.4 The apparent shift of a star as seen from opposite sides of the Earth's orbit is equal to the angular size of the Earth's orbit as seen from the star. Measurement of the size of this shift combined with the known diameter of the Earth's orbit (*AB*) is enough information to calculate the distance to the star.

of this circle as 1 arcsec is to 360°. Since the circumference of a circle of radius r is $2\pi r$, we have

$$\frac{AB}{2\pi r} = \frac{1 \text{ arcsec}}{360°}$$

Since there are 60 arcmin in 1° and 60 arcsec in one arcmin, inserting numbers into this equation gives the result that the distance to the star r is given by the relation

$$r = (206265)(2 \text{ AU})$$

or about 2×10^8 LS, or equivalently about 62×10^{12} km—62 million million km! As it turns out, only the nearest stars are even this close. You can see where the common phrase "astronomical quantities," meaning very large, comes from.

Clearly we need a more convenient unit than the kilometer, or even the light second, to express such distances. One commonly used unit is the **light year (LY)**, defined as the distance that light travels in one year in a vacuum. This unit has the advantage that it is related directly to both the definition of the meter, which is expressed in terms of the speed of light, and the radar measurements that determine the length of the astronomical unit. Another advantage of the light year as a unit is that it emphasizes the fact that as we look out into space, we are also looking back into time. The light that we see from a star 100 LY away left that star 100 years ago. What we study is not the star as it is now but as it was in the past. The light from a distant quasar that reaches our telescopes today left its source before the Earth even existed.

We can calculate the length of the light year very easily by noting that the number of seconds in one year is about 3.15×10^7; thus 1 LY = 3.15×10^7 LS

= 1.0×10^{13} km (approximately). In this text, most distances to stars and even to galaxies and quasars will be given in light years. (For more exact expressions of the light year and other units, see the listing in Appendix 5.)

Astronomers sometimes use an alternative unit for stellar distances called the **parsec (pc)**. One parsec is the distance of a hypothetical star (none exists) with a stellar parallax (half the total angular displacement) of 1 arcsec. The greater the distance of a star, the smaller its parallax. The distance of a star in parsecs (r) is just the reciprocal of its parallax (p) in arcseconds. That is,

$$r = \frac{1}{p}$$

Thus a star with a parallax of 0.1 arcsec would be at a distance of 10 pc, and one with a parallax of 0.05 arcsec would be 20 pc away. Conversion between the two distance units is simple: 1 pc = 3.26 LY, and 1 LY = 0.31 pc.

The Nearest Stars

No known star (other than the Sun) is within 1 LY of the Earth. The stellar neighbors nearest to the Sun are three stars that make up a multiple system. To the naked eye, the system appears as a single bright star, Alpha Centauri, which is only 30° from the south celestial pole and hence not visible from the mainland United States. Alpha Centauri itself is a double star— two stars in mutual revolution, too close together to be separated by the naked eye. The two stars that make up Alpha Centauri have a distance of 4.4 LY. Nearby is the third member of the system, a faint star known as Proxima Centauri. Proxima is slightly closer to us than the other two stars in the system; it has a distance of 4.3 LY.

The nearest star visible to the naked eye from most parts of the United States is the brightest-appearing of all the stars, Sirius. Sirius has a distance of 8 LY. It is interesting to note that light reaches us from the Sun in 8 min and from Sirius in 8 years.

There are more than a thousand stars within a distance of 50 LY, but most are invisible to the unaided eye and actually are intrinsically less luminous than the Sun. Most of the stars visible to the unaided eye, on the other hand, have distances of hundreds or even thousands of LY and are visible not because they are relatively close, but because they are intrinsically very luminous. The nearer stars are described more fully in Chapter 18.

16.3 Motions of Stars

The ancients distinguished between the "wandering stars" (planets) and the "fixed stars," which seemed to maintain permanent patterns in the sky. The stars are, indeed, so nearly fixed on the celestial sphere that the patterns they form—the constellations—look today much as they did when they were first named, more than 2000 years ago. Yet the stars are moving with respect to the Sun, most of them with speeds of many kilometers per second. Their motions are not apparent to the unaided eye in the course of a single human lifetime, but if an ancient observer who knew the sky well—Hipparchus, for example—could return to life today, he would find that several of the stars had noticeably changed their positions relative to the others. After some 50,000 years or so, terrestrial observers will find the handle of the Big Dipper unmistakably bent more than it is now (Figure 16.5). Changes in the positions of the nearer stars can be measured with telescopes after an interval of only a few years.

Proper Motion

The **proper motion** of a star is the rate at which its apparent position changes with respect to other, more distant stars. Proper motion is usually expressed in arcseconds per year. It is a consequence of the intrinsic motion of the star relative to the Sun. Parallax, in contrast, is a measure of the change in apparent direction to a star caused by the orbital motion of the Earth.

The proper motion of a star is almost always an angle that is too small to measure with much precision

in a single year. In an interval of several decades, on the other hand, many stars change their directions by easily detectable amounts. The modern procedure for determining proper motions is to compare the positions of the stars on two different images of the same region of the sky taken a few years apart. Most of the star images on such pictures do not appear to have changed their positions measurably. These are, statistically, the more distant stars that are relatively fixed, even over the time interval spanned by the observations. With respect to these "background" stars, the motions of a few comparatively nearby stars can be observed.

The star with the largest proper motion is Barnard's star, whose position on the celestial sphere changes by 10.34 arcsec each year, enough to move the width of the Moon in just two centuries (Figure 16.6). Barnard's star is the nearest known star beyond the triple system containing Alpha Centauri. Its distance is only 5.9 LY, but since it emits only about 1/2000 as much light as the Sun, it is 25 times too faint to see with the unaided eye. Several hundred stars have proper motions greater than 1 arcsec per year.

Radial Velocity

The *radial velocity* (or line-of-sight velocity), usually measured in kilometers per second, of a star is the speed with which it approaches or recedes from the Sun. Radial velocity can be determined from the Dop-

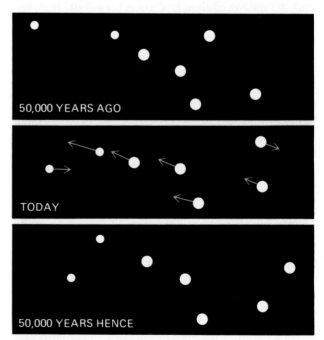

FIGURE 16.5   Appearance of the Big Dipper over 100,000 years.

FIGURE 16.6 Two photographs of Barnard's star, showing its motion over a period of 22 years. *(Yerkes Observatory)*

pler shift of the lines in the spectrum of a star (Section 5.6). W. Huggins made the first radial-velocity determination of a star in 1868. He observed the Doppler shift in one of the hydrogen lines in the spectrum of Sirius and found that this star is approaching the solar system.

Unlike proper motion, which is observable only for the comparatively nearby stars, the radial velocity can be measured for any star that is bright enough for its spectrum to be observed. The radial velocity of a star is only that component of its actual velocity that is projected along the line of sight—that is, that carries the star toward or away from the Sun. Radial velocity is counted as positive if the star is moving away from the Sun and negative if the star is moving toward the Sun.

Since motion of either the star or the observer (or both) produces a Doppler shift in the spectral lines, a knowledge of the radial velocity alone does not enable us to decide whether it is the star or the Sun that is "doing the moving." (Indeed, as we have seen in Chapter 14, it does not even make sense to ask which is moving.) What we really measure, therefore, is the speed with which the distance between the star and Sun is increasing or decreasing—that is, the star's radial velocity with respect to the Sun.

Space Velocity

Radial velocity is a motion of a star along the line of sight. Proper motion is the angular rate of motion produced by the star's motion across, or at right angles to, the line of sight. The radial velocity is expressed in kilometers per second and is independent of distance. The proper motion of a star does not, however, give the star's actual speed in kilometers per second at right angles to the line of sight. The latter is called the **transverse velocity.** To find the transverse velocity of a star, we must know both its proper motion and its distance. A star with a proper motion of 1 arcsec per year, for example, might have

a relatively low transverse velocity and be nearby or a high transverse velocity and be far away.

If we know the distance of a star of a particular proper motion, we can calculate what its transverse velocity must be to produce the observed proper motion. If, in addition, we know its radial velocity—how fast it is moving in the line of sight—we can calculate how fast and in what direction it is moving in space with respect to the Sun. This is called its **space velocity.** By mapping out both the distances and the space velocities of stars, we can determine where the stars are coming from and where they are going, and thus begin to understand the motions of the objects in the Sun's neighborhood.

The Solar Motion

The Sun, a typical star, is in motion, just as the other stars are. In fact, it is a member of our Milky Way Galaxy, a system of a few hundred billion stars. The most luminous stars in the Galaxy lie in a disk, which is flat like a pancake and is rotating. The Sun, partaking of this general rotation of the Galaxy, moves with a speed of about 220 km/s to complete its orbit about the galactic center in a period of about 200 million years.

At first thought, it might seem that the galactic center is the natural reference point for stellar motions. However, our observations of the proper motions and radial velocities of the stars that surround the Sun in space, all in our local neighborhood of the Galaxy, do not give us directly the motions of these stars about the galactic center. The reason is that the stars' orbits around the galactic center and their orbital velocities are both nearly the same as those of the Sun. The motions we observe are merely small differences between the orbital motions of these stars in the Galaxy and that of the Sun.

These small residual motions arise because our neighboring stars' orbits about the galactic center are not absolutely identical to our own. We are overtaking and passing some stars while others are passing us. The slightly different eccentricities and inclinations of our respective orbits bring us closer to some stars and carry us farther from others. We can study these residual motions without knowing anything about the actual motions of stars around the center of the Galaxy. Our situation is analogous to that of a person driving an automobile on a busy highway. All the surrounding cars are going in the same direction and at roughly the same speed, but some are changing lanes and others are passing one another. More or less like the highway traffic, the residual motions of the stars around us seem to be helter-skelter.

We deduce the motion of the Sun with respect to our neighboring stars by analyzing the proper motions and radial velocities of the stars around us. The easiest way to understand how the Sun's motion is measured is to consider the effect it has on the apparent motions of the other stars.

First, consider the radial velocity of stars with respect to the Sun. If we look in the direction toward which the Sun is moving, we find that most of the stars are approaching us because we are moving forward to meet them. The only stars in that direction that are receding from us are those that are moving in the same direction we are going, but at a faster rate, so that they are pulling away from us. The observed radial velocities of all the stars in the direction toward which the Sun is moving do not average to zero, but to -20 km/s, showing that we are moving toward them at about 20 km/s. Similarly, stars in the opposite direction have an average radial velocity of about $+20$ km/s, because we are pulling away from them at that speed.

The Sun's motion also affects the observed proper motions of stars. If the stars were at rest, they would all show a backward drift because of our forward motion, like the telephone posts you pass while driving. As it is, the stars have motions of their own, but only those moving in the same direction we are and at a faster rate appear to have "forward" proper motions—like a few speeding cars passing you on the highway. The rest, by far the majority, appear to drift backward.

William Herschel was the first to attempt to detect the direction of the solar motion from the proper motions of stars. In 1783, he analyzed the proper motions of about a dozen stars and deduced that the Sun was moving in a direction toward the constellation Hercules—a nearly correct result.

Modern analysis of the proper motions and radial velocities of the stars around the Sun has shown that the Sun is moving approximately toward the direction now occupied by the bright star Vega in the constellation of Lyra at a speed of 19.5 km/s (4.14 AU per year). The direction in the sky toward which the Sun is moving is called the *apex* of solar motion.

16.4 Distances to the Stars

Parallax Measurements

Even though the parallaxes of stars are very small indeed, we can measure parallaxes to within a few thousandths of an arcsecond with ground-based telescopes. So far, parallaxes have been measured for nearly 10,000 stars. Only for a fraction of them, however, are the parallaxes large enough (about 0.05 arcsec or more) to be measured with a precision of 10 percent or better. Current efforts that are substituting electronic detectors for photographic plates and computer analysis for hand measurement are expected to permit distances to be determined to 10 percent accuracy for stars with parallaxes as small as 0.02 arcsec—an expansion of this technique from about a distance of 60 LY to one of 160 LY. The next step probably requires going into space. The Hubble Space Telescope, even though its images are imperfect, should be able to measure parallaxes with an accuracy of one-thousandth of an arcsecond or better.

Distances from Stellar Motions

The distances of stars can be estimated from their proper motions as well as from the direct measurement of parallax. The proper motions of stars can be expected to be largest, statistically, for the nearest stars. If, for example, a star is only a few light years away, its proper motion will almost certainly be observable after a few years. The proper motion of a very distant star, on the other hand, may be detectable only after a long time and then only if the star has a very great space velocity. Searches for nearby stars, therefore, are usually conducted by searching for stars of large proper motion. Conversely, remote stars can be identified by their lack of observable proper motions; they serve as standards against which we can measure the parallaxes of the nearby stars to determine their distances.

The proper motion of an individual star does not in itself indicate its distance uniquely. However, *on the average*, stars that are closer to the Sun have larger proper motions, and investigations of such motions do give statistical information about stellar distances. For example, we might identify a large group of intrinsically similar stars from their spectra. If we also wish to know the intrinsic brightness (luminosity) of this class of stars, we might be able to determine this even if none were close enough to determine a reliable parallax. We would simply measure the proper motions of these stars. After enough years had passed, we could use the proper motions to estimate the average distance to the group. While such an approach lacks the rigor of a direct measurement of parallax, it can be extended to much greater distances.

Distances to Clusters of Stars

A special technique has been developed to use the measured space velocities of some stars to determine their distance with a fair degree of accuracy. Many stars are found in loose clusters, bound together gravitationally. One of the nearest such star clusters is called the Hyades, at a distance of about 150 LY, just beyond the reach of parallax measurements. The brightest stars in the Hyades can be seen with the unaided eye, and several dozen members are easily identified with binoculars (Figure 16.7).

When the radial velocities and proper motions of the stars in the Hyades are measured, it is found that they are moving together through space, but that the motion of each star can be traced back to the same point in the sky. It is as if the members of the cluster were diverging from a common point. This is simply the effect of perspective, just as a flight of birds seems to spread out and diverge when it heads toward you. It turns out that the rate at which the Hyades stars diverge from a common point provides a direct measure of the distance to the cluster.

The distance of the Hyades found by this "moving cluster" method is 150 LY. Thus each individual star of the more than 100 in the cluster also is 150 LY away, give or take a few light years owing to the scatter of stars in the cluster. This determination of distance is one of the important rungs in the cosmic distance ladder. It turns out that many of the indirect means of estimating the distance to still more distant

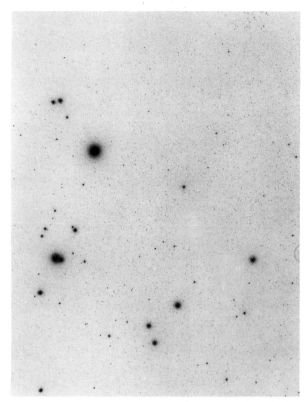

FIGURE 16.7   The Hyades, one of the nearest star clusters (at 150 LY), serves as a benchmark for defining the stellar distance scale. As are many astronomical photographs, this image is a negative, so that the bright stars appear black. *(National Geographic Society–Palomar Observatory Sky Survey)*

groups of stars depend critically on the distance to the Hyades.

Taken together, these measurements of parallaxes and stellar motions yield a picture of the stars within a few hundred light years of the Sun. Beyond this solar neighborhood, only indirect methods can be used to estimate distance, as we shall see in subsequent chapters of this book. But it is well to remember that these indirect techniques are ultimately tied to distances of the stars in the solar neighborhood and to the surveying techniques described in this chapter.

SUMMARY

16.1   Early measurements of length were based on human dimensions, but we now use worldwide standards, such as the metric system, which specify lengths in standard units such as the meter. Distances within the solar system are now determined by timing how long it takes radar signals to travel from the Earth to the surface of a planet and return.

16.2   The shift in position of a nearby star relative to very distant background stars, as viewed from opposite sides of

the Earth's orbit, is called the **parallax** of that star and is a measure of its distance. The units used to measure stellar distance are the **light year (LY)**, which is the distance light travels in one year, and the **parsec (pc)**, which is the distance of a star that has a parallax of 1 arcsec. One parsec = 3.086×10^{13} km = 3.26 LY. The first successful measurements of stellar parallaxes were reported in 1838. The star nearest to the Sun is Alpha Centauri, which is actually a system of three stars at a distance of about 4.4 LY.

ESSAY The Scale of Stellar Distances

In Chapter 2 we introduced a scale model for the solar system in which all dimensions were reduced by a factor of 1 billion (10^9). In this model the Earth was the size of a grape and the Sun was 1.5 m in diameter, with the distance from the Sun to the most distant planets equal to about 5 km. On this same scale the nearest stars are tens of thousands of kilometers away, making our model rather too large to visualize.

Let us therefore shrink our model by another factor of a thousand, so that all dimensions are reduced by 1 trillion (10^{12}). The Sun is now 1.5 mm in diameter, about the size of a mustard seed, and the entire solar system (except the Oort comet cloud) will fit inside a closet. But the nearest star is still 10 km away, on the other side of town.

The space between the stars is extraordinarily empty. Separations between individual stars amount to a million times their diameters. The stars in our model are like mustard seeds scattered many miles apart. Even in star clusters where the stars are much more tightly packed, these seeds are still separated by tens or hundreds of meters. The stellar universe is a big place, and it is a wonderful accomplishment, when you think about it, that we are able to learn so much about the stars by the analyses of the feeble light we collect with our astronomical telescopes.

16.3 Stars move through space. **Proper motion** is the rate at which the apparent position of a nearby star changes with respect to distant stars or galaxies and is measured in arcseconds per year. **Radial velocity** is the speed in kilometers per second with which a star approaches or recedes from the Sun. **Transverse velocity** is the speed of a star at right angles to the line of sight. **Space velocity** is the total velocity of a star in kilometers per second with respect to the Sun. The Sun and most of the nearby stars follow similar orbits around the center of the Galaxy.

16.4 The motions of stars provide clues to their distances. For example, stars with large proper motions are relatively nearby. The distances to nearby clusters can be determined by measuring the radial velocities of member stars and finding the point on the sky toward or from which the stars appear to be moving.

REVIEW QUESTIONS

1. Explain how parallax measurements can be used to determine distances to stars. What would be the advantage of making parallax measurements from Pluto rather than Earth? Is there a disadvantage?

2. Make up a table relating the following units of astronomical distance: kilometer, Earth radius, solar radius, astronomical unit, light year.

3. Define clearly what is meant by radial velocity and proper motion. What is the difference between the transverse velocity of a star and its proper motion? Which of these three quantities—radial velocity, transverse velocity, proper motion—can be determined without knowing the distance to a star? Explain your answer.

4. If a star has a large proper motion, it is likely to be relatively nearby. Explain why.

THOUGHT QUESTIONS

5. Parallaxes are measured in fractions of an arcsecond. One arcsecond equals 1/60 arcmin, which is in turn 1/60°. To get some idea of how big 1° is, go outside at night and find the Big Dipper. The two pointer stars at the end of the bowl of the Dipper are 5.5° apart. The two stars across the top of the bowl are 10° apart. (Ten degrees is also about the width of your fist if you hold it at arm's length and look at it projected against the sky.) Mizar, which is the second star from the end of the handle of the Big Dipper, appears double. The fainter star, Alcor, is about 12 arcmin from Mizar. For comparison, the diameter of the full moon is about 30 arcmin. The belt of Orion is about 3° long.

6. For centuries, astronomers wondered whether comets were true celestial objects, like the planets and stars, or were a phenomenon that occurred in the atmosphere of the Earth. Describe an experiment that would allow you to determine which of these two possibilities is correct.

7. The Sun is much closer to the Earth than are the nearest stars, yet it is not possible to measure the parallax of the Sun accurately by measuring its position directly. Explain why.

8. Show by a diagram how two stars can have the same radial velocity and proper motion but different space motions.

9. Since we observe stars from the Earth, does the Doppler shift we measure in the spectrum of a star indicate directly the radial velocity of the star with respect to the Sun? If not, what kind of correction must be applied?

10. Galaxies and quasars are the most distant objects in the universe. Would you expect to be able to measure their proper motions? Their radial velocities?

11. Parallaxes of stars are sometimes measured relative to the positions of galaxies or quasars. Why is this a good technique?

PROBLEMS

12. A radar astronomer claims that she beamed radio waves to Jupiter and received an echo exactly 48 min later. Do you believe her? Why?

13. Show that a light year contains 9.46 x 10^{12} km.

14. Show that 1 pc equals 3.086×10^{13} km and that it also equals 3.26 LY. Show your calculations.

15. Give the distances to stars having the following parallaxes: (**a**) 0.1 arcsec; (**b**) 0.5 arcsec; (**c**) 0.005 arcsec; (**d**) 0.001 arcsec.

16. Give parallaxes of stars having the following distances: (**a**) 10 pc; (**b**) 3.26 LY; (**c**) 326 LY; (**d**) 10,000 pc.

17. In 50 years a star is seen to change its direction by 100 arcsec. What is its proper motion?

18. Suppose a star at a distance of 33 LY has a radial velocity of 150 km/s. By what percentage does its distance change in 100 years?

Stars differ from one another in their luminosities and masses. This time exposure shows stars near the north celestial pole as they circle the McMath Telescope on Kitt Peak. *(William Livingston/National Optical Astronomy Observatories)*

Annie Jump Cannon (1863–1941), who, during her long career at Harvard College Observatory, personally classified more than 500,000 stellar spectra and arranged them in the following spectral sequence: O, B, A, F, G, K, and M. She received many honors for her work, which laid the groundwork for modern stellar spectroscopy. *(Harvard College Observatory)*

17

ANALYZING STARLIGHT

The stars are suns.

To the astronomer this statement means that stars generate the energy that they emit. They do not shine because they reflect the light from some external source. Most stars produce energy in the same way as the Sun does. Deep in stellar interiors, hydrogen nuclei are fused to form helium, with the small amount of mass lost in the process being converted to energy. Late in their lives, when hydrogen has been depleted, stars are forced to turn to new energy sources. The story of how stars evolve and eventually die as they exhaust the sources of energy available to them will be told in later chapters in this book.

The story of stellar evolution can make sense, however, only after we know what stars are like. Very simple observations are enough to show that not all stars are alike. For example, if you look at the sky you will see that stars appear to have very different brightnesses. Is this property intrinsic to the stars themselves? Do some stars emit much more energy than others? Or do stars merely *appear* to have different brightnesses because some are close by and others are very far away? Some stars appear to be blue-white; others are obviously red. In Chapter 5 we showed that color is a good indication of temperature, so stars can be both hotter and cooler than the Sun. The Sun, of course, is brightest at the wavelengths that the eye sees as yellow.

Other questions about stars can be answered only by means of careful observations with large telescopes. Are stars much larger or smaller than the Sun? Are they made of the same material as the Sun? How massive are they? How do the ages of stars compare with the age of the Sun? Since the stars are suns, it should not surprise you that the same techniques, including spectroscopy, that are used to answer these questions about the Sun can be used to study stars as well.

17.1 The Brightness of Stars

One of the most important characteristics of a star is the total amount of energy that it emits. The total energy emitted per second by a star is its **luminosity.** If we can measure a star's luminosity, we can then determine such other important information as how long that star can continue to shine.

Unfortunately, we cannot determine the luminosity of a star merely by measuring its apparent brightness. As we have seen from the previous chapter, stars are at different distances from us. A star that appears to be bright when we see it in the sky at night could be truly luminous and far away, or it could be intrinsically rather faint but relatively close by. Remember that the light energy that we receive from a star (that is, its apparent brightness) is inversely proportional to the square of its distance (Section 5.1). The Sun, for example, gives us billions of times as much light as any of the other stars, but on the other hand, it is hundreds of thousands of times closer to us than any other star.

To compare the intrinsic energy outputs of other stars, we need to determine two quantities: their apparent brightness—that is, how much of their energy we detect here on Earth—and their distance from us. In Chapter 16 we learned how to measure distances. In this chapter, we now turn to the problem of measuring apparent brightness.

The Magnitude Scale

The branch of observational astronomy that deals with the measurement of stellar brightness and luminosity is called **photometry** (from the Greek *photo*, light, and *-metry*, to measure).

More than 2000 years ago, Hipparchus compiled a catalogue of about a thousand stars. He classified these stars into six categories of brightness, which he called *magnitudes*. The brightest-appearing stars in his catalogue are of the first magnitude; the faintest naked-eye stars are of the sixth magnitude. The other stars are assigned to classes of intermediate-magnitude. This system of stellar magnitudes, which began in ancient Greece, has survived to the present time. Today, however, magnitudes are based on precise measurements of stellar brightness rather than on arbitrary and uncertain eye estimates.

In 1856, after early methods of making accurate measurements of stellar brightness had been developed, Norman R. Pogson proposed the quantitative scale of stellar magnitudes that is now generally adopted. He noted, as did William Herschel before him, that we receive about 100 times as much light from a star of the first magnitude as from one of the sixth. Therefore, a difference of five magnitudes corresponds to a ratio in brightness of 100:1. The physiology of sense perception is such that equal intervals of brightness as perceived by our eyes are in reality equal *ratios* of luminous energy. Pogson proposed, therefore, that the ratio of light flux corresponding to a step of one magnitude be the fifth root of 100, which is about 2.512. Thus, a fifth-magnitude star gives us 2.512 times as much light as one of sixth magnitude, and a fourth-magnitude star, 2.512 times as much light as a fifth, or 2.512 × 2.512 times as much as a sixth-magnitude star. From stars of third, second, and first magnitude, we receive 2.512^3, 2.512^4, and 2.512^5 (= 100) times as much light, respectively, as from a sixth-magnitude star. By assigning a magni-

TABLE 17.1   Magnitude Differences and Light Ratios

Difference in Magnitude	Ratio of Brightness
0.0	1:1
0.5	1.6:1
0.75	2:1
1.0	2.5:1
1.5	4:1
2.0	6.3:1
2.5	10:1
3.0	16:1
4.0	40:1
5.0	100:1
6.0	251:1
10.0	10,000:1
15.0	1,000,000:1
20.0	100,000,000:1
25.0	10,000,000,000:1

tude of 1.0 to the bright stars Aldebaran and Altair, Pogson created a new scale of **apparent magnitudes** that agreed roughly with those in current use at the time.

Table 17.1 gives the approximate ratios of brightness corresponding to several selected magnitude differences. Note that a given ratio of brightness, whether of two bright or of two faint stars, always corresponds to the same magnitude interval. Table 17.2 gives the apparent magnitudes of several astronomical objects. Note that the numerically *smaller*

TABLE 17.2   Some Visual Magnitude Data

Object	Magnitude
Sun	− 26.5
Full Moon	− 12.5
Venus (at brightest)	− 4
Jupiter, Mars (at brightest)	− 2
Sirius	− 1.5
Aldebaran, Altair	1.0
Unaided-eye limit	6.5
Binocular limit	10
15-cm telescope limit	13
4-m CCD limit	26
Hubble Space Telescope	28–29

ESSAY The Stars

One of the most beautiful and awe-inspiring sights in all of nature is the dark night sky filled with stars. The stars that are visible above the horizon change with the seasons and with the time of night. The patterns of the stars, however, are fixed, at least over the time scale of a human life. With star maps it is easy to learn your way around the sky.

All stars are assigned to 1 of 88 different groupings, or *constellations*, of stars. Most of these constellations date back to ancient times, but a few—especially in the Southern Hemisphere—were defined during the past few hundred years. Constellations usually have a story associated with them. There is Orion, the Hunter, permanently separated in the sky from the Scorpion, who caused his death. Andromeda is chained to a rock, offered in sacrifice to atone for the boasts of her mother Cassiopeia. There are dogs, a dragon, a ram, a dolphin, scales, crowns, and crosses. To learn to identify a constellation and to know its story is to make a friend for life.

As you learn the constellations, note that the very brightest stars are not all the same color. An especially good example is found in the winter constellation Orion. In one knee of Orion, we find the bright blue star Rigel. The eastern shoulder, diagonally across from Rigel, is marked by Betelgeuse, the brightest red star in the sky. Binoculars will permit you to see the colors of many fainter stars. Color is an indicator of the temperatures of stars, with red stars having the lowest temperatures.

Stars rise about 4 min earlier every night. Look toward the eastern horizon and note where stars are located at a specific time of night. Wait a week or two to look again, and note the difference in altitude above the horizon. Ancient peoples used the times of rising of specific stars to mark the seasons and the start of a new year.

If you look at very bright stars when they are close to the horizon, you may see them appear to change in brightness and possibly in color as well. This phenomenon is called *twinkling* and is caused by turbulence in the Earth's atmosphere. As small regions in the atmosphere move about, they bend the light from a star, just as a prism does, and cause it to travel in ever-changing directions. The brightness of the star will vary, depending on what fraction of its light is directed precisely toward your eye. The color of the star may also appear to change rapidly if, for example, red light is directed toward your eye and blue light is not, and then, a fraction of a second later, the reverse situation happens.

Bright city lights mask all but the most brilliant stars. The next time you are out in the countryside at night, take time to look at the sky. If it is summertime, you will very likely see the Milky Way arching across the sky, displaying its greatest brilliance toward the south in the direction to the center of the Galaxy. Use binoculars to see dozens of stars in clusters like the Pleiades and the Hyades, both of which are visible in fall. In winter, use binoculars to look at the nebulosity in the sword of Orion, where star formation is vigorously in progress. The Big Dipper rises highest in the northern sky in spring; note that the star in the bend of its handle is a double star.

Learn to know the sky in all seasons of the year—its beauty can bring endless hours of pleasure.

magnitudes are associated with the *brighter* objects. A numerically large magnitude, therefore, refers to a faint object.

A simple equation describes the quantitative relationship between the magnitudes and light flux received from two stars. If m_1 and m_2 are the magnitudes corresponding to stars from which we receive light flux in the amounts l_1 and l_2, the difference between m_1 and m_2 is defined by

$$m_1 - m_2 = 2.5 \log\left(\frac{l_2}{l_1}\right)$$

The so-called first-magnitude stars are not all of the same apparent brightness, nor is a magnitude equal to

1.0 the brightest possible magnitude. As we noted above, magnitude 1 was defined in terms of several of the stars that had traditionally been called first-magnitude stars. The brightest-appearing star, Sirius, sends us about 10 times as much light as the average star of first magnitude and so has a magnitude of -1.5. Several of the planets appear even brighter. Venus, at its brightest, is of magnitude -4.4. The Sun has a magnitude of -26.2. Magnitudes with negative values are, of course, brighter than magnitudes with positive values.

Other Units of Brightness

Although the tradition of expressing the visual brightness of stars in magnitudes goes all the way back to the ancient Greeks, this system has not carried over into newer branches of astronomy. In radio astronomy, for example, no equivalent of the magnitude system has been defined. Rather, the radio astronomer measures directly the amount of energy being collected by the radio telescope and expresses the brightness of each source in terms of the energy reaching the surface of the Earth, for example, in watts per square meter. Similarly, most researchers in the fields of infrared, x-ray, and gamma-ray astronomy use energy units rather than magnitudes to express the results of their measurements.

The Luminosities of Stars

If we can measure the distance to a star, then it is easy for the astronomer to convert from the apparent magnitude, or observed brightness (watts per square meter reaching the Earth), of that star to its true luminosity—the total energy emitted per second at the surface of the star in, say, watts. To do this calculation, it is necessary only to make use of the inverse-square law for light (Section 5.1). Often it is convenient to express the luminosity of other stars in terms of the Sun's luminosity, as when we note that the luminosity of Sirius is 23 times that of the Sun. If we use the symbol L_S to denote the Sun's luminosity, the luminosity of Sirius is thus 23 L_S. We will frequently use this notation in the following chapters of this book.

The luminosity of stars ranges from less than 0.0001 L_S to more than 10^6 L_S. (The luminosity of the Sun is about 4×10^{26} watts.)

Astronomers have, by tradition, used the apparent magnitude of a star at a standard distance as a measure of its intrinsic luminosity. This standard distance is 10 pc (32.6 LY). The idea is to calculate what the apparent magnitude of each star would be if it were viewed from a distance of exactly 10 pc. This calculated magnitude, which depends only on the true luminosity, is called the **absolute magnitude.**

We can use the inverse-square law of light to calculate how luminous all other stars of known distance would appear if they were 10 pc away. Suppose, for example, that a tenth-magnitude star has a distance of 100 pc. If it were only 10 pc away, it would be only one-tenth as far away, and hence 100 times brighter—a difference of five magnitudes. At 10 pc, therefore, it too would appear as a fifth-magnitude star.

The extreme range of absolute magnitudes observed for normal stars is -10 to $+15$, a range of a factor of 10^{10} in luminosity.

Distances of Stars from Their Brightness

We have seen that the luminosity of a star can be calculated if its apparent brightness and its distance are both known. This process can be reversed. Suppose, for example, that the spectrum, color, and other properties of a distant star match those of the Sun exactly. It is then reasonable to conclude that this distant star is likely to have the same luminosity as the Sun, or 1 L_S. A comparison of this known luminosity with the apparent brightness permits the distance to be determined from the inverse-square law.

Most of us have seen examples of this same principle in everyday life. The next time you are driving down the highway at night, notice how the brightness of oncoming headlights increases as a car approaches you. Sometimes at an airport at night you can see the lights of two or three airplanes stacked behind one another in a landing pattern. Again, the increase in brightness of their landing lights with decreasing distance will be very obvious. In other words, if we know the intrinsic brightness of an object, we can use its apparent brightness to estimate a distance. The computation of a star's distance from its apparent brightness and true luminosity is analogous.

17.2 Colors of Stars

Stars have different colors, and all colors do not produce an equal response in the human eye, photographic film, or charge-coupled devices (CCDs). The brightness of a star can depend to some extent, therefore, upon both its color and the detector used to measure it. The human eye, for example, is most sensitive to green and yellow light. It has a lower sensitivity to the shorter wavelengths of blue and violet light and to longer wavelengths of orange and red light. It does not respond at all to ultraviolet or infrared radiation. The eye, in fact, responds roughly to the same kind of light that the Sun emits most intensely. This coincidence is probably not accidental—the eye may have evolved to respond to the kind of light most available on Earth.

Suppose that the total amount of light energy entering a telescope from each of two stars is exactly the same if light of all wavelengths is considered, but that one star emits most of its light in the blue spectral region and the other, in the yellow spectral region. If these stars are observed visually (that is, by looking at them through the telescope), the yellow one will appear brighter because the eye is less sensitive to most of the light emitted by the blue star. If the stars are imaged with an electronic detector that is especially sensitive to blue light, however, the blue star will appear brighter. The difference in brightness of a star as seen with the eye (its visual brightness) and on photographs made with film sensitive to blue light provided astronomers with their first quantitative measure of the colors of stars (Figure 17.1).

Color Indices

One commonly used set of colors for photometry is the U (ultraviolet), B (blue), and V (visual) system. Corresponding wavelengths for these bands are approximately 360 nm, 420 nm, and 540 nm, respectively. The starlight is observed through three filters, each of which transmits only one of these three colors. The results are usually expressed in magnitudes. The difference between any two of these magnitudes—say, between blue and visual magnitudes (B − V)—is called a **color index.** Since the inverse-square law of light applies equally to all wavelengths, the color index of a star would not change if the star's distance were changed.

This idea should make sense to you. After all, you know from everyday experience that the color of an object appears to be the same, apart from effects produced by haze in the Earth's atmosphere, no matter how far away the object is. Color indices of stars, therefore, provide measures of their intrinsic or true colors.

Colors, in turn, indicate the *temperatures* of stars, as we remember from Wien's law (Chapter 5). Hot stars emit more energy in the blue part of the spectrum. Cool stars are brighter at red and infrared wavelengths. By convention, the ultraviolet, blue, and visual magnitudes of the UBV system are adjusted to give a color index of zero to a star with a surface temperature of about 10,000 K. The B − V color indices of stars range from −0.4 for the bluest to more than +2.0 for the reddest. The corresponding range in surface temperature is from about 50,000 K to 2000 K.

17.3 Stars That Vary in Light

Most stars are constant in their luminosity, at least to within a percent or two. Like the Sun, they are in equilibrium, generating a steady flow of energy from their interiors. However, a significant number of stars are seen to vary in brightness.

A graph that shows how the brightness of a variable star changes with time is called a **light curve** of that star (Figure 17.2). The *maximum* is the point of the light curve where the star has its greatest brightness. The *minimum* is the point where it is faintest. If the light variations repeat themselves periodically, the interval between successive maxima is called the *period* of the star. The *amplitude* of the variation is the difference in brightness (usually expressed in magnitudes) between the maximum and minimum. The amplitudes of variable stars range from less than 0.1 magnitude up to several magnitudes.

Pulsating Variables

There is a special class of variable stars that is particularly useful for calculating distances. These stars are

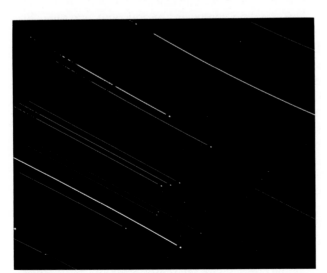

FIGURE 17.1   A time exposure showing the constellation Orion as it rises over Kitt Peak National Observatory. The colors of the various stars are caused by their different temperatures. *(National Optical Astronomy Observatories)*

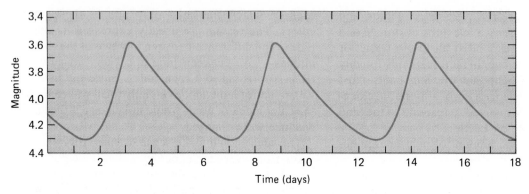

FIGURE 17.2 Light curve (plot of brightness or magnitude as a function of time) of a typical cepheid variable.

the **pulsating variables.** A pulsating variable star actually changes its diameter with time—periodically expanding and contracting, as your chest does when you breathe. The expansion and contraction can be measured by using the Doppler effect to see spectral lines shift toward the blue as the surface of the star moves toward us and then to the red as the surface of the star shrinks back. As it pulsates, the star changes color, indicating that its temperature is also varying. By far the most obvious effect is the changing luminosity of the star, which results in periodic variations in the observed brightness.

The two types of pulsating stars that are most useful for measuring distances are the cepheid variables and the RR Lyrae variables.

Cepheid Variables

Although relatively rare, the cepheid variables are very important in astronomy because, as we shall see, it is possible to determine how far away they are simply by measuring their periods and their apparent brightnesses. **Cepheids** are large yellow pulsating stars named for the prototype and first known star of the group, Delta Cephei. The variability of Delta Cephei was discovered in 1784 by the young English astronomer John Goodricke just two years before his death at the age of 21. The visual magnitude of Delta Cephei varies between 3.6 and 4.3 in a period of 5.4 days. The star rises rather rapidly to maximum light and then falls more slowly to minimum light. The curve in Figure 17.2 represents the light variation of Delta Cephei.

Several hundred cepheid variables are known in our Galaxy. Most cepheids have periods in the range of 3 to 50 days and luminosities about 10,000 times greater than that of the Sun. The amplitudes of cepheids range from 0.1 to 2 magnitudes. Polaris, the North Star, is a small-amplitude cepheid variable that

for a long time varied by one-tenth of a magnitude, or by about 10 percent in visual luminosity, in a period of just under four days. Recent measurements indicate that the amplitude of the pulsation of Polaris is decreasing to zero and that sometime in the future this star will no longer be a pulsating variable. This is just one more piece of evidence that stars really do evolve and change in fundamental ways as they age.

The Period-Luminosity Relation

The importance of cepheid variables lies in the fact that a relation exists between their periods of pulsation (or light variation) and their luminosity. Simply by measuring the period of a cepheid, we can estimate its true luminosity, and since its apparent brightness can be measured, we can calculate how far away it is.

The relation between period and average luminosity was discovered in 1912 by Henrietta Leavitt, an astronomer of the Harvard College Observatory. Some hundreds of cepheid variables had been discovered in the Large and Small **Magellanic Clouds** (Figure 17.3), two great stellar systems that are actually neighboring galaxies (although they were not known to be galaxies in 1912). Leavitt found that the brighter-appearing cepheids always have the longer periods of light variation. The Magellanic Clouds are small galaxies, so it is approximately correct to assume that the stars in each Cloud are all at about the same distance from the Earth. Differences in apparent magnitude therefore correspond directly to differences in luminosity, and the relationship between period and luminosity was readily apparent. In principle, the same conclusion could have been reached by measuring individual parallaxes, and hence luminosities, for cepheids in our own Galaxy. But very few cepheids are close enough for parallaxes to be measured, so for most cepheids in our Galaxy we have no direct way to measure their distances.

FIGURE 17.3   A picture of the Large Magellanic Cloud, which is the galaxy that is nearest to our own Milky Way Galaxy. *(National Optical Astronomy Observatories)*

RR Lyrae Stars

More common than the cepheids, but also less luminous, are the **RR Lyrae variables,** named for RR Lyrae, the best known member of the group. Nearly 5000 of these pulsating variables are known in our Galaxy. Almost all of them are found in the nucleus or the halo of our Galaxy (Chapter 24) or in globular star clusters. The periods of RR Lyrae stars are less than one day; most periods fall in the range from 0.3 to 0.7 day. Their amplitudes never exceed two magnitudes, and most RR Lyrae stars have amplitudes less than one magnitude.

It is observed that the RR Lyrae stars occurring in any particular star cluster all have about the same apparent magnitude. Since they are all at approximately the same distance, it follows that they must also have nearly the same luminosity—about 50 L_S. Figure 17.4 displays the ranges of periods and absolute magnitudes for both the cepheids and the RR Lyrae stars.

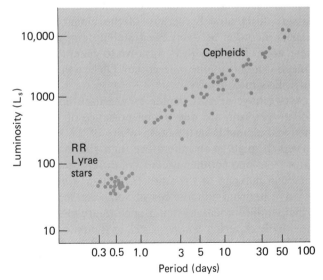

FIGURE 17.4   The period-luminosity relation for cepheid variables. Also shown are the period and luminosity for RR Lyrae stars.

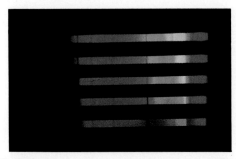

FIGURE 17.5 Several exposures of the spectrum of Vega, which is the fifth brightest star in the sky and has a temperature approximately equal to 9500 K. The strongest lines in the spectrum are produced by the Balmer series of hydrogen. *(National Optical Astronomy Observatories)*

17.4 The Spectra of Stars

As early as 1823, Joseph Fraunhofer observed that stars, like the Sun, have spectra that are characterized by dark lines crossing a continuous band of colors (Figure 17.5). William Huggins, in 1864, first identified some of the lines in stellar spectra with those of known terrestrial elements, showing that the same chemical elements exist in the stars as in the Sun and planets.

Formation of Stellar Spectra

The spectra of different stars differ greatly from one another. In 1863, the Jesuit astronomer Angelo Secchi at the Vatican Observatory classified stars into four groups according to the general arrangement of the dark lines in their spectra. Secchi's scheme subsequently was modified and augmented, until today we recognize seven such principal **spectral classes,** designated by letters of the alphabet. Arranged in order of decreasing stellar temperature, these are O B A F G K M—known to generations of college students by the mnemonic "Oh be a fine (guy) (girl), kiss me."

As we have seen (Chapter 5), the dark lines in a stellar spectrum are due to the presence of various chemical elements in the atmosphere of the star observed. It might seem, therefore, that stellar spectra differ from one another because of differences in the chemical makeup of the stars. Although some stars do have unusual abundances of some elements, the principal differences in stellar spectra are caused by the widely differing *temperatures* in the outer layers of the various stars, not by differences in chemical composition.

Hydrogen, for example, is by far the most abundant element in all stars (except those at advanced stages of evolution—Chapter 21). In the atmospheres of the hottest stars, however, hydrogen atoms are completely ionized. Since the electron and proton are separated, ionized hydrogen can produce no absorption lines. In the atmospheres of the coolest stars, hydrogen is neutral and can produce absorption lines, but in these stars practically all of the hydrogen atoms are in the lowest energy state (unexcited) and can absorb only those photons that can lift an electron from that first energy level to higher ones. The photons so absorbed produce the Lyman series of absorption lines (Chapter 5), which lie in the ultraviolet part of the spectrum and cannot be studied from the ground.

In the visual part of the spectrum, lines of hydrogen show up most strongly in stars with temperatures of about 10,000 K. In a stellar atmosphere with this temperature, many hydrogen atoms are not ionized. Nevertheless, an appreciable number of them are excited to the second energy level, from which they can absorb additional photons and rise to still higher levels of excitation. These photons correspond to the wavelengths of the Balmer series.

At visual wavelengths, therefore, absorption lines due to hydrogen are strongest in the spectra of stars whose atmospheres have temperatures near 10,000 K. They are less conspicuous in the spectra of both hotter and cooler stars, even though hydrogen is roughly equally abundant in all the stars. Similarly, every other chemical element, in each of its possible stages of ionization, has a characteristic temperature at which it is most effective in producing absorption lines in any particular part of the spectrum.

Interpretation of Stellar Spectra

Now that we understand how the temperature of a star determines the physical state of the gases in its outer layers, and thus their ability to produce absorption lines, we need only to observe what patterns of absorption lines are present in the spectrum of a star to estimate its temperature. We can therefore arrange the seven classes of stellar spectra in a continuous sequence in order of decreasing temperature. In the hottest stars (temperatures over 25,000 K), only lines of ionized helium and highly ionized atoms of other elements are conspicuous. Hydrogen lines are strongest in stars with atmospheric temperatures of about 10,000 K. Ionized metals provide the most conspicuous lines in stars with temperatures from 6000 to 8000 K. Lines of neutral metals are the strongest in somewhat cooler stars. In the coolest stars (below 4000 K), absorption bands of some molecules are very strong. The most important among the molecular bands are those due to titanium oxide, a tenacious chemical compound that can exist at the temperatures of the cooler stars. The sequence of spectral types is summarized in Table 17.3 and Figure 17.6.

TABLE 17.3 Spectral Sequence

Spectral Class	Color	Approximate Temperature (K)	Principal Features	Stellar Examples
O	Violet	>28,000	Relatively few absorption lines in observable spectrum. Lines of ionized helium, doubly ionized nitrogen, triply ionized silicon, and other lines of highly ionized atoms. Hydrogen lines appear only weakly.	10 Lacertae
B	Blue	10,000–28,000	Lines of neutral helium, singly and doubly ionized silicon, singly ionized oxygen and magnesium. Hydrogen lines more pronounced than in O-type stars.	Rigel Spica
A	Blue	7500–10,000	Strong lines of hydrogen. Also lines of singly ionized magnesium, silicon, iron, titanium, calcium, and others. Lines of some neutral metals show weakly.	Sirius Vega
F	Blue to white	6000–7500	Hydrogen lines are weaker than in A-type stars but are still conspicuous. Lines of singly ionized calcium, iron, and chromium and lines of neutral iron and chromium are present, as are lines of other neutral metals.	Canopus Procyon
G	White to yellow	5000–6000	Lines of ionized calcium are the most conspicuous spectral features. Many lines of ionized and neutral metals are present. Hydrogen lines are weaker even than in F-type stars. Bands of CH, the hydrocarbon radical, are strong.	Sun Capella
K	Orange to red	3500–5000	Lines of neutral metals predominate. The CH bands are still present.	Arcturus Aldebaran
M	Red	<3500	Strong lines of neutral metals and molecular bands of titanium oxide dominate.	Betelgeuse Antares

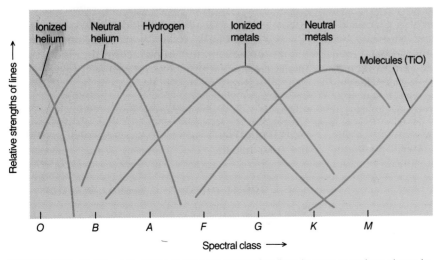

FIGURE 17.6 Relative intensities of different absorption lines in stars at various places in the spectral sequence.

Much of the pioneering work on the interpretation of stellar spectra was carried out at Harvard University in the first decades of the 20th century. The original basis for these studies was a monumental collection of nearly a million photographic spectra of stars, obtained from many years of observations made at Harvard College Observatory in Massachusetts and at its remote observing stations in South America and South Africa. Working with this data base, Annie Jump Cannon developed an empirical scheme for the classification of spectra. She personally measured some 500,000 stars and assigned them to spectral classes. She named the spectral classes with letters of the alphabet in order of the complexity of their spectra, with type A being the simplest. We now understand that spectral classes depend on the temperature of the stars and so always list the types not in order of how many lines are seen but from hottest to coolest—O, B, A, F, G, K, and M.

After Cannon's monumental work, the emphasis shifted from classification to interpretation, culminating in the work of Cecilia Payne, then a doctoral student at Harvard. It was her work, more than any other, that proved that all stars really are composed principally of hydrogen, with the differences in their spectra primarily a consequence of their different temperatures.

Effects of the Size of a Star

Stars come in a wide variety of sizes. At some periods in their evolutionary history, stars can expand to enormous dimensions with correspondingly low densities and atmospheric pressures. Stars of such exaggerated size are called **giants**. Because they are so big, giants are also very luminous. Luckily, for the astronomer who wishes to distinguish such giants from the run-of-the-mill stars (like our Sun), giants can be identified from a study of their spectra.

The pressure in a stellar photosphere affects its spectrum. At the very low densities expected for the extended tenuous photospheres of these giant stars, the pressures are also low, and ionized atoms recombine with electrons more slowly. These low-density gases, therefore, maintain a higher average degree of ionization than do high-density gases of the same temperature. Subtle details in the spectrum of a star of high photospheric pressure thus enable us to distinguish it from the spectrum of a giant star of the same temperature but of low photospheric pressure.

Abundances of the Elements

Dark lines of a majority of the known chemical elements have now been identified in the spectra of the Sun and stars. The lines of all elements are not observable in the spectrum of a star, nor are lines of any one element visible in the spectra of all stars. As we have seen, because of variations among the stars in the temperature and pressure of the photosphere, only certain of the prevailing kinds of atoms are able to produce absorption lines in any one star. The absence of the lines of a particular element, therefore, does not necessarily imply that the element is not present. Only if the physical conditions in the photosphere of a star are such that lines of an element should be visible can we conclude that the absence of observable spectral lines implies low abundance of the element. On the other hand, spectral lines of an element, in the neutral state or in one of its ionized states, certainly imply the presence of that element in the star.

Once due allowance has been made for the prevailing conditions of temperature and pressure in a star's photosphere, analyses of the strengths of absorption lines in its spectrum can yield information regarding the relative abundances of the various chemical elements whose lines appear. Such quantitative analyses have shown that the relative abundances of the different chemical elements in the Sun (Table 13.3) and in most stars are approximately the same.

Hydrogen comprises about three-quarters of the mass of most stars. Hydrogen and helium together make up from 96 to 99 percent of the mass; in some stars they amount to more than 99.9 percent. Among the 4 percent or less of "heavy elements," neon, oxygen, nitrogen, carbon, magnesium, argon, silicon, sulfur, iron, and chlorine are among the most abundant. Generally, but not invariably, the elements of lower atomic weight are more abundant than those of higher atomic weight.

The spectrum of the Sun has been studied more thoroughly than that of any other star, and we know most about the abundances of the elements there. The so-called "cosmic abundances" given in Appendix 18 are primarily solar values. For some rare elements, however, the abundance in the Sun has not been determined from spectral analysis. Our estimate of the amounts of these elements in the universe is based on laboratory measurements of their abundance in primitive meteorites, which are considered representative of unaltered material condensed from the solar nebula.

The Doppler Effect and Stellar Rotation

The radial or line-of-sight velocity of a star can be determined from the *Doppler shift* of the lines in its spectrum (Section 5.6). W. Huggins, an English astronomer who pioneered the use of photography for observing stellar spectra, made the first radial-velocity determination of a star in 1868. He observed the Doppler shift in one of the hydrogen lines in the spectrum of Sirius and found that the star is approaching the solar system.

As a star rotates (unless its axis of rotation happens to be pointed exactly toward the Sun), one of its sides approaches us and the other recedes from us, relative to the star as a whole. For the Sun or a planet, we can observe the light from one edge or the other and measure directly the Doppler shifts that arise from the rotation. A star appears as an unresolved point of light, and we must analyze the light from its entire disk at once. Nevertheless, part of the light from a rotating star, including the spectral lines, is shifted to shorter wavelengths, and part is shifted to longer wavelengths. Each spectral line of the star is a composite of spectral lines originating from different parts of the star's disk, all of which are moving at different speeds with respect to us. The same is true for rotating planets, and radar measurements of the different speeds of various parts of the planetary surface were used to determine the rotational velocities of Venus and Mercury.

The effect produced by a rapidly rotating star is that all its spectral lines are broadened to a characteristic width. The faster the star rotates, the broader is the resulting spectral line. The amount of this rotational broadening of the spectral lines can be measured, and a lower limit to the rate of rotation of the star can be calculated.

Stars are found to rotate at very different speeds. The Sun, with its rotation period of about a month, rotates rather slowly. Many stars rotate in periods of only a day or two. For single stars, hot stars rotate on average more rapidly than cool stars. Members of binary-star systems, however, will generally have been influenced by the tidal effects of their companions, so that their periods of rotation are equal, just as Pluto and its satellite Charon keep the same sides turned toward each other at all times.

Summary of Stellar Spectra

In 1835 the French philosopher Auguste Comte wrote a paper in which he asserted that it would never be possible by any means to study the chemical composition of stars. Yet within a few decades the development of astronomical spectroscopy, together with a rapidly maturing understanding by physicists of the ways atoms absorb and emit radiation, provided the tools to accomplish the "impossible" task.

In this chapter we have seen that spectrum analysis is an extremely powerful technique that allows the astronomer to learn all kinds of things about a star: its detailed chemical composition, the temperature and pressure in its atmosphere, and indirectly its size and luminosity. We can also measure its radial motion and estimate its rotation. It is no wonder that astronomers spend much of their time obtaining and analyzing spectra.

SUMMARY

17.1. For historical reasons, the brightnesses of stars are often expressed in terms of **apparent magnitudes.** If one star is five magnitudes brighter than another, it emits 100 times more energy. The total energy emitted per second by a star is its **luminosity.** Since the **apparent magnitude** of a star depends on its luminosity and distance, determination of apparent magnitude and measurement of the distance to a star provide enough information to calculate its luminosity. Astronomers sometimes specify the luminosity of a star by giving its **absolute magnitude,** which is the brightness the star would appear to have if it were at a distance of 10 pc.

17.2. Stars have different colors, and their colors are an indicator of temperature. The **color index** of a star is the difference in the magnitudes measured at any two different wavelengths. The difference between blue and visual magnitudes, B − V, is one frequently used color index; redder, cooler stars have more positive values of B − V.

17.3. **Cepheids** and **RR Lyrae stars** are two types of **pulsating variable stars. Light curves** of these stars show

that their luminosities vary in a regularly repeating period. Both types of variables follow the **period-luminosity** relation, so their distances can be derived from measurements of their periods.

17.4. The differences in the spectra of stars are principally due to differences in temperature and not composition. The spectra of stars are described in terms of seven **spectral classes.** In order of decreasing temperature, these spectral classes are O, B, A, F, G, K, and M. Spectra of stars of the same temperature but different atmospheric pressure have subtle differences, so spectra can be used to determine whether a star has a large radius and low atmospheric pressure (a **giant** star) or a small radius and high atmospheric pressure. Stellar spectra can be used to determine temperature, pressure, and chemical composition of stars. Measures of shifts of lines produced by the Doppler effect indicate the radial velocity of a star. Broadening of spectral lines by the Doppler effect is a measure of rotational velocity.

REVIEW QUESTIONS

1. What two factors determine how bright a star appears to be when seen in the sky?

2. Explain why color is a measure of the temperature of a star.

3. Suppose you have discovered a new RR Lyrae variable star. What steps would you take to determine its distance?

4. What is the main reason that the spectra of stars are not all identical? Explain.

5. Name at least three characteristics of a star that can be determined by measuring its spectrum. Explain how you would use a spectrum to determine these characteristics.

THOUGHT QUESTIONS

6. Draw a diagram showing how two stars of equal intrinsic luminosity, one of which is blue and the other red, would appear on two CCD images, one taken through a filter that passes mainly blue light and the other through a filter that transmits mainly red light.

7. About how often would you need to observe a cepheid variable star with a period of 50 days in order to determine that period? Explain your reasoning.

8. Star A has lines of ionized helium in its spectrum, and star B has bands of titanium oxide. Which is hotter? Why? The spectrum of another star shows lines of ionized helium and also molecular bands of titanium oxide. What is strange about this spectrum? Can you suggest an explanation?

9. The spectrum of the Sun has hundreds of strong lines of un-ionized iron but only a few lines of helium, which are very weak. A star of spectral type B has very strong lines of helium but very weak iron lines. Does this difference in their spectra mean that the Sun contains more iron and less helium than the B star?

10. What are the approximate spectral classes of stars described as follows?
   a. Balmer lines of hydrogen are very strong; some lines of ionized metals are present.

   b. Strongest lines are those of ionized helium.
   c. Lines of ionized calcium are the strongest in the spectrum; hydrogen lines show with only moderate strength; lines of neutral and ionized metals are present.
   d. Strongest lines are those of neutral metals and bands of titanium oxide.

11. Suppose a spectral line from a star that is not rotating looks very sharp. What would that same line look like in the spectrum of another star that is identical in every way but that is rotating rapidly? Make a sketch to explain your answer.

12. Why is it that only a lower limit to the rate of stellar rotation can be determined from rotational broadening, rather than the actual rotation rate?

13. From the following list, indicate which characteristics of a star can be measured even if its distance is not known: absolute magnitude; apparent magnitude; temperature; spectral type; space motion; proper motion; radial velocity; composition.

PROBLEMS

14. A fifth-magnitude star is about the faintest that can be seen without optical aid unless you have access to a very dark, unpolluted sky. How much fainter is it than a zero-magnitude star? A good pair of binoculars can reveal tenth-magnitude stars. How much fainter are these than stars of zero magnitude?

15. As seen from the Earth, the Sun has an apparent magnitude of about -26. What is the apparent magnitude of the Sun as seen from Saturn, about 10 AU distant?

16. If a star has a color index of $B - V = 2.5$, how many times brighter does it appear in visual light than in blue light, relative to a standard star with $B - V = 0$.

17. What are the approximate spectral classes for stars whose wavelengths of maximum light have the following values (see Section 5.2)?

  a. 290 nm

  b.   50 nm

  c. 600 nm

  d. 1200 nm

  e. 1500 nm

18. Consider a cepheid with a luminosity 10,000 times that of the Sun. If a given telescope and detector can detect a solar-type star out to a distance of only 1000 LY, to what distance can this same telescope observe the cepheid?

19. How many times farther away can a cepheid like that in Exercise 18 be than an RR Lyrae star and still be observed with the same telescope?

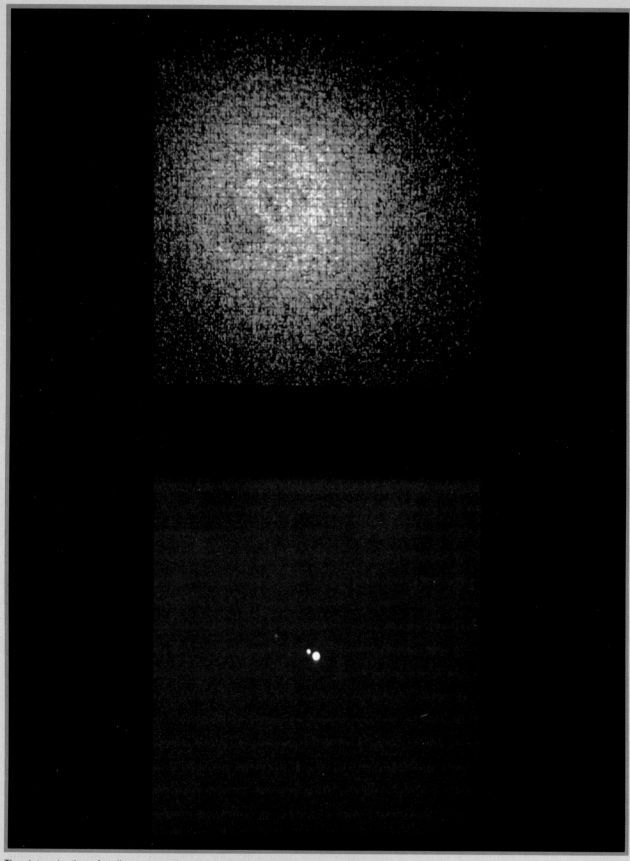

The determination of stellar masses depends on measurement of the gravitational effects of one star on a companion. New techniques make it possible to measure very close binary stars. The top image shows a long time exposure of the fourth-magnitude spectroscopic binary star Sigma Herculis. Atmospheric turbulence produces a blurred image of both stars that is about 2 arcsec in diameter. New techniques can remove this blurring and reveal two stars separated by 0.07 arcsec. *(Anthony Readhead/Palomar Observatory, Caltech)*

Henry Norris Russell (1877–1957),
American astronomer, was a professor
at Princeton University and director of
the observatory. His many interests
included the study of stellar evolution.
Russell and the Danish astronomer
Ejnar Hertzsprung independently
discovered the main sequence of stars,
best illustrated on the famous
Hertzsprung-Russell diagram. *(Princeton
University Archives)*

18

THE STARS: A CELESTIAL CENSUS

The techniques of celestial surveying, of photometry, and of spectroscopy described in the previous two chapters provide ways to study the properties of stars, especially of those stars within a few hundred light years of the Sun. In this chapter we will take a close look at our stellar neighbors.

18.1 The Nearest and Brightest Stars

Astronomers learned a very surprising thing when they measured distances to stars. The brightest stars are not the closest ones. The stars that appear brightest in the nighttime sky are bright not because they are nearby but because they are actually of high intrinsic luminosity. Look at Appendix 14, which gives distances for the 20 brightest stars. Only six of the stars listed are within 32 LY (10 pc) of the Sun. Most of the stars that can be seen without a telescope are hundreds of light years away and are many times more luminous than the Sun. Indeed, among the 6000 stars visible to the unaided eye, at most 50 are intrinsically fainter than the Sun. Figure 18.1 is a histogram showing the distribution of luminosities of the 30 brightest-appearing stars.

If we consider only the brightest stars—those listed in Appendix 14 or plotted in Figure 18.1—we might gain the impression that the luminosity of the Sun is far below average. This is *not* true. Most stars are really much less luminous than the Sun. In fact, most stars are so faint that they can be seen only through a telescope—even if they are very nearby.

Appendix 13 lists the 44 known stars within 5 pc (16 LY) of the Sun. Most of these stars have large proper motion. As we saw in Chapter 16, nearby stars are often discovered because of their large proper motions, which are primarily a reflection of the motion of the Sun through space. Additional nearby stars are discovered from time to time. The total number of stars within 16 LY may even be double the number listed.

The Most Common Stars

The table in Appendix 13 contains a total of 59 stars. Only 3 of the 58 stars other than the Sun are among the 20 brightest-appearing stars: Sirius, Alpha Centauri, and Procyon. We see, then, that most of the nearest stars are intrinsically faint. Stars with luminosities greater than 100 L_S are rare—so rare that not even one is found within 25 LY of the Sun. If we are to find a star of high luminosity, we must search

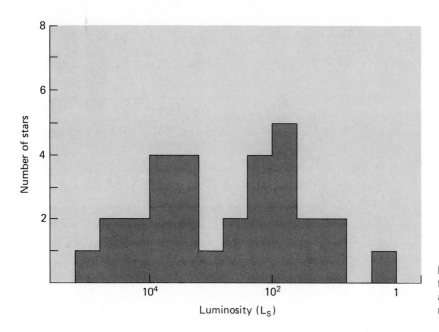

FIGURE 18.1 Distribution in luminosity of the 30 brightest-appearing stars. The units are the numbers of stars per unit of luminosity.

farther from the Sun and explore a larger volume of space.

We can clarify this point with the help of some examples. The Sun would have an apparent magnitude of 4.8 if it were 10 pc (32.6 LY) away and would appear as a very faint star to the unaided eye. Stars much less luminous than the Sun would not be visible at all at that distance. Stars with luminosities in the range from 10^{-2} to 10^{-4} L_S are very common, but a star with a luminosity of 10^{-2} L_S would have to be within 5 LY to be visible to the naked eye. Only Alpha Centauri is closer than this. It is clear, then, that the vast majority of nearby stars, those less luminous than the Sun, do not send enough light across interstellar distances to be seen without a telescope and are therefore very poorly represented in a listing of the apparently brightest stars.

In contrast, consider the highly luminous stars—those 100 times as luminous as the Sun or more. Although they are far less common than fainter stars, they are visible to the unaided eye even out to a distance of 500 LY. A star with luminosity of 10^4 L_S can be seen without a telescope to a distance of 5000 LY (if there is no dimming of light by interstellar dust—see Chapter 19). Such stars are very rare, and we would not expect to find one nearby. The volume of space included within a distance of 5000 LY, however, is enormous, so even though they are intrinsically rare, many such highly luminous stars are readily visible to the unaided eye.

The contrast between these two samples of stars—those that are close to us and those that are readily visible—is an example of a *selection effect*. When a population of objects—stars in this example—includes a great variety of different types, we must be careful what conclusions we draw from an examination of any particular subgroup. Certainly we would be fooling ourselves if we assumed that the stars visible to the unaided eye are characteristic of the general stellar population. Although it requires much more effort to assemble a complete data set for the nearest stars, it is only by doing so that astronomers are able to work out the properties of the vast majority of the stars, which are actually much smaller and fainter than our own Sun.

Of course, the above discussion about what stars are like applies only to the general neighborhood of the Sun. Not all neighborhoods are alike. There are special places—regions of star formation, for example, or the centers of large star clusters—where the characteristics of stars are substantially different from what we see near the Sun.

The Density of Stars in Space

There are at least 59 stars within 16 LY of the Earth, counting the members of binary- and multiple-star systems and the Sun itself. A sphere of radius 16 LY has a volume of about 17,000 cubic LY (1.7×10^4 LY^3). Since this volume of space contains at least 59 stars, the density of stars in space in the neighborhood of the Sun is about one star for every 300 LY^3. The actual stellar density, of course, can be greater than this figure if there are undiscovered stars within

16 LY. At this density we expect a total of nearly 15,000 stars within 100 LY. The mean separation between stars is the cube root of 300, or about 7 LY; thus the Sun and Alpha Centauri, which is 4.4 LY distant, are rather closer than the average among stars in the solar neighborhood. If the matter contained in stars could be spread out evenly over space, and if a typical star has a mass 0.4 times that of the Sun, the mean density of matter in the solar neighborhood would be only about 3×10^{-24} g/cm^3.

The Luminosity Function

Once we measure the luminosities of a large number of stars, we can also determine how many stars in a given volume of space are intrinsically very luminous and how many are faint. Figure 18.2 shows how many stars in the solar neighborhood fall in each successive interval of luminosity. This relationship is called the **luminosity function.** Compare Figure 18.2 with Figure 18.1 and note how unrepresentative of the average stellar population the brightest-appearing stars really are. Figure 18.2 also shows that the Sun is more luminous than the vast majority of stars.

Most of the stellar mass in our neighborhood is contributed by stars that are fainter than the Sun. Most of the starlight from our part of space, however, comes from the relatively few stars of high luminosity. It takes only 10 stars of luminosity 100 L$_S$ to outshine 1000 stars fainter than the Sun, and only 1 star of luminosity 10,000 L$_S$ to outshine 10,000 stars fainter than the Sun.

18.2 Stellar Masses

Many stars are members of double-star systems. For example, among the 59 nearest stars listed in Appendix 13, 28 (roughly one-half) are members of systems containing more than one star. Masses for stars can be calculated from measurements of their orbits around one another, just as the mass of the Sun was derived by measuring the orbits of the planets around it.

Binary Stars

The first **binary, or double, star** was discovered in 1650, less than half a century after Galileo first observed the sky with a telescope. John Baptiste Riccioli, an Italian astronomer, observed that the star Mizar, in the middle of the handle of the Big Dipper, appeared through his telescope as two stars. Since that discovery, thousands of double stars have been catalogued. Although stars most commonly come in pairs, there are also triple and quadruple systems.

If the gravitational forces between stars are like those in the solar system, the orbit of one star about the other must be an *ellipse*. The first to prove that such is the case was Felix Savary. In 1827 he showed that the relative orbit of the two stars in the double system Zeta Ursae Majoris is an ellipse, the stars completing one mutual revolution in a period of 60 years. The orbital motion of another binary star, Kruger 60, is shown in Figure 18.3.

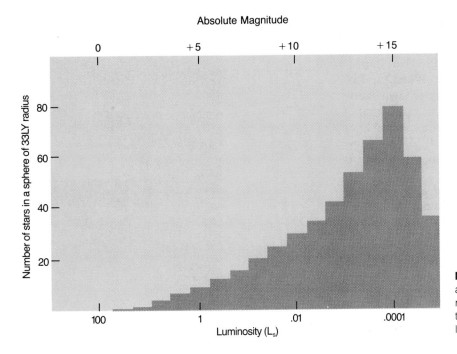

FIGURE 18.2 A luminosity function in astronomy is the distribution of stars with respect to their luminosity. Shown here is the luminosity function of stars in the solar neighborhood.

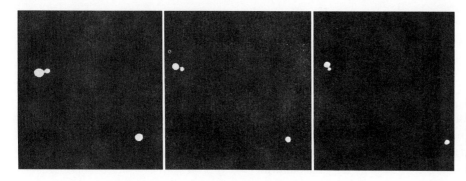

FIGURE 18.3 Revolution of a double star. Three photographs covering a period of about 12 years show the mutual revolution of the components of the nearby star Kruger 60. *(Yerkes Observatory)*

Another class of double stars was discovered by E. C. Pickering at Harvard in 1889. He found that the dark absorption lines in the spectrum of the brighter component of Mizar (the first double star to be discovered) are usually double but that the spacing of the components of the lines varies periodically. At times the lines even become single. He correctly deduced that the brighter component of Mizar, which is called Mizar A, itself is really two stars that revolve about each other in a period of 104 days.

It is not correct to describe the motion of a double-star system by saying that one star orbits the other. Gravitation is a *mutual* attraction. Each star exerts a gravitational force on the other, with the result that both stars orbit a point between them called the **center of mass.** Imagine that the two stars are seated, one at each end of a seesaw. The point at which one would have to place the support to make the seesaw balance is the center of mass.

When one star is approaching us, relative to the center of mass, the other star is receding from us. The radial velocities of the two stars, and therefore the Doppler shifts of their spectral lines, are different. When the composite spectrum of the two stars is observed, each line appears double. When the two stars are both moving across our line of sight, however, they both have the same radial velocity (that of the center of mass of the pair), and the spectral lines of the two stars coalesce (Figure 18.4). A plot showing how the velocities of the stars change with time is called a radial-velocity curve (Figure 18.5).

Stars like Mizar A, which appear as single stars when photographed or observed visually through the

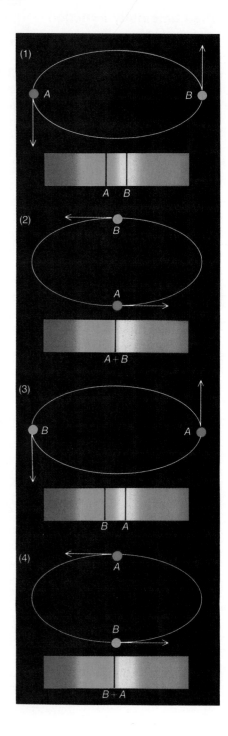

FIGURE 18.4 A schematic drawing of the motions of two stars orbiting each other. When one star approaches the Earth, the other recedes; half a cycle later the situation is reversed. The Doppler shifts of the spectral lines cause the lines to move back and forth. At times, lines from both stars can be seen well separated from each other. When the two stars are moving perpendicular to our line of sight, the two lines are exactly superimposed, and we see only a single spectral line.

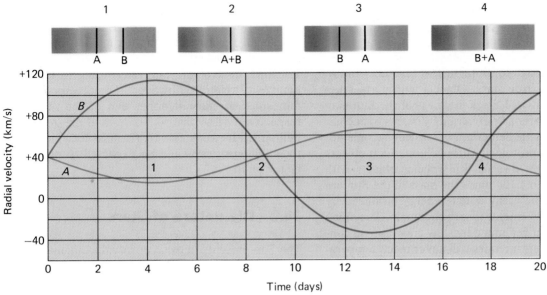

FIGURE 18.5 Radial-velocity curves for a spectroscopic binary system showing how the two components alternately approach and recede from the Earth. The positions on the curve corresponding to the illustrations in Figure 18.4 are marked.

telescope, but which spectroscopy shows really to be double stars, are called **spectroscopic binaries.** Systems that can be observed visually as double stars are called **visual binaries.**

Masses from the Orbits of Binary Stars

We can estimate the masses of double-star systems by using Newton's reformulation of Kepler's third law. The method is similar to that used to calculate the mass of the Sun (Section 2.3). If two objects are in mutual revolution, then the period (P) with which they go around each other is related to the semimajor axis (D) of the orbit of one with respect to the other, according to the equation

$$D^3 = (M_1 + M_2)P^2$$

where D is in astronomical units, P is measured in years, and $M_1 + M_2$ is the sum of the masses of the two stars in units of the mass of the Sun. Thus if we can observe the size of the orbit and the period of mutual revolution of the stars in a binary system, we can calculate the sum of their masses. Most spectroscopic binaries have periods ranging from a few days to a few months, with separations of usually less than 1 AU between their member stars.

The actual analysis of a radial-velocity curve like the one in Figure 18.5 to determine the masses of the stars in a spectroscopic binary is complex, but in principle the idea is simple. The speeds of the stars are measured from the Doppler effect. The speed together with the period (determined from the veloc-

ity curve) tells us the circumference of the orbit and hence the separation of the stars in kilometers or astronomical units. Knowing the period and the separation, we can calculate the sum of the masses of the stars. In addition, the relative orbital speeds of the two stars tell us how much of the total mass each star has. The more massive star has a smaller orbit and hence moves more slowly to get around in the same time. If the binary system is properly oriented and all the steps are carried out carefully, the result is a measurement of the masses of each of the two stars in the system.

In summary, the combination of Doppler measurements of orbital speed and the observed period of the system gives us enough information to calculate the actual linear dimensions in kilometers of the binary-star orbits. With this result and Newton's laws, the determination of mass is straightforward. Note that the method is independent of the distance of the binary system. It works as well for stars 1000 LY away as for those in the solar neighborhood.

The Range of Stellar Masses

What is the largest mass that a star can have? The limit is not known for sure, but searches for massive stars indicate that very few stars have masses greater than about 60 times the mass of the Sun. There is no convincing evidence that there are any stars with masses that significantly exceed about 100 times the mass of the Sun. The rarity of stars with large masses is illustrated by the fact that there are no stars within

30 LY of the Sun that have masses greater than four times the mass of the Sun.

According to theoretical calculations, objects with masses less than 1/12 the mass of the Sun never become hot enough to ignite nuclear reactions and so cannot become true stars. Objects with masses between 1/100 and 1/12 times the mass of the Sun may produce energy for a brief period by means of nuclear reactions involving deuterium but do not become hot enough to fuse protons to form helium. Such objects are called **brown dwarfs.** Still smaller objects with masses less than 1/100 times the mass of the Sun are true planets. They may radiate energy produced by the radioactive elements that they contain. They may also radiate heat generated by slow gravitational contraction, but their interiors will never reach temperatures high enough for nuclear reactions to take place.

The Mass-Luminosity Relation

When we compare the masses and luminosities of stars, it is found that, in general, the more massive stars are also the more luminous. This relation, known as the **mass-luminosity relation,** is shown graphically in Figure 18.6. Each point represents a star of known mass and luminosity. Its horizontal position indicates its mass, given in units of the Sun's mass, and its vertical position indicates its luminosity in units of the Sun's luminosity.

Most stars fall along a line running from the lower left (low mass, low luminosity) corner of the diagram to the upper right (high mass, high luminosity) corner. It is estimated that about 90 percent of all stars obey the mass-luminosity relation illustrated in Figure 18.6.

The range of stellar luminosities is much greater than the range of stellar masses. Luminosities of stars are roughly proportional to their masses raised to the 3.5 power. Most stars have masses between 1/10 and 100 times that of the Sun. According to the mass-luminosity relation, however, the corresponding luminosities of stars at either end of the range are, respectively, less than 0.001 L_S and greater than 1 million L_S. If two stars differ in mass by a factor of 2, their luminosities would be expected to differ by a factor of 10.

18.3 Diameters of Stars

The Sun presents to us an observable angular diameter of about 1/2°. Thus we can calculate directly the Sun's true (linear) diameter, which is 1.39 million km, or about 109 times the diameter of the Earth. Its radius is half this value, or about 700,000 km. The Sun is the only star whose angular diameter is easily resolved. Even through the largest telescopes, apart from the distortions introduced by turbulence in the Earth's atmosphere, stars appear to the eye to be points of light. There are several techniques, however, that the astronomer can use to measure the sizes of stars.

Stellar Diameters of Stars Whose Light Is Blocked by the Moon

One technique, which gives very precise diameters but can be used for only a few stars, is to observe the dimming of light when the Moon passes in front of a star. The technique is to measure the time required

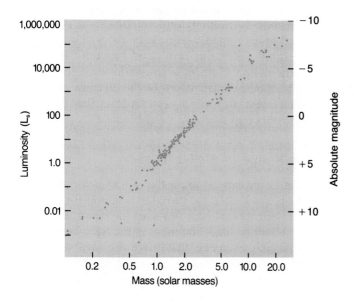

FIGURE 18.6 The mass-luminosity relation. The plotted points show the masses and luminosities of stars for which both of these quantities are known, to an accuracy of 15 to 20 percent. The three points lying below the sequence of points are all white dwarf stars. *(Adapted from data compiled by D.M. Popper)*

for a star's brightness to drop to zero as the edge of the Moon moves across it. Since we know how rapidly the Moon moves in its orbit around the Earth, it is possible to calculate the angular diameter of the star. If the distance to the star is also known, then it is possible to calculate its diameter in kilometers. This method works only for fairly bright stars that happen to lie along the zodiac where the Moon (or much more rarely a planet) can pass in front of them as seen from the Earth.

Eclipsing Binary Stars

A second technique works for those stars that are members of **eclipsing binary systems.** Some double stars are lined up in such a way that when viewed from the Earth each star passes in front of the other during every revolution. When one star blocks the light of the other and prevents it from reaching the Earth, astronomers say that an *eclipse* has occurred. Such double-star systems are called eclipsing binaries.

This type of binary was found very shortly after Pickering discovered that Mizar A is a spectroscopic binary. Radial-velocity observations of the star Algol, in Perseus, showed that it, too, is a spectroscopic binary. The spectral lines of Algol were not observed to be double because the fainter star of the pair gives off too little light compared with the brighter for its lines to be conspicuous in the composite spectrum. Nevertheless, the periodic shifting back and forth of the lines of the brighter star gave evidence that it was revolving about an unseen companion. The lines of both components need not be visible for a star to be recognized as a spectroscopic binary.

The proof that Algol is a double star provided an explanation for observations of Algol's brightness, which changes periodically in a predictable pattern. Normally, Algol is a second-magnitude star, but at intervals of 2d 20h 49m it fades to one-third of its regular brightness. After a few hours, it brightens to normal again. (This effect is easily seen even without a telescope if you know what to look for—try observing Algol if you have access to clear skies without too much light pollution). In 1783, more than a century before the spectroscopic observations of Algol, John Goodricke suggested that its brightness variations might be due to an invisible companion that regularly passes in front of the brighter star and blocks its light. Unfortunately, Goodricke had no way to test this idea, since the equipment available did not permit measurement of Algol's spectrum. Even if it had, the explanation of the Doppler effect for spectral lines came more than 50 years after Goodricke's proposal that Algol is a binary star.

The discovery that Algol is a spectroscopic binary verified Goodricke's hypothesis. The plane in which the stars revolve is turned nearly edgewise to our line of sight, and each star passes in front of the other during every revolution. The eclipse of the fainter star is not very noticeable because the part of it that is covered contributes little to the total light of the system. This second eclipse can, however, be detected by careful measurements. Any binary star will produce eclipses if viewed from the proper direction, near the plane of its orbit. Only a few of these stars, however, are oriented properly to yield eclipses as seen from the particular vantage point of the Earth.

Light Curve of an Eclipsing Binary

During the period of revolution of an eclipsing binary, there are two times when the light from the system diminishes: once when the smaller star passes behind the larger one and once when the smaller star passes in front of the larger one and covers part of it. If the smaller star goes completely behind the larger one, the situation is as illustrated in Figure 18.7. If the

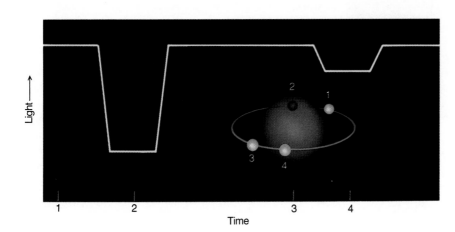

FIGURE 18.7 Schematic light curve of a hypothetical eclipsing binary star with total eclipses (i.e., one star passes directly in front of and behind the other). The numbers indicate parts of the light curve corresponding to various positions of the smaller star in its orbit. In this example, the smaller star is the hotter.

smaller star is never completely hidden behind the larger star, both eclipses are partial.

Each interval during an eclipse when the light from the system is farthest below normal is called a minimum. Both minima are not, in general, equally low in light. The energy emitted from a given area on the surface of a star—say 1 m²—depends on the temperature, as indicated by the Stefan-Boltzmann law (Chapter 5). Hot stars have much greater *surface brightness* (energy emitted per square meter) than do cooler stars. Thus the relative amount of light drop at each minimum depends on the relative surface brightnesses of the two stars, and hence on their temperatures. *Primary minimum* occurs when the *hotter* star is covered (whether or not the hotter star is the larger). *Secondary minimum* occurs when the *cooler* star is covered.

Diameters of Eclipsing Binary Stars

To illustrate how the sizes of the stars are related to the light curve of an eclipsing binary, we may consider a hypothetical binary in which the stars are very different in size, like those illustrated in Figure 18.8. We will assume that the orbit is viewed exactly edge-on. When the small star just starts to pass behind the large star (first contact), the brightness begins to

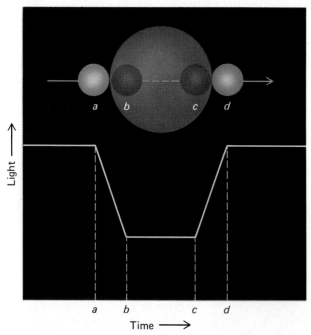

FIGURE 18.8 Contacts in the schematic light curve of a hypothetical eclipsing binary star with central eclipses. From the time intervals between contacts it is possible to estimate the diameters of the two stars.

drop. The total phase of the eclipse begins at second contact, when the small star has gone entirely behind the large one. At the end of totality (third contact), the small star begins to emerge. When the small star has reached last contact, the eclipse is over.

During the time interval between first and second contacts (or between third and last contacts), the small star has moved a distance equal to its own diameter. During the time interval from first to third contacts (or from second to last contacts), the small star has moved a distance equal to the diameter of the large star. If the spectral lines of both stars are visible in the spectrum of the binary, the speed of the small star with respect to the large one is also known. This speed, multiplied by the time intervals from first to second contacts and from first to third contacts, gives, respectively, the diameters of the small and large stars.

In actuality, the orbits generally are not exactly edge-on, and the light from each star may be only partially blocked by the other. Further, binary-star orbits are ellipses, not circles. However, it is a relatively simple geometry problem, at least in principle, to sort out these effects. Then from the depths of the minima and the exact instants of the various phases, it is possible to calculate both the inclination of the orbit and the sizes of the stars in relation to their separation. If the eclipses are partial, the analysis is far more difficult, but even so it can be accomplished.

An eclipsing binary that is also a spectroscopic binary provides a great deal of information, including the orbital speeds of the stars and the linear dimensions of the system. With this knowledge, the shapes of the light curves can be interpreted to yield the sizes of the individual stars. Among the thousands of known eclipsing binaries, however, there are only a few dozen that are so favorably disposed for observation that all the necessary data can be obtained. Only for these few binaries have complete analyses led to fairly reliable values of the diameters of their member stars.

Stellar Diameters from Radiation Laws

For most stars, we must estimate their sizes by using an indirect method based on observations of their brightness and spectrum. The steps are as follows. The luminosity of a star can be obtained from measurements of its apparent brightness and distance. The temperature of a star can be obtained from its color or spectrum. Since stars are fairly good approximations to blackbodies, we can apply the Stefan-Boltzmann law to them (Section 5.2). The energy emitted per square meter of a star is proportional to the fourth power of its temperature. The energy emitted per square meter, multiplied by the total area of the surface of the star in square meters, must be equal to its total luminosity. Since the surface area of a sphere of radius R is $4\pi R^2$, the equation relating luminosity, temperature, and radius is

$$L = 4\pi R^2 \times \sigma T^4$$

Solving for R, we find that

$$R = \sqrt{L/(4\pi\sigma T^4)}$$

A knowledge of a star's temperature and luminosity, therefore, enables us to calculate its size.

Stellar Diameters

The results of the measurements of stellar size confirm that most nearby stars have roughly the size of the Sun—with typical diameters of a million kilometers or so. Faint stars, as might be expected, are generally smaller than more luminous stars. However, there are some dramatic exceptions to this simple generalization.

A few of the very luminous stars, those that are also red in color (indicating relatively low surface temperatures), are truly enormous. These stars are called giants or supergiants. An example is Betelgeuse, the second brightest star in the constellation of Orion and one of the dozen brightest stars in our sky. The distance of Betelgeuse is about 500 LY—not very close—yet its angular diameter in the sky is about 0.1 arcsec, almost big enough to appear as a disk through a large telescope under the best observing conditions. Translated into linear dimensions, this corresponds to a diameter greater than 10 AU, large enough to fill the entire inner solar system almost as far as Jupiter. In Chapter 21, we will look in detail at the evolutionary process that leads to the formation of giant and supergiant stars.

18.4 The H–R Diagram

In 1911, the Danish astronomer Ejnar Hertzsprung compared stars within several clusters by plotting their magnitudes against their colors. In 1913, the American astronomer Henry Norris Russell undertook a similar investigation of stars in the solar neighborhood by plotting the absolute magnitudes of stars of known distance against their spectral classes. These investigations by Hertzsprung and by Russell led to an extremely important discovery concerning the relation between the luminosities and surface temperatures of stars.

Features of the H–R Diagram

The relationship between stellar luminosities and temperatures is best illustrated with the **H–R diagram,** which is simply a plot of stellar color (or temperature) against luminosity. This plot is also frequently called the Hertzsprung-Russell (H–R), diagram in honor of these two astronomers. It is one of the most important and widely used diagrams in astronomy, with applications that extend far beyond the purposes for which it was originally developed nearly a century ago.

The two quantities plotted in the H–R diagram are readily determined for many stars. The *luminosities* can be found from the known distances and the observed apparent magnitudes. Even if the distance is not well determined, an H–R diagram can be plotted for a group of stars all at about the same distance (such as a star cluster) by using apparent or relative brightness. The *surface temperature* of a star is indicated by either its color or its spectral class.

As an example of an H–R diagram, in Figure 18.9 luminosities are plotted against temperatures (and spectral classes and colors) for selected nearby stars for which these data are available. For reasons of convention, temperature increases toward the left and luminosity toward the top. Figure 18.10, a schematic H–R diagram for a large sample of stars, is also shown to make the various features more apparent.

The most significant feature of the H–R diagram is that the stars are not distributed over it at random, as they would be if the stars exhibited all combinations of luminosity and temperature. Instead, we see that the stars cluster into certain parts of the H–R diagram. The majority of stars are aligned along a narrow sequence running from the upper left (hot, highly luminous) part of the diagram to the lower right (cool, less luminous) part. This band of points is called the **main sequence.** It represents a relationship between *temperature* and *luminosity* that is followed by most stars.

A substantial number of stars, however, lie above the main sequence on the H–R diagram, in the upper right (cool, high luminosity) region. These are the **giants.** At the top part of the diagram are stars of even higher luminosity, called **supergiants.** These names are appropriate, since the stars in this part of the H–R diagram do have enormous sizes. Finally, there are stars in the lower left (hot, low luminosity) corner known as **white dwarfs.** To say that a star lies ''on'' or

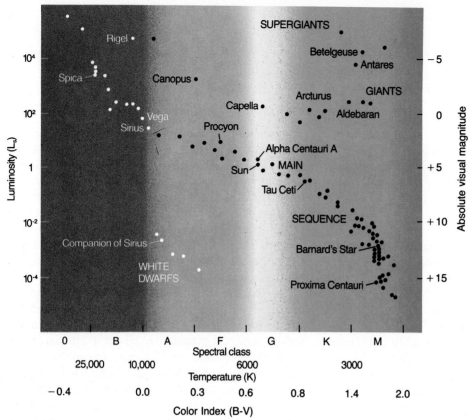

FIGURE 18.9 The H–R diagram for a selected sample of stars. Luminosity is plotted along the vertical axis. Along the horizontal axis, we can plot either temperature or spectral type. Several of the brightest stars are identified by name.

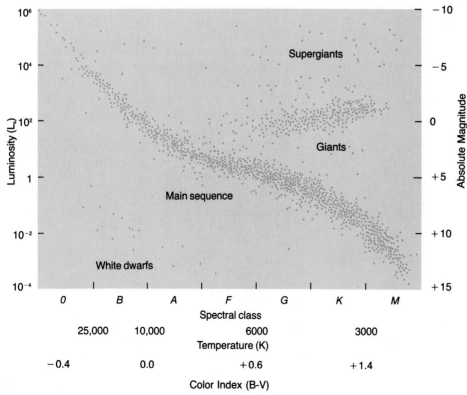

FIGURE 18.10 Schematic H–R diagram for many stars.

"off" the main sequence does not refer to its position in space but only to the point that represents its luminosity and temperature on the H–R diagram.

An H–R diagram that is plotted for stars of known distance (or any other selection criterion) does not show the correct relative proportions of various kinds of stars. The stars in Figure 18.9 were selected because their distances are known. This sample omits many intrinsically faint stars that are nearby but have not had their parallaxes measured. To be truly representative of the stellar population, an H–R diagram should be plotted for all stars within a certain distance (see Problem 15). Unfortunately, our knowledge is reasonably complete only for stars within 10 to 20 LY of the Sun, among which there are no giants or supergiants. It is estimated that about 90 percent of the stars in our part of space are main-sequence stars and about 10 percent are white dwarfs. Fewer than 1 percent are giants or supergiants.

Understanding the Main Sequence

We can use theoretical models of stars to understand the main sequence. The basic structure of stars is really fairly simple to understand, once we get past the complications of computer calculations of models. In fact, for objects obeying the laws of hydrostatic and thermal equilibrium and deriving all their energy from thermonuclear reactions (Chapter 15), astrophysicists have shown that the structure of a star is completely and uniquely determined by just two quantities—its *total mass* and *composition*. This fact provides an interpretation of many features of the H–R diagram.

Imagine a cluster of stars forming from a cloud of interstellar material whose chemical composition is similar to the Sun's. All condensations that become stars will then begin with the same chemical composition and will differ from one another only in their mass. Suppose now that we were to compute a model for each of these stars for the time at which it became stable and derived its energy from nuclear reactions, but before it had time to alter its composition appreciably as a result of these reactions.

The models calculated for these stars would allow us to determine their luminosities, temperatures, and sizes. It turns out that calculations show that the most massive stars are the most luminous, in agreement with the mass-luminosity relationship found empirically for about 90 percent of the nearby stars. We could also plot the results from the models—one point for each star—on the H–R diagram. After doing so, we would find that the most massive stars are the most luminous and the hottest, located at the upper left of the diagram. The least massive are the coolest and have the lowest luminosities, as indicated by their position at the lower right of the plot. The other model stars all lie along a line running diagonally across the diagram.

This diagonal line running across our H–R diagram for model stars lies just where the main sequence is found for real stars. Thus we see that the main sequence is a sequence of stellar masses. It is the locus of points on the H–R diagram representing stars of chemical composition similar to the Sun's and differing from one another only in mass. The fact that most stars lie on the main sequence is an indication that most stars have roughly solar composition. And the actual range in luminosities can be interpreted with our models in terms of the range in stellar masses, from less than 10 percent of the Sun's mass to perhaps 100 times the mass of the Sun. The characteristics of main-sequence stars are listed in Table 18.1.

What about the other stars on the real H–R diagram—the red giants and supergiants and the white dwarfs? These must differ somehow from the majority in internal composition and structure, and in their source of nuclear energy. With the aid of stellar

TABLE 18.1   Characteristics of Main-Sequence Stars

Spectral Type	Mass (Sun = 1)	Luminosity (Sun = 1)	Temperature	Radius (Sun = 1)
O5	40	5×10^5	40,000 K	18
B0	16	2×10^4	28,000 K	7
A0	3.3	80	10,000 K	2.5
F0	1.7	6	7,500 K	1.4
G0	1.1	1.3	6,000 K	1.1
K0	0.8	0.4	5,000 K	0.8
M0	0.4	0.03	3,500 K	0.6

models, it is possible to interpret the location of a star on the H–R diagram—that is, its luminosity and surface temperature—in terms of its interior structure and composition. As we will see in later chapters, we can do even more. We can use the H–R diagram and model calculations to trace out the actual evolution of a star, from its birth to its death.

Stellar Distances from Their Spectra

Examination of the H–R diagram reveals a very important method for determining stellar distances. Suppose, for example, that a star is known to be a spectral class G star on the main sequence. Its luminosity could then be read off the H–R diagram at once. It would have about the same luminosity as the Sun (1 L_S). From this luminosity and the star's apparent brightness, its distance can be calculated.

In general, however, the spectral class alone is not enough to estimate, unambiguously, the intrinsic luminosity of a star. The G star described in the preceding paragraph could have been, for example, a main-sequence star of luminosity 1 L_S, a giant of luminosity 100 L_S, or a supergiant of still higher luminosity. We recall, however (Section 17.4), that pressure differences in the atmospheres of stars of different sizes result in slightly different degrees of ionization for a given temperature. These effects, due to differences in pressure, produce subtle but observable differences in the spectra of stars of the same temperature. It is thus possible to classify a star by its spectrum, not only according to its temperature (spectral class) but also according to whether it is a main-sequence star, a giant, or a supergiant. With both of these items of information, a star's position on the H–R diagram is uniquely determined. Its luminosity, therefore, is also known, and its distance can be calculated.

Extremes of Stellar Luminosities, Diameters, and Densities

The H–R diagram is useful for investigating the extremes in size, luminosity, and density found for stars. The most massive stars are the most luminous ones, at least for main-sequence stars. These stars have luminosities of about 10^4 L_S to 10^5 L_S. A few stars are known that are 10^6 times more luminous than the Sun. These superluminous stars, most of which are at the upper left on the H–R diagram, are very hot stars of spectral types O and B and are very blue. These are the stars that would be the most conspicuous at very great distances in space.

The cool giants and supergiants at the upper right

corner of the H–R diagram are at least a few hundred times as luminous as the Sun (if they are giants) or some thousands of times as luminous (if they are supergiants). These stars also have very much larger diameters than that of the Sun. Indeed, some supergiants are so large that if the solar system could be centered in such a star, the star's surface would lie beyond the orbit of Mars.

It is easy to show that the luminous supergiants must also be very large. Consider, as an example, a red, cool supergiant that has a surface temperature of 3000 K and a luminosity of 10^4 L_S. Although this star is 10,000 times more luminous than the Sun, it has only half its surface temperature. The Stefan-Boltzmann law tells us that the surface brightness is proportional to the fourth power of the temperature (Section 5.2). Thus if the star has half the temperature of the Sun, it emits only $(1/2)^4 = 1/16$ as much light per square meter as the Sun. To have a luminosity of 10,000 L_S, its total surface area must be greater than the Sun's by 160,000 times. Its radius, therefore, is 400 times the Sun's radius.

In contrast, the very common red, cool stars of low luminosity at the lower end of the main sequence are much smaller and more compact than the Sun. An example of such a red dwarf is the star Ross 614B, which has a surface temperature of 2700 K and 1/2000 of the Sun's luminosity. Each square meter of the surface of this star emits only 1/20 as much light as a square meter on the Sun. To have only 1/2000 the Sun's luminosity, the star need have only about 1/100 the Sun's surface area, or 1/10 its radius. A star with such a low luminosity also has a low mass (Ross 614B has a mass about 1/12 that of the Sun) but still would have a mean density about 80 times that of the Sun. Its density must be higher, in fact, than that of any known solid found on the surface of the Earth.

The faint red main-sequence stars are not the stars of the most extreme densities, however. The white dwarfs, at the lower left corner of the H–R diagram, have densities many times greater still.

The White Dwarfs

The first white dwarf stars to be discovered were the companions of much brighter main-sequence stars. The first such object, found in 1914, forms a binary star with Sirius, the brightest-appearing star in the sky. Although it is only 8 LY distant, this faint companion of Sirius (called Sirius B) is quite difficult to see unless you have a rather large telescope.

A good example of a typical white dwarf is the nearby star 40 Eridani B. Its temperature is a relatively hot 12,000 K, but its luminosity is only 1/275 L_S. Calculations show that it has a radius of 0.014 and a volume of 2.5×10^{-6} the Sun's. Its mass, however, is 0.43 times that of the Sun, so its density is about 170,000 times the density of the Sun, or more than 200,000 g/cm^3. A teaspoonful of such material would have a mass of some 50 tons. At such densities, matter cannot exist in its usual state. Although this matter is still gaseous, its atoms are completely stripped of their electrons, and the matter in white dwarfs is said to be **degenerate.**

The British astrophysicist Arthur Eddington described the first known white dwarf this way: "The message of the companion of Sirius, when decoded, ran: 'I am composed of material three thousand times denser than anything you've ever come across. A ton of my material would be a little nugget you could put in a matchbox.' What reply could one make to something like that? Well, the reply most of us made in 1914 was, 'Shut up; don't talk nonsense.'"

As we shall see in Chapter 22, most stars are believed to become white dwarfs near the end of their evolution. Eventually, after many billions of years, white dwarfs radiate away their internal heat, cooling off to become black dwarfs—cold, dense stars no longer shining. White or black dwarfs, however, are not the only possible final evolutionary states for stars. Some stars, as we shall see, evidently become neutron stars, with densities a billion times as great as those of white dwarfs. Still others may collapse to black holes of even greater density. We take up these bizarre objects in later chapters.

SUMMARY

18.1 The stars that appear brightest to our eyes are bright because they are intrinsically very luminous, not because they are the closest stars. Most of the nearest stars are intrinsically so faint that they can be seen only with the aid of a telescope. The luminosity of stars ranges from about 10^{-4} L_S to more than 10^6 L_S. Stars with low luminosity are much more common than stars with high luminosity.

18.2 The masses of stars can be determined by analysis of the orbits of **double stars.** The three types of double stars are **visual binaries, spectroscopic binaries,** and **eclipsing binaries.** Stellar masses range from about 1/12 times the mass of the Sun to (rarely) 100 times the mass of the Sun. The most massive stars are, in most cases, also the most luminous, and this correlation is known as the **mass-luminosity relation.**

18.3 The diameters of stars can be determined by measuring the time it takes an object (the Moon, a planet, or a companion star) to pass in front of it and block its light. Diameters of members of **eclipsing binary systems** can be determined through analysis of orbital motions. Measurement of temperature and total luminosity, combined with the Stefan-Boltzmann law, also makes it possible to calculate the diameter of a star.

18.4 The **H–R diagram** is a plot of stellar luminosity as a function of temperature. Most stars lie on the **main sequence,** which extends diagonally across the H–R diagram from high temperature and high luminosity to low temperature and low luminosity. Main-sequence stars derive their energy from the fusion of hydrogen to helium, and the position of a star along the main sequence is determined by its mass. About 90 percent of the stars in the solar neighborhood lie on the main sequence. Only 10 percent of the stars are **white dwarfs,** and fewer than 1 percent are **giants** or **supergiants.**

REVIEW QUESTIONS

1. How does the intrinsic luminosity of the Sun compare with that of the 30 brightest stars? With that of the stars within 15 LY?

2. Name and describe the three types of binary systems.

3. Describe three ways of determining the diameter of a star.

4. What are the extreme values of the mass, luminosity, temperature, and diameter of stars?

5. Sketch an H–R diagram. Label the axes. Show where cool supergiants, white dwarfs, the Sun, and main-sequence stars are to be found.

THOUGHT QUESTIONS

6. Why do most visual binaries have relatively long periods and most spectroscopic binaries relatively short periods? Under what circumstances could a binary with a relatively long period (over a year) be observed as a spectroscopic binary?

7. Figure 18.7 shows the light curve of a hypothetical eclipsing binary star in which the light of one star is completely blocked by another. What would the light curve look like for a system in which the light of the smaller star is only partially blocked by the larger one? Assume the smaller star is the hotter one. Sketch the relative positions of the two stars that correspond to various portions of the light curve.

8. There are fewer eclipsing binaries than spectroscopic binaries. Explain why. Within 50 LY of the Sun, visual binaries outnumber eclipsing binaries. Why? Which is easier to observe at large distances—a spectroscopic binary or a visual binary?

9. The eclipsing binary Algol drops from maximum to minimum brightness in about 4 h, remains at minimum brightness for 20 min, and then takes another 4 h to return to maximum brightness. Assume that we view this system exactly edge-on, so that one star crosses directly in front of the other. Is one star much larger than the other, or are they fairly similar in size?

10. Consider the following data on five stars:

Star	Apparent Magnitude	Spectrum
1	12	G, main sequence
2	8	K, giant
3	12	K, main sequence
4	15	O, main sequence
5	5	M, main sequence

a. Which is hottest?
b. coolest?
c. most luminous?
d. least luminous?
e. nearest?
f. most distant?
In each case, give your reasoning.

11. Suppose you wanted to search for main-sequence stars with very low mass with a space telescope. Would you design your telescope to detect light in the ultraviolet or in the infrared part of the spectrum. Why?

12. It is possible to construct the equivalent of an H–R diagram for human beings by plotting height against weight. Try doing so for your classmates; obtain additional information for children and babies. Do all combinations of height and weight occur? Can you think of special examples of human beings that deviate from normal relationships?

13. Several very bright stars are identified in Figure 18.9. Select some that are visible at this time of year. What color would you expect each to be? Go outside at night and find these stars. Do they actually have approximately the color that you expected?

14. Approximately 6000 stars are bright enough to be seen without a telescope. Are any of these stars white dwarfs? Use the information given in this chapter and explain your reasoning.

PROBLEMS

15. Plot the luminosity functions of the nearest stars (Appendix 13) and of the 20 brightest stars (Appendix 14). Explain how and why these two luminosity functions differ.

16. Find the combined mass of two stars in a binary system whose period of mutual revolution is two years and for which the semimajor axis of the relative orbit is 2 AU.

17. What is the radius of a star (in terms of the Sun's radius) with the following characteristics:
  a. Twice the Sun's temperature and four times its luminosity?

  b. Eighty-one times the Sun's luminosity and three times its temperature?

18. Two stars, A and B, appear equally bright when their radiation at all wavelengths is measured. They also have identical diameters. Yet, star A has twice the surface temperature of star B. Which is more distant, and by what factor?

19. Verify that a red dwarf with 1/12 the mass of the Sun and 1/10 the radius of the Sun has a density that is 80 times that of the Sun. Calculate the density of the red supergiant described in Section 18.4, which has a mass 50 times that of the Sun and a radius 400 times that of the Sun. The outer parts of such a star would constitute an excellent laboratory vacuum.

20.   Suppose you weigh 70 kg on the Earth. How much would you weigh on the surface of a white dwarf star the same size as the Earth but having a mass 300,000 times larger than the mass of the Earth (nearly the mass of the Sun)?

This region contains some of the most colorful interstellar clouds ever photographed. The bluish nebulae reflect the light of hot stars. The reddish nebulae glow with the light of hydrogen emission. *(Copyright ROE/AAT Board)*

Bengt Georg Daniel Strömgren (1908–1987), Danish astronomer, spent much of his productive career in the United States, especially as Professor and Director of the Yerkes Observatory (University of Chicago). He received many honors for his fundamental work in the study of the structure and evolution of stars, and especially for his pioneering investigation of the physics of the interstellar medium. *(John B. Irwin)*

19

BETWEEN THE STARS: GAS AND DUST IN SPACE

By earthly standards the space between the stars is empty. No laboratory on Earth can produce so complete a vacuum. Yet this "emptiness" contains vast clouds of gas and tiny solid particles. Some of these clouds are visible and are called **nebulae** (Latin for "clouds"). Other clouds can be detected by the energy they emit at infrared or radio wavelengths. Still other clouds make their existence known through their effect on the light that passes through them.

19.1 The Interstellar Medium

The conditions in the tenuous matter between the stars, which astronomers refer to as the **interstellar medium,** vary widely. There are dense clouds with temperatures as low as 10 K and low-density regions with temperatures of a million degrees. The interstellar medium is dynamic. Clouds form, collide, coalesce, and fragment to form stars. Understanding interstellar matter is critical to understanding where and how stars form.

The primary components of the interstellar medium are gas and dust (Figure 19.1). The gas is composed mainly of hydrogen and helium. About 1 percent by mass of interstellar material is in the form of solid material, frozen particles of dust that are sometimes called **interstellar grains.**

Interstellar material is concentrated between the stars in the spiral arms of our own and other galaxies. The density of the interstellar matter in the arms of our Galaxy in the neighborhood of the Sun, for example, is estimated to be 3 to 20 times larger than in the regions between the arms. The gas and dust are not distributed smoothly throughout the spiral arms. Rather, they have a patchy, irregular distribution, being denser in some areas than in others, hence forming "clouds." In the spiral arms, on the average, there is about one atom of gas per cubic centimeter in interstellar space. In addition to the gas, there are in each cubic kilometer a few hundred to a few thousand tiny particles or dust grains, each less than a micrometer (a thousandth of a millimeter) in diameter. In some clouds, the density of gas and dust may exceed the average by as much as a thousand times or more, but even this density is more nearly a vacuum than any attainable on Earth. In contrast, air at sea level contains about 10^{19} molecules per cubic centimeter. Indeed, there is more gas in the Earth's atmosphere in a hypothetical vertical tube with a cross-section of 1 m^2 than would be encountered by extending that same tube from the top of the atmosphere all the way to the edge of the observable universe, which is 10 to 15 billion LY away.

While the density of interstellar matter may be very low, its total mass is substantial. To see why, remember that stars occupy only a tiny fraction of the vol-

FIGURE 19.1   Clouds of luminous gas and opaque dust, such as the nebulosity NGC 3603, are found between the stars. *(Anglo-Australian Telescope Board)*

FIGURE 19.2   NGC 3576 and NGC 3603. Clouds of luminous gas surround compact clusters of very hot stars. These hot stars ionize the hydrogen in the gas clouds. When electrons then recombine with protons and move back down to the lowest energy orbit, emission lines are produced. The strongest line in the visible region is in the red part of the spectrum and is responsible for the color of the photograph. *(Anglo-Australian Telescope Board)*

ume of the Milky Way Galaxy. For example, it takes only about 2 s for light to travel a distance equal to the radius of the Sun but more than four years to travel from the Sun to the nearest star. Although the density of gas and dust surrounding the Sun and lying between it and the nearest stars is small, the volume of space filled by this low-density material is so large that the total mass of gas and dust in the Milky Way Galaxy is equal to about 5 percent of the mass contained in stars. The total mass of interstellar gas and dust in our Galaxy therefore amounts to several billion times the mass of the Sun.

19.2   Interstellar Gas

Some of the most spectacular astronomical photographs (Figure 19.2) show interstellar gas located near hot stars. This gas is heated to temperatures close to 10,000 K by the nearby stars and glows because it emits strong lines of hydrogen. Hydrogen makes up about three quarters of the mass of the interstellar gas, and hydrogen and helium together compose 96 to 99 percent of it.

H II Regions—Gas Near Hot Stars

Interstellar gas near very hot stars is ionized by the ultraviolet radiation from those stars. Since hydrogen is the main constituent of the gas, we often characterize a region of interstellar space according to whether its hydrogen is neutral or ionized. A cloud of ionized hydrogen is called an **H II region.** (Spectroscopists use the Roman numeral I to indicate that an atom is neutral; successively higher Roman numerals are used for each higher stage of ionization. H II refers to hydrogen that has lost its electron.)

Ultraviolet radiation of wavelength 91.2 nm or less can be absorbed by neutral hydrogen. When a hydrogen atom absorbs this radiation, it is stripped of its electron and is ionized (Section 5.5). An appreciable fraction of the energy emitted by the hottest (spectral types O and B) stars lies at wavelengths shorter than 91.2 nm. If such a star is embedded in a cloud of interstellar gas, the ultraviolet radiation from that star ionizes the hydrogen in the gas, converting it into positive hydrogen ions (protons) and free electrons. The detached protons in the gas are continually colliding with free electrons and capturing them, becoming neutral hydrogen again.

The capture of electrons leads to the emission of light by the process of **fluorescence.** As the captured electrons cascade down through the various energy levels of the hydrogen atoms on their way to the lowest energy levels, or ground states, they emit light in the form of emission lines. Lines belonging to all the series of hydrogen (Section 5.5) are emitted—the Lyman series, Balmer series, Paschen series, and so on—but the lines of the Balmer series are most easily observed from the surface of the Earth because our atmosphere blocks the light from the Lyman and most other hydrogen series lines. Color photographs of H II regions appear red because the strongest of the Balmer lines falls in the part of the electromagnetic spectrum that the eye sees as red (Figure 19.2).

When a proton in an H II region captures an electron, light is emitted as the electron cascades down to lower energy levels. The proton then loses that electron again almost immediately by the subsequent absorption of another ultraviolet photon from the star. Thus, although neutral hydrogen is responsible for absorbing and emitting light in H II regions, almost all the hydrogen, at any given time, is in the ionized state.

The interstellar gas, of course, contains other elements besides hydrogen. Many of them are also ionized in the vicinity of hot stars and are capturing electrons and emitting light, just as hydrogen does.

Neutral Hydrogen Clouds

While ionized hydrogen gas is the type of interstellar matter that is most often photographed, observations show that it is not the most abundant form of interstellar gas. The very hot stars required to produce H II regions are rare, and only a small fraction of interstellar matter is close enough to such hot stars to be ionized by them.

Interstellar gas located at large distances from stars does not produce the strong emission lines that make H II regions visible. A cold cloud of gas will, however, produce dark absorption lines in the spectrum of light from a star that lies behind it. The first evidence for absorption by interstellar clouds came from the analysis of a spectroscopic binary star. While most of the lines in the spectrum shifted alternately from longer to shorter wavelengths and back again, as one would expect from the Doppler effect for one star in orbit around another, a few lines in the spectrum did not vary in wavelength. Subsequent work showed that these lines were formed in a cold cloud of gas located between us and the binary star.

The most conspicuous optical interstellar lines are produced by sodium and calcium. Molecular bands produced by CN, CH, and CH$^+$ are seen. Ultraviolet observations made with orbiting telescopes have de-

tected lines of carbon, hydrogen, oxygen, nitrogen, and other elements, of molecular hydrogen, and of CO (carbon monoxide).

The strengths of interstellar lines lead to estimates of the relative abundances of the elements that produce them. For some elements, such estimates are about the same as their relative abundances in the Sun and other stars. For other elements, the relative abundance is noticeably lower. Interstellar gas has especially low abundances of elements that readily condense into solids (notably aluminum, calcium, and titanium, as well as iron, silicon, and magnesium). As we shall see in Section 19.4, it is likely that many of the atoms of these elements have indeed combined to form tiny solid grains of interstellar dust.

Radio Observations of Cold Clouds: The 21-cm Line

Radio observations of the spectral line of hydrogen at a wavelength of 21 cm have provided critical information about the sizes and locations of clouds of cold interstellar gas. A hydrogen atom possesses a tiny amount of angular momentum by virtue of the axial spin of its electron and the electron's orbital motion about the nucleus (proton). In addition, the proton has an axial spin of its own. If the spins of the two particles are in opposite directions, the atom as a whole has a very slightly lower energy than if the two spins are aligned (Figure 19.3). If an atom in the lower energy state (spins opposed) acquires a small amount of energy, the spins of the proton and electron can be aligned, leaving the atom in a slightly *excited state.* If the atom then loses that same amount of energy again, it returns to its ground state. The amount of energy involved is that associated with a photon of 21-cm wavelength.

Neutral hydrogen atoms can be excited by collisions with electrons and other atoms. Such collisions are extremely rare in the sparse gases of interstellar space. An individual atom may wait many years before such an encounter aligns the spin of its proton and electron. Nevertheless, over many millions of years a good fraction of the hydrogen atoms are so excited. An excited atom can then lose its excess energy either by a subsequent collision or by radiating a photon of 21-cm wavelength. An excited atom will wait, on the average, about 10^7 years before emitting a photon and returning to its state of lowest energy. Despite this long wait, there is a definite chance that the atom will radiate away its energy before a second collision can carry away its energy of excitation.

Equipment sensitive enough to detect the 21-cm line of neutral hydrogen became available in 1951. Since that time many other radio lines produced by

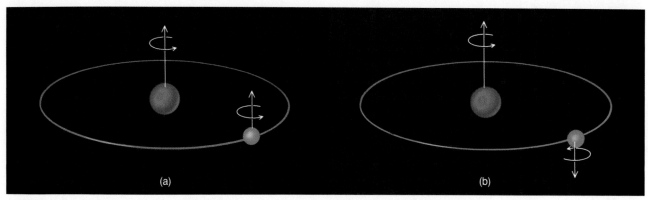

FIGURE 19.3 Formation of the 21-cm line. When the electron in a hydrogen atom is in the orbit closest to the nucleus, the proton and the electron may be spinning either in the same direction *(left)* or in opposite directions *(right)*. When the electron flips over, the atom either gains or loses a tiny bit of energy and either absorbs or emits electromagnetic energy with a wavelength of 21 cm.

both atoms and molecules have been discovered. Observations at 21 cm show that most of the neutral hydrogen in the Galaxy is confined to an extremely flat layer, which is less than 300 LY thick and extends throughout the plane of the Milky Way. Individual cold hydrogen clouds turn out to have temperatures of about 100 K and densities of 50 atoms per cubic centimeter. The diameters of cold clouds range from about 3 to 30 LY, and a light beam that travels 1000 LY through space in the plane of the Galaxy will encounter, on the average, about two of these clouds. About 2 percent of interstellar space is filled with cold clouds. The masses of cold clouds are typically in the range 1 to 1000 times the mass of the Sun, with clouds of low mass being the most common.

Not all of interstellar hydrogen is found in such cold clouds. A comparable amount of mass is found in warm clouds with temperatures of 3000 to 6000 K and typical densities of 0.3 hydrogen atom per cubic centimeter. The hydrogen in these clouds is not ionized. At least 20 percent of the space between stars in the plane of the Milky Way is filled with warm clouds of neutral hydrogen.

Hot Interstellar Gas

Before the launch of astronomical observatories into space, models of the interstellar medium assumed that most of the region between stars was filled with cool hydrogen. Astronomers were therefore surprised, when ultraviolet observations were made above the Earth's atmosphere, to discover interstellar lines at wavelengths of 103.2 and 103.8 nm. These lines are produced by oxygen atoms in the interstellar medium that have been ionized five times. To strip five electrons from their orbits around an oxygen nucleus requires a lot of energy. In fact, these observations imply that the temperature of the interstellar

medium where these atoms occur must be approximately 10^6 K. The density of hydrogen nuclei in these regions is typically a few times 10^{-3} per cubic centimeter, with large variations in density from place to place.

The hot interstellar gas is almost certainly heated by the explosive force of supernovae. Stars nearing the ends of their lives (Chapter 22) explode and send high-temperature gas, moving at velocities of thousands of kilometers per second, out into interstellar space. Astronomers estimate that there is about one supernova explosion every 25 years somewhere in the Galaxy. On the average, the hot gas from a supernova will sweep through any given point in the Galaxy about once every 2 million years. At this rate, the sweeping action is continuous enough to keep most of the space between clouds filled with gas at a temperature of a million degrees.

19.3   A Model of the Interstellar Gas

Table 19.1 summarizes the characteristics of the various types of clouds that populate interstellar space. There are cold, dense clouds in which hydrogen is not ionized. There are clouds so hot that molecules cannot survive and in which atoms are mainly ionized. The challenge for the theoretician is to assemble from the observations a model of the interstellar medium that tells us where we might expect to find the various types of clouds, what the structure of an individual cloud is like, and how the clouds change as time passes.

Structure and Distribution of Interstellar Clouds

One important requirement for a model of the structure of the interstellar medium is that the clouds and

TABLE 19.1   Interstellar Gas

Type of Region	Temperature (K)	Density (number/cm³)	Description
H I: cold clouds	10^2	50	Hydrogen atoms; distributed in clouds with typical diameter of 3–30 LY; fills 2 percent of interstellar space
H I: warm clouds	$3–6 \times 10^3$	0.3	Hydrogen not ionized; fills 20 percent of interstellar space
Hot gas	$10^5–10^6$	10^{-3}	Found well above and below as well as in galactic plane; hydrogen ionized; probably heated by supernova explosions
H II regions	10^4	$10^3–10^4$	Found near hot stars; hydrogen mostly ionized

the gas between the clouds must be at approximately the same pressure. Suppose they were not. If the cloud pressure were higher, the cloud would expand until its pressure matched that of its environment. If, on the other hand, the pressure of the hot gas were greater than that of a cloud embedded in it, the hot gas would compress the cloud and force it to shrink until its pressure became high enough to resist further compression.

The pressure in a gas is proportional to the temperature (T) and to the number of particles per cubic meter in the gas (n). Specifically, the pressure (P) of a gas can be calculated according to the formula

$$P = nkT$$

where k is a constant (and is equal to 1.38×10^{-23} joules deg^{-1}).

If the pressure of a gas cloud is to be equal to the pressure of the intercloud gas that surrounds it, then

$$\frac{n(\text{cloud})}{n(\text{intercloud})} = \frac{T(\text{intercloud})}{T(\text{cloud})}$$

In words, this equation says that if the gas pressures in these two regions are equal, then the region at higher temperature must have fewer particles per cubic centimeter, that is, it must have a lower density.

Figures 19.4 and 19.5 show in a schematic way the most widely accepted model of what interstellar clouds look like based on the requirement that gas pressures must be nearly the same everywhere. Individual clouds are scattered at random throughout the Galaxy. The typical cloud may be a few tens of light years in diameter. The clouds are embedded in gas with a temperature of a million degrees or so, and the

outer portions of the clouds are heated by conduction, that is, by the direct transfer of energy from the 10^6 K gas that surrounds the clouds. The temperature of the outer portion of a typical cloud is about 8000 K. If a cloud is large enough, it can shield its innermost core from being heated, and the core may have a temperature that is 100 times lower and a density correspondingly 100 times higher. Typical values for a cloud core are a temperature of 80 K and a density of 40 hydrogen atoms per cubic centimeter.

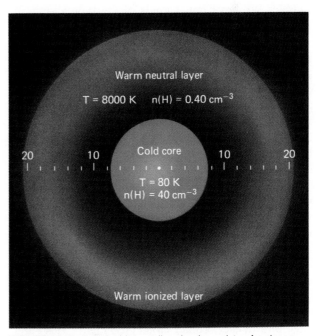

FIGURE 19.4   A typical interstellar cloud consists of a dense, cold core surrounded by a warm envelope. The horizontal scale shows the radius in light years.

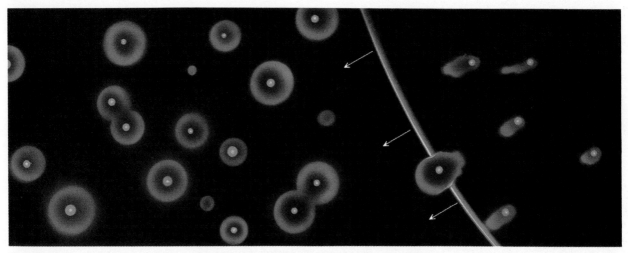

FIGURE 19.5   Interstellar clouds are embedded in hot, low-density gas, which is heated to temperatures as high as 10^6 K by supernova explosions. In the upper right, a supernova remnant is shown sweeping through interstellar space. *(Diagram is taken from work published in* The Astrophysical Journal *by C. McKee and J. Ostriker.)*

Interstellar Matter Around the Sun

The Sun is located in a region where the density of interstellar matter is unusually low. The temperature of interstellar matter in the vicinity of the Sun is about 10^6 K and its density is only 5×10^{-3} hydrogen atoms per cubic centimeter. This region of low-density gas is called the Local Bubble and extends to a distance of about 300 LY from the Sun.

If interstellar space near the Sun contained the normal number of clouds, we would expect to have detected approximately 2000 of them within the Local Bubble. We have not. Clouds of the type shown in Figure 19.4 are conspicuously absent.

We do not know whether conditions within the Local Bubble are typical of large regions of the Galaxy or whether some mechanism blew away all high-density interstellar gas. One possibility is that a supernova explosion in the past 10^5 to 10^7 years swept the region we now see as the Local Bubble nearly clean of interstellar clouds and at the same time heated the small amount of remaining gas to very high temperatures.

While typical cold hydrogen clouds are very rare within the Local Bubble, some clouds of lower density do exist. The Sun itself seems to be inside a cloud with a density of 0.1 hydrogen atom per cubic centimeter and a temperature of 10^4 K. This cloud is so tenuous that it is referred to as Local Fluff. We do see one sizable warm cloud in the direction toward the galactic center but within 60 LY of the Sun. It may be that the Local Fluff is the warm, partially ionized edge of a denser, cooler cloud (see Figure 19.4) and that the Sun is just entering this cloud.

Evolution of Interstellar Clouds

The model of interstellar gas described here presents a picture of the interstellar medium as it appears, on the average, at any given time. The individual clouds do, however, change with time.

We think that clouds are formed initially from the expanding gas around supernovae or hot stars. These clouds are relatively small and do not exceed 100 times the mass of the Sun. The clouds then grow through collisions with other clouds. Ultimately, this process may lead to the formation of giant clouds with diameters as large as 200 LY and masses that exceed 10^5 times the mass of the Sun. It is in these giant clouds that the most vigorous star formation occurs. The newly formed stars then evolve and become supernovae and in the process eject gaseous material, thus starting the cycle over again.

19.4 Cosmic Dust

Figure 19.6 shows a striking example of what is actually a common phenomenon—a dark region on the sky that appears nearly empty of stars. For a long time astronomers debated whether these dark regions were "tunnels" through which we looked beyond the stars of the Milky Way Galaxy into intergalactic space or, alternatively, were dark clouds that obscured the light of the stars beyond. We now know that the latter explanation is correct. Indeed, there are so many clues to the right answer that, with perfect hindsight, it is difficult to understand why astronomers debated the issue for so long.

FIGURE 19.6 A star cluster (NGC 6520) next to a dark cloud of interstellar matter. Old stars in our Galaxy are yellowish in color and form the brightest part of the Milky Way. Superimposed on these background stars in this picture is a dark cloud (Barnard 86), which is visible only because it blocks out the light from the stars beyond. Also in this picture, and possibly associated with the dark cloud, is a small cluster of young blue stars. *(Anglo-Australian Observatory)*

Dark Nebulae

The obscuration of starlight illustrated in Figure 19.6 is produced by a relatively dense cloud or *dark nebula* of tiny solid grains, which are commonly called interstellar dust. Opaque clouds are conspicuous on any photograph of the Milky Way (Figures 24.1 and 24.13). The "dark rift," which runs lengthwise down a long part of the Milky Way and appears to split it in two, is produced by a collection of such obscuring clouds.

While dust clouds are invisible in the optical region of the spectrum, they glow brightly in the infrared. Small dust grains absorb optical and ultraviolet radiation very efficiently. The grains are heated by the absorbed radiation, typically to temperatures between 20 and about 500 K, and reradiate this heat at infrared wavelengths. We can use Wien's law (Section 5.2) to estimate where in the electromagnetic spectrum this radiation will fall. For a temperature of 100 K, this maximum is at about 30 μm (1 μm = 1 micrometer = 10^3 nm), while grains as cold as 20 K will radiate most strongly near 150 μm. The Earth's atmosphere is opaque to radiation at these wavelengths, so emission by interstellar dust is best measured from space.

Observations from above the Earth's atmosphere by IRAS (the Infrared Astronomical Satellite) show that thermal emission from dust is seen throughout the plane of the Milky Way (Figure 19.7). The bright patches of emission have been given the name **infrared cirrus.** The closest infrared cirrus clouds are about 300 LY away.

Reflection Nebulae

The tiny interstellar grains absorb only a portion of the starlight they intercept. At least half of the starlight that interacts with a grain is merely scattered—that is, it is redirected helter skelter in all directions (Figure 19.8). Since neither the absorbed nor the scattered starlight reaches us directly, both absorption and scattering make stars look dimmer. The effects of both processes are termed **interstellar extinction.**

Some dense clouds of dust contain luminous stars within them and scatter enough starlight to become visible. Such a cloud of dust, illuminated by starlight,

FIGURE 19.7 The galactic plane as viewed in the infrared by IRAS. The color coding is such that the warmest dust appears blue and the coldest dust, red. The galactic center is located in the bright region just to the right of center. The wispy emission extending above and below the plane is produced by infrared cirrus. *(NASA)*

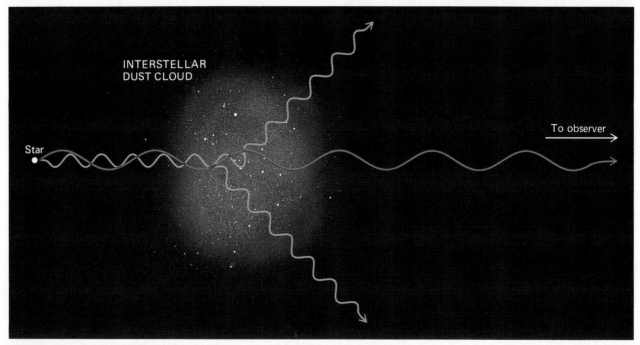

FIGURE 19.8 Interstellar dust scatters blue light more efficiently than red light, thereby making distant stars appear redder.

is called a **reflection nebula.** One of the best known examples is the nebulosity around each of the brightest stars in the Pleiades cluster (Figure 19.9). Blue light is scattered more than red by the dust. A reflection nebula, therefore, usually appears bluer than its illuminating star (Figure 19.10).

Gas and dust are generally intermixed in space, although the proportions are not everywhere exactly the same. The presence of dust is apparent on many photographs of emission nebulae (Figure 19.1). Spectra of H II regions often reveal the faint continuous spectrum (with absorption lines) of the central star, whose light is reflected to us by the dust associated with the gas. Stars cooler than about 25,000 K have so little ultraviolet radiation of wavelengths shorter than 91.2 nm (that is, which can ionize hydrogen) that the reflection nebulae around such stars outshine the emission nebulae. Stars hotter than 25,000 K emit enough ultraviolet energy so that the emission nebulae produced around them generally outshine the reflection nebulae.

Interstellar Reddening

Seventy years ago, astronomers were puzzled by the existence of stars whose spectral lines indicate that they are intrinsically hot and blue, although they are observed to have the red colors of much cooler stars. We know today that the light from these stars is not only dimmed but also **reddened** by interstellar dust.

Most of their violet, blue, and green light has been obscured, but some of their orange and red light, of longer wavelengths, penetrates the obscuring dust and reaches Earth-based telescopes.

We can estimate the total amount by which a star is dimmed from the amount that it is reddened. The extinction of the light from a star increases its apparent color index (the redder the star, the greater the color index—see Section 17.2). The difference between the *observed* color index and the color index that the star *would have* in the absence of obscuration and reddening is called the **color excess.** The $B - V$ color excess, for example, is the amount by which the difference between the blue and visual magnitudes of a star is increased by reddening. In most directions in the Galaxy, the total absorption in the V magnitude is found empirically to be about three times the $B - V$ color excess.

The fact that light is reddened by interstellar dust means that long-wavelength radiation is transmitted more efficiently than short-wavelength radiation. Consequently, if we wish to see farther, we should look at long wavelengths. This simple fact provides one of the motivations for the development of infrared astronomy. In the infrared region at 2 μm (2000 nm), for example, the obscuration is only one-sixth as great as in the visible region (500 nm), and we can therefore study stars that are more than twice as distant before their light is blocked by interstellar dust. This ability to see farther by observing in the

FIGURE 19.9 The Pleiades open star cluster. This cluster contains hundreds of stars and is located about 400 LY from the Sun. The nebulosity is starlight reflected by interstellar dust in a cloud that happens to be passing through the cluster at the present time. *(Anglo-Australian Telescope Board)*

infrared portion of the spectrum represents a major gain for astronomers who are trying to understand the structure of our Galaxy or to probe the galactic center (Chapter 24).

FIGURE 19.10 The Trifid Nebula (M20) in the constellation Sagittarius. In the reddish region, the hydrogen is ionized by nearby hot stars and glows through the process called fluorescence. The red light is produced by the Balmer emission line; the blue light is from a reflection nebula. The Trifid Nebula is about 30 LY in diameter and is about 3000 LY distant from the Sun. *(Kitt Peak National Observatory/National Optical Astronomy Observatories)*

Interstellar Grains

The preceding paragraphs have described the reddening and dimming of starlight by interstellar dust. Observations of these effects reveal something of the nature of the dust particles.

First, we know from the observations that the absorption of light is accomplished by *solid particles* and not by interstellar gas. Except for specific spectral lines, atomic or molecular gas is almost transparent. Consider the Earth's atmosphere. Despite its incredibly high density compared with that of interstellar gas, it is so transparent as to be practically invisible. The quantity of gas that would be required to produce the observed absorption in interstellar space would have to be many thousands of times the amount that can possibly exist. The gravitational attraction of so great a mass of gas would produce effects upon the motions of stars that would be easily detected. Such effects, however, are not observed.

Whereas gas can contribute only negligibly to absorption of light, we know from our everyday experience that tiny solid or liquid particles can be very efficient absorbers. Water vapor in the air is quite invisible. When some of that vapor condenses into tiny water droplets, however, the resulting cloud is opaque. Dust storms, smoke, and smog furnish other familiar examples of the efficiency with which solid particles absorb light.

On the basis of these arguments, we must conclude that widely scattered solid particles in interstellar space are responsible for the observed dimming of starlight. What are these particles made of? And how did they form? The answers to those questions are far from certain, but we can make a few general statements about what the grains must be like.

ESSAY Reddening in the Earth's Atmosphere

We have all seen an example of reddening. The Sun appears much redder at sunset than it does at noon because sunlight is *scattered* by molecules in the Earth's atmosphere. When the Sun is low in the sky, its light must traverse a greater distance through the atmosphere than when the Sun is high in the sky. Over this greater distance there is a higher probability that sunlight will be scattered. Since red light is less likely to be scattered than blue light, the Sun appears more and more red as it approaches the horizon.

Short-wavelength light is also *absorbed* more efficiently by the Earth's atmosphere than is long-wavelength light, and this fact has a practical consequence. It is much easier to get sunburned at noon than in the late afternoon, even though at 4:00 P.M., for example, the Sun feels nearly as hot as at noon. The reason for this is that tans and sunburns are caused primarily by sunlight with wavelengths between 280 and 320 nm. Sunlight at these short wavelengths is so efficiently absorbed by ozone in the Earth's atmosphere that very little of it penetrates the long distance that it must travel to reach the ground early in the morning and late in the afternoon. The heat that we feel, however, is produced mainly by infrared radiation, and these long wavelengths can reach the surface of the Earth even when the Sun is low in the sky.

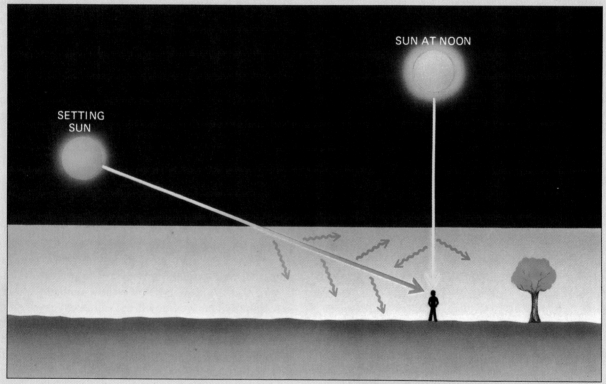

When light from the setting Sun traverses a long path through the Earth's atmosphere, blue light is scattered out of the direct light path, thereby making the Sun look redder.

From measurements of the densities of interstellar dust, that is, of the total amount of dust within a given volume of space, we know that about one of every two atoms heavier than helium must be locked up in the grains. The grains cannot, therefore, be made of rare elements but rather must be composed primarily of the most abundant elements in the universe. After hydrogen and helium, the most abundant elements are oxygen, carbon, and nitrogen. These three elements, along with magnesium, silicon, iron, and perhaps hydrogen, are thought to be the most important components of interstellar dust.

Observations support this line of reasoning. Many heavy elements, including iron, magnesium, and silicon, are less abundant in interstellar gas than they are in the Sun and young stars. These heavy elements are assumed to be missing from the interstellar *gas* because they are condensed into solid particles of interstellar *dust*.

We also know from measurements of both total extinction and reddening that typical individual grains must be slightly smaller than the wavelength of visual light. If the grains were much smaller than the wavelength of light, they would not block it efficiently. For example, a bowling ball, which is much smaller than the wavelength of radio waves, does not keep radio emission from reaching us. On the other hand, if the grains were much larger than the wavelength of optical radiation, then starlight would not be reddened. A bowling ball, which is much larger than the wavelengths of optical radiation, blocks both blue and red light with equal efficiency. A characteristic interstellar dust grain contains 10^6 to 10^9 atoms and has a diameter of 10^{-7} to 10^{-8} m (10 to 100 nm).

Observations of absorption features produced by interstellar material indicate that there are many types of solid particles. Some interstellar grains apparently consist of a core of rock-like material (silicates), including such common minerals as olivine and enstatite, both frequently found in terrestrial igneous rocks and in meteorites. Other grains appear to be nearly pure carbon (graphite). The nuclei of the grains are probably formed in shells of cooling gas ejected by red giants and other stars, including some novae and even possibly supernovae, that are nearing the end of their evolution (Chapter 22). These grain nuclei may then subsequently be incorporated into an interstellar cloud, where they can grow by accreting other atoms. The most widely accepted model pictures the grains as consisting of rocky cores with icy mantles (Figure 19.11). The most common ices are water (H_2O), methane (CH_4), and ammonia (NH_3).

Several processes tend to destroy the grains. If a high-energy ion or photon collides with a grain, it can knock off atoms, and this process is called *sputtering*. If a grain is heated to too high a temperature, atoms can evaporate from its surface. The temperatures required are not very high. In a typical interstellar cloud, the vaporization temperature is 20 K for CH_4, 60 K for NH_3, and 100 K for H_2O. Finally, if two grains collide at a speed of several kilometers per second, both will vaporize. Ultimately, those grains that survive may be incorporated into a new generation of stars, where they will be broken apart by the star's heat into individual atoms to begin the cycle over again.

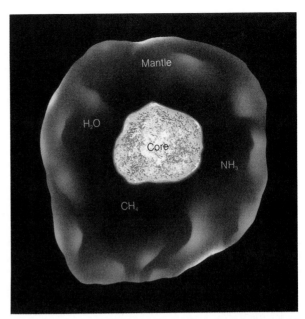

FIGURE 19.11 A typical interstellar grain is thought to consist of a core of rocky material (silicates), graphite, or possibly iron surrounded by a mantle of ices. Typical grain sizes are 10^{-7} to 10^{-8} m.

19.5 Cosmic Rays

In addition to gas and dust, a third class of particles is found in interstellar space. These particles or **cosmic rays** resemble ordinary interstellar gas in terms of composition. In every other respect, however, cosmic rays differ from interstellar gas.

Composition of Cosmic Rays

Cosmic rays are high-speed atomic nuclei, electrons, and positrons. Velocities equal to 0.9 times the speed of light are typical. Most cosmic rays are hydrogen nuclei that have been stripped of their accompanying electron. Helium and heavier nuclei constitute about 9 percent of the cosmic-ray particles. The number of cosmic-ray electrons is only about 2 percent the number of protons. Ten to 20 percent of the cosmic-ray particles with masses equal to the mass of the electron carry positive charge rather than the negative charge that characterizes electrons. Such particles are called **positrons.**

The abundances of various atomic nuclei in cosmic rays mirror the abundances in stars and interstellar gas, with one important exception. The light elements lithium, beryllium, and boron are far more abundant in cosmic rays than in the Sun and stars. These light elements are formed when high-speed cosmic-ray nuclei of carbon, nitrogen, and oxygen collide with protons in interstellar space and break apart.

Cosmic rays reach the Earth in substantial numbers, and we can determine their properties either by capturing them directly or by observing the reactions that occur when cosmic rays collide with atoms in the Earth's atmosphere. The total energy deposited by cosmic rays in the Earth's atmosphere is only about one-billionth the energy received from the Sun but is comparable to the total energy received in the form of starlight. Some of the cosmic rays come to the Earth from the surface of the Sun. Most come from outside the solar system.

Origin of Cosmic Rays

There is a serious problem in identifying the source of cosmic rays. Since light travels in straight lines, we can tell where it comes from simply by looking. Cosmic rays are charged particles, and their direction of motion can be changed by magnetic fields. The paths of cosmic rays are curved both by magnetic fields in interstellar space and by the Earth's own field. Calculations show that cosmic rays may spiral many times around the Earth before entering the Earth's atmosphere, where we can detect them. If an airplane circles an airport many times before landing, it is impossible to determine the direction and city from which it came. So, too, after a cosmic ray orbits the Earth several times, it is impossible to determine where its journey began.

There are a few clues, however, about where cosmic rays might be generated. We know, for example, that magnetic fields in interstellar space are strong enough to keep all but the most energetic cosmic rays from escaping the Galaxy. It therefore seems likely that cosmic rays are produced somewhere inside the Galaxy. The only likely exceptions are the cosmic rays with the very highest energy. Since such cosmic rays move at speeds so great that they are not influenced by interstellar magnetic fields, they could escape our Galaxy. By analogy, they could escape other galaxies as well, so some of the highest energy cosmic rays that we detect may have been created in some distant galaxy.

We can also estimate how far typical cosmic rays travel before they strike the Earth. The light elements lithium, beryllium, and boron hold the key. Since they are formed when carbon, nitrogen, and oxygen strike interstellar protons, we can calculate how long, on average, cosmic rays must spend traveling through space for enough collisions to occur to account for the relative abundances of lithium and the other light elements. It turns out that the required distance is about 30 times around the Galaxy. At speeds of 0.9 times the speed of light, it takes perhaps 3 to 10 million years for the average cosmic ray to travel this distance. Since cosmic rays are only a few million years old and have not been around since the beginning of the universe, they must have been created fairly recently.

The best candidates for a source of cosmic rays are supernova explosions, which mark the deaths of massive stars. There are enough explosions, and the explosions generate enough energy, to account for the observed number of cosmic rays. What we do not know is precisely the mechanism for using the energy involved in these explosions to accelerate protons and other atomic nuclei.

SUMMARY

19.1   About 5 percent of the visible matter in the Galaxy is in the form of gas and dust. The material in the **interstellar medium** is distributed in a patchy way, sometimes forming glowing clouds of gas called **nebulae**. The most abundant elements in the interstellar gas are hydrogen and helium. About 1 percent of the interstellar matter is in the form of solid **interstellar grains.**

19.2   Interstellar gas near hot stars emits light by **fluorescence.** These glowing clouds of ionized hydrogen are called **H II regions** and have temperatures of about 10,000 K. Most hydrogen in interstellar space is not ionized and can best be studied by radio measurements of the 21-cm line. About 2 percent of interstellar space is filled with cold (100 K) clouds with densities of 50 atoms per cubic centimeter. Another 20 percent of the space between stars is filled with warm (3000 to 6000 K) neutral hydrogen clouds at a density of 0.3 atom per cubic centimeter. Some gas has been heated by supernova explosions to temperatures of 10^6 K.

19.3   Observations of interstellar gas are consistent with a model in which this gas is distributed in the form of clouds with cold, high-density cores surrounded by warm, lower density envelopes, all embedded in a hot gas of very low density. The pressure in these three types of regions is approximately the same. The Sun is located at the edge of a low-density (0.1 atom per cubic centimeter; $T = 10^4$ K) cloud called the Local Fluff. The Sun and this cloud are located within the Local Bubble, which is a region extending to about 300 LY from the Sun, where the density of interstellar material is extremely low.

19.4   Interstellar dust grains absorb and scatter starlight. The effects of both processes are referred to as **interstellar extinction** and cause distant stars to appear fainter than they would if no dust grains lay along the path traversed by the starlight. Much of the dust is found in clouds called **infrared cirrus.** Since typical temperatures of the dust are 20 to 500 K, interstellar grains are best observed in the infrared region. Because dust grains scatter blue light more efficiently than red light, stars seen through dust appear **reddened.** The **color excess** provides a quantitative measure of this reddening. Interstellar grains typically have sizes comparable to that of the wavelength of light. They probably have a core of silicates or carbon surrounded by a mantle of such ices as water, ammonia, and methane.

19.5 Cosmic rays are particles that travel through interstellar space at typical speeds of 90 percent the speed of light. The most abundant elements in cosmic rays are the nuclei of hydrogen and helium, but **positrons** are also found. It is likely that many cosmic rays are produced in supernova explosions.

REVIEW QUESTIONS

1. Identify several dark nebulae in photographs in this book. Give the figure numbers of the photographs and specify where the dark nebulae are to be found on them.

2. Why do nebulae near hot stars look red?

3. Describe the characteristics of the various types of interstellar gas clouds.

4. Prepare a table listing the ways in which **(a)** dust and **(b)** gas can be detected in interstellar space.

5. Describe the properties of the dust grains that are found in the space between stars.

6. Why is it difficult to determine where cosmic rays come from?

THOUGHT QUESTIONS

7. Suppose a bright reflection nebula appears yellow. What kind of star probably is producing it?

8. Describe the spectrum of **(a)** starlight reflected by dust; **(b)** a star behind invisible interstellar gas; **(c)** an emission nebula.

9. One way to calculate the size and shape of the Galaxy is to estimate the distances to faint stars from their observed apparent magnitudes and to note the distance at which stars are no longer observable. The first astronomers to try this experiment did not know that starlight is dimmed by interstellar dust. Their estimates for the size of the Galaxy were much too small. Explain why.

PROBLEMS

10. Suppose that the average density of hydrogen gas in our Galaxy is one atom per cubic centimeter. If the Galaxy is a sphere with a diameter of 100,000 LY, how many hydrogen atoms are in the interstellar gas? What is the mass of this quantity of hydrogen?

11. According to the text, there is about one atom of hydrogen in every cubic centimeter of interstellar space and perhaps 10^3 dust grains per 10^9 m^3. If approximately 1 percent of the total mass of the interstellar medium is in the form of dust grains, what is the typical mass of a dust grain?

12. Suppose the density of a typical dust grain is 3 g/cm^3, and its mass is the value found in Problem 11; what is its radius? (Assume the grain is spherical and remember that the volume of a sphere is given by the formula $V = (4/3)\pi R^3$.)

13. A spectral-type A star normally has a color index of 0.0. The blue and visual magnitudes of an A star are observed to be $B = 11.6$ and $V = 10.8$, respectively.

a. What is the color excess of the star?
b. What is the total absorption of its light in visual magnitudes?
c. What is the total absorption of its light in blue magnitudes?

14. The Sun is observed from a distant star to have an apparent blue magnitude of 14.4 and an apparent visual magnitude of 12.8. How far away, approximately, is the star? Assume that the Sun's color index is $B - V = 0.6$ and that its absolute visual magnitude is $+4.8$.

15. According to the model given for the interstellar medium, the cold clouds, warm clouds, and hot gas listed in Table 19.1 should all have about the same pressure. Use the numbers in the table to show that this is true. (Note that the pressure in the H II region is much higher. In fact, H II region are dynamic regions that are not in pressure balance with normal interstellar matter.)

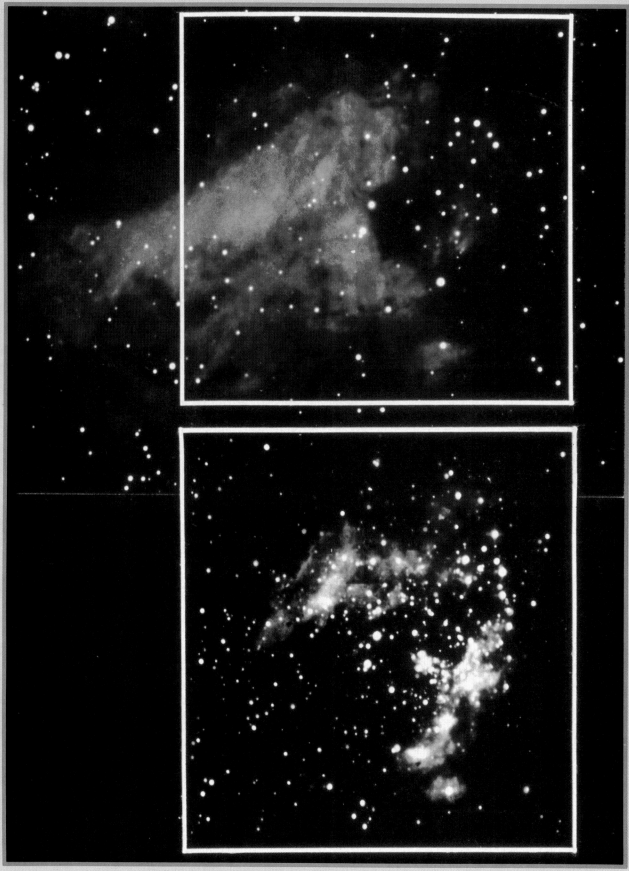

Optical (*top*) and infrared (*bottom*) pictures of M17, a molecular cloud associated with a region of vigorous star formation. Comparison of the optical and infrared images reveals that there is a rich cluster of newly formed stars within the dust cloud. *(Ian Gatley and Charles Lada/ National Optical Astronomy Observatories)*

Frank Drake (b. 1930), an American astronomer, pioneered in the search for extraterrestrial intelligence. His Project Ozma (about 1960) was the first organized attempt to detect radio signals from extraterrestrial civilizations. The project (as expected) did not succeed, but the continuing advocacy by Drake, Carl Sagan, and others has since led to much more realistic and powerful searches for evidence of other intelligent creatures.

THE BIRTH OF STARS

It is natural to think of the stars as fixed, permanent, and unchanging. Yet stars are radiating energy at a prodigious rate, and no source of energy can last forever. For example, deep in the interior of the Sun, 600 million tons of hydrogen are converted to helium every second, with the simultaneous conversion of about 4 million of those tons to energy. At this rate the Sun is able to radiate energy for about 10 billion years.

Stars more massive and more luminous than the Sun exhaust their fuel supply much more rapidly than does the Sun. The most massive stars have only 50 to 100 times the mass of the Sun, yet their luminosities—and correspondingly the rate at which they consume their supply of hydrogen—are a million times greater. Accordingly, these massive stars must exhaust their fuel supply, burn themselves out, and become unobservable in no more than a few million years. The brightest hot star in Orion—Rigel—cannot have been shining when the first human-like creatures walked the Earth.

Astronomers estimate that there are more than 10,000 of these very luminous young stars in our Galaxy—stars with lifetimes that are measured in only millions of years. If, as seems likely, such highly luminous stars have been present throughout the billions of years that our Galaxy has existed, then as these stars die, they must be replaced by new ones. On average, in fact, one new bright star must be formed somewhere in our Galaxy every 500 to 1000 years if the total number of highly luminous stars is to remain approximately constant. And for every such luminous star that is formed, there are many others of more modest mass and luminosity.

As we shall see in this chapter, star formation is a continuous process that is going on *right now*. Stars of all masses, low as well as high, are being formed, and that formation process is taking place in the interiors of clouds of dust and gas, which provide the necessary raw material.

20.1 Star Formation

If we want to find the very youngest stars—stars still in the process of formation—we must look in places where there is plenty of the raw material required to make stars. Stars are made of gas, so we must look in dense clouds of gas (Figure 20.1).

Molecular Clouds: Stellar Nurseries

The most massive clouds—indeed the most massive objects in the Milky Way Galaxy—are the **giant molecular clouds.** Molecular clouds have masses equal to 100 to 1 million times the mass of the Sun, and the diameter of a typical cloud is 50 to 200 LY. Molecular clouds contain both gas and dust, but their interiors

351

FIGURE 20.1 Stars form in clouds of gas and dust. M16 is a cluster of stars that formed about 2 million years ago. The dark areas visible across the face of the nebula are thought to be condensations of material that might one day collapse into yet more stars. *(Anglo-Australian Observatory Board)*

are colder than is typical of most other interstellar clouds. Characteristic temperatures are about 10 K, so most atoms are bound into molecules. Observations show that most stars are born in giant molecular clouds.

More than 60 kinds of molecules have been identified in the giant molecular clouds. Atoms of hydrogen, oxygen, carbon, nitrogen, and sulfur make up molecules of H_2, water, carbon monoxide, ammonia, hydrogen sulfide, and such common organic molecules as formaldehyde and hydrogen cyanide. Relatively heavy molecules, such as HC_9N, are found in some cold clouds, and ethyl alcohol, C_2H_5OH, is quite plentiful—up to one molecule for every cubic meter. The largest of the cold clouds have enough ethyl alcohol to make 10^{28} fifths of 100-proof liquor. Spouses of future interstellar astronauts, however, need not fear that their wives or husbands will become interstellar alcoholics. Even if a spaceship were equipped with a giant funnel 1 km across and could scoop it through such a cloud at the speed of light, it would take about a thousand years to gather up enough alcohol for one standard martini!

The cold interstellar clouds also contain cyanoacetylene (HC_3N) and acetaldehyde (CH_3CHO), generally regarded as starting points for the formation of amino acids necessary for living organisms. The presence of these organic molecules does not, of course, imply the existence of life in space. On the other hand, as we learn more about the processes by which they are produced, we gain an increased understanding of similar processes that must have preceded the beginnings of life on the primitive Earth billions of years ago.

The molecule H_2 and most of the more complex molecules are dissociated by shortwave stellar radiation. These molecules can survive only in places where they are shielded from ultraviolet starlight. The dense, dark giant molecular clouds contain dust, which acts like a thick blanket of interstellar smog and keeps ultraviolet starlight from penetrating the interior of the cloud. That is why these molecules are

FIGURE 20.2 The nearest molecular cloud is in Orion. The extensive network of very faint filaments, which are traceable over most of the constellation of Orion, is optical evidence of a substantial dark cloud of molecular gas and dust. At radio wavelengths, this cloud fills most of the field pictured here. The bright nebula in the lower right contains the Trapezium stars and is a region of active star formation. The Horsehead Nebula appears in the upper left. *(Anglo-Australian Observatory Board)*

force the star to expand. When these two forces are in balance, the star is stable. Major changes in the structure of a star occur when one or the other of these two forces gains the upper hand. Low temperature and high density both work to give gravity the advantage and so facilitate the star formation process.

The Orion Molecular Cloud

The best studied of the stellar nurseries is the Orion region (Figure 20.2). A luminous cloud of dust and gas can be seen with binoculars in the middle of the sword in the constellation of Orion. Associated with the luminous material is a much larger molecular cloud, which is invisible in the optical region of the spectrum. Near the center of the optically bright region of the nebula is the Trapezium cluster of very luminous O-type stars (Figure 20.3). An infrared picture of this same region shows hundreds of stars, invisible on photographs taken at optical wavelengths (Figure 20.4).

The long dimension of the Orion molecular cloud stretches over a distance of about 100 LY. The total quantity of molecular gas is about 200,000 times the mass of the Sun. Star formation began about 12 million years ago at one edge of this molecular cloud near the right-hand (western) shoulder of Orion and has slowly moved through the molecular cloud, leaving

abundant in the interiors of molecular clouds but not elsewhere in the interstellar medium.

Molecular clouds are not smooth but contain clumps or *dense cores* of material that have a very low temperature (10 to 50 K) and a density much higher (10^4 to 10^5 atoms per cubic centimeter) than is typical of most of the rest of the interstellar medium. Both of these conditions—low temperature and high density—are favorable to the star formation process. If a star is to form, it is necessary that a region with a mass comparable to that of a star shrink in radius and increase its density by nearly a factor of 10^{20}, from that typical of a dense core to that of a star. This collapse is brought about by the force of gravity.

The story of stellar evolution is the story of the competition between two forces: *gravity* and *pressure*. As described in Chapter 15, the force of gravity tries to make a star collapse. Internal pressure produced by the motions of the gas atoms tries to

FIGURE 20.3 The central "star" of the three forming the sword of Orion is in fact a group of four stars known as the Trapezium cluster. These stars are easily visible with binoculars and are the brightest members of a substantial cluster hidden by the dust in the nebula. These stars are at a distance of about 1500 LY. *(Anglo-Australian Telescope Board, 1981)*

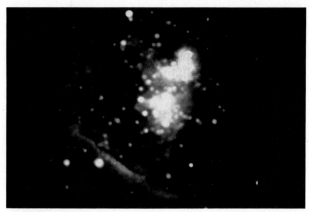

FIGURE 20.4 An infrared picture of the region of the Trapezium stars. Because infrared radiation penetrates the dust, infrared observations at a wavelength of 2.2 μm reveal the cluster of young stars within the molecular cloud. Compare this image with the optical photograph in Figure 20.3; note how many fewer stars are visible in that photograph. *(National Optical Astronomy Observatories)*

FIGURE 20.5 The Rosette Nebula. The cluster of blue stars at the center of the nebula formed less than a million years ago. The gas and dust have been driven away from the bright stars by radiation and intense stellar winds. *(Anglo-Australian Telescope Board)*

behind groups of newly formed stars. The stars in the belt are about 8 million years old, and the stars near the Trapezium are less than 2 million years old.

Star formation is not a very efficient process and uses typically only a few percent—at most perhaps 25 percent—of the gas in a molecular cloud. The leftover material is heated either by stellar radiation or by supernova explosions and blown away into interstellar space. The oldest groups of stars can therefore be easily observed optically because they are no longer shrouded in dust and gas (Figure 20.5).

Because of the correlation between stellar ages and position in the Orion region, we know that star formation has moved progressively through this molecular cloud. While we do not know what caused stars to

begin forming in Orion, there is good evidence that the first generation of stars triggered the formation of additional stars, which led to the formation of still more stars (Figure 20.6). The basic idea is as follows. When a massive star is formed, it emits copious amounts of ultraviolet radiation, which heats the surrounding gas in the molecular cloud. This heating increases the pressure in the gas and causes it to expand. When massive stars exhaust their supply of fuel, they explode, and the energy of the explosion

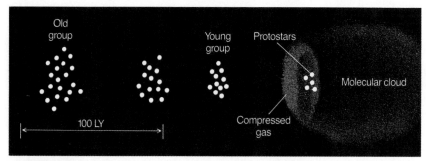

FIGURE 20.6 Schematic diagram showing how star formation can move progressively through a molecular cloud. The oldest group of stars lies to the left of the diagram and has expanded because of the motions of individual stars. Eventually the stars in the group will disperse and will no longer be recognizable as a cluster. The youngest group of stars lies to the right, next to the molecular cloud. This group of stars is only 1 to 2 million years old. The pressure of the hot, ionized gas surrounding these stars compresses the material in the nearby edge of the molecular cloud and initiates the gravitational collapse that will lead to the formation of more stars.

also heats the gas. The hot gases burst into the surrounding cold cloud, compressing the material in it until the cold gas is at the same pressure as the expanding hot gas. In the conditions typical of molecular clouds, the compression is enough to increase the gas density by a factor of 100. At densities this high, stars can begin to form in the compressed gas. This process seems to have occurred not only in Orion but also in many other molecular clouds.

The Birth of a Star

The earliest stages of star formation are still shrouded in mystery. There is almost a factor of 10^{20} difference between the density of a molecular cloud core and that of the youngest stars that can be detected. So far, we have been unable to observe directly what happens within a cloud as material comes together and collapses gravitationally through this range of densities to form a star. We do not know what causes a single large cloud to fragment and form individual stars, nor do we know why multiple-star systems and planets form.

Observations of this stage of stellar evolution are nearly impossible for several reasons. First, the interiors of molecular clouds where stellar births take place cannot be observed with visible light. The dust in these clouds acts like a thick blanket of interstellar smog, which cannot be penetrated by visible radiation (Figure 20.7). It is only with the new techniques of infrared and millimeter radio astronomy that we are able to make any measurements at all. Even so, the time scale for the initial collapse, which is measured in thousands of years, is very short, astronomically speaking. Furthermore, the collapse occurs in a region so small (0.3 LY) that in most cases we cannot resolve it with existing techniques. Accordingly, we have yet to catch a star in the act, so to speak, of its initial collapse. Nevertheless, through a combination of theoretical calculations and the limited observations that are available, astronomers have pieced together a picture of what the earliest stages of stellar evolution are likely to be.

During the period when a condensation of matter in a molecular cloud is contracting to become a true star, we call the object a **protostar.** Since a molecular cloud is more massive than a typical star by a factor of 100,000 or more, many protostars must form within each cloud. The first step in the process of creating stars is the formation within the cloud—through a process that we do not yet understand—of dense cores of material (Figure 20.8a). These cores then attract additional matter because of the gravitational force that they exert on the cloud material that surrounds them. Eventually, the gravitational force be-

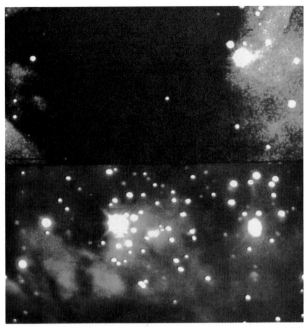

FIGURE 20.7 Images of part of the Orion molecular cloud. The upper image was taken in visible light and shows what appears to be an empty sky in the central region of the picture. This region is not empty but rather contains dust that obscures visible light. The lower image was made with an infrared detector at a wavelength of about 2.2 μm. Electromagnetic radiation at this long wavelength can penetrate the obscuring dust easily. Infrared observations provide a way to detect recently formed stars within the cloud. *(NOAO)*

comes strong enough to overwhelm the pressure exerted by the cold material that forms the dense cores. The material undergoes a rapid collapse, the density of the core increases greatly, and a protostar is formed.

Theory indicates that if the collapsing core is rotating and if its density is strongly peaked toward its center, then a flattened disk will form (Figure 20.8b; Section 20.1). The protostar and disk are embedded in an envelope of dust and gas, which is still falling onto the protostar. This dusty envelope blocks optical light, but infrared radiation can get through. As a result, protostars in this phase of evolution are observable only in the infrared region.

At some point, for reasons that we do not yet understand, observations show that the protostar develops a **stellar wind,** which consists mainly of protons streaming away from the star at velocities of about 200 km/s. This wind eventually sweeps away the obscuring envelope of dust and gas, leaving behind the disk and the protostar, which can now be seen at optical wavelengths (Figure 20.8c and 20.8d). At this point, the protostar is still contracting slowly and has not yet reached the main sequence.

There is observational evidence that most protostars do have disks around them, just as theory predicts. The circumstellar dust around *young stellar*

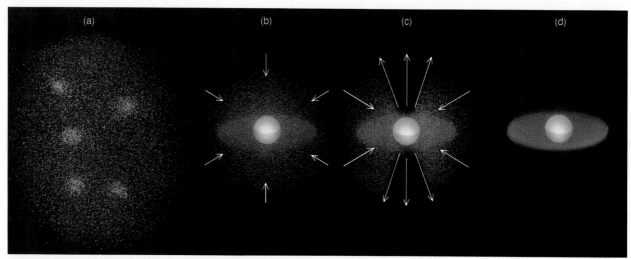

FIGURE 20.8 The formation of a star. (a) Dense cores form within a molecular cloud. (b) A protostar with a surrounding disk of material forms at the center of a dense core, accumulating additional material from the molecular cloud through gravitational attraction. (c) A stellar wind breaks out along the two poles of the star. (d) Eventually this wind sweeps away the cloud material and halts the accumulation of additional material, and a newly formed star surrounded by a disk becomes observable. *(Based on drawings by F. Shu, F. Adams, and S. Lizano)*

objects emits infrared radiation, and analysis of that radiation indicates that in most cases it comes from flattened disks with masses typically in the range of 1 to 10 percent of the mass of the Sun and sizes on the order of 100 AU—only a little more than twice the size of the solar system out to Pluto. In a few cases radio observations have actually detected a rotating disk of gas.

This description of a protostar surrounded by a rotating disk of gas and dust should sound very much like what happened when the Sun and planets formed (Section 12.2). Do the disks around protostars also form planets?

The Formation of Planets

As we have seen, observations suggest that most protostars with masses similar to the mass of the Sun have disks with masses greater than 1 percent of the mass of the Sun. Therefore, most protostars have more than enough material surrounding them to build a planetary system like the one that surrounds our own Sun.

Infrared observations have also been successful in detecting faint disks around a few nearby main-sequence stars (Figure 20.9). The key point, however, is that in every case the disks around main-sequence stars have much less mass—typically 100,000 times less mass—than do the disks around the much younger protostars.

Since protostellar disks start out with much more mass than ends up in the disks around main-sequence stars, it is natural to ask what happened to it. Specifi-

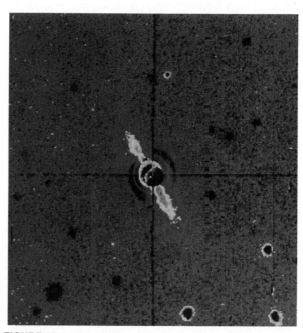

FIGURE 20.9 This optical photograph clearly shows a disk of solid material revolving around the star Beta Pictoris, although the disk is not a planetary system as we know the term. The light of the star itself has been blocked out to show the faint light of the disk more clearly. Evidence of this material was first picked up by IRAS (Infrared Astronomical Satellite), which discovered excess infrared radiation from this star. *(NASA/JPL)*

cally, was the mass that disappeared used to build planets? Unfortunately, we cannot observe planets directly. When the dust surrounding a protostar is distributed in a disk, each individual dust particle is heated by the protostar and radiates its heat in the infrared region. We detect the total radiation from all of the individual particles. If those particles then accrete and form a few planets, we detect only the radiation from the planetary surfaces. The overwhelming majority of the particles are hidden in the interior of the planets where we cannot see them. For this reason, the total radiation from a few planets is much less than the radiation emitted from all of the individual particles that combined to form those planets.

There is indirect evidence, however, that planets do indeed form. The disks surrounding a few main-sequence stars have been thoroughly studied, and the evidence is that the disk particles orbit the stars at large distances. That is, the disk is actually shaped like a 45 rpm record or a compact disk, with very few dust particles in the hole immediately surrounding the central star. One reason for thinking that there is a hole in the disk is that we can actually see the central star. If dust particles extended all the way to the star's surface, they would block the light from the star, making it invisible. Another reason is that these systems do not contain dust at the high temperatures that would be expected if the dust were located very close to the star.

Detailed analysis of the infrared radiation shows that the radius of the central hole is typically about 20 AU. The size of our own solar system is about 40 AU and so is rather similar. The most likely explanation for the fact that the disk does not reach all the way to the surface of the star is that the particles that originally filled the central hole have gradually accreted to form planets.

Observations suggest that it takes only about 10 million years to clear away the dust from the central regions of the disks that surround protostars with masses similar to that of the Sun. If this clearing occurs because the dust is being incorporated into planets, then planets form very quickly relative to the total stellar lifetimes. Remember that the Sun is about 4.5 billion years old and will survive for another 5 billion years.

20.2 The Search for Other Planetary Systems

On the basis of what we know about star formation, it seems likely that there are millions of Jupiters or Neptunes or even Earths surrounding the billions of stars in our Galaxy. Can we prove it? If such planets are indeed out there, can we detect them?

Imaging

If detection implies direct imaging, then no planet—even one as large as Jupiter—could be photographed with current telescopes. The difficulty with direct detection is that any planet will be extremely faint in comparison to its star. Suppose, for example, that you were at a great distance and wished to detect reflected light from the planet Jupiter. Jupiter intercepts and reflects just about one-billionth of the radiation from the Sun, so its apparent brightness in visible light is only 10^{-9} that of the Sun, or a difference of about 22 magnitudes. As seen from Alpha Centauri, the nearest star, the Sun would have an apparent magnitude of about 1 and Jupiter would be magnitude 23.

An astronomer on a planet in the Alpha Centauri system, if equipped with telescopes as large as ours, could detect stars of 23rd magnitude. There would be an additional problem, however. Jupiter would be separated from the Sun by only about 4 arcsec, and its light would be very difficult to detect in the presence of its brilliant companion. If that astronomer wanted to search around several stars for planets—not just around the nearest one—then she would have to observe stars about ten times farther away. In that case, a planet like Jupiter would be magnitude 28, and it would be only 0.4 arcsec from its central star, clearly beyond the limits of detection with techniques available today on Earth.

From the above discussion we conclude that a Jupiter-sized planet orbiting a star might just barely be detected directly in visible light, but this technique will work for only the small number of stars in the immediate solar neighborhood.

Orbital Motion

Alternative approaches to the search for other planetary systems are more promising, since they can be carried out by observing the stars and do not require that the planets themselves actually be seen. Consider a planet in orbit about a star. As described in Chapter 18, both the planet and the star actually revolve about their common center of mass, which is called the barycenter. The sizes of their orbits are inversely proportional to their masses. Suppose the planet, like Jupiter, has a mass about one-thousandth that of the star; the size of the star's orbit is then 10^{-3} that of the planet's.

Let's return to our example of observations of our own system carried out from Alpha Centauri: The diameter of the apparent orbit of Jupiter is 10 arcsec, and that of the orbit of the Sun is 0.010 arcsec, or 10 milli-arcsec. If astronomers there could measure the apparent position of the Sun to this precision, they would see it describe an orbit of diameter 10 milli-arcsec with a period of 12 years, equal to that of Jupiter. From the observed motion and the period, they could deduce the mass and distance of Jupiter using Kepler's laws.

Contemporary astronomical measurements from the Earth's surface have a precision of a few milli-arcseconds, sufficient to detect jovian-mass companions around the nearer low-mass stars. Many decades of positional measurements with small refracting telescopes yielded claims in the 1960s of planetary detection, but all of these claims have since been found to be in error. More recent applications of this technique with higher precision ground-based instruments have turned up nothing. A specially built telescope in space, however, could easily detect a planet with only one-tenth of Jupiter's mass around any of several hundred nearby stars, if such planets are present.

Another approach, which involves the search for tiny changes in radial velocity of the star as it moves around its orbit, has had better luck. To understand how this technique works, consider the Sun. Its radial velocity changes by about 10 m/s with a period of 12 years owing to the gravitational force of Jupiter. A jovian-mass planet at a distance of 1 AU would induce velocity changes of 50 m/s with a period of just one year. Velocity differences of this magnitude in other stars can be detected from measurements of the Doppler effect. A Canadian project carried out at Mauna Kea in Hawaii has attempted this experiment. The observing team found radial-velocity variations for 7 of the 16 stars studied that were consistent with the presence of companions with masses of one to nine times the mass of Jupiter. As this project continues, it should soon be possible to determine with confidence whether the observed effects are really due to orbiting planets.

The search for planets around other stars seems to be a project whose time has come. If there are numerous large planets out there, it is within the capability of modern astronomy to find them. If successful, the search will allow astronomers for the first time to consider the formation of our own planetary system within a broader cosmic context. If the search is carried out with negative results, however, the implication will be almost equally interesting. An absence of planets accompanying single, solar-type stars would challenge our current understanding of the process of star and planet formation and would send astronomers "back to the drawing boards" to develop better theories. A negative result would also raise the profound question of the uniqueness of our own solar system and hence of life in the universe.

20.3 Life in the Universe

It appears likely that many stars will prove to have planets circling them and that we will soon be able to detect some of the more massive of these planets. But is there life on these planets? That is a much more speculative question and one that is much more difficult to answer.

Life in the Solar System

In the quest for life in the universe, it is appropriate to begin with the planets we know best, those in our own solar system. By examining the nature of life on Earth and its possible adaptability to other planets, we can gain a perspective on the environments elsewhere that might also support living things.

The life that we are familiar with developed with our planet, influencing the Earth as it evolved (Chapter 4). While we do not know the details of its origin, life clearly seems to have been the product of the chemical evolution of the early atmosphere and oceans. At this early time, there was no free oxygen, and conditions were at least mildly reducing (that is, dominated chemically by hydrogen rather than oxygen). Temperatures on Earth must have been roughly similar to those today, with at least part of the oceans consisting of liquid water at all seasons and with plentiful sunlight to provide the energy needed for life to grow and diversify.

Laboratory experiments have reproduced many of the early chemical steps along the road to life. The raw materials for organic chemistry were deposited on Earth by impacting comets, and naturally occurring reactions would have resulted in plentiful organic chemicals, enough to make the early oceans a sort of organic "soup." (Similar reactions are taking place today in the atmosphere of Saturn's satellite Titan.) Among the prebiological compounds that must have been present are amino acids, sugars, and proteins. Sometime between 4.0 and 3.8 billion years ago, the Earth had developed self-replicating molecules that could use this organic soup to create copies of themselves, beginning the long evolutionary course leading to plants, pigs, and even people.

Could a similar sequence of events have taken place on other planets in the solar system? In the outer planets and their larger satellites, we see reducing atmospheres and nonbiological organic chemistry that mimic in many ways the conditions on the primitive Earth. Certainly the atmospheres of Titan and the jovian planets contain many of what are believed to be the building blocks of life. Yet there is no evidence that life actually developed. Among the inner planets, Mars has always seemed the most likely abode of life. Data from the Viking landings in 1976, however, show that conditions on Mars are too harsh today for the development of life such as we know it on Earth. The absence of liquid water, and particularly the lethal ultraviolet radiation that reaches the surface unimpeded by ozone or other absorbers in the thin martian atmosphere, are not conducive to biological activity. The possibility remains, however, that Mars may have harbored life billions of years in the past.

We find, then, several planets in our solar system that seem to offer possibilities for the development of life, but only one—the Earth—where living things have proliferated and survived to the present. Does this fact mean that it is very difficult for complex, self-replicating molecules to form? Is life a rare occurrence not only in the solar system but also throughout the Galaxy? We simply do not know. Even without knowing the answer, however, several scientists have begun to search for intelligent life outside the solar system.

The Possibility of Intelligent Life in the Galaxy

The first step in mounting a search for intelligent life is to estimate how likely it is to occur. Although the numerical results are extremely uncertain, the exercise is instructive because it leads to a strategy for possible discovery and communication with other intelligent creatures.

University of California astronomer Frank Drake has pioneered the attempt to estimate the number of potentially communicative civilizations in the Galaxy. Drake's famous equation expresses the number, N, of currently extant civilizations in the Galaxy as the product of seven factors:

$$N = R_s f_p n_p f_b f_i f_c L_c,$$

where R_s is the rate of star formation in the Galaxy, f_p is the fraction of those stars with planetary systems, n_p is the mean number of planets suitable for life per planetary system, f_b is the fraction of those planets suitable for life on which life has actually developed, f_i is the fraction of those planets with life on which intelligent organisms have evolved, f_c is the fraction of those intelligent species that have developed com-

municative civilizations, and L_c is the mean lifetime of those civilizations.

The first three factors are essentially astronomical in nature, the next two are biological, and the last two are sociological. We are able to make some educated estimates regarding the astronomical factors, we may be on shaky ground with the biological ones, and we are almost playing numbers games in trying to estimate values for the last two. Yet some interesting estimates can be made, and limits derived.

The mass of the Galaxy (Chapter 24) is believed to be from 2×10^{11} to 10^{12} solar masses. We do not know the form of much of the mass in the galactic halo, but it is a safe assumption that there are at least 4×10^{11} stars in the Galaxy. Recent estimates indicate that the number of new stars formed per year is about 10. This is also a reasonable guess for the average star formation rate over the past few billion years. Thus we adopt $R_s = 10$ stars per year.

The Sun originated from a cloud of gas and dust—the solar nebula—whose rotation caused it to flatten into a disk from which the planets formed. As discussed above, we expect the similar formation of planetary systems elsewhere to be commonplace around single stars. Let's be very optimistic and assume that all stars form planetary systems, adopting $f_p = 1$.

In our own solar system, at least one planet originally had suitable conditions for the development of life. Other systems may have none, but if we assume that, on the average, a star of the right sort with a planetary system has one suitable planet, we can adopt $n_p = 1$.

Many biologists are of the opinion that given the right kind of planet and enough time, the development of life is inevitable. Let us assume optimistically that they are right. The corresponding value of f_b is thus 1.

Similarly, given the emergence of life, there is a widespread view that with enough time and natural selection a highly intelligent species will certainly evolve. Even were it inevitable that an intelligent species evolve on every planet with life, however, how long should it take? On Earth, it took 4.5×10^9 years. What if we happened to be quick about it, but that the average intelligent species takes, say, 20×10^9 years? Moreover, of the many parallel lines of evolution on Earth, only one (so far) has produced a being with enough intelligence to build a technology. Certainly, one could not rule out a probability as low as 1 percent. For the sake of argument, let us adopt $f_i = 0.01$.

Not all intelligent societies would necessarily develop a technology capable of interstellar communication. We are on the threshold of that capability and possess a natural curiosity about the rest of the universe. It is not certain, however, that this human trait is fundamental to intelligence. Insects, while sometimes highly organized, do not appear to have any curiosity at all. Even if a society were curious, it might have good reason for wishing to have nothing to do with any other civilization. Let us assume that one-tenth of intelligent species form communicative societies; that is, $f_c = 0.1$.

Implications of the Drake Equation

It is generally agreed that the final factor, L_c, is the most uncertain. It is useful, therefore, to leave L_c as an unknown and see what the rest of the equation yields with the numbers we have suggested. If we substitute the optimistic estimates made above into the equation, we find

$$N = 10 \times 1 \times 1 \times 1 \times 0.01 \times 0.1 \times L_c$$
$$= 0.01 \times L_c$$

We can obtain the actual number of communicative civilizations present at any time in the Galaxy by replacing L_c with our estimate of the average lifetime in years.

The only known technology, of course, is our own, and we have only just reached the capability of interstellar communication. Some pessimists have argued that our technology might well end in a few decades. If so, and if we are typical, then L_c might be about 100 years. In rebuttal, some contend that if a communicative society can manage to survive for 100 years, it might well maintain itself for a billion years. We simply do not know.

If $L_c = 100$, then even with our fairly optimistic estimates, the number of communicative civilizations in the Galaxy at any time is only just one—at the present, ourselves! One can make equally plausible arguments for adopting much more pessimistic estimates of the various quantities in the Drake equation, in which case the average number of communicative civilizations drops to 0.0001 or even less, suggesting that there are long intervals in which no one is out there to communicate with. In either case, the search for other civilizations will be fruitless.

On the other hand, if $L_c = 1$ billion years, the optimistic value for N is 10 million—a Galaxy teeming with civilizations with which to communicate. The distance to the nearest such civilization would be expected to be less than 100 LY. While we cannot choose among optimistic and pessimistic estimates, we see the rationale by which estimates are made. These estimates also provide some clues about what strategies are most likely to be successful in the search for extraterrestrial intelligence.

20.4 SETI: The Search for Extraterrestrial Intelligence

There are two ways one might hope to communicate with other intelligent beings. One involves direct contact, and the other relies on communication by means of radio waves. As we shall see, the latter technique is far more likely to be successful.

Direct Contact

The most obvious way to search for other galactic civilizations is to use interstellar travel. From the Drake equation, however, we have seen that the nearest neighboring civilization is expected to be at least a few hundred and probably a thousand or even tens of thousands of light years away. Because nothing can travel faster than light, a visit to another civilization would involve at least hundreds and, more likely, thousands of years.

If we cannot go to them, perhaps we might hope they would come to us. If the nearest civilizations are hundreds of light years away, extraterrestrial visitors cannot already have come to see us as a result of learning about us. Even radio waves that we have inadvertently been emitting into space—our radio and television programs—have only been on their way for a few decades and could have reached only the very nearest stars. It is highly unlikely that anyone there has received them and dispatched spaceships to look us over. If we have been visited, it must have been by random selection by interstellar travelers.

The popular literature is full of accounts of sightings of UFOs, presumably operated by some intelligence, of abductions by aliens, and even of alleged evidence for highly intelligent beings that have visited the Earth and taught people to build such magnificent structures as the pyramids, Easter Island statues, and other marvels. Most scientists are highly skeptical of the extraterrestrial interpretation of reports of lights or erratically accelerating shiny objects in the sky. Hard evidence of objects from space is lacking. Scientists, more than anyone, would delight in finding concrete evidence of alien life—there is so much we have to learn from it! But we still need evidence that can be analyzed by scientists qualified to judge its extraterrestrial origin. Rumors, hearsay, secondhand reports, and eyewitness accounts by lay and inexperienced observers all must be given the benefit of the doubt but still require positive verification before being taken as convincing evidence for life in the universe beyond the Earth.

Radio Contact

The Drake equation tells us that if the lifetime of communicative civilizations is short, then basically we are alone in the Galaxy. This conclusion is valid whatever the particular choice you make of numbers for the various factors in the calculations. Putting it the other way around, a search for extraterrestrial civilizations will succeed only if the average such civilization has a lifetime much longer than our own—perhaps many millions of years longer.

The important conclusion that follows is that any civilization we contact is likely to be very much older, and very much more advanced, than our own. The chance of coming across another civilization like ourselves, just a few decades after its discovery of radio astronomy, is vanishingly small. Thus if we are going to search, it must be with the expectation of discovering a civilization far in advance of our own. It also follows that the best approach is to let them do the talking and to assume that they will be ahead of us in considering the problems of interstellar communication. The proper strategy for SETI, the search for extraterrestrial intelligence, is to listen. There is a very real possibility that we may be able to detect and recognize radio signals from another civilization.

With existing radio astronomy facilities, a number of limited searches for intelligently coded signals have already been made by radio astronomers in the United States and the Soviet Union. To date, there has been no success, but no success would have been expected from such meager efforts. Just what would it take to learn if messages are being beamed in our direction from other civilizations in the Galaxy?

The problem is difficult because we do not know in advance either the location or the nature of any possible broadcast. The entire electromagnetic spectrum is available for potential communication. Most of those who have considered this problem have concluded that radio waves—more specifically, microwaves—offer the most promising part of the electromagnetic spectrum. At these wavelengths, the absorption of energy by the interstellar medium and the emission of competing background radiation from natural sources are both near their minimum values.

All of the early searches for extraterrestrial radio signals were limited to one band or a very few frequency bands, just as we tune a radio to a single station at a time. The fact that so few frequency bands were searched is the primary reason that no success

was expected. The key to an effective SETI program is the development of sensitive receivers that can listen at many bands simultaneously.

The first modern SETI program began in the early 1980s. Financed in part by public contributions to The Planetary Society, this search uses an old 60-ft radio telescope that Harvard University was planning to decommission. The strategy is to point the telescope sequentially at each part of the sky and to measure any radiation received in each of its frequency channels, analyzing the results with sophisticated computer codes that can identify an artificial signal amid the natural babble of the Galaxy. The powerful 8-million channel receiver was paid for by movie-maker Steven Spielberg and placed into operation by The Planetary Society in 1986. A second receiver began operations in the Southern Hemisphere in 1991.

Another approach that is being tried is to search for very narrow-band signals, such as might be produced by an interstellar navigation beacon, in data that are regularly collected with radio telescopes for astronomical purposes. The trick in this technique is to eliminate all of the signals from nearby intelligent life—ourselves! Manmade electronic devices produce several thousand events each day of the type that might also be broadcast by extraterrestrial intelligence. This program has just begun, and it is far too early to estimate its probability of success.

NASA is developing a new generation of sensitive receivers and data processing systems that can be used to carry out a SETI survey of unprecedented sensitivity. This search would make use of antennas in the NASA Deep Space Net (DSN), which is normally used for tracking spacecraft. A number of existing radio telescopes around the world will also be pressed into service. One part of the program would examine 800 solar-type stars at more than a billion separate frequencies. Because long observation times for each source are possible, the targeted search will be literally billions of times more comprehensive than all previous surveys combined. A second part of the program would search the entire sky, but with somewhat less sensitivity than will be used to analyze data from the 800 solar-type stars. The NASA SETI system is scheduled to begin listening on October 12, 1992, the 500th anniversary of Columbus' discovery of the New World.

The consequences of the detection of a message from an advanced civilization are hard to imagine. In part, the significance would depend on our success in deciphering the message. We can hope that any civilization that wishes to be found by the relatively primitive techniques we know of will also have worked out a way to make its message intelligible to us. For one plausible scenario of the detection and decryption of an interstellar signal, read Carl Sagan's novel *Contact*.

Unanswered Questions

The study of star formation is one of the most active areas of astronomical research, and the progress made during the past five years has been nothing less than astounding. Despite the advances, many problems remain unsolved. We do not yet know what process (there may even be more than one) initiates star formation in a molecular cloud. We do not know what determines the masses of the stars that form. We know from the luminosity function that stars of low luminosity, and hence of low mass, are by far the most common, but we do not know why. There are even some star-forming regions where no high-mass stars are formed at all. We do not know what determines whether a collapsing dense core becomes a single or multiple star. We do not know what generates the strong stellar winds that apparently halt the infall of material onto the surface of a newly forming protostar. And, of course, the search for extra–solar system planets and for radio signals from other civilizations continues. These questions will challenge astronomers for years to come.

SUMMARY

20.1 It is possible to estimate the lifetime of a star by determining its total mass and by then calculating how rapidly its hydrogen must be converted to helium in order to account for its luminosity. The most massive stars live only a few million years. A star like the Sun will spend about 10 billion years on the main sequence. **Giant molecular clouds** have masses as large as 10^6 times the mass of the Sun and have typical diameters of 50 to 200 LY. They are the most massive objects in the Galaxy. At the cold temperatures (10 K) that characterize the interiors of these clouds, such molecules as H_2 and CO are abundant; organic molecules are also present. Giant molecular clouds are where most star formation occurs. The best studied molecular cloud is Orion, where star formation began about 12 million years ago and is moving progressively through the cloud. The formation of a star inside a molecular cloud begins with a dense core of material, which accretes matter and collapses owing to gravity. The accumulation of material halts when the **protostar** develops a strong **stellar wind.** It is likely that nearly all protostars are surrounded by a disk containing an amount of mass that may be as large as 10 percent the mass of the Sun—more than enough to form a planetary system.

20.2 Direct searches are under way for planets around nearby stars. Preliminary results suggest that the radial velocities of many solar-type stars vary by a few tens of meters per second. Continued measurements should establish whether these variations are due to the gravitational influence of orbiting planets.

20.3 The Drake equation is used to estimate the probability that there is life elsewhere in our Galaxy. Unfortunately, the uncertainties in the calculation are very large, and the most uncertain factor is the typical lifetime of a technologically advanced society. If the lifetime of a communicative civilization is 10^2 years, then we are probably alone in the Galaxy. If, however, the lifetime is 10^9 years, then there would be 10^7 communicative civilizations, and the nearest might be only a few tens of light years away.

20.4 The search for extraterrestrial intelligence is called SETI. The best way to conduct the search is by listening for radio signals broadcast by other civilizations. Several searches are now under way, and a much more comprehensive NASA survey is set to begin in 1992.

REVIEW QUESTIONS

1. How do we know that the lifetime of an individual star cannot be infinitely long?

2. Why is star formation more likely to occur in cold molecular clouds than in regions where the temperature of the interstellar medium is several hundred thousand degrees?

3. Describe what happens when a star forms. Begin with a dense core of material in a molecular cloud and trace the evolution up to the point at which the newly formed star reaches the main sequence.

4. Is it likely that there are planetary systems around other stars? Describe the evidence.

5. Describe the factors that enter into the Drake equation for estimating the number of communicative civilizations in the Galaxy.

THOUGHT QUESTIONS

6. What arguments would you use to persuade a friend that stars are being born somewhere in the Galaxy in our lifetimes?

7. Look at the four stages of evolution in Figure 20.8. In which stage(s) is the star visible in optical radiation? In which stage(s) is it most nearly in hydrostatic equilibrium? In which stage(s) is it generating energy by converting hydrogen to helium?

8. What are the basic assumptions that underlie the SETI program?

9. Redo the calculation of the Drake equation, putting in values that you think are reasonable. What do you think are the extreme limits for the number of intelligent civilizations of any kind in the Galaxy?

10. What wavelengths of electromagnetic radiation might be suitable for interstellar communication, and why? What wavelengths would certainly *not* be suitable? Why?

11. Suppose we could carry on a two-way radio communication with another civilization. How long would be the minimum time to receive an answer to a transmitted question if that civilization is **(a)** on the Moon? **(b)** on Jupiter? **(c)** on a planet in the Alpha Centauri system? **(d)** on a planet in the Tau Ceti system (see Appendix 13)? and **(e)** at the galactic center?

---

12. If a giant molecular cloud is 100 LY across and has a mass equivalent to 10,000 Suns, what is its average internal density? Suppose that half of the mass of the cloud is made up of dust grains, each with a mass of 10^{-13} kg; how many such grains are there per cubic centimeter in the cloud?

13. The star Rigel has a luminosity of about $4 \times 10^4 \, L_s$. Its mass is uncertain but is probably no more than 50 times the mass of the Sun. Estimate the lifetime of Rigel. (Assume that the lifetime of the Sun is 10 billion years.)

14. Verify the statement in the text that Jupiter would be about 22 magnitudes fainter than the Sun as seen in visible light from a distant star.

15. Determine the angular distance of the Earth from the Sun as seen from distances of **(a)** 1 pc; **(b)** 10 pc; **(c)** 100 pc.

16. Calculate the radial velocity of the Sun as it orbits the center of mass between it and Jupiter. Now repeat the calculation for Saturn and for the Earth.

JILL TARTER

In the case of the NASA program, we are trying to develop the technology that will allow us to remove a number of these limiting assumptions. As a result, the NASA program is at least 200,000 times more powerful than the sum of everything that's been done to date. And, when we consider our ability to discriminate many different types of signals, we find that the overall program is billions of times more comprehensive.

Jill Tarter began her career studying engineering but shifted to astronomy, earning her Ph.D. at Cornell University for a theoretical study of brown dwarf stars. She then moved into radio astronomy, and soon found herself one of a small band of enthusiasts who were developing and advocating the Search for Extraterrestrial Intelligence—SETI. Today she holds joint appointments at the University of California at Berkeley and the NASA Ames Research Center and serves as the Project Scientist for the NASA SETI Project.

Jill, can you tell us what the SETI project is?

SETI is an acronym that stands for "Search for Extraterrestrial Intelligence." As such, it's actually a misnomer, because we don't know how to find evidence of extraterrestrial *intelligence.* The best we can hope to do is to find evidence of extraterrestrial *technology,* and that's what SETI is all about. In our case, for the NASA Microwave Observing Project, we will use existing radio telescopes around the world together with very special computers and signal processing equipment to search the entire sky for microwave signals that have the characteristics that we associate with production by another technology rather than by nature itself.

How does the NASA SETI Project improve upon previous efforts?

The main difference between the NASA Project and all previous radio searches is now we can search in

364

millions of channels simultaneously and we can systematically cover the whole sky. We have had to develop new technology and to get access to many radio telescopes. In past searches, people have generally made a guess on things like frequency. For example, they have picked a "magic" frequency— something that they believe that both sender and transmitter will have decided, "Aha! This is the frequency to use." In this way you can eliminate one whole dimension of the search. Others have picked particular directions, such as the galactic center. These are all just guesses, efforts to whittle down the problem to a size that an individual can handle. The result from that kind of experiment is constrained.

Jill Tarter at work on the SETI Project.

How long will it take to finish the program . . . to achieve this step?

We are going to search a thousand solar-type stars as targets, and we are also going to search the entire sky from 1 to 10 gigahertz to a limiting sensitivity. We are scheduled to begin on Columbus Day 1992, and we should complete our initial search by the year 2000. But that may not be all the searching that needs to be done. What could we do with our technology today, assuming that we had a blank checkbook and no constraints? Well, the search volume that we could cover with unlimited resources is about a million times bigger.

But resources *are* limited. How much money will this project cost?

Resources are definitely limited, especially for something that some people consider speculative or high risk, such as this SETI program. The NASA Microwave Observing Project will cost roughly $100 million over a ten-year period. To get that other factor of a million, we're talking about doing something like erecting an array of telescopes on the far side of the Moon. So, that other factor of a million is going to have a pretty high price tag associated with it. But we might not *need* this additional factor of a million. There's nothing that says we have to do that much searching in order to be successful. The whole idea is to do what we can right now, cheaply and from the surface of the Earth. If it doesn't succeed, then we can face the question of whether to go on to the more expensive project.

You have devoted a lot of energy and effort to your role as the Project Scientist for the NASA SETI Project. In your opinion, what are its odds of success?

Well, obviously nobody's going to work on a project that they don't feel could succeed. I *do* believe that it is probable that there is extraterrestrial intelligent life somewhere else in our Galaxy. The factor of a billion or so in improvement represented by the NASA SETI Project is impressive. But someone who has a very long-term goal like this doesn't think only in terms of the ultimate achievement. In order to get there, we've had to develop some incredibly slick new technologies that push the state-of-the-art in digital signal processing and in signal recognition. So I also get my kicks from the daily triumphs that we have.

The SETI Project also has a short-term payoff. We have a lot of interaction with the public and some people get excited, just as we're excited about this project. When we talk to an audience, we try to get them to think about a universe in which other life forms may exist. Their attitudes may help us as a species to evolve to a more global sense of who we are and what we need to do.

SETI often seems to bring out the best and the worst in people. You've talked about the good part, but you have also been subject to some criticism from the public and politicians about SETI. How have you reacted to this criticism?

Everyone has an opinion about SETI. Most of the time it is positive. The U.S. Government would not support us if public opinion was not generally favorable. But there are people who, for personal or religious reasons, think we're crazy. They either (a) know that the extraterrestrials are already here and they talk to them every Saturday night, or (b) they are absolutely convinced that we are alone in this universe. When confronted by people with these opinions, we roll with the punches. People can think what they want, but they don't have the right to keep the rest of us from knowing.

What will happen if SETI succeeds, if that positive signal comes through?

As we work through this, we always try to keep a critical attitude and be very skeptical. We must be hard nosed and not believe the first signal we find because there may be some other explanation. We need to make sure that we can verify this signal, that someone at another observatory can make an independent confirmation. We must be careful because we have the potential not only of a misunderstanding, but of a deliberate hoax.

I think the initial public reaction is going to be ecstatic. And then, unless there is additional information about the extraterrestrials, like what they look like and what they eat for breakfast, interest will decline. We're not going to be able to answer those questions right away, or perhaps ever. So the discovery is going to slide from the front page to the middle section to the back page of the newspaper. But there will be a change as the discovery becomes integrated into the belief systems of people all over the world.

If the SETI program didn't exist, what do you think you would be doing as an astronomer?

I'd be looking for something else that was hard to find. When I completed my undergraduate degree in engineering physics I was bored and looked around for other, more interesting problems to solve. So I went to graduate school in physics, stumbled onto astrophysics. I *loved* astrophysics. For my thesis I worked on an object that is today still theoretical and is very difficult to find observationally—the brown dwarf star. If I had not become involved with the SETI group, I probably would have continued with the brown dwarf investigations. Technology has come a long way in that field. I expect I might have been associated with using the Hubble Space Telescope to search for brown dwarfs.

How do you think your peers from astronomy feel about the way you focused your efforts on SETI?

I know many of them would not change places with me. Many would find SETI an exciting *avocation,* but they would be unwilling to gamble and make it a *vocation.* Many of the things that I do on a daily basis require skills that in no way resemble what I learned in graduate school. I'm much more of a salesman, and some of my peers are not comfortable with that.

Would you want to change places with some of them?

No, I've got the best job in the world. I really do. SETI is so intellectually satisfying because there is nothing about which I am curious that might not be germane to SETI. I'm beginning to explore biochemistry, geology, sociopolitical situations, cultural evolution. I learn something new every day, and I have a lot of fun doing it. But, most important, I may do something working in SETI that could have a more profound effect on the world than anything I could have done through normal astronomical research. That makes a lot of difference to me.

When I was doing my research on brown dwarfs, I sometimes wondered why the taxpayers should be paying my salary. I didn't know what value my research had to the person on the street, except for the abstract principle about increasing our understanding and knowledge of the universe. When I started working on SETI, that uncertainty went away. Now I'm working on trying to answer a question as old as recorded history: "Are we alone?" The person on the street who's paying my salary can instantly identify with that. I wouldn't change places with anyone.

Optical (blue) and infrared images of NGC 2024 in Orion. The same area of the sky is shown in both images. The optical image shows a dark lane of dust, which obscures the young stars within. Infrared radiation penetrates the dust and permits us to study the characteristics of clusters of very young stars. *(National Optical Astronomy Observatories)*

Fred Hoyle (b. 1915), British astrophysicist and cosmologist, is also known for his science fiction and even for an opera libretto. Hoyle was one of the pioneers in the modern study of stellar evolution, and in the 1950s his deduction about the nature of the carbon nucleus led to the modern theory of nucleosynthesis. *(Floyd Clark, Caltech)*

21

STARS: FROM BIRTH TO OLD AGE

The previous chapter described the formation of stars and planets. How, you should ask, do we know that the description is really correct? And how do we go about figuring out what happens next? After all, the changes we would like to understand require tens of thousands, millions, or even billions of years to take place. In our brief lifetimes, we have only a single snapshot of the Galaxy. Our problem in studying stellar evolution is similar to that of scientists who wish to study the growth and life cycle of humans but are allowed only a few seconds to observe a group of people. Within that few seconds we might see one or two deaths, but nothing else would change. Even the birth of a baby requires much longer than a few seconds. So what do we do?

One approach would be to tabulate the properties of the *group* of humans and look at their distribution. We would see, for example, humans of many different sizes, from 40 cm or so up to about 2 m in length. Some of the variation may reflect differences from one individual to another, but careful examination would probably indicate that there was also an evolutionary process. Small humans become large humans. By noting the numbers of humans in each size range, you could calculate the rate of growth and the fraction of a human lifetime required to grow to full size, even though you never actually measured any change in a specific individual during your brief interval of observation.

Astronomers use a similar technique to study stellar evolution. We see a snapshot of the stellar population, consisting of individual stars at all stages of their life cycle. Some of the differences between stars are due, as we have seen, to their different masses. Others represent changes as a star ages. Through a combination of observation, theory, and clever detective work, we can piece together the life story of a star.

21.1 The H–R Diagram and the Study of Stellar Evolution

One of the best ways to represent a snapshot of a group of stars as they are now is by plotting their properties on an H–R diagram. Even though we do not see dramatic changes in any one star (except for a few violent deaths, as discussed in Chapter 22), we can use the distribution of the stellar population on the H–R diagram, together with stellar models, to follow the evolution of a star.

As a star uses up its nuclear fuel, its luminosity and temperature change. Thus, its position on the H–R diagram, in which luminosity is plotted as a function of temperature, also changes. As a star ages, we must replot it in different places on the diagram. Therefore astronomers often speak of a star *moving* on the H–R diagram or of its evolution *tracing out a path* on the diagram. Of course, the star does not really move at all in a spatial sense. This is just a shorthand way of saying that its temperature and luminosity change as it evolves.

To estimate just how much the luminosity and temperature of a star change as it evolves, we must resort to calculations. In a theoretical study of stellar evolution, we compute a series of *models* for a star, each successive model representing a later point in time. Stars change for a variety of reasons. Protostars, for example, change in size because they are contracting, and main-sequence stars change because they are using up their nuclear fuel. Given a model that represents a star at one stage of its evolution, we can calculate what it will be like at a slightly later time. At each step we find the luminosity and radius of the star, and from these its surface temperature.

Evolutionary Tracks

Let's now use these ideas to follow the evolution of protostars that are on their way to becoming normal main-sequence stars. The evolutionary tracks, that

is, the changes of luminosity and temperature with time, of newly forming stars with a range of stellar masses are shown in Figure 21.1. These young stellar objects are not producing energy by nuclear reactions but derive their energy from gravitational contraction—by the sort of process proposed for the Sun by Helmholtz and Kelvin in the last century (Chapter 15).

Initially, a protostar remains fairly cool, its radius is very large, and its density is very low. It is transparent to infrared radiation, and the heat generated by gravitational contraction can be radiated away freely into space. Because heat builds up slowly inside the protostar, the gas pressure remains low, and the outer layers fall almost unhindered toward the center. Thus the protostar undergoes very rapid collapse, which stops only when the protostar becomes dense and opaque enough to trap the heat released by gravitational contraction.

Stars first become visible only after the stellar wind described in the previous chapter clears away the surrounding dust and gas, and this occurs near the point at which the evolutionary tracks in Figure 21.1

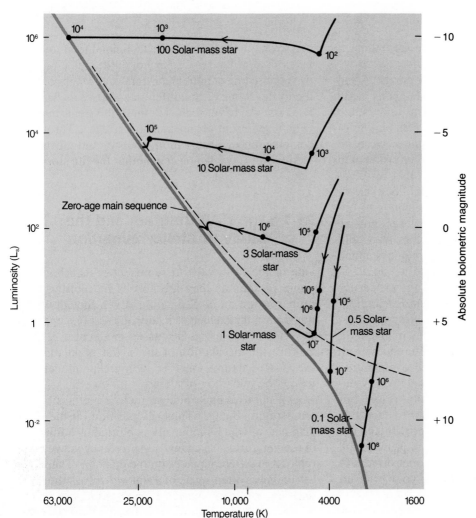

FIGURE 21.1 Theoretical evolutionary tracks of contracting stars or protostars on the H–R diagram. Protostars lying roughly above the dashed line are still surrounded by infalling matter and would be hidden by it.

change from being nearly vertical to being nearly horizontal.

When the central temperature becomes high enough to fuse hydrogen into helium, we say that the star has *reached the main sequence*. It is now approximately in equilibrium, and its rate of evolution slows dramatically. Only the slow depletion of hydrogen as it is transformed into helium in the core gradually changes its properties. The mass of a star determines exactly where it falls on the main sequence. As Figure 21.1 shows, massive stars on the main sequence have high temperatures and high luminosities. Low-mass stars have low temperatures and low luminosities.

Objects of extremely low mass never achieve high enough central temperatures to ignite nuclear reactions. The lower end of the main sequence stops where stars have a mass just barely great enough to sustain nuclear reactions at a sufficient rate to stop gravitational contraction. This critical mass is calculated to be near 1/12 the mass of the Sun.

The Range of Stellar Masses

Since objects with masses less than 1/12 the mass of the Sun never become hot enough to ignite nuclear reactions, they cannot be considered true stars. Objects with masses between 1/100 and 1/12 times the mass of the Sun may produce energy for a brief period by means of nuclear reactions involving deuterium but do not become hot enough to fuse protons to form helium. Such objects are called **brown dwarfs.** Still smaller objects with masses less than 1/100 times the mass of the Sun are true planets; remember that Jupiter has a mass of about 1/1000 times the mass of the Sun. Their interiors will never reach temperatures high enough for nuclear reactions to take place.

At the other extreme, the upper end of the main sequence terminates at the point where the mass of a star would be so high and the internal temperature so great that radiation pressure would dominate. The radiation produced from nuclear reactions would be so extreme that when absorbed by the stellar material it would impart to it a force greater than that produced by gravitation. Such a star could not be stable. The upper limit of stellar mass is thought to be about 100 solar masses.

Evolutionary Time Scales

In general, the pre–main-sequence evolution of a star slows down as the star moves along its evolutionary track toward the main sequence. The numbers labeling the points on each evolution track in Figure 21.1 are the times, in years, required for the embryo stars to reach those stages of contraction. The time

for the whole evolutionary process, however, is highly mass-dependent. Stars of mass much higher than the Sun's reach the main sequence in a few thousand to a million years. The Sun required millions of years; tens of millions of years are required for stars to evolve to the lower main sequence. For all stars, however, we should distinguish three evolutionary time scales:

1. The initial gravitational collapse from interstellar matter is relatively quick. Once the condensation is, say, 1000 AU in diameter, the time for it to reach hydrostatic equilibrium is measured in thousands of years.
2. Pre–main-sequence gravitational contraction is much more gradual. From the onset of hydrostatic equilibrium to the main sequence requires, typically, millions of years. For the stars with masses just barely high enough to ignite hydrogen burning, this phase of evolution can take as long as 100 million years.
3. Subsequent evolution on the main sequence is very slow, and a star changes only as thermonuclear reactions alter its chemical composition. For a star of a solar mass, this gradual process requires billions of years. All evolutionary stages are relatively faster in stars of high mass and slower in those of low mass. Lifetimes for main-sequence stars are listed in Table 21.1.

21.2 Evolution from the Main Sequence to Giants

Once a star has reached the main sequence, it derives its energy almost entirely from the conversion of hydrogen to helium. It remains on the main sequence for most of its "life." Since only 0.7 percent of the hydrogen used up is converted to energy, the star does not change its total mass appreciably. In its central regions, where the nuclear reactions occur, the chemical composition gradually changes as hydrogen is depleted and helium accumulates. This change of

TABLE 21.1	Lifetimes of Main-Sequence Stars	
Spectral Type	Mass (Mass of Sun = 1)	Lifetime on Main Sequence
O5	40	1 million years
B0	16	10 million years
A0	3.3	500 million years
F0	1.7	2.7 billion years
G0	1.1	9 billion years
K0	0.8	14 billion years
M0	0.4	200 billion years

composition forces the star to change its structure, including its luminosity and size. Eventually, the point that represents it on the H–R diagram evolves away from the main sequence. The original main sequence, corresponding to stars of homogeneous chemical composition, is sometimes called the **zero-age main sequence,** where the ''zero'' of time corresponds to the onset of hydrogen fusion.

From Main-Sequence Star to Red Giant

As helium accumulates at the expense of hydrogen in the center of a star, calculations show that the temperature and density in that region must slowly increase. Consequently, the rate of nuclear energy generation increases, and the luminosity of the star gradually rises. This increase amounts to only a few tens of percent over the main-sequence lifetime of a star. Because this change is so small, the main sequence is relatively well defined, with stars of different ages lying along a narrow band in the H–R diagram.

When the hydrogen has been depleted completely in the central part of a star, a core develops containing only helium, ''contaminated'' by whatever small percentage of heavier elements the star had to begin with. Energy can no longer be generated by hydrogen fusion in this helium core, although hydrogen fusion continues farther out in the star. With nothing more to supply heat to the central region of the star, the helium core begins again to contract gravitationally. But as the core contracts, the rest of the star *expands*. These changes result in a substantial and rather rapid readjustment of the star's entire structure, so that the star's luminosity and color no longer correspond to the main sequence. Astronomers say the star ''moves off the main sequence.''

Calculations show that about 10 percent of a star's mass must be depleted of hydrogen before the star evolves away from the main sequence. The more luminous and massive a star, the sooner this happens, ending its term on the main sequence. Because the total rate of energy production in a star must be equal to its luminosity, the core hydrogen is used up first in the very luminous stars. The most massive stars spend only a few million years on the main sequence. A star of 1 solar mass remains there for about 10^{10} years, and a small red star of about 0.4 solar mass has a main-sequence life of some 2×10^{11} years, a value much longer than the age of the universe. Therefore, such low-mass stars would not have had time to complete their main-sequence phase and go on to the next stage of evolution. There exist no post–main-sequence stars that began with a mass less than about 0.8 solar mass. Some post–main-sequence stars do

have lower mass, but only because subsequent to their main-sequence lifetime they have *lost mass*—a not uncommon situation in the late stages of stellar evolution.

As the central helium core contracts, it produces heat energy via the process described by Kelvin and Helmholtz (Section 15.1). This heat is absorbed in the surrounding material, thereby forcing the outer part of the star to expand to enormous proportions. The expansion of the outer layers causes them to cool, and the star becomes red. While the outermost layers cool, the hydrogen in a shell immediately surrounding the stellar core grows ever hotter. The conversion of hydrogen to helium accelerates in this shell, causing most stars to increase in total luminosity. Stars with high luminosity and cool surface temperatures are what we have called red giants. After leaving the main sequence, then, stars move to the upper right portion of the H–R diagram, into the region of the red giants.

Models for Evolution to Red Giants

Figure 21.2, based on theoretical calculations by University of Illinois astronomer Icko Iben, shows the tracks of evolution on the H–R diagram from the main sequence to red giants for stars of several representative masses and with chemical composition similar to that of the Sun. The broad band is the initial or

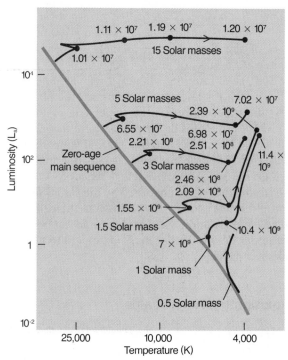

FIGURE 21.2 Predicted evolution of stars from the main sequence to red giants. See text for explanation. *(Based on calculations by I. Iben)*

zero-age main sequence. The numbers along the tracks indicate the times, in years, required for the stars to reach those points in their evolution after leaving the main sequence.

Such models are calculated for single stars, which evolve undisturbed by their neighbors. Remember, however, that more than half of the stars are members of double or multiple systems. If the star has a close companion, some dramatic problems may arise as it evolves off the main sequence.

Suppose, for example, that two stars in a double system are separated by 1 AU, a fairly typical value for spectroscopic binaries. The more massive star of the pair will evolve faster and at some point will begin to expand as it becomes a red giant. But the diameter of a typical red giant can be more than 1 AU, sufficient to engulf the companion star. Before this happens, material from the expanding star will be attracted by the smaller star and will flow toward it, decreasing the mass of the larger star and increasing the mass of the smaller one. In this manner, the evolution of the larger star is arrested in midstride, and it is unable to complete its normal transition to a red giant. At the same time, the mass of the smaller star is increased, and its luminosity goes up, accelerating its own evolution. Depending on the original masses and separation of the two stars, enough mass may be transferred from one to the other to reverse their roles. As you can imagine, such events greatly complicate matters compared with the single-star evolutionary models discussed here.

21.3 Star Clusters

The description of stellar evolution given above is based entirely on calculations of stellar models. No star completes its main-sequence lifetime or its evolu-

tion to a red giant quickly enough for us to observe these structural changes as they happen. Fortunately, nature has provided us a way to test the calculations.

Instead of observing the evolution of a single star, we can look at a group or cluster of stars. If a group of stars is very close together in space and is held together by gravity, it is reasonable to assume that the individual stars in the group all formed nearly at the same time, from the same cloud, and with the same composition. Therefore, we expect that these stars will differ from one another only in their masses and correspondingly in their rates of evolution. Since stars with higher masses evolve more quickly, we can hope to find clusters in which massive stars have already completed their main-sequence phase of evolution and have become red giants, while stars of lower mass in the same cluster are still on the main sequence or are even undergoing pre–main-sequence gravitational contraction.

There are three types of star clusters: globular clusters, open clusters, and stellar associations. Their properties are summarized in Table 21.2. The sizes, absolute magnitudes, and numbers of stars listed in the table for each type of cluster are approximate only and are intended as representative values. Globular clusters contain only very old stars, while open clusters and associations contain young stars.

Globular Clusters

The first type of cluster is the **globular cluster**, so called because of its appearance. About a hundred globular clusters are known in our Galaxy, most of them in a spherical halo surrounding the flat, wheel-like shape formed by the spiral arms. All are very far from the Sun, and some are found at distances of 60,000 LY or more from the galactic plane. One of the most famous globular clusters is called M13, in the

TABLE 21.2 Characteristics of Star Clusters			
	Globular Clusters	**Open Clusters**	**Associations**
Number known in Galaxy	125	~1000	70
Location in Galaxy	Halo and nuclear bulge	Disk (and spiral arms)	Spiral arms
Diameter (LY)	50–300	<50	100–800
Mass (solar masses)	10^4–10^5	10^2–10^3	10^2–10^3?
Number of stars	10^4–10^5	50–10^3	10–100?
Color of brightest stars	Red	Red or blue	Blue
Integrated luminosity of cluster (L_s)	10^4–10^6	10^2–10^6	10^4–10^7
Density of stars (M_s/LY^3)	0.01–30	0.003–0.3	<0.0003
Examples	Hercules Cluster (M13)	Hyades, Pleiades	Zeta Persei, Orion

FIGURE 21.3   The globular cluster M13. *(U.S. Naval Observatory)*

constellation of Hercules, which passes nearly over-head on a summer evening at most places in the United States (Figure 21.3).

A good picture of a typical globular cluster shows it to be a nearly circularly symmetrical system of stars, with the highest concentration of stars near its own center. Most of the stars in the central regions of the cluster are not resolved by ground-based observations as individual points of light but appear as a

nebulous glow. Pictures with the Hubble Space Telescope can resolve the central regions of globular clusters and show hundreds of individual stars (Figure 21.4). The brightest stars in globular clusters are red giants.

The linear diameters of globular star clusters range from 50 LY to more than 100 LY. In one of the nearer globular clusters, more than 30,000 stars have been counted. In addition, there are likely to be many more stars too faint to be seen. If these faint stars are taken into account, most globular clusters must contain hundreds of thousands of member stars. There is plenty of space between the stars, however. If the Earth revolved not about the Sun but about a star in the densest part of a globular cluster, the nearest neighboring stars would be light *months* away and would appear as points of light brighter than any stars in our sky. Many thousands of stars would be scattered uniformly over the sky. The Milky Way would be difficult, if not impossible, to see, and even on the darkest of nights the brightness of the sky would be comparable to faint moonlight.

Open Clusters

Open clusters are found in the disk of the Galaxy, often associated with interstellar matter. Open clusters contain far fewer stars than globular clusters and show little or no strong concentration of stars toward their own centers (Figure 21.5). The stars in these

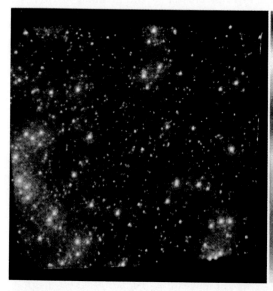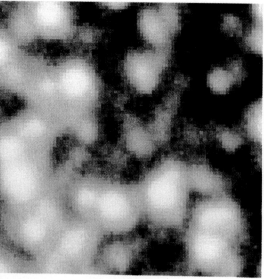

FIGURE 21.4   This pair of images shows a part of the globular star cluster M14, which is at a distance of about 70,000 LY. The image on the right, which was taken at Cerro Tololo Inter-American Observatory in Chile, is a typical ground-based image in which the angular diameters of the stars appear to be about 1.5 arcsec because of blurring by the Earth's atmosphere. The image on the left was taken with the European Space Agency's Faint Object Camera aboard the Hubble Space Telescope. The stellar diameters are about 0.08 arcsec. Hundreds of separate stars can be seen. *(NASA/ESA)*

FIGURE 21.5   The Jewel Box (NGC 4755). This famous cluster of young, bright stars is an open cluster some 8000 LY from the Sun. It was named the Jewel Box from its description by John Herschel as "a casket of variously colored precious stones." *(Anglo-Australian Telescope Board)*

clusters usually appear well separated from one another, even in the central regions (hence the name "open cluster").

Over 1000 open clusters have been catalogued, but many more are identifiable on good search photographs. Yet only the nearest open clusters can be observed, because of interstellar obscuration in the Milky Way plane (Chapter 19). We conclude, therefore, that we see only a small fraction of the open clusters that actually exist in the Galaxy.

Several open clusters are visible to the unaided eye. Most famous among them is the Pleiades (Figure 19.9), which appears as a tiny group of six stars (some people can see more than six) arranged like a dipper in the constellation of Taurus. A good pair of binoculars shows dozens of stars in the cluster, and a telescope reveals hundreds. (The Pleiades is not the Little Dipper; the latter is part of the constellation of Ursa Minor, which also contains the North Star.) The Hyades is another famous open cluster in Taurus, one that has proved to be important in establishing the stellar distance scale (Chapter 16). To the naked eye, the cluster appears as a V-shaped group of faint stars, marking the face of the bull. Telescopes show that the Hyades actually contains more than 200 stars.

Typical open clusters contain several dozen to several hundred member stars. A few open clusters, such as M67, contain more than a thousand stars. Compared with globular clusters, open clusters are small, usually having diameters of less than 30 LY. Bright supergiant stars of high luminosity in some open clusters, however, may cause them to outshine the far richer globular clusters.

Stellar Associations

An **association** appears as a group of several (5 to 50) O stars and B stars scattered over a region of space some 100 to 500 LY in diameter. Because these hot stars are rare, it would be very unlikely for so many of them to exist by chance in so small a volume of space. It is assumed, therefore, that the stars in an association are either physically associated or at least have had a common origin. Stars of other spectral types must also belong to associations, but these more common stars are not conspicuous against the general star field and are not easily observed.

There is evidence that some associations are expanding, probably because most of the gas and dust originally present to bind the system gravitationally has subsequently been blown away. Thus there is no longer sufficient mass in the association to hold all its stars together.

About 70 associations are now catalogued. Like ordinary open clusters, however, they lie in regions occupied by interstellar matter, and many others must be obscured. There are probably several thousand undiscovered associations in our Galaxy.

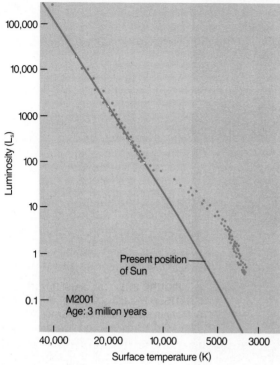

FIGURE 21.6   The H–R diagram of a hypothetical cluster at an age of 3 million years.

FIGURE 21.7   The cluster NGC 2264, which is at a distance of 2500 LY. This region of newly formed stars is a complex mixture of red hydrogen gas, ionized by hot, embedded stars, and dark, obscuring dust lanes. *(Anglo-Australian Telescope Board)*

21.4   Checking out the Theory

Globular clusters are older than open clusters. Because they differ in age, the stars in these two types of clusters are found in different places in the H–R diagram. A valid theory of stellar evolution must be able to explain why the H–R diagrams of globular clusters appear to be so different from those of open clusters.

H–R Diagrams of Young Clusters

What should the H–R diagram be like for a cluster whose stars have recently condensed from an interstellar cloud? After a few million years, the most massive stars should have completed their contraction phase and be on the main sequence, while the less massive ones should be off to the right, still on their way to the main sequence. Figure 21.6 shows the H–R diagram calculated by R. Kippenhahn and his associates at Munich for a hypothetical cluster at an age of 3 million years.

There are real star clusters that fit this description, too. The first to be studied (about 1950) was NGC 2264, a cluster still associated with nebulosity (Figure 21.7). Figure 21.8 shows its H–R diagram. Among the several other star clusters in such an early stage is the one in the middle of the Orion Nebula (Figures 20.3 and 20.4).

After a short time—less than a million years after reaching the main sequence—the most massive stars

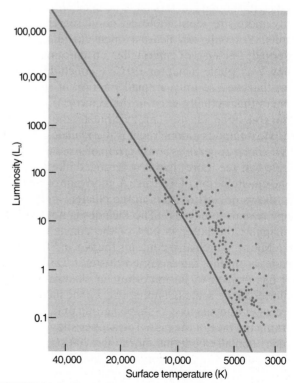

FIGURE 21.8   The H–R diagram of NGC 2264. *(Data by M. Walker)*

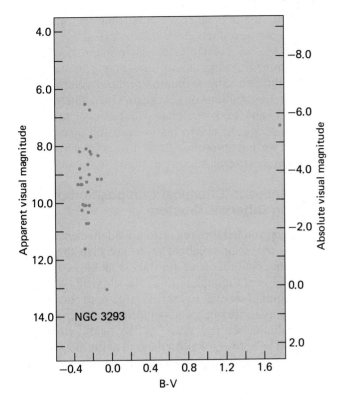

use up the hydrogen in their cores and evolve off the main sequence to become red giants. As more time goes on, stars of successively lower mass leave the main sequence.

Figure 21.9 shows the H–R diagram of the real cluster NGC 3293 (Figure 21.10), which we judge to be a little less than 10 million years old. Note the gap that appears in the H–R diagram for NGC 3293 between the stars near the main sequence and the one red giant in the cluster. In the snapshot of stellar evolution represented by the H–R diagram, a gap does not necessarily represent a locus of temperatures and luminosities that stars avoid. In this particular case, it just represents a domain of temperature and luminosity through which a star moves very quickly as it evolves. We see a gap because at this particular moment we have not caught a star in the process of scurrying across this part of the diagram. This gap, called the *Hertzsprung gap*, is broadest in the H–R diagrams of clusters whose main sequences extend to high luminosities, and it narrows for clusters whose main sequences terminate at successively lower luminosities. The gap disappears in globular clusters and in open clusters that are several billion years old.

H–R Diagrams of Old Clusters

At 4 billion years, stars only a few times as luminous as the Sun begin to leave the main sequence (Figure

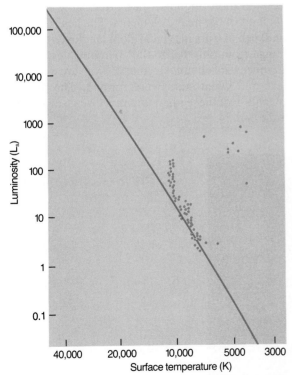

FIGURE 21.9 *Top*, The H–R diagram of NGC 3293. Note the one red giant at the far right of the diagram. Only the brightest stars were observed, so only a portion of the main sequence is shown. *Bottom*, A more complete H–R diagram for M41, a cluster that is only slightly older than NGC 3293. Note that M41 has several red giants.

FIGURE 21.10 The open cluster of stars NGC 3293. All of the stars in this cluster formed at about the same time. The most massive stars, however, exhaust their nuclear fuel more rapidly and hence evolve more quickly than stars of low mass. As stars evolve, they become redder. The bright orange star in NGC 3293 is the member of the cluster that has evolved most rapidly. *(Anglo-Australian Telescope Board)*

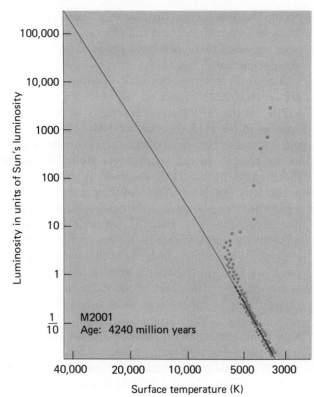

FIGURE 21.11 The H–R diagram of a hypothetical cluster at an age of 4.24 billion years.

21.11). The total main-sequence lifetime for the Sun is expected to be about 10^{10} years. Still older than this are the globular clusters, whose ages are calculated to be between 13 and 15 billion years. The H–R diagram of the globular cluster 47 Tucanae (Figure 21.12) is shown in Figure 21.13. A few open clusters (for example, NGC 188 and M67) are older than the Sun, but they do not approach globular clusters in age.

Note that globular clusters have main sequences that terminate at a luminosity only slightly greater than that of the Sun. Star formation in these systems evidently ceased billions of years ago. Open clusters, on the other hand, are often located in regions of interstellar matter, where star formation can still take place. Indeed, we find open clusters of all ages from less than 1 million to several billion years.

Differences in Chemical Composition of Stars in Different Clusters

Hydrogen and helium, the most abundant elements in stars in the solar neighborhood, are also the most abundant constituents of the stars in all kinds of clusters. The exact abundances of the elements heavier than helium, however, vary from cluster to cluster. In the Sun and most of its neighboring stars, the combined abundance (by mass) of the heavy elements seems to be between 1 and 4 percent of the mass of the star. The strengths of the lines of heavy elements in the spectra of stars in most open clusters show that they, too, have 1 to 4 percent of their matter in the form of heavy elements.

Globular clusters, however, are a different story. Spectra of their brightest stars often show extremely weak lines of the heavy elements. The heavy-element abundance of stars in typical globular clusters is found to range from only 0.1 to 0.01 percent, or even less. Differences in chemical composition are related to where and when stars were formed. The probable explanation of these phenomena is discussed in Chapters 22 and 24.

FIGURE 21.12 The globular cluster 47 Tucanae (NGC 104) is one of the nearest globular clusters and is at a distance of 16,000 LY. This group of old stars has a diameter of about 200 LY. *(National Optical Astronomy Observatories)*

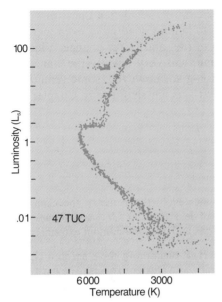

FIGURE 21.13 H–R diagram of the globular cluster 47 Tucanae. *(Data by J. Hesser and collaborators)*

21.5 Nucleosynthesis of Heavy Elements

We shall see in Chapter 28 that we expect hydrogen and helium—but no heavier elements—to have been produced in the first few minutes of the "big bang" that started the expansion of the universe. If this is correct, when our own Galaxy formed, it would have had no carbon, oxygen, nitrogen, sulfur, iron, or any of the other heavy elements that make up the Earth—and us. But if the atoms of which we are composed did not exist when the universe began, where did they come from? They were synthesized by nuclear reactions in stars.

Helium Fusion

Fusion of heavy elements begins in red giants. In these stars the temperature of the central core reaches 10^8 K. At this high temperature, fusion of three helium atoms to form a single carbon nucleus can take place. This process is called the **triple alpha process**—so named because the nucleus of the helium atom is also called an alpha particle. Except in the most massive stars, once this process begins, the entire core is reignited in a flash.

Calculations indicate that when the **helium flash** occurs, the star's surface temperature increases and its luminosity decreases. The point representing the star on the H–R diagram takes on a new position to the left of and somewhat below its place as a red giant. At this time, the core of the star is stable, fusing helium to form carbon. Often a newly formed carbon nucleus is

joined by another helium nucleus to produce a nucleus of oxygen. Surrounding this helium-fusing core is a shell in which hydrogen is fusing to form helium.

As soon as the helium is exhausted in the central regions of a star, however, energy production by means of the fusion of helium to form carbon is shut off. The situation is analogous to that of a main-sequence star when its central hydrogen is used up. Now, however, there is a core of carbon and oxygen surrounded by a shell where the fusion of helium is taking place. Farther out in the star is another shell where hydrogen is fusing to form helium. The star now moves back to the red giant domain on the H–R diagram. In stars with masses similar to that of the Sun, the formation of a carbon-oxygen core marks the end of the generation of nuclear energy at the center of the star.

Nucleosynthesis in Massive Stars

In massive stars, the mass in the outer layers of the star is sufficient to compress the carbon-oxygen core until it becomes hot enough to ignite carbon, which can then form neon, still more oxygen, and finally silicon. After each of the possible sources of nuclear fuel is exhausted, the core contracts until it reaches a temperature high enough to lead to the fusion of still heavier nuclei. In fact, theorists have now found mechanisms whereby virtually all chemical elements of weight up to that of iron can be built up by this **nucleosynthesis** in the centers of the more massive red giant stars, in approximately the relative abundances with which they occur in nature.

When the temperature in the core finally becomes hot enough to fuse silicon to form iron, then energy generation must cease. Up to this point, each fusion reaction has *released* energy. Iron nuclei are, however, so tightly bound that it *requires* energy to fuse iron with any other atomic nucleus. We think today that the most probable place where elements heavier than iron originate is in nuclear reactions that occur in the outbursts of supernovae, which are described in Chapter 22.

Compared with the main-sequence lifetimes of stars, the events that characterize the last stages of stellar evolution pass very quickly. As the star's luminosity increases, its rate of consumption of nuclear fuel goes up rapidly—just at that point in its life when its fuel supply is beginning to run down. It is as if a person, approaching old age, suddenly did everything possible to hasten death, by overeating, overdrinking, smoking, and so on. Soon the plentiful hydrogen that has powered the star for so long is gone, except in cooler layers near the surface where thermonuclear reactions cannot take place.

After hydrogen is exhausted, there are, as we have seen, other sources of nuclear energy, first in the fusion of helium and then of other elements of higher atomic weight. But the energy yield of these reactions is much less than that of the fusion of hydrogen to helium. And to trigger these reactions, the central temperature must be higher than for the fusion of hydrogen to helium, leading to even more rapid consumption of fuel and faster change. Clearly this is a losing game, and very quickly the star will reach its end. In doing so, however, some remarkable things can happen—as we shall see in Chapter 22.

SUMMARY

21.1 The evolution of a star can be described in terms of its changes in temperature and luminosity. These changes can best be followed by plotting them on an H–R diagram. Protostars generate energy through gravitational contraction. The initial gravitational collapse takes several thousand years. After that, a slow contraction continues for, typically, millions of years until the star reaches the **zero-age main sequence** and nuclear reactions begin. The higher the mass of a star, the shorter the time it spends in each stage of evolution. Stars range in mass from about 1/12 to 100 times the mass of the Sun. Objects with masses between that of true planets (1/100 the mass of the Sun) and that of a star are called **brown dwarfs**.

21.2 The fusion of hydrogen to form helium changes the interior composition of a star, which in turn results in changes in temperature, luminosity, and radius. As stars age, they evolve away from the main sequence to become red giants.

21.3. Calculations that show what happens as stars age can be checked by measuring the properties of stars in clusters. The members of a given cluster all formed at about the same time and have the same composition, so that comparison of theory and observations is fairly straightforward. There are three types of star clusters. **Globular clusters** have diameters of 60 to 300 LY, contain hundreds of thousands of old stars, and are distributed in a halo around the Galaxy. **Open clusters** typically contain hundreds of young to middle-aged stars, are located in the plane of the Galaxy, and have diameters of less than 30 LY. **Associations** are found in regions of gas and dust and contain extremely young stars.

21.4 The H–R diagram of the stars in a cluster changes systematically as a cluster evolves. The most massive stars evolve the most rapidly. In the youngest clusters and associations, highly luminous blue stars are on the main sequence; the stars with the lowest masses lie to the right of the main sequence and are still contracting toward it. With passing time, stars of progressively lower mass evolve away from the main sequence. In globular clusters, which have ages typically of 13 to 15 billion years, there are no luminous blue stars at all. The composition of the young stars in open clusters is similar to that of the Sun. The old stars in globular clusters have abundances of elements heavier than helium that are only 0.1 to 0.01 percent, or even less, of those found in the Sun.

21.5 Elements heavier than hydrogen and helium are produced in the interiors of stars by **nucleosynthesis**. The fusion of three helium nuclei produces carbon through the **triple alpha process**. The rapid onset of **helium fusion** in the core of a star is called the helium flash. After hydrogen and helium are exhausted as nuclear fuels, then other elements of higher atomic mass may be fused. Each of these subsequent stages of evolution happens very quickly, and ultimately all stars must use up all of their available supply of energy.

REVIEW QUESTIONS

1. What is the main factor that determines where a star falls along the main sequence?

2. Describe how pressure and gravity influence the structure of a star. What happens when a star exhausts hydrogen in its core and the generation of energy by nuclear fusion of hydrogen to helium stops?

3. Describe the evolution of a star with a mass similar to that of the Sun from the time it is a protostar to the time it becomes a red giant. First give the description in words and then sketch the evolution on an H–R diagram.

4. Suppose you have discovered a new star cluster. How would you go about determining whether it is an open cluster or a globular cluster? List several characteristics that might help you to decide.

5. Explain how an H–R diagram can be used to determine the age of a cluster of stars.

6. How are elements heavier than hydrogen and helium formed?

THOUGHT QUESTIONS

7. Use star charts to identify at least one open cluster that is visible at this time of year. The Pleiades and Hyades are good fall objects, and Praesepe is a good springtime object. Go out and look at these clusters with binoculars and describe what you see. How would you expect the appearance of a globular cluster as seen through binoculars to differ from an open cluster?

8. The H–R diagram for field stars (that is, stars all around us in the sky) shows very luminous main-sequence stars and also various kinds of red giants and supergiants. Explain these features, and interpret the H–R diagram for field stars.

9. In the H–R diagrams for some young clusters, stars of very low and very high luminosity are off to the right of the main sequence, whereas those of intermediate luminosity are on the main sequence. Can you offer an explanation? Sketch an H–R diagram for such a cluster.

10. If the Sun were a member of the cluster NGC 2264, would it be on the main sequence yet? Why?

11. Explain how you could decide whether red giants seen in a star cluster probably had evolved away from the main sequence or were still evolving toward the main sequence.

12. If all the stars in a cluster have the *same age*, how can clusters be useful in studying evolutionary effects?

13. Suppose a star cluster were at such a large distance that it appeared as an unresolved spot of light through the telescope. What would you expect the color of the spot to be if it were the image of the cluster immediately after it was formed? How would the color differ after 10^{10} years. Why?

PROBLEMS

14. Suppose globular clusters have orbits about the galactic center with very high eccentricities—near unity. When a particular globular cluster is at its farthest from the center of the Galaxy, its distance from the center is 3×10^4 LY. What is its period of galactic revolution? (*Hint:* 1 LY = 6.3×10^4 AU. Assume that the mass of the Galaxy is 10^{12} solar masses.)

15. From the data of Table 21.2, estimate the average mass of the stars in each of the three different cluster types. Is the average mass of stars in an open cluster likely to be larger or smaller than the average mass in a globular cluster?

16. What is the density in solar masses per cubic light year of the following clusters: **(a)** a globular cluster 50 LY in diameter containing 10^5 stars? **(b)** a stellar association of 100 solar masses and 60 LY in radius?

    Answer: (a) 1.5 solar masses per cubic light year; **(b)** 0.0003 solar masses per cubic light year.

17. A main-sequence star of color index 0 has an absolute magnitude of about +1. In the H–R diagram of a certain cluster, it is noted that stars of 0 color index have apparent magnitudes of about +6. How distant is the cluster? (Ignore interstellar absorption.)

18. Suppose a star spends 10^{10} years on the main sequence and uses up 10 percent of its hydrogen. Then it quickly becomes a red giant with a luminosity 100 times as great as the luminosity it had on the main sequence and remains a red giant until it exhausts the rest of its hydrogen. How long a time would it be a red giant? Ignore helium fusion and other nuclear reactions and assume that the star brightens from main sequence to red giant almost instantaneously.

19. Stars exist that are as much as a million times more luminous than the Sun. Consider a star of mass 2×10^{32} kg and luminosity 4×10^{32} watts. Assume that the star is 100 percent hydrogen, all of which can be converted to helium, and calculate how long it can shine at its present luminosity. There are about 3×10^7 s in a year.

20. Perform a similar computation for a typical star less massive than the Sun, such as one whose mass is 1×10^{30} kg and whose luminosity is 4×10^{25} watts.

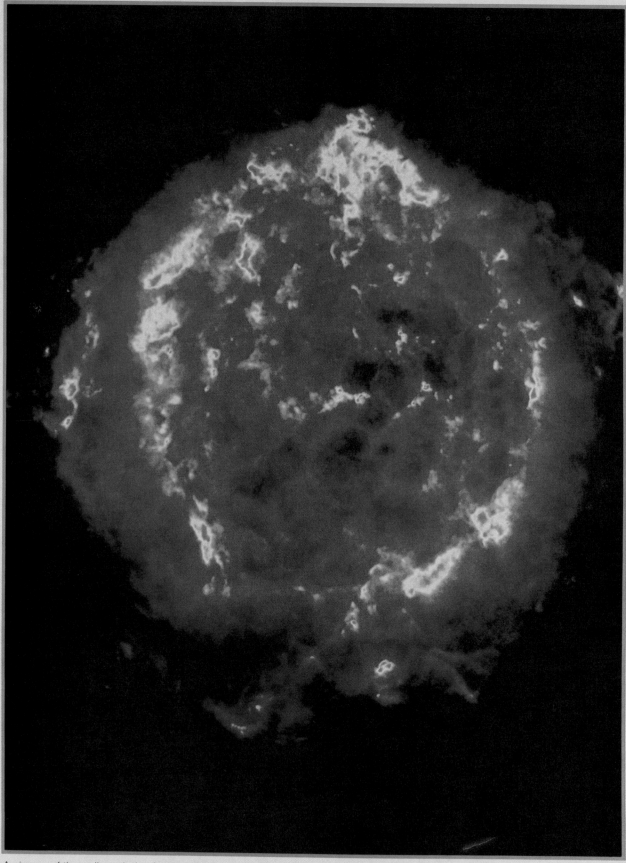

An image of the radio emission from the expanding gas produced by the explosion of
Supernova Cassiopeia A. This supernova exploded in the latter half of the 17th century.
(National Radio Astronomy Observatory/AUI)

Subrahmanyan Chandrasekhar (b. 1910) was born in India and educated at Madras and Cambridge Universities. He has spent most of his career at the University of Chicago, where he has made fundamental contributions in almost every area of astrophysics, including the physical theory of white dwarfs. Chandrasekhar received the Nobel prize in 1983.

22

STARS: DEATH WITH STYLE

Before dawn on February 24, 1987, Ian Shelton, a Canadian astronomer working at an observatory in Chile, pulled a photographic plate from the developer. Two nights earlier he had begun a survey of the Large Magellanic Cloud, a small galaxy that is the Milky Way's nearest neighbor in space. Shelton planned to study variable stars in the Cloud, but the plates from the first two nights did not have sharp images. It was difficult to guide long exposures with the telescope that Shelton was using, which was more than 50 years old. The plate from the third night was a success. The images were sharp. Shelton examined the Tarantula Nebula, a region of bright glowing gas where star formation is occurring. Nearby, where there should have been only faint stars, he saw a large spot on his plate. Concerned at first that his photograph was flawed, Shelton went outside to look at the Large Magellanic Cloud—and saw that a very bright new object had actually appeared in the sky (Figure 22.1).

What Shelton had discovered was an exploding star, a **supernova.** Now known as SN 1987A, since it was the first supernova discovered in 1987, this brilliant newcomer to the southern sky gave astronomers for the first time an opportunity to study in detail the death of a star. This process is fundamental to shaping the evolution of galaxies and to creating the chemical elements of which we ourselves are made.

Sooner or later all stars must die, since eventually every star exhausts its store of nuclear energy. Without an internal energy source, a dying star can only collapse until it attains an enormous density. We know of three possible high-density end states for stars: *white dwarfs, neutron stars,* and *black holes.* Which of these three a star becomes depends on only one thing—its mass at the time it collapses.

As a star contracts toward one of these endpoints, its structural changes are sometimes accompanied by loss of mass. This mass loss may occur gently, in the form of a stellar wind, but often it is explosive and can destroy most or all of the star. Any major loss of mass will influence the star's evolution, and the final state of the object after collapse may be very different from what we would have expected from the initial mass of the star on the main sequence. We thus must be careful in the following discussions to distinguish between the initial, main-sequence mass of a star and the mass of the final collapsed object.

22.1 White Dwarfs

Most stars complete their evolution by becoming white dwarfs. White dwarfs (Section 18.4) are the endpoint of stars that begin their main-sequence evolution with masses up to about eight times that of the

FIGURE 22.1 Before and after pictures of the field around Supernova 1987A in the Large Magellanic Cloud. The difference in image quality between these pictures is an effect of the Earth's atmosphere, which was steadier when the plates used to make the presupernova picture were taken. *(Anglo-Australian Telescope Board)*

Sun. When such a star exhausts its nuclear fuel, it shrinks under the pressure of its overlying layers until its internal pressure becomes great enough to support its own weight. Equilibrium is established only when the star has reached an enormous density, typically about a million times the density of water.

Structure of White Dwarfs

White dwarf stars are far more dense than any substance with which we are familiar on Earth. Their peculiar structure is a result of certain rules that govern the behavior of electrons. According to these rules, which have been verified by studying the behavior of electrons in the laboratory, no two electrons can be in the same place at the same time doing the same thing. We specify the *place* of an electron by its precise position in space and specify what it is doing by its momentum and the way it is spinning.

Imagine the electrons in the interior of a star. The temperature inside a star is so high that the atoms are

stripped of virtually all their electrons. If the temperature is high enough and the density low enough, as it is in normal stars, the electrons will be moving rapidly. It is very unlikely that any two of them will be in the same place moving in exactly the same way at the same time. But what happens when a star exhausts its store of nuclear energy?

What happens is that the star begins to shrink. In the absence of a source of heat, the stellar interior starts to cool, and its pressure will no longer be high enough to resist the crushing mass of the outer layers of the star. The collapse will halt only when the electrons are crowded so closely together that further collapse would require two or more electrons to violate the rule against occupying the same place and having the same momentum. Such a gas is said to be **degenerate.** The electrons in a degenerate gas resist being crushed more closely together and so exert a tremendous new kind of pressure that halts the gravitational collapse.

White dwarfs are thus stars whose electrons have become degenerate. Given the mass of a star, we can calculate how small it can be before the degenerate electrons stop its collapse, and hence what the radius of the final white dwarf will be. A white dwarf of mass like that of the Sun has a diameter about that of the Earth (several thousand kilometers). The theoretically calculated relation between the masses and the radii of white dwarfs is shown in Figure 22.2. Note that the larger the mass of the star, the smaller is its radius.

According to these calculations, which were first carried out by astrophysicist S. Chandrasekhar, a white dwarf with a mass of about 1.4 times that of the

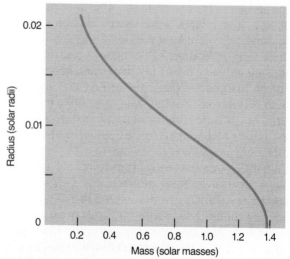

FIGURE 22.2 Theoretical relation between the masses and the radii of white dwarf stars.

Sun would have a radius of zero! As we shall see, what this result means is that even the pressure exerted by degenerate electrons is not enough to halt the collapse of stars that are more massive than 1.4 solar masses when they reach this stage of evolution. The maximum mass that a star can have and still become a white dwarf—1.4 times the mass of the Sun—is called the **Chandrasekhar limit**. Stars with masses exceeding this limit must collapse still further. They are the ones that become supernovae.

White dwarfs have hot interiors—tens of millions of degrees Kelvin. At those temperatures and at the high densities of white dwarf stars, any remaining hydrogen would undergo violent fusion into helium, making the stars many times more luminous than what is actually observed. Consequently, white dwarfs can have no hydrogen in their interiors. For all but the white dwarfs with the lowest mass, a similar argument shows that helium is also absent, since any helium would have been converted to oxygen and carbon. In fact, most white dwarfs are probably composed primarily of carbon and oxygen.

A white dwarf is a star that has exhausted its nuclear fuel. Since it can no longer contract, its only energy source is the heat represented by the motions of the atomic nuclei in its interior. The light by which we see a white dwarf comes from this internal stored heat. Gradually, however, the white dwarf must radiate away its heat, so it must slowly fade. After many billions of years, that heat will be gone, the nuclei will cease their motion, and the white dwarf will no longer shine—a final cold body with the mass of a sun and the size of a planet.

22.2 Mass Loss

What kinds of stars eventually end their lives as white dwarfs? The critical determining factor is *mass*. White dwarfs can be no more massive than the Chandrasekhar limit of 1.4 times the mass of the Sun. Yet, somewhat surprisingly, observations indicate that stars that have masses larger than 1.4 times the mass of the Sun at the time they are on the main sequence also complete their evolution by becoming white dwarfs. White dwarfs have been found in young open clusters—clusters so young that only stars with masses greater than five times the mass of the Sun have had time to exhaust their supply of nuclear energy and complete their evolution to the white dwarf stage. In the Pleiades, for example, stars with masses four to five times the mass of the Sun are still on the main sequence. Yet this cluster also has at least one white dwarf, and the star that has become this dwarf must have had a main-sequence mass that exceeded five solar masses.

If stars that initially have masses as large as five times the mass of the Sun (or, according to theoretical calculations, as large as eight solar masses) are to become white dwarfs, then somehow they must get rid of enough matter so that their total mass at the time nuclear energy generation ceases is less than 1.4 solar masses. Let us look at one of the ways in which stars can lose mass.

Planetary Nebulae

Stars with masses no greater than a few times the mass of the Sun lose mass by forming **planetary nebulae**. Planetary nebulae are shells of gas ejected from, and expanding about, certain extremely hot stars (Figure 22.3). The name is derived from the fact that a

FIGURE 22.3 The Helix Nebula, NGC 7293, is at a distance of about 400 LY and is the planetary nebula nearest the Sun. On photographs the Helix has a diameter about the same as that of the full Moon. The greenish color is produced by emission lines of ionized oxygen; the red color is due to nitrogen and hydrogen. *(Anglo-Australian Telescope Board)*

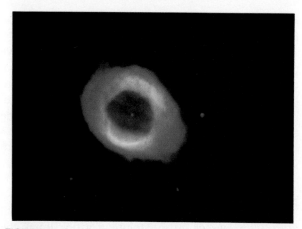

FIGURE 22.4   The Ring Nebula. The image is coded in such a way that red corresponds to emission from singly ionized nitrogen, green corresponds to emission from doubly ionized oxygen, and blue corresponds to emission from ionized helium. *(Courtesy Bruce Balick, University of Washington)*

few planetary nebulae bear a superficial resemblance, when viewed through a small telescope, to a planet. Actually, they are thousands of times larger than the entire solar system and have nothing whatever to do with planets. Planetary nebulae can be seen because they absorb ultraviolet radiation from their central stars and emit this energy as visible light.

The most famous planetary nebula is the Ring Nebula in Lyra (Figure 22.4). It is typical of many planetary nebulae in that, although actually a hollow shell of material emitting light, it appears as a ring. The explanation is that near the center we are looking through the thin dimensions of the front and rear parts of the shell, while along its periphery our line of sight encounters a long path through the glowing material. Similarly, a soap bubble often appears to be a thin ring. Altogether, about a thousand planetary nebulae have been catalogued in our own Galaxy. Doubtless there are many distant ones that have escaped detection, so there must be some tens of thousands. From spectroscopy we calculate that the shells must have masses of 10 to 20 percent that of the Sun. The shells typically expand about their parent stars at speeds of 20 to 30 km/s.

A typical planetary nebula has a diameter of about 1 LY. If we assume that the gas shell has always expanded at the speed with which it is now enlarging about its parent star, we can determine that the shells have been ejected within the past 50,000 years. After about 100,000 years, the shell is so enlarged that it is too thin and tenuous to be seen. When we take account of the relatively short time over which planetary nebulae exist, we find that they are very common and that an appreciable fraction of all stars must sometime evolve through the planetary nebula phase.

The central stars of planetary nebulae have surface temperatures as hot as 100,000 K and are among the hottest stars known. Despite their high temperatures, the central stars of planetary nebulae do not have exceedingly high luminosities—some emit little more total energy than does the Sun. They must, therefore, be stars of small size. Thus a planetary nebula may be the last ejection of matter by a star before it collapses to a white dwarf.

Almost certainly, planetary nebulae originate from red giants. We have already seen (Section 21.5) that red giants have small, dense cores. As a core contracts, it reaches a density so high that the electrons become degenerate, so it is really like a small white dwarf at the center of a red giant star. At some point in the giant star's evolution, the outer envelope detaches and is ejected as a planetary nebula, leaving behind a white dwarf as its core. We do not know if all white dwarfs must originate with the formation of a planetary nebula, but it is suspected that this scenario is a common one.

Consequences of Mass Loss

As we have seen, many stars manage to qualify for white-dwarfhood by ejecting their outer layers into space. Very significantly, however, the material they shed is not the same as that from which they were formed. The nuclear reactions by which they shine alter the chemical composition of the gases inside stars.

We saw in Chapter 21 that stars convert hydrogen into helium. Moreover, at least some stars in some stages of their evolution are building up helium into carbon and heavier elements. Thus, inside stars, lighter elements are gradually being converted into heavier ones. The technical term for this process is **nucleosynthesis.** During the red giant phase of evolution, material from the central regions is dredged up and mixed with the outer layers of the star. When stars nearing the end of their evolution eject these outer layers into the interstellar medium, that matter is richer in heavy elements than was the material from which the stars formed. In other words, a gradual *enrichment* of the heavy-element abundance in interstellar matter is taking place. The heavy-element abundance in stars that are forming from interstellar matter now is thus higher than in those that formed in the past. The fact that the oldest known stars (those in globular clusters) are the stars with the lowest known abundance of heavy elements provides evidence to support this scenario.

We have reason to think that the first generation of stars began their evolution with a composition of nearly pure hydrogen and helium and that all the other elements were synthesized in the hot centers of stars at advanced stages of their evolution. Stars such as the Sun, in whose outer layers heavy elements are observed spectroscopically, thus have to be of the second (or higher) generation. *They formed from matter that was once part of other stars.* It is a grand concept that the planets (and we ourselves!) are composed of atoms that were synthesized in earlier generations of stars—that we are literally made of "stardust."

22.3 Evolution of the Sun

With the information we have been assembling on stellar evolution, we can look again at the evolution of our own Sun and its influence on the planets. Figure 22.5 uses an H–R diagram to summarize our current ideas on the evolution of a star of about one solar mass. In its early stages, the star contracts and moves to the left (toward higher temperatures), reaching the main sequence near the present location of the Sun. The time required for the Sun to contract to the main sequence was probably a few tens of millions of years.

From Main-Sequence Star to Red Giant

Since the Sun reached the main sequence, it has increased somewhat in luminosity, probably by about 30 to 50 percent. Interestingly, the surface temperature of the Earth has not varied much over this same interval of time. Apparently, changes in our planet's atmosphere have compensated approximately for the increasing luminosity of the Sun. During this main-sequence interval of 4 to 5 billion years, the Sun has depleted much of the hydrogen at its very center, but a pure helium core has not yet had time to form. It is not certain how much more time the Sun has before starting to evolve to the red giant stage, but a good guess is that it has lived out about half of its main-sequence life. We can probably look forward to at

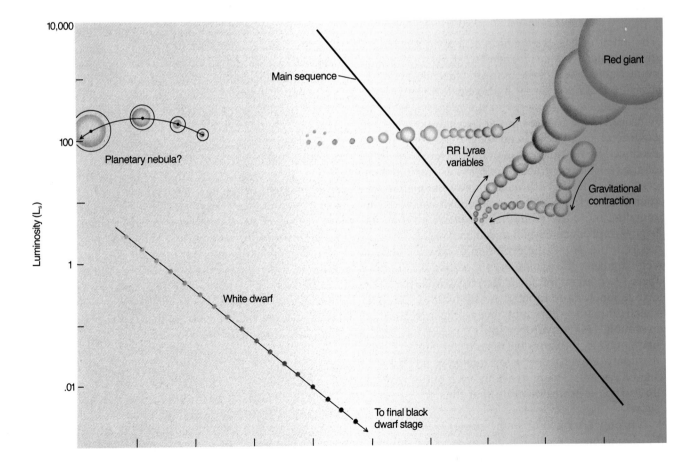

FIGURE 22.5 Summary of the evolutionary track on the H–R diagram for a star with a mass equal to 1.2 times the mass of the Sun. The Sun, which is now in its main-sequence phase of evolution, will go through the phases shown in the diagram.

least another 5 billion years before the Sun's structure undergoes large changes.

Sometime in the future the Sun will leave the main sequence and evolve to a red giant. Thus about 5 billion years from now, the Sun will expand until its photosphere reaches nearly to the orbit of Mars. The Earth, well inside the Sun and exposed to temperatures of thousands of degrees, will gradually vaporize. The gases in the greatly distended outer layers of the Sun will be very tenuous, but they should still offer enough resistance to the partially vaporized Earth to slow it in its orbital motion. Calculations suggest that the Earth will spiral inward toward the very hot interior of the Sun, reaching its final end about 10,000 years after being swallowed by the Sun.

22.4 Evolution of Massive Stars

Stars of masses up to about eight times the mass of the Sun probably end their lives as white dwarfs. But stars can have masses as large as 100 times the mass of the Sun. What is their fate?

Nuclear Fusion of Heavy Elements

The initial stages of evolution of a massive star are quite similar to what happens to stars with masses comparable to that of the Sun. On the main sequence, massive stars are fusing hydrogen to form helium, converting mass to energy in the process. After the hydrogen in the core of the star is exhausted, the core contracts, since there is no longer enough pressure produced by the generation of heat to resist the force of gravity. The contraction releases gravitational energy and so heats both the core and the matter surrounding it. Hydrogen fusion continues in a shell surrounding the core, while helium fusion in the core begins to produce carbon and then oxygen.

Up to this point, the evolution of massive stars is very similar to the evolution of stars like the Sun, but there is one important difference. Massive stars evolve much more rapidly. A star like the Sun may spend 10 billion years in its main-sequence phase of evolution, fusing hydrogen in its core. A star with a mass of 60 to 100 times the mass of the Sun exhausts its supply of hydrogen in a few million years. Massive stars have much more available fuel, but they use it up at such a prodigious rate that their lifetimes are short.

It is only after helium in the core is exhausted that the evolution of a massive star takes a very different course from that of solar-type stars. In a massive star, the weight of the outer layers is sufficient to force the core of carbon and oxygen to contract, heating both the core and its surroundings. Helium fusion begins in

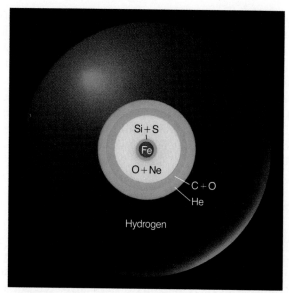

FIGURE 22.6 Just before its final gravitational collapse, a massive star resembles an onion. The iron core is surrounded by layers of silicon, sulfur, oxygen, neon, carbon mixed with some oxygen, helium, and finally hydrogen.

a shell surrounding the carbon-oxygen core. Outside this shell there is a second shell in which hydrogen is fusing to form helium. The carbon-oxygen core continues to shrink until it becomes hot enough to ignite carbon, which can then form neon, still more oxygen, and finally silicon. After each of the possible sources of nuclear fuel is exhausted, the core contracts until it reaches a temperature high enough to lead to the fusion of still heavier nuclei. When the temperature in the core finally becomes hot enough to fuse silicon to form iron, energy generation ceases. Up to this point, each fusion reaction has *released* energy. Iron nuclei are, however, so tightly bound, that energy must be *absorbed* to fuse iron with any other atomic nucleus.

At this stage of its evolution, a massive star resembles an onion, with an iron core and, at progressively larger distances from the center, shells of decreasing temperature in which nuclear reactions involving nuclei of progressively lower mass—silicon, oxygen, neon, carbon, helium, and finally hydrogen—are taking place (Figure 22.6).

Explosion!

What happens next to a star with an iron core? The computations become very complicated, but we can trace the events in a schematic way. In effect, a massive star builds a white dwarf in its center where no nuclear reactions are taking place. For stars that begin their evolution with masses of at least 12 times that of the Sun, this white dwarf is made of iron. For stars with initial masses in the range 8 to 12 times the mass of the Sun, the white dwarf that forms the core is made of oxygen, neon, and magnesium because the star never gets hot enough to form elements as heavy as iron. Whatever its composition, the white dwarf embedded in the center of the star is supported against further gravitational collapse by degenerate electrons.

While no energy is being generated within the white dwarf core of the star, fusion does still occur in shells surrounding the core. As the core accretes the ashes of the fusion reactions going on in the surrounding shell, the mass of the core grows. Ultimately, it is pushed over the Chandrasekhar limit of 1.4 times the mass of the Sun. That is, it becomes so massive that the force exerted by degenerate electrons is no longer great enough to resist gravity. The electrons merge with protons inside the nuclei of iron and other atoms to produce neutrons. The removal of electrons eliminates the main source of support for the core, and it collapses.

This collapse occurs very rapidly. In less than a second, the core, which originally was approximately the same diameter as the Earth, collapses to a diameter that is less than 100 km. The speed with which material falls inward reaches one-fourth the speed of light. The collapse halts only when the density of the core reaches the density of an atomic nucleus. In effect, the matter in the core has merged to form a single gigantic nucleus.

This nuclear material strongly resists further compression, abruptly halting the collapse. The shock of the abrupt jolt generates waves throughout the outer layers of the star and causes the star to blow off those outer layers in a violent **supernova** explosion.

Table 22.1 summarizes the discussion so far about what happens to stars of different initial masses. The mass limits that correspond to various outcomes may change somewhat as models improve. It is also not certain whether or not stars with masses in the range four to eight times the mass of the Sun form white

dwarfs. The nuclear reactions in these stars may go one step beyond helium fusion to reactions involving carbon, which may occur explosively and produce supernovae. If the uncertainties are kept in mind, Table 22.1 does provide a useful guide to what theorists think is the ultimate result of stellar evolution.

22.5 Supernovae

Supernovae were discovered long before astronomers realized that these spectacular cataclysms marked the death of stars. Five supernovae in our own Galaxy have been observed in the past 1000 years, all before the invention of the telescope. The Chinese reported the temporary appearance of "guest stars" in A.D. 1006, 1054, and 1181. The two remaining galactic supernovae occurred in 1572 (Figure 22.7) and 1604. The latter was observed in considerable detail by Johannes Kepler. From the historical records, from studies of the remnants of supernova explosions in our own Galaxy, and from analyses of supernovae in other galaxies, we estimate that, on average, one supernova explosion occurs somewhere in the Milky Way Galaxy every 25 to 100 years.

At their maximum brightness, the most luminous supernovae have about 10 billion times the luminosity of the Sun. For a brief time, a supernova may outshine the entire galaxy in which it appears. After maximum brightness, the star fades in light until it

TABLE 22.1 Fate of Stars of Different Masses	
Initial Mass (Mass of Sun = 1)	**Final Evolutionary State**
< 0.01	Planet
0.01 < M < 0.08	Brown Dwarf
0.08 < M < 0.25	Helium white dwarf
0.25 < M < 8	Carbon-oxygen white dwarf
8 < M < 12	Oxygen-neon-magnesium white dwarf
12 < M < 40	Supernova; neutron star
40 < M	Supernova; black hole

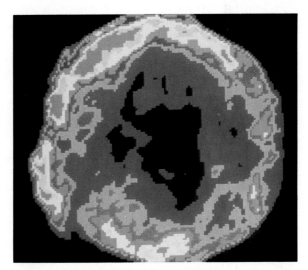

FIGURE 22.7 A map of the radio emission from the expanding shell of gas ejected by the supernova explosion observed by Tycho Brahe in 1572. The different colors correspond to different intensities of radio emission, produced by extremely energetic electrons gyrating in a magnetic field. This was a Type I supernova, and no central source has been found. Apparently nothing remains of the star that exploded. *(National Radio Astronomy Observatory/AUI)*

disappears from telescopic visibility within a few months or years after its outburst. Supernovae eject material at the time of their outbursts. The velocities of ejection may reach 10,000 km/s.

Types of Supernovae

The available evidence suggests that there are two distinct types of supernovae. A Type II supernova is the type that has been described here, and it marks the death of a massive star. Supernovae can also occur in a binary system that initially contains a white dwarf and a nearby companion. The intense gravitational force exerted by the white dwarf attracts matter from the companion star. The mass of the white dwarf, which initially must have been less than 1.4 times the mass of the Sun, begins to build up, and eventually it may exceed the Chandrasekhar limit. At this point, the white dwarf must begin to collapse. As it does so, it heats up, new nuclear reactions begin, and the energy released is so great that it completely disrupts the star. Gases are blown out into space at velocities of several thousand kilometers per second. No central star remains behind (Figure 22.7). The explosion completely destroys the white dwarf. Such supernovae are called Type I.

Observations are consistent with this model, in that Type I supernovae occur primarily in types of galaxies (ellipticals) or regions of galaxies (away from spiral arms) where there are large numbers of old, low-mass stars. In contrast, Type II supernovae appear in regions where there are large numbers of young, massive stars.

Supernova 1987A

The 1987 explosion of a Type II supernova in the Large Magellanic Cloud at long last provided astronomers with an opportunity to test their calculations of how stars die (Figure 22.8). Also, for the first time, we know what the star was like before it exploded.

By combining theory and observation, astronomers have reconstructed the life story of the star that became SN 1987A. Formed about 10 million years ago, the star originally had a mass about 20 times that of the Sun. For 90 percent of its life it lived quietly on the main sequence, converting hydrogen to helium. At this time, its luminosity was about 60,000 times the luminosity of the Sun, and it was spectral type O. The temperature in its core was about 40 million K, and the central density was about 5 g/cm³, or about five times the density of water.

FIGURE 22.8 A photograph of SN 1987A and the region of the Large Magellanic Cloud where it occurred. The large cloud of gas is called the Tarantula Nebula and is a region of active star formation. *(National Optical Astronomy Observatories)*

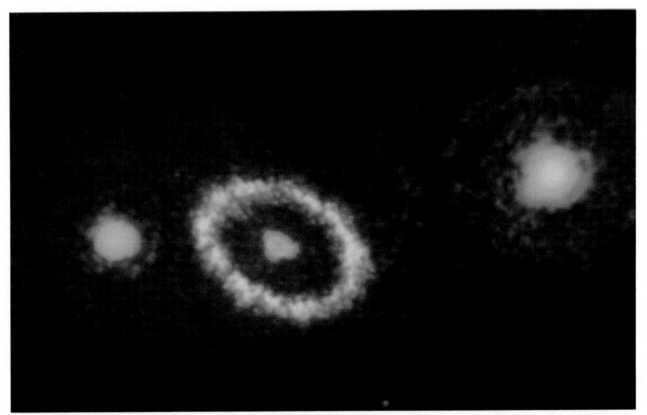

FIGURE 22.9 This image, which was taken with the Hubble Space Telescope, shows (in yellow) a ring of stellar material ejected by the star that ultimately became SN 1987A. This material was ejected long before the supernova explosion. Within 100 more years, the expanding debris from the supernova, which appears in red in this image, will plow into the ring and tear it apart. The blue stars to the left and right of the ring are not associated with the supernova. *(NASA)*

By the time hydrogen was exhausted at the center of the star, a helium core of about six times the mass of the Sun had developed, and hydrogen fusion was proceeding in a shell surrounding this core. The core contracted and grew hotter, until it reached a temperature of 170 million K and a density of 900 g/cm^3, at which time helium began to fuse to form carbon and oxygen. The surface of the star expanded to a radius of about 10^8 km, or about the distance from the Earth to the Sun. The star's luminosity nearly doubled, to 100,000 times the luminosity of the Sun, and the star became a red supergiant. While it was a red supergiant, the star lost some mass. This mass has now been detected by observations with the NASA Hubble Space Telescope (Figure 22.9).

Helium fusion lasted for only about 1 million years, forming a core of carbon and oxygen with a mass about four times the mass of the Sun. When helium was exhausted at the center of the star, the core contracted again, the surface of the star also decreased in radius, and the star became a blue supergiant with a luminosity still about equal to 100,000 L$_s$, which was what it was when it exploded. When the contracting core reached a temperature of

700 million K and a density of 150,000 g/cm^3, nuclear reactions began to convert carbon to neon, sodium, and magnesium. This phase lasted only about 1000 years.

The core, having exhausted carbon as a fuel, again contracted and heated—this time to a temperature of 1.5 billion K and a density of 10^7 g/cm^3—at which point the conversion of neon to oxygen and magnesium, and then oxygen to silicon and sulfur, lasted for several years. At temperatures of 3.5 billion K and densities of 10^8 g/cm^3, the silicon began to melt into a sea of helium, neutrons, and protons, which then combined with some of the remaining silicon and sulfur nuclei to form iron.

When the silicon is converted to iron, and the mass of the core exceeds 1.4 times the mass of the Sun, the collapse begins. In the case of SN 1987A, the collapse lasted only a few tenths of a second, and the velocity of the collapse in the outer portion of the iron core reached 70,000 km/s. The outer shells of neon, helium, and hydrogen, however, do not know about the collapse. Information travels through the star at the speed of sound and cannot reach the surface in the few tenths of a second required for the collapse to

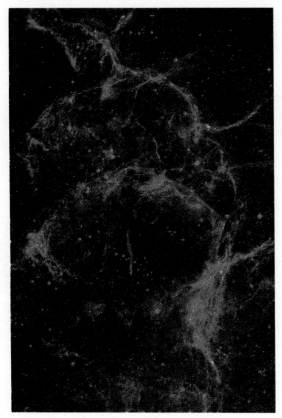

FIGURE 22.10 About 120 centuries ago, an inconspicuous star in the constellation Vela became a supernova. A portion of its expanding shell, which now covers 6° of the sky, is shown here. Ultimately, the shell expanding around SN 1987A will thin out and look like this. *(Anglo-Australian Telescope Board)*

great, and the core rebounds. Infalling material runs into the brick wall of the rebounding core and is blown outward. The material that is ejected in the ensuing explosion is rich in heavy elements. Studies confirm that the composition is what would be predicted from the models of the composition of the stellar interior immediately prior to the explosion. This shell of gas will expand and disperse into interstellar space, and the atoms in it may someday be incorporated into a new generation of stars (Figure 22.10). Table 22.2 summarizes the steps that led inexorably to the explosion that was SN 1987A. The data are taken from detailed calculations by S. Woosley (University of California at Santa Cruz) and his collaborators.

Brightness Variations of SN 1987A

Figure 22.11 shows how the brightness of SN 1987A changed with time. In a single day, the star soared in brightness by a factor of about 1000, from an apparent magnitude of about 12 to a magnitude of 5. The star then continued to increase slowly in brightness until it was about the same apparent magnitude as the stars in the Little Dipper. Up until about day 40 after the outburst, the energy being radiated away was produced in the explosion itself.

When the supernova explodes, the force of the blast hurls most of the material surrounding the iron core into space. As the shock waves pass violently through the layers containing oxygen and silicon, more heavy elements are formed. One of these newly formed elements is radioactive nickel, with an atomic mass of 56. Nickel-56 is unstable and changes spontaneously with a half-life of about 6 days to cobalt-56, which in turn decays with a half-life of about 77 days to iron-56, which is stable. Gamma rays are emitted when these radioactive nuclei decay, and those gamma rays, which are absorbed in the overlying gas and re-emitted at visible wavelengths, are responsible

occur. The surface layers hang suspended, much like a cartoon character that dashes off the edge of a cliff and hangs momentarily in space before he realizes that he no longer is held up by anything.

The collapse of the core continues until the densities rise to several times that of an atomic nucleus. The resistance to further collapse becomes very

TABLE 22.2 Evolutionary Phases That Preceded SN 1987A

Phase	Central Temperature (K)	Central Density (g/cm³)	Duration of Phase
Hydrogen fusion	40×10^6	5	9×10^6 years
Helium fusion	170×10^6	900	10^6 years
Carbon fusion	700×10^6	150,000	10^3 years
Neon fusion	1.5×10^9	10^7	Several years
Oxygen fusion	2.1×10^9	10^7	Several years
Silicon fusion	3.5×10^9	10^8	Days
Core collapse	200×10^9	2×10^{14}	Tenths of a second

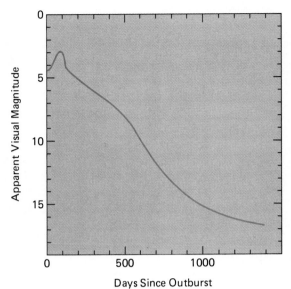

FIGURE 22.11 A graph that shows how the brightness of SN 1987A changed with time. *(Courtesy N. Suntzeff/CTIO)*

for virtually all of the radiation detected from SN 1987A after day 40. From the total amount of radiation seen, astronomers can estimate how much nickel was produced. The total mass of nickel-56 turns out to be about 7 or 8 percent of the mass of the Sun, and this mass is in agreement with theoretical predictions for a supernova explosion like 1987A. Some gamma rays also escape directly without being absorbed. These gamma rays have been detected at the wavelengths expected for the decay of radioactive nickel and cobalt, again confirming the theory of what is going on.

The radioactive nuclei produced in the explosion will all ultimately decay into stable elements and will cease to produce energy in the form of gamma rays. With no new source of energy, SN 1987A will slowly fade away. By the beginning of 1991, it was observable only with very large telescopes.

Neutrinos from SN 1987A

Brilliant as SN 1987A was in the southern skies, less than 1/10 of 1 percent of the energy of the explosion appeared as optical radiation. One of the predictions of models of supernovae is that a large number of neutrinos should be ejected from the core of the star at the time of the collapse. When the collapse occurs, the electrons merge with protons to form neutrons, and this reaction also releases neutrinos, which escape from the star at the speed of light. The energy carried away by these neutrinos is truly astounding. In the first second, the total luminosity of the neutrinos is 10^{46} watts, which exceeds the luminosity of all the stars in all the galaxies in the part of the universe that we can observe. And the supernova

generated this energy in a volume less than 50 km in diameter! Supernovae are by far the most violent events in the universe.

One of the most exciting results from observations of SN 1987A is that astronomers detected the neutrinos at the right time and in the expected quantity, thereby obtaining strong confirmation that the theoretical calculations of what happens when a star explodes are actually correct. The neutrinos were detected by two instruments, which might be called "neutrino telescopes," about 3 h before the brightening of the star was first observed in the optical region of the spectrum. Both "telescopes," one in Japan and the other under Lake Erie, consist of several thousand tons of purified water surrounded by several hundred detectors that are sensitive to light. Incoming neutrinos interact with the water to produce positrons and electrons, which move rapidly through the water and emit deep blue light.

The Japanese system detected 11 neutrino events over an interval of 13 s, and the instrument beneath Lake Erie measured 8 events at the same time. Since the neutrino telescopes are located in the Northern Hemisphere, and the supernova occurred in the Southern Hemisphere, the neutrinos detected had already passed through the Earth and were on their way back out into space when they were captured!

Only a few neutrinos were detected because the probability that they will interact with matter is very, very low. It is estimated that the supernova actually released 10^{58} neutrinos. About 50 billion of these neutrinos passed through every square centimeter on the Earth, and about a million people experienced a neutrino interaction within their bodies. Of course, this interaction had absolutely no biological effect and went completely unnoticed.

Since the neutrinos come directly from the heart of the supernova, their energies provide a measure of how hot the star was at the time of the explosion. The central temperature was about 200 billion K.

22.6 Pulsars and the Discovery of Neutron Stars

What's left now of the star that exploded to become SN 1987A? According to theory, there should be an object with a density approximately equal to that of an atomic nucleus composed essentially entirely of neutrons at the center of the gaseous cloud that has been blown out into space. In the case of SN 1987A, this **neutron star** has not yet (July, 1991) been detected. As the cloud of gas surrounding SN 1987A expands and disperses, astronomers will continue to search for the evidence of the neutron star that they are confident is there. Fortunately, from observations of the remnant of the much older Crab supernova, where the cloud has thinned enough so that we can observe to its center, we know exactly what to look for.

The Crab Nebula

The best studied example of the remnant of a supernova explosion is the Crab Nebula in Taurus, a chaotic, expanding mass of gas (Figure 22.12). The Doppler shifts of spectral lines formed in the center of the Crab Nebula show the gases there to be moving toward us at speeds up to 1450 km/s. If we assume that the nebula has always expanded at this same rate, we can derive its age by calculating how long it would take for the nebula to reach its present size. It turns out that both the location and the computed time of formation of the Crab Nebula are in good agreement with the occurrence of the supernova of 1054. The Crab Nebula, therefore, must be the material ejected during that stellar explosion.

Observations of the center of the Crab Nebula have demonstrated conclusively that supernova explosions leave behind a neutron star. The first neutron star to be discovered, however, did not lie in the Crab, and initially it presented astronomers with a mystery.

The Discovery of Neutron Stars

In 1967, Jocelyn Bell, a graduate research student at Cambridge University, was studying distant radio sources with one of the Cambridge radio telescopes. In the course of her investigation, Bell made a remarkable discovery—one that won her advisor, Antony Hewish, the Nobel prize in physics, because his analysis of the object (and other similar ones) revealed the first evidence for neutron stars.

What Bell had found, in the constellation of Vulpecula, was a source of rapid, sharp, intense, and extremely regular pulses of radio radiation, the pulses arriving exactly every 1.33728 s. For a time there was speculation that they might be signals from an intelligent civilization. Radio astronomers half-jokingly dubbed the source "LGM," for "little green men," and withheld announcement pending more careful study. Soon, however, three additional similar sources were discovered in widely separated directions in the sky. When it became apparent that this type of source was fairly common, astronomers concluded that it was highly unlikely that they were signals from other civilizations. By now hundreds of such sources have been discovered. They are called **pulsars,** for pulsating radio sources.

The pulse periods of different pulsars range from a little longer than 1/1000 s to nearly 10 s. One pulsar is in the middle of the Crab Nebula, and it has a pulse

FIGURE 22.12 The Crab Nebula in the constellation Taurus. This nebula is the remnant of a supernova explosion, which was seen on Earth in A.D. 1054. It is located some 6300 LY away and is approximately 6 LY in diameter, still expanding outward. This composite of images shows hydrogen (red) and sulfur (blue) emission; different colors thus signify different ionization conditions and chemical compositions in various portions of the nebula. *(National Optical Astronomy Observatories)*

period of 0.033 s. The period is observed to be very slowly increasing, showing that pulsars evolve, pulsing gradually more slowly as they age. The source of the pulses is what appears to be a star at the center of the nebula. In addition to pulses of radio energy, there are pulses of optical and x-ray radiation from the Crab as well (Figure 22.13).

The energy emitted by pulsars is not small. The Crab pulsar emits considerably more energy than the Sun does. Yet the energy that we detect from the Crab pulsar is not constant, but arrives in pulses or bursts that occur 30 times each second. What type of object can emit such bursts of energy?

One clue comes from measurements of the masses of pulsars. Four pulsars that emit x rays are also members of binary-star systems for which enough information is available to calculate their masses using Kepler's laws. These four x-ray pulsars have masses in the range of 1.4 to 1.8 times that of the Sun. This is consistent with the masses that theorists predict that neutron stars should have.

Their masses, combined with other evidence, prove that pulsars are, in fact, *spinning neutron stars*—just what would be expected from the theory of evolution of very massive stars. Neutron stars are the densest objects in the universe. The force of gravity at the surface of a neutron star is 10^{11} times the force of gravity at the surface of the Earth. The interior of a neutron star is composed of about 95 percent neutrons with a small number of protons and electrons mixed in. In effect, a neutron star is a giant atomic nucleus with a mass about 10^{57} times the mass

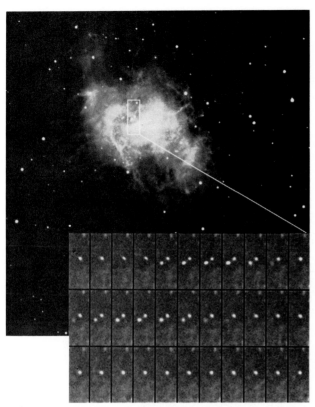

FIGURE 22.13 A series of photographs of the central part of the Crab Nebula taken by Nigel Sharp at Kitt Peak National Observatory. Note the star that seems to blink on and off; it is the pulsar, which has a period of 0.033 s. *(National Optical Astronomy Observatories)*

star. At the surface of the neutron star, protons and electrons escape but are caught up in this spinning field and accelerated nearly to the speed of light, where they emit energy over a broad range of the electromagnetic spectrum. The pulsar radiation is, however, confined to a narrow, rotating beam like a lighthouse beacon. As the rotation carries first one and then the other magnetic pole of the star into our view, we see a rapidly varying emission—the pulses that make the object a pulsar (Figure 22.14).

Table 22.3 compares the characteristics of a typical white dwarf and a typical neutron star.

Evolution of Pulsars

From observations of the several hundred pulsars discovered so far, astronomers have concluded that one new pulsar is born somewhere in the Galaxy every 30 to 100 years. Calculations suggest that the typical lifetime of a pulsar is about 10 million years.

The energy radiated by a pulsar robs the star of rotational energy. Thus, theory predicts that the rotating neutron stars should gradually slow down and that the pulsars should slowly increase their periods. Indeed, as we have already mentioned, the Crab pulsar is actually observed to be increasing the interval between its pulses. According to present ideas, the Crab pulsar is rather young and has a short period (we know it is only about 900 years old), while the other, older pulsars have already slowed to longer periods. Pulsars thousands of years old have lost too much energy to emit appreciably in the visible and x-ray wavelengths and are observed only as radio pulsars; their periods are a second or more.

Only 3 of more than 400 pulsars discovered so far are embedded in visible clouds of gas. The lifetime of a pulsar is about 100 times longer than the length of time required for the expanding gas to disperse into interstellar space. This fact explains why we find only about 1 percent of the pulsars to be still surrounded by the gas ejected during the supernova explosion.

One other possible end state for a star is that of a black hole, which involves one of the more bizarre predictions of general relativity theory. We shall discuss black holes in the next chapter.

of a proton. There are approximately 100 million neutron stars in our Galaxy.

A neutron star itself is so small—typically about 10 km in diameter—that it cannot be observed. What we see, instead, is pulses of x rays, light, and radio waves, all produced by the magnetic field embedded in the rapidly spinning neutron star. The rapid spin is another example of the conservation of angular momentum. Even if the star begins with a very slow rotation when it is on the main sequence, it speeds up in the process of collapse until it spins in just a fraction of a second.

Any magnetic field that existed in the original star is highly compressed if the core collapses to a neutron

TABLE 22.3 Properties of Typical White Dwarfs and Neutron Stars

	White Dwarf	Neutron Star
Mass (mass of Sun = 1)	1.0 (always <1.4)	1.5 (always <3)
Radius	5000 km	10 km
Density	5×10^5 g/cm³	10^{14} g/cm³

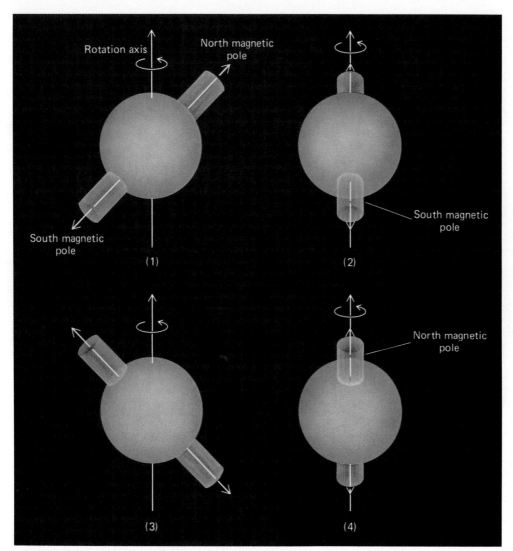

FIGURE 22.14 A diagram showing how emission at the magnetic poles of a neutron star can give rise to pulses of emission as the star rotates. This model requires that the magnetic poles not be located in the same place as the rotation poles.

SUMMARY

22.1 Stars that begin their lives with masses less than about eight times the mass of the Sun end their evolution as **white dwarfs.** A typical white dwarf has a mass about like that of the Sun and a diameter comparable to that of the Earth. The pressure exerted by **degenerate electrons** keeps white dwarfs from contracting to still smaller diameters.

22.2 During the course of their evolution, stars may shed their outer layers and lose a significant fraction of their initial mass. Stars with masses up to eight (and possibly as much as 12) times the mass of the Sun can lose enough mass to become white dwarfs, which have masses less than the **Chandrasekhar limit,** which is 1.4 solar masses. **Planetary nebulae** are shells of gas ejected by stars. The mass that is lost is enriched with heavy elements synthesized by nuclear fusion in the stellar interior.

22.3 The Sun required a few tens of millions of years to contract and become a main-sequence star. It will spend about 10 billion years as a main-sequence star, converting hydrogen to helium in its core. Astronomers estimate that the Sun is now about 5 billion years old. In another 5 billion years it will expand to become a red giant, with its radius extending nearly to the orbit of Mars. The Earth will be vaporized.

22.4 In a massive star, hydrogen fusion in the core is followed by several other fusion reactions involving heavier elements. Just before it exhausts all sources of energy, a massive star has an iron core surrounded by shells of (in order of increasing distance from the center and decreasing temperature) silicon, oxygen, neon, carbon, helium, and hydrogen.

22.5 When the mass of the iron core of a star exceeds the Chandrasekhar limit, the core collapses until its density exceeds that of an atomic nucleus. The core rebounds and transfers energy outward, blowing off the outer layers of the star in a **supernova** explosion. Studies of SN 1987A, including the detection of neutrinos, have confirmed theoretical calculations of what happens during the explosion. This type of explosion of a massive star is called a Type II supernova. Supernova explosions can also occur when a white dwarf star accretes enough matter from a nearby companion so that its total mass exceeds 1.4 solar masses. The white dwarf then explodes as a Type I supernova.

22.6 At least some supernovae leave behind a rotating **neutron star,** which is called a **pulsar.** Pulsars emit pulses of radiation at regular intervals. Their periods are approximately in the range of 0.001 to 10 s and are related to the period of the neutron star. Pulsars lose energy as they age, the rotation slows, and their periods increase.

REVIEW QUESTIONS

1. What is a white dwarf? Describe its structure.

2. Describe the evolution of a star with a mass two times that of the Sun, from the main-sequence phase of its evolution until it becomes a white dwarf.

3. Describe the evolution of a massive star up to the point at which it becomes a supernova. How does the evolution of a massive star differ from that of the Sun? Why?

4. What is a supernova?

5. What is the evidence that a supernova can produce a neutron star?

6. Describe the basic characteristics of pulsars.

THOUGHT QUESTIONS

7. You observe an expanding shell of gas through a telescope. What measurements would you make to determine whether you have discovered a planetary nebula or the remnant of a supernova explosion?

8. Arrange the following stars in order of age:
a. A star with no nuclear reactions going on in the core, which is made primarily of carbon and oxygen
b. A star of uniform composition from center to surface; the star contains hydrogen but has no nuclear reactions going on in the core
c. A star that is fusing hydrogen to form helium in its core
d. A star that is fusing helium to carbon in the core and hydrogen to helium in a shell around the core
e. A star that has no nuclear reactions going on in the core but is fusing hydrogen to form helium in a shell around the core

9. Would you expect to find any white dwarfs in the Orion Nebula?

10. Suppose no stars had ever formed that were more massive than about two times the mass of the Sun. Would life as we know it have been able to develop?

11. Would you be more likely to observe a Type II supernova (the explosion of a massive star) in a globular cluster or in an open cluster? Why?

12. Use contemporary sources to determine whether or not a neutron star has been detected at the position of SN 1987A.

13. Show the evolutionary track in the H–R diagram for the star that exploded as SN 1987A. Begin when the star was on the main sequence and end just before the explosion.

PROBLEMS

14. The gas shell of a particular planetary nebula is expanding at the rate of 20 km/s. Its diameter is 1 LY. Find its age. For this calculation, assume that there are 3×10^7 s/yr and 10^{13} km/LY.

15. Prepare a chart or diagram that exhibits the relative sizes of a typical red giant, the Sun, a typical white dwarf, and a neutron star of mass equal to the Sun's. You may have to be clever to devise such a diagram.

16. Suppose the central star of a planetary nebula is 16 times as luminous as the Sun and 20 times as hot (about 110,000 K). Find its radius, in terms of the Sun's. Compare this radius with that of a typical white dwarf.

The Keyhole Nebula in the constellation Carina surrounds the peculiar variable star Eta Carinae. Astronomers believe that Eta Carinae is a massive supergiant star that sheds an amount of material equal to the mass of the Sun every 100 to 1000 years. Astronomers are trying to learn whether stars like this can shed enough mass to avoid becoming a black hole when they finally exhaust their sources of nuclear energy. *(National Optical Astronomy Observatories)*

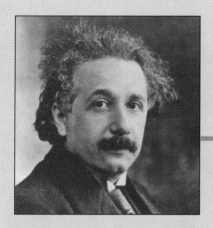

Albert Einstein (1879–1955) received the Nobel prize in 1921, not for his theory of relativity but for the photoelectric effect. At that time his ideas on relativity were still at the frontier. Einstein believed that such seemingly diverse areas of physics as mechanics and electromagnetic phenomena—and even gravitation— were guided in the same way by underlying principles.

23

GENERAL RELATIVITY: CURVED SPACETIME AND BLACK HOLES

Most stars end their lives as white dwarfs and neutron stars. A very massive star, however, may collapse to one of the strangest objects ever predicted by theory—a black hole. To understand what a black hole is, we must first learn something about general relativity.

The theory of general relativity is one of the major intellectual achievements of the 20th century. Until recently, however, this theory had little impact either on our daily lives or on scientific research. For a long time there were only three tests of general relativity, none of them precise enough to provide compelling support for the theory. In the past two decades, however, general relativity has become essential in understanding pulsars, quasars, and other astronomical objects and events.

23.1 Principle of Equivalence

The fundamental insight that led to the formulation of general relativity is deceptively simple. Galileo noted that all bodies, despite their different masses, if dropped together, fall to the ground at the same rate. According to Newton's law of gravitation, the Earth pulls on a more massive object with a greater force than it does on a less massive one. So then why doesn't the more massive one fall faster? The reason is that according to Newton's second law, the more

massive an object, the larger the force *required* to accelerate it. Since the greater force exerted by the Earth on the more massive object is exactly equal to the greater force required to accelerate it, the two objects fall at the same rate of speed. In fact, *any* two freely falling bodies, independent of their composition and internal structure, will follow identical paths. This similarity of behavior of all types of objects is an **equivalence principle.**

Einstein broadened this equivalence principle and as a consequence reached sweeping conclusions about the very fabric of space and time. The basic assumption of Einstein's equivalence principle is that life in a freely falling laboratory is indistinguishable from, and hence equivalent to, life with no gravity. Similarly, life in a stationary laboratory in a gravitational field is indistinguishable from life in an accelerating laboratory far from any gravitational force.

Gravity or Acceleration?

To explore the implications of this idea, let's consider as an example the case of a foolhardy boy and girl who simultaneously jump into a bottomless chasm from opposite sides of its banks (Figure 23.1). If we ignore air friction, then while they fall they accelerate downward at the same rate and feel no external force acting on them. They can throw a ball back and forth between them, aiming always in a straight line, as if

FIGURE 23.1 The brave couple playing catch as they descend into a bottomless abyss. Since the boy, girl, and ball all fall at the same velocity, it appears to them that they can play catch by throwing the ball in a straight line between them. Within their world, there appears to be no gravity.

there were no gravitation. The ball, falling along with them, moves directly between them.

It's very different on the surface of the Earth. Everyone knows that a ball, once thrown, falls to the ground. Thus, in order to reach its target (the catcher) the ball must be aimed upward so that it follows a parabolic arc—falling as it moves forward—until it is caught at the other end.

Because our freely falling boy, girl, and ball are all falling at the same rate and in the same direction, we could enclose them in a large box falling with them. No one inside that box is aware of any gravitational force. Nothing falls to the ground, or anywhere else, but moves in a straight line in the most simple, natural way, obeying Newton's laws. By having our box fall with the boy and girl, we have removed the force of gravitation. We have defined an environment—or in mathematical terms, we have selected a *coordinate system*—that is accelerating at just the right rate to compensate for gravitation. Here is the principle of equivalence: A force of gravitation is equivalent to an acceleration of the coordinate system of the observer.

Einstein himself pointed out how our weight seems to be reduced in an elevator when it accelerates from

a stop to a rapid downward velocity. Similarly, our weight seems to increase in an elevator that is increasing its upward velocity. Stand on a scale in an elevator and see what happens! In a *freely falling* elevator, with no air friction, we would lose our weight altogether.

This idea is not hypothetical. Astronauts in the Space Shuttle orbiting the Earth live in just such an environment (Figure 23.2). The Shuttle in orbit is, of course, falling freely around the Earth. While in free fall the astronauts live in a magical world where there seem to be no gravitational forces. One can give a wrench a shove, and it moves at constant speed across the orbiting laboratory. One can lay a pencil in midair, and it remains there, as if no force is acting on it.

Appearances are misleading. There *is* a force. Neither the Shuttle nor the astronauts are *really* weightless, for they continually fall around the Earth, pulled by its gravity. But since all fall together—Shuttle, astronauts, wrench, and pencil—within the Shuttle all gravitational forces appear to be absent.

Thus the Shuttle provides an excellent example of the principle of equivalence—how local effects of

FIGURE 23.2 When the Space Shuttle is in orbit, everything stays put or moves uniformly because there is no apparent gravitation acting inside the spacecraft. *(NASA)*

gravitation can be removed by a suitable acceleration of the coordinate system. To the astronauts it is as if they are far off in space, remote from all gravitating objects.

Suppose the astronauts *actually were* in remote space and activate the engines of their ship, producing acceleration. The ship would then push up against their feet, giving the impression of a gravitational tug. If one were to drop a small coin and a hammer, the floor of the ship would move up to meet both objects

at the same time. To the astronauts, though, it would seem that the hammer and coin fell to the floor together. In other words, an acceleration of one's local environment produces exactly the same effect as a gravitational attraction. The two are indistinguishable—again, the equivalence principle.

Trajectories of Light and Matter

Einstein postulated that the equivalence principle is a fundamental fact of nature. If so, then there must be *no* way in which an astronaut, at least by experiments within a spacecraft, can distinguish between weightlessness in remote interstellar space and free fall in a gravitational field about a planet like the Earth. Experiments must have the same result whatever the objects involved and whether or not the test involves falling objects, interactions of atoms, or light.

But how about light? If the equivalence principle really applies to light, then the consequences are profound. Suppose that astronauts shine a beam of light along the length of their ship, which is in a free-fall orbit about a planet. The ship is falling away from a straight-line path. If the beam of light follows a straight line, which from everyday experience we believe it does, then won't the light strike above its target (Figure 23.3)?

Not so, according to Einstein. If the principle of equivalence is correct, there must be *no* way of knowing whether one is accelerated. Hence the light beam *must fall with the ship* if that ship is in orbit about a gravitating body (Figure 23.3). Is light actually bent from its straight-line path by the force of gravity? Einstein preferred to think that light always follows the shortest path—but that path may not be straight. On the Earth, we know that the shortest distance between two points is not a straight line but follows the arc of a *great circle* (the equator is a great circle). Suppose we define **spacetime** to be a system of coordinates within which we can specify the time and place of any event. If spacetime is curved, then the shortest path between two points may be curved.

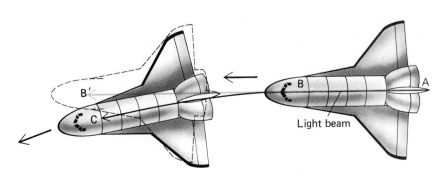

Light beam

FIGURE 23.3 In a spaceship moving to the left (in this figure) in its orbit about a planet, light is beamed from the rear, *A*, toward the front, *B*. We might expect the light to strike at *B′*, above the target in the ship, which has fallen out of its straight path in its orbit about the planet. Instead, the light, bent by gravity, follows the curved path and strikes at *C*.

By examining the consequences of the equivalence principle, Einstein concluded that we live in a *curved* spacetime and that gravitation is equivalent to, and indistinguishable from, curvature of spacetime. A massive body distorts space and time, and the path of a second body passing near such a mass is controlled by these distortions.

To understand this concept, we will explore in more detail what is meant by spacetime and by curvature of spacetime.

23.2 Spacetime

There is nothing mysterious about four-dimensional spacetime. Imagine yourself in the rear seats at an outdoor concert at the Hollywood Bowl. The sound from the orchestra in the shell, hundreds of feet away, takes a substantial fraction of a second to reach you, and the players seem to be behind the beat of the conductor. When a piece is finished, you first hear the applause from people near you, and slightly later from the front of the amphitheater. Because of the finite travel time of sound, all people do not hear the same note of music at the same time.

Events in Spacetime

Light also has a finite speed, so we never see an instantaneous snapshot of events around us (as we saw in Chapter 14). The speed of light is so great that within a single room we obtain *effectively* an instantaneous snapshot, but this is certainly not the case astronomically. We see the Moon as it was just over a second ago, and the Sun as it was about 8 min ago. At the same time, we see the stars by light that left them years ago, and the other galaxies as they were millions of years in the past. We do not observe the world about us at an instant in time, but rather we see different things about us as different *events* in spacetime.

Relatively moving observers do not even agree on the order of events. As an example, suppose that I am riding exactly in the middle of a boxcar on a train moving at uniform speed. Suppose you are standing on the ground beside the tracks and that, just at the instant that we are abreast of each other, two lightning bolts strike the ends of my boxcar and the ground at points *A* and *B* exactly below the ends of the boxcar (Figure 23.4).

You will receive the light signals from the lightning bolts that struck at *A* and *B* at the same time. Since these signals will have traveled identical distances at the speed of light to reach you, you will conclude that

the strikes at *A* and *B* occurred simultaneously. That is not what I see. When you receive the signals from *A* and *B*, I have already moved forward (Figure 23.4). The signal from the lightning strike at the front of the boxcar will already have swept past me on its way to you; the signal from the strike at the back of the boxcar will not yet have reached me. Therefore, I must conclude that the front of the boxcar was struck first. We disagree. Two events that are simultaneous to one observer will not appear to be simultaneous to a second observer who is moving with respect to the first.

At this point, you may wish to ask, "Well, who is really right?" According to relativity, both are correct—in their own worlds or their own *reference frames*. Furthermore, no one reference frame is to be preferred to any other. Space and time are inextricably connected. We need to describe the universe not just in terms of three-dimensional space, but in terms of four-dimensional spacetime.

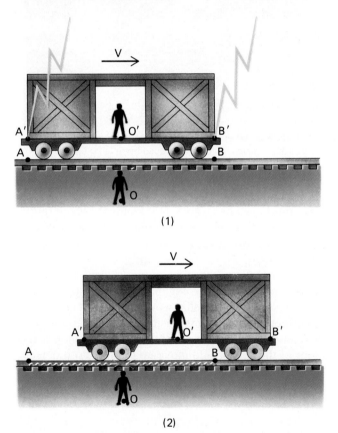

(1)

(2)

FIGURE 23.4 An experiment demonstrating that events that appear to one observer to happen simultaneously can appear to another observer to happen at different times. When the boxcar passes the observer beside the tracks, lightning bolts strike the ends of the boxcar and points *A* and *B* on the ground. The stationary observer will see these events simultaneously. The observer on the train will think that the front of the boxcar was struck first.

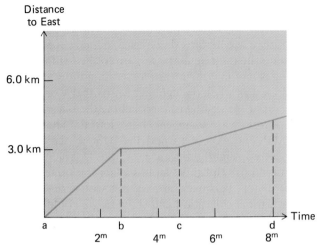

FIGURE 23.5 The progress of a motorist traveling east across town. Distance traveled is plotted along the vertical axis. The time that has elapsed since the motorist left his starting point is plotted along the horizontal axis.

Visualizing Spacetime

We can easily represent the spatial positions of objects in two dimensions on a flat sheet of paper (for example, the map of a city). To plot three dimensions on a page, the draftsperson uses projections. Architectural drawings of a home generally show three projections: a floor plan and two different elevations—say, the house as seen from the east and from the north—to give all necessary information. By the use of perspectives, we can also give an impression of a three-dimensional view.

A two-dimensional projection is used to illustrate four-dimensional spacetime. Figure 23.5, for example, shows the progress of a motorist driving to the east across town. How much time has elapsed since he left home is shown on the horizontal axis, and how far he has traveled eastward is shown on the vertical axis. From *a* to *b* he drove at a uniform speed. From *b* to *c* he stopped for a traffic light and made no progress, and from *c* to *d* he drove more slowly because of increased traffic.

Figure 23.6 shows a rather conventional two-dimensional representation of spacetime. Time increases upward in the figure, and one of the three spatial dimensions is shown horizontally. If we measure time in years and distance in light years, light goes one unit of distance in one unit of time, and so flows along diagonal lines as shown. "Here and now" is at the origin of the diagram. At this instant we can receive information of a past event along such a line as *AO*. In this case the messenger was going slower than light, so she covered less distance than light would in the same time. Because nothing can go

faster than light, we cannot, right now, know of something happening at point *B* in spacetime, for the message along *BO* would have to travel faster than light. We will have to wait until we are at *C* in the future, before a light or radio beam can get us the word along path *BC*.

Curvature of Spacetime

To understand what is meant by curvature of spacetime, let's consider a familiar analogy. A simple Mercator-type map, with lines of constant latitude running horizontally and lines of constant longitude running vertically, is fine for showing a small area of the Earth—say, a single city—without noticeable distortion. But such a map cannot show a large area of the curved Earth without distortion. Everyone knows how distorted and enlarged countries near the poles appear on the usual flat world maps. We cannot map the Earth with ordinary (Euclidean) plane geometry.

Indeed, if we travel far enough in a straight line on the surface of the Earth, we end back at our starting point. Our path is a great circle (the equator is such a great circle). More generally, if we take into account the slight polar flattening of the Earth, as well as effects of such irregularities as mountains, our "straight-line" path is called a **geodesic,** which means "Earth divider."

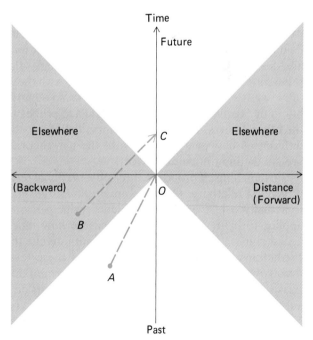

FIGURE 23.6 A spacetime diagram. In this kind of diagram, the time scale along the vertical axis and the distance scale along the horizontal axis are chosen in such a way that travel at the speed of light is represented by a line that makes a 45° angle with respect to the space and time axes.

Einstein showed how to find spacetime coordinates within which all objects move as they would if there were no forces. In a small local region, where a gravitational field is uniform, those coordinates are the ones used in plane geometry. A city map, for example, makes use of plane geometry. But to describe paths of objects over a large region, where the gravitational field varies, the geometry used to describe spacetime must be curved, just as we must use curved geometry to describe a large area of the spherical Earth.

The distribution of matter determines the nature of a gravitational field, so it is the distribution of matter that determines the geometry of curved spacetime. Within this curved spacetime, everything moves in the simplest possible way. In analogy with Earth geometry, the paths of light and material objects in spacetime are called geodesics.

23.3 Tests of General Relativity

Relativity is different from Newtonian theory in several ways. First, the signals that govern gravitational interactions are not instantaneous but travel with the speed of light. Matter and energy are equivalent, so that not only matter itself but also energy contributes to gravitation—that is, to the geometry of spacetime. Energy (light, for example) and mass are both affected by that geometry. Where speeds are low compared with that of light and where the gravitational field is relatively weak—and both of these conditions are met throughout most of the solar system—the predictions of general relativity must agree with those of Newton's theory, which has served us so admirably in our technology and in guiding space probes to the other planets. In familiar territory, therefore, the differences between predictions of the two theories are subtle and difficult to detect.

Einstein himself proposed three observational tests of general relativity. One, the *gravitational redshift*, is actually a test of the equivalence principle as it applies to light. The other two measure how much spacetime is curved and so test quantitatively the predictions of general relativity. One of these tests involves observation of the *deflection of starlight* that passes close to the Sun, and the second depends on a subtle effect in the *motion of the planet Mercury*. With the advent of spacecraft, another and more precise test of general relativity has become possible.

The Gravitational Redshift

Let us consider an experiment with light in a freely falling laboratory (the Einstein elevator). Suppose we shine a light beam—say, a laser beam of a precise frequency—upward from floor to ceiling. Now the laboratory accelerates downward, gaining speed. By the time the light beam travels up to the ceiling, that ceiling is moving downward faster than the source on the floor was when the light left it. In other words, the receiver at the ceiling is approaching the place where the source was when the light left it. Therefore, wouldn't we expect to find the light at the ceiling blueshifted slightly because of the Doppler effect (Section 5.6)? But this would violate the principle of equivalence, because the blueshift would reveal our downward acceleration and show us we could not be weightless in free space. Therefore, Einstein postulated, there must be a redshift, due to the light's moving upward against gravity, that exactly compensates for the Doppler shift that would otherwise be observed. If so, that gravitational redshift should be observed in radiation climbing upward in a gravitational field—even at the surface of the Earth.

This idea was verified experimentally in the 1960s, at the Jefferson Physical Laboratory at Harvard University. Radioactive cobalt, which is a source of gamma rays, was placed in the basement of the building. The detector was a layer of cobalt placed at the top of the building, 20 m above the source. If there were no gravitational redshift, the gamma rays traveling upward from the basement should have been absorbed by the cobalt at the top of the building. Remember that a substance absorbs electromagnetic radiation at exactly the same wavelengths at which it emits radiation (Section 5.5). In fact, the gamma rays were not absorbed. In traveling upward against the Earth's gravitation, they suffered a gravitational redshift, which changed their wavelengths. To absorb them, the detector had to be moved slowly to produce a Doppler shift to compensate for the Earth's gravitational redshift. The speed of the detecting cobalt needed to make it absorb the gamma rays from the emitting cobalt in the basement was so slow that it would require a full year to travel the 20-m gap between emitter and detector. That speed produced a Doppler shift that agreed with the value needed to compensate for Einstein's predicted redshift to within 1 percent.

Far higher accuracy has been attained recently in the near-Earth environment with space-age technology. In the mid-1970s, a hydrogen maser, which is a device that produces a microwave radio signal at a particular wavelength, was carried by a rocket to an altitude of 10,000 km. This maser was used to detect the radiation from a similar maser on the ground. That radiation showed a gravitational redshift due to the Earth's field that confirmed the relativity predictions to within a few parts in 10,000.

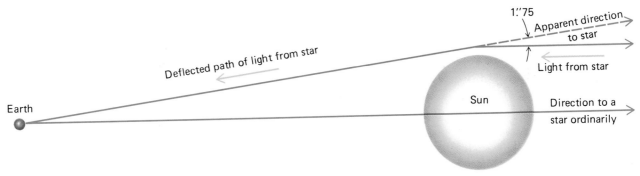

FIGURE 23.7 Starlight passing near the Sun is deflected slightly, so that it no longer travels in a straight line.

Deflection of Starlight

The strength of the gravitational acceleration at the surface of the Sun is 28 times its value at the surface of the Earth. Since spacetime is curved in regions where the gravitational field is strong, we would expect to find that light passing near the Sun would appear to follow a curved path (Figure 23.7). Einstein calculated from general relativity theory that starlight just grazing the Sun's surface should be deflected by an angle of 1.75 arcsec.

Stars cannot be seen or photographed near the Sun in bright daylight, but with difficulty they can be photographed close to the Sun at times of total solar eclipses. In a paper published during World War I, Einstein suggested an eclipse observation to look for the deflection of light passing near the Sun. A single copy of that paper, passed through neutral Holland, reached the British astronomer Arthur S. Eddington. The next suitable eclipse was on May 29, 1919. The British organized two expeditions to observe it, one on the island of Principe, off the coast of West Africa, and the other in Sobral, in North Brazil. Despite some problems with the weather, both expeditions obtained successful photographs of stars near the Sun. Measures of their positions were then compared with measurements on photographs of the same stars taken at other times of the year when the Sun was elsewhere in the sky. The stars seen near the Sun were indeed displaced, and to the accuracy of the measurements, which was about 10 percent, the shifts were consistent with the predictions of relativity. It was a triumph that made Einstein a world celebrity.

The measurements made in 1919 were good enough to distinguish between Newton's theory of gravity, which predicts no deflection of starlight, and Einstein's theory of gravity, which does predict a deflection. Nevertheless, 10 percent accuracy is hardly a convincing demonstration that a theory is completely correct. Far higher accuracy has been obtained recently at radio wavelengths. Simultaneous observations of a source of radio waves with two telescopes far apart (a radio interferometer—see Chapter 6) can pinpoint the direction of the source very precisely. Observations of remote astronomical radio sources show that the difference in the directions to any two of them depends on their positions relative to the Sun. Measurements of the apparent shifts in direction as the Sun moves through the sky confirm the predictions made by Einstein to within 1 percent.

Advance of Perihelion of Mercury

According to relativity, the energy and momentum associated with the motion of a body, and even its gravitational energy, all contribute to its effective mass, and hence to the force of gravitation on it. Mercury has a highly elliptical orbit, so that it is only about two-thirds as far from the Sun at perihelion as it is at aphelion. As required by Kepler's second law, Mercury moves fastest when nearest the Sun, which is also the time when the gravitational force exerted by the Sun is greatest. According to relativity, these effects combine to produce a very tiny additional push on Mercury, over and above that predicted by Newtonian theory, at each perihelion. The result of this effect is to make the major axis of Mercury's orbit rotate slowly in space. Each successive perihelion occurs in a slightly different direction as seen from the Sun (Figure 23.8). The prediction of relativity is that the direction of perihelion should change by 43 arcsec per century.

The gravitational effects (perturbations) of the other planets on Mercury also produce an advance of its perihelion. According to Newtonian theory, the gravitational force exerted by Venus will cause Mercury's perihelion to advance by 277 arcsec per century, Jupiter contributes another 153 arcsec, the Earth accounts for 90 arcsec, and Mars and the remaining planets are responsible for an additional 10

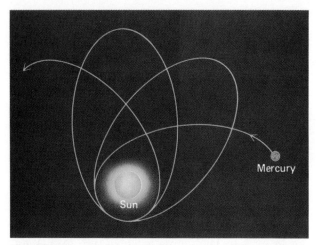

FIGURE 23.8 The major axis of the orbit of a planet, such as Mercury, rotates in space because of various perturbations. In the case of Mercury, the amount of the rotation is larger than can be accounted for by the gravitational forces exerted by the other planets.

arcsec. The total of these contributions amounts to about 531 arcsec per century. In the last century, however, it was observed that the actual advance is 574 arcsec per century. The discrepancy was first pointed out by Leverrier, codiscoverer of Neptune. In analogy with Neptune, it was assumed that an undiscovered planet was responsible. The hypothetical planet, which was supposed to be closer to the Sun than is Mercury, was even named for the god Vulcan. Vulcan, of course, never materialized, and that 43-arcsec anomaly was subsequently explained by relativity. The relativistic advance of perihelion

can also be observed in the orbits of several asteroids that come close to the Sun.

Time Delay of Light

According to general relativity, if light from a distant source passes very near the edge of the Sun, it will take longer for that light to reach the Earth than one would expect on the basis of Newton's law of gravity. The smaller the distance between the ray of light and the edge of the Sun at closest approach, the longer the delay in the arrival time. Half of the delay is a result of the gravitational redshift that occurs when light climbs out of a strong gravitational field. The other half of the delay is a consequence of the curvature of spacetime.

To see why curvature of spacetime causes a time delay, imagine a spacetime formed of a rubber sheet (Figure 23.9). Imagine the Sun as a heavy ball that causes the rubber sheet to sag in the middle. In this spacetime, light can travel only along the surface of the rubber sheet. When a light ray travels near the Sun, it must pass into the depression and therefore cover a longer distance than it would if the Sun were not present and there were no curvature of spacetime in this region.

The time delay in light that passes near the Sun is very small. For example, when Mars is beyond the Sun at superior conjunction, the delay in the arrival time of a radar signal sent from the Earth to Mars and reflected back again amounts to only 250 millionths of a second according to general relativity. The total

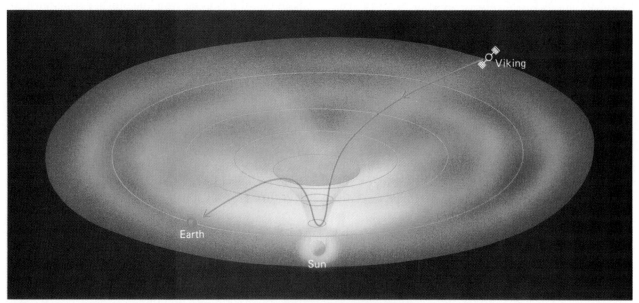

FIGURE 23.9 Radio signals from the Viking lander on Mars are delayed when they pass near the Sun, where spacetime is curved relatively strongly. In this picture, spacetime is pictured as a two-dimensional rubber sheet.

time required for the round trip is 42 min. In 250 millionths of a second, light travels 75 km, so measurement of the delay can be estimated by determining the difference in the true distance to Mars, which is known from its orbit, and its apparent distance as measured from the time required for a signal to reach the Earth from Mars.

The most accurate measurements of time delay were made by using signals sent from the Viking spacecraft that landed on Mars. The predictions of general relativity turned out to be correct within the accuracy of the observations, which was 1 part in 1000.

23.4 Black Holes

Let's do a thought experiment—that is, an experiment that is just in our heads, since it is impossible to do this experiment in a laboratory. We already know that a rocket must be launched from the surface of the Earth at a very high velocity if it is to escape the pull of the Earth's gravity. In fact, any object—rocket, bullet, ball—that is thrown into the air with a velocity that is less than 11 km/s will fall back to the Earth's surface. Only those objects launched with a velocity greater than 11 km/s can escape into space.

For the Sun, the escape velocity is higher yet—618 km/s. Now imagine that we begin to compress the Sun and force it to shrink in diameter. When the Sun reaches the diameter of a neutron star (less than 100 km), the velocity required to escape the gravitational pull of the shrunken Sun is about half the speed of light. Suppose we continue to compress the Sun to a smaller and smaller diameter. Ultimately, the escape velocity will exceed the speed of light. If light is composed of particles, one might expect it, too, just like rockets, to feel the influence of gravity and so be unable to escape. If no light can escape, the object is invisible.

In modern terminology, we call such an object a **black hole,** a name suggested by the American scientist John Wheeler in 1969. The idea that such objects might exist is, however, not a new one. Cambridge professor and amateur astronomer John Michell wrote a paper in 1783 about the possibility that stars might exist for which the escape velocity exceeds that of light. In 1796, the French mathematician Pierre Simon, Marquis de Laplace, wrote about similar objects, which he termed "dark bodies."

Behavior of Light Near a Black Hole

A rocket moving away from the Earth is slowed down by the Earth's gravitational attraction. Light, however, always moves at the same velocity, so photons are not slowed down when they move away from a strong source of gravity. Calculations based on Newton's theory of gravity cannot tell us what will happen to light when it attempts to escape from a black hole. The theory of general relativity is required to calculate what actually happens when the gravitational force becomes so large.

What the calculations show is that when the star becomes so dense that the escape velocity equals the speed of light, then everything, including light itself, is trapped inside. Nothing at all can escape through that surface where the escape velocity equals the speed of light. That surface is called the **event horizon,** and its radius is the Schwarzschild radius, named for Karl Schwarzschild, an astronomer who was a member of the German army in World War I and died of an illness on the Russian front in 1916. Schwarzschild was the first to use the equations of general relativity to work out what happens to light in such a situation. The event horizon is the boundary of the black hole. All that is inside is hidden forever from us.

The size of the event horizon is proportional to the mass of the collapsed object. For a black hole of 1 solar mass, the event horizon is about 3 km in radius; thus the entire black hole is about one-third the size of a neutron star of that same mass. The event horizons of larger black holes—if they exist—have greater radii. For example, if a globular cluster of 100,000 stars could collapse to a black hole, it would be 300,000 km in radius, a little less than half the radius of the Sun. If the entire Galaxy could collapse to a black hole, it would be only about 10^{12} km in radius—about 0.1 LY. On the other hand, for the Earth to become a black hole, it would have to be compressed to a radius of only 1 cm—about the size of a golf ball. A typical asteroid, if crushed to a small enough size to be a black hole, would have the dimensions of an atomic nucleus!

It happens that the correct size of the event horizon can be calculated by naively (and incorrectly) assuming that Newton's laws apply. The formula for the velocity of escape from a spherical body such as a planet or a black hole is $\sqrt{(2GM/R)}$, where M and R are the mass and radius of the spherical body, respectively. In the case of a black hole, the escape velocity is equal to c, the speed of light. Therefore, for a black hole

$$c = \sqrt{\frac{2GM}{R}}$$

If we solve for R, we find that the radius of the event horizon of a black hole of mass M is

$$R = 2GM/c^2$$

One Way to Make a Black Hole

Theory tells us what a black hole must be like. But do black holes exist? And if so, how were they created?

The most plausible way to manufacture a black hole is through the collapse of a star that has used up all of its store of nuclear energy. As we have seen, however, only massive stars can end their lives as black holes. Stars with initial main-sequence masses of six to eight times the mass of the Sun or less apparently complete their evolution by becoming white dwarfs. At the time that nuclear burning ceases, the mass of the core of these stars is less than 1.4 times the mass of the Sun, and electron degeneracy can halt the collapse. Stars with masses of 12 to about 100 times the mass of the Sun are thought to burn nuclear fuel until a core of iron and nickel is created. When this core forms, no more energy can be produced by the fusion of atomic nuclei, and the core begins to collapse. If the mass of this core is less than two to three times the mass of the Sun, then the formation of a core of densely packed neutrons is able to halt the collapse abruptly, and the resulting shock waves blow off the outer layers of the star and produce a Type II supernova explosion.

The critical question, and one that theory has not yet been able to answer, is this: Are the cores of some stars so massive that they exceed the upper limit on the mass of a neutron star (two to three times the mass of the Sun)? If so, then nothing that we now know of can halt the collapse, and a black hole will be formed.

If theory cannot yet tell us what conditions produce a black hole rather than a neutron star, then do observations provide any evidence that black holes actually do exist? And how would we go about looking for an object that we cannot see? The answer is that we can look for the gravitational effects of a black hole on a nearby star. As stars collapse into black holes, they leave behind their gravitational fields. If a black hole is a member of a double-star system, then we may be able to detect the black hole by studying the orbital motion of its companion.

Candidates for Black Holes

To find a black hole, we must first find a star whose motion (determined from the Doppler shift of its spectral lines) shows it to be a member of a binary-star system with a companion of mass too high to be a white dwarf or a neutron star. Second, that companion star must not be visible, for a black hole, of course, gives off no light. But being invisible is not enough, for a relatively faint star might be unseen next to the light of a brilliant companion. Therefore, we must also have evidence that the unseen star, of mass too high to be a neutron star, is a collapsed object—one of extremely small size.

Modern space astronomy provides a means to determine if a candidate object is collapsed. If matter falls toward or into a small object of high gravity (and possibly a black hole), the matter is accelerated to high speed. Near the event horizon of a black hole, matter is moving at velocities that approach the speed of light. Internal friction can heat it to very high temperatures—up to 100 million K or more. Such hot matter emits radiation in the form of x rays. What we are looking for, then, are x-ray sources associated with binary stars with invisible companions of high mass. We cannot prove that such a system contains a black hole, but at present we have no other theory for what the invisible massive companion can be.

We can easily understand the origin of the infalling gas that produces the x-ray emission. Stars in close binary systems can exchange mass, especially as one of the members evolves into a red giant and overflows into the gravitational field of the smaller companion. Suppose that one star in such a double-star system has evolved to a black hole and that the second star has become so large that its outer layers pass through the point of no return between the stars and fall toward the black hole. The mutual revolution of the giant star and black hole causes the material from the former to flow not directly into the black hole but, because of conservation of angular momentum, to spiral around it. The infalling gas collects in a disk of matter called an **accretion disk.** In the inner part of the accretion disk, the matter is revolving about the black hole so fast that internal friction heats it up to temperatures where it emits x rays.

Another way to form an accretion disk in a binary-star system is from material ejected from the companion of the black hole as a stellar wind. Some of that ejected gas will flow closely enough to the black hole to be captured by it into the disk (Figure 23.10). Such a case, we think, is the binary system containing the first x-ray source discovered in Cygnus—Cygnus X-1. The visible star is spectral type B.

Measurements of the Doppler shifts of the B star's spectral lines show that it has an unseen companion of mass nearly ten times that of the Sun. The companion is therefore too massive to be either a white dwarf or a neutron star. If the companion is a small collapsed object, then it must be a black hole. The x rays from it strongly suggest that it is indeed a small collapsed object, for we have thought of no other way to produce those x rays than in gas heated by an infall toward a tiny massive object.

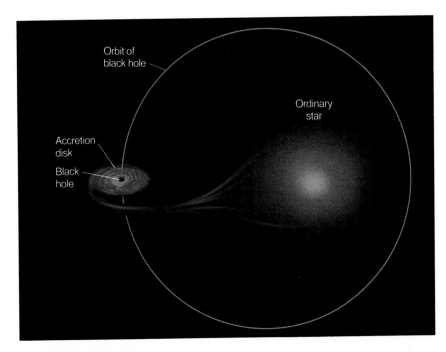

FIGURE 23.10 Mass lost from a giant star through a stellar wind streams toward a black hole and swirls around it before it finally falls in. In the inner portions of the accretion disk, the matter is revolving so fast that internal friction heats it to very high temperatures and x rays are emitted.

There are a few other binary systems, one of which is in the Large Magellanic Cloud, that meet all the conditions to be prime candidates for black holes.

Myths About Black Holes

Much of the modern folklore about black holes is misleading. One idea is that black holes are monsters that go about sucking things up with their gravity. Actually, the gravitational attraction at a large distance from a black hole is the same as that around any other star (or object) of the same mass. Even if another star, or a spaceship, were to pass one or two solar radii from a black hole, Newton's laws would give an excellent account of what would happen to it. It is only very near the surface of a black hole that its gravitation is so strong that Newton's laws break down. For a black hole of the mass of the Sun, light would have to come within 4.5 km of its center to be trapped. A solar-mass black hole, remember, is only 3 km in radius—a very tiny target. A star would be far, far safer to us as an interloping black hole than it would have been in its former stellar dimensions.

A Trip into a Black Hole

Still, it is interesting to contemplate a trip into a black hole. Suppose that the invisible companion of the star associated with Cygnus X-1 is a black hole of ten solar masses. What would you see if a daring astronaut bravely flies into it in a spaceship?

At first he darts away from you as though he were approaching any massive star. However, when he nears the event horizon of Cygnus X-1—some 30 km in radius and presumably near the center of the accretion disk—things change. The strong gravitational field around the black hole makes his clocks run more slowly as seen by you. Signals from him reach you at greatly increased wavelengths because of the gravitational redshift. As he approaches the event horizon, his time slows to a stop—as seen by you—and his signals are redshifted through radio waves to infinite wavelength. He fades from view as he seems to you to come to a stop, frozen at the event horizon.

All matter falling into a black hole appears to an outside observer to stop and fade at the event horizon, frozen in place, and taking an infinite time to fall through it—including the matter of a star itself that is collapsing into a black hole. For this reason, black holes are sometimes called *frozen stars.*

This, however, is only as you, well outside the black hole, see things. To the astronaut, time goes at its normal rate, and he crashes right on through the event horizon, noticing nothing special as he does so—except for enormously strong tidal forces that rip him apart. At least this is true of ten-solar-mass black holes. If there were a very massive black hole, say, thousands of millions of solar masses, its event horizon would be large enough that its tidal forces are not severe. An astronaut, in principle, could survive a trip into it.

But in no case is there an escape. Once inside, astronaut, light, and everything else are doomed to remain hidden forever from the universe outside. Moreover, current theory predicts that after entering the black hole, the astronaut races irreversibly to the center, to a point of zero volume and infinite density.

Mathematicians call such a point a **singularity**. At this point, the laws of physics as we now know them break down, and we can make no predictions about what will happen to matter at such extraordinary densities. We can never know what goes on inside the event horizon, but we know of no reason why black holes themselves should not exist.

SUMMARY

23.1.   The **equivalence principle** is the foundation of general relativity. According to this principle, there is no way that anyone, by experiment within a laboratory, spaceship, or other local environment, can distinguish between an acceleration of that environment and an external gravitational force. By considering the consequences of this principle, Einstein concluded that we live in a curved **spacetime.**

23.2   The universe and events within it are described in terms of a spacetime that is characterized by three spatial dimensions and a time dimension. In this four-dimensional spacetime, events that appear to one observer to occur at the same time may not appear to be simultaneous to another observer. The shortest path between two points of the curved surface of the Earth lies along a great circle or **geodesic.** In analogy, the paths of light and material objects in spacetime are also called geodesics. The distribution of matter determines the curvature of spacetime.

23.3   At low speeds and in weak gravitational fields, the predictions of general relativity agree with the predictions of Newton's theory of gravity. General relativity predicts, and Newtonian theory does not, that light climbing up out of a gravitational field will be redshifted; that starlight will be deflected when it passes near the Sun; and that the position where Mercury is at perihelion should change by 43 arcsec per century even if there were no other planets in the solar system to perturb its orbit. Furthermore, light that passes near the edge of the Sun will take longer to reach the Earth than would be expected on the basis of Newton's law of gravity. All of these predictions have been verified by experiment.

23.4   Theory suggests that stars with stellar cores more massive than three times the mass of the Sun at the time they exhaust their nuclear fuel and collapse will become **black holes.** Observations of binary stars that emit x rays suggest that a few black holes do exist. The surface surrounding a black hole where the escape velocity equals the speed of light is called the **event horizon.** Nothing, not even light, can escape through the event horizon from the black hole.

REVIEW QUESTIONS

1.   Restate the principle of equivalence in your own words.

2.   If general relativity offers the best description of what happens in the presence of gravity, why do physicists still make use of Newton's equations describing gravitational forces?

3.   Draw a diagram showing the progress in spacetime of an automobile traveling northward. For the first hour, in city traffic, it goes only 30 km/h. Then, in the country, for 3 h it goes 90 km/h. Finally, because the driver is late, he drives the car for 1 h at 140 km/h.

4.   Einstein's general theory of relativity made predictions about the outcome of several experiments that had not yet been carried out at the time that the theory was first published. Describe three experiments that verified the predictions of the theory.

5.   If a black hole emits no radiation, how can a scientist hope to verify that such strange objects actually exist?

THOUGHT QUESTIONS

6.   Consider a bucket nearly full of water. A spring is attached to the middle of the inside of the bottom of the bucket, and at the other end of the spring is a cork. The cork, trying to float to the top of the water, stretches the spring somewhat. Now suppose the bucket, with its water, spring, and cork, is dropped from a high building so that it remains upright as it falls. What happens to the spring and cork? Explain your answer thoroughly.

7.   A monkey hanging from a branch of a tree sees a hunter aiming a rifle directly at him. The monkey then sees a flash, telling him that the rifle has been fired. The monkey, reacting quickly, lets go of the branch and drops, so that the bullet will pass harmlessly over his head. Does this act save the monkey's life? Why?

8.   Some of the Skylab astronauts exercised by running around the inside wall of their cylindrical vehicle. How could they stay against the wall while running, rather than float aimlessly inside the Skylab? What physical principles are involved?

9.   Make up a new example of a geodesic in spacetime and show it on a spacetime diagram.

10.   Why would we not expect x rays from a disk of matter about an ordinary star?

11.   Look elsewhere in this book for the necessary data and indicate what the final stage of evolution—white dwarf, neutron star, or black hole—will be for each of the following kinds of stars:
   a. Spectral type O main-sequence star
   b. B main-sequence star
   c. A main-sequence star
   d. G main-sequence star
   e. M main-sequence star

12.   Which is likely to be more common in our Galaxy— white dwarfs or black holes?

PROBLEMS

13.   What would be the radius of a black hole with the mass of the planet Jupiter?

14.   Suppose that the Earth were collapsed to the size of a golf ball, becoming a small black hole. **(a)** What would be the revolution period of the Moon around it, at a distance of 400,000 km? **(b)** What would be the revolution period of a spacecraft orbiting at a distance of 6000 km? **(c)** What would be the revolution period of a miniature spacecraft orbiting at a distance of 0.1 m? **(d)** For this mini-spaceship, calculate its orbital speed and compare it with the speed of light.

15.   If the Sun could suddenly collapse to a black hole, how would the period of the Earth's revolution about it differ from what it is now?

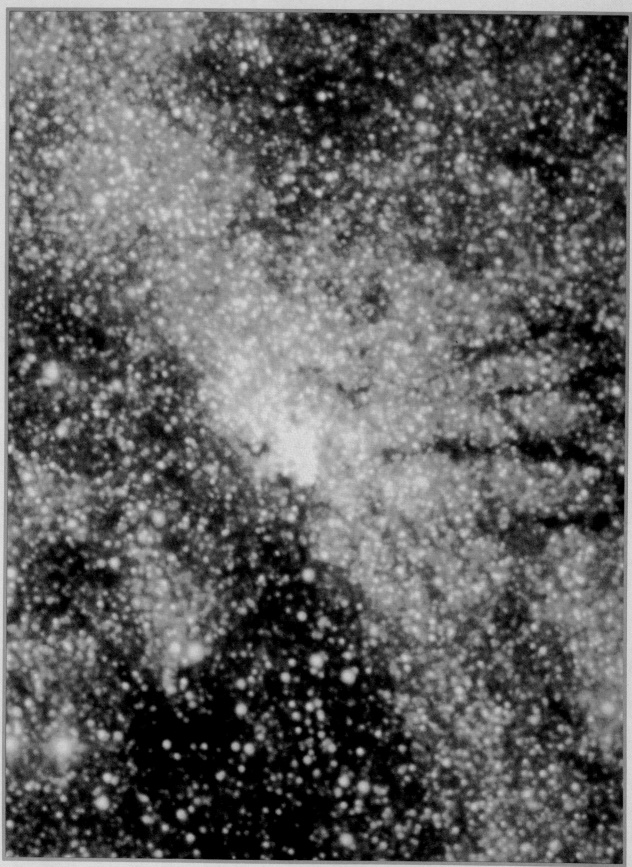

An infrared image of the central 150 LY of our Galaxy. The bright central region is the nucleus of the Galaxy, a region that some astronomers believe harbors a black hole. In optical images, the center of the Galaxy is completely obscured by dust. The color photograph is a composite of three images made through 1.2-μm, 1.65-μm, and 2.2-μm filters, represented by blue, green, and red, respectively, in the final image. *(Ian Gatley/ National Optical Astronomy Observatories)*

Harlow Shapley (1885–1972) began his career as a newspaper reporter. In his twenties he enrolled in the University of Missouri, where he searched through the catalogue for a suitable major and found astronomy under the A's. After earning his bachelor's degree, he went to Princeton for his Ph.D. Subsequently, at Mount Wilson, his study of globular clusters revealed the true extent of our Galaxy. *(Yerkes Observatory)*

24

THE MILKY WAY GALAXY

One of the most striking features in a truly dark sky is a band of faint white light that stretches from one horizon to the other (Figure 24.1). Because of its appearance, this band of light is called the Milky Way. In the Northern Hemisphere, the band is brightest in the region of the constellation Cygnus and is best viewed in the summer. In the Southern Hemisphere, the Milky Way is even brighter—so bright, in fact, that Indians in South America gave names to various portions of it just as northern astronomers gave constellation names to conspicuous groupings of stars. Unfortunately, the Milky Way is not bright enough to be seen from urban areas with their artificial lighting, and so many city dwellers have not seen the Milky Way.

In 1610, Galileo made the first telescopic observations of the Milky Way and discovered that it is composed of a multitude of individual stars. We now know that the Sun is located within a disk-shaped system of stars. The Milky Way is the light from nearby stars that lie more or less in the plane of the disk. We call this great stellar system, which includes all of the individual stars that we can see except with the largest telescopes, the **Milky Way Galaxy** or, more simply, just the **Galaxy.**

24.1 The Architecture of the Galaxy

William Herschel showed in 1785 that the stellar system to which the Sun belongs has the shape of a wheel or disk. Herschel used a telescope to count the numbers of stars that he saw in various directions. He found that the numbers of stars were about the same in any direction around the Milky Way, a result that seemed to show that the Sun was near the center of the Galaxy (Figure 24.2). We now know that Herschel was right about the shape of the Galaxy but wrong about the location of the Sun. Interstellar dust absorbs the light from stars and restricts optical observations in the plane of the Milky Way to stars within about 6000 LY. Herschel was able to observe only a tiny fraction of the system of stars that surrounds us.

Overview of the Galaxy

With modern instrumentation, astronomers can make observations at radio and infrared wavelengths, and electromagnetic radiation at these long wavelengths penetrates the dust easily. On the basis of these new observations, astronomers now picture the Galaxy as

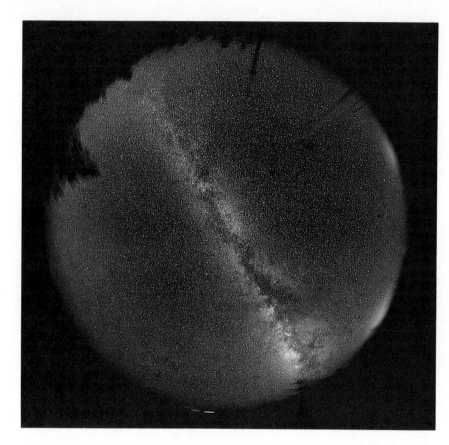

FIGURE 24.1 This full-sky view of the Milky Way was taken from the summit of Mount Graham, which is a mountain 3200 m high in southeastern Arizona. Note the dark rift through the center of the Milky Way. Around the horizon in a counterclockwise direction beginning from the bottom, one can see the lights of Willcox, Tucson, and Phoenix and the silhouettes of test towers and fir trees. The brightest point-like object in the lower left of the picture is the planet Jupiter. *(Roger Angel, Steward Observatory/University of Arizona)*

a thin, circular **disk** of luminous matter distributed across a region about 100,000 LY in diameter, with a thickness of about 1000 LY. Dust and gas, the raw material from which stars form, are fairly closely confined to the galactic disk. The disk of the Galaxy is apparently embedded in a **halo** of largely nonluminous or dark matter that extends to a distance of at least 150,000 LY from the galactic center. The globular clusters are also distributed in a much larger spherical halo around the galactic center. The Sun orbits the center of the Galaxy at a distance of nearly 28,000 LY.

Within 3000 LY of the galactic center, the stars are no longer confined to the disk but rather form a spheroidal **nuclear bulge** of old stars. In ordinary visible radiation it is possible to observe stars in the

FIGURE 24.2 A copy of a diagram by William Herschel, showing a cross-section of the Milky Way system based on counting stars in various directions. The large circle shows the location of the Sun.

ESSAY   Light Pollution

Bright lights have robbed most city dwellers of the pleasure of looking at the stars. Many people have never seen the Milky Way. "Light pollution" has brightened the night sky and made it difficult or impossible to see the constellations, watch for meteor showers, or identify the occasional passing comet. Artificial lights are also a problem for many major observatories, making it impossible to detect faint stars and galaxies against the bright sky background. Indeed, Mount Wilson Observatory was recently closed in part because of increasing skyglow from Los Angeles Basin.

Lights, you may say, are the price we must pay for living in large cities. Lights are essential for safety and security.

Fortunately, there is a solution that provides high-quality lighting while minimizing the amount of light pollution. The key is first of all to direct the light to where it is most useful. That means that outdoor lighting should be shielded so that no light is emitted directly upward, where it does no good, but down toward streets and sidewalks and sports fields, where it is needed. Engineers have designed lighting fixtures that shield street lights in such a way that no light is projected upward.

The second step is to choose the proper source of light. Figure 24A shows the spectra of several types of light sources. Note that the low-pressure sodium lamp emits light at only a few discrete wavelengths. The rest of the spectrum is free of contaminating light, and this is the kind of light that astronomers prefer. Fortunately, low-pressure sodium lights are also extremely energy-efficient, are the most cost-effective form of street lighting, and are the best light for visual acuity, since they emit most of their energy in the yellow part of the spectrum, where the eye is most sensitive.

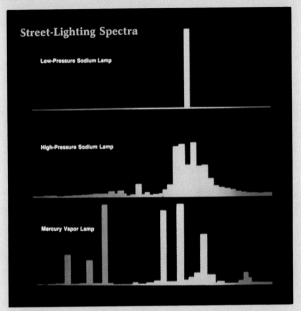

FIGURE 24A Comparison of the spectra of three forms of street lighting. Mercury vapor lights are the most common, but much of the energy is emitted in the red and blue regions of the spectrum, at wavelengths to which the eye is not very sensitive. In contrast, low-pressure sodium lamps emit all of their energy in a very narrow wavelength region where the eye is very sensitive. This type of lighting leaves most of the spectrum free of radiation that would contaminate observations of faint stars and galaxies.

Check the lights in your own city. What type of lights are they? Are they shielded? Notice the city lights next time you are in a plane at night. Do you see the lights themselves? If so, then some of the light is being emitted directly upward, where it does no good. Or do you see only the light reflected from the pavement of the streets? In this case, the lighting is properly shielded. If you live in or near a large city, check the sky. What part of it is contaminated with artificial light? In what directions is it brightest? What is the faintest star you can see? Now drive away from the city and notice how the appearance of the sky changes. How far do you have to go before the sky appears to be truly dark?

FIGURE 24.3 Although the nucleus of our Galaxy is completely hidden by dust at visible wavelengths, there are regions where the obscuration is low and we can see into the cloud of old, population II stars that are concentrated toward the center and form the nuclear bulge. This is a picture of such a region, which is called Baade's window. The density of stars in this photograph peaks at a distance of about 28,000 LY. The brightest star in the picture is Gamma Sagittarii, a deep yellow star that appears in the foreground at a distance of about 100 LY and can be seen with the unaided eye. *(Anglo-Australian Telescope Board)*

bulge only in directions where the obscuration by interstellar dust is unusually low (Figure 24.3). The first picture that actually succeeded in showing the bulge as a whole was taken at infrared wavelengths from a satellite orbiting the Earth (Figure 24.4).

A schematic diagram of the Galaxy is shown in Figure 24.5. The Sun lies far from the galactic center. The young stars in the disk of the Galaxy are concentrated in a series of **spiral arms.** In its overall characteristics, the Milky Way Galaxy resembles the Andromeda Galaxy (M31), which is at a distance of about 2.3 million LY (Figure 24.6).

Globular Clusters and the Center of the Galaxy

This picture of the Galaxy is one of the major achievements of 20th-century astronomy. Until early in this century, astronomers accepted Herschel's conclusion that the Galaxy was centered approximately at the Sun and extended only a few thousand light years from it. The shift from the "heliocentric" to the "galactocentric" view of the Milky Way Galaxy, as well as the first knowledge of its true size, came about largely through the efforts of Harlow Shapley and his investigation of the distribution of globular clusters. Because of their brilliance, and the fact that they are not confined to the central plane of the Galaxy, where they would otherwise be largely obscured by interstellar dust, globular clusters can be observed (with telescopes) to very large distances.

Most globular clusters contain at least a few RR Lyrae variable stars (Section 17.3), whose absolute magnitudes are known. The distance to an RR Lyrae star in a globular cluster, and hence to the cluster

FIGURE 24.4 This image presents a view of the Milky Way Galaxy obtained by an experiment aboard the Cosmic Background Explorer Satellite (COBE). This image was taken in the near-infrared part of the spectrum and permits us to see, for the first time, the bulge of old stars that surround the center of our Galaxy. Redder colors correspond to regions where there is more dust, and this dust forms a thin disk of material. In optical regions of the spectrum, this dust absorbs so much radiation that we cannot see the bulge of old stars. *(NASA).*

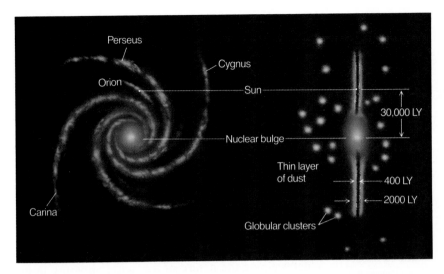

FIGURE 24.5 Schematic representation of the Galaxy. The Sun is located on the inside edge of the short Orion spur.

itself, can therefore be calculated from its observed apparent magnitude. Shapley used this technique to obtain distances to globular clusters.

In 1917, Shapley used the distances and directions of 93 globular clusters to map out their three-dimensional distribution in space. He found that the clusters form a roughly spheroidal system. The center of that spheroidal system is not at the Sun, however, but at a point in the middle of the Milky Way in the direction of Sagittarius, and at a distance of some

FIGURE 24.6 The spiral galaxy in Andromeda (M31) looks very much like our own Galaxy. Note the bulge of older yellowish stars in the center, the bluer and younger stars in the outer regions, and the dust in the disk that blocks some of the light from the bulge. *(Caltech/Palomar Observatory)*

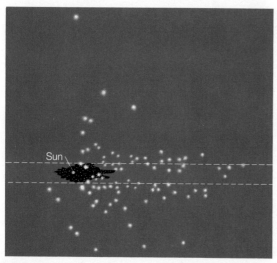

FIGURE 24.7 A copy of a diagram by Shapley, showing the distribution of globular clusters in a plane perpendicular to the Milky Way and containing the Sun and the center of the Galaxy. Herschel's diagram (Figure 24.2) is shown centered on the Sun, approximately to scale.

25,000 to 30,000 LY (Figure 24.7). Shapley then made the bold—and correct—assumption that the system of globular clusters is centered upon the center of the Galaxy. Today this assumption has been verified by many pieces of evidence, including the observed distributions of globular clusters in other spiral galaxies.

The Galactic Halo

The globular clusters are distributed in a sphere centered on the center of the Galaxy. A sparse "haze" of individual stars—not members of clusters but far outnumbering the cluster stars—also exists in the region outlined by the globular clusters. This haze of stars and clusters forms the galactic halo, a region whose volume exceeds that of the main disk of the Galaxy by many times. Individual RR Lyrae stars have been found in significant numbers as far away as 30,000 to 50,000 LY on either side of the galactic plane, which shows that the halo must have an overall thickness of at least 100,000 LY.

Interstellar Matter

In addition to stars, the disk of the Galaxy contains interstellar gas and dust. Observations at 21 cm show that the neutral atomic hydrogen in the Galaxy is confined to an extremely flat layer, which is only about 400 LY thick. In the plane of the Galaxy, this hydrogen extends well beyond the Sun to a distance of about 80,000 LY from the center of the Galaxy.

Dust, too, is confined to the disk, with the highest concentrations occurring in the spiral arms. The thickness of the dust layer is also about 400 LY.

There is very little emission from dust lying outside the Sun's orbit around the center of the Galaxy.

The most massive molecular clouds are found in the spiral arms. In many cases, individual clouds have gathered into large complexes containing a dozen or more discrete clumps. Since the large molecular clouds and molecular cloud complexes are the sites where star formation occurs, most young stars are also to be found in spiral arms. That is why the spiral arms shine so brilliantly in photographs of spiral galaxies (Figure 24.8).

24.2 Spiral Structure of the Galaxy

Studies of the 21-cm line have played a key role in determining the nature of the spiral structure in our own Galaxy. While the interpretation of the measurements is somewhat controversial, it appears likely that the Galaxy has four major spiral arms, with some smaller spurs (Figure 24.5). The Sun appears to be near the inner edge of a short arm or spur called the *Orion arm*, which is about 15,000 LY long and contains such conspicuous features as the Cygnus Rift (the great dark nebula in the summer Milky Way) and the Orion Nebula. More distant, and therefore less conspicuous, are the *Sagittarius-Carina* and *Perseus arms*, which are located, respectively, about 6000 LY inside and outside the Sun's position relative to the galactic center. Both of these arms, and the Cygnus arm, are about 80,000 LY long. The fourth arm is unnamed and is difficult to detect because emission from it is confused with strong emission from the central regions of the Galaxy, which lie between it and us.

Formation and Permanence of Spiral Structure

At the Sun's distance from its center, the Galaxy does not rotate like a solid wheel. Individual stars obey Kepler's third law. Remember that Pluto takes longer than the Earth to complete one full circuit around the Sun. In just the same way, stars in larger orbits do not keep abreast of those in smaller ones but trail behind. This effect is called *differential galactic rotation*.

It is not surprising that much of the interstellar material in our Galaxy is concentrated into elongated features that resemble spiral arms. No matter what the original distribution of the material might be, the differential rotation of the Galaxy would be expected to form it into spirals. Figure 24.9 shows the development of spiral arms from two irregular blobs of interstellar matter, as the portions of the blobs closest to the galactic center move fastest, while those farther away trail behind.

FIGURE 24.8 The spiral galaxy NGC 2997. The two spiral arms, which appear to originate in the yellow nucleus, are peppered with bright red blobs of ionized hydrogen that are similar to regions of star formation in our own Milky Way Galaxy. The hot blue stars that generate most of the light in the arms of the galaxy form within these hydrogen clouds. *(Anglo-Australian Telescope Board)*

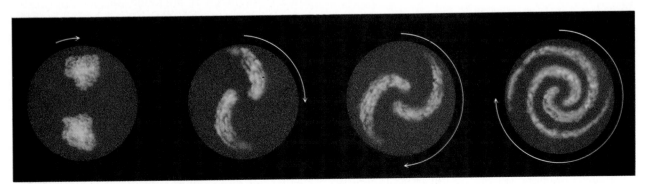

FIGURE 24.9 Hypothetical formation of two spiral arms from irregular clouds of interstellar material based simply on the rotation of the Galaxy, which is more rapid for objects closer to the center.

It is harder to understand, however, why the arms are not wound more tightly than they are. At the Sun's distance from the center of the Galaxy, the Galaxy rotates once in about 2×10^8 years. Its total age, however, is believed to be about 10^{10} years. In this time, a point near the Sun has made at least 50 revolutions. With so many turns, we would expect the spiral arms to be wound very tightly and to lie very much more closely together than they do.

How can spiral structure last for the lifetime of a galaxy? To answer this question, astronomers have calculated how stars and gas clouds would move if they had circular paths about the galactic center and were influenced both by the gravitational fields produced by the Galaxy as a whole and by the matter forming the spiral arms themselves. The calculations show that objects should slow down slightly in the regions of the spiral arms and linger there longer than elsewhere in their orbits. Thus a wave of higher density builds up where the spiral arms are. The regions of higher density rotate more slowly than does the actual material in the Galaxy, so that the stars, gas, and dust pass slowly through the spiral arms. This theory for the formation of spiral arms is referred to as the **spiral density wave** model.

As a good analogy for what happens, suppose you are driving on a freeway with three lanes. Suppose that there are cars that are moving unusually slowly in all three lanes ahead of you. Traffic behind these three cars will be forced to slow down, and the density of cars will increase. Some individual cars may manage to get past the three slowly moving cars, but others will take their place in the traffic jam. Viewed from high above the freeway, the point of maximum density of cars would appear to move along the freeway at the same speed as that of the three slowly moving cars. The place of maximum density would thus be moving more slowly than the cars either well in front of the traffic jam or behind it. In just the same way, stars slow down when they pass through the spiral arms, which are the places of maximum density.

As gas and dust clouds approach the inner boundaries of an arm and encounter the higher density of slowly moving matter, they collide with it. It is here, where the shock of the collision occurs, that theory predicts star formation is most likely to take place. We know from our own Galaxy that the youngest stars are found in the spiral arms. In some other galaxies, where we can view the spiral arms face-on, we see young stars, along with the densest dust clouds, near the inner boundaries of spiral arms, just as theory predicts (Figure 24.10). Spiral density waves have also been observed directly in the rings of Saturn (Section 10.5).

There may, however, be other ways, unrelated to spiral density waves, to produce elongated structures that look like spiral arms. For example, star formation may move progressively through molecular clouds and produce extended regions of young stars that mimic spiral arms (Section 20.1). We have one fairly satisfactory theory—the spiral density wave theory—for explaining the structure of the Milky Way Galaxy. We should be aware, however, that there may be other ways to develop spiral structure, and we cannot yet be certain about how our Galaxy evolved.

24.3 Stellar Populations in the Galaxy

Striking correlations are found between the velocities of stars and their other characteristics, including age, chemical composition, velocities, and location in the Galaxy. Some classes of stars, which are referred to as a group as **population I** stars, are found only in regions of interstellar matter. Such stars are restricted to the disk, are especially concentrated in the spiral arms, and are described as low-velocity objects. Examples are bright supergiants, main-sequence stars of high luminosity (spectral classes O and B), and young open star clusters. Molecular clouds are found in similar places and also have low velocities.

The distributions of some other classes of objects, which are collectively referred to as **population II** stars, show no correlation with the location of spiral arms. These objects are found throughout the disk of the Galaxy, with the greatest concentration toward the nuclear bulge. They also extend into the galactic halo. Examples are planetary nebulae, since most planetary nebulae are formed by old stars with masses similar to that of the Sun, and RR Lyrae variables. The globular clusters, which are also classified as population II, are found almost entirely in the halo and central bulge of the Galaxy. Objects in this category typically have high velocities relative to the Sun.

Discovery of Two Types of Stellar Populations

The terms "population I" and "population II" were first applied to different classes of stars by Walter Baade, of Mount Wilson Observatory, from work carried out during World War II. As a German national, Baade was not allowed to become involved in war research, so he was able to make a great deal of

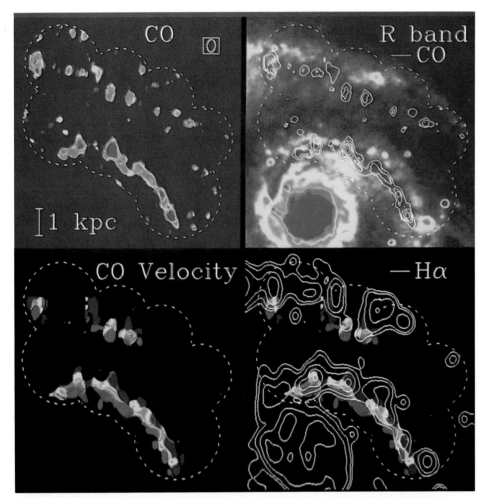

FIGURE 24.10 Measurements of M51, the Whirlpool Galaxy, demonstrate that spiral density waves lead to star formation. All four panels show the northwest part of M51. The upper left picture shows the locations of molecular clouds, which are detected by measurements of carbon monoxide (CO). Two segments of spiral arms are clearly seen. In this picture, the molecular gas is moving counterclockwise. In the upper right picture, contour lines show the locations of the same molecular clouds relative to the visible spiral arms. Note that the clouds lie in dusty regions along the inner part of each spiral arm. In the lower pictures, the molecular clouds of highest velocity have been color-coded red; those of lower velocity are blue. As the gas moves from red areas to blue, it slows by 20 to 30 km/s. The contours in the right-hand plot show H II regions (ionized hydrogen gas); these are regions containing massive hot stars, so star formation has occurred recently here. Note that star formation occurs downstream (that is, in a counterclockwise direction) from the molecular clouds. This spatial relationship shows that star formation occurs *after* the molecular clouds have been slowed down and compressed by their passage through regions of higher density. *(Courtesy S.N. Vogel, University of Maryland)*

use of the Mount Wilson telescopes. Aided by reduced sky brightness resulting from the wartime blackout of Los Angeles, Baade was able to distinguish the two populations of stars in the nearby Andromeda Galaxy. He was impressed by the similarity of the stars in the nuclear bulge of the Andromeda Galaxy to those in the globular clusters and the halo of our own Galaxy, and by the differences between all of these stars and those found in the spiral arms. On

this basis, he referred to all the stars in the halo, nuclear bulge, and globular clusters as population II. The bright blue stars in the spiral arms were called population I.

Today we can interpret the phenomenon of different stellar populations in the light of stellar evolution. Thus, population I comprises stars of many different ages, including some that were recently formed or are still forming from gas and dust in the galactic disk.

Population II, on the other hand, consists entirely of old stars, formed early in the history of the Galaxy.

Chemical Composition

Measurements show that there are differences in the chemical compositions of population I and population II stars. Nearly all stars appear to be composed mostly of hydrogen and helium, but the abundance of the heavier elements is not the same in all stars. In the Sun and in other population I stars, the heavy elements (elements heavier than hydrogen and helium) account for about 1 to 4 percent of the total stellar mass. Population II stars in the outer galactic halo and in globular clusters have much lower abundances of the heavy elements—often less than one-tenth or even one-hundredth that of the Sun.

The abundance of heavy elements in stars varies systematically with the time that stars were born. Old population II stars, which were formed 10 billion or more years ago, have a lower abundance of heavy elements than do the Sun and other population I stars. This result indicates that the heavy-element content of the Galaxy has increased over time. As discussed in earlier chapters, we expect heavy elements to be created in stars. These heavy elements are returned to the interstellar medium when stars die, only to be recycled by becoming part of a new generation of stars.

Since Baade's pioneering work, we have learned that the notion that all stars can be characterized as either old, with low abundances of elements heavier than hydrogen and helium, or young and rich in heavy elements is an oversimplification. The stars in the nuclear bulge of the Galaxy are all fairly old. Their mean age is in the range of 11 to 14 billion years, and none are younger than about 5 billion years, yet the abundance of heavy elements in these stars is about twice that of the Sun. Astronomers think that star formation in the nuclear bulge occurred very rapidly shortly after the Milky Way Galaxy formed, so even stars 11 to 14 billion years old were enriched with heavy elements expelled in supernova explosions of the first generations of massive stars.

Completely the opposite situation occurs in the Small Magellanic Cloud. Even the youngest stars in this galaxy are deficient in heavy elements, presumably because star formation has occurred so slowly that there have been, so far, relatively few supernova explosions. Therefore, the abundance of a star depends not only on when it forms but also on the star formation history of the parent galaxy prior to the time that the star began its gravitational collapse from interstellar matter.

High- and Low-Velocity Stars

The majority of the stars near the Sun are population I stars and move nearly parallel to the Sun's path about the galactic nucleus. Their speeds with respect to the Sun are generally less than 40 or 50 km/s. For this reason, population I stars are sometimes also referred to as **low-velocity** stars. The radial velocities of nearby gas clouds are also low. The gas clouds, like the Sun, move in roughly circular orbits about the galactic nucleus. Population I objects do not move very far from the plane of the Milky Way Galaxy.

Population II objects, including globular clusters, typically have speeds relative to the Sun in excess of 80 km/s and are called **high-velocity** stars. They move along elongated elliptical orbits that cross the disk of the Galaxy at rather large angles (Figure 24.11). These orbits are rather like the orbits of comets revolving about the Sun in the solar system.

The term "high velocity" or "low velocity" refers to the speed of an object *with respect to the Sun* and has nothing to do with its motion in the Galaxy. Most high-velocity stars lag behind the Sun in its motion about the galactic center and hence are actually revolving about the Galaxy with speeds *less* than those of the low-velocity stars near the Sun.

24.4 The Mass of the Galaxy

Nearby external galaxies can be used to deduce the motion of the Sun in the Galaxy. The Large Magellanic Cloud, for example, has a radial velocity of recession of about 270 km/s, while the Andromeda Galaxy has a velocity of approach of about 300 km/s. When measurements of other galaxies and globular clusters are combined, they indicate that these velocities are caused not by the motion of Andromeda and the Large Magellanic Cloud but by the motion of the Sun. The Sun is moving in the general direction away from the Large Magellanic Cloud and toward the Andromeda Galaxy.

More precisely, the Sun is moving in the direction of the constellation Cygnus, with a speed of about 220 km/s. This direction lies in the Milky Way and is about 90° from the direction of the galactic center, which shows that the Sun's orbit is nearly circular and lies in the main plane of the Galaxy. As viewed from the north side of the galactic plane, the orbital motion of the Sun is clockwise.

The period of the Sun's revolution about the nucleus, the *galactic year*, can be found by dividing the circumference of the Sun's orbit by its speed; it comes out roughly 200 million (2×10^8) of our terrestrial years. We can observe, therefore, only a

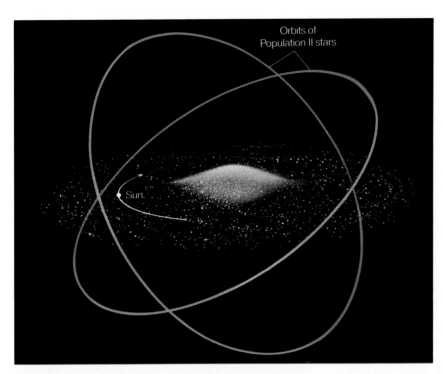

FIGURE 24.11 The orbits of stars in the Galaxy. The Sun and other population I stars orbit in the plane of the Galaxy. Population II stars have orbits that move high above and below the plane.

"snapshot" of the Galaxy in rotation. We do not actually see stars traverse appreciable portions of their orbits.

An Estimate of the Mass of the Galaxy

We can make an estimate of the mass of the inner part of the Galaxy (the part that lies inside the Sun's orbit) with an application of Kepler's third law (as modified by Newton). Assume the Sun's orbit to be circular and the Galaxy to be roughly spherical, so we can treat it as though its mass internal to the Sun were concentrated at a point at the galactic center. If the Sun is 28,000 LY from the center, its orbit has a radius of 1.8×10^9 AU (there are 6.3×10^4 AU in 1 LY). Since its period is 2×10^8 years, we have

$$\text{mass}_{\text{Galaxy}} = \frac{(1.8 \times 10^9)^3}{(2 \times 10^8)^2} = 10^{11} \text{ solar masses}$$

More sophisticated calculations based on complicated models give a similar result.

It must be emphasized that this is only the mass contained in the volume inside the Sun's orbit. It is a good estimate for the total mass of the Galaxy if and only if no more than a small fraction of its mass is to be found beyond the radius that marks the Sun's distance from the galactic center. For many years, astronomers thought this assumption was reasonable, since the number of bright stars and the amount of luminous matter drop dramatically at distances more than about 30,000 LY from the galactic center.

A Galaxy of Mostly Invisible Matter

In science, reasonable assumptions can turn out to be wrong. Observations now show that while there is relatively little luminous matter lying beyond 30,000 LY, there must be a lot of nonluminous, invisible, dark matter at large distances from the galactic center.

We can understand how astronomers reached this conclusion by remembering that according to Kepler's third law, objects orbiting at large distances from a massive object move more slowly in their orbits than do objects closer to that central mass. We have already seen an example of this idea in the case of the solar system. The outer planets move more slowly in their orbits than do the ones close to the Sun.

There are a few objects, including gas clouds and globular clusters, that lie well outside the luminous boundary of the Milky Way. If all of the mass of our Galaxy were concentrated within the luminous region, then these very distant objects should travel around their galactic orbits at lower speeds than, for example, the Sun does. In just the same way, the velocity of Pluto in its orbit is slower than the velocity of the Earth.

In fact, globular clusters and other objects at large distances from the luminous boundary of the Milky Way Galaxy are *not* moving more slowly than the Sun. As we look outward to objects between 30,000 and 150,000 LY from the center of the Galaxy, we

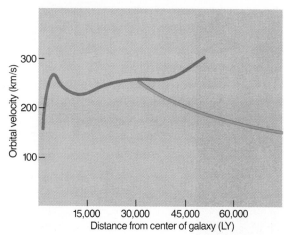

FIGURE 24.12   The rotation curve of carbon monoxide (CO) gas in the Milky Way Galaxy is shown in red. The blue curve shows what the rotation curve would have looked like if all of the matter in the Galaxy were located inside a radius of 30,000 LY.

find that their orbital velocities remain constant at about 250 km/s (Figure 24.12). The only way that this can happen is if there is a huge amount of additional mass beyond the *visible* boundary of the Milky Way Galaxy that can hold these objects in their high-speed orbits. This matter is, except for its gravitational force, entirely invisible and undetectable!

Studies of the motions of very remote globular clusters show that the total mass of the Galaxy out to a radius of 150,000 LY is about 10^{12} solar masses, ten times greater than the amount of matter inside the solar orbit. Theoretical arguments suggest that this **dark matter** is distributed in a spherical halo around the Galaxy. But what is it?

Since this matter is invisible, it cannot be in the form of ordinary stars. It cannot be gas in any form. If it were neutral hydrogen, its 21-cm radiation would have been detected. If it were ionized hydrogen, it would emit visible radiation. If it were molecular hydrogen, absorption bands would have been observed in ultraviolet spectra of objects lying beyond the Galaxy. The halo cannot consist of interstellar dust, since dust in the required quantities would block the light from distant galaxies. The dark matter cannot be black holes of stellar mass or neutron stars, or the accretion of interstellar matter onto such objects would produce more x rays than are observed. Also, formation of black holes and neutron stars is preceded by supernova explosions, which scatter heavy elements into space to be incorporated into subsequent generations of stars. If the halo mass consisted of stellar-sized black holes and neutron stars, then the young stars that are shining today would have to contain much larger abundances of heavy elements than they actually do.

The possibilities that remain to account for the dark matter in the Galaxy are low-mass objects such as brown dwarfs, white dwarfs that formed from an early generation of stars and that have now cooled and ceased to shine, black holes that have masses a million times the mass of the Sun, or exotic subatomic particles of a type not yet detected on Earth. As we shall see, dark matter is a major constituent of other galaxies as well (Chapter 27).

Stop a moment to think how startling this conclusion is. About 90 percent of the mass in our Galaxy is invisible, and we do not even know what it is made of!

24.5   The Nucleus of the Galaxy

The center of the Galaxy lies in the direction of the constellation Sagittarius. We cannot see the nucleus in visible light or in the ultraviolet region, because those wavelengths are absorbed by the intervening interstellar dust (Figure 24.13). In the optical region

FIGURE 24.13   This wide-angle picture covers over 50° of the sky in the direction of the center of the Milky Way. The galactic center is totally obscured at optical wavelengths. *(Anglo-Australian Telescope Board 1980)*

FIGURE 24.14 Infrared radiation can penetrate the dust that lies between us and the center of the Milky Way Galaxy. This photograph at a wavelength of 2200 nm shows the cluster of cool stars at the center of the Galaxy. *(National Optical Astronomy Observatories)*

of the spectrum, light from the central region of the Galaxy is dimmed by a factor of a trillion (10^{12}). High-energy x rays and gamma rays, however, force their way through the interstellar medium and can be recorded by instruments orbiting above the Earth's atmosphere. Also, infrared and radio radiation, whose wavelengths are long compared with the sizes of the interstellar grains, flows around the dust particles and reaches us from the center of the Galaxy (Figure 24.14). The very bright radio source in that region, known as Sagittarius A, was the first cosmic radio source discovered.

Astronomers have used observations at all of these wavelengths to try to determine what lies at the center of the Milky Way. The question of most interest to astronomers is whether or not there is a black hole at the center of the Galaxy. A great deal of effort has gone into trying to answer this question, and the jury is still out.

The Central Few Light Years

What do these observational techniques tell us about the structure of the nucleus of the Milky Way Galaxy? The center of the Galaxy is surrounded by a clumpy and rather irregular disk of molecular clouds. These clouds form a donut-shaped ring that has an inner diameter of about 10 LY. The outer boundary of the ring extends to a distance of at least 25 LY from the galactic center. The individual molecular clouds have sizes in the range typically 0.5 to 1.5 LY (Figure 24.15). The ring appears to be rotating about the

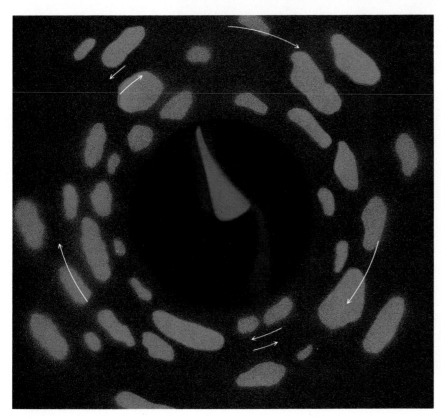

FIGURE 24.15 Schematic drawing of the central 20 LY of the Galaxy. Streamers of ionized gas are falling into the center of the Galaxy. Surrounding the central regions is a ring of dust, and beyond this lie individual clouds of gas.

galactic center, but the motions of the individual clumps are turbulent, and the clouds must sometimes collide with one another.

It is possible to measure the gas in the ring spectroscopically and so derive its velocity from the Doppler effect. To force the material in the ring to follow a circular orbit around the nucleus of the Galaxy, mass must be concentrated in the nucleus in sufficient quantity to exert the required gravitational force. From the observed rotational velocity of the dust ring, we estimate that the total mass required is 2 to 5 million times the mass of the Sun.

Measurements show that the density of gas and dust is 10 to 100 times higher in the ring than in the region interior to the ring. That is, the ring surrounds a nearly empty cavity (empty except, as we shall see, for stars and possibly a black hole). Within this cavity, there is very little dust or cold gas. There are some streamers of hot, ionized gas in the cavity (Figure 24.16), but the total mass of the ionized gas in the central few light years of the Galaxy is only about 100 times the mass of the Sun. The most widely accepted idea is that this hot gas is falling from the ring toward the center of the Galaxy. As clouds in the ring collide,

the velocity of some of the material in the clouds is slowed to the point that the material can no longer remain in orbit. This material then spirals inward toward the galactic center.

The cavity and the surrounding ring cannot last forever. Over a period of only 100,000 years, material falling inward will blur the sharp edge that now separates the ring from its nearly empty interior. Over about that same time interval, collisions between the clouds will tend to eradicate the individual clumps and produce a smooth ring of material. In the past 100,000 years, some remarkable event must have cleared out the cavity and severely disturbed the ring of gas. One possibility is an explosion of some kind at the center of the Galaxy. According to this hypothesis, the explosion blew material outward and cleared the cavity of dust and gas.

The measurements of the amount of gas in the cavity still do not account for most of the mass—a few million times the mass of the Sun—in the center of the Galaxy. There are also stars inside the ring (Figure 24.17). These stars are relatively close together. The average distance between stars is perhaps 1/300 the average distance between stars in the vicinity of

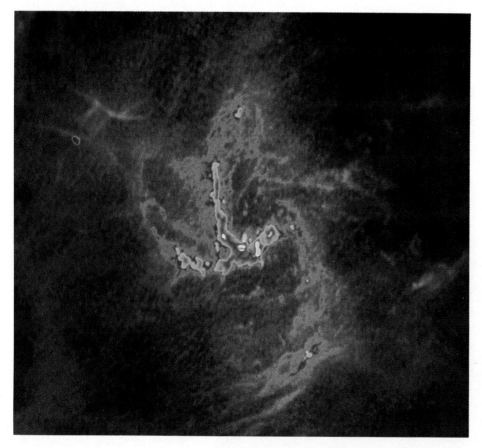

FIGURE 24.16 An image of a region approximately 10 LY across at the center of our Galaxy shows radio emission from hot gas. The most likely interpretation is that this gas is falling into the nucleus. The emission is strongest (red color) where the gas is most dense. The mass of this gas is about 100 times the mass of the Sun. The bright red region in the middle is the compact radio source at the center of the Galaxy. *(National Radio Astronomy Observatory/AUI)*

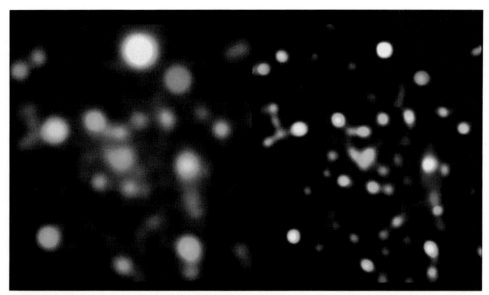

FIGURE 24.17   Infrared image of the stars in a region 2 LY across surrounding the galactic center. The picture shows the original image on the left and a computer-enhanced version, on the right, of the center of the Milky Way. The observations were made at 1.65 μm, 2.2 μm, and 3.45 μm, which are represented in the color image by blue, green, and red, respectively. The different colors of the stars are indicative of age, composition, and absorption by obscuring gas and dust clouds. *(Darren DePoy and Nigel Sharp/National Optical Astronomy Observatories)*

the Sun. But even the mass of the stars is not enough to add up to the 2 to 5 million solar masses that we are looking for. Is this mass hidden inside a black hole?

The Central Energy Source— A Black Hole?

A black hole itself emits no electromagnetic radiation. We cannot hope to *see* a black hole in the center of the Galaxy even if one is there. We must use circumstantial evidence, as they say in mystery stories, to decide whether or not a black hole is the only plausible explanation for the kinds of things we *do* see.

The most direct way to prove that there is a black hole in the center of the Galaxy is to show that there is a large amount of mass within a very tiny volume. We can measure the velocities of stars and gas that lie very close to the galactic center and then determine what gravitational force is exerted by the mass concentrated at the center. These measurements indicate that the amount of mass located within 1 to 2 LY of the center of the Galaxy is equal to 3 to 4 million times the mass of the Sun.

If this mass is actually contained within a black hole, then the size of the black hole must be much smaller than 1 to 2 LY. A black hole with a mass of about 3 million solar masses would have a radius similar to that of our own Sun. Unfortunately, we cannot see fine enough detail with existing telescopes to measure the velocities of gas and stars that are only a few tens of astronomical units from the center of the Galaxy. Therefore, we do not yet know if the mass at the center of the Galaxy is as highly concentrated as we would expect for a black hole.

There is indirect evidence, however, that the mass may indeed be very highly concentrated. There is an intense radio source, which is called Sagittarius A*, at or very near what we believe is the exact center of the Galaxy. Measurements with the VLA radio telescope show that the diameter of this radio source is no larger than the diameter of Jupiter's orbit (10 AU), and it may be even somewhat smaller.

If there were a black hole at the center of the Galaxy, it could account for the intense radio emission from Sagittarius A*. Matter—gas, dust, and even perhaps stars—is attracted by the gravitational force of a black hole. Material spirals in toward the black hole and forms an *accretion disk* of material around it. As the material spirals ever closer to the black hole, it accelerates and heats through compression to millions of degrees. This hot matter then would be the source of the radio emission from the galactic center.

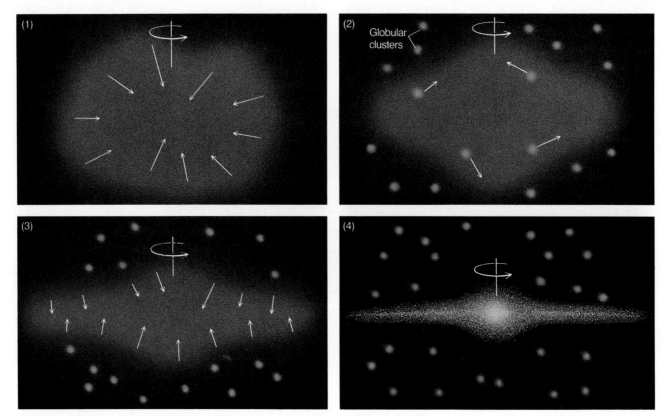

FIGURE 24.18 The Galaxy probably formed from an isolated, rotating cloud of gas that collapsed owing to gravity. Halo stars and globular clusters either formed prior to the collapse or were formed elsewhere and attracted to the Galaxy early in its history by gravity. Stars in the disk formed late, and the gas from which they were made was contaminated with heavy elements produced in early generations of stars.

Another argument in favor of thinking that Sagittarius A* has a high mass is that its proper motion is zero. Ordinary stars in the galactic center change their positions by measurable amounts because of the gravitational tugs and pulls of other nearby stars. We could be very unlucky, and Sagittarius A* could just happen to be moving exactly toward us. It seems more likely, however, that Sagittarius A* is so much more massive than ordinary stars that it is not measurably moved by their gravitational influence.

Suppose there is a black hole of several million solar masses in the center of the Galaxy. Where did the mass come from? At the present time, matter from such sources as colliding gas clouds in the ring is falling into the galactic center at the rate of about one solar mass per 1000 years. If matter had been falling in at the same rate for about 5 billion years, and we have no idea whether or not this is a reasonable assumption, then it would have easily been possible to accumulate the matter needed to form a black hole with a mass of several million solar masses.

The density of stars near the galactic center is such that we would expect a star to pass near the black hole and be disrupted by it every few thousand years. As this material falls into the black hole, there should be a brilliant outburst. Perhaps such an outburst drove gas and dust out of the galactic center a few tens of thousands of years ago, leaving behind the cavity and the clumpy ring of molecular clouds that we now see. Between outbursts, according to this idea, the black hole should be fairly quiescent, as it is now.

So scientists find a great deal of circumstantial evidence that there is a black hole in the center of the Galaxy. But a black hole may not be the only possibility. Gamma rays and radio emission can, for example, also be produced in an accretion disk around a neutron star. Over the next few years, astronomers will turn their new observational tools—gamma-ray, x-ray, infrared, and radio telescopes—to the galactic center to try to prove or disprove the existence of a black hole at the heart of our own Galaxy. As we shall see in later chapters, there is evidence that black holes exist in other galaxies and in quasars as well.

24.6 The Formation of the Galaxy

One of the least well understood events in the evolution of the universe is the formation of galaxies. The flattened shape of our own Galaxy suggests that the basic process was similar to the way in which the Sun and solar system formed (Section 12.2).

According to the traditional model, the Galaxy formed from a single rotating cloud. Since the oldest stars that we know—stars in the halo and in globular clusters—are distributed in a sphere centered on the nucleus of the Galaxy, we can assume that the protogalactic cloud was spherical. Because the oldest stars in the halo have ages of 13 to 15 billion years, we estimate that the collapse of the gas cloud occurred about that long ago. As the cloud collapsed, it formed a thin disk, just as was true for the early solar system. The stars that were formed far from the disk before the cloud collapsed did not participate in the collapse but continue to orbit in the halo (Figure 24.18).

More modern models include the possibility that the Galaxy may have been assembled from smaller bodies of matter. According to these new ideas, some of the halo stars and globular clusters, as well as stars in the central bulge of the Galaxy, formed in independent clumps or subunits. These clumps then assembled, drawn together by gravity into a roughly spherical structure. The gas associated with these subunits, already enriched in heavy elements from earlier generations of stars, then collapsed to form a disk.

Gravitational forces within the disk caused the gas in the disk to fragment into clouds or clumps with masses like those of star clusters. These individual clouds then fragmented further to form individual stars. Since the oldest stars in the disk are nearly as old as the stars in the halo, the collapse was rapid (astronomically speaking) and required perhaps no more a few hundred million years.

Some low-mass members of the earliest generations of stars are still shining. The abundance of elements heavier than helium in these very old stars is 1 percent (or less) of the abundance of heavy elements in the Sun. More massive members of these earliest generations of stars have already exhausted their store of nuclear energy and ceased to shine. These stars produced heavy elements in their interiors that were then recycled into the interstellar medium and subsequently incorporated into new generations of stars, which accordingly have a higher content of heavy elements.

SUMMARY

24.1 The Sun is located in the outskirts of the **Milky Way Galaxy.** The **Galaxy** consists of a **disk,** which contains dust, gas, and young stars; a **nuclear bulge,** which contains old stars; and a spherical **halo,** which contains very old stars, including the members of globular clusters and RR Lyrae variables. Analysis by Shapley of the distribution of globular clusters gave the first indication that the Sun is not located at the center of the Galaxy. Radio observations at 21 cm show that atomic hydrogen is confined to a flat disk, which has a thickness of only 400 LY near the Sun. Atomic hydrogen extends beyond the Sun to a distance of at least 80,000 LY from the galactic center. Dust is found in the same locations as atomic hydrogen inside the Sun's orbit; there is very little dust outside the Sun's orbit. The most massive molecular clouds, where star formation is active, are concentrated in the spiral arms.

24.2 Studies of atomic hydrogen show that the Galaxy has four main **spiral arms.** The Galaxy does not rotate as a solid body. The stars within it follow orbits that obey Kepler's laws, so stars closer to the galactic center complete their orbits more quickly than do the more distant stars. The **spiral density wave** theory is one way to account for the spiral arms. Calculations show that the gravitational forces within the Galaxy will cause stars and gas clouds to slow down in the vicinity of the spiral arms, thereby leading to higher densities of material. When molecular clouds attempt to pass through these regions of higher density, star formation is triggered.

24.3 The properties of stars are closely related to their positions in the Galaxy. Old stars are referred to as **population II** stars and are found in the halo, in globular clusters, and in the nuclear bulge. Young **population I** stars are found in the disk and are especially concentrated in the spiral arms. The Sun is a member of population I. Population I stars, which formed after previous generations of stars had produced heavy elements and ejected them into the interstellar medium, have higher metal abundances than do population II stars.

24.4 The mass of the Galaxy can be determined by measuring the orbital velocities of stars or interstellar matter. The Sun revolves completely around the center of the Galaxy in about 200 million years. The total mass of the Galaxy is about 10^{12} solar masses, and about 90 percent of this mass consists of **dark matter** that emits no electromagnetic radiation and can be detected only because of the gravitational force it exerts on visible stars and interstellar matter. This dark matter is mostly located in the halo; candidates for what the dark matter might be include brown dwarfs, white dwarfs, massive black holes, or some type of subatomic particle not yet detected on Earth.

24.5 The **nucleus** of the Galaxy contains a compact radio source. Measurements of the velocities of stars and gas show that the central 1 to 2 LY of the Galaxy contain a mass that is 2 to 5 million times the mass of the Sun. A massive black hole can explain these observations, but it is not yet certain that a black hole is the only possibility.

24.6 The Galaxy formed about 13 to 15 billion years ago. Models suggest that the stars in the halo and globular clusters formed first, while the Galaxy was spherical. The gas, somewhat enriched in heavy elements by the first generation of stars, then collapsed from a spherical distribution to a disk-shaped distribution. Stars are still forming today from the gas and dust that remain in the disk. Star formation occurs most rapidly in the spiral arms, where the density of interstellar matter is highest.

REVIEW QUESTIONS

1. Explain why we see the Milky Way as a faint band of light stretching across the sky.

2. Describe where in the Galaxy you would expect to find globular clusters, molecular clouds, and atomic hydrogen.

3. Describe several characteristics that distinguish population I stars from population II stars.

4. Describe the orbits of population I and population II stars around the center of the Galaxy.

5. Describe the evidence that indicates that there may be a black hole at the center of our Galaxy.

6. Explain why the abundances of heavy elements in stars correlate with their positions in the Galaxy.

THOUGHT QUESTIONS

7. Suppose the Milky Way were a band of light extending only halfway around the sky (that is, in a semicircle). What, then, would you conclude about the Sun's location in the Galaxy? Give your reasoning.

8. The globular clusters revolve around the Galaxy in highly elliptical orbits. Where would the clusters spend most of their time? (Think of Kepler's laws.) At any given time, would you expect most globular clusters to be moving at high or low speeds with respect to the center of the Galaxy? Why?

9. Consider the following five kinds of objects: (1) open cluster, (2) giant molecular cloud, (3) globular cluster, (4) a group of O and B stars, (5) planetary nebula.
   a. Which one or ones are found only in spiral arms?
   b. Which one or ones are found only in the parts of the Galaxy that are *not* in the spiral arms?
   c. Which are thought to be very young?
   d. Which are thought to be very old?
   e. Which have stars that are of the highest temperatures?

10. Where in the Galaxy do you suppose undiscovered globular clusters may exist?

11. Why does star formation occur primarily in the disk of the Galaxy?

12. Where in the Galaxy would you expect to find supernovae of Type II, which are the explosions of massive stars? Where would you expect to find supernovae of Type I, which involve the explosions of white dwarfs?

13. According to Chapter 22, astronomers believe that there is one supernova explosion somewhere in our Galaxy every 25 to 100 years, yet we have not seen a galactic supernova in nearly 400 years. Is it likely that we would be able to observe *every* galactic supernova explosion? Why or why not?

14. Suppose stars evolve without losing mass. Once matter is incorporated into a star, it remains there forever. How would the appearance of the Galaxy be different from what it is now? Would there be population I and population II stars? What other differences would there be?

PROBLEMS

15. Suppose the mean mass of a star in the Galaxy were only one-third of a solar mass. Use the value for the mass of the Galaxy found in the text to find how many stars the system contains.

16. Assume that the Sun orbits the center of the Galaxy at a speed of 220 km/s and at a distance that, for convenience, we will assume to be 30,000 LY from the center. **(a)** Calculate the circumference of the Sun's orbit, assuming it to be approximately circular. **(b)** Calculate the Sun's period, the "galactic year." **(c)** Use Newton's formulation of Kepler's third law to calculate the mass of the Galaxy inside the orbit of the Sun. **(d)** It is estimated that the total mass of the Galaxy is ten times that within the Sun's orbit. If this mass were to collapse inside the orbit of the Sun, what would be the length of the galactic year? **(e)** In this case, what would be the Sun's new orbital speed?

17. Construct a rotation curve, that is, a curve like Figure 24.12, for the solar system by using the orbital velocities of the planets. How does this curve differ from the rotation curve for the Galaxy? What does this curve tell you about where most of the mass in the solar system is concentrated?

18. Suppose that all of the mass in our Galaxy lies inside the orbit of the Sun, which we will assume to be circular with a radius of 30,000 LY; assume the Sun's orbital velocity is 220 km/s. Use Kepler's third law to find the orbital speed of a cloud of atomic hydrogen at a distance of 60,000 LY. In other words, verify the data in Figure 24.12. What would be the orbital speed of a globular cluster at a distance of 150,000 LY?

19. Calculate the rotation speed of the gas ring at the galactic center if the mass inside the ring is 2×10^6 solar masses and the radius of the ring is 25 LY.

20. Calculate how much mass would fall into the black hole at the center of the Galaxy in a billion years at the current rate of infall of one solar mass per 1000 years.

Observations at many wavelengths are used to study galaxies. Three separate sets of observations have been used to make this image of the Whirlpool Galaxy. Optical radiation, which is shown in red, highlights the younger stars, as well as the dust. The radio emission shown in green is due to thermal emission from H II regions and synchrotron emission from electrons moving at high speed in regions where magnetic fields are present. Radio observations of the line of neutral hydrogen at 21 cm, which are shown in blue, give the distribution of cold neutral hydrogen gas. *(NRAO/AUI)*

Walter Baade (1893–1960), born in Westphalia, joined the staff of the Mount Wilson Observatory in 1930. He discovered the two populations of stars, expanded the distance scale and age of the universe, and was an early investigator of supernovae and radio sources. He is still recognized as one who used large telescopes to the very best advantage. *(Caltech)*

GALAXIES

The "analogy [of the nebulae] with the system of stars in which we find ourselves . . . is in perfect agreement with the concept that these elliptical objects are just [island] universes—in other words, Milky Ways. . . ."

So wrote Immanuel Kant (1724–1804) concerning the faint patches of light that telescopes revealed in large numbers. Unlike the true gaseous nebulae that populate the Milky Way (Chapter 24), the nebulous-appearing luminous objects referred to by Kant are found in all directions in the sky except where obscuring clouds of interstellar dust intervene. Despite Kant's speculation that these patches of light are actually systems like our own Milky Way Galaxy, the weight of astronomical opinion rejected this idea for nearly two centuries. The proof in 1924 that our Galaxy is not unique and central in the universe is one of the great advances in cosmological thought.

25.1 Galactic or Extragalactic?

Faint star clusters, glowing gas clouds, dust clouds reflecting starlight, and galaxies all appear as faint, unresolved luminous patches when viewed visually with telescopes of only moderate size. Since the true natures of these various objects were not known to early observers, all of them were called "nebulae."

Nebula (plural *nebulae*) literally means "cloud." Today, we usually reserve the word nebula for the true gas or dust clouds.

Catalogues of Nebulae

One of the earliest catalogues of nebulous-appearing objects was prepared in 1781 by the French astronomer Charles Messier. Messier was a comet hunter, and as an aid to himself and others he made a list of 103 objects that might be mistaken for comets. Because Messier's list contains some of the most conspicuous star clusters, nebulae, and galaxies in the sky, these objects are often referred to by their numbers in his catalogue—for example, M31, the great galaxy in Andromeda.

By 1908 nearly 15,000 nebulae had been catalogued and described. Some had been correctly identified as star clusters and others as gaseous nebulae (such as the Orion nebula). The nature of most of them, however, still remained unexplained. If they were nearby, with distances comparable to those of observable stars, they would have to be luminous clouds of gas within our Galaxy. If, on the other hand, they were very remote, far beyond the foreground stars of the Galaxy, they could be systems of thousands of millions of stars. In this case, they would be galaxies in their own right or, as Kant had described them, ''island universes.'' To determine which idea is right, it was necessary to measure the distances to at least some of the nebulae.

The Solution to the Controversy

An end to controversy was brought about by the discovery of variable stars in some of the nearer nebulae in 1923 and 1924. Edwin Hubble, working with the 100-in. (2.5-m) telescope at the Mount Wilson Observatory, analyzed the light curves of variables he had discovered in M31, M33, and NGC 6822 and found that they were cepheids. Although cepheid variables are supergiant stars, the ones studied by Hubble appeared very faint—near magnitude 18. Those stars, therefore, and the systems in which they were found, must lie far beyond the boundaries of the Milky Way Galaxy. The ''nebulae'' had been established as galaxies (Figure 25.1).

25.2 The Extragalactic Distance Scale

One of the most important, difficult, and controversial problems in modern observational astronomy is the measurement of the distances to galaxies. Galax-

ies are far too remote to display parallaxes or proper motions. To determine distances to galaxies, we must therefore resort to a multistep process. The traditional approach to determining distances goes roughly as follows. First, we derive distances to individual nearby stars in our own Galaxy by measuring parallaxes and proper motions. With knowledge of the luminosities of these nearby stars, we can then determine distances to clusters, which contain stars similar to those with luminosities already known from their parallaxes or proper motions. Once we measure the distance to a cluster, we know the intrinsic luminosities of every star within the cluster. Fortunately, clusters contain some stars, including cepheid variables, that are much more luminous than any of the nearby stars for which we can obtain parallaxes by direct measurement. These stars are so luminous, in fact, that ones just like them can be detected in other galaxies. Since we can measure the apparent brightnesses of these stars and already know their intrinsic luminosities from studies of stars in clusters in our own Galaxy, we can use the inverse-square law for the propagation of light (Section 5.1) to determine the distances to the galaxies to which they belong. Any object whose intrinsic luminosity is known is referred to as a **standard candle.** Luminous stars can

FIGURE 25.1 The nearby spiral galaxy M83. This galaxy is about 10 million LY away and has a diameter of 30,000 LY. *(Copyright Anglo-Australian Telescope Board, 1977)*

ESSAY The Scale of the Universe

In Chapter 16 we discussed a model for the universe in which all dimensions were reduced by a factor of 1 trillion (10^{12}). In this model the Sun is the size of a mustard seed, and the nearest star is 10 km distant. In order to provide a more comprehensible model for our own Galaxy, let us now shrink our model by another thousandfold, thus reducing all dimensions by a factor of 10^{15}.

In this new model universe, the diameter of the Sun is only a little more than the wavelength of light, and the typical spacing between stars in the solar neighborhood is a few meters. Most of the stars visible to the unaided eye would fit into a volume about 5 km on a side—still much larger than any building. To contain our entire Galaxy, we would need a space about the volume of the Moon. Clearly, our model universe needs to shrink still more in order for us to visualize it.

Suppose we take as a reduction factor 10^{18}. The Sun is now reduced to the size of a single atom, the naked-eye stars occupy an ordinary room, and the Galaxy fits into a space about 1 km across—it would drop nicely into Meteor Crater in Arizona. The nearest other galaxies similar to our own are a few kilometers away—about like the spacing of farming villages in England or Germany, where the settlements are within walking distances of the fields. In our model, you could walk to the Andromeda Galaxy in a couple of hours, and a day hike would take you across the Local Group of galaxies. But to reach the nearest large cluster of galaxies—the Virgo Cluster—you would have to find some better form of transportation, since it is several hundred kilometers distant.

We could continue to shrink our model, but the illustrations given above are sufficient to show the scale of the universe of galaxies. Perhaps more impressive than the distances themselves is the fact that the light we see from these galaxies left them long ago. As we survey far beyond the Milky Way, we look back farther and farther, probing the depths of time as well as of space.

---

be used as standard candles for measuring extragalactic distances.

Individual stars can be detected only in relatively nearby galaxies. At larger distances, we must use objects that are even brighter than the brightest stars as standard candles. Globular clusters, H II regions, and supernovae have all proved useful. At the greatest distances of all, we use entire galaxies as standard candles for determining distances to clusters of galaxies.

The Distances to Galaxies

Some of the more important ways by which we estimate distances to galaxies are the following:

Cepheids Cepheid variables gave Hubble the first clue to the remote nature of galaxies and are still one of the best ways to estimate extragalactic distances. If we can recognize a cepheid in a galaxy, we can find its luminosity or absolute magnitude from its period through the period-luminosity relation (Section 17.3). Comparison of their known luminosities and observed brightnesses enables us to find their distances and hence the distances to the galaxies in which they occur, with the inverse-square law of light. Thus, the cepheids can serve as standard candles. The most luminous cepheids have luminosities of about $2 \times 10^4 \, L_s$, which makes them supergiant stars. Even so, they can be detected in only about 30 of the nearest galaxies—even with the world's largest telescopes.

Brightest Stars The most luminous stars are even brighter than cepheids and can be seen to greater distances. Thus, once calibrated in those galaxies whose distances are known from observations of cepheids, the brightest stars can be used as standard candles. Young, high-mass supergiant stars range in luminosity up to $10^6 \, L_s$ and so can extend the distance scale to more than six times the distance to which cepheids can be seen.

Planetary Nebulae A new technique depends on measurements of the brightnesses of planetary nebulae. It is possible to construct a luminosity function for planetary nebulae, just we have already done for stars (Figure 18.2). Measurements show that the luminosity function for planetaries drops sharply to zero at high luminosities. Furthermore, the relative number of planetaries in each range of intrinsic luminosity is the same for all types of galaxies. To determine the distance to a galaxy, one must find a sample of planetaries within the galaxy and measure their luminosity function. This technique can be used in galaxies as far away as 45 to 60 million LY with existing telescopes.

Novae Novae can be recognized in nearby galaxies, and from their light curves we know what their approximate luminosities are. The brightest novae, however, even at maximum light, do not outshine the brightest stars. To measure distances farther out in the universe, we need brighter standard candles.

Globular Clusters Although globular clusters range considerably in total luminosity (because they differ in their numbers of stars), we sometimes recognize many such clusters in one galaxy. If we assume that the brightest of them is like the brightest globular cluster in our own Galaxy (intrinsic luminosity of 10^6 L_s), that object becomes a standard candle. Distances determined this way are rather uncertain and do not extend the distance scale beyond that determined by the brightest individual young stars. The method is useful, however, because globular clusters are often seen in galaxies without a young population of supergiant stars.

Supernovae Because of their high luminosities (up to 10^{10} L_s), supernovae can be seen in very remote galaxies and would seem to be ideal standard candles. The only problem is that supernovae differ considerably among themselves in luminosity at maximum light. Some progress is being made in calibrating different kinds of supernovae, and in the future they may well hold the key to the extragalactic distance scale. The difficulty at present is that supernovae occur rarely in any one galaxy, and few supernovae have appeared in galaxies with well-known distances.

21-cm Line Width A very powerful technique of distance determination, called the Tully-Fisher technique after the two astronomers who developed it, was introduced in the late 1970s. A typical spiral galaxy contains a great deal of neutral hydrogen gas in revolution about its center. The 21-cm radiation from this hydrogen in different parts of the galaxy, moving at different speeds in our line of sight, therefore displays a range of Doppler shifts, so that the entire 21-cm line radiation from the galaxy is observed as a broad band. After a simple correction for the tilt of the plane of the galaxy to our line of sight, the width of that 21-cm line gives a measure of the maximum rotational velocity in the galaxy. That rotational velocity is correlated with the galaxy's mass, and we might expect the luminosity to be correlated with the mass as well, and hence with the 21-cm line width. Indeed, it is. The correlation was first noted between the line width and the total visual light from a galaxy. The relation is even better if the galaxy's infrared luminosity is used instead, because infrared radiation is less affected by interstellar dust. This method of distance determination, then, is to observe the 21-cm line width, thereby learning the galaxy's infrared luminosity, then to observe the galaxy's apparent infrared brightness, and finally to calculate the distance from the inverse-square law.

Total Light of Galaxies If all galaxies were identical, they would all emit the same total amount of light, and the magnitude of a galaxy as a whole would indicate its distance. Galaxies range enormously in total luminosity. Nevertheless, some types of galaxies display a relatively small range of luminosities. For those galaxies, rough distances can be estimated from their total apparent magnitudes. Unfortunately, most galaxies do not have distinguishing characteristics that enable us to estimate their true luminosities. For example, appearance alone is usually not enough to distinguish a dwarf elliptical galaxy from a giant elliptical galaxy. We can tell which galaxies are highly luminous and which ones are not only if we see a collection of them of various brightnesses, side by side in a cluster. Thus we can use the apparent brightness of the brightest members (say, the average of the brightest five) in a large cluster to estimate the distance to the cluster.

These techniques are summarized in Table 25.1, which gives the type of galaxy for which the specific technique is useful (galaxy types are described in Section 25.4), the range of distances over which the technique can be applied, and the reliability of the distance estimates derived with each technique.

Accuracy of Extragalactic Distances

With so many ways of finding distances to galaxies, one might think that the distance scale is well settled. Unfortunately, it is not. The various standard candles are difficult to calibrate, and the measurements are difficult to make. Experts differ in their judgment of the proper interpretation of the observations, the calibration of the standard candles, and what standards are the most reliable. These differences in judgment translate to a difference in the distances of remote galaxies of a factor of 2.

Most of the uncertainty in assigning distances is caused by an uncertainty in the luminosities of the standard candles. Because of this uncertainty, we may not be able to determine whether or not a given

TABLE 25.1   Methods for Estimating the Distance to Galaxies

Method	Reliability	Galaxy Type for Which Method Is Useful	Approximate Distance Range over Which Method Is Useful (millions of LY)
Cepheids	Very	Spirals, irregulars	0–50
Brightest stars	Moderate	Spirals, irregulars	0–150
Planetary nebulae	Very	All	0–70
Novae	Very	Sc, irregulars	0–150
Globular clusters	Moderate	All	0–150
Supernovae	Moderate	All	0–600
21-cm line width	Very	Spirals, irregulars	0–75
Total light of galaxies	Low	Spirals, irregulars	0–300
Brightest galaxy in cluster	Very	Ellipticals in clusters of galaxies	70–13,000
Radial velocities	Very	All	300–13,000

galaxy is at a distance of 2 billion or 4 billion LY. We can, however, easily determine that the standard candles in one galaxy are, for example, four times fainter than in another and that one galaxy must be twice as far away as the other. As a result, the *relative* distances of galaxies are known with much higher accuracy than their *actual* distances.

Perhaps an uncertainty of whether a particular galaxy is 2 billion or 4 billion LY away may seem to be pretty poor precision in an exact science like astronomy, where we know the distances between the planets in the solar system to better than one part in a million. But uncertainty is common at the frontier of research. At least we are far beyond the controversy of whether or not the "nebulae" are extragalactic.

25.3   The Expanding Universe

Early in the present century astronomers began to collect evidence that all galaxies (except for a few of the nearest ones) are moving away from us. This apparent expansion of everything away from us has many implications for the structure and history of the universe, as we shall further discuss in Chapter 28. In this section, we will describe how the observational evidence was obtained and how it is used to help determine the distances to remote galaxies and clusters of galaxies.

The Evidence: Slipher's Observations

The universe is *expanding*. This fundamental observation underlies all of modern cosmological thought. Curiously, this fact was discovered as a by-product of the search for distant solar systems.

In 1894, Percival Lowell established an observatory in Flagstaff, Arizona, to study the planets and to search for life in the universe. Lowell thought that the spiral nebulae might be solar systems in the process of formation—like the solar nebula (Chapter 12). He therefore asked one of the observatory's young astronomers, Vesto M. Slipher, to photograph the spectra of some of the spiral nebulae to see if their spectral lines suggest chemical compositions like those expected for newly forming planets.

The Lowell Observatory's major instrument was a 24-in. refracting telescope, which was not at all well suited to observations of faint spiral nebulae. With technology available in those days, exposure times were 20 to 40 h long! Slipher's first spectrum, in 1912, was of the Andromeda Galaxy, M31, which was not then known to be a galaxy. Its spectrum did not suggest new planets but did display a Doppler shift of its absorption lines that indicated a motion of that object toward the Earth of about 300 km/s. We know today that most of that speed is caused by the revolution of the Sun about the center of our own Galaxy, carrying us roughly in the direction of M31. Over the next 20 years, Slipher photographed spectra of more than 40 additional nebulae. Only a few, now known to be very nearby and recognized as members of our Local Group, are approaching us. Slipher found the overwhelming majority to be receding at speeds as large as 1800 km/s.

The Hubble Law

The profound implications of Slipher's work became apparent only during the 1920s, when Hubble found ways of estimating the distances of the spiral nebulae. Hubble carried out the key observations in collabora-

FIGURE 25.2   Milton Humason. *(Courtesy of the Archives, Caltech)*

tion with a remarkable man: Milton Humason (Figure 25.2). Humason began his astronomical career by driving a mule train up the trail on Mount Wilson to the observatory. In those early days, supplies had to be brought up that way; even astronomers hiked up to the mountaintop for their turns at the telescope. Humason became interested in the work of the astronomers and took a job as janitor at the observatory. After a time he became a night assistant, helping the astronomers run the telescope and take data. Eventually, he made such a mark that he became a full astronomer at the observatory.

By the late 1920s Humason was collaborating with Hubble by photographing the spectra of faint galaxies with the 100-in. telescope. In 1931 Hubble and Humason jointly published their classic paper in *The Astrophysical Journal* (Figure 25.3). In this paper, they compared distances and velocities of remote galaxies moving away from us at speeds as high as 20,000 km/s. Furthermore, Hubble and Humason showed that the velocities of recession of galaxies are proportional to their distances from us. Written as an equation, the relationship between velocity and distance is

$$v = H \cdot d$$

where v is the recession speed, d is the distance, and H is a number called the **Hubble constant.** This equation, expressing the law of **redshifts,** is now known as the **Hubble law.**

The relative distances to clusters of galaxies are known fairly well. To the accuracy of the observations, remote clusters of galaxies show the same proportionality between velocity and distance, that is, the same value of the Hubble constant, as nearby galaxies. For many years, estimates of the value of the Hubble constant have been in the range of 15 to 30 km/s per million light years. The most recent work appears to be converging on a value near 25 km/s per million light years. In other words, if H is 25 km/s per million light years, a galaxy moves away from us at a speed of 25 km/s for every million light years of its distance. As an example, a galaxy that is 100 million LY away will be moving away from us at a speed of 2500 km/s.

Expansion Requires a Velocity-Distance Relation

The fact that galaxies obey the Hubble law shows beyond doubt that the universe is expanding uniformly. A uniformly expanding universe is one that is expanding at the same rate everywhere. In such a universe, we and all other observers within it, no matter where they are located, *must* observe a proportionality between the velocities and distances of remote galaxies.

To see why, imagine a ruler made of flexible rubber, with the usual lines marked off at each centimeter. Now suppose someone with strong arms grabs each end of the ruler and slowly stretches it, so that, say, it doubles in length in 1 min (Figure 25.4). Consider an intelligent ant sitting on the mark at 2 cm—intentionally not at either end or in the middle. He measures how fast other ants, sitting at the 4-, 7-, and 12-cm marks, move away from him as the ruler

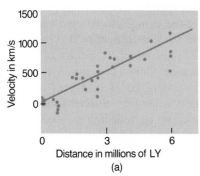

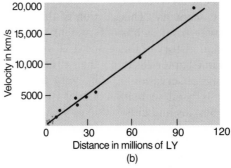

FIGURE 25.3   (a) Hubble's original velocity-distance relation, adapted from his 1929 paper in the Proceedings of the National Academy of Sciences. (b) Hubble and Humason's velocity-distance relation, adapted from their 1931 paper in *The Astrophysical Journal*. The red dots at the lower left are the points in the diagram in the 1929 paper (a). Comparison of the two graphs shows how rapidly the determination of distances and redshifts of galaxies progressed in the two years between these publications.

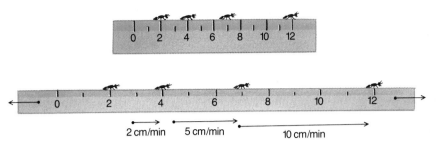

FIGURE 25.4 Stretching a ruler. See text for explanation.

stretches. The one at 4 cm, originally 2 cm away, has doubled its distance in 1 min; it has moved 2 cm/min. Similarly, the ones at 7 cm and 12 cm, which were originally 5 and 10 cm distant, have had to move away at 5 and 10 cm/min, respectively, to reach their current distances of 10 cm and 20 cm. All ants move at speeds proportional to their distance.

Now repeat the analysis, but put the intelligent ant on some other mark—say, on 7 or 12—and you'll find that in all cases, as long as the ruler stretches uniformly, this ant finds that every other ant moves away at a speed proportional to its distance.

For a three-dimensional analogy, look at the raisin bread in Figure 25.5. The cook has put too much yeast in the dough, and when he sets the bread out to rise, it doubles in size during the next hour and all the raisins move farther apart. Some representative distances from one of the raisins (chosen arbitrarily, but not at the center) to several others are shown in the figure. Since each distance doubles during the hour, each raisin must move away from the one selected as origin at a speed proportional to its distance. The same is true, of course, no matter which raisin you start with. But the analogy must not be carried too far; in the bread it is the expanding dough that carries the raisins apart, but in the universe no pervading medium is presumed to separate the galaxies.

From the foregoing, it should be clear that if the universe is uniformly expanding, all observers everywhere, including us, must see all other objects moving away from them at speeds that are greater in proportion to their distances. As Hubble and Humason showed, that is precisely what observers on Earth do see.

As telescopes have grown larger and detectors have become more sensitive, it has become possible to observe more and more remote galaxies with greater and greater speeds of recession (Figure 25.6). The current (July 1991) record-holder as the most distant galaxy so far discovered is 4C 41.17, which is moving away from us at 92 percent of the speed of light. At this high velocity, lines that are so far into the ultraviolet region that they cannot be observed from the ground are shifted into the yellow and red part of the spectrum.

The expansion of the universe does not imply that the galaxies and clusters of galaxies themselves are expanding. The raisins in our raisin bread analogy do not grow in size as the loaf expands. Similarly, gravitation holds galaxies and clusters together, and they simply separate without changing in size as the universe expands, just as do the raisins in the bread.

25.4 Types of Galaxies

Galaxies differ a great deal among themselves, but the majority of optically bright galaxies fall into two general classes: spirals and ellipticals. A minority are classed as irregular.

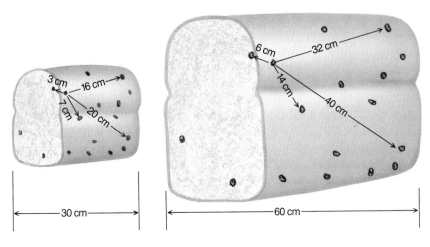

FIGURE 25.5 Expanding raisin bread. See text for explanation.

FIGURE 25.6 A distant cluster of galaxies in Hydra, with a radial velocity of about 20 percent that of light. *(National Optical Astronomy Observatories)*

FIGURE 25.7 The nearby spiral galaxy in Andromeda, which is similar in size and structure to the Milky Way Galaxy. Its distance is about 2.4 million LY, and its diameter is about 100,000 LY. This galaxy contains over 300 billion stars. *(National Optical Astronomy Observatories)*

Spiral Galaxies

Our own Galaxy and the Andromeda Galaxy (Figure 25.7), which is believed to be much like it, are typical large **spiral galaxies.** Like our Galaxy (Chapter 24), a spiral consists of a nucleus, a disk, a halo, and spiral arms. Interstellar material is usually spread throughout the disks of spiral galaxies. Bright emission nebulae are present, and absorption of light by dust is also often apparent, especially in those systems turned almost edge-on to our line of sight (Figure 25.8). The spiral arms contain the young stars, which include luminous supergiants. These bright stars and the emission nebulae make the arms of spirals stand out like the arms of a Fourth-of-July pinwheel (Figure 25.1). Open star clusters can be seen in the arms of nearer spirals, and globular clusters are often visible in their halos. In M31, for example, more than 200 globular clusters have been identified. Spiral galaxies contain both young and old stars.

Perhaps a third or more of spiral galaxies display conspicuous bars running through their nuclei. The

FIGURE 25.9 NGC 4650, a barred spiral of type SBa in the constellation Centaurus. *(National Optical Astronomy Observatories)*

spiral arms of such a system usually begin from the ends of the bar, rather than winding out directly from the nucleus. These are called barred spirals (Figure 25.9). Some astronomers believe that almost all spirals, including probably the Milky Way Galaxy, contain at least a weak bar. Studies of the rotations of some barred spirals show that their inner parts (out to the ends of the bars) are rotating approximately as solid bodies.

In both normal and barred spirals, we observe a gradual transition of morphological types. At one extreme, the nuclear bulge is large and luminous, the arms are faint and tightly coiled, and bright emission nebulae and supergiant stars are inconspicuous. At the other extreme are spirals in which the nuclear bulges are small—almost lacking—and the arms are loosely wound, or even wide open. In these latter galaxies, there is a high degree of resolution of the arms into luminous stars, star clusters, and emission nebulae. Our Galaxy and M31 are both intermediate between these two extremes. Photographs of spiral galaxies, illustrating this transition of types, are shown in Figures 25.10 and 25.11. All spirals rotate in the sense that their arms trail, as does our own Galaxy.

Spiral galaxies range in diameter from about 20,000 to more than 100,000 LY, and the atomic hydrogen in the disks often extends to far greater diameters. From the limited observational data available, their masses are estimated to range from 10^9 to 10^{12} times the mass of the Sun. The total luminosities of most spirals fall in the range of 10^8 to 10^{11} L_s. Our Galaxy and M31 are relatively large and massive, as spirals go.

FIGURE 25.8 NGC 253, one of the dustiest of the spiral galaxies. Much of the interior structure of this galaxy is obscured by dust, but two spiral arms and many blueish clusters of stars can be seen around the outer edge. This galaxy is only 10 million LY distant. It appears elongated because we view it nearly edge-on. *(Anglo-Australian Telescope Board)*

Elliptical Galaxies

Elliptical galaxies are spherical or ellipsoidal systems that consist almost entirely of old stars. They contain no trace of spiral arms. Their light is dominated by red stars (population II), and in this respect, ellipti-

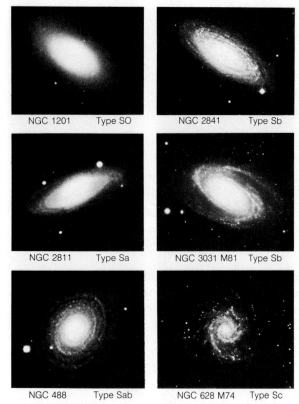

FIGURE 25.10   Types of spiral galaxies. *(Palomar Observatory, Caltech)*

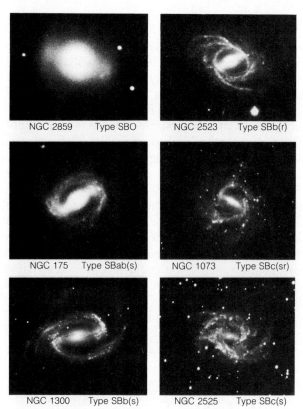

FIGURE 25.11   Types of barred spirals. *(Palomar Observatory, Caltech)*

cals resemble the nuclear bulge and halo components of spiral galaxies. In the larger nearby ellipticals, many globular clusters can be identified. Dust and emission nebulae are not conspicuous in elliptical galaxies, but many ellipticals do contain a small amount of interstellar matter. X-ray data indicate that 1 to 2 percent of the total mass of ellipticals may be in the form of gas at a temperature that exceeds a million degrees.

Elliptical galaxies show various degrees of flattening, ranging from systems that are approximately spherical to those that approach the flatness of spirals (Figure 25.12). The rare giant ellipticals (for example, M87—Figure 25.13) reach luminosities of 10^{11} L_s. The

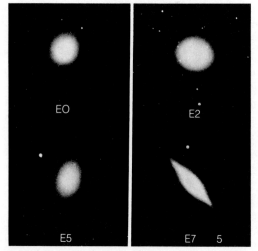

FIGURE 25.12   Types of elliptical galaxies. *(Palomar Observatory, Caltech)*

FIGURE 25.13   The giant elliptical galaxy M87, which has an active nuclear region that is a prominent source of radio and x-ray emission. *(Copyright Anglo-Australian Telescope Board)*

FIGURE 25.14 Leo II, a dwarf elliptical galaxy. *(R. Schild, Center for Astrophysics)*

mass of the stars in giant ellipticals is typically at least 10^{12} times the mass of the Sun. The diameters of these large galaxies extend over at least several hundred thousand light years and are considerably larger than the largest spirals.

Elliptical galaxies range all the way from the giants, which we have just described, to dwarfs, which we think are the most common kind of galaxy. An example of a dwarf elliptical is the Leo II system, shown in Figure 25.14. There are so few bright stars in this galaxy that even its central regions are transparent. However, the total number of stars (most of which are too faint to show in Figure 25.14) is probably at least several million. The luminosity of this typical dwarf is approximately 10^6 L_s, about equal to the luminosity of the brightest known individual stars or to that of the brightest globular clusters.

Intermediate between the giant and dwarf elliptical galaxies are systems such as M32 and NGC 205, two companions of M31. They can be seen in the photograph of M31 (Figure 25.7); NGC 205 is the one that is farther from M31.

Irregular Galaxies

As many as 25 percent of all galaxies fall into the class of **irregular galaxies,** a term that includes a wide variety of sizes and shapes. They do not show circular symmetry but rather have an irregular or chaotic appearance. Many irregulars appear to be undergoing relatively intense star formation activity, with bright young star clusters and clouds of ionized gas. Irregular galaxies contain stars of both population I and population II.

The two best-known irregular galaxies are the Large and Small Magellanic Clouds (Figure 25.15 and 25.16), our nearest extragalactic neighbors. Although they are invisible from the United States and Europe, these two systems are prominent from the Southern Hemisphere, where they look like wispy clouds detached from the Milky Way. They are only about one-tenth as distant as the Andromeda spiral. The Large Cloud contains the 30 Doradus complex, which is one of the largest and most luminous groupings of supergiant stars and associated gas known in any galaxy. Supernova 1987A (Section 22.5) occurred near 30 Doradus.

The Small Magellanic Cloud is greatly elongated and considerably less massive than the Large Cloud. This narrow wisp of material is six times longer than wide and is pointed directly toward our Galaxy, stretching from about 150,000 LY out to 250,000 LY. Apparently the Small Cloud is the victim of a near-collision with the Large Cloud that took place some 200 million years ago. It is now being pulled apart by the gravitation of the Milky Way.

Classification of Galaxies

Of the several classification schemes that have been suggested for galaxies, one of the earliest and simplest, and the one most used today, was invented by

FIGURE 25.15 The Large Magellanic Cloud, a satellite of our own Galaxy, is visible to the naked eye from the Southern Hemisphere. The large red nebula (the Tarantula) is the site of active star formation and contains many young supergiant stars. *(National Optical Astronomy Observatories)*

FIGURE 25.16 The Small Magellanic Cloud. This dwarf irregular galaxy is a satellite of our own Milky Way Galaxy. *(National Optical Astronomy Observatories)*

Hubble during his study of galaxies in the 1920s. Hubble's scheme consists of three principal classification sequences: ellipticals, spirals, and barred spirals. The irregular galaxies form a fourth class of objects in Hubble's classification.

The ellipticals are classified according to their degree of flattening or ellipticity. Hubble denoted the spherical galaxies by E0, and the most highly flattened by E7. The classes E1 through E6 are used for galaxies of intermediate ellipticity. Hubble's classification of elliptical galaxies is based on the appearance of their images, not upon their true shapes. An E7 galaxy, for example, must really be a relatively flat elliptical galaxy seen nearly edge-on, but an E0 galaxy could be one of any degree of ellipticity, seen face-on.

Hubble classed the normal spirals as S and the barred spirals as SB. Lowercase letters a, b, and c are added to denote the extent of the nucleus and the tightness with which the spiral arms are coiled. For example, Sa and SBa galaxies are spirals and barred spirals in which the nuclei are large and the arms tightly wound. Sc and SBc are spirals of the opposite extreme. Our Galaxy and M31 are classed as Sb.

Some galaxies have the disk shape of spirals but no trace of spiral arms. Hubble regarded these as galaxies of a type intermediate between spirals and ellipticals and classed them S0.

Hubble's classification scheme for all but irregular galaxies is illustrated in Figure 25.17, in which the morphological forms are sketched and labeled, with the three principal sequences joined at S0. Figure 25.17 provides an easy way to remember the shapes of galaxies, but it does not represent an evolutionary sequence. It is true that as spirals consume their gas, star formation will stop, and the spiral arms will grad-ually become less conspicuous. Over long periods, spirals will therefore begin to look more like S0 galaxies. Collisions of spiral galaxies in the centers of dense clusters of galaxies may play a role in forming the most massive elliptical galaxies. It is likely, however, that most of today's elliptical galaxies have always been ellipticals. The most important characteristics of the different kinds of galaxies are summarized in Table 25.2.

25.5 Properties of Galaxies

The diameter of a galaxy in light years can be calculated from its angular size once the distance to the galaxy is known, just as we calculate the diameter of the Sun or of a planet. Also, if we know the distance to a galaxy, we can apply the inverse-square law of light and calculate its total luminosity from the amount of light we receive from it. Thus our knowledge of the diameters and luminosities of galaxies is dependent on the accuracy of our estimates of the distances to galaxies. It also follows that an error in the extragalactic distance scale will lead to systematic errors in the derived sizes and luminosities.

Masses of Galaxies

We determine the masses of galaxies, like those of other astronomical bodies, by measuring their gravitational influences on other objects or on the stars within them. Internal motions in galaxies provide the most reliable methods of measuring their masses. The procedure for spiral galaxies is to observe the rotation of a galaxy from the Doppler shifts of radio lines of neutral hydrogen (H) or carbon monoxide (CO) and then to compute its mass with the help of Kepler's

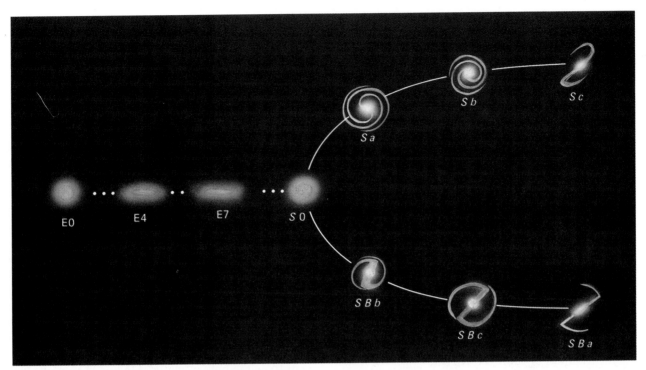

FIGURE 25.17 Hubble's classification scheme for galaxies.

third law. For example, such observations of M31, the Andromeda Galaxy, show it to have a mass (within the main visible part of the galaxy, out to a distance of 100,000 LY from the center) of about 4×10^{11} solar masses, which is about the same as the mass of our own Galaxy. (The total mass of M31 is higher than 4×10^{11} solar masses because we have not included the material that lies more than 100,000 LY from the center. Like our own Galaxy, Andromeda appears to have a large amount of dark matter that lies beyond the luminous boundary.)

Elliptical galaxies are not highly flattened and are not in rapid rotation. Nevertheless, the velocities of the stars in such a galaxy depend on its gravitational attraction for them, and hence on its mass. The spectrum of a galaxy is a composite of the spectra of its many stars, whose different motions produce different Doppler shifts. The lines in the composite spectrum, therefore, are broadened, and the amount by which they are broadened indicates the range of speeds with which the stars are moving with respect to the center of mass of the galaxy. The range of speeds depends, in turn, on the force of gravity that holds the stars within the galaxies. With information about the speeds, it is therefore possible to calculate the mass of the galaxy.

TABLE 25.2   Characteristics of Galaxies of Different Types

	Spirals	Ellipticals	Irregulars
Mass (solar masses)	10^9 to 10^{12}	10^5 to 10^{13}	10^8 to 10^{11}
Diameter (thousands of LY)	20–200	5–500	5–30
Luminosity (solar units)	10^8 to 10^{11}	10^6 to 10^{11}	10^7 to 2×10^9
Absolute visual magnitude	−15 to −22.5	−9 to −23	−13 to −20
Population content of stars	Old and young	Old	Old and young
Composite spectral type	A to K	G to K	A to F
Interstellar matter	Both gas and dust	Almost no dust; little gas	Much gas; some are deficient in dust; others contain large quantities of dust
Mass-light ratio	2–20	100	1

Mass-Light Ratio

A shorthand way of characterizing a galaxy is by giving the ratio of its mass, in units of the solar mass, to its light output, in units of the solar luminosity. For the Sun, of course, this ratio would be unity. Galaxies, however, are not composed entirely of stars that are like the Sun. The overwhelming majority of stars are less luminous than the Sun, and usually these stars contribute most to the mass of a system, without accounting for very much light. Thus, the **mass-light ratio** is generally greater than one. Galaxies in which star formation is still occurring tend to have mass-light ratios in the range of 1 to 10, while in galaxies consisting mostly of an older stellar population, the ratio is 10 to 20.

But the above comments refer only to the inner, more conspicuous parts of galaxies. In Chapter 24 we discussed the evidence for invisible matter in the outer disk and halo of our own Galaxy, extending much farther from the galactic center than do the more conspicuous bright stars. Recent measurements of the rotations of outer parts of nearby galaxies, such as the Andromeda spiral, suggest that they, too, have extended distributions of dark matter around the visible disk of stars and dust. This largely invisible matter adds to the mass of the galaxy while contributing nothing to its luminosity, resulting in high mass-light ratios. If dark, invisible matter is present in a galaxy, its mass-light ratio can be as high as 100. The mass-light ratios measured for various types of galaxies are given in Table 25.2.

The outer parts of at least some galaxies, therefore, contain matter that has not yet been identified by its light but only by means of its gravitational influence. The anomaly is sometimes called the missing mass problem. In fact, the mass is there, it is the light that is missing.

What is this dark matter? Extensive searches have been made for both hot and cold gas, and neither is present in sufficient quantity to account for the dark matter. The most likely explanation seems to be that the dark matter in galaxies is composed of very massive collapsed objects (e.g., black holes), star-like objects that are not massive enough to burn hydrogen (brown dwarfs), or massive neutrinos or exotic subnuclear particles (see Chapter 28).

Whatever the composition of the dark matter, these measurements of other galaxies support the conclusion reached already from studies of the rotation of our own Galaxy—namely, that probably 90 percent or more of all the material in the universe cannot be observed directly in any part of the electromagnetic spectrum. The light that we see from galaxies does not trace the bulk of material that is present in space and so may give us a very misleading picture of the large-scale structure of the universe. An understanding of the properties and distribution of this invisible matter is crucial, however. Through the gravitational force that it exerts, dark matter probably plays a dominant role in the formation of galaxies. As we shall see in Chapter 28, it may also determine the ultimate fate of the universe.

SUMMARY

25.1. Faint star clusters, clouds of glowing gas, dust clouds reflecting starlight, and galaxies all appear as faint patches of light in telescopes of the quality available at the beginning of the 20th century. It was only when the discovery of cepheid variables in the Andromeda Galaxy was announced in 1924 that it became firmly established that there are other galaxies similar to the Milky Way in size and content.

25.2. Astronomers determine the distances to galaxies by measuring the apparent magnitudes of objects whose intrinsic luminosities are known and then using the inverse-square law for light. Such objects are known as **standard candles.** Some useful standard candles are cepheids, planetary nebulae, novae, supernovae, and globular clusters. The rotational velocities of spiral galaxies are correlated with their absolute magnitudes and are very useful indicators of distance.

25.3. The universe is expanding. Observations show that the lines in the spectra of distant galaxies are **redshifted,** and their velocities of recession are proportional to their distances from us. The relationship between velocity and distance is known as the **Hubble law.** The rate of recession, which is known as the **Hubble constant,** is approximately 25 km/s per million light years. We are not at the center of this expansion; an observer in any other galaxy would see the same expansion that we do.

25.4. The majority of bright galaxies are either **spirals** or **ellipticals.** Spiral galaxies contain both old and young stars, as well as interstellar matter, and have typical masses in the range of 10^9 to 10^{12} solar masses. Our own Galaxy is a large spiral. Ellipticals are spheroidal or elliptical systems that consist almost entirely of old stars, with very little interstellar matter. Elliptical galaxies range in size from giant ellipticals, which are more massive than any spiral, down to dwarf ellipticals, which have masses of only about 10^6 solar masses. A small percentage of galaxies are classified as **irregular.** The Milky Way's nearest neighbors in space are the Large and Small Magellanic Clouds, which are both irregular galaxies.

25.5. The masses of spiral galaxies are determined from measurements of their rates of rotation. The masses of elliptical galaxies are estimated from analyses of the random motions of the stars within them. Galaxies are characterized by their **mass-light** ratios. The luminous parts of galaxies with active star formation have mass-light ratios typically in the range of 1 to 10; the luminous parts of elliptical galaxies, which contain only old stars, have mass-light ratios of typically 10 to 20. The mass-light ratios of whole galaxies, including their halos, are as high as 100, indicating that a great deal of dark matter is present.

REVIEW QUESTIONS

1. What is a galaxy? Describe the main distinguishing features of spiral, elliptical, and irregular galaxies.

2. Make up a chart of distance indicators for galaxies, explain how each indicator can be used to derive distances, and show the distance range over which each can be applied.

3. What is the Hubble law and what does it mean?

4. What does it mean to say that the universe is expanding?

5. What are the three main types of galaxies? Describe the characteristics of each.

6. Explain what the mass-light ratio is and why it is smaller in regions of star formation in spiral galaxies than it is in the central regions of elliptical galaxies.

THOUGHT QUESTIONS

7. Why can we not determine distances to galaxies by the same method we use to measure the parallaxes of stars?

8. Starting with the determination of the size of the Earth, outline all the steps one has to go through to obtain the distance to a remote cluster of galaxies.

9. Why can the redshifts in the spectra of galaxies *not* be explained by the absorption of their light by intergalactic dust?

10. Suppose that the Milky Way Galaxy were truly isolated and that there were no other galaxies within 100 million LY. Suppose that galaxies are observed in larger numbers at distances greater than 100 million LY. Why would it be more difficult to determine accurate distances to those galaxies than if there were also galaxies relatively close by?

11. Classify the following galaxies according to Hubble type: **(a)** a galaxy that is chaotic in appearance, with no symmetry and no resolved stars; **(b)** a galaxy with an elliptical image whose major axis is twice its minor axis; **(c)** a galaxy with very tightly wound spiral arms and a large nucleus.

12. Where might the gas and dust (if any) in an elliptical galaxy come from?

PROBLEMS

13. Plot the velocity-distance relation for the raisins in the bread analogy from the numbers given in Figure 25.5.

14. Repeat Problem 13, but use some other raisin for a reference and measure the distances with a ruler. Is your new plot the same as the last one?

15. Suppose a supernova explosion occurred in a galaxy at a distance of 10^8 LY. If we are only now detecting it, how long ago did the supernova actually occur? According to the Hubble law, what is the radial velocity for this galaxy?

16. A cluster of galaxies is observed to have a radial velocity of 60,000 km/s. Find the distance to the cluster.

17. Consider the possibility that the Milky Way and Andromeda are in circular orbit about each other. From their separations and masses as given in this chapter, calculate their period of revolution. Compare this result with the age of the universe (10 to 15 billion years).

18. On the basis of results of the previous question, is Andromeda likely to be useful in determining the Hubble constant?

19. Calculate the mass-light ratio for a globular cluster with a luminosity of 10^6 L_s and 10^5 stars. Do the same for a superluminous star of 100 solar masses with the same luminosity of 10^6 L_s.

The Very Large Array (VLA) of radio telescopes, which is located in New Mexico, is used to map the radio emission from quasars. This picture shows the array at night illuminated by moonlight. Part of the constellation Orion can be seen. *(NRAO/AUI)*

QUASARS AND ACTIVE GALAXIES

E. Margaret Burbidge (b. 1919) has held a variety of posts in British and American astronomy, including the Directorship of the Royal Observatory in Greenwich and the Presidency of the American Association for the Advancement of Science. Her research has concentrated on stellar nucleosynthesis, galaxies, and especially the nature of quasars.

Through the first half of the 20th century, astronomers viewed the universe as a rather placid place. They assumed that galaxies formed billions of years ago and then evolved slowly as the populations of stars within formed, aged, and died. That picture has completely changed. We now see that the universe is shaped by violent events, including cataclysmic explosions of supernovae, collisions of whole galaxies, and the prodigious outpouring of energy as matter interacts in the environment surrounding massive black holes. One of the keys to our changed view of the nature of the universe was the discovery of quasars.

26.1 Quasars

If the Sun were typical among stars as a radio emitter, we would not expect to observe strong radio emission from stars. Astronomers were, therefore, surprised when in 1960 two radio sources were identified with what appeared to be stars. There seemed to be no chance that the identifications were in error, because the precise positions of the radio sources were pinned down by noting the exact instants they were occulted by the Moon. By 1963 the number of such "radio stars" had increased to four (Figure 26.1). They were especially perplexing objects because their optical spectra showed emission lines that at first could not be identified with known chemical elements.

Redshifts

The breakthrough in determining what these radio stars actually are came in 1963. Maarten Schmidt, at Caltech's Palomar Observatory, was puzzling over the spectrum of one of the objects, which was named 3C 273 because it was the 273rd entry listed in the third Cambridge catalogue of radio sources. Schmidt recognized that the emission lines in the spectrum had the same spacing between them as the Balmer lines of hydrogen, but the lines of 3C 273 were shifted far to the red of the wavelengths where the Balmer lines are normally located. Indeed, these lines were at such long wavelengths that if the redshifts were attributed to the Doppler effect, 3C 273 was receding from us at a speed of 45,000 km/s, or about 15 percent the speed of light! This is an astounding speed for something that looks like a star. Any true star moving at more than a few hundred kilometers per second would be able to overcome the gravitational force of the Galaxy and escape completely from it.

After Schmidt's proposed identification of the lines in 3C 273, the emission lines in other star-like radio sources were re-examined to see if they, too, might be well-known lines with large redshifts. Such proved to

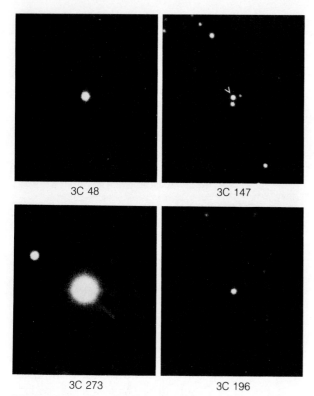

3C 48   3C 147

3C 273   3C 196

FIGURE 26.1 Quasi-stellar radio sources photographed with the 5-m Hale telescope. *(Palomar Observatory, Caltech)*

be the case, but the other objects were found to be receding from us at even greater speeds. Evidently, they could not be nearby stars. They merely look like stars because they are so far away that they have a very small apparent size. They were given the name *quasi-stellar radio sources,* or simply quasi-stellar objects (abbreviated QSO). Later, similar objects were found that are not sources of strong radio emission. Today they are all designated by the term **quasar,** and about 90 percent of all known quasars are not radio sources. Some astronomers think that radio emission is a temporary phase in the evolution of quasars.

Thousands of quasars have now been discovered, and spectra are available for a representative sample. All the spectra show large to very large redshifts. The largest redshift measured to date (July 1991) corresponds to a relative shift in wavelength of $\Delta\lambda/\lambda = 4.9$. The Lyman α line of hydrogen, which has a laboratory wavelength of 121.5 nm in the ultraviolet portion of the spectrum, is shifted all the way through the visible region to 700 nm! At such high velocities the simple Doppler formula (Section 5.6) must be modified to take into account the effects of special relativity. If we apply the relativistic form of the formula for the Doppler shift (Problem 12), we find that a redshift

of 4.9 corresponds to a velocity of more than 94 percent the speed of light.

What are the quasars? They are at the distances of galaxies, but they are certainly not normal galaxies. Galaxies contain stars, so the spectra of normal galaxies have absorption lines, just as do the spectra of stars. Quasar spectra are dominated by emission lines. Accounting for the energy emitted by quasars presents another fundamental problem. The difficulty is that if the quasars obey the Hubble law and are at the distances that correspond to their redshifts, then they are more luminous than the brightest galaxies. Furthermore, as we shall see, they must generate this enormous amount of energy in a volume of space that is no larger in diameter, and is sometimes much smaller, than the distance from the Sun to the nearest star.

Luminosities

We can determine the distance to a galaxy if we know its redshift. Let us assume for the moment that quasars also obey the Hubble law and that we can estimate their distances accurately by measuring their velocities. If this assumption is true, then quasars are *extremely* luminous compared with ordinary galaxies. In visible light most are far more energetic than the brightest elliptical galaxies. Quasars also emit energy at x-ray wavelengths, and many are radio sources as well. Some quasars have total luminosities as large as 10^{14} L_S, or 10 to 100 times the brightness of the brighter elliptical galaxies.

Finding a mechanism to produce this much energy would be difficult under any circumstances. But the quasars present an additional problem. Quasars vary in luminosity on time scales of months, weeks, or even, in some cases, days. This variation is irregular, evidently at random, and can amount to a few tens of percent. Since quasars are highly luminous, a change in brightness by, for example, a factor of 2 means an extremely large amount of energy is released rather suddenly—equivalent to 10^{14} L_S or to the total conversion of about ten Earth masses per minute from mass into energy. Moreover, because the fluctuations occur in such short times, the part of a quasar responsible for the light (and radio) variations must be smaller than the distance light travels in a month or so.

To see why this must be so, consider a cluster of stars 10 LY in diameter (Figure 26.2) at a very large distance from Earth. Suppose every star in this cluster brightens simultaneously and remains bright. We would first see the light from stars on the near side; five years later we would see light from stars at the

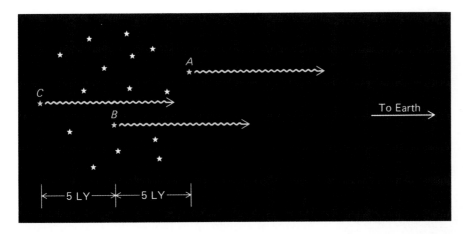

FIGURE 26.2 A diagram showing why light variations from a large region in space appear to last for an extended period as viewed from Earth. Suppose all the stars in this cluster brighten simultaneously and instantaneously. In this example, star *A* will appear to the observer to brighten five years before star *B*, which in turn, appears to brighten five years earlier than star *C*.

center. It would be ten years before we detected light from stars on the far side. Even though this cluster brightened instantaneously, from Earth it would appear that ten years elapsed before maximum brightness was reached. In other words, if an object brightens suddenly, it will seem to us to brighten over a period that is equal to the time it takes light to travel across the object from its far side.

In general, the time scale for significant changes in brightness sets an upper limit on the size of the region that brightened. Since quasars vary on time scales of months, the region where the energy is generated can be no larger than a few light months. Some quasars vary on even shorter time scales, and for them the energy must be generated in a region that is even smaller. The challenge, then, is to devise a power source that can generate more energy than an entire galaxy in a volume of space that, in some cases, is no larger than our solar system.

26.2 Origin of the Redshifts of Quasars

It was the great difficulty of devising a physical model to account for this flood of energy that led some astronomers to suggest that the redshifts of the quasars are not due to the Doppler effect. The spectral lines in galaxies are shifted to the red because the universe is expanding. These astronomers argued that rather than being a consequence of the expansion of the universe, the redshift of the spectral lines in quasars was produced by some physical mechanism that Earth-bound scientists had not previously observed. If this hypothesis were correct, then the measured redshift could not be used to estimate the distances of quasars. Since there were at the time no alternative methods for estimating their distances, the quasars could be assumed to be close enough to

us so that their energy output was within the range of that associated with normal galaxies.

Associations of Objects with Different Redshifts?

As support for this point of view, some astronomers, most notably Halton Arp, who is now at the European Southern Observatory, have sought evidence for physical associations between high-redshift quasars and low-redshift normal galaxies. If two objects are physically associated, they must be at the same distance. If they also have very different redshifts, then we would be forced to conclude that redshifts are not always a reliable indicator of distance. Logic therefore demands that there must be some physical process other than the Doppler effect that can produce very large redshifts.

There are indeed many cases in which quasars appear close to galaxies on the sky. It is always possible, however, that these are chance superpositions of two objects that are really at very different distances. There remain too few examples of an apparent association between a quasar and a galaxy with discordant redshifts to convince most astronomers that the redshifts of quasars are not cosmological in origin.

Quasars in Clusters of Galaxies

In fact, several astronomers have turned this argument around and have searched for clusters of galaxies in the vicinity of quasars. If redshifts can be measured and distances derived for these normal types of objects, and if the redshifts turn out to be the same as that of the nearby quasar, then we would have compelling evidence that the quasar also obeys the Hubble law. This task is not easy observationally because normal galaxies are fainter than quasars and are therefore more difficult to detect. Nevertheless,

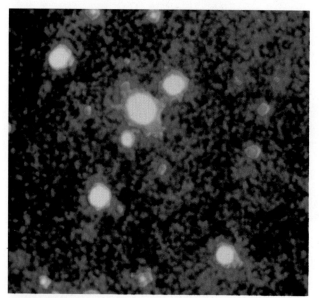

FIGURE 26.3 Quasar 3C 275.1, the first to be found at the center of a cluster of galaxies, appears as the brightest object near the center of this image. The quasar nucleus is surrounded by a gas cloud that is elliptical in shape. Its redshift indicates that this quasar is 7 billion LY away and that the light that we now observe left the quasar more than 2 billion years before our solar system formed. *(National Optical Astronomy Observatories)*

studies to date show that quasars are often surrounded on the sky by small clusters of galaxies, and the cluster galaxies exhibit the same redshift as the quasar (Figure 26.3). It is highly improbable that the apparent velocities of quasar and galaxy would coincide unless the two objects were physically associated and at the same distance.

There have been other observations as well that support the hypothesis that quasars are located at the distances indicated by their redshifts. One key result is the discovery that many relatively nearby quasars ($\Delta\lambda/\lambda = 0.5$) are not true point sources but rather are embedded in a faint, fuzzy-looking patch of light. The colors of this fuzz are like those of spiral galaxies. In a few cases, spectra have been obtained and indicate that the light of the fuzz is derived from stars, demonstrating that quasars are located in galaxies.

26.3 Active Galaxies

When quasars were first discovered, it was thought that they were much more luminous than galaxies. They are indeed more luminous than normal galaxies, but we have now found bona fide galaxies—albeit peculiar ones—that fill in the luminosity gap. These peculiar galaxies share many of the properties of the quasars, although to a less spectacular degree. Members of this class of galaxy are referred to as **active**

galaxies. Since the production of abnormal amounts of energy occurs in their centers, they are said to have **active galactic nuclei.** In effect, active galaxies have mini-quasars embedded in their nuclei.

Seyfert Galaxies

Seyfert galaxies, which are spirals with star-like nuclei (Figure 26.4), are one type of active galaxy. Like quasars, Seyferts have strong, broad emission lines, which indicate that there are gas clouds near their nuclei and that these clouds are moving at high velocity. The width of the lines indicates that the gas is moving at speeds up to thousands of kilometers per second.

One of the earliest images taken with the Hubble Space Telescope (HST) shows the nuclear region of the Seyfert galaxy NGC 1068 (Figure 26.5). The galaxy itself is a barred spiral galaxy, but it contains an extraordinarily bright nucleus. The HST image, freed from the blurring of the Earth's atmosphere, shows the clouds of hot gas in the very center of the galaxy. These clouds are ionized by radiation emitted from

FIGURE 26.4 The Seyfert galaxy NGC 1566, which is at a distance of about 50 million LY, appears on this photograph to be a normal spiral. However, it has a very luminous nucleus, which has many of the characteristics of a quasar, though it is much less energetic. The active region at the center of NGC 1566 has recently been found to vary on a time scale of less than a month, which indicates that it is extremely compact. Spectra show that hot gas near the tiny nucleus is moving at an abnormally high velocity, suggesting that it may be in orbit around a massive black hole. *(Anglo-Australian Telescope Board)*

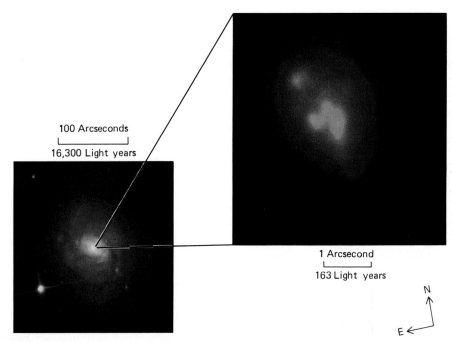

100 Arcseconds

16,300 Light years

1 Arcsecond

163 Light years

N

E

FIGURE 26.5 An optical picture of the Seyfert galaxy NGC 1068 shows the bright nucleus at the center of a spiral galaxy. The inset is a picture taken with the Hubble Space Telescope (HST), which shows clouds of ionized gas at the very center of this galaxy. *(NASA)*

the nucleus in a direction perpendicular to the plane of the galaxy (Figure 26.6).

Some Seyferts, again like quasars, are radio or x-ray sources or both, and all emit strongly in the infrared region. Some show brightness variations over a period of a few months, so as we concluded for the quasars, the region from which the radiation comes can be no more than a few light months across.

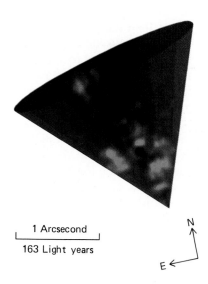

1 Arcsecond

163 Light years

N

E

FIGURE 26.6 The HST image of the clouds in the center of NGC 1068 has been computer-processed to show more detail. The cone was artificially added to the image to illustrate how radiation is beamed from the hidden nucleus. *(NASA)*

The visual luminosities of Seyferts are about normal for spiral galaxies, but when account is taken of their infrared emission, their total luminosities are found to be about 100 or so times normal.

The Seyfert and other peculiar galaxies tend to be more luminous than normal galaxies but less luminous than quasars. Their bright but point-like nuclei indicate that enormous amounts of energy are being emitted from a small region at their centers. The crucial point about Seyferts and other active galaxies is that a significant fraction of their power output comes from a source other than individual stars.

The Seyfert properties can be recognized easily only in relatively nearby galaxies. It is quite possible that 1 or 2 percent of all spiral galaxies have these active nuclei. Alternatively, it is possible that all spiral galaxies (even our own?) have these properties 1 or 2 percent of the time.

N-Type Galaxies and BL Lac Objects

Intermediate between the quasars and such galaxies as M87 or the Seyfert galaxies is a class known as N-type galaxies. N galaxies have small nuclei that are very bright compared with the main parts of those galaxies. Often they appear as stellar images superimposed on faint wispy or nebulous backgrounds. Their

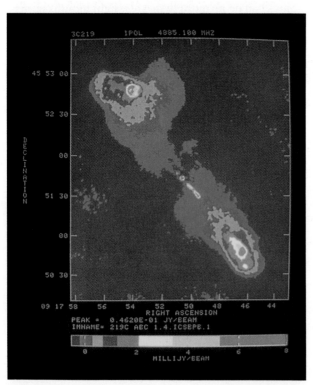

FIGURE 26.7 This radio image shows the extended lobes of radio emission and a short jet emanating from the core of the elliptical radio galaxy 3C 219. *(NRAO/AUI)*

bright nuclei indicate that enormous amounts of energy are being emitted from those regions.

Objects of another class believed to be related to Seyfert galaxies and quasars are the BL Lac objects, named for the prototype, BL Lacertae. BL Lac is a stellar-appearing object that shows large, irregular variations in luminosity. Like other BL Lac objects, it has no spectral lines, but it is a strong radio source.

In 1974, a spectrum was obtained at Palomar of the light passing through a ring-shaped aperture centered on BL Lac itself, but blocking out light from its central image. The source of faint light passing through the annular opening surrounding BL Lac proved to have a spectrum like that of a normal galaxy with a radial velocity of 21,000 km/s. BL Lac (and presumably other objects in its class) is evidently the brilliant nucleus of a distant galaxy.

Elliptical Galaxies

Finally, it has been known since 1948 that many giant elliptical galaxies that appear comparatively normal in the optical region of the spectrum are powerful emitters of radio energy. Some, M87 (Figure 25.13) being one example, emit thousands of times as much radio energy as is typical of bright galaxies. In some radio galaxies, the bulk of the radio emission comes from small regions within them, while in some others—the core-halo sources—there are bright sources in the nucleus of the galaxy surrounded by larger extended regions of radio emission. In about three-quarters of the radio galaxies, the radio source is double, with most of the radiation coming from extended regions on opposite sides of the galaxy (Figure 26.7). Typically, the two emitting regions are far larger than the galaxy itself and are centered a few hundred thousand light years away from it. Often radio observations reveal two well-delineated jets of radio radiation pointing away from the galaxy to the large extended sources. These jets can be more than a million light years long (Figure 26.8).

Presumably, ionized gases are shot out along the jets into the radio "clouds" by an extremely intense source of energy in the nucleus of the galaxy. It is thought that the ionized gas eventually collides with neutral gas and slows to a stop, defining the sometimes rather sharp outer edges of the emitting regions. Similar jets are seen in more than half of all quasars (Figure 26.9).

Another interesting structure displayed by some radio galaxies is that of the head-tail source. These are probably associated only with galaxies in clusters or at least in regions of space pervaded by intergalactic gas. In head-tail radio galaxies, the two opposing radio lobes fold back along themselves, so

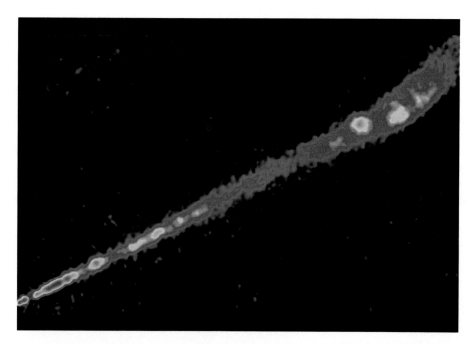

FIGURE 26.8 The radio jet associated with NGC 6251. This unusually long jet is over 300,000 LY in length. The flow direction is from lower left to upper right, and the jet expands as it moves away from the nucleus of the galaxy. In this and the radio images that follow, the color coding represents intensity, with blue being the lowest intensity. *(NRAO/AUI)*

that the radio galaxy resembles a comet (Figure 26.10). We think this is a result of the drag of intergalactic gas through which the radio galaxy is moving.

Some elliptical galaxies that are strong sources of radio emission have star-like nuclei. An interesting question is whether or not even normal ellipticals,

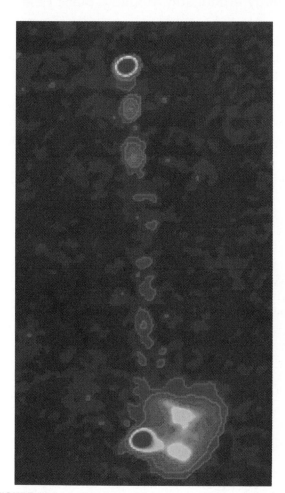

FIGURE 26.9 A radio image of the quasar 1007 + 417 taken with the VLA. The quasar is located at the bright point near the top of the picture. A jet leads to a lobe of radio emission at the bottom of the picture. *(NRAO/AUI)*

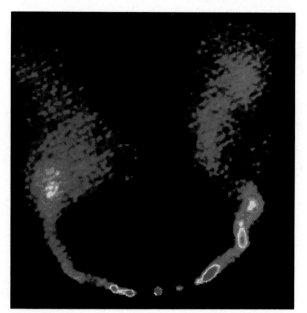

FIGURE 26.10 A radio image of NGC 1265. The two jets extend about 60,000 LY through space. The galaxy itself is moving through the Perseus cluster of galaxies at a speed of about 2000 km/s. Particles ejected from the galaxy into the two jets are swept backward by the pressure of the intergalactic gas within the cluster to form the "U" shape. *(NRAO/AUI)*

like active galaxies and quasars, have strong concentrations of light at their centers. As long as astronomers were forced to work from the ground, blurring by the Earth's atmosphere made it impossible to answer this question.

One of the first images taken by the HST was of an apparently normal elliptical galaxy (Figure 26.11). There is a *very* strong concentration of light, and therefore of stars, at the center of this galaxy. This observation suggests that the cores of many galaxies may be far more densely populated by stars than had been expected. What causes this concentration? Is it a black hole? Astronomers do not yet know the answer to this question.

Most astronomers now view quasars as simply the most extreme example of galaxies with active galactic nuclei. As we have seen, even our own Galaxy has a compact source of energy at its center, and broad emission lines indicate that high-velocity gas is present there as well. While the energy emitted by this compact source is tiny relative to that generated by a quasar, the Galaxy may represent the low-energy extreme of a continuum of levels of activity in compact nuclei in galaxies.

26.4 Black Holes—The Power Behind the Quasars?

The observations of quasars and of all the various types of galaxies that are unusually active emitters of

optical, x-ray, and radio radiation suggest that what these objects have in common is a compact source of enormous energy, evidently buried in the nucleus of a galaxy. Many models have been offered to account for this energy source, including stellar collisions in dense galactic cores, supermassive stars, extraordinarily powerful supernovae, and others.

The most widely accepted model at the present time is that quasars, and presumably other types of active galaxies as well, derive their energy output from an enormous black hole at the center of what would otherwise be a normal galaxy. The black hole must be very large—perhaps a billion solar masses. Given such a massive black hole, relatively modest amounts of additional material—only about ten solar masses per year—falling into the black hole would be adequate to produce as much energy as a thousand normal galaxies and could account for the total energy of a quasar.

Energy Production Around a Black Hole

A black hole itself can, of course, radiate no energy. The energy comes from material very close to the black hole. The idea is that the black hole captures matter—stars, dust, and gas—which is orbiting around in the dense nuclear regions of the galaxy. This material then spirals in toward the black hole and forms an accretion disk of material around it. As the material spirals ever closer to the black hole, it accel-

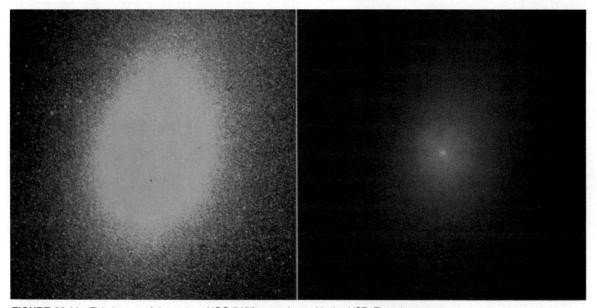

FIGURE 26.11 This image of the galaxy NGC 7457 was taken with the HST. The picture on the left shows the central portion of the galaxy. Note that the density of stars increases toward the galactic center. The picture on the right is of the same galaxy, but the contrast has been adjusted to reveal a surprisingly high concentration of stars exactly at the galaxy's core. Stars in the core are crammed together at least 30,000 times more densely than they are in our own stellar neighborhood. It is not yet known whether or not a black hole is responsible for holding this tight core together gravitationally. *(NASA)*

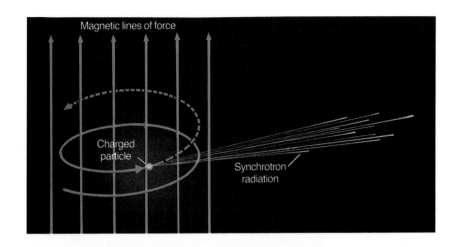

FIGURE 26.12 The emission of synchrotron radiation by a charged particle moving at nearly the speed of light in a magnetic field.

erates and heats through compression to millions of degrees. This hot matter can radiate prodigious amounts of energy as it falls into the black hole.

A number of the phenomena that we observe can be explained naturally in terms of this model. First and foremost, it can produce the amount of energy that is actually observed to be emitted by quasars and active galactic nuclei. Detailed calculations show that about 10 percent of the rest mass of matter falling into a black hole is converted to energy. Remember that during the entire course of the evolution of a star like the Sun, only a tiny fraction of its rest mass will be converted to energy by nuclear fusion. Infall into a black hole is a very efficient way to produce energy.

Since the black hole is also fairly compact in terms of its circumference, the emission produced by infalling matter comes from a small volume of space. As we recall, this condition is required to explain the fact that quasars vary on a time scale of weeks to months.

The radio radiation from quasars is in the form of *synchrotron radiation*. When a charged particle enters a magnetic field, the field compels it to move in a circular or spiral path around the lines of force. The particle is thus accelerated and radiates energy (Figure 26.12). If the speed of the particle is nearly that of light, the energy it radiates is called synchrotron radiation. This terminology was chosen because particles radiate in the same way when they are accelerated to these speeds in a laboratory synchrotron.

We find many astronomical examples in which energetic electrons are spiraling through magnetic fields and are emitting synchrotron radiation. We do not yet, in all cases, know the origin of these electrons nor the mechanisms that give them their great speeds. The compact radio source in the nucleus of the Galaxy (Section 24.5) is one source of synchrotron radiation.

In the case of quasars, the synchrotron radiation produces emission not only at radio wavelengths but also in the visible and x-ray regions of the spectrum.

Some quasars, however, have excessive infrared radiation, probably caused by warm dust re-emitting radiation that is absorbed at shorter wavelengths.

Quasars also have emission lines (with which we measure their redshifts). These must originate from ionized gas at not too high a temperature. The emission lines certainly do not come from the same region as the x-ray and synchrotron radiation. At the temperatures required to produce x-ray emission, the gas would be completely ionized and no atomic emission lines could be produced. The strengths of the emission lines vary on times scales of months, however, so they cannot be spread throughout the galaxy in which the quasar is found. Models suggest that the broad emission lines are formed in relatively dense clouds within about 0.5 LY of the black hole. The broadening of the lines is produced by the Doppler effect, but it is not known whether the motions of the gas are caused by turbulence, rotation around the black hole, expansion, or contraction. It is also not known whether these clouds may play a role in providing fuel for the black hole.

Radio Jets

All galaxies emit radio waves. The supernova remnants in the Milky Way Galaxy, for example, are sources of radio emission. Quasars and other active galactic nuclei are, however, much stronger sources of radio waves, emitting 10^8 times (or even more) as much radio energy as do normal galaxies such as our own. Since the radio radiation is synchrotron emission, strong magnetic fields and electrons moving at speeds approaching that of light must be present.

As we have seen, quasars and other active galaxies also emit jets that extend far beyond the limits of the parent galaxy. Observations have traced these jets to within 3 to 30 LY of the parent quasar or galactic nucleus. The black hole and accretion disk are, of course, much smaller than 1 LY, but it is presumed that the jets originate from the vicinity of the black hole.

Why are energetic particles ejected into jets, or often into two oppositely directed jets, rather than in all directions? It may be that the accretion disk around the black hole is dense enough to prevent radiation from escaping in all but the two directions perpendicular to the disk (Figure 26.13). The basic idea behind the formation of jets is that matter in the accretion disk will move inward toward the black hole. Some of this matter will not actually fall into the black hole but will feed the jets. That is, some infalling matter will be accelerated by the intense radiation pressure in the vicinity of the black hole and will be

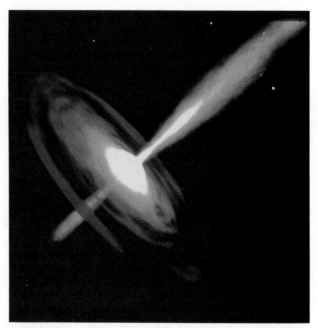

FIGURE 26.14 Radio jets associated with galaxies and quasars are powered by material falling into a massive, spinning black hole. The high pressures and temperatures generated in the accretion disk surrounding the black hole cause some of the infalling gas to be ejected along the direction of the black hole's spinning axis to create the galactic jet. *(Artist's concept by Dana Berry, STScI)*

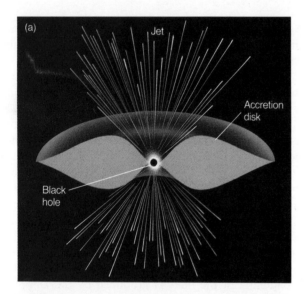

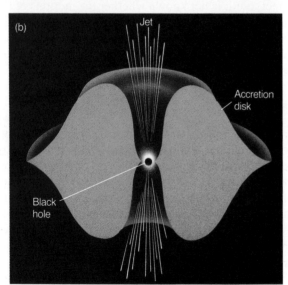

FIGURE 26.13 Schematic drawings of two accretion disks around large black holes. (a) A thin accretion disk. (b) A "fat" disk, of the type needed to account for channeling outflow of hot material into narrow jets oriented perpendicular to the disk.

blown out into space along the rotation axis of the black hole in a direction perpendicular to the plane of the accretion disk (Figure 26.14). The detailed mechanism for converting the energy associated with infall into an outward flowing jet remains a matter of controversy for theorists.

Figure 26.15 summarizes what we think we know about the structure of quasars and of the regions surrounding them.

Evolution of Quasars

If matter in the accretion disk is continually being depleted by falling into the black hole or being blown out from the galaxy in the form of jets, then a quasar can continue to radiate only as long as there is gas available to replenish the accretion disk. Where does this matter come from? One possibility is that very dense star clusters form near the centers of galaxies. These stars might then supply the fuel, either through gas that is lost during the normal course of stellar evolution by means of stellar winds and supernova explosions or because the tidal forces exerted by the black hole are strong enough to tear the stars apart. An alternate source of fuel may come from collisions of galaxies. That is, if two galaxies collide and merge, then gas and dust from one galaxy may come close

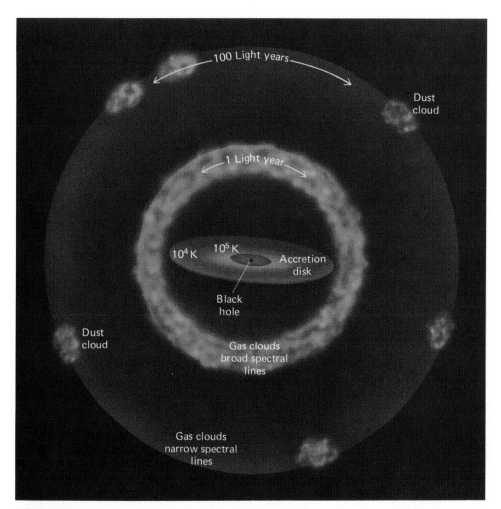

FIGURE 26.15 A diagram showing the various components that make up a quasar. Each circle represents a region 100 times larger in diameter than the one inside it. At the center of a galaxy is a massive black hole. The black hole is surrounded by an accretion disk, which has its highest temperatures closest to the black hole. Within a few light years of the black hole, there are rapidly moving gas clouds that produce broad absorption lines. At larger distances, gas clouds produce sharp spectral lines.

enough to the black hole in the other to be devoured by it and so provide the necessary fuel.

Observations show that very bright quasars were much more common a few billion years ago than they are now. That is, there are many more quasars at great distances, where we are seeing the universe as it was several billion years back in time, than there are nearby. Many astronomers believe that quasars are newly formed galaxies passing through some violent stage of formation. After this stage of formation, the quasar becomes quiescent. One possible explanation is that as quasars age they simply run out of fuel—that as time passes, all of the gas, dust, and stars available to fuel the black hole are consumed by it. Indeed, many of the relatively nearby, still-active quasars appear to be embedded in galaxies that have recently been involved in collisions with other galaxies. Gas and dust from this second galaxy have ap-

parently been swept up by a dormant black hole and so have provided the new source of fuel required to rekindle it.

This model suggests that there should be low-level quasar-like activity in some nearby galaxies. That is, black holes should exist in the nuclei of nearby galaxies but should be producing only low levels of activity either because they have relatively small masses or because they have already accreted nearly all of the matter available in their vicinity. Emission lines like those seen in quasars and Seyferts but of much lower intensity are indeed seen in the nuclei of many otherwise normal nearby galaxies. As we have already seen (Section 24.5), there is also some evidence that a black hole of at least modest size may lie hidden at the center of our own Milky Way Galaxy.

Of course, these ideas are still speculative. We know we have a small, compact source of enormous energy at the heart of every quasar. It seems plausible to associate that source of energy with a giant black hole, but this subject is at the frontier of research in astrophysics. New observations and theories may substantially modify the picture presented here during the next few years.

26.5 Gravitational Lenses

The general theory of relativity predicts that light will be deflected in the vicinity of a strong gravitational field. Quasars offer a chance to test this theory. If light from a distant quasar passes nearby an intervening galaxy on its way to Earth, then gravitational lensing may cause us to see two or more quasars when in fact there is only one.

The First Gravitational Lens

In 1979, astronomers D. Walsh, R. F. Carswell, and R. J. Weymann of the University of Arizona noticed that a pair of quasars, separated by only 6 arcsec and known collectively as 0957 + 561 (the numbers give their coordinates in the sky), are remarkably similar in appearance and spectra (Figure 26.16). They are both about 17th magnitude, and both have a redshift ($\Delta\lambda/\lambda$) of 1.4. The astronomers suggested that the two quasars might actually be only one and that we are seeing two images produced by an intervening object, acting as a gravitational lens (Section 23.3).

We now know that there is an 18th-magnitude galaxy that lies in the same direction as one of the quasars. In fact, the galaxy turns out to be a member of a cluster of galaxies, which has a redshift of 0.39 and thus is much closer than the quasar. The geometry and estimated mass of the galaxy are correct to produce the gravitational lens effect. A schematic of the lens is shown in Figure 26.17.

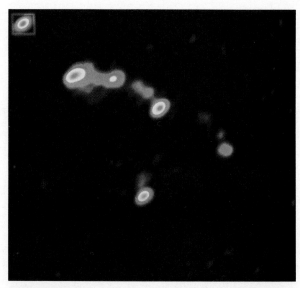

FIGURE 26.16 A radio image of the double quasar 0957 + 561. A massive galaxy acting as a gravitational lens forms multiple images of a quasar lying far beyond it. The two images of the quasar are the bright point-like objects just above and below the center of the picture. The weak blue image just above the lower image of the quasar is the galaxy that forms the gravitational lens. Several other radio sources are seen above and to the left of the double quasar. *(National Radio Astronomy Observatory/AUI)*

Other Gravitational Lenses

There is fairly convincing evidence for several other gravitational lenses, and searches are under way to discover still more. The search is difficult. If theoretical calculations are correct, the light from only about one quasar in a thousand will pass closely enough to a galaxy so that the galaxy can act as a gravitational lens and produce a double image of the background quasar.

Gravitational lenses can produce not only double images but also multiple images (Figure 26.18) and even arcs (Figure 26.19) and rings. Images produced

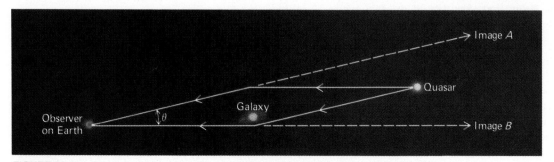

FIGURE 26.17 Gravitational lens associated with the double quasar 0957 + 561. Two light rays from the quasar are shown being bent in passing a foreground galaxy, and they arrive together at Earth. We thus see two images of the quasar, in directions A and B. This simple schematic does not reflect the subtle complexities produced by the finite size of the galaxy or by the cluster of which it is a member, nor does it reflect the relativity theory itself, but it does illustrate how we can observe multiple images of one object. The angular separation, θ, of the two images at Earth is greatly exaggerated, being only 6 arcsec in reality.

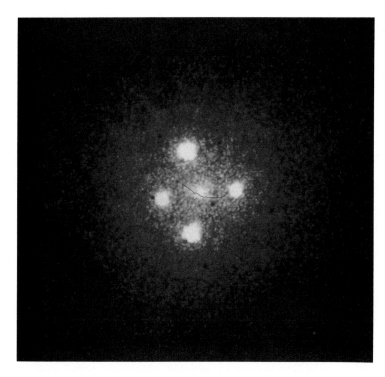

FIGURE 26.18 This image of a gravitational lens, which is referred to as Einstein's cross, was taken with the HST. The four concentrations of light at the ends of the cross bars are four images of a single distant quasar. The quasar is at a distance of approximately 8 billion LY. A galaxy at a distance of 400 million LY serves as the gravitational lens. The diffuse central object is the bright core of this intervening galaxy.

by point-like gravitational lenses can appear much brighter than the actual source would appear to be in the absence of lensing. Some of the brightest quasars and BL Lac objects may owe their apparently high luminosities to enhancement by gravitational lensing.

Galaxies may not be the only gravitational lenses. As we have already seen, a large fraction of the material in the universe is not luminous. This dark matter may also act as a gravitational lens. If this dark matter is in the form of point-like objects (e.g., black holes), the separation of the two images of the background quasar will be too small to be detected with optical telescopes. If, however, the black holes are massive (about a million times the mass of the Sun), the images may be resolvable with radio telescopes used as interferometers. Searches for such double images are under way and, if successful, may tell us what the dark matter is made of.

FIGURE 26.19 Gravitational lenses can produce very complex images. In this photograph, the blue arc is the image of a distant galaxy smeared out by the gravitational lensing effects of the mass within the foreground cluster of galaxies. *(National Optical Astronomy Observatories)*

SUMMARY

26.1. **Quasars** were first discovered because of their strong radio emission. Optical spectroscopy of the star-like sources of the radio emission has shown that the quasars have redshifts ranging from 15 to (so far) nearly 95 percent of the speed of light. If quasars are at the distances implied by their redshifts, then they are 10 to 100 times the luminosity of the brighter normal galaxies. This tremendous energy output is generated in a small volume—in some cases, in a region no larger than our own solar system.

26.2. Some quasars are members of small groups or clusters. Others have fuzz around them that has the spectrum of a normal galaxy. In such cases, the quasars clearly obey the Hubble law, which relates velocity or redshift to distance. A few astronomers have continued to argue that there are other cases in which a quasar is physically associated with a galaxy of a different redshift and that some as yet unknown physical process produces the redshift. The arguments for such associations are not widely accepted.

26.3. Most astronomers now view quasars as the most extreme example of a class of peculiar galaxies that generate large amounts of energy in a small **active galactic nucleus.** Examples of **active galaxies** include Seyferts, N galaxies, and BL Lac objects. Even some giant elliptical galaxies are strong radio sources. In many cases, there are jets of radio emission extending to large radio-emitting regions located on either side of the galaxy or quasar. These peculiar galaxies have luminosities intermediate between those of normal galaxies and quasars.

26.4. Both active galactic nuclei and quasars are thought to derive their energy from material falling toward, and forming a hot accretion disk around, a massive (up to 10^9 solar masses) black hole. Quasars were much more common billions of years ago than they are now, and astronomers speculate that quasars mark some violent stage in the formation of galaxies. Quasar activity can apparently be retriggered by collisions between galaxies, which provide a new source of fuel to feed the black hole.

26.5. Observations of quasars give evidence that the gravitational lensing effects predicted by general relativity actually do occur. Gravitational lenses can produce a variety of types of images of quasars, including double or multiple images and even arcs and rings. Analysis of lensed quasars may provide clues to the nature of the dark matter.

REVIEW QUESTIONS

1. Describe some differences between quasars and normal galaxies.

2. Describe the arguments that support the idea that at least some quasars are at the distance indicated by their redshifts.

3. In what ways are active galaxies like quasars and different from normal galaxies?

4. Describe the process by which a black hole can explain the energy radiated by quasars.

5. What is a gravitational lens? Show with a diagram how it is possible to see two quasars in the sky when in reality there is only one.

THOUGHT QUESTIONS

6. Suppose you observe a star-like object in the sky. How would you determine whether it is actually a star or a quasar?

7. Why don't any of the methods for establishing distances to galaxies, which were described in Chapter 25, work for quasars?

8. Do quasars obey the Hubble law, which relates redshift and distance? Describe the evidence.

9. One of the early hypotheses to explain the high redshifts of quasars was that these objects had been ejected at very high velocities from galaxies. This idea was rejected, because no quasars have been found that have large blue shifts. Explain why we would expect to see quasars with both blueshifted and redshifted lines if they were ejected from nearby galaxies.

10. If we see a double image of a quasar produced by a gravitational lens and can obtain a spectrum of the galaxy that is acting as a gravitational lens, then we can put limits on the distance to the quasar. Explain how.

PROBLEMS

11. Rapid variability in quasars indicates that the region in which the energy is generated must be small. Show why this is true. Specifically, suppose that the region in which the energy is generated is a transparent sphere 1 LY in diameter. Suppose that in 1 s this region brightens by a factor of 10 and remains bright for two years, after which it returns to its original luminosity. Draw its light curve as viewed from Earth.

12. If special relativity is taken into account, the exact formula for the change in wavelength due to the Doppler effect is given by the equation $\Delta\lambda/\lambda = [\sqrt{(1 + v/c)}/\sqrt{(1 - v/c)}] - 1$. Show that if the velocity v is much less than the speed of light c, then this equation is equivalent to the equation for the Doppler effect given in Section 5.6.

13. Assume that its redshift is a Doppler shift and use the relativistic form of the Doppler equation given in Problem 12 to verify that a quasi-stellar source with a redshift of $\Delta\lambda/\lambda = 3.53$ has a radial velocity of 91 percent the speed of light.

14. The greatest redshift measured to date for any quasar is $\Delta\lambda/\lambda = 4.9$. What is the observed wavelength of its Lyman α line of hydrogen, which has a laboratory or rest wavelength of 121.6 nm? Would this line be observable with a ground-based telescope in a quasar with zero redshift? Would it be observable from the ground in a quasar with a redshift of $\Delta\lambda/\lambda = 4.9$?

15. In what circumstances would a gravitational lens produce an image of a quasar that was a ring, rather than two or more points of light?

The nearby elliptical galaxy Centaurus A (NGC 5128) is crossed by a broad, irregular dust lane. The dark band is believed to be the remains of a dusty spiral galaxy, which is being disrupted and absorbed by the giant elliptical. *(National Optical Astronomy Observatories)*

Edwin Powell Hubble (1889–1953) left a career in law to study astronomy. In 1919 he joined the staff of the Mount Wilson Observatory, where he was the first to show that the "nebulae" are actually galaxies like our own. He was also the first to show that the large-scale structure of the universe is homogeneous. In 1929 he gave the first evidence for the expansion of the universe. *(Caltech)*

27

STRUCTURE AND EVOLUTION OF THE UNIVERSE

Celestial objects rarely journey through space alone. The Earth is but one of nine planets orbiting the Sun. The Sun itself seems somewhat unusual in that, as far as we know, it is a single star. Most stars are at least double, and many are members of either galactic or globular clusters. All of the stars and clusters in the Milky Way Galaxy are gravitationally bound, orbit a common center, and will complete their evolution in close proximity. Does this cosmic togetherness persist on still larger scales? Are most galaxies to be found in clusters of galaxies? What is the structure of the universe as a whole? How are galaxies distributed in space? Are there as many in one direction of the sky as in any other? And if we count fainter and fainter galaxies, presumably farther and farther away, do we find that their numbers increase in the way they should if galaxies are distributed uniformly in depth?

27.1 Distribution of Galaxies in Space

The first person to try to answer these questions was Edwin Hubble. Hubble had at his disposal what were then the world's largest telescopes—the 100-in. (2.5-m) and 60-in. (1.5-m)—reflectors on Mount Wilson. Although those telescopes can probe to great depths,

they can do so only in small fields of view. To photograph the entire sky with the 100-in. telescope would take not just a lifetime, but thousands of years. So instead, Hubble sampled the sky in many regions. In the 1930s Hubble photographed 1283 sample areas or fields with the telescopes, and on each photograph he carefully counted the numbers of galaxy images.

Hubble's Survey of Faint Galaxies

The results of Hubble's survey are shown in Figure 27.1, which is a map of the sky shown in what are called *galactic coordinates (Appendix 6):* The Milky Way, across the middle of the plot, defines the galactic equator, and the top and bottom of the map—the galactic poles—are 90° away from the Milky Way. The empty sectors at the lower right and left are the parts of the sky too far south to observe from Mount Wilson. Each symbol represents one of the regions of the sky surveyed by Hubble, and the size of the symbol indicates the relative number of galaxies he could observe in that area.

The first obvious thing to notice is that we do not see galaxies in the direction of the Milky Way. The obscuring clouds of dust in our Galaxy hide what lies in those directions. Hubble called this part of the sky the *zone of avoidance*. Near the Milky Way, the

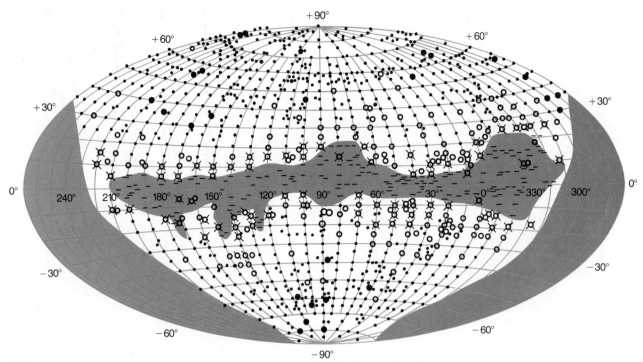

FIGURE 27.1 Distribution of galaxies according to Hubble's survey. Note that galaxies are not observed along the galactic equator because of obscuration by dust in the plane of the Milky Way.

counts of galaxies are below average and are denoted by open circles. The farther away we look from the direction of the Milky Way, the less obscuring foreground dust lies in our line of sight and the more galaxies we see. From the counts of galaxies in different directions, Hubble determined that light is dimmed by about 0.25 magnitude in the blue region of the spectrum in traversing a half-thickness of the Galaxy at the Sun's position. Having derived how foreground dust dims the light from galaxies, Hubble could correct his counts to allow for the effect.

After such correction, Hubble found that on the large scale the distribution of galaxies is **isotropic,** which means that if we look at a large enough area of the sky, we find as many galaxies in one direction as in any other. Moreover, Hubble found that the numbers of galaxies increase with faintness about as we would expect if they were distributed uniformly in depth.

These findings of Hubble were enormously important, for they indicated, at least relative to the precision of his data, that *the universe is isotropic and homogeneous*—the same in all directions and at all distances. In other words, his results indicate not only that the universe is about the same everywhere but also that the part we can see around us, aside from small-scale local differences, is representative of the whole. This idea of the uniformity of the universe is called the **cosmological principle** and is the

starting assumption for nearly all theories of cosmology (Chapter 28).

Hubble's survey was in two dimensions only. He simply counted the numbers of galaxies that he could see in particular directions. He did not know how far away those galaxies were—distant, highly luminous galaxies could not be distinguished from nearby faint galaxies. To understand what the three-dimensional structure of the universe is like, it is necessary to measure the distances of galaxies as well as their positions on the sky. This requires spectroscopy of each individual galaxy, which is an extremely time-consuming task, so we have detailed data for only a few regions. Nevertheless, these results are both surprising and puzzling, as we shall see.

The Local Group

The region of the universe for which we have the most detailed information is, as you would expect, our own local neighborhood. It turns out that our own Galaxy is a member of a small group of galaxies, which is called the Local Group. It is spread over about 3 million LY and contains at least 26 members. There are 3 large spiral galaxies (our own, the Andromeda Galaxy, and M33), at least 9 dwarf irregulars, 2 intermediate ellipticals, and 12 known dwarf ellipticals. Appendix 16 gives the properties of the galaxies that are generally accepted to be members of

FIGURE 27.2   The Local Group.

the Local Group. Figure 27.2 is a plot of the Local Group.

The average of the motions of all the galaxies in the Group indicates that its total mass is about 5×10^{12} solar masses. Although this mass estimate is rather uncertain, it implies that extensive amounts of dark matter must be present in the Local Group.

The Neighboring Groups and Clusters

Well beyond the edge of the Local Group we find other small groups of galaxies (Figure 27.3). At a distance of several tens of millions of light years, however, a group like the Local Group, seen in projection against the background of very many more distant galaxies, would not be noticed. At large distances, we recognize only very rich clusters that have

enough densely packed member galaxies to stand out against the background.

The nearest moderately rich cluster is the famous Virgo Cluster, a system with thousands of members (Figure 27.4). It contains a concentration of mostly

FIGURE 27.3   Many galaxies are found in small groups. This group, which is known as Stephan's Quintet, contains five prominent galaxies. Included in the group are one spiral, two peculiar barred spirals, and two elliptical galaxies. *(National Optical Astronomy Observatories)*

FIGURE 27.4   The central region of the Virgo Cluster of galaxies. The nearest large cluster (50 million LY distant), Virgo, with its hundreds of bright galaxies, is the dominant feature of the Local Supercluster of galaxies. *(Anglo-Australian Telescope Board)*

FIGURE 27.5 The central part of the Coma Cluster of galaxies. *(National Optical Astronomy Observatories)*

elliptical galaxies that includes M87, which is both a radio and an x-ray source (Figure 25.13). Within several degrees of M87 are many spirals as well as ellipticals and, associated with the brightest galaxies, very many dwarfs, like the dwarf ellipticals in the Local Group. All of the galaxies in the Virgo Cluster are too remote for us to observe their cepheids, and distances estimated by some of the other techniques described in Section 25.2 are quite uncertain. There is a growing consensus that the distance to the concentration of galaxies around M87 is about 50 million LY. At this distance, its linear diameter is about 5 million LY. Even this large system, however, would not be recognizable in remote parts of the universe.

A good example of a cluster that is much larger than the Virgo Cluster is the Coma Cluster, which has a linear diameter of at least 10 million LY and thousands of observable galaxies (Figure 27.5). The cluster is centered on two giant ellipticals, whose luminosities are about 4×10^{11} L_S. The Coma Cluster contains more and more members at magnitudes that are successively fainter. There is every reason to expect dwarf elliptical galaxies to be present; if so, the total number of galaxies in the cluster might be tens of thousands. The mass of this cluster is about 4×10^{15} solar masses.

Rich clusters like Coma usually show marked spherical symmetry and high central concentration. They contain few, if any, spiral galaxies in the cluster core (Figure 27.6) but rather have a membership dominated by ellipticals and those galaxies that resemble spirals but without spiral arms or obvious evidence of interstellar matter (those classed as S0—Section 25.4). Clusters of galaxies, particularly rich clusters like Coma, are usually sources of x rays. The x rays from a cluster are thermal radiation from gas at a temperature of 10^7 to 10^8 K. The gas is located between the galaxies.

There may be a significant relation between the presence of hot gas and of S0 galaxies in a cluster. X-ray emission lines of heavy elements such as iron have been observed at such intensity as to suggest that the abundance of heavy-elements in the hot gas is similar to that in the Sun, rather than the matter being all or nearly all hydrogen and helium, as current theories predict for the primordial matter from which the clusters formed. This suggests that at least some of the x-ray–emitting gas must have undergone nucleosynthesis in stellar interiors. This processed matter was then ejected into interstellar space within the cluster galaxies by such mechanisms as supernova outbursts. Finally, this material was swept from the galaxies by collisions between them and by their moving through the intracluster gas, or possibly by internal processes such as stellar winds and supernova explosions (Figure 27.7). Such sweeping of interstellar matter from galaxies stops star formation in them, and the spiral arms gradually disappear, leaving the galaxies as type S0. The swept gas is hot because the galaxies collide with one another or pass

FIGURE 27.6 Cluster of galaxies in Fornax. Most of the galaxies in this cluster are elliptical galaxies, which are composed of rather old, yellowish stars with little or no gas and no evidence of star formation. This photograph illustrates an important property of ellipticals—namely, that they are gregarious and tend to be found in regions of high concentrations of other galaxies. Spirals usually are isolated or lie on the outskirts of clusters. There is one spiral in the corner of the picture. *(Anglo-Australian Telescope Board)*

through intracluster gas at speeds up to thousands of kilometers per second.

27.2 Galaxy Encounters and Collisions

The sweeping of gas from galaxies is only one example of how the dense environment in the centers of clusters can modify the structure and evolution of galaxies. Because the spacing between galaxies in these regions is small relative to their sizes, collisions of whole galaxies are not unusual.

Interacting Galaxies

The first evidence that galaxy collisions occur came from analyses of strange-appearing pairs of galaxies interacting with each other (Figure 27.8). We can now understand many of these in terms of gravitational tidal effects. We might guess that two galaxies that happen to pass close to each other would pull matter out of each toward the other. Calculations show that bridges of matter can indeed form between the galaxies. "Tails" of material also string out away from each galaxy in a direction opposite to that of the

other. Because of the rotation of the galaxies, the tails and bridges can take on unusual shapes, especially when account is taken of the fact that the orbital

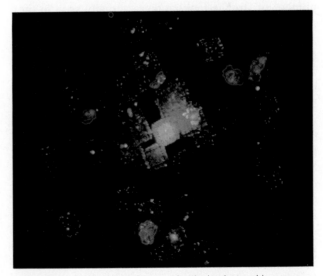

FIGURE 27.7 The Virgo Cluster of galaxies imaged in x-rays (blue) and hydrogen radio emission (red). The x-ray data were obtained with the Einstein Observatory. Note that the x-ray and hydrogen emissions arise from different parts of the cluster. It seems likely that galaxies near the center of the cluster have been stripped of their hydrogen gas. *(National Radio Astronomy Observatory/AUI)*

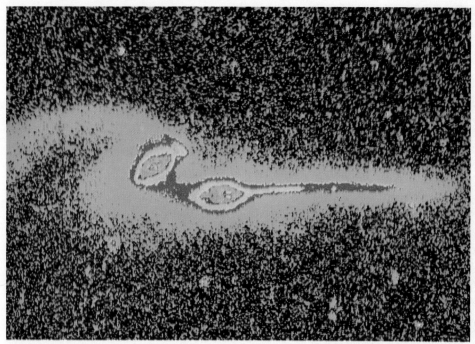

FIGURE 27.8 NGC 4676AB (the Mice). A classic system of colliding galaxies that have produced narrow tails as a consequence of the interaction. In this computer-processed image, the different colors correspond to different intensities. *(W. Keel and R. Kennicutt/National Optical Astronomy Observatories)*

motions of the galaxies can lie in a plane at any angle to our line of sight. Models of interacting galaxies have been calculated that mimic the appearances of a number of strange-looking pairs actually seen in the sky (compare Figures 27.9 and 27.10).

The interstellar matter in galaxies is much more affected by galaxy interactions than are the stars. Interstellar gas clouds are large and are likely to experience direct impacts with other clouds. These violent collisions compress the gas in the clouds, and the increased density can lead to vigorous star formation (Figure 27.10). In some interacting galaxies, the star formation is so intense that all of the available gas will be exhausted in only a few million years, so that the burst of star formation is clearly only a temporary phenomenon (Figure 27.11). Bursts of star formation are very rare in isolated galaxies.

Galactic Mergers and Cannibalism

If galaxies collide with slow enough relative speed, they may ultimately merge. Calculations show that some parts of slowly colliding galaxies can be ejected, while the main masses become binary (or multiple) systems with small orbits about each other. Such a newly formed binary galaxy, surrounded by a mutual envelope of stars and possibly interstellar matter, may eventually coalesce into a single large galaxy. This process is especially likely in the collisions of the most massive members of a cluster of galaxies, which tend to have the lowest relative speeds and to be concentrated toward the center. Mergers may convert spirals to giant ellipticals.

Other processes within clusters can also affect the morphology and evolution of galaxies. While we use the term **merger** to refer to the interaction of two galaxies of comparable size, the swallowing of a small galaxy by one that is much larger is described as **galactic cannibalism.** Two mechanisms are relevant. The first is *tidal stripping*. If a small galaxy approaches a large one too closely, then its self-gravity may be inadequate to retain the stars and gas in its outer regions. The tidal forces of the larger galaxy will dominate and will rip stars away from the galaxy of lower mass. The physics is the same as that discussed in Section 10.4, which considered how rings might form around planets.

A large galaxy can swallow, or cannibalize, the dense core of a smaller galaxy through a second mechanism, which is referred to as *dynamical friction*. The basic idea is that if the core of the smaller galaxy is moving rapidly through the envelope of

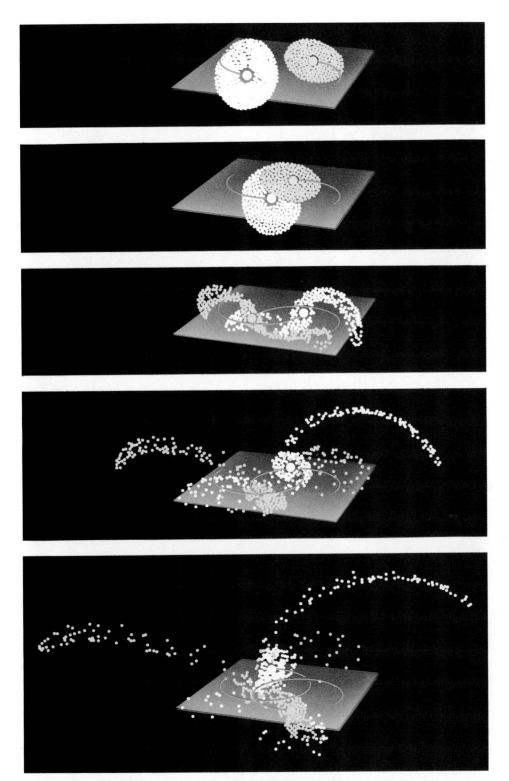

FIGURE 27.9 A sequence of five frames from a computer-produced motion picture that simulates the tidal distortion of two interacting galaxies. In this computer run, the initial conditions were chosen to see if the strange appearance of the pair of galaxies NGC 4038 and NGC 4039 could be accounted for in terms of tidal effects. Compare the last two frames with the photograph of the galaxies in Figure 27.10. *(Alar Toomre, MIT)*

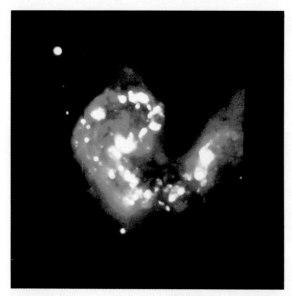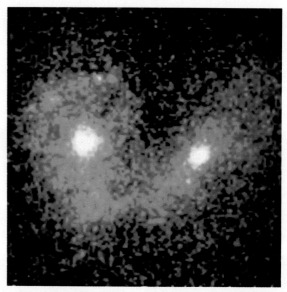

FIGURE 27.10 Comparison of optical (*left*) and infrared (*right*) images of the colliding galaxies NGC 4038 and NGC 4039. The optical image shows bright regions where hot, young stars are forming and dark patches of dust. The infrared image shows the distribution of old red stars in these two colliding galaxies. The nuclei of the two galaxies are evident, and the distribution of old stars seems relatively undistorted by the collision. *(National Optical Astronomy Observatories)*

stars of the larger galaxy, it will lose energy and decelerate while the stars in the larger galaxy will accelerate. This process causes the smaller galaxy to slow and spiral into the massive one.

In a cluster of galaxies, the most massive and most luminous galaxies are usually found near the center. These galaxies are often much more luminous than typical galaxies, and they frequently have more than one nucleus. These galaxies have acquired their unusually high luminosities by swallowing nearby galaxies of lower mass. The multiple nuclei are the remnants of their victims.

FIGURE 27.11 Optical (*left*) and infrared (*right*) images of M51, the Whirlpool Galaxy. The outlying arm, which reaches to the companion galaxy, is much brighter in the optical than in the infrared region. This difference in color suggests that most of the stars in this arm are hot, young stars and that the formation of these stars was stimulated by the interaction of the two galaxies. *(National Optical Astronomy Observatories)*

27.3 The Large-Scale Distribution of Matter

After astronomers discovered clusters of galaxies, they naturally wondered whether there were still larger structures in the universe. Are there clusters of clusters of galaxies? The answer is yes, and it is these very large-scale structures of matter, which contain one or more clusters of galaxies, that we call **superclusters.** Between the superclusters, there are great voids where few, if any, galaxies can be found.

The existence of superclusters and voids may seem to violate our earlier conclusion that the universe is homogeneous. After all, how can the universe be the same everywhere if some regions contain galaxies and others are apparently empty? Most astronomers still believe that the universe is homogeneous, but only on a very large scale—that is, only for regions large enough to include a number of superclusters and voids. This idea can be understood through an analogy. The residents of your neighborhood may be fairly typical of the residents of the city as a whole in which you live. But the people within the particular house in which you live may not be typical in number, age, or other characteristics. Just as we must take a fairly large sample of the population to obtain a representative group of people, so must we consider a fairly large volume of space in order to find within it the characteristics and kinds of objects that are typical of other large volumes of space. Astronomers are continuing to survey even larger volumes of space to try to *prove* that one region of the universe is typical of all other regions of comparable size.

The Local Supercluster

The best studied of the superclusters is the one that includes the Milky Way Galaxy. The most prominent grouping of galaxies within the Local Supercluster is the Virgo Cluster. The Milky Way Galaxy lies in the outskirts of the Local Supercluster. The diameter of the Local Supercluster is at least 60 million LY, and its mass probably is about 10^{15} solar masses.

Perhaps the most startling fact revealed by detailed study of the Local Supercluster is that space is mostly empty (Figure 27.12). Most of the galaxies are concentrated into individual clusters, and these clusters occupy only about 5 percent of the total volume of space contained within the boundaries of the Local Supercluster. A major problem for any theory of the formation of galaxies and of large-scale structure in the universe is to explain why galaxies are so closely clumped and why most of the universe is devoid of luminous matter.

Voids

The Local Supercluster is not unique in being mostly empty. Surveys in other directions in the sky yield the same result (Figure 27.13). The density of galaxies in the empty regions, or voids, is at least five times lower than it is in the well-populated regions. The largest structure seen so far in the universe is a sheet of galaxies that is at least 500 million LY long, 200 million LY high, and about 15 million LY thick. Referred to as the "Great Wall," this sheet of galaxies is about 250 million LY from our own Galaxy. The mass of the Great Wall is estimated to be 2×10^{16} solar masses, a factor of 10 greater than the mass of the Local Supercluster. Furthermore, the Great Wall extends over the entire region of space covered by the survey. Studies of larger volumes of space may lead to the discovery of even larger structures.

Knowledge of exactly *how* galaxies are distributed may provide clues to *why* they are distributed that

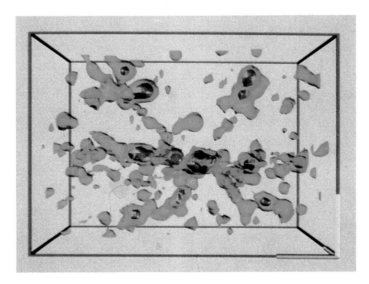

FIGURE 27.12 The distribution of galaxies in the Local Supercluster, a volume of space approximately 150 million LY across. Galaxies are found in clumps and small groups, while much of the volume of space shown contains no galaxies at all. *(Brent Tully, University of Hawaii)*

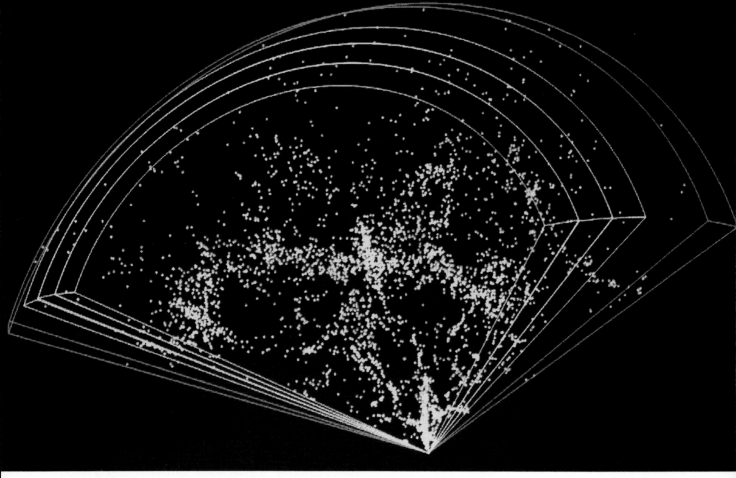

FIGURE 27.13 A three-dimensional "slice of the universe." Each dot represents the position of a galaxy in space. The Milky Way lies at the apex of the slice; the outside rim of the slice lies 650 million LY from the Milky Way. Note the concentration of galaxies in narrow bands or lanes with large voids between them. An analogous distribution would be obtained if we took a slice through a collection of bubbles of various sizes. The Great Wall is the band of galaxies stretching from left to right across the middle of the picture. *(Margaret Geller/Harvard-Smithsonian Center for Astrophysics)*

way. We know that the universe, only a few hundred thousand years after it was formed, was extremely smooth (Chapter 28). The challenge for the theoretician is to find a way to turn that featureless universe into the complex one that we see today, with dense regions containing many galaxies and nearly empty regions containing almost none.

For example, on the basis of data like those plotted in Figure 27.13, some astronomers argued that galaxies were distributed on the walls of bubble-like structures and that the matter originally in the interiors of these bubbles had been somehow cleaned out and pushed toward the walls. But how? The typical random velocities of galaxies today are only about 300 km/s. In the entire lifetime of the universe—that is, in 10 to 15 billion years—a galaxy moving with that velocity would travel a distance of only about 10 to 15 million LY. This distance is small relative to the 150-million-LY diameters of the largest voids. An alterna-

tive is to suggest that there were giant explosions during the first billion years of the universe that swept the gas into bubble-like shells and that galaxies subsequently formed in these shells. Unfortunately, despite considerable ingenuity, astronomers have yet to devise a way to produce explosions energetic enough to account for the largest voids.

While we do not have a good explanation for why the voids exist, this discussion should serve to illustrate how knowledge of the large-scale distribution of matter can help us to choose the correct theory for the events that occurred when galaxies were forming.

27.4   Formation of Galaxies

As we look off into space, we look back in time, for we see remote objects as (and where) they were far in the past, when light left them to begin its long journey across space to reach our telescopes. Remote objects, therefore, are in a sense historical documents in the universe, even though we may have difficulty in interpreting their message.

The Ages of Galaxies

One starting point for all theories of galaxy formation is the fact that most galaxies are very old indeed, and there are several observations that lead to this conclusion. For example, there are stars in globular clusters in our own Galaxy that are 13 to 15 billion years old. Therefore, the Milky Way must be at least this old.

The universe itself is not significantly older than the globular cluster stars. The age of the universe is derived from the observed rate of expansion. As we have seen (Section 25.3), galaxies are moving farther and farther apart. If we project this expansion back in time, we find that all of the galaxies were very close together sometime between 10 and 15 billion years ago. The major uncertainty in the age of the universe is caused by uncertainties in estimates of how far away galaxies are from us at the present time. It appears that the globular cluster stars in the Milky Way Galaxy must have formed during the first 2 to 3 billion years after the expansion of the universe began.

There is other evidence that galaxies formed very early and have been around for billions of years. For example, if we look at galaxies that are several billion light years away, we are seeing light that left those galaxies several billion years ago. We are, in effect, looking back in time and seeing those galaxies as they were when they were much younger than our own Galaxy is now. The most distant galaxies for which we have some information on composition emitted the light that we now observe when the universe was only about half its present age. Yet some of these galaxies have about the same luminosity and colors, and hence about the same stellar content, as do nearby and much older galaxies. This similarity of galaxies that span half of the age of the universe suggests that at least some galaxies were fully formed and quite mature several billion years ago.

We can probe still further back in time, and still closer to the beginning of the universe, by observing quasars, which are much brighter than normal galaxies and can be seen at larger distances. Astronomers find that the composition of the gas in distant quasars is very much like the composition of the gas in our own Galaxy. Specifically, the gas in quasars contains not only hydrogen and helium but also heavier elements such as carbon, nitrogen, and oxygen. These heavy elements were not present when the universe began but rather were manufactured in the first generations of stars that evolved within newly formed galaxies. Since quasars contain large amounts of heavy elements, at least one generation of stars had already completed its evolution even before the light that we now see was emitted. Given the distance to quasars, this means that some galaxies must have formed when the universe was less than 20 percent as old as it is now.

The Masses of Galaxies

The mass of the Milky Way Galaxy, excluding the invisible matter that makes up the large, dark halo, is about 10^{11} times the mass of the Sun. Most other spiral galaxies have masses within about a factor of 100 of this same value. The largest ellipticals are somewhat more massive than the largest spirals, but any theory of galaxy formation must explain why galaxies that are very much more massive do not occur.

Another important characteristic of galaxies is that small galaxies, which are also the fainter galaxies, are more common than large ones. In this respect, galaxies are much like stars; remember that there are many more faint stars with low masses than bright stars with large masses.

Environmental Influence

In the case of human beings there has been continuing controversy over the question of nature versus nurture. How many of our characteristics are determined by the genes we inherit, and to what extent are we a product of our environment? Similar questions can be asked about galaxies, and observations of galaxies in clusters suggest that the surrounding environment may be the dominant factor in determining what types of galaxies are formed.

Observations show that 80 to 90 percent of the galaxies in the high-density environments in the centers of clusters of galaxies are ellipticals and disk-shaped galaxies that have very little gas, no spiral arms, and no recent star formation. Conversely, isolated galaxies found in regions outside clusters or groups of galaxies, where the density of material is low, are mostly spirals.

Star Formation in Galaxies

The evolution of spiral and elliptical galaxies differs in a fundamental way. In spirals, star formation is a continuous process that is still occurring today. In elliptical galaxies, even the youngest stars are older than the Sun. Since there is very little dust or gas in ellipticals, star formation cannot take place in the present era.

Where did the gas and dust go? Much of it must have been consumed very rapidly in the formation of the first generations of stars. But star formation alone would not be efficient enough to consume *all* of the gas and dust. In any case, as stars evolve, they lose mass via stellar winds, by forming planetary nebulae, or by exploding. In the process, they inject dust and gas into the space between the stars, replenishing the interstellar material.

It must be that gas and dust are somehow efficiently *removed* from elliptical galaxies. One possibility is that the gas is swept out. As we have seen, ellipticals occur in clusters of galaxies. In these clusters, gas is present between the galaxies, as we know from x-ray observations. As an elliptical galaxy orbits about within a cluster, it moves rapidly (typical velocities are 1000 km/s) through the gas that lies within the cluster but outside the galaxies. This intergalactic gas bombards whatever small amount of gas may lie within the boundaries of the elliptical and drives the gas from the galaxy (see Figure 27.7). The pressure of the intergalactic gas is too small to affect the motions of the stars in the galaxy.

Spiral galaxies are able to retain their gas and dust because they lie isolated in regions of space where the distances to other galaxies are fairly large and the density of intergalactic gas is too low to sweep them clean. The Milky Way and Andromeda Galaxies are examples.

27.5 Evolution of Galaxies—The Theories

The universe is expanding. At the beginning, as we shall see in Chapter 28, the universe was very smooth. Matter was distributed uniformly. Now, however, the universe is certainly no longer smooth. Matter has clumped into stars. Stars themselves are not found everywhere in space but clump together to form galaxies. Even the galaxies are not uniformly distributed but have congregated to form clusters and superclusters with great voids between. The challenge for the theorist is to understand how an initially smooth (or nearly smooth) distribution of matter in the universe gives rise to the complex structure that we now see.

Top-Down or Bottom-Up

There are many ideas about how structure might have formed. Here we will look at the two possibilities that have been explored in the most detail. Top-down

theories assume that large structures—that is, supercluster-sized concentrations of matter—formed first and then fragmented to form galaxies. Bottom-up theories hypothesize that small structures formed first and then merged to build larger ones. If this bottom-up picture is correct, galaxies formed first and then gradually assembled to build first clusters and then superclusters of galaxies.

Both top-down and bottom-up theories assume that the universe was not initially absolutely smooth, but rather contained small fluctuations in density. As the universe expanded, the regions of higher density accumulated additional mass because they exerted a slightly larger than average gravitational force on surrounding material.

As in the case of star formation, the fate of these regions of higher density depended on the balance between pressure and gravity. If the gravitational force exceeded the pressure force, then each individual region would ultimately stop expanding in diameter with the expansion of the universe. It would then begin to collapse to form a cluster of stars, a galaxy, a cluster of galaxies, or even a supercluster of galaxies.

Unfortunately, the calculations of what happened are very difficult, and it is unclear what is the most likely size of the first high-density regions to collapse. There are two possibilities. The typical initial condensations may have been very large, with total masses equal to 10^{15} times the mass of the Sun, which is the mass contained within a supercluster. Alternatively, they may have been rather small and contained only 10^6 times the mass of the Sun, which is about the mass of a large globular cluster. Intermediate-sized condensations are not likely to have been formed initially.

Superclusters First?

Top-down theories calculate the consequences if only the very large-scale density fluctuations were able to collapse. Initially, gas clouds with masses of about 10^{15} times that of the Sun began to collapse. The collapse was irregular, and the structure formed was a pancake-shaped blob. Within the pancake, many individual regions of high density formed, and these too began to collapse. Calculations show that stable structures could be formed only for masses less than 10^{12} times that of the Sun and with diameters less than 300,000 LY— just the size of galaxies (Figure 27.14). Larger fragments were extremely diffuse and were destroyed in collisions with other fragments before stars could form within them.

This top-down model has the advantage that it can explain why galaxies are no larger than they are observed to be, but it has one fatal flaw. It takes a long

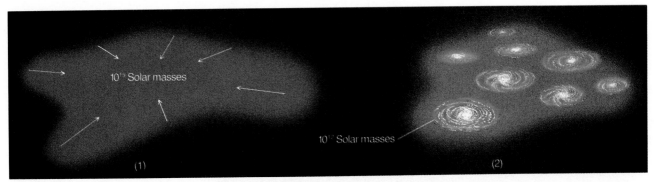

FIGURE 27.14 Schematic diagram showing how galaxies might have formed if large-scale supercluster structures formed first and then fragmented to form galaxies.

time for the original pancake to collapse and fragment into still smaller structures the size of galaxies. This model predicts that galaxies should still be forming in the present era. While astronomers have found a few nearby galaxies that may be young, it is clear that most galaxies formed billions of years ago. The top-down model is not compatible with the existence of galaxies and quasars when the universe was only 20 percent of its present age, and thus it conflicts with the observations.

Bottom-Up?

The alternative approach is to assume that small-scale structures formed first. This idea is the one that is most widely accepted as of this writing, but it will likely be modified as both theory and observation improve.

The basic assumption of the bottom-up models is that the first regions of higher density that collapsed had masses about 10^6 times that of the Sun, or about the size of a large globular cluster. Their collapse began when the universe was no more than 1 or 2 percent of its present age. As time passed, regions containing ever larger mass, possibly as large as the mass of a giant elliptical galaxy, began to collapse as well. Galaxies were formed either from the collapse of single galaxy-sized clouds or through the merger of several smaller structures. Clusters of galaxies formed as individual galaxies congregated, drawn together by their mutual gravitational attraction. First a few galaxies came together to form a group, much like our own Local Group. Then the groups began to combine to form clusters and eventually superclusters, like the Local Supercluster (Figure 27.15).

This model explains in a natural way why there are more small galaxies than large ones. The collapsing clouds are initially gaseous and collide and merge to form galaxies. The more collisions and mergers that

occur, the larger the galaxy that finally emerges. Since it is more likely that a given cloud will experience a few collisions and mergers than a large number, very large galaxies should be rare. A detailed theory of galaxy-building through this process, however, has not yet been worked out.

Giant elliptical galaxies are spherical and are found in the regions of highest density. It is likely that they were formed through the collision and merger of many fragments. Any collision of two systems of gas and stars will tend to stir up the orbits of the individual stars within each system and also to strip away from the outer regions matter that is not strongly held to the system by gravity. The result of many collisions is to build spherical systems that do not have extended disks.

According to this theory, spiral galaxies are formed in relatively isolated regions. A single cloud of gas collapses undisturbed to a disk, in which stars are then formed. A spiral might, through collisions with smaller systems, acquire some of the stars that populate its halo and its nuclear bulge. These stars are distributed in a spherical fashion, just as are the stars in ellipticals, which are also built through mergers. As the isolated cloud collapses to form a spiral, it may leave behind some fragments that become dwarf galaxies. Many spirals, including the Milky Way Galaxy, are surrounded by a swarm of small galaxies.

The bottom-up model predicts that there should be very few young galaxies in the present era, and indeed young nearby galaxies appear to be very rare. This model also predicts that clusters and superclusters should still be in the process of forming. Observations of the motions of galaxies in clusters, and the fact that clusters of galaxies often have very irregular rather than spherical shapes, suggest that they are still in the process of accumulating their member galaxies.

Many astronomers would label the ideas presented here as speculative, and it is clear that we are on very

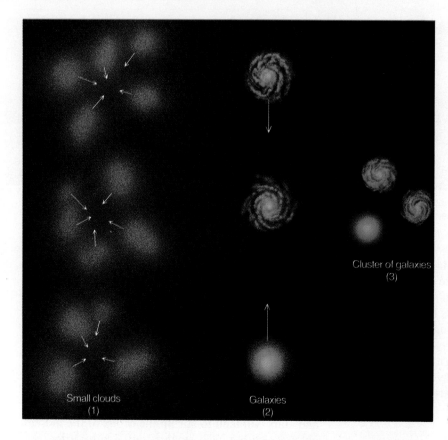

Small clouds
(1)

Galaxies
(2)

Cluster of galaxies
(3)

FIGURE 27.15 Schematic diagram showing how galaxies might have formed if small clouds formed first and then congregated to form galaxies and then clusters of galaxies.

uncertain ground when we attempt to describe the formation of galaxies. Both observational and theoretical efforts to understand this process are being vigorously pursued, and it is quite likely that the ideas presented here will be greatly modified by future research.

27.6 A Universe of (Mostly) Dark Matter?

Astronomers quite naturally focus their attention on studying the objects in the universe that they can *see*. Galaxies, stars, and interstellar gas and dust all emit electromagnetic radiation and can be studied with telescopes. It may very well be, however, that all this luminous matter is less than 10 percent (and possibly even less than 1 percent) of the matter in the universe. Studies of luminous matter may be missing the most important constituent of the universe! It may be that astronomers are, in effect, studying the flotsam and jetsam carried on the surface of the ocean, and ignoring the ocean itself.

Since the dark matter does not emit electromagnetic radiation, we can detect its influence only through the gravitational force that it exerts on other objects. In this sense, the dark matter resembles black holes. Indeed, black holes are one possible contributor to the dark matter, but there are other possibilities as well.

Evidence that dark matter exists comes from studies of *motions*—motions of individual stars in galaxies and motions of whole galaxies within clusters of galaxies. By studying the orbit of an object and applying Newton's law of gravity, astronomers can calculate the mass that must be present to hold the object in its orbit. When these calculations are made, the mass turns out to be much larger than we would expect from the number of visible stars and the amount of gas and dust that we can see.

Several experiments all yield this result. The orbits of halo stars and globular clusters indicate that the disk of the Galaxy is surrounded by a spherical envelope of dark matter. The extent of this envelope is uncertain but could reach to a distance of 300,000 LY from the center of the Galaxy. The total mass of this envelope of dark matter is about five times the mass of the luminous material in the Galaxy. The rotation curves of other spirals yield a similar result for the ratio of dark to luminous matter. While less certain, there is evidence that elliptical galaxies have massive dark halos as well.

It is possible to estimate the mass of a cluster of galaxies in a similar way. If we measure the velocities with which the member galaxies are moving, we can then calculate what the total mass in the cluster must be to keep the individual galaxies from flying off into space. The estimates of the amount of dark matter are less certain than the values derived from analyzing the rotation of stars in individual galaxies. It appears likely, however, that the amount of dark matter contained within giant clusters of galaxies is at least equal to the fraction of dark matter in individual galaxies and could be as much as ten times larger.

We cannot search for dark matter in the voids that separate clusters of galaxies, since so far we have found no luminous objects whose motions we can study. As a result, we simply do not know what the gravitational force is in the voids. We cannot, therefore, know whether the voids are filled with invisible matter.

Section 25.5 described the use of the mass-light ratio to characterize the matter in galaxies or clusters of galaxies. Remember that for systems containing mostly old stars, the mass-light ratio is typically 10 to 20. Mass-light ratios of 100 or more are a signal that a substantial amount of dark matter is present. Table 27.1 summarizes the results of measurements of mass-light ratios for various classes of objects. Very large mass-light ratios are found for all systems of galaxy size and larger, and this result indicates that dark matter is present in all of these types of objects.

What *is* the dark matter? If there is only five to ten times as much dark matter as luminous matter, then the dark matter could consist of normal particles—protons and neutrons. These protons and neutrons are not assembled into stars, or we would see them. Neither can the dark matter be in the form of dust and gas, or again we could detect it. The protons and neutrons could be in the form of black holes, brown dwarfs, or white dwarfs, and they might perhaps appear in different forms in different places. The dark

TABLE 27.1   Mass-Light Ratios

Type of Object	Mass-Light Ratio
Visible matter in Milky Way Galaxy	2
Total Matter in Milky Way Galaxy	10
Binary and small groups of galaxies	50–150
Clusters of galaxies	250–300

matter in giant clusters of galaxies might not be assembled into the same types of objects as the dark matter in the halo of the Milky Way and other spiral galaxies. However, if there is 100 times more dark than bright matter, as some theories of the universe require, then it may be that the dark matter is composed of some new type of particle that we have not yet detected on Earth. This possibility will be discussed in the next chapter.

Dark matter also helps to explain one observation that the bottom-up theory does not account for. According to that theory, matter should be distributed fairly uniformly throughout the universe. As we have seen, however, there are large voids in space where galaxies have not formed. One solution is to assume that dark matter *is* distributed uniformly. The assumption is then made that for some reason galaxy formation is *biased,* so that only some matter turns into galaxies and other matter does not. If this idea is correct, then the voids would not be empty at all but rather would be filled with failed galaxies that for some reason never formed stars. Astronomers are now trying to develop convincing theories for why galaxy formation should be so strongly biased that it occurs in some places and not others.

SUMMARY

27.1. Counts of galaxies in various directions can be used to show that the universe on the large scale is *homogeneous* (the same at all distances) and *isotropic* (the same in all directions). The uniformity of the universe is referred to as the **cosmological principle.** The Milky Way Galaxy is a member of the Local Group, which contains 26 member galaxies. There are richer clusters of galaxies that have thousands of members. Rich clusters of galaxies often contain hot (10^7 to 10^8 K) x-ray–emitting gas. This gas has been enriched in heavy elements through nucleosynthesis in stel-

lar interiors; the enriched material was then ejected into the intergalactic medium.

27.2. Galaxies in clusters are close enough together that collisions are likely. These collisions change the orbits of the member stars, leading to galaxies with distorted shapes and long tails; trigger star formation through compression of interstellar clouds; may lead to the **merger** of galaxies of comparable size; or result in **galactic cannibalism,** in which a small galaxy is swallowed by a much larger one.

27.3. Clusters of galaxies often group together with other clusters to form large-scale structures called **superclusters,** which can extend over distances of 300 million LY or more. Clusters and superclusters fill only a small fraction of space; much of the universe is devoid, or nearly so, of luminous matter.

27.4. Galaxies were formed when the universe was no more than 1 or 2 billion years old. Low-mass galaxies are much more common than high-mass galaxies. Elliptical galaxies tend to be found in the centers of dense clusters of galaxies, while spiral galaxies tend to be relatively isolated from other galaxies. These characteristics provide important constraints on models of the formation of galaxies.

27.5. The challenge for theories of galaxy formation is to show how an initially smooth distribution of matter can develop the structure—galaxies and galaxy clusters—that we see today. Calculations show that the first condensations of matter are likely to have contained the mass either of a supercluster of galaxies or of a globular cluster. Observations seem to favor the initial condensation of globular cluster–sized masses, which then congregate to form galaxies and clusters of galaxies.

27.6. The visible matter in the universe does not exert a large enough gravitational force to hold stars in their orbits within galaxies or to hold galaxies in their orbits around other galaxies. There is at least 5 to 10 times, and perhaps as much as 100 times, more dark matter than luminous matter. Astronomers do not know yet whether the dark matter is made of ordinary matter—protons and neutrons, for example—or of some totally new type of particle not yet detected on Earth.

REVIEW QUESTIONS

1. What is the cosmological principle? What would the impact be on studies of the origin and evolution of the universe if this principle were not valid?

2. Describe how galaxies are distributed in space.

3. What is the evidence that galaxies are nearly as old as the universe?

4. Describe two possible ways in which galaxies might form. Which possibility seems more likely? Why?

5. What is the evidence that a large fraction of the matter in the universe is invisible?

THOUGHT QUESTIONS

6. If galaxies are distributed at random in space, statistical theory enables us to predict how many fields should contain a certain number of galaxies each. Hubble found that a far larger number of fields had too few galaxies, and also that a far larger number of fields had too many galaxies, than would be expected for a random distribution. Explain why this result suggests that galaxies tend to be clustered.

7. Would an indefinite hierarchy of clustering (clusters upon clusters, without limit) be consistent or inconsistent with the cosmological principle? Why?

8. Observations show that the number of quasars is vastly greater at faint magnitudes and large redshifts than would be predicted for a uniform distribution in space. What does this tell us about the evolution of quasars? Explain clearly and completely why the cosmological principle is violated if quasars are not temporary phenomena at an earlier epoch of the history of the universe.

9. Assume for the sake of illustration that the dust in our Galaxy all lies in a relatively thin, flat disk, with the Sun in the central plane of that disk. Under these circumstances, show by a diagram that the obscuration of distant galaxies is less and less at greater and greater directions from the Milky Way (that is, at greater and greater galactic latitudes).

10. Why do we know less about the formation of galaxies than about the formation of stars?

11. Given the ideas presented here about how galaxies form, would you expect to find a giant elliptical galaxy in the Local Group? Why or why not? Is there a giant elliptical in the Local Group?

12. Can an elliptical galaxy evolve into a spiral?

13. Suppose you were to develop a theory to account for the evolution of the population of New York City. Would your theory most closely resemble a bottom-up or a top-down theory as we have applied those terms to galaxy evolution?

PROBLEMS

14. Suppose on one survey you count galaxies to a certain limiting faintness. On a second survey you count galaxies to a limit that is four times fainter.

   a. To how much greater distance does your second survey probe?
   b. How much greater is the volume of space you are reaching in your second survey?
   c. If galaxies are distributed homogeneously, how many times as many galaxies would you expect to count on your second survey?

15. If galaxies have typical velocities of 300 km/s with respect to their neighbors, calculate how long it would take for a galaxy to move across

   a. The Local Group
   b. The Virgo Cluster
   c. The Local Supercluster
   d. A typical void

Compare these times with the age of the universe, which is about 15 billion years.

16. Suppose we imagine superclusters to be flat (or stringy) and to have an average thickness of 50 million LY in the line of sight and to be separated from one another by voids 150 million LY across. How many superclusters would lie overlapping in projection in a typical line of sight out to a distance of 10 billion LY (about as far as we could hope to see galaxies)?

17. Theory suggests that elliptical galaxies may form by accumulating small condensations of matter, but we do not suggest that stars grow by accumulating matter from other stars. Why is it more likely that one galaxy will collide with another than that two stars will collide? (*Hint:* Compare the spacing between stars in the solar neighborhood with their diameters. Then compare the diameter of galaxies in the Local Group with their separations.)

MARGARET GELLER

Margaret Geller is Professor of Astronomy at Harvard University and Senior Scientist at the Harvard-Smithsonian Center for Astrophysics. She works in extragalactic astronomy, and her recent research contributions include the largest studies of the three dimensional distribution of galaxies, examination of the properties of individual systems of galaxies, and estimates of the average density of matter in the universe.

Geller was encouraged to become a scientist by her father, Seymour Geller, a solid state chemist. She did her undergraduate work in physics at the University of California, Berkeley and was only the second woman to receive a Ph.D. in physics from Princeton University. Geller has appeared frequently on television science and news programs, including The Infinite Voyage, A Galactic Odyssey, Smithsonian World, The Astronomers, *and the* MacNeil/Lehrer NewsHour. *In July 1990, she was awarded a MacArthur Foundation Fellowship. She is a member of the American Academy of Arts and Sciences.*

Margaret Geller is best known for her work on the distribution of galaxies.

What do you think is your most important scientific contribution?

The most important contribution is the one John Huchra and I have made together. We've found, by measuring redshifts of galaxies, that galaxies are located on very thin surfaces, which surround or nearly surround big dark regions that we call "voids." And these coherent patterns in the universe are very common. They're really not explained by any current theory.

When you started out to do the survey, did you have any idea that it would turn out to be as surprising and significant as it did?

We started in about 1980, when there was already a more limited redshift survey available. The results of this shallow survey were important but certainly not striking.

This survey didn't see the structures because the volume of space it covered was smaller than the structures. You could put that whole survey inside one of the big voids discovered in the new survey. When we started out to extend the survey, we began by studying individual clusters of galaxies because that seemed a well-defined thing to do, but in 1980–81, Bob Kirshner and his colleagues discovered a huge void in the direction of the constellation of Boötes. The void is a big region—150 million light years across—with a very low density of galaxies inside it. And I think we, like everybody else, were very skeptical about this observation and thought, ''Well, this can't be. Maybe it's the only region like that in the universe.'' About four years later, we finally realized that we had the wherewithal to test whether regions like this were a common feature of the universe.

The best way to look for these structures is to do a carefully designed survey. The way I usually explain it is: suppose that you wanted to know whether the Earth had continents and oceans but you were able to look at only one part in 10^5 of it—an area the size of the state of Rhode Island. If you choose just one small area, it would most likely land in the ocean, and you wouldn't learn much about what the Earth is really like. On the other hand, if you choose a long thin path—say a great circle around the Earth—it would probably intersect both continents and oceans. It would even tell you that both are big and of comparable size. Now, granted, there are a few great circles that might pass only through oceans because the oceans are connected and the continents aren't, but those are not common.

The great circle on the Earth is analogous to the slice that we took of the universe. We actually expected to set a limit on the number and size of the structures, not to find the structures themselves. John and I thought studying this slice was a good thesis project. We were so sure that we were not going to detect any structure that we didn't even encourage our graduate stu-

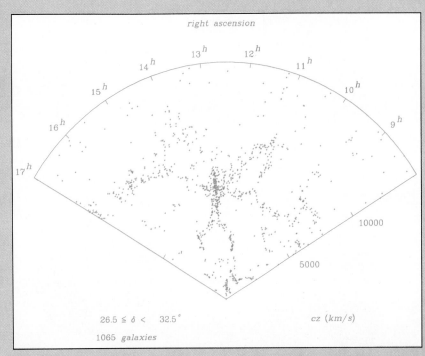

The map of the distribution of galaxies in space published by V. Lapparent, M. Geller, and J. Huchra. The galaxies appear to be distributed in elongated structures that surround empty regions where there are almost no galaxies. These empty regions have typical diameters of about 100 million light years. More extensive surveys have resulted in the map shown in Figure 27.13. *(Harvard Smithsonian Center for Astrophysics)*

dent, Valerie Lapparent, to plot up her data along the way. And so, she and John and the technicians who make the observations put all their data together, and it was reduced here at Cambridge. It wasn't until the end of the summer of 1985 that Valerie plotted it up. And then we saw the data. It was surprising to everyone! Nobody expected such a striking pattern.

If you looked at the universe at some other time in the past or the future, would these patterns in the distribution of galaxies look the same as they do now?

One of the great discoveries of this century is that the universe evolves. We expect that if we were to look at the universe in the past, we would learn something about the evolution of the large structures in the distribution of galaxies. One of the exciting things about the large telescopes being built on the ground now is that we will be able to do galaxy

surveys at large redshift. We expect to be able to look at the evolution of these structures directly. It's sort of like having a time machine and looking into the past, just as the geologic record on the surface of the Earth allows us to explore the history of the Earth.

We essentially have two pictures of the universe. We have one when the universe was only 100,000 years old, and we have one now when the universe is 10 to 20 billion years old. The picture we have of the early universe [when it was only 100,000 years old] is the cosmic background radiation (see Chapter 28); this radiation is remarkably smooth. This smooth background tells us that at the early epochs in the universe, normal matter was very uniformly distributed. Now we look out and see that at the present time galaxies are arranged in these big patterns. What happened in between? It's like going to a movie theater. You arrive at the beginning but then have to leave right away. You return to see the end, and you have to guess everything that happened.

Four of the five galaxies in the group known as Stephan's quintet. This true-color picture was constructed by combining images in red, green, and blue light. *(W. Schoening and N. Sharp/NOAO)*

How did you get started working on this problem in the first place?

I was a graduate student at Princeton. I was a student of Jim Peebles, and he is interested in this area. I started working on analyses of the way galaxies like the Milky Way are distributed on large scales. Then I came to the Center for Astrophysics, and John Huchra came about a year later. Slowly but surely we moved toward doing these kinds of problems. In general, our goal is to look for physical questions which can be addressed by taking large, well-defined samples of data. We do this for a large range of problems in the large-scale structure area. We've been working together for 15 years on this and related issues.

When did you decide to become an astronomer?

I decided when I went to graduate school. And even after that, I had

many doubts. As late as 1980, I had lots of doubts.

Why did you have doubts?

I wasn't really sure that I enjoyed it, basically. I wasn't really doing the kind of problem that I thought I'd like to be doing. It didn't excite me as much as I felt that I wanted to be excited by whatever I spent my life doing. As a student, I did well in the sciences. I started in mathematics, then went into physics. I was going to go into solid state physics, but I was advised by Charles Kittel, who is a famous solid state theorist, that I should look for a field that would be exciting 10 years after I got my Ph.D., when I was mature as a scientist. This advice seemed very strange advice to me at the time, but now I give the same advice to my students. And I was very lucky, because I did pick a field that really opened up at about that time.

Was there any particular individual who influenced your decision to become either a scientist or an astronomer?

My father is a scientist and was certainly a strong influence. Had he not been such a strong influence, I would have gone into the arts, and in fact I am now doing some things in this area.

What kinds of things?

Well, Boyd Estus, a filmmaker, and I have made a video *Where the Galaxies Are* about the survey. Reactions to the video have been extremely positive. It seems to appeal to people all the way from Nobel laureates to kids. I'm interested in the link between the arts and science. I've always been interested in design. Now I find I'm having some chance to do something with that interest. Boyd and I are beginning to make another film, which is more about the people who do the redshift survey and how people become scientists. It'll be about six of us, two students, two postdocs, John and me. We'll tell each person's story and how they got here in a way that I hope will be very general. I'd like to show some of the humor and the warmth of these people while telling the story very simply of what a redshift survey really is. I've always, all my life, had very broad interests, and I think I could easily have done something other than science.

Would you recommend a career in astronomy?

I think that it is really a wonderful thing to have a career where you can be creative. I think today that there are some serious frustrations, having to do with how much of your time you have to spend getting funds and so on. I also think there are issues today which are rather different from when I made the decision.

Could you elaborate on that?

I think that when I grew up, during the Sputnik era, we were a country with great imagination, with the attitude that we could lead broadly. Now in Washington you hear the statement that we can't lead in everything. My attitude is that that's a sure prescription for not leading in anything. And we have to make choices, etc., etc., etc., and things are becoming very constrained. Science is being done more by committee than by individuals. I must say that I think the lead of the U. S. in many areas in science is rapidly eroding. Although I think science is exciting intrinsically, I do not think that it's anywhere near as attractive to be a scientist as it was in this country.

I've had a wonderful time in science. It's been very exciting. I've had a very rich, exciting life. I've met a lot of people, and science has opened many doors . . . even opened doors to do other things. Part of the reason I can explore other skills that I have is because I've done something in science.

It's exciting to discover something nobody's seen before. After all, we're the first three people to ever see that slice of the universe. You see something like that and you wonder, "Why me?" It's sort of like being Columbus. You wonder how those people felt, and you think you share something with them, and every piece that you see—every piece of the puzzle that you see—it's like that.

On the other hand, I think that there are many frustrations in any profession. I think it's rather sad that there's been this loss of imagination in this country. It's very important that we worry about our social problems and social issues, but I think that we give lip service to things that are really of fundamental importance, like education. I think that's really serious, that we're really forgetting that you could cure all your social problems but if you don't also invest in creative things for the future, what are you living for?

One of the reasons that I became a scientist was that I perceived that this was the country where, if you worked in science, you could really be on the forefront. My job has allowed me to follow my dream, which, for me is the reason I do science.

The open dome of the Mayall 4-m telescope frames the setting Sun. Much of the observing time on this telescope is devoted to observational cosmology. *(National Optical Astronomy Observatories)*

28

THE BIG BANG

Cosmology, which is the study of the organization and evolution of the universe, is one of the most fascinating and rapidly changing fields of modern research. Through a combination of theory and observation, physics and astronomy, scientists now think that they can trace the evolution of the universe back to within tiny fractions of a second of the instant when it began. It is important to recognize, however, that the ideas that we will discuss in this chapter lie on the frontiers of science. Our description of the evolution of the universe, particularly during the first second, may seem as primitive to future generations of scientists as ancient Greek ideas seem to us now.

28.1 The Expanding Universe

The universe is *expanding*. This fundamental fact is the basis for modern cosmological models. As we saw in Section 25.3, the expansion of the universe was discovered about 60 years ago, and this discovery was one of the great events in modern science.

A Universe with a Beginning

If we were to make a movie of the expanding universe and run it backward, we would find that all of the matter in the universe was once concentrated in an infinitesimally small volume. We call this time the *beginning of the universe,* and from the rate of ex-

pansion, we calculate that this beginning occurred between 10 and 15 billion years ago.

As telescopes have grown larger (Figure 28.1) and astronomical detectors have become more sensitive, it has become possible to observe more and more remote galaxies with greater and greater speeds of recession (Figure 28.2). One of the most distant galaxies detected so far is shown in Figure 28.3.

There is no way to *prove* that the redshifts of these distant objects might not be due to some cause other than the expansion of the universe. If, however, the redshifts are not the result of expansion, then some new physical principles unknown to modern scientists are at work. It is much simpler, and characteristic of the scientific method, to adopt the more straightforward explanation—that the redshifts measure velocities of recession—especially as no reasonable alternative has been devised.

General Relativity and the Expanding Universe

Theoretical calculations show that the universe *must* be either expanding or contracting. A static universe is not stable. According to Newton's law of gravity, all objects exert a mutual attraction on one another. If the universe is of finite size, it is obvious that even if galaxies were initially stationary, they would inevitably begin to move more closely together, unable to

FIGURE 28.1 The world's largest telescopes devote much of their observing time to the study of distant objects to determine what the universe was like when it was young. This picture shows the dome of the Mayall 4-m telescope. The telescope is the largest at Kitt Peak National Observatory, which is located near Tucson, Arizona. *(National Optical Astronomy Observatories)*

resist the pull of gravity. Newton was not able to prove whether or not an *infinite* universe could be static.

In 1917, Einstein used his new general theory of relativity to prove that static universes could not exist, even if they were infinite. Since astronomers at that time had not discovered the expansion of the universe, Einstein altered his equations with the introduction of a new term, called the **cosmological constant.** The cosmological constant represents a repulsion that can balance gravitational attraction over large distances and permit a static universe.

There is, however, no evidence for such a repulsion in nature. Einstein included the cosmological constant to force his model of the universe to conform to the widely held notion that the universe is neither expanding nor contracting. When Einstein learned that subsequent observations demonstrated that the universe is expanding, he is reported to have said that the introduction of the cosmological constant was "the biggest blunder of my life."

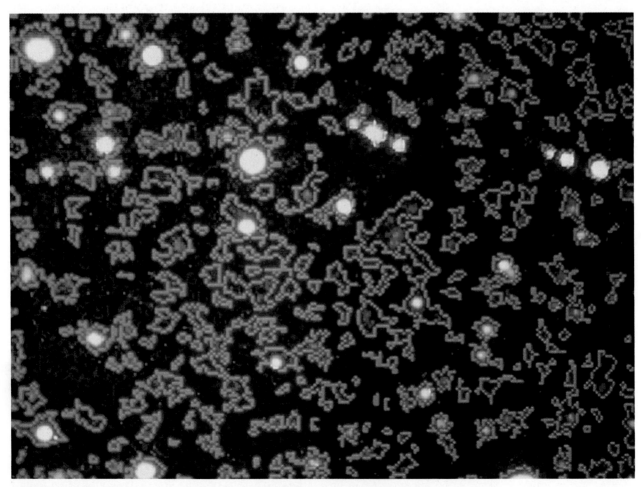

FIGURE 28.2 Studies of distant galaxies and quasars provide clues to the early evolution of the universe. Nearly every object in this image is a galaxy. The faintest galaxies were probably no more than 1 to 2 billion years old when the light we now see left them. *(National Optical Astronomy Observatories)*

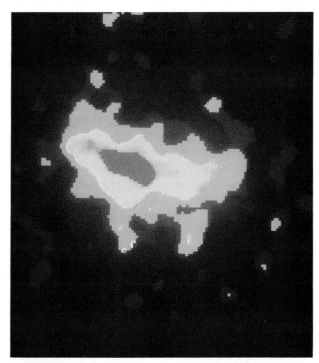

FIGURE 28.3 A photograph of the most distant galaxy known. The different colors represent different levels of brightness. The galaxy is obviously irregular in shape, but it is so faint that we know very little more about its morphology, stellar content, or composition. Light left this galaxy when the universe was only about 20 percent of its present age. *(Courtesy Ken Chambers)*

The Extent of the Observable Universe

In all directions we see galaxies and clusters of galaxies. At greater and greater distances they appear ever fainter, and beyond a few billion light years we detect only the greatest giants among the galaxies—the most luminous members of great clusters. The current record redshift belongs to a radio galaxy whose lines are redshifted by an amount $\Delta\lambda/\lambda = 3.8$. With the assumption that the Hubble constant H is equal to 25

km/s per million light years, this redshift corresponds to a distance of about 10 billion LY.

Quasars can be seen to somewhat greater distances. The current record redshift for a quasar is $\Delta\lambda/\lambda = 4.9$, which corresponds to a velocity of 95 percent of the speed of light. We are seeing this object as it was only a little over 1 billion years after the beginning of the universe. At these large redshifts, which translate to very large distances, even the quasars seem to thin out. This thinning out is real. At the distances of the farthest observed quasars, we are looking back to a time when the universe had only about 10 percent of its present age. At a still earlier time, quasars evidently did not exist, or at least were exceedingly rare. Perhaps galaxies had not formed yet or were only recently formed and had not yet had time to produce massive black holes.

The actual distance to which we can see depends on the Hubble constant and on how much the expansion is slowed over time by the mutual gravitational attraction of galaxies. Table 28.1 summarizes how far we can see for $H = 25$ km/s per 10^6 LY on the assumption that the expansion of the universe is not slowing down (Section 28.2).

At great distances, then, and in all directions, luminous objects disappear, but it is a boundary in *time*, not *space*. Many people think this means we must be seeing the "place" where the universe began. But there is no special location or site of creation. As far as we know, the universe is and always was everywhere. It is infinite in its spatial extent, but not in its age. We see only a *finite* part of the *infinite* universe, but what we can see is limited by time, not by distance.

28.2 Cosmological Models

Let us now see what cosmological theory predicts in detail for the past and future of the universe. Com-

TABLE 28.1	Velocity–Distance Relationship for $H = 25$ km/s per 10^6 LY in a Nearly Empty Universe	
Velocity (speed of light = 1)	**Distance (billions of light years)**	**Least Luminous Observable Objects**
0.001	0.01	Ordinary galaxies
0.40	4.3	Clusters of galaxies
0.60	6.5	Radio galaxies
0.78	8.4	Radio galaxies
0.88	9.8	Radio galaxies/quasars
0.93	10.6	Quasars

Go outside at night and look at the sky. How would you describe it? No doubt one of the first characteristics you would mention is the fact that the *sky is dark at night*. This turns out to be a very significant observation—one that has implications for our understanding of cosmology.

The fact that the sky is dark has puzzled many scientists over the centuries, including the German physician and amateur astronomer Wilhelm Olbers. In 1826, Olbers argued that if the universe is uniform and infinite, both standard assumptions in many cosmologies, and populated by eternal and unchanging stars, then we should see an infinite number of stars. In fact, in every direction we look, our line of sight, if we extend it far enough, should run into a star. As a good analogy, imagine that you are standing in a forest. No matter what direction you look in, provided that the forest is large enough, you will see a tree.

So, too, with stars in an infinite universe. Indeed, it is possible to show mathematically that in an infinite and eternal universe filled with stars like the Sun, every point in the sky should appear to us to be as bright as the surface of the Sun. It would seem as if we were living inside a fiery, hot furnace that poured down on the Earth 180,000 times as much energy as the Sun does. The fact that the sky is dark is often referred to as *Olber's paradox*.

Why, then, is the sky not blazing with heat and light? Scientists have offered two classes of explanations. The first class hypothesizes that the universe is infinite, that the stars do emit the expected amount of light, but that, for some reason, light does not reach the Earth. Suggestions for stopping the light have included reddening by intergalactic dust and the redshifting of light because of the expansion of the universe. Quantitatively, none of the suggestions has worked out.

The second class of explanations assumes that for some reason the number of bright stars is not infinite. Early scientists speculated that perhaps the universe itself is not infinite, but if it is not, then we have the problem of what lies outside the universe. Before there was a good understanding of stellar evolution, some scientists suggested that perhaps many stars were dark and radiated no energy.

It turns out that the number of observable bright stars is indeed not infinite, but for a reason that became apparent only in this century. *The sky is dark at night because the universe is young.* It takes time for light to traverse the distance from a star or a system of stars—that is, from a galaxy—to us. Accordingly, we can see a galaxy only if it is near enough so that light emitted by it has had time to reach us during the finite lifetime of the universe. Because the universe is not infinitely old, we do not in fact see an infinite number of stars or galaxies.

Olber's paradox offers an excellent example of how a very simple observation may have profound implications. The challenge for the scientist is to look at the world with fresh eyes and see that things we take for granted may offer important clues about the world in which we live.

---

mon to most cosmological theories is the *cosmological principle*—the assumption that on very large scales the universe is homogeneous—and we will make that assumption here. We shall also adopt the best description of gravitation yet found, that of *general relativity*. For simplicity and in common with nearly all cosmologists today, we assume that the *cosmological constant* is zero. Of course, the universe may not be, in reality, described by so simple a model as the best theory we know today. The question is, can we explain what we see with these assumptions? If so, then there is no need to look for a more complicated model.

The models that we will consider here all presume that the universe began at a particular finite time in the past and that the universe is evolving today. These models make solid, testable predictions about such observable quantities as the age of the universe and the abundance of helium relative to hydrogen. As we shall see, these predictions appear to be consistent with what is actually observed.

The Age of the Universe

As the universe expands, galaxies and clusters of galaxies separate from one another. Thus if we extrapolate backward in time, we would find them coming together, until some time in the distant past when all matter was crowded to an extreme density—a condition that marks a unique beginning of the universe, or at least of that universe we can know about. At that beginning, the universe suddenly began its expansion with a phenomenon called the **big bang.** We return in the next section to the physical conditions of the universe at and just after the big bang.

The total amount of matter in the universe creates gravitation, whereby all objects pull on all other objects (including light). This mutual attraction must slow the expansion, which means that in the past the expansion must have been at a greater rate than it is today. How much greater depends on the importance of gravitation in decelerating the expansion. At the extreme, if the total mass-energy density has always been low enough that gravitation is ineffective (an essentially "empty" universe), the deceleration would be zero. Only in that case would the universe have been expanding at the present rate ever since the big bang.

That extreme of a nearly empty universe corresponds to the greatest age of the universe (since the big bang) consistent with the observed expansion. In universes containing more mass, the rate of expansion would be slowed by gravity. The expansion would have been faster in the past, and the galaxies would have reached their present separations in a smaller time. Consequently, we can obtain an estimate of the *maximum age* of the universe by asking how long it would take for distant galaxies, always moving away from us at their present rates, to have reached those distances.

Call the maximum possible age T_0. So far, we have expressed the Hubble law in words only, but it is very convenient also to write it as the simple equation

$$v = H \cdot d$$

where v is the radial velocity of a galaxy of distance d. But velocity is just distance divided by time; that is, $v = d/T_0$. Hence combining these two expressions we have

$$\frac{d}{T_0} = H \cdot d$$

or

$$T_0 = \frac{1}{H}$$

We see, then, that the maximum age of the universe is just the *reciprocal of the Hubble constant.* If we use consistent units and adopt a Hubble constant of 25 km/s per million light years, then this time is approximately 13 billion years (Problem 19). For another way to derive this same result, see Problem 20.

Maximum and Minimum Ages

The age of the universe that is calculated in this way depends, of course, on the present rate of expansion—the value of the Hubble constant. Radial velocities of galaxies can be determined from the redshifts of their spectral lines, so the accuracy of H depends on the accuracy of our measurements of the distances to galaxies. If distance estimates were in error by a factor of 2, the Hubble constant would also be wrong by a factor of 2. The estimate of the age of the universe would, as a result, also be wrong by a factor of 2 (Figure 28.4). Recently proposed values for the Hubble constant range from about 15 to 30 km/s per million light years (50 to 100 km/s per million parsecs). These values give maximum ages for the universe of about 20 to 10 billion years, respectively. Our choice of value for the Hubble constant (25 km/s per 10^6 LY) gives $1/H$ as 13 billion years.

These calculated ages are maximum values. The true age of the universe must be less than $1/H$ because the expansion has to be decelerating. As we shall see below, the age for a universe that is decelerating just enough to halt its expansion eventually is two-thirds of the maximum age. Thus the range of acceptable ages for a Hubble constant of 25 km/s per million light years is between 9 and 13 billion years. That is our best present estimate of the age of the universe—the time since it all started with the big bang.

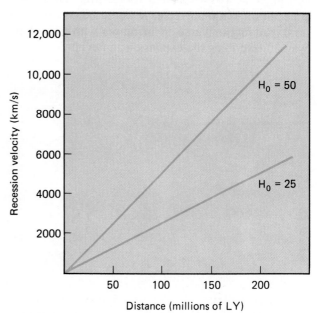

FIGURE 28.4 Suppose we measure a velocity of 5000 km/s for a distant galaxy. If that galaxy is actually at a distance of 200 million LY, then the Hubble constant is 25 km/s per million light years. But suppose we estimate the distance incorrectly by a factor of 2 and think the galaxy is at a distance of 100 million LY. Then we will think the Hubble constant is 50 km/s per million light years and that the age of the universe is only half its true value.

The best alternative method of estimating the age of the universe is from calculations of the evolution of the oldest stars we know, namely, the members of globular clusters. Current models yield ages in the range of 13 to 15 billion years and so are in approximate agreement with the age derived from determination of the Hubble constant. If H does eventually prove to be as large as 30 km/s per million light years, however, then we would be faced with a problem, since the oldest stars would be older than the universe itself—and that, of course, cannot be! We would then have to reconsider both our models for stellar evolution and our basic cosmological model.

The Scale of the Universe

By assumption (the cosmological principle), the universe is always homogeneous if we consider a large enough volume of space. Consequently, the expansion rate must be uniform (the same in all directions), so that the universe must undergo a uniform change in scale or size with time. It is customary to represent that scale by R, which is a function of time. The actual value of the scale is arbitrary—we could think of it as being the distance between any two representative galaxies, since R changes in the same way everywhere. R plays the same role as the "scale of miles" on a terrestrial map. It tells us by how much the universe has expanded (or contracted) at any time. For a static universe, R would be constant. In an expanding universe, R increases with time. One way of describing the expansion and evolution of the universe is to describe the change in R since the universe began.

One approach is to use the equations of general relativity to provide a mathematical description of R. But even without general relativity we can understand in a general way how the scale of the universe changes with time. In a uniformly expanding (nearly zero mass) universe, R simply grows in direct proportion to elapsed time. But in a more massive universe, there is a gravitational deceleration, and the rate of change of R decreases as the expansion is slowed.

The degree of deceleration depends on very uncertain observational parameters: the *Hubble constant* and the *mean density of matter* in space. By "mean density" we mean the mass of matter (including the equivalent mass of energy) that would be contained in each unit of volume (say, 1 m³) if all of the stars, galaxies, and other objects were taken apart, atom by atom, and if all of those particles, along with the light and other energy, were distributed throughout all space with absolute uniformity.

The behavior of R—the rate of change of distance in the expanding universe—is shown for several possible models in Figure 28.5. Time increases to the right and the scale length, R, upward in the figure. At the present (marked along the time axis), R is increasing at the rate indicated by the Hubble constant. The straight dashed line corresponds to the nearly empty universe with no deceleration. In this case, R has been increasing uniformly with time since the big bang about 15 billion years ago. This initial time, the reciprocal of the Hubble constant, is labeled T_0 in the figure. The other curves represent varying amounts of

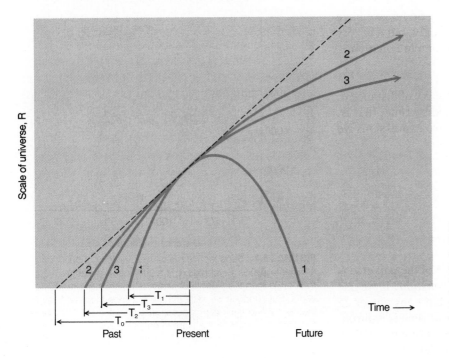

FIGURE 28.5 A plot of $R(t)$, the scale of the universe, against time for various cosmological models. Curve 1 represents a closed universe, curve 2 represents an open universe, and curve 3 is for a universe with a critical density. The dashed line is for an empty universe, which is one in which the expansion is not slowed by gravity.

deceleration, although they all correspond to the observed rate of expansion at the present. The curves that include deceleration must, of course, start from the big bang at shorter times in the past.

If the mean density of the universe is high enough and the deceleration is above a critical value, R is given by curve 1. We see that in this case the universe stops expanding some time in the future and begins contracting. Eventually, the scale drops to zero, with what Princeton physicist John A. Wheeler calls the "big crunch." In this case, the universe is said to be *closed,* for it cannot expand forever. It is tempting to speculate that another big bang might follow the "crunch," giving rise to a new expansion phase, and ensuing contraction, perhaps oscillating between successive big bangs indefinitely in the past and future. Such speculation is sometimes referred to as the *oscillating theory* of the universe. It is not really a theory, however, because we know of no mechanism that can produce another big bang. General relativity (and other theories) predict, instead, that at the crunch the universe will collapse into a universal black hole. (Of course, we have no complete theory for the first big bang either!) In any case, the oscillating theory is a speculation on a possible variation of the closed model of the universe. Alternatively, if the mean density is too low (curve 2 in Figure 28.5), gravitation is never important enough to stop the expansion, and the universe expands forever. In this case, the universe is said to be *open.*

At a **critical value** of the density (curve 3), the universe can just barely expand forever. This density marks the boundary between the families of open and closed universes. The age of the "critical density" universe is exactly $2/3\ T_0$. Open universes have ages between $2/3\ T_0$ and T_0, and closed universes ages of less than $2/3\ T_0$.

The various possibilities for the evolution of the universe are analogous to those for a rocket launched from the Earth. If the rocket has just the critical velocity of escape from the Earth (about 11 km/s), it barely escapes the Earth on a parabolic orbit. This situation corresponds to the critical density universe. At lower and higher launch velocities, the rocket falls back to Earth or escapes completely from the Earth—analogous to closed and open universes, respectively.

If the universe is closed in the sense that material bodies cannot separate forever, then neither can light or other radiation. The closed universe is *unbounded,* for no edge can ever be observed, but it is still *finite.* On the other hand, if the universe is open, its extent is *infinite.*

The critical density universe, separating the open and closed ones, has a particularly appealing geome-try in terms of general relativity. In it, light travels in straight lines. This universe is said to be flat, and it can be described in terms of ordinary plane or Euclidean geometry that we all were taught in high school.

Summary of Models of the Universe

The possibilities described in the previous paragraphs are summarized in Table 28.2. For this table, the Hubble constant, H, has been assumed to equal 25 km/s per million light years.

Note particularly the value of the critical density. This value, the density at which the universe is just barely closed, is about 10^{-29} g/cm^3. If the Hubble constant is larger than 25 km/s per million light years, the critical density is also larger, perhaps up to 2×10^{-29} g/cm^3. With a small Hubble constant, it could be as small as 0.5×10^{-29} g/cm^3. But some value near 10^{-29} is the density we should look for if we expect the universe eventually to stop its expansion.

The Future of the Universe

There are several observational tests by which we hope to be able to distinguish between the evolving cosmological models. One is the determination of the mean density of matter in space. We can estimate the mean density from the number of galaxies and clusters we observe out to a given distance and from a knowledge of the masses of these objects. There is considerable uncertainty in the masses of clusters of galaxies, and we do not know how much dark matter (if any) may exist in the space between galaxies. Nevertheless, such estimates indicate a mean density less than 10^{-30} g/cm^3. This is below the critical density by a factor of at least 5 and probably more and suggests an open universe. The estimates are too uncertain, however, to be sure of the conclusion.

As we will see in the following section, the production of deuterium in the early universe is very sensitive to the density of the universe within the first few

TABLE 28.2	Evolving Relativistic Cosmological Models for which $H = 25$ km/s per Million Light Years	
Kind of Universe	**Age (T) (units 10^9 years)**	**Mean Density (ρ) (g/cm^3)**
Closed	$T < 9$	$\rho > 10^{-29}$
Open	$9 < T < 13$	$\rho < 10^{-29}$
Flat	$T = 9$	$\rho = 10^{-29}$

minutes of the expansion. The proportion of deuterium in interstellar space is thought to be a measure of that formed shortly after the big bang, because any deuterium inside a star is rather quickly converted to helium. It is very difficult to detect deuterium in space, but careful measures show that the ratio of deuterium to hydrogen is probably in the range of 10^{-4} to 10^{-5}. From this, a crude estimate of the density of the early universe can be inferred, and from that knowledge it is possible to predict what present-day density would result. The calculation suggests a present density of about 10^{-31} g/cm^3, again pointing to an open universe.

Other Cosmologies

The discussion above is based on particular assumptions—the cosmological principle, conservation of mass-energy, and general relativity theory with zero cosmological constant. Any or all of these assumptions *could* be wrong, and if so, there is an infinity of other possibilities.

Another model of cosmology, which received wide attention in the 1950s, is the *steady-state theory* of Hermann Bondi, Thomas Gold, and Fred Hoyle. The steady state is based on a generalization of the cosmological principle called the *perfect cosmological principle,* in which it is assumed that on the large scale the universe is not only the same everywhere, but for all time. In a steady-state universe, mass-energy is not conserved. As the universe expands and matter would otherwise thin out, new matter is continuously being created to keep the mean density the same at all times.

The steady-state theory predicts at what rate matter is created to ensure that there will always be the same admixture of young and old stars and galaxies. The creation would occur so gradually (presumably as individual atoms coming into being here and there) that we would not notice it. The steady-state universe is infinite and eternal and has much philosophical appeal. But we have already seen (Chapter 26) that quasars are more common at large redshifts than nearby, so it appears that there were more quasars when the universe was young. This evolution of quasars violates the perfect cosmological principle, which requires the universe to be the same at all times. Moreover, we shall see subsequently that there is direct evidence that the universe has evolved from a hot, dense state, strongly supporting the idea of a big bang.

28.3 The Beginning

The Belgian cosmologist Georges Lemaître was probably the first to propose a specific model for the big bang itself. He envisioned all the matter of the universe starting in one great bulk he called the primeval atom. The primeval atom broke into tremendous numbers of pieces, each of them further fragmenting, and so on, until what was left were the present atoms of the universe, created in a vast nuclear fission. In a popular account of his theory he wrote, ''The evolution of the world could be compared to a display of fireworks just ended—some few red wisps, ashes and smoke. Standing on a well-cooled cinder we see the slow fading of the suns and we try to recall the vanished brilliance of the origin of the worlds.''

Today we know much more about nuclear physics, and the primeval fission model cannot be correct. Yet Lemaître's vision inspired more modern work and in some respects was quite prophetic.

In the 1940s the American physicist George Gamow suggested a universe with the opposite kind of beginning—nuclear fusion. He worked out the details with Ralph Alpher, and they published the results in 1948. (They added the name of physicist Hans Bethe to their paper, so that the coauthors would be Alpher, Bethe, and Gamow, a pun on the first three letters of the Greek alphabet: alpha, beta, and gamma.) Gamow's universe started with fundamental particles that built up the heavy elements by fusion in the big bang. His ideas were close to our modern view, except that the conditions in the primordial universe were not right for atoms to fuse to carbon and beyond, and only hydrogen and helium should have been formed in appreciable abundances. (The heavier elements formed later in stars, as described in Chapter 21.)

The Standard Model of the Big Bang

The modern theory for the evolution of the early universe is called the *standard model of the big bang.* Three simple ideas hold the key to tracing the changes that occurred during the first few minutes after the universe began.

The first essential piece of information is that the universe cools as it expands (Figure 28.6). In the first fraction of a second, the universe was unimaginably hot. By the time 0.01 s had elapsed, the temperature had dropped to 100 billion (10^{11}) K. After about 3 min, the temperature had reached about 1 billion (10^9) K, still some 70 times hotter than the interior of the Sun. After 700,000 years, the temperature was down to a mere 3000 K, and the universe has continued to cool since that time.

All of these temperatures but the last are derived from theoretical calculations, since obviously no one was there to measure them directly. As we shall see, however, we have actually detected the feeble glow of radiation emitted at a time when the universe was only about 700,000 years old.

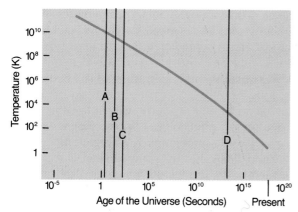

FIGURE 28.6 Standard model of the early universe showing how the temperature varies with time. The vertical line labeled *A* designates approximately the time at which neutrinos stop interacting with matter. *B* denotes the time when positrons and electrons annihilate one another. Helium synthesis occurs at time *C*, and the universe becomes transparent to radiation at time *D*.

The second step in understanding the evolution of the universe is to realize that at very early times it was so hot that collisions of photons could produce particles of matter. Detailed calculations with the theories of modern physics allow us to reconstruct the complex interactions between photons and a vast number of subatomic particles that were present in the fireball at that time.

The third basic concept of this standard theory is that the energy of the photons depends only on the temperature in the fireball. The higher the temperature, the higher the energy of a typical photon, and the more massive the particles that can be produced by the collision of two such photons. Thus as the fireball cooled, the makeup of both its photons and its material particles changed in a predictable way.

Keeping these three ideas in mind, we will now trace the evolution of the universe from the time that it was about 0.01 s old and had a temperature of about 10^{11} K. Why not begin at the very beginning? The reason is that when the universe was less than about 0.01 s old, the kinds of particles present are so unfamiliar to us that the theory is on shaky ground. Only at temperatures below 10^{11} K can physicists calculate the evolution of the fireball with some confidence, using data obtained from large particle accelerators on Earth.

The universe, 0.01 s after the beginning, consists of a dense soup of matter and radiation. Each particle collides rapidly with other particles. There are protons and neutrons and electrons and positrons, as well as a sea of exotic particles that will later play a role as "dark matter." The picture is of a seething cauldron of a universe, with photons colliding and interchanging energy, often meeting in such a violent impact that the photons themselves are destroyed and

leave in their wake an electron-positron pair, of neutrons being converted to protons and protons to neutrons through collisions of particles. It is much too hot for protons and neutrons to combine to form heavier atomic nuclei.

By the time the universe is a little more than 1 s old, and the temperature has dropped to a mere 10^{10} K, the density has dropped to the point where neutrinos no longer interact with matter but simply travel freely through space. In fact, these neutrinos should now be all around us. Since they have been traveling through space unimpeded and hence unchanged since the universe was 1 s old, observations of them would offer one of the best tests of the big bang model. Unfortunately, the very characteristic that makes them so useful, the fact that they interact so weakly with matter that they have survived unaltered for all but the first second of time, also renders them undetectable, at least with present techniques. Perhaps someday someone will devise a way to capture these messengers from the past.

The universe continues to expand and cool. At 3 min and 46 s, a big event takes place: Protons and neutrons begin to be bound into stable atomic nuclei as the temperature falls to 9×10^8 K. The first step in building atomic nuclei is the collision of a neutron and proton to form deuterium (heavy hydrogen), and essentially all of the neutrons are used up by this reaction. Collisions then convert nearly all the deuterium to helium, which has an atomic mass of 4 (2 protons and 2 neutrons). There is no stable particle of mass 5, however, and we believe that very few heavier elements are formed at this early stage of the universe. Instead, heavy elements are predominantly produced later deep in the interiors of stars.

The model predicts that when the helium is created, there will be 1 helium atom for every 12 hydrogen atoms. In units of mass, this means that about 25 per cent of the matter in the universe should be helium and about 75 percent should be hydrogen. Some deuterium also survived, but only a small amount—probably less than 1 part in 10,000. The actual amount of deuterium formed depends critically on the density of the fireball. If the density was fairly high, most of the deuterium would have been built up into helium.

It is a striking success of the standard model that the predicted ratio of hydrogen to helium—3 to 1 by mass—is just the ratio observed in the Sun, Jupiter, stars, and interstellar matter. A small enhancement of the helium must have resulted from nucleosynthesis in stars, to be sure, but by far most of the helium must be primordial—especially in the outer layers of old stars. Hence the agreement between the predicted and observed abundance of helium must be regarded as a second triumph for the big bang theory (the first being the expansion of the universe). We come to a third successful prediction in the next subsection.

For the next few hundred thousand years, the fireball was like a stellar interior—hot and opaque, with radiation being scattered from one particle to another. By about 700,000 years after the big bang, the temperature had dropped to about 3000 K and the density of atomic nuclei to about 1000 per cubic centimeter. Under these conditions, the electrons and nuclei combined to form stable atoms of hydrogen and helium. With no free electrons to scatter photons, the universe became transparent, and matter and radiation no longer interacted; subsequently each evolved in its separate way.

One billion years after the big bang, stars and galaxies had probably begun to form, but we are not sure of the precise mechanisms. Certainly, however, deep in the interiors of stars, matter was reheated, stars began to shine, nuclear reactions were ignited, and the gradual synthesis of the heavier elements began.

The fireball must not be thought of as a localized explosion—like an exploding superstar. There were no boundaries and no site of the explosion. It was everywhere. The fireball still exists, in a sense. It has expanded greatly, but the original matter and radiation are still present and accounted for. The stuff of our bodies came from material in the fireball.

Discovery of the Cosmic Background Radiation

What happened to the radiation released when the universe became transparent at the tender age of just under a million years? That question was first considered in the late 1940s by Alpher and Robert Herman, both associates of Gamow. They realized that just before the universe became transparent it must have been radiating like a blackbody at a temperature of 3000 K. If we could have seen that radiation just after neutral atoms formed, it would have resembled radiation from a reddish star. But that was at least 10 billion years ago, and in the meantime the scale of the universe has increased a thousandfold. The light emitted by the once hot gas in our part of the universe is now billions of light years away.

To observe that glow of the early universe, we must look out to such great distance—10 to 15 billion LY— that we see the universe as it was when it was only about 700,000 years old. Now those remote parts of the universe, because of its expansion, should be receding from us at a speed within two parts in a million of that of light. The radiation would be redshifted to wavelengths a thousand times those at which it was emitted.

When a blackbody approaches us, the Doppler shift shortens the wavelengths of its light and causes it to mimic a blackbody of higher temperature. When a blackbody recedes, it mimics a cooler blackbody. Alpher and Herman predicted that the glow from the fireball should now be at radio wavelengths and should resemble the radiation from a blackbody at a temperature of only 5 K—just a few degrees above absolute zero. But there was no way at the time they published their conclusion to observe such radiation from space, so the prediction was forgotten.

In the mid-1960s, however, the same idea occurred independently to Princeton physicist Robert H. Dicke, who realized that microwave radio telescopes might detect the dying glow of the big bang. He and his Princeton colleagues began construction of a suitable microwave receiver on the roof of the Princeton biology building. They were not, however, the first to observe the radiation.

Unknown to them, a few miles away in Holmdel, New Jersey, Arno Penzias and Robert Wilson of the Bell Laboratories were using a delicate microwave horn antenna to make careful measures of the absolute intensity of radio radiation coming from certain places in the Galaxy. They were plagued by some unexpected background noise in the system that they could not get rid of. They checked everything and eliminated the Galaxy as a source, as well as the Sun, the sky, the ground, and even the equipment.

At one point they realized that a couple of pigeons had made their home in the antenna and nested up near the throat of the horn, where it was warmer. Penzias and Wilson could chase the birds away while they observed, but they found that the birds left, as Penzias puts it, a layer of white, sticky, dielectric substance coating the inside of the horn. That substance would radiate, producing radio interference. They disassembled the horn and cleaned it, and the unwanted noise did go down somewhat, but it did not go away completely.

Finally, Penzias and Wilson decided that they had to be detecting radiation from space. Penzias mentioned it in a telephone conversation with another radio astronomer, Bernard Burke, who was aware of the Princeton work. Burke got Penzias and Wilson in touch with Dicke, and it was soon realized that the predicted glow from the primeval fireball had been observed.

Penzias and Wilson received the Nobel prize for their work in 1978. And perhaps almost equally fitting, just before his death in 1966, Lemaître learned about the discovery of his "vanished brilliance."

Properties of the Cosmic Background Radiation

The faint glow of radio radiation is now called the **cosmic background radiation (CBR)**. Since the early work by Penzias and Wilson, the CBR has been very thoroughly studied throughout the entire radio spectrum, with observations from ground-based radio telescopes, with instruments carried aloft in balloons, with a receiver in a U2 reconnaissance airplane, and with detectors flown in spacecraft.

The most accurate measurements have been made with a satellite orbiting the Earth. This satellite, which is named the Cosmic Background Explorer (COBE), was launched by NASA in 1989 (Figure 28.7) The first data received showed that the CBR closely matches that expected from a blackbody with a temperature of 2.735 K (Figure 28.8). This is exactly the result that one would expect if the CBR is indeed redshifted radiation emitted by a hot gas shortly after the universe began.

One of the things the CBR shows, therefore, is that the universe has evolved from a hot, uniform state. This provides direct observational support for an evolving universe, as opposed, for example, to a steady-state universe, which would not be filled with radiation from such a cosmic fireball.

At any given wavelength, the CBR is extremely *isotropic* on the small scale. Observations made by COBE show that any fluctuations in its intensity are less than a few parts in 10^5. The uniformity of the radiation tells us that at any age of less than a million years the universe had to be extremely uniform in temperature and, as it turns out, in density as well. But even if astronomers have not yet detected them, there must have been at least some density variations in order to allow matter to clump gravitationally to form stars and galaxies.

While small-scale fluctuations associated with galaxy formation have not yet been seen, observations show that the CBR has a slightly higher temperature in one direction than in the exact opposite direction in

FIGURE 28.7 Observations with the Cosmic Background Explorer (COBE), which is shown here as it leaves the launch pad, have provided the most accurate measurements of the cosmic background radiation (CBR). *(NASA)*

the sky. This is because of our own motion through space. If you approach a blackbody, its radiation is all Doppler-shifted to shorter wavelengths and resembles that from a slightly hotter blackbody. If you move away from it, the radiation appears like that from a slightly cooler blackbody. Such an effect has been searched for and observed in the microwave background. It represents the motion of the Sun with respect to the universe as a whole.

The direction of the Sun's motion is roughly toward the constellation Leo, but a good part of the motion is the Sun's revolution about the center of the Galaxy. When we take into account galactic rotation, the Gal-

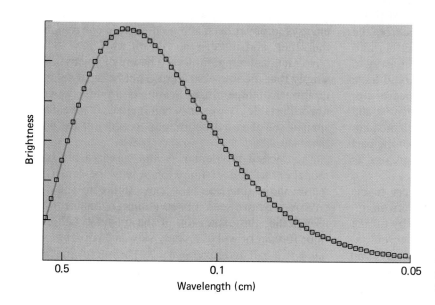

FIGURE 28.8 Measurements of the variation of intensity with wavelength by COBE show that the CBR follows the pattern expected for a blackbody with a temperature of 2.735 K.

axy's motion in the Local Group, and the motion of the Local Group toward the center of the Virgo Supercluster, we find that the Local Group as a whole has an additional motion with a speed of about 500 km/s in the general direction of the constellation Hydra.

The motion of the Local Group toward Hydra is in addition to the uniform expansion of the universe as a whole. But what can cause the Local Group to have such a high velocity toward Hydra? One possibility is that it is due to the gravitational attraction of an unusually large concentration of galaxies, and possibly dark matter as well, that is pulling the Local Group toward Hydra. Astronomers have dubbed this mass concentration the "Great Attractor." To account for the motion of the Local Group, theory says the Great Attractor must be an approximately spherical region of the universe, about 250 million LY in diameter, in which the density of matter is very much larger than the average for the universe as a whole. Its center is about 125 million LY from us. The mass of the Great Attractor, including both luminous and dark matter, is estimated to be 3×10^{16} solar masses, equivalent to tens of thousands of galaxies. Observations of galaxies in this region of the sky are now being made in an attempt to confirm the theoretical calculations.

A Look Back to the Big Bang

Since the CBR comes from the time when the fireball first became transparent, it is at the farthest point in space and time to which we can now observe. If we could see that radiation visually, it would be as if it were coming from an opaque wall. No radiation from a more distant source could ever reach us—for that source would have to lie further back in time, where it would be behind that opaque wall.

As time goes on, we can see farther and farther away, and more galaxies would come into view (if we had a large enough telescope), for as the time from the big bang becomes greater, we are looking further into the past to see the fireball, and hence further away in space (Figure 28.9). At earlier times, we could have seen (had we been here) only relatively nearer objects, and there were fewer galaxies between us and the threshold provided by the fireball itself. Thus, not only does the universe expand with time, but the part of it accessible to observation becomes greater as well.

On Earth, the microwave radiation is very feeble compared with, say, sunlight. But far off in intergalactic space, that radio background is by far the most intense radiation around. The observed radiation comes equally from all directions and gives no direction to a "center" of the universe. The universe, its "center," and its origin are all around us. As we

have noted before, there is a boundary to the observable universe, but it is a boundary in time, not in space.

28.4 The Inflationary Universe

The big bang model is successful in explaining the relationship between velocity and distance that is observed for galaxies; it accounts for the CBR; and it explains why about 25 percent of the mass of the universe is in the form of helium. There are, however, several important characteristics of the universe that the simple big bang model cannot explain, and we shall consider two in detail.

Problems with the Standard Big Bang Model

The first problem for which the big bang offers no explanation is the *uniformity* of the universe. The CBR is the same, no matter which direction we look, to an accuracy of at least 1 part in 10,000. There is, however, a maximum distance that light can have traveled since the time the universe began. This distance is called the *horizon distance*, because any two objects separated by this distance cannot ever have been in contact. No information, no physical process can propagate faster than the speed of light. One region of space separated by more than the horizon distance from another lies completely out of sight; throughout the history of the universe it has been truly beyond the horizon.

If we measure the CBR in two opposite directions in the sky, we are observing regions that were separated by more than 90 times the horizon distance at the time the CBR was emitted (Figure 28.10). We can see both, but they can never have seen each other. Why, then, are their temperatures so precisely the same? According to the standard big bang model, energy has never been able to flow from one region to the other, and there is no reason why they should have identical temperatures. The only explanation is simply that the universe started out being absolutely uniform in temperature. Scientists are always very uncomfortable, however, when they must appeal to a special set of initial conditions to account for what they see.

The second problem with the standard big bang model is that it does not explain why the density of matter in the universe is so close to the critical density. As we have seen, observations are unable to tell us whether the expansion of the universe will continue forever or will ultimately slow and perhaps even come to a halt and reverse itself. The interesting fact, however, is that the universe is so nearly balanced between these two possibilities that we cannot yet

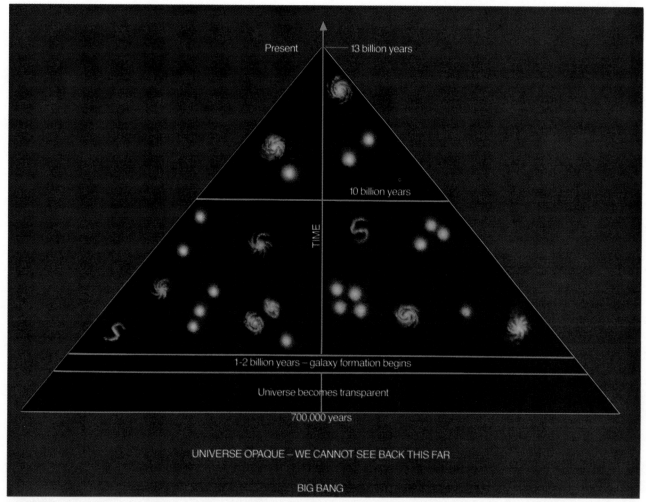

FIGURE 28.9 As we look to larger and larger distances and further back into time, we see more and more galaxies. Ultimately, we see back to the point at which the universe became transparent, which occurred about 700,000 years after the expansion began. Since the universe was opaque before this time, we cannot directly observe earlier eras.

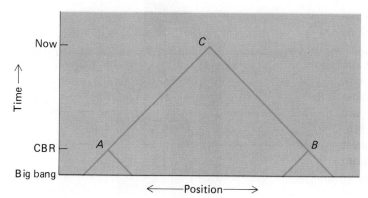

FIGURE 28.10 In this two-dimensional spacetime diagram, we are located at position *C*. The bottom line corresponds to the beginning of the universe. Light travels at a 45° angle to the vertical. Since no signal can travel faster than light, we can observe only those parts of the universe that lie within the large triangle. *A* and *B* are locations in spacetime when the CBR was emitted. We are just now receiving light from *A* and *B*. Note that in the time elapsed between the beginning of the universe and the time that the CBR was emitted no information could have traveled from *A* to *B*. Since no information could have been exchanged between *A* and *B*, it is difficult to explain why they were at exactly the same temperature when the CBR was emitted.

TABLE 28.3   The Forces of Nature

Force or Interaction	Examples of Interacting Particles	Relative Strength Now	Range	Important Applications
Gravitation	All	10^{-38}	Whole universe	Motions of planets, stars, galaxies
Electromagnetic	All charged	10^{-2}	Whole universe	Atoms, molecules, electricity
Weak	Electrons	10^{-5}	10^{-17} m	Radioactive decay
Strong	Protons, neutrons	1	10^{-15} m	Nuclear forces

determine which is correct. There could have been, after all, so little matter that it would be obvious that the universe is open and that the expansion will continue forever. Alternatively, there could have been so much matter that the universe would be clearly and unambiguously closed. Instead, the amount of matter present is within a factor of 5 to 10 of the value that corresponds to precise balance between these two situations.

To find the answers to these problems, we must make a digression and talk about the forces acting on the tiniest particles in the universe. Then we will return to discussing the grand picture of how the universe might have evolved.

Grand Unified Theories

In the terminology of elementary particle physics, an *interaction* is any process that affects the elementary particles that make up the material universe. Such processes include the creation and annihilation of particles, radioactive decay, and absorption or scattering of energy. There are four types of interactions or forces that describe all known physical processes. These four are gravity, electromagnetism, the *weak interactions,* and the *strong interactions* (Table 28.3).

Although the force of gravity is the one that most obviously affects our everyday lives and appears strong to us, the force of gravity between two elementary particles, say, two protons, is by far the weakest of the forces. Electromagnetism, which includes both magnetic and electric forces, holds atoms together and produces the electromagnetic radiation that we use to study the universe. The weak interactions are in fact much stronger than gravity but act only over very small distances—distances comparable to a hundredth the size of an atomic nucleus. The weak interaction is involved in radioactive decay and in reactions that result in the production of neutrinos. The strong interaction is also effective only over nuclear dimensions and is the force that holds protons and neutrons together in an atomic nucleus.

In exploring the fundamental nature of these four forces, physicists have developed so-called *grand unified theories (GUTs).* In these theories, the strong, weak, and electromagnetic forces are not three independent forces, but rather different manifestations or aspects of what is in fact a single force. The theories predict that at high enough temperatures, there would in fact be only one force. At lower temperatures, however, a mechanism causes this single force to look like three different forces (Figure 28.11). The mechanism is given the name *spontaneous symmetry breaking.* Unfortunately, the temperatures at which the three forces are predicted to become one force are so high that they cannot be reached in any terrestrial

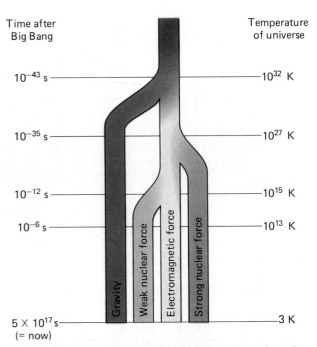

FIGURE 28.11   The strength of the four forces depends on the temperature of the universe. This diagram shows that at very early times when the temperature of the universe was very high, all four forces resembled one another and were indistinguishable. As the universe cooled, the forces took on separate and distinctive characteristics.

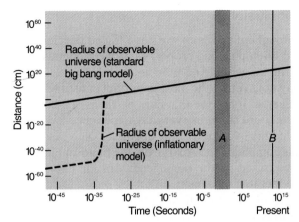

FIGURE 28.12 Radius of the observable universe as a function of time for the standard big bang model (solid line) and for the inflationary model (dashed line). The two models are the same for all times after 10^{-30} s. Electrons, positrons, and the lightest atomic nuclei are formed during the time interval labeled A. The universe becomes transparent to radiation at the time designated B.

laboratory. Only the early universe at times earlier than 10^{-35} s was hot enough so that there was one force instead of three.

Some forms of the GUTs predict, in fact, that a remarkable event occurred about the time that the universe was 10^{-35} s old and spontaneous symmetry breaking was taking place. The equations of general relativity, combined with the special state of matter at that time, predict that gravity could briefly have been a repulsive force. We know that now in our own time gravity is an attractive force that slows the expansion of the universe, but for a brief instant near 10^{-35} s after the expansion began, gravity could actually have accelerated the expansion. It is as if the cosmological constant were, for an instant of time, not equal to zero.

A model universe in which this rapid early expansion occurs is called an **inflationary universe.** The inflationary universe is identical to the big bang universe for all time after the first 10^{-30} s. Prior to that time, associated with the spontaneous symmetry breaking, there was a brief period of extraordinarily rapid expansion or inflation during which the scale of the universe increased by a factor of as much as 10^{75}; this increase is 10^{50} times more than predicted by standard big bang models (Figure 28.12). As the universe expanded, its temperature dropped below the critical value at which all three forces behave in a symmetrical fashion. In the cooler, asymmetrical universe, the nuclear forces dominate the electromagnetic force and continue to do so in our world today.

Prior to the inflation, all of the universe that we can now see was causally connected. That is, the horizon distance was large enough to include all of the universe that we can now observe. There was adequate time for the observable universe to homogenize itself and come to the same temperature.

Besides accounting for the fact that we measure the CBR to be the same in all directions, GUTs make some predictions about the nature of the universe. For example, for reasonable choices for the parameters in the theory, it is possible to produce an excess of matter over antimatter, and we do now live in a universe that contains far more matter than antimatter. (If the amounts were equal, the matter and antimatter would have annihilated each other and we wouldn't be here!) The theory also predicts that protons should not have an infinite lifetime but should decay, on average, after a period that is in excess of 10^{30} years. Fortunately, there are so many protons in our world that it should be possible to detect a few protons that decay sooner than this average value, and experiments are being conducted to try to do so. So far, no decays have been detected, so some of the simpler forms of GUTs have already been ruled out.

One solid prediction of the inflationary universe model is that the mean density of matter in the universe should precisely equal the critical density. (The only way to avoid this result is to assume that some special conditions occurred during the early expansion of the universe. Of course, the desire to avoid an appeal to special initial conditions was one of the original motivations for devising the inflationary theory.) As we have seen, observations indicate that the amount of matter that we have so far found in galaxies and in clusters of galaxies is a factor of 5 to 10 below the critical density. Table 28.4 summarizes the amount of matter that is estimated to be present in various types of astronomical objects. Visible stars in total contribute only 1/1000 of the mass required to reach critical density. Even if we add to the stars and interstellar matter in galaxies the invisible dark matter that is detected through its gravitational influence

TABLE 28.4	Amounts of Mass Present in Astronomical Objects of Various Sizes	
Object	**Ratio of Measured Density to Critical Density**
Stars	0.001
Individual galaxies	0.01
Binary galaxies and small groups of galaxies	0.01–0.1
Rich clusters and superclusters of galaxies	0.2

on luminous objects, we are only up to 20 percent of the critical density.

For their ideas about the inflationary universe to be correct, theorists must hope that observers have somehow failed to notice at least 80 percent of the total mass of the universe. How should astronomers go about looking for it? If the required additional matter is indeed present, then it must not emit light or we would have seen it. It cannot lie within galaxies, or even within the Local Supercluster, or we would have detected its gravitational effects. And it cannot be in the form of protons and neutrons or combinations thereof, including black holes, or there would be far less deuterium in the universe.

What is this dark matter? Theory has produced a number of candidates—new types of particles in addition to the protons, neutrons, electrons, and other subatomic particles that we have already detected. And now the hunt is on—in huge accelerators where exotic new particles might be produced, deep in underground mines where scientists are trying to trap elusive dark matter particles just as they once succeeded in capturing neutrinos, and in university laboratories around the world. Scientists are taking a big gamble in devoting a substantial portion of their careers to searching for particles that may not even exist. The gamble is high but so is the reward. Detection of dark matter and determination of its characteristics would revolutionize cosmological thought in the same way as did the discoveries of the expansion of the universe and of the CBR.

Conclusion

Throughout this book we have traced the fascinating and often puzzling properties of the luminous matter in the universe. And now we find that visible matter may not even be the most important constituent of the universe. It may be that dark particles of a kind completely unknown in everyday experience have dominated the evolution of the universe. Do we even need this complication? Apart from the philosophical problems raised by the homogeneity of the universe and the fact that its density differs by only a factor of 5 to 10 from the critical density, observations are still entirely consistent with an open universe with an age of 10 to 15 billion years. Is a model that requires that most of the matter in the universe be invisible really an improvement?

The GUTs that have been devised to account for the universe are beautiful and elegant, and very often in science beauty and simplicity have been a guide to truth. But the ultimate test must be whether the theories make verifiable predictions and whether those predictions are validated in laboratory experiments. The new cosmological theories are driving a host of developments in observational and experimental astrophysics. Over the next 20 years, we should know whether current models contain truth as well as beauty. Whatever the ultimate judgment, it is quite clear that we will be able to probe much closer to the beginning of our universe than we would have thought possible only 20 years ago.

SUMMARY

28.1 **Cosmology** is the study of the organization and evolution of the universe. The universe is expanding, and this fact is the starting point for modern cosmological theories. From the rate of expansion of the universe, we can estimate that all of the matter in the universe was concentrated in an infinitesimally small volume 10 to 15 billion years ago. The beginning of the expansion is what astronomers mean when they talk about the beginning of the universe.

28.2 The most widely accepted model of the universe is the **big bang** model. The factor that controls the evolution of the universe is the density of matter (and energy). If the density is high, then the rate of expansion will slow and possibly even reverse direction so that the galaxies all come together again (a *closed universe*). Observations suggest that the density is actually so low that the expansion will continue forever (an *open universe*).

28.3 The universe cools as it expands. The energy of photons is determined by their temperature, and calculations show that in the hot early universe photons had so much energy that when they collided with one another they

could produce material particles. As the universe expanded and cooled, protons and neutrons formed first, then electrons and positrons. Fusion reactions then produced helium nuclei. Finally, the universe became cool enough to form hydrogen atoms. At this time the universe became transparent to radiation. Scientists have detected the **cosmic background radiation (CBR)** from the hot early universe. Measurements with the COBE satellite show that the CBR now has a temperature of 2.735 K.

28.4 The big bang model does not explain why the CBR has the same temperature in all directions. Neither does it explain why there was originally more matter than antimatter nor why the density of the universe is so close to the **critical density.** New theories, which involve a period of very rapid expansion, or **inflation,** when the universe was 10^{-35} s old, are being developed to try to explain these observations. One prediction of these new theories is that the density of the universe should be exactly equal to the critical density. This prediction can be true only if at least 80 percent of the matter in the universe is invisible and so far undetected.

REVIEW QUESTIONS

1. Astronomers of today can actually observe galaxies as they were billions of years ago. Explain.

2. Describe three possible futures for the universe. What property of the universe determines which of these possibilities is the correct one?

3. Which formed first in the early universe—protons and neutrons or electrons and positrons? Why?

4. Which formed first—hydrogen nuclei or hydrogen atoms? Why?

5. Describe at least two characteristics of the universe that are explained by the big bang model.

6. Describe two properties of the universe that are not explained by the big bang cosmology.

THOUGHT QUESTIONS

7. What is the most useful probe of the early evolution of the universe—a giant elliptical galaxy or an irregular galaxy like the Large Magellanic Cloud? Why?

8. What are the advantages and disadvantages of using quasars to probe the early history of the universe?

9. Consider the plot of radial velocities against the distances of remote clusters of galaxies. Are the measured distances and radial velocities that are plotted the *present* values of these quantities for the clusters? In each case, why? If not, try to describe how the diagram might differ if we did plot the present-day values of cluster distances and velocities.

10. Suppose the universe will expand forever. Describe what will become of the radiation from the primeval fireball. What will the future evolution of galaxies be like?

11. Some theorists argue that the universe is at just the critical density. Do the observations support this hypothesis?

12. Summarize the evidence for the existence of dark matter in the universe.

13. In this text we have discussed many motions of the Earth as it travels through space with the Sun. Describe as many of these motions as you can.

14. There are a variety of ways of estimating the ages of various objects in the universe. Describe some of these ways and indicate how well they agree with one another and with the age of the universe itself as estimated by its expansion.

15. In the 19th century both geology and biology were based on the idea that evolution was a slow process. In the 20th century, we have come to the conclusion that violent events have played a significant role in shaping the evolution of the universe and everything in it, including the evolution of life on Earth. Discuss some of these violent events, starting with the big bang and including some events that have affected the Earth directly.

16. Since the time of Copernicus, each revolution in astronomy has moved humans farther from the center of the universe. Now it appears that we may not even be made of the most common form of matter. Trace the changes in scientific thought about the central nature of the Earth, the Sun, and our Galaxy on a cosmic scale.

17. Construct a time line for the universe and indicate when various significant events, from the beginning of the expansion to the formation of the Sun to the appearance of humans on Earth, occurred.

PROBLEMS

18. The Andromeda Galaxy is approaching the Sun at a velocity of about 300 km/s. Does this indicate that the universe is not expanding? Compare this velocity with that of the Sun in its orbit around the center of the Galaxy. Suppose Andromeda is orbiting the Milky Way Galaxy with a period of 10 billion years. What velocity would it have?

19. Show that if $H = 25$ km/s per million light years, then the maximum age of the universe is approximately 13 billion years.

20. It is possible to derive the age of the universe, given the value of the Hubble constant and the distance to a galaxy. Consider a galaxy at distance of 400 million LY, receding from us at a velocity v. If the Hubble constant is 25 km/s per million light years, what is its velocity? How long ago was that galaxy right next door to our own Galaxy if it has always been receding at its present rate? Express your answer in years. Since the universe began when all galaxies were very close together, this number is an estimate for the age of the universe.

Histories of Astronomy

Berendzen, R., Hart, R., and Seeley, D., *Man Discovers the Galaxies*. New York: Neale Watson Academic Publications, 1976. (Excellent history of the development of modern extragalactic astronomy.)

Boorstin, D. J., *The Discoverers*. New York: Random House, 1983. (Brilliant essays on the development of astronomy, timekeeping, and geography.)

Cooper, H. S. F., *Imaging Saturn: The Voyager Flights to Saturn*. New York: Holt, Rinehart and Winston, 1982. (Eyewitness account of the Voyager exploration of Saturn.)

Cooper, H. S. F., *The Search for Life on Mars*. New York: Harper and Row, 1980. (Eyewitness account of the Viking program, especially the Martian biology experiments.)

Ferris, T., *Coming of Age in the Milky Way*. New York: William Morrow and Co., 1988. (The story of humankind's effort to comprehend the scale of the universe and our place within it. The book ranges from Aristotle's crystalline spheres to the latest theories of cosmology and includes biographical descriptions of some of the key individuals in the development of astronomy.)

Ferris, T., *The Red Limit*, 2nd ed. New York: Morrow, Quill, 1983. (Superbly written history of modern cosmology and the astronomers who played a role in it.)

Grosser, M., *The Discovery of Neptune*. New York: Dover, 1979. (An engrossing account of the discovery of new planets in the solar system.)

Hoskin, M., *Stellar Astronomy: Historical Studies*. Buckinghamshire, England: Science History Publications Ltd., 1982. (Account of the development of stellar astronomy by a leading historian of astronomy.)

Hoyt, W. G., *Lowell and Mars*. Tucson, AZ: University of Arizona Press, 1976. (Excellent account of the martian canal controversy.)

Hoyt, W. G., *Planets X and Pluto*. Tucson, AZ: University of Arizona Press, 1980. (Excellent account of the discovery of Pluto.)

Koestler, A., *The Sleepwalkers*. New York: Macmillan, 1959. (A famous and very interesting history of the beginnings of modern science.)

Krupp, E. C., *Echoes of the Ancient Skies*. New York: Harper and Row, 1983. (Excellent discussion of archaeoastronomy for the scientifically literate layperson.)

Mailer, N., *Of a Fire on the Moon*. New York: Little, Brown, and Co., 1970. (Interesting account of the Apollo program by a major contemporary writer, with emphasis on the reasons so many Americans enthusiastically supported this effort.)

Osterbrock, D., *Eye on the Sky: Lick Observatory's First Century*. Berkeley, CA: University of California Press. (The story of one of the first major science institutions in this country and of the strong personalities that shaped its development.)

Smith, R. W., *The Space Telescope: A Study of NASA, Science, Technology, and Politics*. Cambridge, England: Cambridge University Press, 1989. (A detailed history of how the Hubble Space Telescope came to be. Interesting insights into how big science projects are actually carried out.)

Tucker, W. H., *The Star Splitters (NASA SP-466)*. Washington, DC: U.S. Government Printing Office, 1984. (History of the beginnings of high-energy astronomy, with detailed accounts of the NASA High-Energy Observatories.)

Tucker, W. H., and Giacconi, R., *The X-ray Universe*. Cambridge, MA: Harvard University Press, 1985. (Insider's account of the history of x-ray astronomy, with emphasis on the highly successful Einstein Observatory.)

Van Helden, A., *Measuring the Universe*. Chicago: University of Chicago Press, 1985. (A survey of human views of the scale of the universe from the time of Aristarchus to that of Halley.)

Washburn, M. L., *Distant Encounters: The Exploration of Jupiter and Saturn*. New York: Harcourt Brace Jovanovich, 1983. (A journalist's account of the Voyager program.)

Washburn, M. L., *Mars at Last*. New York: Putnam, 1977. (Well-written journalist's account of the Viking program.)

Telescopes, Instrumentation, and the Electromagnetic Spectrum

Field, G. B., and Chaisson, E. J., *The Invisible Universe*. Boston: Birkhauser, 1985. (Semipopular book on the new regions of the electromagnetic spectrum and the space observatories that astronomers would like to build to explore them.)

Henbest, N., and Marten, M., *The New Astronomy*. New York: Cambridge University Press, 1983. (A beautiful collection of x-ray, UV, IR, and radio images of astronomical objects.)

Rowan-Robinson, M., *Cosmic Landscape*. New York: Oxford University Press, 1979. (Beautifully written popular account of the various regions of the electromagnetic spectrum and what each tells us of the universe, from planets to the big bang.)

Tucker, W., and Tucker, K., *The Cosmic Inquirers: Modern Telescopes and Their Makers*. Cambridge, MA: Harvard University Press, 1986. (An inside view of major new telescopes—the Very Large Array, Einstein, Gamma Ray Observatory, Infrared Astronomy Satellite, and Hubble Space Telescope—with emphasis on the persons who struggled to make them a reality.)

Planetary System

Beatty, J. K. (ed.), *The New Solar System*, 3rd ed. Cambridge, MA: Sky Publishing Corp., 1991. (Well-edited semipopular descriptions written by leading planetary scientists, illustrated in color.)

Brandt, J. C., and Chapman, R. D., *Introduction to Comets*. Cambridge, England: Cambridge University Press, 1981. (Introductory text on comets for the advanced undergraduate.)

Carr, M. H., *The Surface of Mars*. New Haven, CT: Yale University Press, 1981. (The most comprehensive and authoritative book on Mars, beautifully illustrated with Viking photographs, written at a semitechnical level.)

Chapman, C. R., *Planets of Rock and Ice*. New York: Charles Scribner's Sons, 1982. (An excellent introduction to the terrestrial planets and satellites, well written and authoritative, aimed at the scientifically literate layperson.)

Chapman, C.R., and Morrison, D., *Cosmic Catastrophes*. New York: Plenum, 1989. (An account of changing concepts of the role of catastrophes in shaping the history of the Earth, other planets, and the evolution of life.)

Cooper, H. S. F., *Moon Rocks*. New York: Dial Press, 1970. (A journalist's account of the Apollo program, first published in the New Yorker.)

Cortright, E. M., *Apollo Expeditions to the Moon (NASA SP-350)*. Washington, DC: U.S. Government Printing Office, 1975. (Beautiful illustrations, well captioned.)

Dodd, R. T., *Thunderstones and Shooting Stars*. Cambridge, MA: Harvard University Press, 1986. (An outstanding introduction to meteorites and their significance.)

Elliot, J., and Kerr, R., *Rings*. Cambridge, MA: MIT Press, 1984. (Collaboration between a research scientist and a science writer that discusses the history of ring studies and current knowledge for the scientifically literate layperson.)

Frazier, K., *Solar System*. Alexandria, VA: Time-Life Books, 1985. (Very well written and illustrated introduction for the layperson.)

French, B. M., *The Moon Book*. New York: Penguin Press, 1977. (Readable introduction to the Moon, written for the layperson.)

Goldsmith, D., *Nemesis: The Death Star and Other Theories of Mass Extinction*. New York: Walker and Co., 1985. (Well-written, semipopular account of the current debate on the astronomical causes of mass extinctions.)

Hartmann, W., Miller, R., and Lee, P., *Out of the Cradle: Exploring the Frontiers Beyond Earth*. New York: Workman, 1984. (Beautiful collection of paintings and essays on the exploration of the planetary system.)

Hutchison, R., *The Search for Our Beginnings*. New York: Oxford University Press, 1983. (Highly readable introduction to meteoritics and the origin of the solar system.)

Lamb, H. H., *Climate, History, and the Modern World*. London and New York: Methuen, 1982. (A description of what is known about the history and causes of long-term changes in the Earth's climate.)

Littmann, M., *Planets Beyond: Discovering the Outer Solar System*. New York: Wiley Science Editions, 1988. (Contemporary overview of the outer planets by a science educator, much of it from a historical viewpoint.)

Morrison, D., *Voyage to Saturn (NASA SP-451)*. Washington, DC: U.S. Government Printing Office, 1982. (The Voyager encounters with Saturn and their results; written for the layperson.)

Morrison, D., and Owen, T., *The Planetary System*. Reading, MA: Addison-Wesley, 1988. (College text in planetary science.)

Morrison, D., and Samz, J., *Voyage to Jupiter (NASA SP-439)*. Washington, DC: U.S. Government Printing Office, 1982. (The Voyager encounters with Jupiter and their results; written for the layperson.)

Raup, D. M., *The Nemesis Affair: A Story of the Death of Dinosaurs and the Ways of Science*. New York: W. W. Norton, 1986. (Outstanding account of the theory that periodic mass extinctions are due to collisions of comets with the Earth, written by a leading paleontologist and major participant in this exciting and controversial debate.)

Schneider, S. H., and Londer, R., *The Coevolution of Climate and Life*. San Francisco: Sierra Club Books, 1984. (A modern discussion of weather, climate, human society, and planetary evolution written for the interested layperson.)

Sullivan, W. S., *Landprints*. New York: Times Books, 1985. (Highly readable account of geology for the layperson, written by a leading science journalist.)

Whipple, F. L., *The Mystery of Comets*. Washington, DC: Smithsonian Institution Press, 1985. (Discussion of comets for the layperson, written by one of the leading cometary experts.)

Sun, Stars, and Interstellar Medium

Cohen, M., *In Darkness Born: The Story of Star Formation*. Cambridge, England: Cambridge University

Press, 1988. (A summary of the theory of star formation and of the observations of regions where star formation is currently active.)

Frazier, K., *Our Turbulent Sun*. Englewood Cliffs, NJ: Prentice-Hall, 1982. (Good popular-level account of the Sun, with particular attention to solar-terrestrial relations and possible effects of the Sun on Earth's climate and history.)

Friedlander, M. W., *Cosmic Rays*. Cambridge, MA: Harvard University Press, 1989. (A well-written and easy-to-read description of cosmic rays—what they are and where they come from.)

Friedman, H., *Sun and Earth*. New York: Scientific American Books, 1985. (A discussion of the Sun, its variations, and its effects on the Earth. Describes results from recent experiments in space.)

Giovanelli, R. G., *Secrets of the Sun*. Cambridge, England: Cambridge University Press, 1984. (A detailed description by one of the leading solar physicists of solar activity, including sunspots, the sunspot cycle, the chromosphere, the corona, and the role of magnetic fields.)

Greenstein, G., *Frozen Star*. New York: Charles Scribner's Sons, 1983. (An eloquently written introduction to pulsars and black holes, which also portrays the scientists involved in this area of research.)

Kaler, J. B., *Stars and Their Spectra*. Cambridge, England: Cambridge University Press, 1989. (A detailed introduction to the properties of stars. In nontechnical language, this book provides a textbook-style description of stars of all spectral types. It also discusses spectroscopy and the evolution of stars.)

Kippenhahn, R., *100 Billion Suns* (trans. J. Steinberg). New York: Basic Books, 1983. (Excellent popular account of stellar evolution.)

Malin, D., and Murdin, P., *Colours of the Stars*. New York: Cambridge University Press, 1984. (A beautifully produced coffee table–style book of astronomical images made with modern photographic techniques.)

Marschall, L. A., *The Supernova Story*. New York: Plenum Press, 1988. (An easy-to-read but thorough and scientifically accurate discussion of ancient and modern observations and ideas about supernovae. The book of choice for anyone who wishes to know about SN 1987A.)

Mitton, S., *Daytime Star: The Story of Our Sun*. New York: Charles Scribner's Sons, 1983. (Well-written discussion of the Sun for the layperson.)

Noyes, R. W., *The Sun, Our Star*. Cambridge, MA: Harvard University Press, 1982. (A well-presented picture of the Sun, including chapters on climate and solar energy.)

Sullivan, W., *Black Holes*. Garden City, NY: Doubleday, 1979. (Well-written account for the layperson by a leading science journalist.)

Washburn, M., *In the Light of the Sun*. New York: Harcourt Brace Jovanovich, 1981. (Well-written popular account of the Sun and modern astrophysical ideas.)

Wentzel, D.G., *The Restless Sun*. Washington, DC: Smithsonian Institution Press, 1989. (A broad overview of the Sun, including solar activity and solar terrestrial relations. A good first book to read to explore in more detail the topics described in this text.)

Relativity and Modern Physics

Carrigan, R. A., and Trower, W. P. (eds.), *Particle Physics in the Cosmos*. New York: W. H. Freeman and Co., 1989. (Articles from *Scientific American* that describe particle physics and its relevance to such astrophysical topics as the formation of galaxies and the origin and fate of the universe.)

Davies, P., *Other Worlds*. New York: Simon and Schuster, 1980. (Popular discussion of quantum theory and cosmology.)

Davies, P. C. W., *The Forces of Nature*. Cambridge, England: Cambridge University Press, 1979. (An excellent semipopular account of modern physics.)

Einstein, A., *Relativity: The Special and General Theory*. New York: Crown Publishers, 1961. (A classic popular account, by Einstein himself, of his epoch-making new physics.)

Feynman, R., *The Character of Physical Law*. Cambridge, MA: MIT Press, 1965. (A series of popular lectures by one of the most articulate theoretical physicists of our time; an excellent insight into modern physics.)

Gardner, M., *The Relativity Explosion*. New York: McGraw-Hill, 1966. (A very readable description of the meaning of relativity for the layperson.)

Hawking, S. W., *A Brief History of Time*. New York: Bantam Books, 1988. (An introduction to some of the most important scientific ideas about the cosmos, with thoughtful and thought-provoking ideas about what it means to search for the ultimate beginnings of our universe.)

Kaufmann, W. J., III., *The Cosmic Frontiers of General Relativity*. Boston: Little, Brown and Co., 1977. (A layperson's account of relativity theory and some of the possible esoteric consequences of it.)

Sciama, D. W., *The Physical Foundations of General Relativity*. Garden City, NY: Doubleday, 1969. (A description of the physical basis of Einstein's theory for the person with only a smattering of algebra.)

Will, C. M., *Was Einstein Right?* New York: Basic Books, 1986. (An authoritative but fun-to-read book that describes modern experimental tests of general relativity.)

Galaxies and Cosmology

Barrow, J., and Silk, J., *The Left Hand of Creation: Origin and Evolution of the Expanding Universe*. New York: Basic Books, 1983. (Authoritative account of the origin and evolution of the universe, with a good discussion of the relationship between subatomic physics and cosmology.)

Bartusiak, M., *Thursday's Universe*. New York: Times Books, 1986. (A first-rate introduction to ideas about the evolution of the universe and to the scientists and facilities that have contributed to major breakthroughs.)

Chaisson, E., *Cosmic Dawn*. Boston: Little, Brown, and Co., 1981. (An eloquent introduction to the development of "particles, galaxies, stars, planets, life, and culture.")

Cohen, N., *Gravity's Lens: Views of the New Cosmology*. New York: John Wiley & Sons, 1988. (Brief descriptions of recent research in such areas as the big bang, cosmic background radiation, gravitational lenses, and quasars, written in a light and entertaining style.)

Cornell, J., *Bubbles, Voids, and Bumps in Time: The New Cosmology*. Cambridge, England: Cambridge University Press, 1989. (Six essays on such subjects as dark matter, the distribution of galaxies, and inflationary cosmology by leading researchers. An outstanding introduction to modern ideas about the origin and evolution of the universe.)

Ferris, T., *Galaxies*. San Francisco: Sierra Club Books, 1981. (A magnificently illustrated coffee table–style book with informative text.)

Harrison, E. R., *Darkness at Night*. Cambridge, MA: Harvard University Press, 1987. (A combination of history and science is used to find the profound implications of the answer to a simple question: Why is the sky dark at night?)

Hodge, P., *Galaxies*. Cambridge, MA: Harvard University Press, 1986. (Comprehensive introduction to the galaxies, written for the nonastronomer.)

Hubble, E., *The Realm of the Nebulae*. New Haven, CT: Yale University Press, 1936; also New York: Dover, 1958. (A classic book for the educated layperson by one of the great astronomers of our time, describing his exploration of the extragalactic universe.)

Parker, B., *Creation*. New York: Plenum Press, 1988. (The story of the early universe and of the scientists who have contributed to developing our modern views.)

Shipman, H. L., *Black Holes, Quasars, and the Universe*, 2nd ed. Boston: Houghton Mifflin, 1980. (One of the best popular or semipopular accounts of black holes, relativity, and active galactic nuclei available.)

Silk, J., *The Big Bang*. New York: W. H. Freeman and Co., 1989. (This revised and updated edition describes the evolution of the universe from the beginning to the present. Written by one of the leading experts, this book explores modern ideas about the creation and evolution of the universe and includes a detailed discussion of the formation of galaxies. The language is nontechnical, but careful reading is required to appreciate the full range and implications of the ideas presented.)

Trefil, J. S., *The Dark Side of the Universe*. New York: Charles Scribner's Sons, 1988. (A discussion of theoretical work on dark matter and its implications for such problems as the formation of galaxies. Entertainingly written, with complex ideas explained in terms of vivid analogies to phenomena in everyday life.)

Trefil, J. S., *The Moment of Creation*. New York: Charles Scribner's Sons, 1983. (Excellent discussion of modern cosmological thinking for the interested layperson, incorporating elementary particle physics, grand unified theories, and the inflationary universe.)

Tucker, W., and Tucker, K., *The Dark Matter*. New York: William Morrow, 1988. (A clear presentation of the evidence for dark matter on all scales from galaxies to superclusters. Possible explanations of what the dark matter is are examined, and implications for the future evolution of the universe are explored.)

Wagoner, R., and Goldsmith, D., *Cosmic Horizons: Understanding the Universe*. New York: W. H. Freeman and Co., 1983. (Clearly written guide to the latest theories and observations about the origin, structure, and evolution of the cosmos.)

Weinberg, S., *The First Three Minutes*, 2nd ed. New York: Basic Books, 1988. (Although somewhat dated, this remains one of the best popular accounts of modern cosmology, lucidly written by a Nobel Laureate physicist who has been at the forefront in the study of the basic forces of nature.)

Life in the Universe

Goldsmith, D., and Owen, T., *The Search for Life in the Universe*. Menlo Park, CA: Benjamin/Cummings, 1980. (Comprehensive and well-written text, covering all of astronomy from the perspective of the search for life; a good successor to Shklovskii and Sagan.)

McDonough, Thomas R., *The Search for Extraterrestrial Intelligence*. New York: Wiley, 1987. (Good, modern account of SETI.)

Sagan, C., *The Cosmic Connection*. New York: Anchor Press–Doubleday, 1973. (Essays by America's best-known scientist that establish a perspective on our place in the universe.)

Shklovskii, I. S., and Sagan, C., *Intelligent Life in the Universe*. New York: Dell, 1966. (The classic work that investigates extraterrestrial life in a thoughtful and authoritative manner—dated but still fascinating.)

Pseudoscience

Abell, G. O., and Singer, B. (eds.), *Science and the Paranormal*. New York: Charles Scribner's Sons, 1981. (A collection of critical essays on allegedly paranormal topics by well-known scientists and science writers.)

Culver, R. B., and Ianna, P. A., *The Gemini Syndrome: Star Wars of the Oldest Kind*. Tucson, AZ: Pachart, 1979. (An excellent account of astrology.)

Frazier, K. (ed.), *Paranormal Borderlands of Science*. Buffalo, NY: Prometheus Books, 1981. (A collection of articles from *The Skeptical Inquirer* that scrutinize a variety of claims of the paranormal.)

Gardner, M., *Fads and Fallacies in the Name of Science*. New York: Dover, 1957. (Fascinating look at pseudoscience.)

Gardner, M., *Science: Good, Bad, and Bogus*. Buffalo, NY: Prometheus Books, 1981. (A collection of essays from the famous philosopher, mathematician, and science writer.)

Klass, P. J., *UFOs Explained*. New York: Vintage Books, 1976. (A revealing expose of the UFO phenomenon by the foremost skeptical investigator of the subject.)

Radner, D., and Radner, M., *Science and Unreason*. Belmont, CA: Wadsworth, 1982. (A splendid little book on how to distinguish pseudoscience from science.)

Randi, J., *Flim-Flam*. New York: Lippincott and Crowell, 1980. (World-famous magician exposes frauds, some of which duped scientists as well as the public.)

Sheaffer, R., *The UFO Verdict: Examining the Evidence*. Buffalo, NY: Prometheus Books, 1981. (A skeptical survey of the evidence for extraterrestrial spaceships by a well-known UFO investigator.)

Standon, A., *Forget Your Sun Sign*. Baton Rouge, LA: Legacy Publishing Co., 1977. (Skeptical discussion of astrology.)

Star Atlases, Sky Guides, and Sky Lore

Allen, R. H., *Star Names*. New York: Dover, 1963. (An exhaustive reference on star names and their origins.)

Menzel, D. H., and Pasachoff, J. M., *A Field Guide to the Stars and Planets*, 2nd ed. Boston: Houghton Mifflin, 1983. (An all-purpose guide to the sky for those who have binoculars or telescopes or who just want to identify the constellations.)

Minnaert, M., *The Nature of Light and Color in the Open Air*. New York: Dover, 1954. (The definitive book on the origin of optical phenomena in the sky, including rainbows, halos, shadow bands, and hundreds of others.)

Moore, P., *Exploring the Night Sky with Binoculars*. Cambridge, England: Cambridge University Press, 1986. (Describes what can be seen in the sky with only binoculars. Instrucions are given for finding many prominent clusters of stars.)

Motz, L., and Nathanson, C., *The Constellations: An Enthusiast's Guide to the Night Sky*. New York: Doubleday, 1988. (For each constellation, this book gives a description, the mythology, the location of significant objects, and the astronomical background needed to understand why they are significant.)

Popular Journals on Astronomy

Astronomy, published monthly by Astromedia Corporation, 441 Mason Street, P.O. Box 92788, Milwaukee, WI 53202.

Mercury, published bimonthly by the Astronomical Society of the Pacific, 390 Ashton Avenue, San Francisco, CA 94112.

The Planetary Report, published bimonthly by The Planetary Society, 65 N. Catalina Avenue, Pasadena, CA 91106.

Sky and Telescope, published monthly by Sky Publishing Corporation, 49 Bay State Road, Cambridge, MA 02139.

Popular articles on astronomy also appear frequently in *Scientific American* and *Science News*.

Career Information

Information about a career in astronomy is available from the Executive Officer of the American Astronomical Society, at Suite 300, 2000 Florida Avenue, N.W., Washington, DC 20009.

APPENDIX 2
Glossary

absolute magnitude Apparent magnitude a star would have at a distance of 10 pc.

absolute zero A temperature of $-273°C$ (or 0 K), where all molecular motion stops.

absorption spectrum Dark lines superimposed on a continuous spectrum.

accelerate To change velocity; to speed up, slow down, or change direction.

acceleration of gravity Numerical value of the acceleration produced by the gravitational attraction on an object at the surface of a planet or star.

accretion Gradual accumulation of mass, as by a planet forming by the building up of colliding particles in the solar nebula or gas falling into a black hole.

accretion disk A disk of matter falling in toward a massive object; the disk shape is the result of conservation of angular momentum.

active galactic nucleus A galaxy is said to have an active nucleus if unusually violent events are taking place in its center, emitting in the process very large quantities of electromagnetic radiation. Seyfert galaxies and quasars are examples of galaxies with active nuclei.

active region Areas on the Sun where magnetic fields are concentrated; sunspots, prominences, and flares all tend to occur in active regions.

albedo The fraction of incident sunlight that a planet or minor planet reflects.

alpha particle The nucleus of a helium atom, consisting of two protons and two neutrons.

altitude Angular distance above or below the horizon, measured along a vertical circle, to a celestial object.

altitude-azimuth mount A mounting for a telescope, one axis of which permits motion in altitude and the other axis of which permits the telescope to move in azimuth (horizontally).

amplitude The range in variability, as in the light from a variable star.

angular diameter Angle subtended by the diameter of an object.

angular momentum A measure of the momentum associated with motion about an axis or fixed point.

antapex (solar) Direction away from which the Sun is moving with respect to the local standard of rest.

Antarctic Circle Parallel of latitude 66°30′S; at this latitude the noon altitude of the Sun is 0° on the date of the summer solstice.

antimatter Matter consisting of antiparticles: antiprotons (protons with negative rather than positive charge), positrons (positively charged electrons), and antineutrons.

aperture The diameter of an opening, or of the primary lens or mirror of a telescope.

apex (solar) The direction toward which the Sun is moving with respect to the local standard of rest.

aphelion Point in its orbit where a planet is farthest from the Sun.

apogee Point in its orbit where an Earth satellite is farthest from the Earth.

apparent magnitude or brightness A measure of the observed light flux received from a star or other object at the Earth.

Arctic Circle Parallel of latitude 66°30′N; at this latitude the noon altitude of the Sun is 0° on the date of the winter solstice.

array (interferometer) A group of several telescopes that is used to make observations at high angular resolution.

association A loose cluster of young stars whose spectral types, motions, or positions in the sky indicate that they have probably had a common origin.

asteroid An object orbiting the Sun that is smaller than a major planet, but that shows no evidence of an atmosphere or of other types of activity associated with comets. Also called a minor planet.

asteroid belt The region of the solar system between the orbits of Mars and Jupiter in which most asteroids are located. The main belt, where the orbits are generally the most stable, extends from 2.2 to 3.3 AU from the Sun.

asteroid family A group of asteroids that have similar orbital characteristics.

astrology The pseudoscience that deals with the supposed influences on human destiny of the configurations and locations in the sky of the Sun, Moon, and planets; a primitive religion having its origin in ancient Babylonia.

astronomical unit (AU) Originally meant to be the semimajor axis of the orbit of the Earth; now defined as the semimajor axis of the orbit of a hypothetical body with the mass and period that Gauss assumed for the Earth. The semimajor axis of the orbit of the Earth is 1.000000230 AU.

atmospheric refraction The bending, or refraction, of light rays from celestial objects by the Earth's atmosphere.

atom The smallest particle of an element that retains the properties that characterize that element.

atomic mass unit   *Chemical*: one-sixteenth of the mean mass of an oxygen atom. *Physical*: one-twelfth of the mass of an atom of the most common isotope of carbon. The atomic mass unit is approximately the mass of a hydrogen atom, 1.67×10^{-27} kg.

atomic number   The number of protons in each atom of a particular element.

atomic weight   The mean mass of an atom of a particular element in atomic mass units.

aurora   Light radiated by atoms and ions in the ionosphere, mostly in the magnetic polar regions.

autumnal equinox   The intersection of the ecliptic and celestial equator where the Sun crosses the equator from north to south.

azimuth   The angle along the celestial horizon, measured eastward from the north point, to the intersection of the horizon with the vertical circle passing through an object.

Balmer lines   Emission or absorption lines in the spectrum of hydrogen that arise from transitions between the second (or first excited) and higher energy states of the hydrogen atom.

bands (in spectra)   Emission or absorption lines, usually in the spectra of chemical compounds or radicals, so numerous and closely spaced that they coalesce into broad emission or absorption bands.

bar   A force of 100,000 newtons acting on a surface area of 1 square meter is equal to 1 bar. The average pressure of the Earth's atmosphere at sea level is equal to 1.013 bars.

barred spiral galaxy   Spiral galaxy in which the spiral arms begin from the ends of a "bar" running through the nucleus rather than from the nucleus itself.

barycenter   The center of mass of two mutually revolving bodies.

baryons (and antibaryons)   The heavy atomic nuclear particles, such as protons and neutrons.

basalt   Igneous rock, composed primarily of silicon, oxygen, iron, aluminum, and magnesium, produced by the cooling of lava. Basalts make up most of Earth's oceanic crust and are also found on other planets that have experienced extensive volcanic activity.

big bang theory   A theory of cosmology in which the expansion of the universe is presumed to have begun with a primeval explosion.

billion   In the United States and France and in this text, one thousand million (10^9); in Great Britain and Germany, originally, one million million (10^{12}). Usage in Great Britain is changing to conform to usage in the United States.

binary star   A double star; two stars revolving about each other.

binding energy   The energy required to separate completely the constituent parts of an atomic nucleus.

black dwarf   A presumed final state of evolution for a star, in which all of its energy sources are exhausted and it no longer emits radiation.

black hole   A hypothetical body whose velocity of escape is equal to or greater than the speed of light; thus no radiation can escape from it.

blackbody   A hypothetical perfect radiator, which absorbs and re-emits all radiation incident upon it.

Bohr atom   A particular model of an atom, invented by Niels Bohr, in which the electrons are described as revolving about the nucleus in circular orbits.

brown dwarf   An object intermediate in size between a planet and a star. The approximate mass range is from about twice the mass of Jupiter up to the lower mass limit for self-sustaining nuclear reactions, which is 0.08 solar mass. Also called infrared dwarf.

caldera   A volcanic crater, often resulting from the partial collapse of the summit of a shield volcano.

canals (on Mars)   Supposed long, narrow, straight, dark lines first reported on Mars by Schiaparelli and believed by Lowell and others to be the work of intelligent Martians. The martian canals (not to be confused with martian channels) were later shown to have been an illusion.

carbonaceous meteorite   A primitive meteorite made primarily of silicates but often including chemically bound water, free carbon, and complex organic compounds. Also called carbonaceous chondrites.

carbon-nitrogen-oxygen (CNO) cycle   A series of nuclear reactions in the interiors of stars involving carbon as a catalyst, by which hydrogen is transformed to helium.

Cassegrain focus   An optical arrangement in a reflecting telescope in which light is reflected by a second mirror to a point behind the primary mirror.

CBR   See *cosmic background radiation (CBR)*.

CCD   See *charged-coupled device (CCD)*.

cD galaxy   A supergiant elliptical galaxy frequently found at the center of a cluster of galaxies.

celestial equator   A great circle on the celestial sphere 90° from the celestial poles; the circle of intersection of the celestial sphere with the plane of the Earth's equator.

celestial mechanics   The branch of astronomy that deals with the motions and gravitational influences of the members of the solar system.

celestial meridian   An imaginary line on the celestial sphere passing through the north and south points on the horizon and through the zenith.

celestial poles   Points about which the celestial sphere appears to rotate, intersections of the celestial sphere with the Earth's polar axis.

celestial sphere   Apparent sphere of the sky; a sphere of large radius centered on the observer. Directions of objects in the sky can be denoted by the position of those objects on the celestial sphere.

center of gravity   Center of mass.

center of mass   The mean position of the various mass elements of a body or system, weighted according to their distances from that center of mass; that point in an isolated system that moves with constant velocity, according to Newton's first law of motion.

cepheid variable   A star that belongs to a class of yellow supergiant pulsating stars. These stars vary periodically in brightness, and the relationship between their periods and luminosities is useful in deriving distances to them.

Chandrasekhar limit   The upper limit to the mass of a white dwarf (equals 1.4 times the mass of the Sun).

charged-coupled device (CCD)   An array of electronic detectors of electromagnetic radiation, used at the focus of a telescope (or camera lens). A CCD acts like a photographic plate of very high sensitivity.

chemical condensation sequence   The calculated chemical compounds and minerals that would form in a cooling gas of cosmic composition, presented as a function of the temperature in the gas; used to infer the composition of grains that formed in the solar nebula at different distances from the protosun.

chondrite   A primitive stony meteorite that contains small spherical particles called chondrules.

chromosphere   That part of the solar atmosphere that lies immediately above the photospheric layers.

circular (satellite) velocity   The critical speed that a revolving body must have in order to follow a circular orbit.

circumpolar regions   Portions of the celestial sphere near the celestial poles that are either always above or always below the horizon.

climate   The weather conditions (temperature, precipitation, seasonal variations) averaged over a long enough span of time to eliminate most random variations—usually representing an average over a 10- to 20-year period.

cluster of galaxies   A system of galaxies containing several to thousands of member galaxies.

color excess   The amount by which the color index of a star is increased when its light is reddened in passing through interstellar absorbing material.

color index   Difference between the magnitudes of a star or other object measured in light of two different spectral regions, for example, blue minus visual $(B - V)$ magnitudes.

color-magnitude diagram   Plot of the magnitudes (apparent or absolute) of the stars in a cluster against their color indices.

coma (of comet)   The diffuse gaseous component of the head of a comet.

comet   A small body of icy and dusty matter that revolves about the Sun. When a comet comes near the Sun, some of its material vaporizes, forming a large head of tenuous gas, and often a tail.

compound   A substance composed of two or more chemical elements.

conduction   The transfer of energy by the direct passing of energy or electrons from atom to atom.

conic section   The curve of intersection between a circular cone and a plane; these curves can be ellipses, circles, parabolas, or hyperbolas.

conservation of angular momentum   The law that angular momentum is conserved in the absence of any force not directed toward or away from the point or axis about which the angular momentum is referred—that is, in the absence of a torque.

constellation   A configuration of stars named for a particular object, person, or animal; or the area of the sky assigned to a particular configuration.

contacts (of eclipses)   The instants when certain stages of an eclipse begin.

continental drift   A gradual drift of the continents over the surface of the Earth due to plate tectonics.

continuous spectrum   A spectrum of light composed of radiation of a continuous range of wavelengths or colors rather than only certain discrete wavelengths.

convection   The transfer of energy by moving currents of a fluid containing that energy.

core (of a planet)   The central part of a planet, consisting of higher density material, often metallic.

corona (of Galaxy)   A region lying above and below the plane of the Galaxy that contains hot gas, which emits x rays.

corona (of Sun)   Outer atmosphere of the Sun.

coronagraph   An instrument for observing the chromosphere and corona of the Sun outside of eclipse.

coronal hole   A region in the Sun's outer atmosphere where visible coronal radiation is absent.

cosmic background radiation (CBR)   The microwave radiation coming from all directions that is believed to be the redshifted glow of the big bang.

cosmic rays   Atomic nuclei (mostly protons) that are observed to strike the Earth's atmosphere with exceedingly high energies.

cosmological constant   A term that arises in the development of the equations of general relativity, which represents a repulsive force in the universe. The cosmological constant is usually assumed to be zero.

cosmological principle   The assumption that, on the large scale, the universe at any given time is the same everywhere.

cosmology   The study of the organization and evolution of the universe.

crater   A circular depression (from the Greek word for cup), generally of impact origin. The rarer volcanic craters are usually identified as such (see *caldera*); crater by itself is used in this text to refer to an impact crater.

crescent moon   One of the phases of the Moon when its elongation is less than 90° from the Sun and it appears less than half full.

critical density   In cosmology, the density that provides enough gravity to bring the expansion of the universe just to a stop after infinite time.

crust (of Earth)   The outer layer of the Earth.

dark matter   Nonluminous mass, whose presence can be inferred only because of its gravitational influence on luminous matter. Dark matter may constitute as much as 99 percent of all the mass in the universe. The composition of the dark matter is not known.

dark nebula   A cloud of interstellar dust that obscures the light of more distant stars and appears as an opaque curtain.

daylight saving time   A time 1 hr more advanced than standard time, usually adopted in spring and summer to take advantage of long evening twilights.

deceleration parameter (q_0)   A quantity that characterizes the future evolution of the various models of the universe based on general relativity.

declination   Angular distance north or south of the celestial equator to some object, measured along an hour circle passing through that object.

degenerate gas   A gas in which the allowable states for the electrons have been filled; it behaves according to different laws from those that apply to "perfect" gases.

density   The ratio of the mass of an object to its volume.

deuterium   A "heavy" form of hydrogen, in which the nucleus of each atom consists of one proton and one neutron.

differential galactic rotation   The rotation of the Galaxy, not as a solid wheel, but so that parts adjacent to one another do not always stay close together.

differential gravitational force   The difference between the respective gravitational forces exerted on two bodies near each other by a third, more distant body.

differentiation (geological)   A separation or segregation of different kinds of material in different layers in the interior of a planet.

diffraction   The spreading out of light in passing the edge of an opaque body.

diffraction grating   A system of closely spaced equidistant slits or reflecting strips that, by diffraction and interference, produce a spectrum.

disk (of planet or other object)   The apparent circular shape that a planet (or the Sun or Moon) displays when seen in the sky or viewed telescopically.

disk of Galaxy   The central disk or "wheel" of our Galaxy, superimposed on the spiral structure.

dispersion   Separation, from white light, of different wavelengths being refracted by different amounts.

Doppler effect   Apparent change in wavelength of the radiation from a source due to its relative motion in the line of sight.

Drake equation   An equation used to estimate the number of civilizations in the Galaxy.

dust tail (of a comet)   A cometary tail, usually broad, somewhat curved, and yellow-white, made up of dust grains released from the nucleus of the comet.

dwarf (star)   A main-sequence star (as opposed to a giant or supergiant).

Earth-approaching asteroid   An asteroid with an orbit that crosses the Earth's orbit or that will at some time cross the Earth's orbit as it evolves under the influence of the planets' gravity.

east point   The point on the horizon 90° from the north point (measured clockwise as seen from the zenith).

eccentricity (of ellipse)   Ratio of the distance between the foci to the major axis.

eclipse   The cutting off of all or part of the light of one body by another; in planetary science, the passing of one body into the shadow of another. (See *occultation, transit.*)

eclipsing binary star   A binary star in which the plane of revolution of the two stars is nearly edge on to our line of sight, so that the light of one star is periodically diminished by the other passing in front of it.

ecliptic   The apparent annual path of the Sun on the celestial sphere.

effective temperature   See *temperature (effective).*

ejecta   Material excavated from an impact crater; includes the ejecta blankets surrounding lunar craters, crater rays, and (in the case of the Earth) dust released into the atmosphere by an impact.

electromagnetic force   One of the four fundamental forces or interactions of nature; the force that acts between charges and binds atoms and molecules together.

electromagnetic radiation   Radiation consisting of waves propagated through the building up and breaking down of electric and magnetic fields; these include radio, infrared, light, ultraviolet, x rays, and gamma rays.

electromagnetic spectrum   The whole array or family of electromagnetic waves.

electron   A negatively charged subatomic particle that normally moves about the nucleus of an atom.

electron volt   The kinetic energy acquired by an electron that is accelerated through an electric potential of 1 volt; 1 electron volt is 1.60207×10^{-19} joule.

element   A substance that cannot be decomposed, by chemical means, into simpler substances.

elementary particle   One of the basic particles of matter. The most familiar of the elementary particles are the proton, neutron, and electron.

ellipse   A conic section: the curve of intersection of a circular cone and a plane cutting completely through the cone.

elliptical galaxy   A galaxy whose apparent photometric contours are ellipses and that contains no conspicuous interstellar material.

ellipticity   The ratio (in an ellipse) of the major axis minus the minor axis to the major axis.

emission line   A discrete bright spectral line.

emission nebula   A gaseous nebula that derives its visible light from the fluorescence of ultraviolet light from a star in or near the nebula.

emission spectrum   A spectrum consisting of emission lines.

energy level (in an atom or ion)   A particular level, or amount, of energy possessed by an atom or ion above the energy it possesses in its least energetic state.

epicycle   A circular orbit of a body in the Ptolemaic system, the center of which revolves about another circle (the deferent).

equator   A great circle on the Earth, 90° from its poles.

equatorial mount   A mounting for a telescope, one axis of which is parallel to the Earth's axis, so that a motion of the telescope about this axis can compensate for the Earth's rotation.

equinox   One of the intersections of the ecliptic and celestial equator.

equivalence principle   Principle that a gravitational force and a suitable acceleration are indistinguishable within a sufficiently local environment.

escape velocity   The velocity a body must achieve to break away from the gravity of another body and never return to it.

eucrite meteorite   One of a class of basaltic meteorites believed to have originated on the asteroid Vesta.

event   A point in four-dimensional spacetime.

event horizon   The surface through which a collapsing star is hypothesized to pass when its velocity of escape is equal to the speed of light, that is, when the star becomes a black hole.

excitation   The process of imparting to an atom or an ion an amount of energy greater than that it has in its normal or least-energy state.

exclusion principle   See *Pauli exclusion principle*.

extinction   Attenuation of light from a celestial body produced by the Earth's atmosphere, or by interstellar absorption.

extragalactic   Beyond our own Milky Way Galaxy.

eyepiece   A magnifying lens used to view the image produced by the objective of a telescope.

family (of asteroids)   A group of asteroids with similar orbital elements, indicating a probable common origin in a collision sometime in the past.

fall (of meteorites)   Meteorites seen in the sky and recovered on the ground.

fault   In geology, a crack or break in the crust of a planet along which slippage or movement can take place, accompanied by seismic activity.

field   A mathematical description of the effect of forces, such as gravity, that act on distant objects. According to this concept, a given mass produces a gravitational field in the space surrounding it, which produces a gravitational force on objects within that space.

find (of meteorites)   A meteorite that has been recovered but was not seen to fall.

filtergram   A photograph of the Sun (or part of it) taken through a special filter that transmits only a small range of wavelengths.

fireball   A spectacular meteor.

fission   The breakup of a heavy atomic nucleus into two or more lighter ones.

flare   A sudden and temporary outburst of light from an extended region of the solar surface.

fluorescence   The absorption of light of one wavelength and re-emission of it at another wavelength; especially the conversion of ultraviolet into visible light.

flux   The rate at which energy or matter crosses a unit area of a surface.

focal length   The distance from a lens or mirror to the point where light converged by it comes to a focus.

focus (of ellipse)   From any point on an ellipse, the sum of the distances to two fixed points inside the ellipse is a constant. Each of these two points is a focus of the ellipse.

focus (of telescope)   Point where the rays of light converged by a mirror or lens meet.

forbidden lines   Spectral lines that are not usually observed under laboratory conditions because they result from atomic transitions that are highly improbable.

force   That which can change the momentum of a body; numerically, the rate at which the body's momentum changes.

Fraunhofer line   An absorption line in the spectrum of the Sun or of a star.

Fraunhofer spectrum   The array of absorption lines in the spectrum of the Sun or a star.

free-free transition   An atomic transition in which the energy associated with an atom or ion and a passing electron changes during the encounter, but without capture of the electron by the atom or ion.

frequency   Number of vibrations per unit time; number of waves that cross a given point per unit time (in radiation).

full moon   That phase of the Moon when it is at opposition (180° from the Sun) and its full daylight hemisphere is visible from the Earth.

fusion   The building up of heavier atomic nuclei from lighter ones.

galactic cannibalism   The process by which a larger galaxy strips material from a smaller one.

galactic cluster   An "open" cluster of stars located in the spiral arms or disk of the Galaxy.

galaxy   A large assemblage of stars; a typical galaxy contains millions to hundreds of thousands of millions of stars.

Galaxy   The galaxy to which the Sun and our neighboring stars belong; the Milky Way is light from remote stars in the Galaxy.

Galilean satellite   Any of the four largest of Jupiter's satellites, discovered by Galileo.

gamma rays   Photons (of electromagnetic radiation) of energy higher than those of x rays; the most energetic form of electromagnetic radiation.

geocentric   Centered on the Earth.

geodesic   The path of a body in spacetime.

geology   The study of the Earth's crust and surface and of the processes that influence them; by extension, similar studies of any solid planetary object.

giant (star)   A star of large luminosity and radius.

giant molecular cloud   Large, cold interstellar clouds, with diameters of tens of parsecs and typical masses of 10^5 solar masses; found in the spiral arms of galaxies, these clouds are where massive stars form.

gibbous moon   One of the phases of the Moon in which more than half, but not all, of the Moon's daylight hemisphere is visible from the Earth.

globular cluster   One of about 120 large star clusters that form a system of clusters centered on the center of the Galaxy.

grand unified theories (GUTs)   Physical theories that attempt to describe the four interactions (forces) of nature as different manifestations of a single force.

granite   The type of igneous silicate rocks that make up most of the continental crust of the Earth.

granulation   The "rice-grain"–like structure of the solar photosphere; granulation is produced by upwelling currents of gas that are slightly hotter, and therefore brighter, than the surrounding regions, which are flowing downward into the Sun.

gravitation   The mutual attraction of material bodies or particles.

gravitational constant, G   The constant of proportionality in Newton's law of gravitation; in metric units G has the value 6.672×10^{-11} newton·m²/kg².

gravitational energy   Energy that can be released by the gravitational collapse, or partial collapse, of a system.

gravitational lens   A configuration of celestial objects, one of which provides one or more images of the other by gravitationally deflecting its light.

gravitational redshift The redshift of electromagnetic radiation caused by a gravitational field. The slowing of clocks in a gravitational field.

gravitational waves Oscillations in spacetime, produced by changes in the distribution of matter.

great circle Circle on the surface of a sphere that is the curve of intersection of the sphere with a plane passing through its center.

greenhouse effect The blanketing of infrared radiation near the surface of a planet by, for example, carbon dioxide in its atmosphere.

Greenwich meridian The meridian of longitude passing through the site of the old Royal Greenwich Observatory, near London; origin of longitude on the Earth.

ground state The lowest energy state of an atom.

H See *Hubble constant*.

H I region Region of neutral hydrogen in interstellar space.

H II region Region of ionized hydrogen in interstellar space.

half-life The time required for half of the radioactive atoms in a sample to disintegrate.

halo (of galaxy) The outermost extent of our Galaxy or another, containing a sparse distribution of stars and globular clusters in a more or less spherical distribution.

head (of comet) The main part of a comet, consisting of its nucleus and coma.

heavy elements In astronomy, usually those elements of greater atomic number than helium.

Heisenberg uncertainty principle A principle of quantum mechanics that places a limit on the precision with which the simultaneous position and momentum of a body or particle can be specified.

helio- Prefix referring to the Sun.

heliocentric Centered on the Sun.

helium flash The nearly explosive ignition of helium in the triple-alpha process in the dense core of a red giant star.

Hertz A unit of frequency: one cycle per second. Named for Heinrich Hertz, who first produced radio radiation.

Hertzsprung gap A V-shaped gap in the upper part of the Hertzsprung-Russell diagram where few stable stars are found.

Hertzsprung-Russell (H-R) diagram A plot of absolute magnitude against temperature (or spectral class or color index) for a group of stars.

highlands (lunar) The older, heavily cratered crust of the Moon, covering 83 percent of its surface and composed in large part of anorthositic breccias.

high-velocity star (or object) A star (or object) with high space motion relative to the Sun; generally an object that does not share the orbital velocity of the Sun about the galactic nucleus.

homogeneous Having a consistent and even distribution of matter that is the same everywhere.

horizon (astronomical) A great circle on the celestial sphere 90° from the zenith.

horizontal branch A sequence of stars on the Hertzsprung-Russell diagram of a typical globular cluster of approximately constant absolute magnitude (near $M_v = 0.8$).

horoscope A chart showing the positions along the zodiac and in the sky of the Sun, Moon, and planets at some given instant and as seen from a particular place on Earth—usually corresponding to the time and place of a person's birth.

Hubble constant Constant of proportionality between the velocities of remote galaxies and their distances. The Hubble constant is thought to lie in the range of 15 to 30 km/s per million LY.

Hubble law The law of the redshifts. The radial velocities of remote galaxies are proportional to their distances from us.

hydrostatic equilibrium A balance between the weights of various layers, as in a star or the Earth's atmosphere, and the pressures that support them.

hypothesis A tentative theory or supposition, advanced to explain certain facts or phenomena, which is subject to further tests and verification.

igneous rock Any rock produced by cooling from a molten state.

image The optical representation of an object produced by light rays from the object being refracted or reflected, as by a lens or mirror.

impact basin Large features in the lunar crust formed by the impact of projectiles up to 100 km in diameter.

inclination (of an orbit) The angle between the orbital plane of a revolving body and some fundamental plane—usually the plane of the celestial equator or of the ecliptic.

index of refraction A measure of the refracting power of a transparent substance; specifically, the ratio of the speed of light in a vacuum to its speed in the substance.

inertia The property of matter that requires a force to act on it to change its state of motion; momentum is a measure of inertia.

inertial system A system of coordinates that is not itself accelerated, but that either is at rest or is moving with constant velocity.

inflationary universe A theory of cosmology in which the universe is assumed to have undergone a phase of very rapid expansion during the first 10^{-30} s. After this period of rapid expansion, the big bang and inflationary models are identical.

infrared cirrus Patches of interstellar dust, which emit infrared radiation and look like cirrus clouds on the images of the sky produced by the Infrared Astronomy Satellite.

infrared radiation Electromagnetic radiation of wavelength longer than the longest (red) wavelengths that can be perceived by the eye, but shorter than radio wavelengths.

interference A phenomenon of waves that mix together such that their crests and troughs can alternately reinforce and cancel one another.

International Date Line An arbitrary line on the surface of the Earth near longitude 180° across which the date changes by one day.

interstellar dust   Tiny solid grains in interstellar space, thought to consist of a core of rock-like material (silicates) or graphite surrounded by a mantle of ices. Water, methane, and ammonia are probably the most abundant ices.

interstellar gas   Sparse gas in interstellar space.

interstellar lines   Absorption lines superimposed on stellar spectra, produced by the interstellar gas.

interstellar medium or interstellar matter   Interstellar gas and dust.

inverse-square law (for light)   The amount of energy (light) flowing through a given area in a given time (flux) decreases in proportion to the square of the distance from the source of energy or light.

ion   An atom that has become electrically charged by the addition or loss of one or more electrons.

ion tail (of comet)   See *plasma tail (of a comet)*.

ionization   The process by which an atom gains or loses electrons.

ionosphere   The upper region of the Earth's atmosphere in which many of the atoms are ionized.

irons (meteorites)   One of the three main types of meteorites, typically made of about 90 percent iron and 9 percent nickel, with traces of other elements.

irregular galaxy   A galaxy without rotational symmetry; neither a spiral nor an elliptical galaxy.

irregular satellite   A planetary satellite with an orbit that is retrograde, or of high inclination or eccentricity.

irreversible process   A process (for example, a chemical change) that cannot be reversed, for instance, because of the loss of one or more of the chemical products. The dissociation of water vapor in a planetary atmosphere is usually irreversible because the hydrogen escapes and thus cannot recombine with oxygen to form water again.

isotope   Any of two or more forms of the same element, whose atoms all have the same number of protons but different numbers of neutrons.

isotropic   The same in all directions.

joule   The metric unit of energy; the work done by a force of 1 newton (N) acting through a distance of 1 m.

jovian planet   Any of the planets Jupiter, Saturn, Uranus, and Neptune.

Kepler's laws   Three laws, discovered by J. Kepler, that describe the motions of the planets.

kinetic energy   Energy associated with motion; the kinetic energy of a body is one-half the product of its mass and the square of its velocity.

laser   An acronym for *l*ight *a*mplification by *s*timulated *e*mission of *r*adiation; a device for amplifying a light signal at a particular wavelength into a coherent beam.

latitude   A north-south coordinate on the surface of the Earth; the angular distance north or south of the equator measured along a meridian passing through a place.

law   A statement of order or relation between phenomena that, under given conditions, is presumed to be invariable.

law of areas   Kepler's second law: the radius vector from the Sun to any planet sweeps out equal areas in the planet's orbital plane in equal intervals of time.

law of the redshifts   The relation between the radial velocity and distance of a remote galaxy: The radial velocities are proportional to the distances of galaxies.

leap year   A calendar year with 366 days, inserted approximately every 4 years to make the average length of the calendar year as nearly equal as possible to the tropical year.

light   Electromagnetic radiation that is visible to the eye.

light curve   A graph that displays the time variation in light or magnitude of a variable or eclipsing binary star.

light year   The distance light travels in a vacuum in one year; 1 LY = 9.46×10^{12} km, or about 6×10^{12} mi.

line broadening   The phenomenon by which spectral lines are not precisely sharp but have finite widths.

line profile   A plot of the intensity of light versus wavelength across a spectral line.

linear diameter   Actual diameter in units of length.

Local Group   The cluster of galaxies to which our Galaxy belongs.

local standard of rest   A coordinate system that shares the average motion of the Sun and its neighboring stars about the galactic center.

Local Supercluster   The supercluster of galaxies to which the Local Group belongs.

longitude   An east-west coordinate on the Earth's surface; the angular distance, measured east or west along the equator, from the Greenwich meridian to the meridian passing through a place.

low-velocity star (or object)   A star (or object) that has low space velocity relative to the Sun; generally an object that shares the Sun's high orbital speed about the galactic center.

luminosity   The rate of radiation of electromagnetic energy into space by a star or other object.

luminosity class   A classification of a star according to its luminosity for a given spectral class.

luminosity function   The relative numbers of stars (or other objects) of various luminosities or absolute magnitudes.

luminous flux or luminous energy   Light.

lunar   Referring to the Moon.

lunar eclipse   An eclipse of the Moon.

Lyman lines   A series of absorption or emission lines in the spectrum of hydrogen that arise from transitions to and from the lowest energy states of the hydrogen atoms.

Magellanic Clouds   Two neighboring galaxies visible to the naked eye from southern latitudes.

magma   Mobile, high-temperature molten state of rock, usually of silicate mineral composition and with dissolved gases and other volatiles.

magnetic field   The region of space near a magnetized body within which magnetic forces can be detected.

magnetic pole   One of two points on a magnet (or the Earth) at which the greatest density of lines of force emerge. A compass needle aligns itself along the local lines of force on the Earth and points more or less toward the magnetic poles of the Earth.

magnetosphere   The region around a planet in which its intrinsic magnetic field dominates the interplanetary field carried by the solar wind; hence, the region within which charged particles can be trapped by the planetary magnetic field.

magnifying power or magnification   The number of times larger (in angular diameter) an object appears through a telescope than with the naked eye.

magnitude   A measure of the amount of light flux received from a star or other luminous object.

main sequence   A sequence of stars on the Hertzsprung-Russell diagram, containing the majority of stars, that runs diagonally from the upper left to the lower right.

major axis (of ellipse)   The maximum diameter of an ellipse.

major planet   A jovian planet.

mantle (of Earth)   The greatest part of the Earth's interior, lying between the crust and the core.

mare (pl. maria)   Latin for "sea"; name applied to the dark, relatively smooth features that cover 17 percent of the Moon.

mass   A measure of the total amount of material in a body; defined either by the inertial properties of the body or by its gravitational influence on other bodies.

mass defect   The amount by which the mass of an atomic nucleus is less than the sum of the masses of the individual nucleons that compose it.

mass extinction   The sudden disappearance in the fossil record of a large number of species of life, to be replaced by new species in subsequent layers. Mass extinctions are indications of catastrophic changes in the environment, such as might be produced by a large impact on the Earth.

mass-light ratio   The ratio of the total mass of a galaxy to its total luminosity, usually expressed in units of solar mass and solar luminosity. The mass-light ratio gives a rough indication of the types of stars contained within a galaxy and whether or not substantial quantities of dark matter are present.

mass-luminosity relation   An empirical relation between the masses and luminosities of many (principally main-sequence) stars.

Maunder minimum   The interval from 1645 to 1715 when solar activity was very low.

Maxwell's equations   A set of four equations that describe the fields around magnetic and electric charges, and how changes in those fields produce forces and electromagnetic radiation.

mean density of matter in the universe   The average density of the universe if all of its matter and energy could be smoothed out to absolute uniformity.

mean solar day   Average length of the apparent solar day.

mean solar time   Mean solar time is based on the mean solar day, which is defined to have a duration equal to the average length of an apparent solar day.

meridian (celestial)   The great circle on the celestial sphere that passes through an observer's zenith and the north (or south) celestial pole.

meridian (terrestrial)   The great circle on the surface of the Earth that passes through a particular place and the north and south poles of the Earth.

Messier catalogue   A catalogue of nonstellar objects compiled by Charles Messier in 1787.

metamorphic rock   Any rock produced by the physical and chemical alteration (without melting) of another rock that has been subjected to high temperature and pressure.

metastable level   An energy level in an atom from which there is a low probability of an atomic transition accompanied by the radiation of a photon.

meteor   The luminous phenomenon observed when a meteoroid enters the Earth's atmosphere and burns up; popularly called a "shooting star."

meteor shower   Many meteors appearing to radiate from a common point in the sky caused by the collision of the Earth with a swarm of meteoritic particles.

meteorite   A portion of a meteoroid that survives passage through the atmosphere and strikes the ground.

meteorite fall   The occurrence of a meteorite striking the ground.

meteoritics   The study of meteorites and of the formation of the solar system.

meteoroid   A meteoritic particle in space before any encounter with the Earth.

meteorology   The study of planetary atmospheres, weather, and climate.

Michelson-Morley experiment   The classic (1887) experiment by A. A. Michelson and E. W. Morley, in which they tried to measure the speed of the Earth in space by timing the speed of light in different directions. The failure of their experiment was later explained by special relativity.

micrometeorite   A meteoroid so small that, on entering the atmosphere of the Earth, it is slowed quickly enough that it does not burn up or ablate but filters through the air to the ground.

micron   Old term for micrometer (10^{-6} meter).

microwave   Shortwave radio wavelengths.

Milky Way   The band of light encircling the sky, which is due to the many stars and diffuse nebulae lying near the plane of the Galaxy.

minerals   The solid compounds (often primarily silicon and oxygen) that form rocks.

minor planet   See *asteroid*.

model atmosphere (or photosphere)   The result of a theoretical calculation of the run of temperature, pressure, density, and so on, through the outer layers of the Sun or a star.

molecule   A combination of two or more atoms bound together; the smallest particle of a chemical compound or substance that exhibits the chemical properties of that substance.

momentum   A measure of the inertia or state of motion of a body; the momentum of a body is the product of its mass and velocity. In the absence of a force, momentum is conserved.

monochromatic   Of one wavelength or color.

nadir   The point on the celestial sphere 180° from the zenith.

nebula   Cloud of interstellar gas or dust.

neutrino   A fundamental particle that has little or no rest mass and no charge but that does have spin and energy.

neutron A subatomic particle with no charge and with mass approximately equal to that of the proton.

neutron star A star of extremely high density composed almost entirely of neutrons.

new moon Phase of the Moon when its longitude is the same as that of the Sun.

Newtonian focus An optical arrangement in a reflecting telescope, in which a flat mirror intercepts the light from the primary before it reaches the focus and reflects it to a focus at the side of the telescope tube.

Newton's laws The laws of mechanics and gravitation formulated by Isaac Newton.

nongravitational force The force that acts on comets to change their orbits, due not to the gravitational influence of the other members of the planetary system, but rather to the rocket effect of gases escaping from the cometary nucleus.

nonthermal radiation See *synchrotron radiation*.

nova A star that experiences a sudden outburst of radiant energy, temporarily increasing its luminosity by hundreds to thousands of times.

nuclear Referring to the nucleus of the atom.

nuclear bulge Central part of our Galaxy.

nuclear transformation Transformation of one atomic nucleus into another.

nucleosynthesis The building up of heavy elements from lighter ones by nuclear fusion.

nucleus (of atom) The heavy part of an atom, composed mostly of protons and neutrons, and about which the electrons revolve.

nucleus (of comet) The solid chunk of ice and dust in the head of a comet.

nucleus (of galaxy) Central concentration of matter at the center of a galaxy.

oblate spheroid A solid formed by rotating an ellipse about its minor axis.

oblateness A measure of the "flattening" of an oblate spheroid; numerically, the ratio of the difference between the major and minor diameters (or axes) to the major diameter (or axis).

occultation The passage of an object of large angular size in front of a smaller object, such as the Moon in front of a distant star or the rings of Saturn in front of the Voyager spacecraft.

Oort comet cloud The spherical region around the Sun from which most "new" comets come, representing objects with aphelia at about 50,000 AU, or extending about a third of the way to the nearest other stars.

opacity Absorbing power; capacity to impede the passage of light.

open cluster A comparatively loose or "open" cluster of stars, containing from a few dozen to a few thousand members, located in the spiral arms or disk of the Galaxy; sometimes referred to as a galactic cluster.

optical double star Two stars at different distances nearly lined up in projection so that they appear close together, but that are not really gravitationally associated.

optics The branch of physics that deals with light and its properties.

orbit The path of a body that is in revolution about another body or point.

oscillation A periodic motion; in the case of the Sun, a periodic or quasi-periodic expansion and contraction of the whole Sun or some portion of it.

oxidizing In chemistry, referring to conditions in which oxygen dominates over hydrogen, so that most other elements form compounds with oxygen. In very oxidizing conditions, such as are found in the atmosphere of the Earth, free oxygen gas (O_2) or even atomic oxygen (O) is present.

ozone A heavy molecule of oxygen that contains three atoms rather than the more normal two. Designated O_3.

parabola A conic section of eccentricity 1.0; the curve of the intersection between a circular cone and a plane parallel to a straight line in the surface of the cone.

parallax An apparent displacement of an object due to a motion of the observer.

parallax (stellar) An apparent displacement of a nearby star that results from the motion of the Earth around the Sun; numerically, the angle subtended by 1 AU at the distance of a particular star.

parent In referring to the process of radioactive decay, the name given to the radioactive isotope that is destroyed in the decay process to produce a new, or daughter, isotope.

parent molecules The original molecules (for example, H_2O, CO_2, CO, CH_4) that dissociate to form the radicals (C_2, CN, OH) actually observed in comets.

Pauli exclusion principle Quantum mechanical principle by which no two particles of the same kind can have the same position and momentum.

peculiar velocity The velocity of a star with respect to the local standard of rest; that is, its space motion, corrected for the motion of the Sun with respect to our neighboring stars.

perfect cosmological principle The assumption that, on the large scale, the universe appears the same from every place and at all times.

perfect gas An "ideal" gas that obeys the perfect gas laws.

perfect gas laws Certain laws that describe the behavior of an ideal gas: Charles' law, Boyle's law, and the equation of state for a perfect gas.

perfect radiator Blackbody; a body that absorbs and subsequently re-emits all radiation incident upon it.

periastron The place in the orbit of a star in a binary-star system where it is closest to its companion star.

perigee The place in the orbit of an Earth satellite where it is closest to the center of the Earth.

perihelion The place in the orbit of an object revolving about the Sun where it is closest to the Sun's center.

period A time interval; for example, the time required for one complete revolution.

period-luminosity relation An empirical relation between the periods and luminosities of cepheid variable stars.

perturbation The disturbing effect, when small, on the motion of a body as predicted by a simple theory, produced by a third body or other external agent.

photometry The measurement of light intensities.

photon A discrete unit of electromagnetic energy.

photon sphere A surface surrounding a black hole, of radius about 1.4 times that of the event horizon, where a photon can have a closed circular orbit.

photosphere The region of the solar (or a stellar) atmosphere from which continuous radiation escapes into space.

pixel An individual picture element in a detector; for example, a particular silicon diode in a CCD.

plage A bright region of the solar surface observed in the monochromatic light of some spectral line.

Planck's constant The constant of proportionality relating the energy of a photon to its frequency.

planet Any of the nine largest bodies revolving about the Sun, or any similar non–self-luminous bodies that may orbit other stars.

planetarium An optical device for projecting on a screen or domed ceiling the stars and planets and their apparent motions in the sky.

planetary nebula A shell of gas ejected from, and enlarging about, a certain kind of extremely hot star that is nearing the end of its life.

planetary system A term used in this text to refer to all of the solar system except the Sun: the planets, their satellites, rings, comets, asteroids, meteoroids, dust, and the solar wind.

planetesimals The hypothetical objects, from tens to hundreds of kilometers in diameter, that formed in the solar nebula as an intermediate step between tiny grains and the larger planetary objects we see today. The comets and some asteroids may be leftover planetesimals.

plasma A hot ionized gas.

plasma tail (of a comet) A cometary tail, usually narrow and bluish in color, extending straight away from the Sun and consisting of plasma streaming away from the head under the influence of the solar wind. Also called an ion tail.

plate tectonics The motion of segments or plates of the outer layer of the Earth over the underlying mantle.

polar axis The axis of rotation of the Earth; also, an axis in the mounting of a telescope that is parallel to the Earth's axis.

Population I and II Two classes of stars (and systems of stars), classified according to their spectral characteristics, chemical compositions, radial velocities, ages, and locations in the Galaxy.

positron An electron with a positive rather than negative charge; an antielectron.

postulate An essential prerequisite to a hypothesis or theory.

potential energy Stored energy that can be converted into other forms; especially gravitational energy.

precession (of Earth) A slow, conical motion of the Earth's axis of rotation, caused principally by the gravitational torque of the Moon and Sun on the Earth's equatorial bulge.

precession of the equinoxes Slow westward motion of the equinoxes along the ecliptic that results from precession.

pressure Force per unit area; expressed in units of atmospheres or pascals.

prime focus The point in a telescope where the objective focuses the light.

prime meridian The terrestrial meridian passing through the site of the old Royal Greenwich Observatory; longitude 0°.

primitive In planetary science and meteoritics, an object or rock that is little changed, chemically, since its formation, and hence representative of the conditions in the solar nebula at the time of formation of the solar system. Also used to refer to the chemical composition of an atmosphere that has not undergone extensive chemical evolution.

primitive meteorite A meteorite that has not been greatly altered chemically since its condensation from the solar nebula; called in meteoritics a chondrite (either ordinary chondrite or carbonaceous chondrite).

primitive rock Any rock that has not experienced great heat or pressure and therefore remains representative of the original condensates from the solar nebula— never found on any object large enough to have undergone melting and differentiation.

principle of equivalence Principle that a gravitational force and a suitable acceleration are indistinguishable within a sufficiently local environment.

principle of relativity Principle that all observers in uniform relative motion are equivalent; the laws of nature are the same for all, and no experiment can reveal an absolute motion or state of rest.

prism A wedge-shaped piece of glass that is used to disperse white light into a spectrum.

prominence A phenomenon in the solar corona that commonly appears like a flame above the limb of the Sun.

proper motion The angular change per year in the direction of a star as seen from the Sun.

proton A heavy subatomic particle that carries a positive charge; one of the two principal constituents of the atomic nucleus.

proton-proton cycle A series of thermonuclear reactions by which nuclei of hydrogen are built up into nuclei of helium.

protoplanet (or -star or -galaxy) The original material from which a planet (or a star or galaxy) condensed.

pulsar A variable radio source of small angular size that emits radio pulses in very regular periods that range from 0.03 to 5 s.

pulsating variable A variable star that pulsates in size and luminosity.

q_0 See *deceleration parameter* (q_0).

quantum efficiency The ratio of the number of photons incident on a detector to the number actually detected.

quantum mechanics The branch of physics that deals with the structure of atoms and their interactions with one another and with radiation.

quarter moon Either of the two phases of the Moon when its longitude differs by 90° from that of the Sun; the Moon appears half full at these phases.

quasar A stellar-appearing object of very high redshift, presumed to be extragalactic and highly luminous; an active galactic nucleus.

RR Lyrae variable   One of a class of giant pulsating stars with periods less than one day.

radar   The technique of transmitting radio waves to an object and then detecting the radiation that the object reflects back to the transmitter; used to measure the distance to, and motion of, a target object.

radial velocity   The component of relative velocity that lies in the line of sight.

radial velocity curve   A plot of the variation of radial velocity with time for a binary or variable star.

radiant (of meteor shower)   The point in the sky from which the meteors belonging to a shower seem to radiate.

radiation   A mode of energy transport whereby energy is transmitted through a vacuum; also the transmitted energy itself.

radiation pressure   The transfer of momentum carried by electromagnetic radiation to a body that the radiation impinges upon.

radio astronomy   The technique of making astronomical observations in radio wavelengths.

radio galaxy   A galaxy that emits greater amounts of radio radiation than average.

radio telescope   A telescope designed to make observations in radio wavelengths.

radioactive dating   The technique of determining the ages of rocks or other specimens by the amount of radioactive decay of certain radioactive elements contained therein.

radioactivity (radioactive decay)   The process by which certain kinds of atomic nuclei naturally decompose, with the spontaneous emission of subatomic particles and gamma rays.

red giant   A large, cool star of high luminosity; a star occupying the upper right portion of the Hertzsprung-Russell diagram.

reddening (interstellar)   The reddening of starlight passing through interstellar dust, caused because dust scatters blue light more effectively than red.

redshift   A shift to longer wavelengths of the light from remote galaxies; presumed to be produced by a Doppler shift.

reducing   In chemistry, referring to conditions in which hydrogen dominates over oxygen, so that most other elements form compounds with hydrogen. In very reducing conditions free hydrogen (H_2) is present and free oxygen (O_2) cannot exist.

reflecting telescope   A telescope in which the principal optical component (objective) is a concave mirror.

reflection   The return of light rays by an optical surface.

reflection nebula   A relatively dense dust cloud in interstellar space that is illuminated by starlight.

reflectivity   Measure of the brightness of an object relative to a sphere of equal size made of a perfectly reflecting diffuse white material (also known as the geometric albedo).

refracting telescope   A telescope in which the principal optical component (objective) is a lens or system of lenses.

refraction   The bending of light rays passing from one transparent medium (or a vacuum) to another.

regular satellites   Planetary satellites that have orbits of low or moderate eccentricity in approximately the plane of the planet's equator.

relative orbit   The orbit of one of two mutually revolving bodies referred to the other body as origin.

relativistic particle (or electron)   A particle (electron) moving at nearly the speed of light.

relativity   A theory formulated by Einstein that describes the relations between measurements of physical phenomena by two different observers who are in relative motion at constant velocity (the special theory of relativity) or that describes how a gravitational field can be replaced by a curvature of spacetime (the general theory of relativity).

resolution   The degree to which fine details in an image are separated, or resolved.

resolving power   A measure of the ability of an optical system to resolve, or separate, fine details in the image it produces; in astronomy, the angle in the sky that can be resolved by a telescope.

resonance   An orbital condition in which one object is subject to periodic gravitational perturbations by another, most commonly arising when two objects orbiting a third have periods of revolution that are simple multiples or fractions of each other.

rest mass   The mass of an object or particle as measured when it is at rest in the laboratory.

retrograde (rotation or revolution)   Backward with respect to the common direction of motion in the solar system; counterclockwise as viewed from the north, and going from east to west rather than from west to east.

retrograde motion   An apparent westward motion of a planet on the celestial sphere or with respect to the stars.

revolution   The motion of one body around another.

rift zone   In geology, a place where the crust is being torn apart by internal forces, generally associated with the injection of new material from the mantle and with the slow separation of tectonic plates.

right ascension   A coordinate for measuring the east-west positions of celestial bodies; the angle measured eastward along the celestial equator from the vernal equinox to the hour circle passing through a body.

rotation   Turning of a body about an axis running through it.

runaway greenhouse effect   A process whereby the heating of a planet leads to an increase in its atmospheric greenhouse effect and thus to further heating, thereby quickly altering the composition of its atmosphere and the temperature of its surface.

Russell-Vogt theorem   The theorem that the mass and chemical composition of a star determine its entire structure if it derives its energy entirely from thermonuclear reactions.

satellite   A body that revolves about a planet.

Schwarzschild radius   See *event horizon*.

science   The attempt to find order in nature or to find laws that describe natural phenomena.

scientific method   A specific procedure in science: (1) the observation of phenomena or the results of experiments; (2) the formulation of hypotheses that describe these phenomena and that are consistent with the body of knowledge available; (3) the testing of these hypotheses by noting whether or not they adequately predict and describe new phenomena or the results of new experiments.

sedimentary rock   Any rock formed by the deposition and cementing of fine grains of material.

seeing   The unsteadiness of the Earth's atmosphere, which blurs telescopic images.

seismic waves   Vibrations traveling through the Earth's interior that result from earthquakes.

seismology (geology)   The study of earthquakes and the conditions that produce them and of the internal structure of the Earth as deduced from analyses of seismic waves.

seismology (solar)   The study of small changes in the radial velocity of the Sun as a whole or of small regions on the surface of the Sun. Analyses of these velocity changes can be used to infer the internal structure of the Sun.

semimajor axis   Half the major axis of a conic section.

separation (in a visual binary)   The angular separation of the two components of a visual binary star.

SETI   The search for extraterrestrial intelligence, usually applied to searches for radio signals from other civilizations.

Seyfert galaxy   A galaxy belonging to the class of those with active galactic nuclei; one whose nucleus shows bright emission lines; one of a class of galaxies first described by C. Seyfert.

shepherd satellite   Informal term for a satellite that is thought to maintain the structure of a planetary ring through its close gravitational influence—originally, the two Saturn satellites, Prometheus and Pandora, that orbit just inside and outside the F ring.

shower (meteor)   Many meteors, all seeming to radiate from a common point in the sky, caused by the encounter by the Earth of a swarm of meteoroids moving together through space.

sidereal period   The period of revolution of one body about another with respect to the stars.

sidereal time   The local hour angle of the vernal equinox.

sidereal year   Period of the Earth's revolution about the Sun with respect to the stars.

sign (of zodiac)   Astrological term for any of 12 equal sections along the ecliptic, each of length 30°. Starting at the vernal equinox, and moving eastward, the signs are Aries, Taurus, Gemini, Cancer, Leo, Virgo, Libra, Scorpio, Sagittarius, Capricorn, Aquarius, and Pisces.

simultaneity   The occurrence of two events at the same time. In relativity, absolute simultaneity is seen not to have meaning, except for two simultaneous events occurring at the same place.

singularity   A theoretical point of zero volume and infinite density to which any object that becomes a black hole must collapse, according to the general theory of relativity.

SNC meteorite   One of a class of basaltic meteorites now believed by many planetary scientists to be impact-ejected fragments from Mars.

solar activity   Phenomena of the solar atmosphere: sunspots, plages, and related phenomena.

solar antapex   Direction away from which the Sun is moving with respect to the local standard of rest.

solar apex   The direction toward which the Sun is moving with respect to the local standard of rest.

solar eclipse   An eclipse of the Sun by the Moon, caused by the passage of the Moon in front of the Sun. Solar eclipses can occur only at the time of new moon.

solar motion   Motion of the Sun, or the velocity of the Sun, with respect to the local standard of rest.

solar nebula   The cloud of gas and dust from which the solar system formed.

solar seismology   The study of pulsations or oscillations of the Sun in order to determine the characteristics of the solar interior, including composition and rotation rate.

solar system   The system of the Sun and the planets, their satellites, the minor planets, comets, meteoroids, and other objects revolving around the Sun.

solar time   A time based on the Sun; usually the hour angle of the Sun plus 12 hr.

solar wind   A radial flow of plasma leaving the Sun.

solstice   Either of two points on the celestial sphere where the Sun reaches its maximum distances north and south of the celestial equator.

space velocity or space motion   The velocity of a star with respect to the Sun.

spacetime   A system of one time and three spatial coordinates, with respect to which the time and place of an event can be specified; also called spacetime continuum.

spectral class (or type)   A classification of a star according to the characteristics of its spectrum.

spectral line   Radiation at a particular wavelength of light produced by the emission or absorption of energy by an atom.

spectral sequence   The sequence of spectral classes of stars arranged in order of decreasing temperatures of stars of those classes.

spectrometer   An instrument for obtaining a spectrum; in astronomy, usually attached to a telescope to record the spectrum of a star, galaxy, or other astronomical object.

spectroscopic binary star   A binary star in which the components are not resolved optically, but whose binary nature is indicated by periodic variations in radial velocity, indicating orbital motion.

spectroscopic parallax   A parallax (or distance) of a star that is derived by comparing the apparent magnitude of the star with its absolute magnitude as deduced from its spectral characteristics.

spectroscopy   The study of spectra.

spectrum   The array of colors or wavelengths obtained when light from a source is dispersed, as in passing it through a prism or grating.

speed   The rate at which an object moves without regard to its direction of motion; the numerical or absolute value of velocity.

spicule A jet of rising material in the solar chromosphere.

spiral arms Arms of interstellar material and young stars that wind out in a plane from the central nucleus of a spiral galaxy.

spiral density wave A mechanism for the generation of spiral structure in galaxies; a density wave interacts with interstellar matter and triggers the formation of stars. Spiral density waves are also seen in the rings of Saturn.

spiral galaxy A flattened, rotating galaxy with pinwheel-like arms of interstellar material and young stars winding out from its nucleus.

spring tide The highest tidal range of the month, produced when the Moon is near either the full or the new phase.

sputtering The process by which energetic atomic particles striking a solid alter its chemistry and eject additional atoms or molecular fragments from the surface.

standard candle An astronomical object of known luminosity; such an object can be used to determine distances.

standard time The local mean solar time of a standard meridian, adopted over a large region to avoid the inconvenience of continuous time changes around the Earth.

star A self-luminous sphere of gas.

star cluster An assemblage of stars held together by their mutual gravitation.

statistical parallax The mean parallax for a selection of stars, derived from the radial velocities of the stars and the components of their proper motions that cannot be affected by the solar motion.

steady state (theory of cosmology) A theory of cosmology embracing the perfect cosmological principle and involving the continuous creation of matter.

Stefan-Boltzmann law A formula from which the rate at which a blackbody radiates energy can be computed; the total rate of energy emission from a unit area of a blackbody is proportional to the fourth power of its absolute temperature.

stellar evolution The changes that take place in the sizes, luminosities, structures, and so on, of stars as they age.

stellar model The result of a theoretical calculation of the run of physical conditions in a stellar interior.

stellar parallax The angle subtended by 1 AU at the distance of a star; usually measured in seconds of arc.

stellar wind The outflow of gas, sometimes at speeds as high as hundreds of kilometers per second, from a star.

stony meteorite A meteorite composed mostly of stony material.

stony-iron meteorite A type of meteorite that is a blend of nickel-iron and silicate materials.

stratosphere The layer of the Earth's atmosphere above the troposphere (where most weather takes place) and below the ionosphere.

strong nuclear force or strong interaction The force that binds together the parts of the atomic nucleus.

subduction zone In terrestrial geology, a region where one crustal plate is forced under another, generally associated with earthquakes, volcanic activity, and the formation of deep ocean trenches.

submillimeter (part of the electromagnetic spectrum) That part of the electromagnetic spectrum with wavelengths of a few hundred micrometers.

summer solstice The point on the celestial sphere where the Sun reaches its greatest distance north of the celestial equator.

Sun The star about which the Earth and other planets revolve.

sunspot A temporary cool region in the solar photosphere that appears dark by contrast against the surrounding hotter photosphere.

sunspot cycle The semiregular 11-year period with which the frequency of sunspots fluctuates.

supercluster A large region of space (more than 100 million LY across) where matter is concentrated into galaxies, groups of galaxies, and clusters of galaxies; a cluster of clusters of galaxies.

supergiant A star of very high luminosity.

supernova An explosion that marks the final stage of evolution of a star. A Type I supernova is thought to occur when a white dwarf accretes enough matter to exceed the Chandrasekhar limit, collapses, and explodes. A Type II supernova is thought to mark the final collapse of a massive star.

surface gravity The weight of a unit mass at the surface of a body.

surveying The technique of measuring distances and relative positions of places over the surface of the Earth (or elsewhere); generally accomplished by triangulation.

synchrotron radiation The radiation emitted by charged particles being accelerated in magnetic fields and moving at speeds near that of light.

synodic period The interval between successive occurrences of the same configuration of a planet; for example, between successive oppositions or successive superior conjunctions.

tail (of a comet) See *dust tail (of a comet)* and *plasma tail (of a comet)*.

tangential (transverse) velocity The component of a star's space velocity that lies in the plane of the sky.

tectonic Activity and motion that result from expansion and contraction of the crust of a planet.

telescope An optical instrument used to aid in viewing or measuring distant objects.

temperature (Celsius; formerly centigrade) Temperature measured on scale where water freezes at 0° and boils at 100°.

temperature (color) The temperature of a star as estimated from the intensity of the stellar radiation at two or more colors or wavelengths.

temperature (effective) The temperature of a blackbody that would radiate the same total amount of energy that a particular object, such as a star, does.

temperature (excitation) The temperature of a star as estimated from the relative strengths of lines in its spectrum that originate from atoms in different stages of excitation.

temperature (Fahrenheit) Temperature measured on a scale where water freezes at 32° and boils at 212°.

temperature (ionization) The temperature of a star as estimated from the relative strengths of lines in its spectrum that originate from atoms in different stages of ionization.

temperature (Kelvin) Absolute temperature measured in Celsius degrees.

temperature (kinetic) A measure of the mean energy of the molecules in a substance.

temperature (radiation) The temperature of a blackbody that radiates the same amount of energy in a given spectral region as does a particular body.

terrestrial planet Any of the planets Mercury, Venus, Earth, Mars, and sometimes Pluto or the Moon.

theory A set of hypotheses and laws that have been well demonstrated to apply to a wide range of phenomena associated with a particular subject.

thermal energy Energy associated with the motions of the molecules in a substance.

thermal equilibrium A balance between the input and outflow of heat in a system.

thermal radiation The radiation emitted by any body or gas that is not at absolute zero.

thermodynamics The branch of physics that deals with heat and heat transfer among bodies.

thermonuclear energy Energy associated with thermonuclear reactions or that can be released through thermonuclear reactions.

thermonuclear reaction A nuclear reaction or transformation that results from encounters between nuclear particles that are given high velocities (by heating them).

tidal force A differential gravitational force that tends to deform a body.

tidal heating Generation of heat in a planetary object through repeated tidal stresses from a larger nearby object, probably important for the Moon early in its history and responsible today for the high level of volcanic activity on Io.

tidal stability limit The distance—approximately 2.5 planetary radii from the center—within which differential gravitational forces (or tides) are stronger than the mutual gravitational attraction between two adjacent orbiting objects. Within this limit, fragments are not likely to accrete or assemble themselves into a larger object. Also called the Roche limit.

tide Deformation of a body by the differential gravitational force exerted on it by another body; in the Earth, the deformation of the ocean surface by the differential gravitational forces exerted by the Moon and Sun.

ton (metric) One thousand kilograms.

total eclipse An eclipse of the Sun in which the Sun's photosphere is entirely hidden by the Moon, or an eclipse of the Moon in which it passes completely into the umbra of the Earth's shadow.

transition region The region in the Sun's atmosphere where the temperature rises very rapidly from the relatively low temperatures that characterize the chromosphere to the high temperatures of the corona.

triple-alpha process A series of two nuclear reactions by which three helium nuclei are built up into one carbon nucleus.

Trojan asteroid One of a large number of asteroids that share Jupiter's orbit about the Sun, but either preceding or following Jupiter by 60°.

Tropic of Cancer Parallel of latitude 23.5° N.

Tropic of Capricorn Parallel of latitude 23.5° S.

tropical year Period of revolution of the Earth about the Sun with respect to the vernal equinox.

troposphere Lowest level of the Earth's atmosphere, where most weather takes place.

turbulence Random motions of gas masses, as in the atmosphere of a star.

21-cm line A spectral line of neutral hydrogen at the radio wavelength of 21 cm.

ultraviolet radiation Electromagnetic radiation of wavelengths shorter than the shortest (violet) wavelengths to which the eye is sensitive; radiation of wavelengths in the approximate range 10 to 400 nm.

umbra The central, completely dark part of a shadow.

uncertainty principle See *Heisenberg uncertainty principle*.

universe The totality of all matter and radiation and the space occupied by same.

variable star A star that varies in luminosity.

vector A quantity that has both magnitude and direction.

velocity A vector that denotes both the speed and the direction a body is moving.

velocity of escape The speed with which an object must move in order to enter a parabolic orbit about another body (such as the Earth), and hence move permanently away from the vicinity of that body.

vernal equinox The point on the celestial sphere where the Sun crosses the celestial equator passing from south to north.

very-long-baseline interferometry (VLBI) A technique of radio astronomy whereby signals from telescopes thousands of kilometers apart are combined to obtain very high resolution with interferometry.

visual binary star A binary star in which the two components are telescopically resolved.

volatile materials Materials that are gaseous at fairly low temperatures. This is a relative term, usually applied to the gases in planetary atmospheres and to common ices (H_2O, CO_2, and so on), but it is also sometimes used for elements such as cadmium, zinc, lead, and rubidium that form gases at temperatures up to 1000 K. (These are called volatile elements, as opposed to refractory elements.)

volume A measure of the total space occupied by a body.

watt A unit of power.

wavelength The spacing of the crests or troughs in a wave train.

weak nuclear force or weak interaction The nuclear force involved in radioactive decay. The weak force is characterized by the slow rate of certain nuclear reactions—such as the decay of the neutron, which occurs with a half-life of 11 min.

weather The state of a planetary atmosphere—its composition, temperature, pressure, motion, and so on—at a particular place and time.

weight A measure of the force due to gravitational attraction.

white dwarf A star that has exhausted most or all of its nuclear fuel and has collapsed to a very small size; such a star is near its final stage of evolution.

Wien's law Formula that relates the temperature of a blackbody to the wavelength at which it emits the greatest intensity of radiation.

winter solstice Point on the celestial sphere where the Sun reaches its greatest distance south of the celestial equator.

Wolf-Rayet star One of a class of very hot stars that eject shells of gas at very high velocity.

x rays Photons of wavelengths intermediate between those of ultraviolet radiation and gamma rays.

x-ray stars Stars (other than the Sun) that emit observable amounts of radiation at x-ray frequencies.

year The period of revolution of the Earth around the Sun.

Zeeman effect A splitting or broadening of spectral lines due to magnetic fields.

zenith The point on the celestial sphere opposite to the direction of gravity; or the direction opposite to that indicated by a plumb bob.

zero-age main sequence Main sequence for a system of stars that have completed their contraction from interstellar matter and are now deriving all their energy from nuclear reactions, but whose chemical composition has not yet been altered by nuclear reactions.

zodiac A belt around the sky 18° wide centered on the ecliptic.

zodiacal light A faint illumination along the zodiac, which is due to sunlight that has been reflected and scattered by interplanetary dust.

zone of avoidance A region near the Milky Way where obscuration by interstellar dust is so heavy that few or no exterior galaxies can be seen.

APPENDIX 3
Power of Ten Notation

In astronomy and other sciences, it is often necessary to deal with very large or very small numbers. For example, the Earth is 150,000,000,000 m from the Sun, and the mass of the hydrogen atom is 0.00000000000000000000000000167 kg. Instead of writing and carrying so many zeros, the numbers are usually written as figures between 1 and 10 multiplied by the appropriate power of 10. For example, 150,000,000,000 is 1.5×10^{11}, and the mass of the hydrogen atom given above is written simply as 1.67×10^{-27} kg. Additional examples are given below:

one hundredth	=	0.01 =	10^{-2}
one tenth	=	0.1 =	10^{-1}
one	=	1 =	10^{0}
ten	=	10 =	10^{1}

one hundred	=	100 =	10^{2}
one thousand	=	1000 =	10^{3}
one million	=	1,000,000 =	10^{6}
one billion	=	1,000,000,000 =	10^{9}

The powers-of-ten notation is not only compact and convenient, it also simplifies arithmetic. To multiply two numbers expressed as powers of ten, you need only add the exponents. And to divide, you subtract the exponents. Following are several examples:

$$100 \times 100,000 = 10^{2} \times 10^{5} = 10^{2+5} = 10^{7}$$
$$0.01 \times 1,000,000 = 10^{-2} \times 10^{6} = 10^{6-2} = 10^{4}$$
$$1,000,000 \div 1000 = 10^{6} \div 10^{3} = 10^{6-3} = 10^{3}$$
$$100 \div 1,000,000 = 10^{2} \div 10^{6} = 10^{2-6} = 10^{-4}$$

APPENDIX 4
Units

In the American system of measure (originally developed in England), the fundamental units of length, mass, and time are the yard, pound, and second, respectively. There are also, of course, larger and smaller units, which include the ton (2240 lb), the mile (1760 yd), the rod (16½ ft), the inch (1/36 yd), the ounce (1/16 lb), and so on. Such units are inconvenient for conversion and arithmetic computation.

In science, therefore, it is more usual to use the metric system, which has been adopted in virtually all countries except the United States. The fundamental units of the metric system are

length: 1 meter (m)
mass: 1 kilogram (kg)
time: 1 second (s)

A meter was originally intended to be 1 ten-millionth of the distance from the equator to the North Pole along the surface of the Earth. It is about 1.1 yd. A kilogram is about 2.2 lb. The second is the same in metric and American units.

The most commonly used quantities of length and mass of the metric system are the following:

Length

1 km	= 1 kilometer	= 1000 meters	= 0.6214 mile
1 m	= 1 meter	= 1.094 yards	= 39.37 inches
1 cm	= 1 centimeter	= 0.01 meter	= 0.3937 inch
1 mm	= 1 millimeter	= 0.001 meter	= 0.1 cm
1 μm	= 1 micrometer	= 0.000 001 meter	= 0.0001 cm
1 nm	= 1 nanometer	= 10^{-9} meter	= 10^{-7} cm

also: 1 mile = 1.6093 km
       1 inch = 2.5400 cm

Mass

1 metric ton	= 10^6 grams	= 1000 kg	= 2.2046×10^3 lb
1 kg	= 1000 grams	= 2.2046 lb	
1 g	= 1 gram	= 0.0022046 lb	= 0.0353 oz
1 mg	= 1 milligram	= 0.001 g	

also: 1 lb = 0.4536 kg
       1 oz = 28.3495 g

Three temperature scales are in general use:

1. Fahrenheit (F); water freezes at 32°F and boils at 212°F.

2. Celsius or centigrade* (C); water freezes at 0°C and boils at 100°C.

3. Kelvin or absolute (K); water freezes at 273 K and boils at 373 K.

All molecular motion ceases at −459°F = −273°C = 0 K. Thus, Kelvin temperature is measured from this lowest possible temperature, called *absolute zero*. It is the temperature scale most often used in astronomy. Kelvins are degrees that have the same value as centigrade or Celsius degrees, since the difference between the freezing and boiling points of water is 100 degrees in each.

On the Fahrenheit scale, water boils at 212 degrees and freezes at 32 degrees; the difference is 180 degrees. Thus, to convert Celsius degrees or Kelvins to Fahrenheit, it is necessary to multiply by 180/100 = 9/5. To convert from Fahrenheit to Celsius degrees or Kelvins, it is necessary to multiply by 100/180 = 5/9.

Example 1: What is 68°F in Celsius and in Kelvins?

$$68°F − 32°F = 36°F \text{ above freezing.}$$

$$\frac{5}{9} \times 36° = 20°;$$

thus,

$$68°F = 20°C = 293 \text{ K.}$$

Example 2: What is 37°C in Fahrenheit and in Kelvins?

$$37°C = 273° + 37° = 310 \text{ K;}$$

$$\frac{9}{5} \times 37° = 66.6 \text{ Fahrenheit degrees;}$$

thus,

$$37°C \text{ is } 66.6°F \text{ above freezing}$$

or

$$37°C = 32° + 66.6° = 98.6°F.$$

---

*Celsius is now the name used for centigrade temperature; it has a more modern standardization but differs from the old centigrade scale by less than 0.1°.

APPENDIX 5
Some Useful Constants

Physical Constants

speed of light	c	$= 2.9979 \times 10^8$ m/s
constant of gravitation	G	$= 6.672 \times 10^{-11}$ N m^2/kg^2
Planck's constant	h	$= 6.626 \times 10^{-34}$ joules
mass of hydrogen atom	m_H	$= 1.673 \times 10^{-27}$ kg
mass of electron	m_e	$= 9.109 \times 10^{-31}$ kg
charge of electron	e	$= 4.803 \times 10^{-10}$ eu
Rydberg constant	R	$= 1.0974 \times 10^7$ per m
Stefan-Boltzmann constant	σ	$= 5.670 \times 10^{-8}$ joule/m^2·deg^4
constant in Wien's law	$\lambda_{max}T$	$= 2.898 \times 10^{-3}$ m·deg
electron volt (energy)	eV	$= 1.602 \times 10^{-19}$ joules
energy equivalent of 1 ton TNT	E	$= 4.3 \times 10^9$ joules

Astronomical Constants

astronomical unit	AU	$= 1.496 \times 10^{11}$ m
light year	LY	$= 9.461 \times 10^{15}$ m
parsec	pc	$= 3.086 \times 10^{16}$ m
sidereal year	yr	$= 3.158 \times 10^7$ s
mass of Earth	M_E	$= 5.977 \times 10^{24}$ kg
equatorial radius of Earth	R_E	$= 6.378 \times 10^6$ m
obliquity of ecliptic	ϵ	$= 23°\ 27'$
surface gravity of Earth	g	$= 9.807$ m/s^2
escape velocity of Earth	v_E	$= 1.119 \times 10^4$ m/s
age of Earth	A_E	$= 4.55 \times 10^9$ yr
mass of Sun	M_S	$= 1.989 \times 10^{30}$ kg
equatorial radius of Sun	R_S	$= 6.960 \times 10^8$ m
luminosity of Sun	L_S	$= 3.83 \times 10^{26}$ watts
solar constant (at Earth)	S	$= 1.37 \times 10^3$ watts/m^2
Hubble constant	H	$= 75 \pm 15$ km/s/mpc
Age of "empty" universe	1/H	$= 1.3 \times 10^{10}$ yr
Critical density of universe	ρ	$= 10^{-29}$ g/cm^3 (approximately)

APPENDIX 6
Astronomical Coordinate Systems

Several astronomical coordinate systems are in common use. In each of these systems the position of an object in the sky, or on the celestial sphere, is denoted by two angles. These angles are referred to a *reference plane,* which contains the observer, and a *reference direction,* which is a direction from the observer to some arbitrary point lying in the reference plane. The intersection of the reference plane and the celestial sphere is a great circle, which defines the "equator" of the coordinate system. At two points, each 90° from this equator, are the "poles" of the coordinate system. Great circles passing through these poles intersect the equator of the system at right angles.

One of the two angular coordinates of each coordinate system is measured from the equator of the system to the object along the great circle passing through it and the poles. Angles on one side of the equator (or reference plane) are reckoned as positive; those on the opposite are negative. The other angular coordinate is measured along the equator from the reference direction to the intersection of the equator with the great circle passing through the object and the poles.

The system of terrestrial latitude and longitude provides an excellent analogue. Here the plane of the terrestrial equator is the fundamental plane, and the Earth's equator is the equator of the system; the North and South terrestrial Poles are the poles of the system. One coordinate, the *latitude* of a place, is reckoned north (positive) or south (negative) of the equator along a meridian passing through the place. The other coordinate, *longitude,* is measured along the equator to the intersection of the equator and the meridian of the place from the intersection of the equator and the Greenwich meridian. The direction (from the center of the Earth) to this latter intersection is the reference direction. Terrestrial longitude is either east or west (whichever is less), but the corresponding coordinate in celestial systems is generally reckoned in one direction from 0 to 360° (or, equivalently, from 0 to 24^h).

The following table lists the more important astronomical coordinate systems and defines how each of the angular coordinates is defined.

System	Reference Plane	Reference Direction	"Latitude" Coordinate	Range	"Longitude" Coordinate	Range
Horizon	Horizon plane	North point (formerly the south point was used by astronomers)	Altitude, h; toward the zenith (+) toward the nadir (−)	±90°	Azimuth, A; measured to the east along the horizon from the north point	0 to 360°
Equator	Plane of the celestial equator	Vernal equinox	Declination, δ; toward the north celestial pole (+) toward the south celestial pole (−)	±90°	Right ascension, α or R.A.; measured to the east along the celestial equator from the vernal equinox	0 to 24^h
Ecliptic	Plane of the Earth's orbit (ecliptic)	Vernal equinox	Celestial latitude, β; toward the north ecliptic pole (+) toward the south ecliptic pole (−)	±90°	Celestial longitude, λ; measured to the east along the ecliptic from the vernal equinox	0 to 360°
Galactic	Mean plane of the Milky Way	Direction to the galactic center	Galactic latitude, b; toward the north galactic pole (+) toward the south galactic pole (−)	±90°	Galactic longitude, l; measured along the galactic equator to the east from the galactic center	0 to 360°

APPENDIX 7
Elements of an Orbit

The elements of an orbit are those numbers needed to specify both the nature of the orbit and the location of an object in its orbit at any time. Two numbers (usually eccentricity and semimajor axis) are needed to describe the size and shape of the orbit. Three other numbers are needed to specify the orbit's orientation in space. The final two elements give the orbital period and specify where the object is at some particular time, so that its location at other times can be computed.

The table below lists the set of elements that is conventional for describing orbits of objects revolving about the Sun. However, other equivalent data can also specify an orbit and may be more convenient for calculation, depending on the exact problem addressed.

Name	Symbol	Definition
Semimajor axis	a	Half of the distance between the points nearest the foci on the conic that represents the orbit (usually measured in astronomical units).
Eccentricity	e	Distance between the foci of the conic divided by the major axis.
Inclination	i	Angle of intersection between the orbital planes of the object and of the Earth.
Longitude of the ascending node	Ω	Angle from the vernal equinox (where the ecliptic and celestial equator intersect with the Sun crossing the equator from south to north), measured to the east along the ecliptic plane, to the point where the object crosses the ecliptic traveling from south to north (the ascending node).
Argument of perihelion	ω	Angle from the ascending node, measured in the plane of the object's orbit and in the direction of its motion, to the perihelion point (its closest approach to the Sun).
Time of perihelion passage	T	One of the precise times that the object passed the perihelion point.
Period	P	The sidereal period of revolution of the object about the Sun.

APPENDIX 8
Some Nuclear Reactions of Importance in Astronomy

Given here are the series of thermonuclear reactions that are most important in stellar interiors. The subscript to the left of a nuclear symbol is the atomic number; the superscript to the left is the atomic mass number. The symbols for the positive electron (positron) and electron are e^+ and e^-, respectively, for the neutrino is ν, and for a photon (generally of gamma-ray energy) is γ.

1. The Proton-Proton Chains

(Important below 15×10^6 K)

There are three ways the proton-proton chain can be completed. The first (a_1, b_1, c_1) is the most important, but depending on the physical conditions in the stellar interior, some energy is released by one or both of the following alternatives: a_1, b_1, c_2, d_2, e_2, and a_1, b_1, c_2, d_3, e_3, f_3.

(a_1) $^1_1H + ^1_1H \rightarrow ^2_1H + e^+ + \nu$

(b_1) $^2_1H + ^1_1H \rightarrow ^3_2He + \gamma$

(c_1) $^3_2He + ^3_2He \rightarrow ^4_2He + 2^1_1H$

or (c_2) $^3_2He + ^4_2He \rightarrow ^7_4Be + \gamma$

(d_2) $^7_4Be + e^- \rightarrow ^7_3Li + \nu$

(e_2) $^7_3Li + ^1_1H \rightarrow 2^4_2He$

or (d_3) $^7_4Be + ^1_1H \rightarrow ^8_5B + \gamma$

(e_3) $^8_5B \rightarrow ^8_4Be + e^+ + \nu$

(f_3) $^8_4Be \rightarrow 2^4_2He$

2. The Carbon-Nitrogen Cycle

(Important above 15×10^6 K)

(a) $^{12}_6C + ^1_1H \rightarrow ^{13}_7N + \gamma$

(b) $^{13}_7N \rightarrow ^{13}_6C + e^+ + \nu$

(c) $^{13}_6C + ^1_1H \rightarrow ^{14}_7N + \gamma$

(d) $^{14}_7N + ^1_1H \rightarrow ^{15}_8O + \gamma$

(e) $^{15}_8O \rightarrow ^{15}_7N + e^+ + \nu$

(f) $^{15}_7N + ^1_1H \rightarrow ^{12}_6C + ^4_2He$

3. The Triple-Alpha Process

(Important above 10^8 K)

(a) $^4_2He + ^4_2He \rightarrow ^8_4Be + \gamma$

(b) $^4_2He + ^8_4Be \rightarrow ^{12}_6C + \gamma$

APPENDIX 9
Orbital Data for the Planets

Planet	Semimajor Axis		Sidereal Period		Mean Orbital Speed (km/s)	Orbital Eccentricity	Inclination of Orbit to Ecliptic (°)
	AU	10^6 km	Tropical Years	Days			
Mercury	0.3871	57.9	0.24085	87.97	47.9	0.206	7.004
Venus	0.7233	108.2	0.61521	224.70	35.0	0.007	3.394
Earth	1.0000	149.6	1.000039	365.26	29.8	0.017	0.0
Mars	1.5237	227.9	1.88089	686.98	24.1	0.093	1.850
(Ceres)	2.7671	414	4.603		17.9	0.077	10.6
Jupiter	5.2028	778	11.86		13.1	0.048	1.308
Saturn	9.538	1427	29.46		9.6	0.056	2.488
Uranus	19.191	2871	84.07		6.8	0.046	0.774
Neptune	30.061	4497	164.82		5.4	0.010	1.774
Pluto	39.529	5913	248.6		4.7	0.248	17.15

Adapted from *The Astronomical Almanac* (U.S. Naval Observatory), 1981.

APPENDIX 10
Physical Data for the Planets

Planet	Diameter (km)	Diameter (Earth = 1)	Mass (Earth = 1)	Mean Density (g/cm³)	Rotation Period (days)	Inclination of Equator to Orbit (°)	Surface Gravity (Earth = 1)	Velocity of Escape (km/s)
Mercury	4878	0.38	0.055	5.43	58.6	0.0	0.38	4.3
Venus	12,104	0.95	0.82	5.24	−243.0	177.4	0.91	10.4
Earth	12,756	1.00	1.00	5.52	0.997	23.4	1.00	11.2
Mars	6794	0.53	0.107	3.9	1.026	25.2	0.38	5.0
Jupiter	142,800	11.2	317.8	1.3	0.41	3.1	2.53	60
Saturn	120,540	9.41	94.3	0.7	0.43	26.7	1.07	36
Uranus	51,200	4.01	14.6	1.2	−0.72	97.9	0.92	21
Neptune	49,500	3.88	17.2	1.6	0.67	29	1.18	24
Pluto	2200	0.17	0.0025	2.0	−6.387	118	0.09	1

APPENDIX 11
Satellites of the Planets

Planet	Satellite Name	Discovery	Semimajor Axis (km × 1000)	Period (days)	Diameter (km)	Mass (10²⁰ kg)	Density (g/cm³)
Earth	Moon	—	384	27.32	3476	735	3.3
Mars	Phobos	Hall (1877)	9.4	0.32	23	1×10^{-4}	2.0
	Deimos	Hall (1877)	23.5	1.26	13	2×10^{-5}	1.7
Jupiter	Metis	Voyager (1979)	128	0.29	20	—	—
	Adrastea	Voyager (1979)	129	0.30	40	—	—
	Amalthea	Barnard (1892)	181	0.50	200	—	—
	Thebe	Voyager (1979)	222	0.67	90	—	—
	Io	Galileo (1610)	422	1.77	3630	894	3.6
	Europa	Galileo (1610)	671	3.55	3138	480	3.0
	Ganymede	Galileo (1610)	1070	7.16	5262	1482	1.9
	Callisto	Galileo (1610)	1883	16.69	4800	1077	1.9
	Leda	Kowal (1974)	11,090	239	15	—	—
	Himalia	Perrine (1904)	11,480	251	180	—	—
	Lysithea	Nicholson (1938)	11,720	259	40	—	—
	Elara	Perrine (1905)	11,740	260	80	—	—
	Ananke	Nicholson (1951)	21,200	631 (R)	30	—	—
	Carme	Nicholson (1938)	22,600	692 (R)	40	—	—
	Pasiphae	Melotte (1908)	23,500	735 (R)	40	—	—
	Sinope	Nicholson (1914)	23,700	758 (R)	40	—	—
Saturn	Unnamed	Voyager (1985)	118.2	0.48	15?	3×10^{-5}	—
	Pan	Voyager (1985)	133.6	0.58	20	3×10^{-5}	—
	Atlas	Voyager (1980)	137.7	0.60	40	—	—
	Prometheus	Voyager (1980)	139.4	0.61	80	—	—
	Pandora	Voyager (1980)	141.7	0.63	100	—	—
	Janus	Dollfus (1966)	151.4	0.69	190	—	—
	Epimetheus	Fountain, Larson (1980)	151.4	0.69	120	—	—
	Mimas	Herschel (1789)	186	0.94	394	0.4	1.2
	Enceladus	Herschel (1789)	238	1.37	502	0.8	1.2
	Tethys	Cassini (1684)	295	1.89	1048	7.5	1.3
	Telesto	Reitsema et al. (1980)	295	1.89	25	—	—
	Calypso	Pascu et al. (1980)	295	1.89	25	—	—
	Dione	Cassini (1684)	377	2.74	1120	11	1.4
	Helene	Lecacheux, Laques (1980)	377	2.74	30	—	—
	Rhea	Cassini (1672)	527	4.52	1530	25	1.3
	Titan	Huygens (1655)	1222	15.95	5150	1346	1.9
	Hyperion	Bond, Lassell (1848)	1481	21.3	270	—	—
	Iapetus	Cassini (1671)	3561	79.3	1435	19	1.2
	Phoebe	Pickering (1898)	12,950	550 (R)	220	—	—
Uranus	Cordelia	Voyager (1986)	49.8	0.34	40?	—	—
	Ophelia	Voyager (1986)	53.8	0.38	50?	—	—
	Bianca	Voyager (1986)	59.2	0.44	50?	—	—
	Cressida	Voyager (1986)	61.8	0.46	60?	—	—
	Desdemona	Voyager (1986)	62.7	0.48	60?	—	—
	Juliet	Voyager (1986)	64.4	0.50	80?	—	—
	Portia	Voyager (1986)	66.1	0.51	80?	—	—
	Rosalind	Voyager (1986)	69.9	0.56	60?	—	—
	Belinda	Voyager (1986)	75.3	0.63	60?	—	—
	Puck	Voyager (1985)	86.0	0.76	170	—	—

Satellites of the Planets (Continued)

Planet	Satellite Name	Discovery	Semimajor Axis (km × 1000)	Period (days)	Diameter (km)	Mass (10²⁰ kg)	Density (g/cm³)
	Miranda	Kuiper (1948)	130	1.41	485	0.8	1.3
	Ariel	Lassell (1851)	191	2.52	1160	13	1.6
	Umbriel	Lassell (1851)	266	4.14	1190	13	1.4
	Titania	Herschel (1787)	436	8.71	1610	35	1.6
	Oberon	Herschel (1787)	583	13.5	1550	29	1.5
Neptune	Naiad	Voyager (1989)	48	0.30	50	—	
	Thalassa	Voyager (1989)	50	0.31	90	—	
	Despina	Voyager (1989)	53	0.33	150	—	
	Galatea	Voyager (1989)	62	0.40	150	—	
	Larissa	Voyager (1989)	74	0.55	200	—	
	Proteus	Voyager (1989)	118	1.12	400	—	
	Triton	Lassell (1846)	355	5.88 (R)	2720	220	2.1
	Nereid	Kuiper (1949)	5511	360	340	—	—
Pluto	Charon	Christy (1978)	19.7	6.39	1200	—	—

APPENDIX 12
Total Solar Eclipses from 1972 through 2030

Date	Duration of Totality (*min*)	Where Visible
1972 July 10	2.7	Alaska, Northern Canada
1973 June 30	7.2	Atlantic Ocean, Africa
1974 June 20	5.3	Indian Ocean, Australia
1976 Oct. 23	4.9	Africa, Indian Ocean, Australia
1977 Oct. 12	2.8	Northern South America
1979 Feb. 26	2.7	Northwest U.S., Canada
1980 Feb. 16	4.3	Central Africa, India
1981 July 31	2.2	Siberia
1983 June 11	5.4	Indonesia
1984 Nov. 22	2.1	Indonesia, South America
1987 March 29	0.3	Central Africa
1988 March 18	4.0	Philippines, Indonesia
1990 July 22	2.6	Finland, Arctic Regions
1991 July 11	7.1	Hawaii, Mexico, Central America, Brazil
1992 June 30	5.4	South Atlantic
1994 Nov. 3	4.6	South America
1995 Oct. 24	2.4	South Asia
1997 March 9	2.8	Siberia, Arctic
1998 Feb. 26	4.4	Central America
1999 Aug. 11	2.6	Central Europe, Central Asia
2001 June 21	4.9	Southern Africa
2002 Dec. 4	2.1	South Africa, Australia
2003 Nov. 23	2.0	Antarctica
2005 April 8	0.7	South Pacific Ocean
2006 March 29	4.1	Africa, Asia Minor, U.S.S.R.
2008 Aug. 1	2.4	Arctic Ocean, Siberia, China
2009 July 22	6.6	India, China, South Pacific
2010 July 11	5.3	South Pacific Ocean
2012 Nov. 13	4.0	Northern Australia, South Pacific
2013 Nov. 3	1.7	Atlantic Ocean, Central Africa
2015 March 20	4.1	North Atlantic, Arctic Ocean
2016 March 9	4.5	Indonesia, Pacific Ocean
2017 Aug. 21	2.7	Pacific Ocean, U.S.A., Atlantic Ocean
2019 July 2	4.5	South Pacific, South America
2020 Dec. 14	2.2	South Pacific, South America, South Atlantic Ocean
2021 Dec. 4	1.9	Antarctica
2023 April 20	1.3	Indian Ocean, Indonesia
2024 April 8	4.5	South Pacific, Mexico, East U.S.A.
2026 Aug. 12	2.3	Arctic, Greenland, North Atlantic, Spain
2027 Aug. 2	6.4	North Africa, Arabia, Indian Ocean
2028 July 22	5.1	Indian Ocean, Australia, New Zealand
2030 Nov. 25	3.7	South Africa, Indian Ocean, Australia

Star	Distance (LY)	Radial Velocity (km/s)	Spectra of Components A	B	C	Visual Magnitudes of Components A	B	C	Visual Luminosities of Components (L_S) A	B	C
Sun			G2V			−26.8			1.0		
Proxima Centauri*	4.3	−16	M5V			+11.05			5.8×10^{-5}		
α Centauri	4.4	−22	G2V	K0V		−0.01	+1.33		1.4	0.44	
Barnard's Star	5.9	−108	M5V			+9.54			4.4×10^{-4}		
Wolf 359	7.7	+13	M8V			+13.53			1.7×10^{-5}		
Lalande 21185	8.2	−84	M2V			+7.50			5.2×10^{-3}		
Luyten 726-8	8.5	+30	M5.5V	M5.5V		+12.45	+12.95		6.3×10^{-5}	4.0×10^{-5}	
Sirius	8.6	−8	A1V	wd		−1.46	+8.68		2.3	1.9×10^{-3}	
Ross 154	9.5	−4	M4.5V			+10.6			4.0×10^{-4}		
Ross 248	10.2	−81	M6V			+12.29			$+1 \times 10^{-4}$		
ε Eridani	10.7	+16	K2V			+3.73			0.30		
Ross 128	10.8	−13	M5V			+11.10			3.3×10^{-4}		
Luyten 789-6	10.8	−60	M6V			−12.18			1.20×10^{-4}		
61 Cygni	11.1	−64	K5V	K7V		+5.22	+6.03		0.076	0.036	
ε Indi	11.2	−40	K5V			+4.68			0.13		
τ Ceti	11.3	−16	G8V			+3.50			0.44		
Procyon	11.4	−3	F5IV-V	wd		+0.37	+10.7		7.6	5.2×10^{-4}	
BD + 59°1915	11.5	+5	M4V	M5V		+8.90	+9.69		2.8×10^{-3}	1.4×10^{-3}	
BD + 43°44	11.6	+17	M1V	M6V		+8.07	+11.04		6.3×10^{-3}	4.0×10^{-4}	
CD − 36°15693	11.7	+10	M2V			+7.36			0.012		
G51-15	11.9		MV			+14.8			1.3×10^{-5}		
Luyten 725-32	12.4		M5V			+11.5			3.0×10^{-4}		
BD + 5°1668	12.4	+26	M5V			+9.82			1.3×10^{-3}		
CD − 39°14192	12.6	+21	M0V			+6.67			0.025		
Kapteyn's Star	12.7	+245	M0V			+8.81			4.0×10^{-3}		
Kruger 60	12.8	−26	M3V	M4.5V		+9.85	+11.3		1.4×10^{-3}	4.0×10^{-4}	
Ross 614	13.4	+24	M7V	?		+11.07	+14.8		4.8×10^{-4}	1.6×10^{-5}	
BD − 12°4523	13.7	−13	M5V			+10.12			1.2×10^{-3}		
Wolf 424	13.9	−5	M5.5V	M6V		+13.16	+13.4		8.3×10^{-5}	6.9×10^{-5}	
v. Maanen's Star	14.1	+54	wd			+12.37			1.6×10^{-4}		
CD − 37°15492	14.5	+23	M3V			+8.63			5.8×10^{-3}		
Luyten 1159-16	14.7		M8V			+12.27			2.3×10^{-4}		
BD + 50°1725	15.0	−26	K7V			+6.59			0.040		
CD − 46°11540	15.1		M4V			+9.36			3.3×10^{-3}		
CD − 49°13515	15.2	+8	M3V			+8.67			6.3×10^{-3}		
CD − 44°11909	15.3		M5V			+11.2			6.3×10^{-4}		
BD + 68°946	15.3	−22	M3.5V			+9.15			4.0×10^{-3}		
G158 − 27	15.4		MV			+13.7			6.3×10^{-5}		
G208-44/45	15.5		MV	MV		+13.4	+14.0		8.3×10^{-5}	4.8×10^{-5}	
Ross 780	15.6	+9	M5V			+10.7			1.6×10^{-3}		
40 Eridani	15.7	−43	K0V	wd	M4.5V	+4.43	+9.53	+11.17	0.33	3.0×10^{-3}	6.9×10^{-4}
Luyten 145-141	15.8		wd			+11.44			5.2×10^{-4}		
BD + 20°2465	16.1	+11	M4.5V			+9.43			3.3×10^{-3}		
70 Ophiuchi	16.1	−7	K1V	K5V		+4.2	+6.0		0.44	0.83	
BD + 43°4305	16.3	−2	M4.5V			+10.2			1.7×10^{-3}		

*Proxima Centauri is sometimes considered an outlying member of the α Centauri system.

Adapted from data supplied by the U.S. Naval Observatory.

APPENDIX 14

The Twenty Brightest Stars

Star	Right Ascension (1950)		Declination (1950)		Distance* (pc)	Proper Motion (arcsec/yr)	Spectra of Components			Visual Magnitudes of Components			Luminosities of Components		
	(h)	(m)	°	'			A	B	C	A	B	C	A	B	C
Sirius	6	42.9	−16	39	2.7	1.33	A1V	wd		−1.46	+8.7		4×10^1	2×10^{-3}	
Canopus	6	22.8	−52	40	30	0.02	F0Ib-II			−0.72			2×10^3		
α Centauri	14	36.2	−60	38	1.3	3.68	G2V	K0V		−0.01	+1.3		2	8×10^1	
Arcturus	14	13.4	+19	27	11	2.28	K2IIIp			−0.06			1×10^2		
Vega	18	35.2	+38	44	8.0	0.34	A0V			+0.04			9×10^1		
Capella	5	13.0	+45	57	14	0.44	GIII	M1V	M5V	+0.05	+10.2	+13.7	2×10^2	1×10^{-1}	3×10^{-3}
Rigel	5	12.1	−8	15	250	0.00	B8 Ia	B9		+0.14	+6.6		8×10^4	2×10^2	
Procyon	7	36.7	+5	21	3.5	1.25	F5IV-V	wd		+0.37	+10.7		9	6×10^{-4}	
Betelgeuse	5	52.5	+7	24	150	0.03	M2Iab			+0.41v			1×10^5		
Achernar	1	35.9	−57	29	20	0.10	B5V			+0.51			5×10^2		
β Centauri	14	00.3	−60	08	90	0.04	B1III	?		+0.63	+4		9×10^3	2×10^2	
Altair	19	48.3	+8	44	5.1	0.66	A7IV-V			+0.77			1×10^2		
α Crucis	12	23.8	−62	49	120	0.04	B1IV	B3		+1.39	+1.9		8×10^3	5×10^3	
Aldebaran	4	33.0	+16	25	16	0.20	K5III	M2V		+0.86	+13		2×10^2	2×10^{-2}	
Spica	13	22.6	−10	54	80	0.05	B1V			+0.91v			6×10^3		
Antares	16	26.3	−26	19	120	0.03	MIb	B4eV		+0.92v	+5.1		5×10^4	2×10^2	
Pollux	7	42.3	+28	09	12	0.62	KOIII			+1.16			6×10^1		
Fomalhaut	22	54.9	−29	53	7.0	0.37	A3V	K4V		+1.19	+6.5		2×10^2	2×10^{-1}	
Deneb	20	39.7	+45	06	430	0.00	A2Ia			+1.26			8×10^4		
β Crucis	12	47.8	−59	24	150	0.05	B0.5IV			+1.28v			1×10^4		

*Distances of the more remote stars have been estimated from their spectral types and apparent magnitudes, and are only approximate.

Note: Several of the components listed are themselves spectroscopic binaries. A "v" after a magnitude denotes that the star is variable, in which case the magnitude at median light is given. A "p" after a spectral type indicates that the spectrum is peculiar. An "e" after a spectral type indicates that emission lines are present. When the luminosity classification is rather uncertain, a range is given.

Type of Variable	Kind of Star	Peak Absolute Magnitude	Peak Luminosity (L_S)	Period (days)	Description	Example
Cepheids (type I)	F and G supergiants	−2 to −5	10^3 to 10^4	3 to 50	Regular pulsation as a stage in the late evolution of moderately massive stars of solar-type composition. Period-luminosity relation exists.	δ Cep
Cepheids (type II)	F and G supergiants	−3	about 10^3	5 to 30	Regular pulsation as a stage in the late evolution of moderately massive stars depleted in metals. Period-luminosity relation exists.	W Vir
RR Lyrae	A and F giants	0	about 50	0.5 to 1	Very regular, small-amplitude pulsations as a stage in the late evolution of stars depleted in metals.	RR Lyr
Long-period	M red giants	−1 to −5	10^2 to 10^4	80 to 600	Large-amplitude, semiperiodic variations in evolved, luminous red giants that are losing mass. Much of luminosity is in infrared.	o Ceti (Mira)
T Tauri	Young stars G to M	0 to +8	$\frac{1}{20}$ to 50	irregular	Rapid and irregular variations of young stars still embedded in gas and dust from their formation.	T Tau
Novae	O to A binaries	−5 to −8	10^4 to 10^5	—	Eruptive event with ejection of shell due to explosive hydrogen fusion in the atmosphere of one of two binaries exchanging mass. Star brightens in a few days by as much as 10,000 times.	GK Per
Supernovae (type I)	White dwarfs in binary systems	−15 to −20	10^8 to 10^{10}	—	Catastrophic disruption of white dwarf that accretes mass from its binary companion until it collapses. Star brightens in a few days by 10^{10} or more can outshine an entire galaxy at maximum.	—
Supernovae (type II)	Massive red supergiants	−15 to −18	10^8 to 10^9	—	Catastrophic ejection of most of the mass from the collapsing stellar core of a massive, evolved star. Star brightens in a few days by 10^6 or more, can outshine an entire galaxy at maximum.	SN1987A

APPENDIX 16

The Local Group of Galaxies

Galaxy	Type	Right Ascension (1980) h m	Declination (1980) °	Visual Magnitude (m)	Distance (kpc)	Distance (1000 LY)	Diameter (kpc)	Diameter (1000 LY)	Absolute Magnitude (M_v)	Radial Velocity (km/s)	Mass (Solar Masses)
Our Galaxy	Sb	— —	—	—	—	—	30	100	(−21)	—	2×10^{11}
Large Magellanic Cloud	Irr I	5 26	−69	0.9	48	160	10	30	−17.7	+276	2.5×10^{10}
Small Magellanic Cloud	Irr I	0 51	−73	2.5	56	180	8	25	−16.5	+168	
Ursa Minor system	E4 (dwarf)	15 8.6	+67		70	220	1	3	(−9)		
Sculptor system	E3 (dwarf)	0 58.9	−33	8.0	83	270	2.2	7	−11.8		(2 to 4 × 10^6)
Draco system	E2 (dwarf)	17 19.9	+57		100	330	1.4	4.5	(−10)		
Carina system	E3 (dwarf)	6 41.1	−50		(170)	(550)	1.5	4.8	(−10)		
Fornax system	E3 (dwarf)	2 38.9	−34	8.4	250	800	4.5	15	−13.6		(1.2 to 2 × 10^7)
Leo II system	E0 (dwarf)	11 12.4	+22		230	750	1.6	5.2	−10.0	+39	(1.1 × 10^6)
Leo I system	E4 (dwarf)	10 7.4	+12	12.0	280	900	1.5	5	−10.4	+39	
NGC 6822	Irr I	19 43.8	−14	8.9	460	1500	2.7	9	−14.8	−32	
NGC 147	E6	0 32.0	+48	9.73	570	1900	3	10	−14.5		
NGC 185	E2	0 37.8	+48	9.43	570	1900	2.3	8	−14.8	−305	
NGC 205	E5	0 39.2	+41	8.17	680	2200	5	16	−16.5	−239	
NGC 221 (M32)	E3	0 41.6	46	8.16	680	2200	2.4	8	−16.5	−214	
IC 1613	Irr I	1 2.1	+1	9.61	680	2200	5	16	−14.7	−238	
Andromeda galaxy (NGC 224; M31)	Sb	0 41.6	+41	3.47	680	2200	40	130	−21.2	−266	3×10^{11}
And I	(dwarf)	0 44.6	+37	(14)	(680)	(2200)	0.5	1.6	(−11)		
And II	E0 (dwarf)	1 15.2	+33	(14)	(680)	(2200)	0.7	2.3	(−11)		
And III	E3 (dwarf)	0 34.2	+36	(14)	(680)	(2200)	0.9	0.9	(−11)		
NGC 598 (M33)	Sc	1 32.7	+30	5.79	720	2300	17	60	−18.9	−189	8×10^9

The Messier Catalogue of Nebulae and Star Clusters

M	NGC or (IC)	Right Ascension (1980) h	m	Decli- nation (1980) °	'	Apparent Visual Magnitude	Description
1	1952	5	33.3	+22	01	8.4	"Crab" nebula in Taurus; remains of SN 1054
2	7089	21	32.4	−0	54	6.4	Globular cluster in Aquarius
3	5272	13	41.2	+28	29	6.3	Globular cluster in Canes Venatici
4	6121	16	22.4	−26	28	6.5	Globular cluster in Scorpius
5	5904	15	17.5	+2	10	6.1	Globular cluster in Serpens
6	6405	17	38.8	−32	11	5.5	Open cluster in Scorpius
7	6475	17	52.7	−34	48	3.3	Open cluster in Scorpius
8	6523	18	02.4	−24	23	5.1	"Lagoon" nebula in Sagittarius
9	6333	17	18.1	−18	30	8.0	Globular cluster in Ophiuchus
10	6254	16	56.1	−4	05	6.7	Globular cluster in Ophiuchus
11	6705	18	50.0	−6	18	6.8	Open cluster in Scutum Sobieskii
12	6218	16	46.3	−1	55	6.6	Globular cluster in Ophiuchus
13	6205	16	41.0	+36	30	5.9	Globular cluster in Hercules
14	6402	17	36.6	−3	14	8.0	Globular cluster in Ophiuchus
15	7078	21	28.9	+12	05	6.4	Globular cluster in Pegasus
16	6611	18	17.8	−13	47	6.6	Open cluster with nebulosity in Serpens
17	6618	18	19.6	−16	11	7.5	"Swan" or "Omega" nebula in Sagittarius
18	6613	18	18.7	−17	08	7.2	Open cluster in Sagittarius
19	6273	17	01.4	−26	14	6.9	Globular cluster in Ophiuchus
20	6514	18	01.2	−23	02	8.5	"Trifid" nebula in Sagittarius
21	6531	18	03.4	−22	30	6.5	Open cluster in Sagittarius
22	6656	18	35.2	−23	56	5.6	Globular cluster in Sagittarius
23	6494	17	55.8	−19	00	5.9	Open cluster in Sagittarius
24	6603	18	17.3	−18	26	4.6	Open cluster in Sagittarius
25	(4725)	18	30.5	−19	16	6.2	Open cluster in Sagittarius
26	6694	18	44.1	−9	25	9.3	Open cluster in Scutum Sobieskii
27	6853	19	58.8	+22	40	8.2	"Dumbbell" planetary nebula in Vulpecula
28	6626	18	23.2	−24	52	7.6	Globular cluster in Sagittarius
29	6913	20	23.3	+38	27	8.0	Open cluster in Cygnus
30	7099	21	39.2	−23	16	7.7	Globular cluster in Capricornus
31	224	0	41.6	+41	10	3.5	Andromeda galaxy
32	221	0	41.6	+40	46	8.2	Elliptical galaxy; companion to M31
33	598	1	32.7	+30	33	5.8	Spiral galaxy in Triangulum
34	1039	2	40.7	+42	43	5.8	Open cluster in Perseus
35	2168	6	07.5	+24	21	5.6	Open cluster in Gemini
36	1960	5	35.0	+34	05	6.5	Open cluster in Auriga
37	2099	5	51.1	+32	33	6.2	Open cluster in Auriga
38	1912	5	27.3	+35	48	7.0	Open cluster in Auriga
39	7092	21	31.5	+48	21	5.3	Open cluster in Cygnus
40		12	21	+59			Close double star in Ursa Major
41	2287	6	46.2	−20	43	5.0	Loose open cluster in Canis Major
42	1976	5	34.4	−5	24	4	Orion nebula
43	1982	5	34.6	−5	18	9	Northeast portion of Orion nebula

The Messier Catalogue of Nebulae and Star Clusters (Continued)

M	NGC or (IC)	Right Ascension (1980) h	m	Declination (1980) °	′	Apparent Visual Magnitude	Description
44	2632	8	39	+20	04	3.9	Praesepe; open cluster in Cancer
45		3	46.3	+24	03	1.6	The Pleiades; open cluster in Taurus
46	2437	7	40.9	−14	46	6.6	Open cluster in Puppis
47	2422	7	35.7	−14	26	5	Loose group of stars in Puppis
48	2548	8	12.8	−5	44	6	"Cluster of very small stars"
49	4472	12	28.8	+8	06	8.5	Elliptical galaxy in Virgo
50	2323	7	02.0	−8	19	6.3	Loose open cluster in Monoceros
51	5194	13	29.1	+47	18	8.4	"Whirlpool" spiral galaxy in Canes Venatici
52	7654	23	23.3	+61	30	8.2	Loose open cluster in Cassiopeia
53	5024	13	12.0	+18	16	7.8	Globular cluster in Coma Berenices
54	6715	18	53.8	−30	30	7.8	Globular cluster in Sagittarius
55	6809	19	38.7	−30	59	6.2	Globular cluster in Sagittarius
56	6779	19	15.8	+30	08	8.7	Globular cluster in Lyra
57	6720	18	52.8	+33	00	9.0	"Ring" nebula; planetary nebula in Lyra
58	4579	12	36.7	+11	55	9.9	Spiral galaxy in Virgo
59	4621	12	41.0	+11	46	10.0	Spiral galaxy in Virgo
60	4649	12	42.6	+11	40	9.0	Elliptical galaxy in Virgo
61	4303	12	20.8	+4	35	9.6	Spiral galaxy in Virgo
62	6266	16	59.9	−30	05	6.6	Globular cluster in Scorpius
63	5055	13	14.8	+42	07	8.9	Spiral galaxy in Canes Venatici
64	4826	12	55.7	+21	39	8.5	Spiral galaxy in Coma Berenices
65	3623	11	17.9	+13	12	9.4	Spiral galaxy in Leo
66	3627	11	19.2	+13	06	9.0	Spiral galaxy in Leo; companion to M65
67	2682	8	50.0	+11	53	6.1	Open cluster in Cancer
68	4590	12	38.4	−26	39	8.2	Globular cluster in Hydra
69	6637	18	30.1	−32	23	8.0	Globular cluster in Sagittarius
70	6681	18	42.0	−32	18	8.1	Globular cluster in Sagittarius
71	6838	19	52.8	+18	44	7.6	Globular cluster in Sagittarius
72	6981	20	52.3	−12	38	9.3	Globular cluster in Aquarius
73	6994	20	57.8	−12	43	9.1	Open cluster in Aquarius
74	628	1	35.6	+15	41	9.3	Spiral galaxy in Pisces
75	6864	20	04.9	−21	59	8.6	Globular cluster in Sagittarius
76	650	1	41.0	+51	28	11.4	Planetary nebula in Perseus
77	1068	2	41.6	−0	04	8.9	Spiral galaxy in Cetus
78	2068	5	45.7	0	03	8.3	Small emission nebula in Orion
79	1904	5	23.3	−24	32	7.5	Globular cluster in Lepus
80	6093	16	15.8	−22	56	7.5	Globular cluster in Scorpius
81	3031	9	54.2	+69	09	7.0	Spiral galaxy in Ursa Major
82	3034	9	54.4	+69	47	8.4	Irregular galaxy in Ursa Major
83	5236	13	35.4	−29	31	7.6	Spiral galaxy in Hydra
84	4374	12	24.1	+13	00	9.4	Elliptical galaxy in Virgo
85	4382	12	24.3	+18	18	9.3	Elliptical galaxy in Coma Berenices
86	4406	12	25.1	+13	03	9.2	Elliptical galaxy in Virgo
87	4486	12	29.7	+12	30	8.7	Elliptical galaxy in Virgo
88	4501	12	30.9	+14	32	9.5	Spiral galaxy in Coma Berenices
89	4552	12	34.6	+12	40	10.3	Elliptical galaxy in Virgo
90	4569	12	35.8	+13	16	9.6	Spiral galaxy in Virgo

The Messier Catalogue of Nebulae and Star Clusters (Continued)

M	NGC or (IC)	Right Ascension (1980) h	m	Declination (1980) °	′	Apparent Visual Magnitude	Description
91	omitted						
92	6341	17	16.5	+43	10	6.4	Globular cluster in Hercules
93	2447	7	43.7	−23	49	6.5	Open cluster in Puppis
94	4736	12	50.0	+41	14	8.3	Spiral galaxy in Canes Venatici
95	3351	10	42.9	+11	49	9.8	Barred spiral galaxy in Leo
96	3368	10	45.7	+11	56	9.3	Spiral galaxy in Leo
97	3587	11	13.7	+55	07	11.1	"Owl" nebula; planetary nebula in Ursa Major
98	4192	12	12.7	+15	01	10.2	Spiral galaxy in Coma Berenices
99	4254	12	17.8	+14	32	9.9	Spiral galaxy in Coma Berenices
100	4321	12	21.9	+15	56	9.4	Spiral galaxy in Coma Berenices
101	5457	14	02.5	+54	27	7.9	Spiral galaxy in Ursa Major
102	5866(?)	15	05.9	+55	50	10.5	Spiral galaxy (identification in doubt)
103	581	1	31.9	+60	35	6.9	Open cluster in Cassiopeia
104*	4594	12	39.0	−11	31	8.3	Spiral galaxy in Virgo
105*	3379	10	46.8	+12	51	9.7	Elliptical galaxy in Leo
106*	4258	12	18.0	+47	25	8.4	Spiral galaxy in Canes Venatici
107*	6171	16	31.4	−13	01	9.2	Globular cluster in Ophiuchus
108*	3556	11	10.5	+55	47	10.5	Spiral galaxy in Ursa Major
109*	3992	11	56.6	+53	29	10.0	Spiral galaxy in Ursa Major
110*	205	0	39.2	+41	35	9.4	Elliptical galaxy (companion to M31)

*Not in Messier's original (1781) list; added later by others.

Element	Symbol	Atomic Number	Atomic Weight* (Chemical Scale)	Number of Atoms per 10^{12} Hydrogen Atoms
Hydrogen	H	1	1.0080	1×10^{12}
Helium	He	2	4.003	8×10^{10}
Lithium	Li	3	6.940	2×10^{3}
Beryllium	Be	4	9.013	3×10^{1}
Boron	B	5	10.82	9×10^{2}
Carbon	C	6	12.011	4.5×10^{8}
Nitrogen	N	7	14.008	9.2×10^{7}
Oxygen	O	8	16.0000	7.4×10^{8}
Fluorine	F	9	19.00	3.1×10^{4}
Neon	Ne	10	20.183	1.3×10^{8}
Sodium	Na	11	22.991	2.1×10^{6}
Magnesium	Mg	12	24.32	4.0×10^{7}
Aluminum	Al	13	26.98	3.1×10^{6}
Silicon	Si	14	28.09	3.7×10^{7}
Phosphorus	P	15	30.975	3.8×10^{5}
Sulfur	S	16	32.066	1.9×10^{7}
Chlorine	Cl	17	35.457	1.9×10^{5}
Argon	Ar(A)	18	39.944	3.8×10^{6}
Potassium	K	19	39.100	1.4×10^{5}
Calcium	Ca	20	40.08	2.2×10^{6}
Scandium	Sc	21	44.96	1.3×10^{3}
Titanium	Ti	22	47.90	8.9×10^{4}
Vanadium	V	23	50.95	1.0×10^{4}
Chromium	Cr	24	52.01	5.1×10^{5}
Manganese	Mn	25	54.94	3.5×10^{5}
Iron	Fe	26	55.85	3.2×10^{7}
Cobalt	Co	27	58.94	8.3×10^{4}
Nickel	Ni	28	58.71	1.9×10^{6}
Copper	Cu	29	63.54	1.9×10^{4}
Zinc	Zn	30	65.38	4.7×10^{4}
Gallium	Ga	31	69.72	1.4×10^{3}
Germanium	Ge	32	72.60	4.4×10^{3}
Arsenic	As	33	74.91	2.5×10^{2}
Selenium	Se	34	78.96	2.3×10^{3}
Bromine	Br	35	79.916	4.4×10^{2}
Krypton	Kr	36	83.80	1.7×10^{3}
Rubidium	Rb	37	85.48	2.6×10^{2}
Strontium	Sr	38	87.63	8.8×10^{2}
Yttrium	Y	39	88.92	2.5×10^{2}
Zirconium	Zr	40	91.22	4.0×10^{2}
Niobium (Columbium)	Nb(Cb)	41	92.91	2.6×10^{1}
Molybdenum	Mo	42	95.95	9.3×10^{1}
Technetium	Tc(Ma)	43	(99)	—
Ruthenium	Ru	44	101.1	68
Rhodium	Rh	45	102.91	13
Palladium	Pd	46	106.4	51
Silver	Ag	47	107.880	20
Cadmium	Cd	48	112.41	63

The Chemical Elements (Continued)

Element	Symbol	Atomic Number	Atomic Weight* (Chemical Scale)	Number of Atoms per 10^{12} Hydrogen Atoms
Indium	In	49	114.82	7
Tin	Sn	50	118.70	1.4×10^2
Antimony	Sb	51	121.76	13
Tellurium	Te	52	127.61	1.8×10^2
Iodine	I(J)	53	126.91	33
Xenon	Xe(X)	54	131.30	1.6×10^2
Cesium	Cs	55	132.91	14
Barium	Ba	56	137.36	1.6×10^2
Lanthanum	La	57	138.92	17
Cerium	Ce	58	140.13	43
Praseodymium	Pr	59	140.92	6
Neodymium	Nd	60	144.27	31
Promethium	Pm	61	(147)	—
Samarium	Sm(Sa)	62	150.35	10
Europium	Eu	63	152.0	4
Gadolinium	Gd	64	157.26	13
Terbium	Tb	65	158.93	2
Dysprosium	Dy(Ds)	66	162.51	15
Holmium	Ho	67	164.94	3
Erbium	Er	68	167.27	9
Thulium	Tm(Tu)	69	168.94	2
Ytterbium	Yb	70	173.04	8
Lutecium	Lu(Cp)	71	174.99	2
Hafnium	Hf	72	178.50	6
Tantalum	Ta	73	180.95	1
Tungsten	W	74	183.86	5
Rhenium	Re	75	186.22	2
Osmium	Os	76	190.2	27
Iridium	Ir	77	192.2	24
Platinum	Pt	78	195.09	56
Gold	Au	79	197.0	6
Mercury	Hg	80	200.61	19
Thallium	Tl	81	204.39	8
Lead	Pb	82	207.21	1.2×10^2
Bismuth	Bi	83	209.00	5
Polonium	Po	84	(209)	—
Astatine	At	85	(210)	—
Radon	Rn	86	(222)	—
Francium	Fr(Fa)	87	(223)	—
Radium	Ra	88	226.05	—
Actinium	Ac	89	(227)	—
Thorium	Th	90	232.12	1
Protactinium	Pa	91	(231)	—
Uranium	U(Ur)	92	238.07	1
Neptunium	Np	93	(237)	—
Plutonium	Pu	94	(244)	—
Americium	Am	95	(243)	—
Curium	Cm	96	(248)	—
Berkelium	Bk	97	(247)	—
Californium	Cf	98	(251)	—
Einsteinium	E	99	(254)	—
Fermium	Fm	100	(253)	—
Mendeleevium	Mv	101	(256)	—
Nobelium	No	102	(253)	—

*Where mean atomic weights have not been well determined, the atomic mass numbers of the most stable isotopes are given in parentheses.

Constellation (Latin name)	Genitive Case Ending	English Name or Description	Abbre- viation	Approximate Position α h	δ °
Andromeda	Andromedae	Princess of Ethiopia	And	1	+40
Antila	Antilae	Air pump	Ant	10	−35
Apus	Apodis	Bird of Paradise	Aps	16	−75
Aquarius	Aquarii	Water bearer	Aqr	23	−15
Aquila	Aquilae	Eagle	Aql	20	+5
Ara	Arae	Altar	Ara	17	−55
Aries	Arietis	Ram	Ari	3	+20
Auriga	Aurigae	Charioteer	Aur	6	+40
Boötes	Boötis	Herdsman	Boo	15	+30
Caelum	Caeli	Graving tool	Cae	5	−40
Camelopardus	Camelopardis	Giraffe	Cam	6	+70
Cancer	Cancri	Crab	Cnc	9	+20
Canes Venatici	Canum Venaticorum	Hunting dogs	CVn	13	+40
Canis Major	Canis Majoris	Big dog	CMa	7	−20
Canis Minor	Canis Minoris	Little dog	CMi	8	+5
Capricornus	Capricorni	Sea goat	Cap	21	−20
Carina*	Carinae	Keel of Argonauts' ship	Car	9	−60
Cassiopeia	Cassiopeiae	Queen of Ethiopia	Cas	1	+60
Centaurus	Centauri	Centaur	Cen	13	−50
Cepheus	Cephei	King of Ethiopia	Cep	22	+70
Cetus	Ceti	Sea monster (whale)	Cet	2	−10
Chamaeleon	Chamaeleontis	Chameleon	Cha	11	−80
Circinus	Circini	Compasses	Cir	15	−60
Columba	Columbae	Dove	Col	6	−35
Coma Berenices	Comae Berenices	Berenice's hair	Com	13	+20
Corona Australis	Coronae Australis	Southern crown	CrA	19	−40
Corona Borealis	Coronae Borealis	Northern crown	CrB	16	+30
Corvus	Corvi	Crow	Crv	12	−20
Crater	Crateris	Cup	Crt	11	−15
Crux	Crucis	Cross (southern)	Cru	12	−60
Cygnus	Cygni	Swan	Cyg	21	+40
Delphinus	Delphini	Porpoise	Del	21	+10
Dorado	Doradus	Swordfish	Dor	5	−65
Draco	Draconis	Dragon	Dra	17	+65
Equuleus	Equulei	Little horse	Equ	21	+10
Eridanus	Eridani	River	Eri	3	−20
Fornax	Fornacis	Furnace	For	3	−30
Gemini	Geminorum	Twins	Gem	7	+20
Grus	Gruis	Crane	Gru	22	−45
Hercules	Herculis	Hercules, son of Zeus	Her	17	+30
Horologium	Horologii	Clock	Hor	3	−60
Hydra	Hydrae	Sea serpent	Hya	10	−20
Hydrus	Hydri	Water snake	Hyi	2	−75
Indus	Indi	Indian	Ind	21	−55
Lacerta	Lacertae	Lizard	Lac	22	+45

The Constellations (Continued)

Constellation (Latin name)	Genitive Case Ending	English Name or Description	Abbre-viation	α h	δ °
Leo	Leonis	Lion	Leo	11	+15
Leo Minor	Leonis Minoris	Little lion	LMi	10	+35
Lepus	Leporis	Hare	Lep	6	−20
Libra	Librae	Balance	Lib	15	−15
Lupus	Lupi	Wolf	Lup	15	−45
Lynx	Lyncis	Lynx	Lyn	8	+45
Lyra	Lyrae	Lyre or harp	Lyr	19	+40
Mensa	Mensae	Table Mountain	Men	5	−80
Microscopium	Microscopii	Microscope	Mic	21	−35
Monoceros	Monocerotis	Unicorn	Mon	7	−5
Musca	Muscae	Fly	Mus	12	−70
Norma	Normae	Carpenter's level	Nor	16	−50
Octans	Octantis	Octant	Oct	22	−85
Ophiuchus	Ophiuchi	Holder of serpent	Oph	17	0
Orion	Orionis	Orion, the hunter	Ori	5	+5
Pavo	Pavonis	Peacock	Pav	20	−65
Pegasus	Pegasi	Pegasus, the winged horse	Peg	22	+20
Perseus	Persei	Perseus, hero who saved Andromeda	Per	3	+45
Phoenix	Phoenicis	Phoenix	Phe	1	−50
Pictor	Pictoris	Easel	Pic	6	−55
Pisces	Piscium	Fishes	Psc	1	+15
Piscis Austrinus	Piscis Austrini	Southern fish	PsA	22	−30
Puppis*	Puppis	Stern of the Argonauts' ship	Pup	8	−40
Pyxis* (= Malus)	Pyxidus	Compass on the Argonauts' ship	Pyx	9	−30
Reticulum	Reticuli	Net	Ret	4	−60
Sagitta	Sagittae	Arrow	Sge	20	+10
Sagittarius	Sagittarii	Archer	Sgr	19	−25
Scorpius	Scorpii	Scorpion	Sco	17	−40
Sculptor	Sculptoris	Sculptor's tools	Scl	0	−30
Scutum	Scuti	Shield	Sct	19	−10
Serpens	Serpentis	Serpent	Ser	17	0
Sextans	Sextantis	Sextant	Sex	10	0
Taurus	Tauri	Bull	Tau	4	+15
Telescopium	Telescopii	Telescope	Tel	19	−50
Triangulum	Trianguli	Triangle	Tri	2	+30
Triangulum Australe	Trianguli Australis	Southern triangle	TrA	16	−65
Tucana	Tucanae	Toucan	Tuc	0	−65
Ursa Major	Ursae Majoris	Big bear	UMa	11	+50
Ursa Minor	Ursae Minoris	Little bear	VMi	15	+70
Vela*	Velorum	Sail of the Argonauts' ship	Vel	9	−50
Virgo	Virginis	Virgin	Vir	13	0
Volans	Volantis	Flying fish	Vol	8	−70
Vulpecula	Vulpeculae	Fox	Vul	20	+25

*The four constellations Carina, Puppis, Pyxis, and Vela originally formed the single constellation, Argo Navis.

INDEX

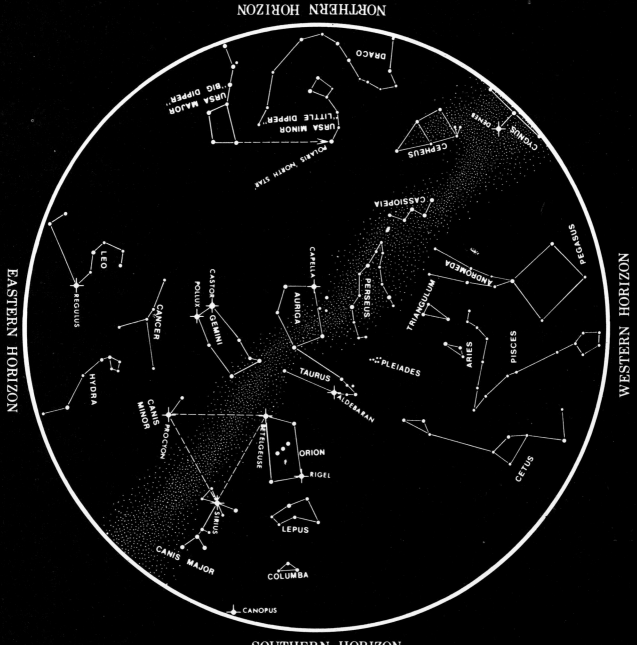

NORTHERN HORIZON

EASTERN HORIZON

WESTERN HORIZON

SOUTHERN HORIZON

THE NIGHT SKY IN JANUARY

Latitude of chart is 34°N, but it is practical throughout the continental United States.

To use: Hold chart vertically and turn it so the direction you are facing shows at the bottom.

Chart time (Local Standard):

10 p.m. First of month

9 p.m. Middle of month

8 p.m. Last of month

THE NIGHT SKY IN FEBRUARY

Latitude of chart is 34°N, but it is practical throughout the continental United States.

To use: Hold chart vertically and turn it so the direction you are facing shows at the bottom.

Chart time (Local Standard):

10 p.m. First of month

9 p.m. Middle of month

8 p.m. Last of month

Star Chart from *GRIFFITH OBSERVER*, Griffith Observatory, Los Angeles

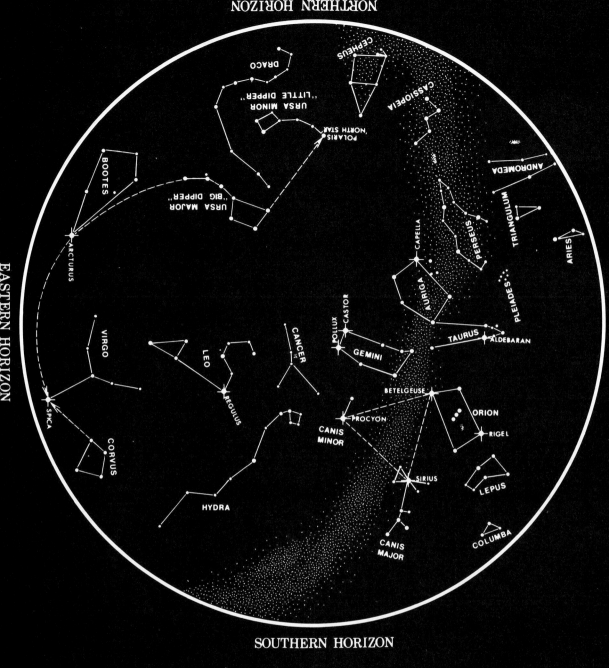

SOUTHERN HORIZON

THE NIGHT SKY IN MARCH

Latitude of chart is 34°N, but it is
practical throughout the continental
United States.

To use: Hold chart vertically and turn
it so the direction you are facing
shows at the bottom.

Chart time (Local Standard):

10 p.m. First of month

9 p.m. Middle of month

8 p.m. Last of month

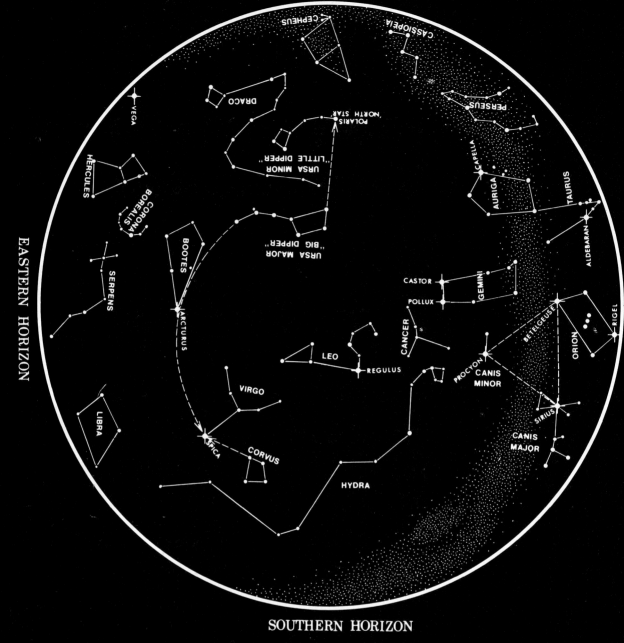

SOUTHERN HORIZON

THE NIGHT SKY IN APRIL

Latitude of chart is 34°N, but it is
practical throughout the continental
United States.

To use: Hold chart vertically and turn
it so the direction you are facing
shows at the bottom.

Chart time (Local Standard):

10 p.m. First of month

9 p.m. Middle of month

8 p.m. Last of month

Star Chart from *GRIFFITH OBSERVER*, Griffith Observatory, Los Angeles

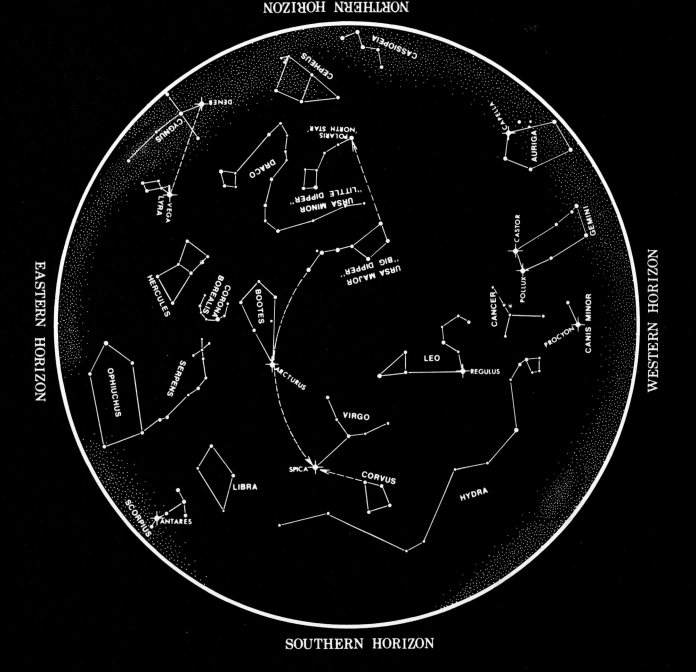

THE NIGHT SKY IN MAY

Latitude of chart is 34°N, but it is
practical throughout the continental
United States.

To use: Hold chart vertically and turn
it so the direction you are facing
shows at the bottom.

Chart time (Local Standard):

10 p.m. First of month
9 p.m. Middle of month
8 p.m. Last of month

Star Chart from *GRIFFITH OBSERVER*, Griffith Observatory, Los Angeles

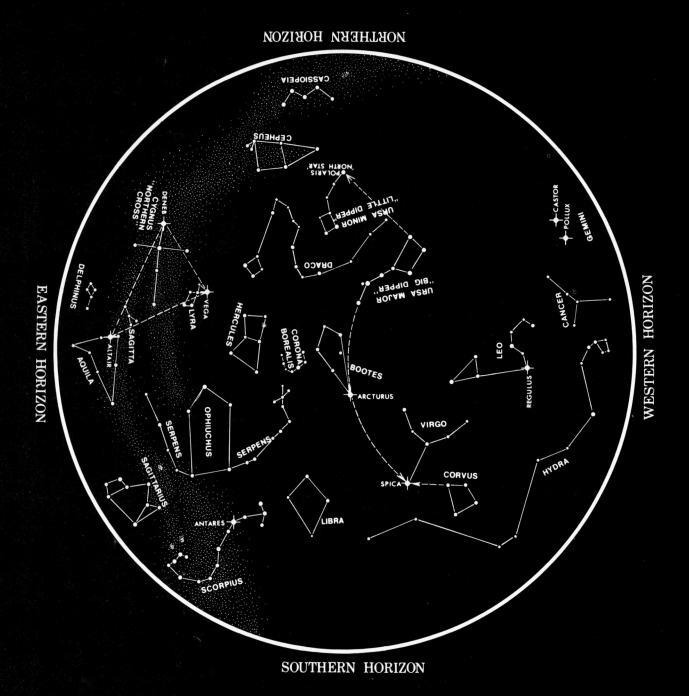

NORTHERN HORIZON

SOUTHERN HORIZON

THE NIGHT SKY IN JUNE

Latitude of chart is 34°N, but it is
practical throughout the continental
United States.

To use: Hold chart vertically and turn
it so the direction you are facing
shows at the bottom.

Chart time (Local Standard):

10 p.m. First of month

9 p.m. Middle of month

8 p.m. Last of month

Star Chart from *GRIFFITH OBSERVER*, Griffith Observatory, Los Angeles

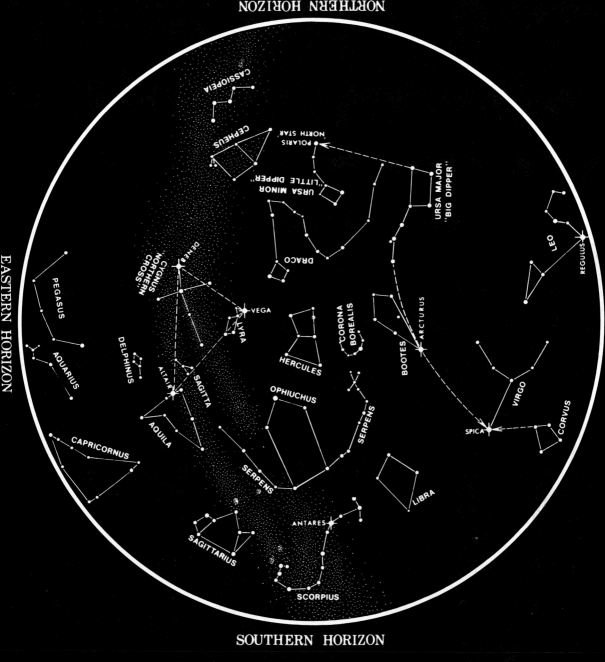

NORTHERN HORIZON

EASTERN HORIZON

SOUTHERN HORIZON

THE NIGHT SKY IN JULY

Latitude of chart is 34°N, but it is
practical throughout the continental
United States.

To use: Hold chart vertically and turn
it so the direction you are facing
shows at the bottom.

Chart time (Local Standard):

10 p.m. First of month

9 p.m. Middle of month

8 p.m. Last of month

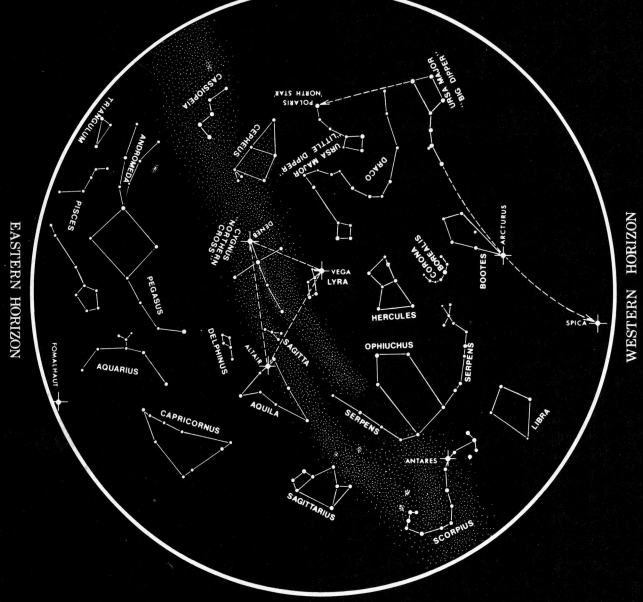

SOUTHERN HORIZON

THE NIGHT SKY IN AUGUST

Latitude of chart is 34°N, but it is practical throughout the continental United States.

To use: Hold chart vertically and turn it so the direction you are facing shows at the bottom.

Chart time (Local Standard):

10 p.m. First of month

9 p.m. Middle of month

8 p.m. Last of month

Star Chart from *GRIFFITH OBSERVER*, Griffith Observatory, Los Angeles

NORTHERN HORIZON

EASTERN HORIZON

WESTERN HORIZON

SOUTHERN HORIZON

THE NIGHT SKY IN SEPTEMBER

Latitude of chart is 34°N, but it is practical throughout the continental United States.

To use: Hold chart vertically and turn it so the direction you are facing shows at the bottom.

Chart time (Local Standard):

10 p.m. First of month

9 p.m. Middle of month

8 p.m. Last of month

Star Chart from *GRIFFITH OBSERVER*, Griffith Observatory, Los Angeles

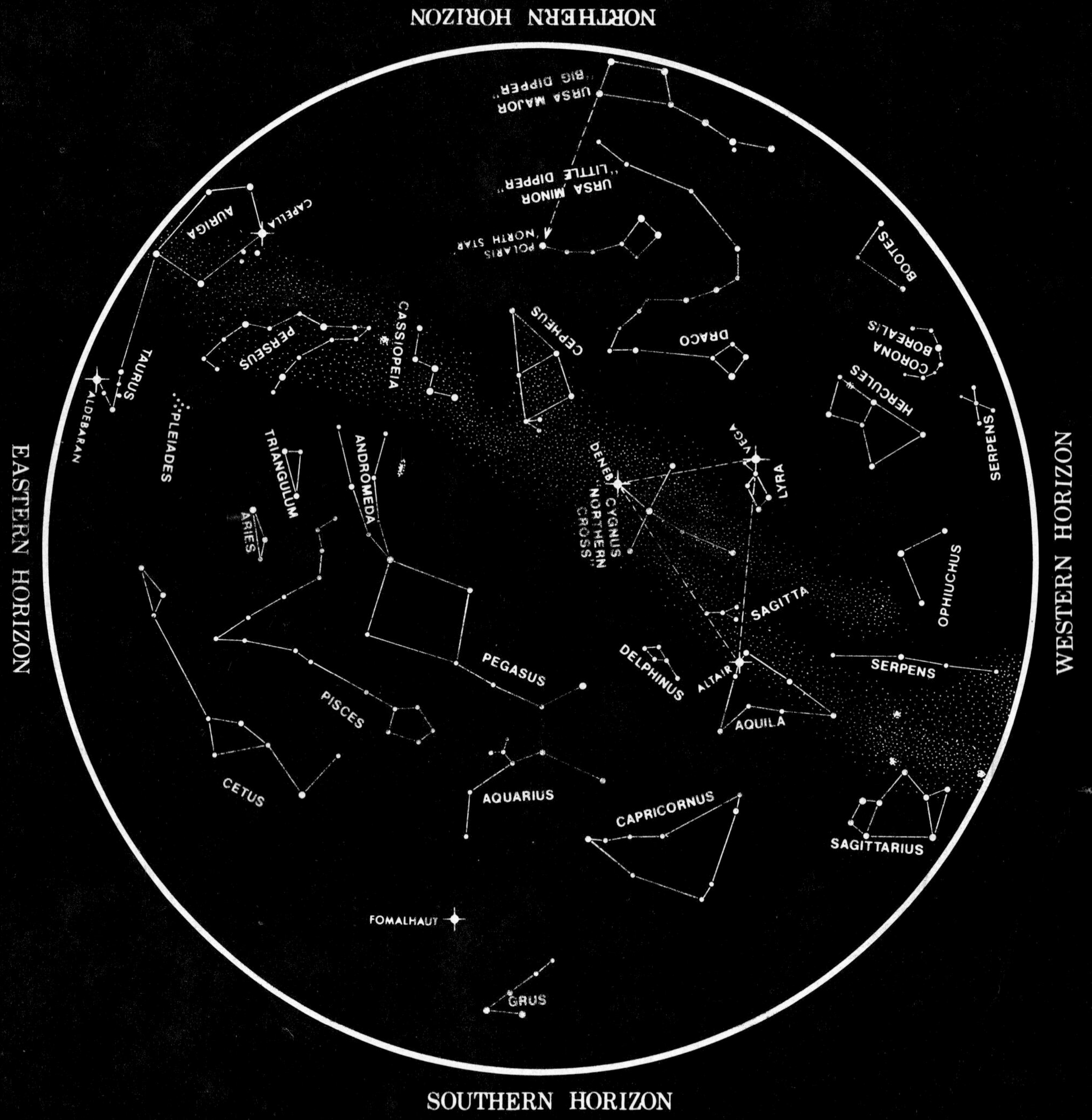

THE NIGHT SKY IN OCTOBER

Latitude of chart is 34°N, but it is
practical throughout the continental
United States.

To use: Hold chart vertically and turn
it so the direction you are facing
shows at the bottom.

Chart time (Local Standard):

10 p.m. First of month

9 p.m. Middle of month

8 p.m. Last of month

Star Chart from *GRIFFITH OBSERVER*, Griffith Observatory, Los Angeles

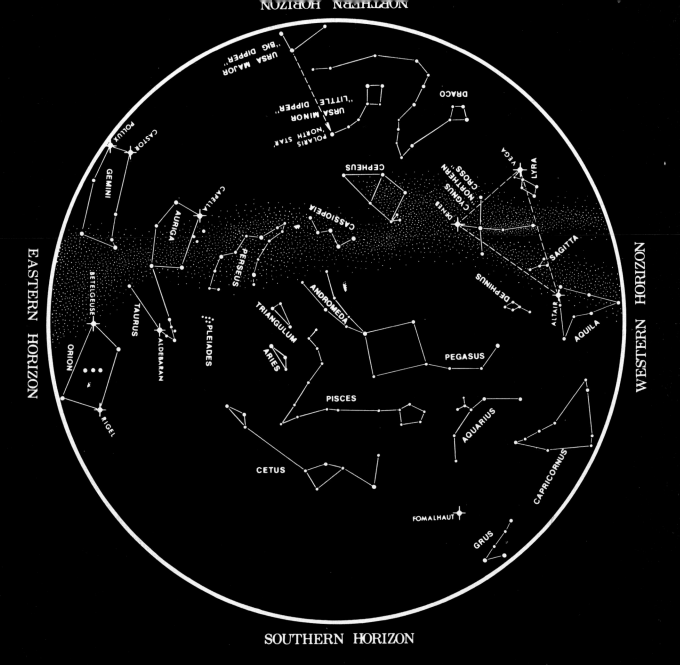

THE NIGHT SKY IN NOVEMBER

SOUTHERN HORIZON

Latitude of chart is 34°N, but it is
practical throughout the continental
United States.

To use: Hold chart vertically and turn
it so the direction you are facing
shows at the bottom.

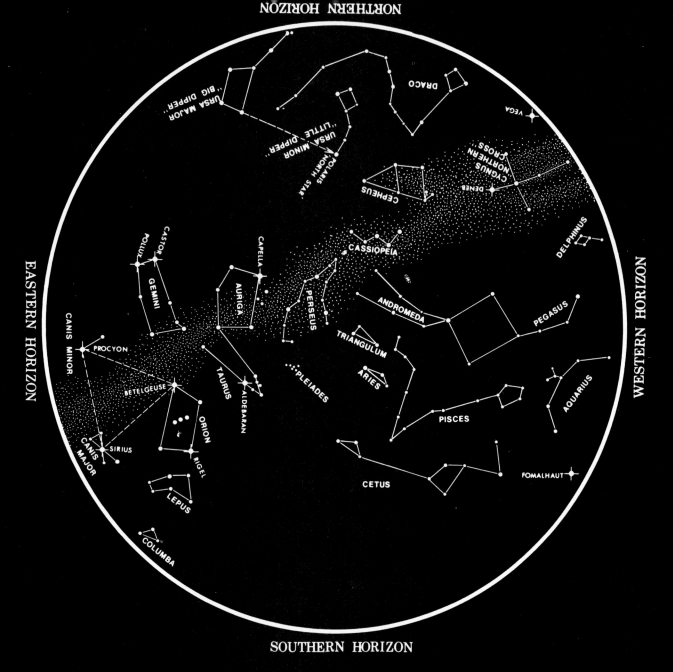

THE NIGHT SKY IN DECEMBER

Latitude of chart is 34°N, but it is
practical throughout the continental
United States.

To use: Hold chart vertically and turn
it so the direction you are facing
shows at the bottom.

Chart time (Local Standard):

10 p.m. First of month

9 p.m. Middle of month

8 p.m. Last of month